Turmeric

The Genus *Curcuma*

Turmeric

The Genus *Curcuma*

Edited by
P. N. Ravindran
K. Nirmal Babu
K. Sivaraman

CRC PRESS

Boca Raton   London   New York   Washington, D.C.

FIRST INDIAN REPRINT, 2012

This book contains information obtained from authentic and highly regarded sources. Reprinted material is quoted with permission, and sources are indicated. A wide variety of references are listed. Reasonable efforts have been made to publish reliable data and information, but the author and the publisher cannot assume responsibility for the validity of all materials or for the consequences of their use.

Neither this book nor any part may be reproduced or transmitted in any form or by any means, electronic or mechanical, including photocopying, microfilming, and recording, or by any information storage or retrieval system, without prior permission in writing from the publisher.

Direct all inquiries to CRC Press LLC, 2000 N.W. Corporate Blvd., Boca Raton, Florida 33431.

© 2007 Taylor & Francis Group, LLC CRC Press is an imprint of Taylor & Francis Group

Trademark Notice: Product or corporate names may be trademarks or registered trademarks, and are used only for identification and explanation, without intent to infringe.

Visit the CRC Press Web site at www.crcpress.com

Printed and bound in India by
Replika Press Pvt. Ltd.

ISBN 10 : 0-8493-7034-5
ISBN 13 : 978-0-8493-7034-2

FOR SALE IN SOUTH ASIA ONLY.

This book is dedicated to
The ARYA VAIDYA SALA (AVS), KOTTAKKAL
Kerala, India
For preserving the ancient traditions of Ayurveda in its pristine glory;
For being a trailblazer in nurturing Ayurveda to suit the modern world;
For extending its healing touch to millions of ailing people across the globe;
For being in the forefront of Ayurvedic practice, teaching, and research;
For being a national leader in the area of medicinal plant research.

Preface to the Series

There is increasing interest in industry, academia, and the health sciences in medicinal and aromatic plants. In passing from plant production to the eventual product used by the public, many sciences are involved. This series brings together information that is currently scattered through an ever-increasing number of journals. Each volume gives an in-depth look at one plant genus, about which an area specialist has assembled information ranging from the production of the plant to market trends and quality control.

Many industries are involved, such as forestry, agriculture, chemical, food, flavor, beverage, pharmaceutical, cosmetic, and fragrance. The plant raw materials are roots, rhizomes, bulbs, leaves, stems, barks, wood, flowers, fruits, and seeds. These yield gums, resins, essential (volatile) oils, fixed oils, waxes, juices, extracts, and spices for medicinal and aromatic purposes. All these commodities are traded worldwide. A dealer's market report for an item may say "drought in the country of origin has forced up prices."

Natural products do not mean safe products, and account of this has to be taken by the above industries, which are subject to regulation. For example, a number of plants that are approved for use in medicine must not be used in cosmetic products.

The assessment of "safe to use" starts with the harvested plant material, which has to comply with an official monograph. This may require absence of, or prescribed limits of, radioactive material, heavy metals, aflatoxin, pesticide residue, as well as the required level of active principle. This analytical control is costly and tends to exclude small batches of plant material. Large-scale, contracted, mechanized cultivation with designated seed or plantlets is now preferable.

Today, plant selection is not only for the yield of active principle, but for the plant's ability to overcome disease, climatic stress, and the hazards caused by mankind. Such methods as *in vitro* fertilization, meristem cultures, and somatic embryogenesis are used. The transfer of sections of DNA is giving rise to controversy in the case of some end uses of the plant material.

Some suppliers of plant raw material are now able to certify that they are supplying organically farmed medicinal plants, herbs, and spices. The Economic Union directive CVO/EU No. 2092/91 details the specifications for the *obligatory* quality controls to be carried out at all stages of production and processing of organic products.

Fascinating plant folklore and ethnopharmacology lead to medicinal potential. Examples are the muscle relaxants based on the arrow poison, curare, from species of *Chondrodendron*, and the antimalarials derived from species of *Cinchona* and *Artemisia*. The methods of detection of pharmacological activity have become increasingly reliable and specific, frequently involving enzymes in bioassays and avoiding the use of laboratory animals. By using bioassay-linked fractionation of crude plant juices or extracts, compounds can be specifically targeted which, for example, inhibit blood platelet aggregation, or have antitumor, or antiviral, or any other required activity. With the assistance of robotic devices, all the members of a genus may be readily screened. However, the plant material must be fully authenticated by a specialist.

The medicinal traditions of ancient civilizations such as those of China and India have a large armamentarium of plants in their pharmacopoeias that are used throughout Southeast Asia. A similar situation exists in Africa and South America. Thus, a very high percentage of the world's population relies on medicinal and aromatic plants for their medicine. Western medicine is also responding. Already in Germany all medical practitioners have to pass an examination in phytotherapy before being allowed to practice. It is noticeable that medical, pharmacy, and health-related schools throughout Europe and the United States are increasingly offering training in phytotherapy.

Multinational pharmaceutical companies have become less enamored of the single compound, magic-bullet cure. The high costs of such ventures and the endless competition from "me-too" compounds from rival companies often discourage the attempt. Independent phytomedicine companies have been very strong in Germany. However, by the end of 1995, 11 (almost all) had been acquired by the multinational pharmaceutical firms, acknowledging the lay public's growing demand for phytomedicines in the Western world.

The business of dietary supplements in the Western world has expanded from the health store to the pharmacy. Alternative medicine includes plant-based products. Appropriate measures to ensure their quality, safety, and efficacy either already exist or are being answered by greater legislative control by such bodies as the U.S. Food and Drug Administration and the recently created European Agency for the Evaluation of Medicinal Products, which is based in London.

In the United States, the Dietary Supplement and Health Education Act of 1994 recognized the class of phytotherapeutic agents derived from medicinal and aromatic plants. Furthermore, under public pressure, the U.S. Congress set up an Office of Alternative Medicine, which in 1994 assisted the filing of several Investigational New Drug (IND) applications, required for clinical trials of some Chinese herbal preparations. The significance of these applications was that each Chinese preparation involved several plants and yet was handled as a *single* IND. A demonstration of the contribution to efficacy of *each* ingredient of *each* plant was not required. This was a major step toward more sensible regulations in regard to phytomedicines.

My thanks are due to the staff of Taylor & Francis who have made this series possible and especially to the volume editors and their chapter contributors for the authoritative information.

Dr. Roland Hardman

Preface

Among the crops used by humankind, the history of spices is perhaps the most adventurous, the most fascinating, and the most romantic. In the misty distant past, when the primitive man was roaming around the forests in search of food and shelter, he might have tested and tasted many roots and leaves and might have selected those that were aromatic and spicy as of special value and used them to propitiate his primitive gods to save him from the raging storm, thunder, lightning, and rain. Out of the misty darkness of that distant past, the early civilizations blossomed when man settled down and started practicing agriculture. In all civilizations, the aromatic plants were given special status, and many were probably used as offerings to gods. Gradually, man might have started using them for curing various illnesses and, in the course of time, spices and aromatic plants acquired magical associations about their properties. Among all the civilizations, it was in the Indian and Chinese where that profound knowledge gradually evolved in the use of plants and plant products for the treatment of human ailments.

From the dawn of human civilization, spices were sought after as eagerly as gold and precious stones. Discovery of the spice land was one of the aims of all circumnavigations and the great explorations that the period of Renaissance witnessed. One such navigational venture in search of the famed land of spices and ivory reached the ancient Malabar Coast of India on May 20, 1498. Vasco da Gama discovered the sea route to India. The decades that followed witnessed the Portuguese establishment of trade relations with the Malabar Coast and, subsequently, they emerged as powerful players in the game of power politics of the region. The landing of Gama also witnessed the transition from Medieval to the Modern India and the rising of the global imperialism and colonial power struggles. The Portuguese, the Dutch, French, and finally the British established their supremacy over the spices trade in the decades that followed.

Turmeric has been valued as a source of medicine and color in the whole of South Asia, from ancient times. Probably man would have been attracted to this plant due to its attractive color and in due course, it acquired many religious and sociocultural associations. For the ancient people of India, turmeric was the "*Oushadhi* — the medicinal herb," and possibly it might have played a great role in the day-to-day life of ancient Indians as a wound healer, as a medicine for stomach ache, flatulence, poison, etc., for dyeing clothes and yarns, and for worshipping their gods and goddesses. This plant has acquired great importance in the present-day world with its antiaging, anticancer, anti-Alzheimer's, antioxidant, and a variety of other medicinal properties. This volume is the first comprehensive monographic treatment on turmeric. It covers all aspects of turmeric — botany, genetic resources, crop improvement, chemistry, biotechnology, production technology, post harvest and processing, pharmacology, medicinal uses, traditional uses, and its use as a spice and a flavorant. There is also a chapter on related economically important species.

The book comprises 15 chapters, each written by experts in their respective field and each chapter having an exhaustive bibliography. Experts from Japan, India, United States, and United Kingdom collaborated in the production of this monograph.

The editors of this volume have extensive experience in genetic resource collection, conservation, botany, breeding, biotechnology, and agronomy of turmeric. As the former national coordinator for spices research under the Indian Council of Agricultural Research, the senior editor had the opportunity of associating with various turmeric research programs being carried out in India — the only country having a strong research and development program in turmeric. The editors tried to collect and collate as much information as possible about turmeric in this volume and with over 2,156 citations this becomes a very valuable database. This was not an easy task, and the editors would like to place on record their gratitude to the various chapter authors for their tremendous efforts. During the preparation of this volume, Dr. Roland Hardman has been a source of inspiration.

We hope that this book will be invaluable to all those who are involved in the production, processing, marketing, and use of turmeric. It is further hoped that this book will kindle interest in the minds of the readers and will act as a catalyst for more research on the possible uses and value addition of this wonderful gift of nature — turmeric.

The Editors

Acknowledgments

The editors express their deep gratitude to the contributors of various chapters of this monograph who found time to collaborate in its production.

The editors are thankful to Dr. Roland Hardman, the series editor of Medicinal and Aromatic Plants: Industrial Profiles, for giving us the opportunity to edit this volume. As in the case of the earlier volumes on black pepper, cardamom, cinnamon, cassia, and ginger, Dr. Hardman provided us with constant help, encouragement, and updated literature searches.

We are thankful to Dr. Minoo Divakaran for her help at all stages of the preparation of this volume. Ms. Jana Skornickova supplied the photographs included in Chapter 1 (Fig. 1.1), for which we thank her. We wish to express our thanks to Dr. Remashree for contributing the diagrams 2.4 and 2.6, to Mr. Praveen for the photograph 2.7, and to Sudhakaran for the diagrams 2.1 to 2.3 included in Chapter 2, one of which was also used in the cover design. We are also thankful to Nishanth for typing work.

In the preparation of this volume, we made use of the published information from many research workers, of course with full citation. Some of them are not with us now, but their contributions will continue to be remembered and studied by the students, through this book, for many a decade to come. We acknowledge with gratitude the contributions of all turmeric workers and salute them in reverence.

We acknowledge with gratitude the direct and indirect help (including photographs) given by our colleagues and coworkers at the Indian Council of Agricultural Research, Indian Institute of Spices Research, Calicut, All India Coordinated Research Project on Spices and its various centers, and Spices Board, Cochin, in compiling this volume.

The senior editor thanks the management of Centre for Medicinal Plants Research (AVS), Kottakkal and his colleagues for giving unstinted support during the preparation of this monograph. The editors are grateful to P.K. Warier, managing trustee of AVS, for giving permission to dedicate this book to the AVS.

The editors express their sincere gratitude to Taylor & Francis, Boca Raton, Florida, U.S., and its commissioning editor John Sulzycki, for giving them the opportunity to edit this very important monograph on turmeric.

We are indebted to the members of our families whose help, understanding, and consideration sustained us during preparation of this work. We thank our well wishers and all those who helped us in the compilation of this volume.

The Editors

P.N. Ravindran, Ph.D. is one of the foremost spice experts in the world. He was formerly director of the Indian Institute of Spices Research (IISR), ICAR; national coordinator for Spices Research, ICAR; coordinating director of the Centre for Medicinal Plants Research (CMPR), Kottakkal; and principal scientist and head of the Division of Crop Improvement and Biotechnology, IISR, Calicut. The world's largest spices gene bank and the national conservatories for spices were established at the IISR under his leadership during the period of 1984 to 1998. He has published more than 250 research and review papers and edited earlier monographs in this series on MAAP-Industrial Profiles including black pepper, cardamom — the genus *Elettaria*; cinnamon and cassia — the genus *Cinnamomum*; and ginger — the genus *Zingiber.* He also compiled *Advances in Spices Research: History and Achievements of Spices Research in India since Independence* (Agrobios, 2006). In addition, he co-edited *Biotechnology of Horticultural and Plantation Crops* (Malhothra, 2000). He was the founder and founding secretary and later president of the Indian Society for Spices; founder, founding editor and later chief editor of the international journal, *Journal of Spices and Aromatic Crops.* He is at present involved in guiding and monitoring the research activities on medicinal plants at the CMPR. His current interests also include the sociocultural history of medicinal plants and sacred plants used in religious rituals.

K. Nirmal Babu, Ph.D. received his doctorate from the University of Calicut and did post-doctoral work at the University of California, Davis, USA. He has done pioneering work in the areas of genetic resources conservation, including *in vitro* conservation, crop improvement, and biotechnology of spices. He is the co-editor of the MAPP international monographs on the genus *Cinnamomum* and the genus *Zingiber* and also the volume *Advances in Spices Research.* He has published more than 130 research and review papers in the areas of biotechnology, molecular characterization, crop improvement, and germplasm conservation, He is currently working in biotechnology and the molecular biology of major spices at the Division of Crop Improvement and Biotechnology, Indian Institute of Spices Research, Calicut, Kerala, India.

K. Sivaraman, Ph.D. obtained his doctoral degree from the Tamil Nadu Agricultural University, Coimbatore, India where he specialized in turmeric-based cropping systems. In a career spanning 28 years, he has contributed significantly in the field of agronomy and farming system management with specific reference to spices and coconut. He has been involved in the development of agro-technologies for spices at the Indian Institute of Spices Research at Calicut. Later, as director of the Directorate of Arecanut and Spices Development at Calicut under the Ministry of Agriculture, Government of India, he has been involved in implementing and monitoring development programs on spices and medicinal and aromatic plants. Presently he is involved in the study of long-term organic farming with particular reference to sugarcane at Sugarcane Breeding Institute (ICAR), Coimbatore, where he is a principal scientist. He has published more than 60 scientific papers and three books including *Cropping Systems in the Tropics — Principles and Management, Indian Spices and their Utilization,* and *Spices and Herbs in Coconut Based Intensive Farming Systems in India.*

Contributors

K.M. Abdulla Koya
Indian Institute of Spices Research
Calicut, Kerala, India

K.V. Balakrishnan
R & D Divison
Synthite Industrial Chemicals Ltd
Kerala, India

B. Bharat Aggarwal
Cytokine Research Laboratory
Department of Experimental Therapeutics
The University of Texas M.D. Anderson
  Cancer Center
Houston, Texas

Chitra Sundaram
Cytokine Research Laboratory
Department of Experimental Therapeutics
The University of Texas M.D. Anderson
  Cancer Center
Houston, Texas

S. Devasahayam
Indian Institute of Spices Research
Calicut, Kerala, India

N.P. Dohroo
Division of Vegetable Crops
Dr Y.S. Parmer University of Horticulture
  and Forestry
Himachal Pradesh, India

Sethi Gautam
Cytokine Research Laboratory
Department of Experimental Therapeutics
The University of Texas M.D. Anderson
  Cancer Center
Houston, Texas

S.P. Geetha
Centre for Medicinal Plants Research
Kottakkal, Kerala, India

Haruyo Ichikawa
Cytokine Research Laboratory
Department of Experimental Therapeutics
The University of Texas M.D. Anderson
  Cancer Center
Houston, Texas

D. Indra Bhatt
Cytokine Research Laboratory
Department of Experimental Therapeutics
The University of Texas M. D. Anderson
  Cancer Center
Houston, Texas

Seok Ahn Kwang
Cytokine Research Laboratory
Department of Experimental Therapeutics
The University of Texas M.D. Anderson
  Cancer Center
Houston, Texas

Lutfun Nahar
Londonderry, Northern Ireland
United Kingdom

M.S. Madan
Indian Institute of Spices Research
Calicut, Kerala, India

D. Minoo
Indian Institute of Spices Research
Calicut, Kerala, India

Seeram Navindra
UCLA Center for Human Nutrition
David Geffen School of Medicine
Los Angeles, California

K. Nirmal Babu
Indian Institute of Spices Research
Calicut, Kerala, India

K.V. Peter
Kerala Agricultural University
Vellanikkara, Kerala, India

K. Praveen
Indian Institute of Spices Research
Calicut, Kerala, India

K.S. Premavalli
Division of Food Preservation
Defence Food Research Laboratory
Mysore, Karnataka, India

P.N. Ravindran
Centre for Medicinal Plants Research
Kottakkal, Kerala, India

T. Rehse
Department of Biology
Duke University
Durham, North Carolina
and
Department of Botany
University of Calicut
Kerala, India

R. Remadevi
Department of Dravyaguna Vigyana
Vidyaratnam P.S. Varier Ayurveda College
Kottakkal, Kerala, India

M. Sabu
Department of Botany
University of Calicut
Kerala, India

Satyajit D. Sarker
School of Biomedical Sciences
University of Ulster at Coleraine
Londonderry, Northern Ireland
United Kingdom

Santosh K. Sandur
Cytokine Research Laboratory
Department of Experimental Therapeutics
The University of Texas M.D. Anderson
  Cancer Center
Houston, Texas

Shishir Shishodia
Department of Biology
Texas Southern University
Houston, Texas

K.N. Shiva
Indian Institute of Spices Research
Calicut, Kerala, India

K. Sivaraman
Sugarcane Breeding Institute
Coimbatore, Tamil Nadu, India

J. Skornickova
Department of Botany
University of Calicut
Kerala, India
and
The Herbarium
Singapore Botanic Gardens
National Park Road
Singapore, India

V. Sumathi
Indian Institute of Spices Research
Calicut, Kerala, India

E. Surendran
Department of Kaya Chikitsa
Vaidyaratnam P.S. Varier Ayurveda College
Kottakkal, Kerala, India

Kimura Takeatsu
Department of Kampo Pharmaceutical Science
Nihon Pharmaceutical University
Saitama, Japan

P.A. Valsala
College of Horticulture
Kerala Agricultural University
Vellanikkara, Kerala, India

Contents

1 Turmeric — The Golden Spice of Life

P.N. Ravindran

CONTENTS

1.1 INTRODUCTION

Turmeric is known as the "golden spice" as well as the "spice of life." It has been used in India as a medicinal plant, and held sacred from time immemorial. Turmeric has strong associations with the sociocultural life of the people of the Indian subcontinent. This "earthy herb of the Sun" with the orange-yellow rhizome was regarded as the "herb of the Sun" by the people of the Vedic period. No wonder the ancients regarded turmeric as the *Oushadhi*, the healing herb, the most outstanding herb, the one herb above all others (Jager, 1997). Turmeric has at least 6000 yr of documented history of its use as medicine and in many socioreligious practices.

Turmeric is probably a native of Southeast Asia, where many related species of *Curcuma* occur wildly, though turmeric itself is not known to occur in the wild. Turmeric is cultivated most extensively in India, followed by Bangladesh, China, Thailand, Cambodia, Malaysia, Indonesia, and Philippines. On a small scale, it is also grown in most tropical regions in Africa, America, and Pacific Ocean Islands. India is the largest producer, consumer, and exporter of turmeric.

The name turmeric has originated from the Medieval Latin name *terramerita,* which became *terre merite* of French, meaning deserved earth or meritorious earth, a name by which powdered turmeric was known in commerce. Ancient Indians had given many names for turmeric, each one denoting a particular quality as listed below;

Ranjani	Denotes that which gives color
Mangal prada	Bringing luck
Krimighni	Killing worms, antimicrobial

Mahaghni	Indicates antidiabetic properties
Anestha	Not offered for sacrifice or *homa*
Haridra	Indicating that it is dear to *Hari* (Lord Krishna)
Varna-datri	That gives color, indicating its use as enhancer of body complexion
Hemaragi	Having golden color
Bhadra	Denotes auspicious or lucky
Pavitra	Holy
Hridayavilasini	Giving delight to heart, charming
Shobhna	Brilliant, indicating the brilliant color

In the Sanskrit language, turmeric has about 55 synonyms that are associated with its religious or medicinal uses. In English, turmeric was also known as yellow root and Indian saffron. The names of turmeric and related species in a few languages are given in Table 1.1.

Herbal experts consider turmeric as one of the greatest gifts of Mother Nature because it is endowed with a variety of curative properties. In all South Asian countries, turmeric has been in use from ancient time as a spice, food preservative, coloring agent, and cosmetic and in the traditional systems of medicine (*Ayurveda, Sidha, Unani,* and *Tibetan*). In the past, turmeric as well as its wild relatives such as *C. aromatica* (*Vanaharidra* — Jungle turmeric), was used to dye clothes (cotton, silk, and wool), though the color degrades rapidly in presence of sunlight. In modern times, the coloring matter of turmeric (curcumin) is used as a safe food color in cheese, spices, mustard, cereal products, pickles, potato flakes, soups, ice creams, yogurt, etc. Studies have indicated that curcumin is nontoxic to humans even at a dose of 8000 mg/d taken continuously (Cheng, 2001). The medicinal uses of turmeric and curcumin are indeed diverse, ranging from cosmetic face cream to the prevention of Alzheimer's disease. Turmeric is also qualified as the queen of natural Cox-2 inhibitors (Duke, 2003). Recent researches on turmeric are focused on its antioxidant, hepatoprotective, anti-inflammatory, anticarcinogenic, and antimicrobial properties, in addition to its use in cardiovascular and gastrointestinal disorders.

1.2 AREA, PRODUCTION, AND TRADE

India is the world's largest producer, consumer, and exporter of turmeric. The annual production is about 635,950 t from an area of 175,190 ha (2002 to 2003). This is about 5.5 and 20.6% of area and production of spices in India, respectively (2002 to 2003). In the last 30 yr, the area, production, and productivity of turmeric exhibited an increasing trend, and the production has moved up at an annual growth rate of 7.6% and area at 2.8%. The productivity also doubled during this period.

Indian turmeric is regarded as the best in the world market because of its high curcumin content. All over the world, the Aleppey Finger Turmeric with over 6% curcumin is preferred for curcumin extraction. India is the major exporter of turmeric, exporting to over 100 countries throughout the world. UAE is the major importer of turmeric from India, followed by the U.S., Japan, U.K., Iran, Singapore, Sri Lanka, and South Africa. Turmeric export registered a growth rate of 7.56, 17.19, and 8.95% in terms of quantity, value, and unit value, respectively, during the period 1981–1982 to 2000–2001. The International Trade Centre, Geneva, has estimated an annual growth rate of 10% in the world demand for turmeric.

In India, the state of Andhra Pradesh is the largest producer of turmeric, with an area of 64,100 ha and a production of 346,400 t. The other major producing states are Tamil Nadu, Orissa, Karnataka, West Bengal, and Maharashtra. The productivity of turmeric in Andhra Pradesh is estimated at 5404 kg/ha, which is 33% higher than the national average. The names of major turmeric production centers and areas in India are given in Table 1.2.

TABLE 1.1
List of Common Names of Turmeric and Some of the Related Species of *Curcuma*

Botanical Name	Languages	Common Names
Curcuma longa	Bengali	Haldi, pitras
(turmeric)	Burmese	Hansanwen, sanae, tarum
	Cambodia	Banley, pauly, romiet
	Cantonese	Wong kewng, yuet kau
	Chinese	Chiang husang, kiang husang, yuchin
	Dualao	Quinamboy
	English	Turmeric, Indian saffron
	French	*Curcuma*, Saffron de India, sochet des Indes, souchet, souchet long, souchet odorant, teri-merit
	German	Glbwurzel, kurkuma
	Gujarati	Halada
	Hamsa	Ganjamau
	Hebrew	Kurkum
	Hindi	Haldi
	Ilocano	Culiago, cuming
	Italian	*Curcuma*
	Java	Kumir, kuing, warangan, koeneng temen, kunyit, kunir bentis
	Kannada	Arishina
	Konkani	Halad, ollod, ollodi
	Malacca	Kunyit
	Malagasy	Tamo tamo
	Malaya	Watkam, wang keong, kunyit
	Malayalam	Mannal, manjal, manjalkua
	Marathi	Haled
	Modjakerto	Kumirbantis
	Mundari	Hatusassang
	Oriya	Haldi
	Panpangan	Angari, culalo
	Persian	Darzardi, zharachobabi, tardhubah
	Portugese	Acafrao da india
	Punjabi	Haldar, halja
	Sanskrit	Ameshta, bahula, bhadra, dhirgharaja, gandaplashika, gauri, gharshani, haldi, haridra, harita, hemaragi, hemaragini, hridvilasini, jayanti, jwarantika, kanchani, kaveri, krimighna, kshamada, kshapa, lakshmi, mangalaprada, mangalya, mehagni, nisha, nishakhya, nishawa, pavitra, pinga, pinja, pita, patavaluka, pitika, rabhangavasa, ranjani, ratrimanika, shifa, shiva, shobhana, shyama, soughagouhaya, suvarna, suvarnavarna, tamasini, umavara, vauragi, varavarnini, varnadatri, varnini, vishagni, yamini, yohitapriya, yuvati
	Sinhalese	Kaha
	Spanish	*Curcuma*
	Tagalog	Dialo
	Tamil	Manjal
	Telugu	Pasupu, pampi
	Visayan	Calanag, calavaga
	Zambales	Lisangoy
Curcuma amada	Arabic	Daruhaldi
	Bengali	Amada
	English	Mango ginger
	Gujarati	Ambahaldi

Continued

TABLE 1.1 *(Continued)*
List of Common Names of Turmeric and Some of the Related Species of *Curcuma*

Botanical Name	Languages	Common Names
	Konkani	Ambahaldi
	Malayalam	Mangayincchi
	Marathi	Ambehaldi
	Naguri	Bundu sasang
	Persian	Darchula
	Tamil	Mangaincchi
	Telugu	Mamidiallam
	Urdu	Ambahaldi
Curcuma	English	East Indian arrowroot, narrow-leaved turmeric, wild arrowroot
angustifolia	Marathi	Tikkar, tavakhira, tavakula
	Kannada	Kovehittu, kuve gidda
	Gujarati	Tavakhara
	Hindi	Tavakhira, tikhur
	Persian	Thavakhira
	Sanskrit	Gavayodhbhava, tawakshira, godhumaja, pavakshira, pisthika, yavaja, talakshira, talasambuta
Curcuma	Arabic	White jadvar
aromatica	Bengali	Benhaldi
	Marathi	Ambehalad, ranhald, sholi
	Mundari	Hattumundu sasang, mundu sasang
	Burmese	Kiasanvin
	Kannada	Kasturi arishina
	English	Cochin turmeric, wild turmeric
	Gujarati	Kapur katchali, van halda
	Hasada	Birsang
	Hindi	Banhaldi, banharidra, jangali haldi
	Konkani	Ran halad
	Malayalam	Anakkuva, kattumanjal
	Naguri	Bandusasang
	Portugese	Zedoaria amaralla
	Sadani	Bon haldi
	Sanskrit	Aranya haridra, sholi, sholikka, vana haridra, vana ristha
	Sinhalese	Dudda kaha, wal kaha
	Tamil	Kasturi manjal, kattumanjal
	Telugu	Kasturi pasupu, tella kasturi pasupu
Curcuma	Bengali	Ekangi, kathura, sati, savi
zedoaria	Burmese	Thannuvan
	Cambodia	Prateal, vong preah a tit
	Dutch	Rhonde zedoar
	English	Zedoary
	French	Zedoare, zedoaire buylpeaux, zedoire
	German	Zedoarwurgil, zittuver
	Gujarati	Kacchuri
	Hindi	Kaucchura, kali haldi
	Italian	Zedoac
	Kannada	Kachura
	Java	Tamoclama
	Malayalam	Kacholam, kacchuri kizhangu, pulakizhangu, manjakkuva

Continued

TABLE 1.1 *(Continued)*
List of Common Names of Turmeric and Some of the Related Species of *Curcuma*

Botanical Name	Languages	Common Names
	Marathi	Kacchura, kacchuri, narkacchuura
	Persian	Kazur, urkelkafur
	Urdu	Kacchura
	Portugese	Zedoaria
	Russian	Tzitvar
	Telugu	Katchili gaddalu, kachhooram
	Tamil	Katchili kilangu, pulakilangu
	Sanskrit	Dravida, durlabha, gandhamulaka
	Sinhalese	Haru kaha

Source: From Velayudhan et al. (1999). www.plantnames.unimelb.edu.au/sorting/curcuma.html, 2005

TABLE 1.2
Area and Production of Turmeric in the Major Turmeric Producing Regions in India

State	Area (ha)	Production (t)	Productivity (kg/ha)
Andhra Pradesh	64,100	346,400	5,404
Karnataka	6,800	28,900	4,250
Kerala	3,700	7,900	2,135
Maharashtra	6,900	8,700	1,261
Orissa	28,100	61,200	2,178
Tamil Nadu	28,300	150,000	5,300
West Bengal	13,300	20,000	1,504

1.3 HISTORY

The earliest reference about turmeric can be seen in *Atharvaveda* (Ca. 6000 yr B.P.), in which turmeric is prescribed to charm away jaundice. It was also prescribed in the treatment of leprosy. Reference to turmeric has also been made in the *Yajnavalkyasamhita* (composed, Ca. 4000 yr B.P.) at the time of the epic Ramayana. Turmeric was listed as a coloring plant in an Assyrian herbal dating about 2600 yr B.P. Marco Polo, in 1280 A.D., mentioned turmeric as growing in the Fokien region of China (Rosengarten, 1969). Evidences indicate that turmeric was under cultivation in India from ancient times, but whether the turmeric that they used was *C. longa* or some other species having yellow rhizome, is not known. Garcia de Orta (1563) described turmeric under the name *Crocus indicus*. Fluckiger and Hanbury (1879) wrote "several varieties of turmeric, distinguished by the names of the countries or districts in which they are produced are found in the English market; although they present differences that are sufficiently appreciable to the eye of the experienced dealer, the characters of each sort are scarcely so marked or so constant as to be recognizable by mere verbal descriptions." Linschoten (1596), while describing with the utmost details the trade in Cochin makes no mention of turmeric.

Sopher (1964) states that "the distribution and uses of turmeric in domestic sites outside India, especially in Celebes, the Moluccas, and Polynesia indicates their antiquity and suggests an early cultural connection between the people of these areas and the indigenous pre-Aryan cultivators of

India ... the indigenous use of turmeric in magical rites intended to produce fertility then became an important part of the established Hindu ceremony and as such was taken to the Hinduized Kingdoms of Southeast Asia." The use of turmeric by the Betsiko people of the Malagasy Republic suggests that the introduction was of Malayo-Polynesian origin (Purseglove et al., 1981). Burkill (1936), quoting Hammerstein, states that turmeric reached East Africa in the eighth century and West Africa in the 13th century; it is used in the latter area only as a dye. It was introduced into Jamaica by Edwards in 1783, where it has become naturalized (Purseglove et al., 1981).

The original home of turmeric is shrouded in mystery. Although several species of *Curcuma* are natives of India, some of which might have been used as *haridra* (turmeric), there is little evidence to indicate that *C. longa* is a native of India. All the earlier writers speak of turmeric only as cultivated (e.g., Roxburgh, 1810; Watt, 1872). Ainslie (cited by Watt, 1872) remarked, "*Curcuma longa* grows wild in Cochin China, and is there called as *Kuong huynh*. Loureiro gives us a long list of the medicinal virtues of turmeric in lepra, jaundice, and other disorders." Watt (1872) remarked, "Although there are Sanskrit names for the plant and also names for it in most of the languages of India, the suggestion may be offered that it is most probably a Chinese or Cochin Chinese species, which may have superscribed some of the indigenous curcumas formerly in use and which bore the names now given to this plant, just as the true arrowroot plant in rapidly displacing the indigenous East India Species." Darzell and Gibson (Watt, 1872) have treated this plant as introduced into Bombay. Purseglove et al. (1981) stated that turmeric was domesticated in Southeast Asia, where it might have originated as a triploid hybrid.

In all probability, it seems that the true turmeric (*C. longa*) came to India from the ancient regions of Cochin China (present day Vietnam) or China either through the movement of the ancient tribal people during their migration to the Northeast region of India, or through the Buddhist monks and ancient travellers who reached India during the post-Buddha era. To the ancients turmeric was not a spice but was a dye and a remedy for many ailments. The travellers might have been carrying turmeric rhizomes as a remedy for two of the most common ailments that they were usually subjected to — wounds and stomach troubles. Gradually, turmeric became popular in India and in course of time replaced the indigenous types that were in use. It possibly might have been introduced into cooking for preservation of food products and subsequently used to impart color to the dishes. For such uses, the other *Curcuma* species might not have been preferred due to their very bitter taste. Taste, color, and medicinal property all merged in *C. longa*, which in due course acquired magical associations. In course of time, turmeric became associated with many traditions and myths in the centuries that followed. Sopher (1964) writes, "the wild *Curcuma* from which *Curcuma domestica* evolved may first have attracted attention as an incidental source of food, but the important property that became the object of conscious selection was the yellowish color of turmeric. As a quickly growing plant with a strikingly colored rhizome, turmeric acquired magical properties, some apparently associated with the fertility of the earth ... Attitudes and practices expressing these ideas would be disseminated together with the human dispersal of the plant over a wide area."

The earliest account of cultivation of turmeric is that of Roxburgh (1810). The Agricultural department of Bengal published information on the system of cultivation (Ridley, 1912). The system of cultivation and processing after harvest has not changed much over the countries.

1.4 TURMERIC AS A DYE PLANT

The yellow color of turmeric is due to the presence of a group of compounds known as curcuminoids, of which the most important one is curcumin. Earlier in India, turmeric was used as a dye. Many studies were carried out on the dye properties of turmeric by the British, with an objective of using it as a source of commercial dye. Buck (cited by Watts, 1872) observed that the dye given by turmeric is a dull yellow color, it is fleeting, and except in dyeing the commoner sort of cloth, seldom used, except in combination. The action of alkali on turmeric changes its color to red. In spite of the many efforts by the European dye experts, the color could not be rendered permanent. For color extraction, hard rhizomes were preferred. The dye was employed mainly in coloring cotton fabrics and in calico printing. Alum was used to purify the color and to destroy all shades of red. The dyers of Calcutta produced a brilliant yellow, known as *basanti rang* by mixing turmeric with carbonate of soda and lemon or limejuice. Turmeric was used to produce green shades in combination with indigo. The fabric was first dyed with indigo and then dipped in a solution of turmeric. Turmeric was also used to sharpen or brighten other colors, such as the *Singahar* (*Nyctanthes arbortristis*), lac dye, *Al* (*Morinda tinctoria*), safflower (*Carthamus tinctorius*), and *toon* (*Cedrela toona*).

The Indian calico printers used to prepare a dye by mixing turmeric with the rind of pomegranate and alum. Turmeric was used for dyeing wool and silk for the production of compound shades such as olives and browns. With mordants, turmeric gives other shades. If the wool is mordanted with aluminium or tin, the color is more brilliant, and with the latter becomes more orange. Potassium dichromate and ferrous sulphate mordants produce olive and brown, respectively. Curcumin forms a complex with boric acid — borocurcumin — and from this complex, rosocyanin, a coloring pigment can be obtained.

Presently, turmeric is not used for any commercial coloring. However, both turmeric and curcumin are used widely for imparting yellow color to a variety of food products.

1.5 SOCIOCULTURAL ASSOCIATIONS WITH TURMERIC

The ancients, while searching for food and shelter might have been attracted to this plant because of its deep orange color and its potential as a wound-healing agent, which might have been a boon to them. Gradually they might have used it because of its amazing effect in curing their stomach problems and a variety of other illnesses. Eventually, magical properties might have been associated with this plant, which gradually assumed great importance in their religious and sociocultural traditions. As an auspicious item, turmeric became associated with many Hindu customs and traditions. Recently Remadevi and Ravindran (2005) summarized the myths and traditions associated with turmeric, on which the following discussion is based.

The Hindus, both tribal and civilized, consider turmeric as sacred and auspicious. It is associated with several rituals from ancient period and the tradition still goes on. In some tribal communities in Tamil Nadu, Andhra Pradesh, and in the Northeast regions of India, a piece of turmeric tied to a thread, dyed yellow with turmeric powder, is used as the nuptial string (*mangalsutra*). Even now, in the village and urban Hindu community, this practice is very much prevalent; rich people use gold chain also along with the natural yellow rhizome, but the poor depend on turmeric alone. Turmeric is also used as an amulet and a piece of turmeric tied on the hand is believed to keep away evil spirits.

Turmeric is used to give yellow coloration to clothes, threads, etc. In olden days, in Kerala, the new clothes given to children during Onam festival were yellow colored, the exact reason for which it is unknown, but it might be due to the attractive power of yellow color and its association with Lord Krishna, who was *Peethamber dhari* (wearing only yellow clothes) and with goddess Durga. Narayana Guru, a social revolutionary and the uplifter of the Thiyya (Ezhava) community of Kerala, selected yellow as the color of his dress and made it the color of the order of *Sanyasins* that he established. Earlier, such clothes were prepared by dying them with turmeric. It is not the antiseptic property or the medicinal property that made people go for such practices; one probable reason might have been the association of turmeric with the goddess Durga and snake worship. Turmeric is also associated with many after-death rites. The dead body of married women or the woman who performs *Sati* (widow burning herself in the funeral pyre of the husband was a practice prevalent among certain tribes in India during the Medieval times) is taken to the funeral pyre and clothed in a robe dyed with turmeric. In Tamil Nadu, application of turmeric over a dead body is a custom and is intended for cleansing.

Stevens (1920) described the use of turmeric in marriage customs of the Tribe *Dandasis* (or *Sansis*) of Punjab. The headman or a respected elder of the community places a nut-cutter with rice and an areca nut between the united hands of the couple and ties their hands with seven turns of a thread already dyed with turmeric. The parents of the bride and the bridegroom pour turmeric water from a conch shell or from a leaf over their hands seven times, thereby concluding the marriage ceremony. In the *Tareya* tribe of Tamil Nadu, turmeric is smeared on the doors of all the households invited for a marriage ceremony. Fawcett (1892) lists the items that are to be taken in ceremonial procession when a Basian (a tribal community in Bellary district of Karnataka) girl gets married. The list contains turmeric together with rice, coconuts, betel leaves, betel nuts, banana, gold *tali*, silver bangles, and toe rings.

Dubois (1806) mentions, "The nuptial bath called *Nalangu* is accompanied by making the bride and groom sit on a raised platform facing east. Married women then rub their heads with sesame oil followed by smearing the exposed parts of their bodies with powdered *kumkum* (turmeric powder mixed with lime)." The *Dharmasutras*, which originated during the Buddhist era (600 B.C. to 500 A.D.), also describes the marriage ceremonies during that period in northern India. Marriage ceremonies of those days were divided into five parts. In the first part, known as the *Mandapam* ceremony, the important ritual is the *gatraharidra*, which is smearing the body with turmeric paste to generate sex desire in the new couples.

Turmeric is also associated intimately with many other day-to-day customs of the people of India. There is a custom related to the "ceremonial bath" of brides on the fourth day of marriage and the girls on the fourth day of their first menstruation, when their bodies are smeared with turmeric paste in oil and then they are bathed in turmeric water first and then in ordinary water. Among the Tamil, Kannada, and Telugu communities, women put yellow markings with turmeric paste on their forehead, cheeks, and neck as a custom on auspicious days. Dymock et al. (1867) quote a well known *Vaidya* (traditional medical practitioner) from Bombay, that in former times, married women used to apply turmeric daily in the evening; a practice still being followed in many villages. After finishing the household chores, married women dip their hands in turmeric water and pass them lightly over their cheeks. The mistress of the house performs the same for other married friends who visit at that time. The guest is detained until the lamps are lighted. They believe that goddess *Lakshmy* (goddess of health and prosperity) may visit the house during that time.

Even now, among many village folks in the rural community of India, during marriage ceremonies, five married women or five virgins anoint the bride with turmeric and oil. The bride puts on a robe dyed with turmeric, until the day of marriage. Turmeric and oil are sent to the bridegroom, and the bridegroom applies turmeric on his body. The marriage contract is stained or spotted with turmeric. A portion of the wall is daubed with turmeric and small lines of *kumkum* (a mixture of turmeric and lime). The bridegroom ties a thread around the bride's wrist, to which is attached a piece of turmeric and betel nut. At the end of the ceremonies, the bridal party plays with turmeric water, dashing it over one another. In the Karnataka region, turmeric is reported to be used for external application all over the body of the bride and bridegroom on the previous day of marriage for the fertility of the future couple.

Until the first half of the 20th century, the womenfolk in the wealthy families of Kerala used to smear their bodies with a mixture of turmeric and sandal paste almost daily before taking bath. Turmeric is perhaps thus the first ever known cosmetic. This knowledge spread to Rome (from the Malabar Coast with which Rome was maintaining an active trade relationship), and eventually turmeric was exported to Rome along with other spices. It soon became the most important beauty aid for the Roman women.

In some communities, the association with turmeric starts at birth. In such cases, the newborn, immediately after birth, is given a mixture of turmeric and coconut milk (yellow milk) as the first intake. Turmeric powder mixed with water was used during the occasion of adoption, in olden days. Turmeric water is of great importance in many occasions. During the worship of goddess *Kali*, ablutions with turmeric water are considered sacred.

Turmeric has gone so deep into the cultural heritage of the Indian community that a separate icon, known as *haridra ganesha* (Turmeric *Ganesha*) came into existence. In such worship, people (especially the tribal and rural people of Tamil Nadu and Andhra Pradesh) in some occasions worship *Ganesha* in the form of turmeric rhizome, which indicates how strong the association is between turmeric and the people of the region. Similarly, Telugu people make a cone with turmeric powder and worship it as goddess *Gowrie* during the *Dasara* festival. They throw rice or jowar (*Sorghum* grains) colored yellow with turmeric to bless the newly wed couples.

Turmeric is treated as *Mangalaprada* — which brings prosperity. The term *Mangala* also is a synonym of turmeric (in Sanskrit), which means auspicious. *Mangala devata* is goddess Sree Parvathy. Because of the above-mentioned associations, turmeric is regarded as dear to goddess Lakshmy and Parvathy. This may be the reason for considering turmeric as an essential substance in marriage ceremonies and other auspicious occasions.

In Durga (a form of goddess Parvathy) temples, a mixture of turmeric and lime is commonly used. In *Kali* (or *Durga*) worship, *gurusi* (or *kuruthi* — a form of worship in which in olden times animal sacrifice was in vogue) is practiced using turmeric water mixed with lime, which gives a deep red color similar to blood. The turmeric–limewater mix is used daily in most *Durga* temples during worship. The dark red liquid may also symbolize the victory of good over evil, and the myth of *Durga* slaying the evil demon king Darika.

There is a school of thought that the auspicious or sacred association attributed to turmeric is the direct outcome of Sun worship in one form or the other. The idea of festivity connected with the color yellow, through its association with the Sun, has given turmeric an erotic significance. This is another reason why yellow is the chief color at weddings and in the relation between sexes. Apart from the custom of smearing the body with turmeric at weddings, garments dyed or marked at the corners with turmeric color are considered lucky and to possess protective powers.

Turmeric is mentioned in Ramayana as one of the eight ingredients of *Arghya*, a respectful oblation made to gods and venerable men. All the religious customary practices in the temples are associated with turmeric and turmeric powder. In rituals like *Homa* and *Pooja*, as well as in snake worship, elaborate *tantric* designs (*kalam*) are prepared with turmeric powder, rice powder, etc. *Kalams* are intricate designs of gods, goddesses, serpents, or demons, prepared by using natural dyes (Figure 1.1). Turmeric powder is used for yellow color; turmeric mixed with lime is used for creating the red color. Such *kalams* are, in fact, beautiful creations of art.

Turmeric has played an important role in curing disease too. When smallpox and chickenpox were prevalent, turmeric powder and neem leaves were the antiseptics used for the healing of eruptions on the skin. Before taking a bath, the patients were invariably smeared with turmeric paste and then bathed in water boiled with neem leaves. Ash of turmeric rhizome was applied to the skin for healing the eruptions. In such practices, it was not only the antiseptic property that was made use of, but also the strong belief of the association of goddess *Durga* with the disease such as smallpox and chickenpox.

1.6 IMPORTANT RELATED SPECIES

There are many species of *Curcuma*, which are related to turmeric and are economically useful, mostly as herbal or tribal medicine. The main ones are the following:

1.6.1 C. AMADA ROXB. (MANGO GINGER)

C. amada Roxb is endemic to South Asia, which is found wild in many parts of Northeast and in the hills of South India. Cultivated for its edible rhizome, it has the flavor of green mango and is

FIGURE 1.1 Elaborate *tantric* designs (*Kalam*) prepared with turmeric, rice powder, etc. during religious ceremonies (a) depicting snake gods and (b) depicting goddess Kali.

used as a vegetable. It is also used in preparation of pickles, chutney, etc., and in traditional and tribal medicines.

1.6.2 C. *ANGUSTIFOLIA* ROXB. (WILD OR INDIAN ARROWROOT)

C. angustifolia Roxb is a native of India found wild in Bengal, northeast regions, and western costal plains and hills. It is also abundant in Madhya Pradesh, Chattisgarh, Orissa, and Andhra Pradesh, and in the hills of Tamil Nadu and Kerala. It was an article of trade in the past. The tribals of Orissa and northeastern India cultivate this species as a source of arrowroot starch, which is used as food and in the Indian system of medicines. The flour, when boiled with milk, forms an excellent diet for patients and children. The tribal people extract the arrowroot flour by grating the rhizomes on a stone. The liberated flour is washed down to a vessel of water where it sediments. The flour is washed repeatedly with water and then dried in the sun. This flour was an article of export in the past; in 1869 to 1870, 3729 cwt of the flour valued at Rs. 14,152 was exported from Madras; and from Travancore (the southern part of present day Kerala state), the average annual export of this arrowroot was around 250 candies (1 candy = 200 kg) (Watt, 1872).

1.6.3   *C. AROMATICA* SALISB.

C. aromatica is the *vanharidra* mentioned in the classical Ayurvedic texts. It is found distributed from China southwards to Sri Lanka. Ainslie (quoted by Watt, 1872) recorded, "the native women prize it much ... used externally, which gives a particular lively tinge to their naturally dark complexions, and a delicious fragrance to their whole frame." *C. aromatica* holds an important place in native perfumery and cosmetics. It grows wild in many parts of India, and is cultivated in Andhra Pradesh and Orissa. Fresh rhizome is yellowish and gives out a strong camphoraceous smell. Dymock (quoted by Watt, 1872) reports, "like turmeric, its principal use is as a dyeing agent." Rhizome was valued as medicine, being regarded as a tonic and carminative, and was a valued toiletry and cosmetic item.

1.6.4   *C. CAESIA* ROXB. (BLACK TURMERIC, BLACK ZEDOARY)

C. caesia Roxb. is a native of northeast India extending up to the present day Bastar region of Chattisgarh. The rhizome has a deep bluish black or grayish black color. It is used in native medicines.

1.6.5   *C. ZEDOARIA* (CHRISTM.) ROSC. (THE LONG AND THE ROUND ZEDOARY)

This species occurs mainly in the northeastern and west coastal regions of India, extending to the hills. It has also been cultivated in earlier times for extracting arrowroot powder and for the production of *Abir* (a dye used by Hindus during the *Holi* festival). Rhizome is pale yellow or colorless. It is ground to a powder, which is purified by repeated washing and dried. Dried powder is mixed with a decoction of sappan wood (*Cesalpinia sappan*) when the red color is obtained. However, this old practice has long been discontinued with the advent of the aniline dyes in the market. It is used in traditional and local medicines as a stimulant and carminative. This was reported to have cosmetic properties.

Species such as *Curcuma aeruginosa*, *Curcuma caulina*, *Curcuma leucorrhiza*, *Curcuma pseudomontana*, and *Curcuma rubescens* are also used as sources of arrowroot powder and in local and tribal medicines. *Curcuma zanthorrhiza* was in use earlier as a dye, and now as a cosmetic, often as a substitute for *C. aromatica*.

1.7   R&D EFFORTS IN TURMERIC

Perhaps, India is the only country where there is a strong R&D base for turmeric. However, research on turmeric, especially in the area of pharmacology, is being pursued by many workers in many countries (U.S., U.K., France, Japan, Thailand, etc.).

The first ever research on turmeric in India was initiated at Udayagiri in Orissa in 1944 under the Imperial Council for Agriculture Research. However, organized research programs were initiated in independent India during the first Five-Year Plan. Based on a recommendation of the Spices Enquiry Committee (1953), turmeric research was started in Kandaghat (Punjab), Targaon (Maharashtra), and Thodupuzha and Ambalavayal (Kerala). A scheme for turmeric research was initiated in 1955 at Andhra Pradesh (at Peddapalem). However, the real impetus for turmeric research was received with the organization of the All India Coordinated Spices and Cashew Improvement Project. In 1975, research programs were started in two centers, Coimbatore (Tamil Nadu Agriculture University — TNAU) and Pottangi (Orissa University of Agricultural and Technology, High Altitude Research Station).

From 1975 onward, work in the areas of germplasm collection, evaluation, conservation, and development of agrotechnology have been taken up. Jagtial, under the Andhra Pradesh Agricultural University became another important center for turmeric research. The conservation and evaluation work was taken up by the All India Cashew and Spices Improvement Project at Central Plantation Crops Research Institute (CPCRI), Kasaragod where its headquarters was located. It was also at this time, Central Food Technology Research Institute (CFTRI) in Mysore started some research programs in the areas of postharvest technology and quality studies in turmeric. Subsequently improved curing methods and technology for processing, oleoresin extraction, and extraction of curcumin were developed. Later, the Indian Council of Agricultural Research felt the need for intensifying research on spices and established a Regional Station of the CPCRI at Calicut, Kerala. This center became the National Research Centre for Spices in 1986, and then Indian Institute of Spices Research in 1995. Research programs on turmeric, such as germplasm collection, conservation, evaluation, varietal improvement, crop production, and crop protection, were initiated in the research center at Calicut (Kozhikode). In 1986, the original coordinated project was split into independent projects for cashew and spices, and the new All India Coordinated Research Project on Spices (AICRPS) came into existence with headquarters at the National Research Centre for Spices at Calicut. Under the AICRPS, turmeric research was initiated at the Agricultural Universities at Orissa (Pottangi), Himachal Pradesh (Solan), Uttar Pradesh (Kumarganj), Bihar (Dholi), Chattisgarh (Raigarh), and West Bengal (Pundibari).

The biological effects of turmeric and its coloring matter, curcumin (curcuminoids), are being researched upon during the past 50 yr or so, and many research publications came out (for reviews: Ammon and Wahl, 1991; Chattopadhyay et al., 2004; Joe et al., 2004; Khanna, 1999, etc. and the other chapters in this volume). A wide spectrum of biological actions has been reported for turmeric and curcumin.

The present monograph is the first attempt to collect and collate the information so far generated on this golden spice and medicinal plant. All the aspects, botany, chemistry, agronomy, plant protection, post harvest technology, pharmacology, uses, etc., are dealt in detail in this volume in separate chapters. There are also chapters on the role of curcumin in modern medicine, and the importance of turmeric in the traditional medicines. Scientists from Japan, India, U.K., and U.S. collaborated in the production of this first comprehensive monograph on turmeric, the golden spice and the spice of life.

REFERENCES

Ainslie. *Materia Indica* (Quoted by Watt, 1872).

Ammon, H.P.T. and Wahl, M.A. (1991) Pharmacology of *Curcuma longa*. *Planta Med* 57:1–7.

Anonymous. (2005) Sorting Curcuma names. www.plantnames.unimelb.edu.au/sorting/curcuma.html, accessed on 29 Dec, 2005.

Burkill, T.H. (1936) *A Dictionary of Economic Products of the Malay Peninsula.* Kualalampur: Ministry of Agri. & Cooperation (Reprint).

Chattopadhyay, I., Biswas, K., Bandyopadhyay, U. and Banerjee, R.K. (2004) Turmeric and Curcumin: biological actions and medicinal applications. *Curr. Sci.,* 87:44–53.

Cheng, A.L. (2001) Phase I clinical trial of Curcumin, a chemopreventive agent, in patients with high-risk or pre-malignant lesions. *Anticancer Res.,* 21(4B):2895–2900.

Darzell and Gibson (Cited from Watt, 1872)

Dubois, J.A. (1806) *Hindu Manners, Customs, and Ceremonies.* Oxford: Clarendon Press, (Translated at Ed. H.K.Beauchamp).

Duke, J.A. (2003) *CRC Handbook of Medicinal Spices.* Boca Raton, FL: CRC Press.

Dymock, W., Warden, C.J.H. and Hooper, D. (1867) *Phamacographia Indica.* Part. VI. Bombay: Educational Society's Press, Byculla, (Reprint).

Fawcett (1892) Quoted from Mahindru (1982).

Fluckiger, E.A. and Hanbury, D. (1879) *Pharmacographia: A History of the Principal Drugs of Vegetable Origin met within Great Britain and British India.* London: Macmillan.

Garcia de Orta (1563) Cited from Watt (1872).

Jager, P. de (1997) *Turmeric.* Vidyasagar Pub., California, USA. p. 67.

Khanna, N.M. (1999) Turmeric–nature's precious gift. *Curr. Sci.,* 76:1351–1356.

Linschoten J.H. Van. (1596) *Voyage of John Huygen Van Linschoten in India.* Vol II. (Quoted from Watt, 1872).

Mahindru, S.H. (1982) *Spices in Indian Life.* Delhi: S. Chand & Sons.

Mahinkumar, Y., and Lokesh, B.R. (2004) Biological properties of curcumin–cellular and molecular mechanisms of action. *Crit. Rev. Food Sci. Nutr.,* 44:97–111.

Purseglove, J.W., Brown, E.G., Green, C.L., and Robbins, S.R.J. (1981) *Spices* Vol.2. London: Longman.

Remadevi, R. and Ravindran, P.N. (2005) Turmeric: myths and traditions. *Spice India,* 18(9):11–17.

Ridley, H.N. (1912) *Spices.* London: Macmillan.

Ravindran, P.N. (2006) Spices: definition, classification, history, properties, uses, and role in Indian life. In: Ravindran, P.N., Nirmal Babu, K., Shiva, K.N., and Kallupurackal, J.A., eds. *Advances in Spices Research,* pp. 1–42, Agrobios, India.

Rosengarten, F. Jr. (1969) *The Book of Spices.* Philadelphia, PA: Livingston Pub. Co.

Roxburgh, W. (1810) *Asiatic Res,* XI, 335.

Roxburgh, W. (1820) *Flora Indica,* Serampore.

Sopher, D.E. (1964) Indigenous uses of turmeric (*Curcuma domestica*) in Asia and Oceania. Anthropos, 59:93–127.

Stevens, S. (1920) *The Rites of Twice Born.* Vol.II. London: Oxford University Press.

Velayudhan, K.C., Muralidharan, V.K., Amalraj, V.A., Gautam, P.L., Mandal, S., and Kumar, D. (1999) Curcuma Genetic Resources. National Bureau of Plant Genetic Resources, Regional Station, Trichur, Kerala.

Watt, G. (1872) *A Dictionary of the Economic Products of India.* Vol. II. New Delhi: Today and Tomorrows Pub (Reprint, 1972).

2 Botany and Crop Improvement of Turmeric

P.N. Ravindran, K. Nirmal Babu, and K.N. Shiva

CONTENTS

2.1 INTRODUCTION

Curcuma, a very important genus in the family Zingiberaceae, consists of about 110 species, distributed in tropical Asia and the Asia–Pacific region. The greatest diversity of the genus occurs in India, Myanmar, and Thailand, and extends to Korea, China, Australia, and the South Pacific. Many species of *Curcuma* are economically valuable, the most important being *Curcuma longa*, known as turmeric commercially. In spite of its economic importance, the genus is poorly understood, botanically and chemically.

In addition to *C. longa*, the genus includes other economically important species such as *C. aromatica*, used in medicine and in toiletry articles; *C. kwangsiensis*, *C. ochrorhiza*, *C. pierreana*, *C. zedoaria*, *C. caesia*, etc., used in folk medicines of the Southeast Asian nations; *C. alismatifolia*, *C. roscoeana*, etc., having floricultural importance; *C. amada*, used as a vegetable in a variety of culinary preparations, pickles, and salads; and *C. zedoaria*, *C. malabarica*, *C. pseudomontana*, *C. montana*, *C. decipiens*, *C. angustifolia*, *C. aeruginosa*, etc., used in the production of arrowroot powder.

This chapter summarizes the available information on botany, genetic resources, and crop improvement aspects of turmeric.

2.2 TAXONOMY

2.2.1 *CURCUMA* L.

Linn. Sp. Pl. 2, 1753; Gen. Pl. ed. 5.3:1754; Roxburgh, Asiat. Res. 11:338, 1810; Fl. Indica. 1:20, 1820; Bentham in Bentham & Hooker f. Gen. Pl. 3:643, 1883; Dalzell & Gibson, Bombay Fl. 274, 1861; Baker in Hooker f. Fl. Brit. India 6:209, 1890; Trimen, Handb. Fl. Ceylon 4:240, 1898; Schuman in Engler, Pflanzer. 4(46):99, 1904; Cooke, Fl. Pres. Bombay 2:729, 1907; Fyson, Fl. Nilgiri & Pulney Hill Tops, 1:408, 1915; Fl. S. Indian Hill Stations 2:597, 1932; Fischer. Rec. Bot. Sur. India 9:177, 1921; in Gamble, Fl. Pres. Madras 8:1481, 1928; Ridley, Fl. Malay Penin. 4:253, 1924; Holttum. Gard. Bull. Singapore 13:65, 1950.

Holttum (1950) gives the following description for the genus *Curcuma*:

> Rhizome fleshy, complex, the base of each aerial stem consisting of an erect, ovoid or ellipsoid structure (Primary tuber), ringed with the bases of old scale leaves, bearing when mature several horizontal or curved rhizomes, which are again branched. Roots fleshy, many of them bearing ellipsoid tubers. Leaf shoots bearing a group of leaves surrounded by bladeless sheaths, the leaf sheaths forming a pseudostem; total height of leafy shoots 1 to 2 m. Leaf blades usually more or less erect, often with a purple-flushed strip on either side of the midrib; size and proportional width varying from the outermost to the innermost (uppermost) leaf. Petioles of outermost leaf short or none, of inner leaves fairly long, channeled. Ligule forming a narrow up-growth across the base of the petiole, its ends joined to thin edges of the sheath, the ends in most species simply decurrent, rarely raised as prominent auricles. Inflorescence either terminal on the leaf shoot, the scape enclosed by the leaf

sheaths; or on a separate shoot from the base of the leaf shoot, the scape covered by rather large bladeless sheaths. Bracts large, very broad, each joined to those adjacent to it for about half of its length, the basal parts thus forming enclosed pockets, the free ends more or less spreading, the whole forming a cylindrical spike; uppermost bracts usually larger than the rest and differently colored, a few of them sterile (the group is called coma). Flowers in cincinni of two to seven, each cincinnus in the axil of a bract. Bracteoles thin, elliptic with the sides inflexed, each one at right angles to the last, quite enclosing the buds but not tubular at the base. Calyx short, unequally toothed and split nearly half way down one side. Corolla tube + staminal tube tubular at the base, the upper half cup shaped, the corolla lobes inserted on the edges of the cup, and the lip, staminodes and stamen just above them. Corolla lobes thin, translucent white or pink to purplish, the dorsal one hooded and ending in a hollow hairy point. Staminodes elliptic-oblong, their inner edges folded under the hood of the dorsal petal. Labellum obovate, consisting of a thickened yellow middle band which points straight towards or in some what reflexed, its tip slightly cleft, and thinner pale (white or pale yellow) side-lobes up-curved and overlapping the staminodes. Filament of stamen short and broad, constricted at the top, anther versatile, the filament joined to its back, the pollen sacs parallel, with usually a curved spur at the base of each; connective some times protruded at the apex into a small crest. Stylodes cylindrical, 4 to 8 mm long. Ovary trilocular; fruit ellipsoid, thin-walled, dehiscing and liberating the seeds in the mucilage of the bract pouch; seeds ellipsoid; with a lacerate aril of few segments which are free to the base.

The genus *Curcuma* was established by Linnaeus (1753). The generic epithet is derived from the Arabic word *karkum,* meaning yellow, referring to the yellow color of the rhizome, and *Curcuma* is the latinized version (Purseglove et al., 1981; Sirirugsa, 1999). The earliest description of turmeric is found in Rheede's *Hortus Malabaricus,* which described it under the local name *Manjella kua,* which was later established (Burt, 1977) as the lectotype of *Curcuma.* He further reinstated the name of *C. longa* L. and *C. domestica* Val. as its synonym.

The genus *Curcuma* consists of about 110 species (Table 2.1) distributed chiefly in South and Southeast Asia. Baker (1890–1892) described 27 species in the *Flora of British India.* He subdivided the genus into three sections: exantha, mesantha, and hitcheniopsis. The section exantha comprises 14 species, including turmeric and other economically important species such are *C. anguistifolia* Roxb. (Indian arrow root), *C. aromatica* Salisb, and *C. zedoaria* Rose.

In yet another study, the genus *Curcuma* is subdivided into two subgenera: *Eucurcuma* and *Paracurcuma. Eucurcuma* consists of three sections: (i) tuberosa (sessile root tuber present); (ii) nontuberosa (sessile root tuber absent); and (iii) stolonifera (stoloniferous tuber present) (Velayudhan et al., 1999b). However, morphologically, it is not correct to consider the underground-branched rhizome as sessile tubers. Mongaly and Sabu (1993) revised the South Indian *Curcuma.*

TABLE 2.1
Distribution of the *Curcuma* Species

Geographic Area	*Curcuma* Species (Approximate)
Bangladesh	16–20
China	20–25
India	40–45
Cambodia, Vietnam, and Laos	20–25
Malaysia	20–30
Nepal	10–15
The Philippines	12–15
Thailand	30–40
Total	100–110

2.2.2 C. LONGA L.

C. longa Linn. Sp. Pl. 1:2, 1753; Koenig in Retzius, Obs. Bot. 3:72, 1983; Roxburgh, Asiat. Res. 11:340, 1810; Fl. Indica 1:32, 1820; Dalzell and Gibson, Bombay Fl. Suppl. 87, 1861; Baker in Hooker f. Fl. Brit. India 6:214, 1890; Schuman in Engler, Pflanzer. 4(46):108, 1904; Fischer in Gamble, Fl. Pres. Madras 8:1483, 1928; Burtt and Smith, Notes Roy. Bot. Gard. Edinburgh 31:185, 1972; in Dassanayaka, Rev. Hand b. Fl. Ceylon 4:500, 1983; Burtt, Note, Roy. Bot. Gard. Edinburgh 35:209, 1977; Lectotype: Manjella Kua Rheede, Hort. Malab. 11:21, t. 11, 1692.

   C. domestica Valeton. Bull. Jard. Bot. Buitenzerg, 2 Ser. 27:31, 1918; Redley, Fl. Malay Peninsula, 4:254, 1924; Holttum, Gard. Bull, Singapore, 13:68, 1950.

   The following description of the species is based mainly on Holttum (1950). Primary tuber ellipsoid, c.5 by 2.5 cm; emitting many rhizomes 5 to 8 cm long, 1.5 cm thick, straight or a little curved, bearing secondary branches, the whole forming a dense clump; color inside deep orange, outside yellowish orange; root tubers usually absent. Leaf stem up to about 1 m; leaf blade rarely over 50 cm; usually 30 by 7 to 8 cm; wholly green. Petiole thin, abruptly broadened at the sheath. Ligule lobes small (1 mm); sheath near ligules with ciliate edges (Figure 2.1). Inflorescence apical on the leaf shoot, 10 to 15 cm long, 5 to 7 cm wide. Coma bracts white or white streaked with green, grading to light green bracts lower down; bracts adnate for less than half their length, elliptic, lanceolate, acute, length 5 to 6 cm. Bracteoles up to 3.5 cm long (Figure 2.2 and Figure 2.3). Flowers 5 to 5.5 cm long; petals white, staminodes and lip creamy-white with yellow median band on the lip. Calyx truncate, 1 cm long, minutely pubescent. Corolla tube 2.5 cm long, white, glabrous, lobes unequal, dorsal lobe larger, 1.5 × 1.7 cm, concave, white hooded, hood hairy; lateral lobes linear, 1.5 × 1.2 cm; white, glabrous. Labellum ca. 2.2 × 2.5 cm, trilobed, middle lobe emarginate; lateral staminodes linear, 1.5 × 0.8 cm; tip slightly incurved, included within the dorsal corolla lobe. Style long, filiform, stigma bilipped. Epigynous glands two, 5 mm long. Filaments united to anther about the middle of the pollen sacs; spurs very large, broad, diverging, a little curved with the thin apex always recurved outwards. Ovary 5 mm, tricarpellary, syncarpous; ovules many on axial placenta (Figure 2.2 and Figure 2.3), pubescent toward the tip, fruiting absent or extremely rare.

2.2.3 RELATED SPECIES

2.2.3.1 C. amada Roxb.

C. amada Roxb. Asiat. Res. 11:341, 1810, Fl. Indica 1:33, 1820.

   The epithet *amada* is derived from Bengali, meaning mango, referring to the characteristic taste of the rhizome resembling green mango. This species resembles *C. longa* in its external morphology, but its rhizome is creamy-yellow and not bright orange as in turmeric. Rhizome large, 4.5 × 3.4 cm, creamy-yellow inside, white toward the periphery, sessile tubers thick, 5 to 10 × 2 to 3 cm, cylindrical or ellipsoidal, branched, root tubers absent. Leafy shoot 70 to 100 cm tall, leaves four to five, petiole 5 to 10 cm; lamina 45 to 60 × 14 to 15 cm; lower ones smaller; oblong-lanceolate, lower surface puberulous, upper glabrous, tip hairy. Inflorescence lateral (very rarely central) produced early in the season, spike 12 to 18 cm long, peduncle 20 to 22 cm; fertile bracts 15 to 18, coma bracts 5 to 8 numbers, ca. 6 × 2.5 cm, fused at the base only, light violet. Fertile bract ca. 4 × 3.5 cm; lower two-thirds fused to form a pouch, green. Each bract subtends a cincinnus of four to five flowers, flowers large and 4 to 5 cm long. Corolla tube funnel shaped, pale yellow, minutely pubescent, lobes unequal; labellum ca. 18 × 1.5 cm, three-lobed, pale yellow with a median dark yellow band, glabrous. Lateral staminodes ca. 10.5 × 6.9 cm, apex slightly incurved, pale yellow. Stamen white, basal spur 1 mm long, slightly convergent, glabrous. Style long, filiform, stigma closely appressed with anther lobes. Epigynous glands two. Ovary trigonal, 3 mm long, tricarpellory, syncarpous, with many ovules in axile placenta, densely hairy. Fruiting not reported.

FIGURE 2.1 Line diagram of turmeric plant.

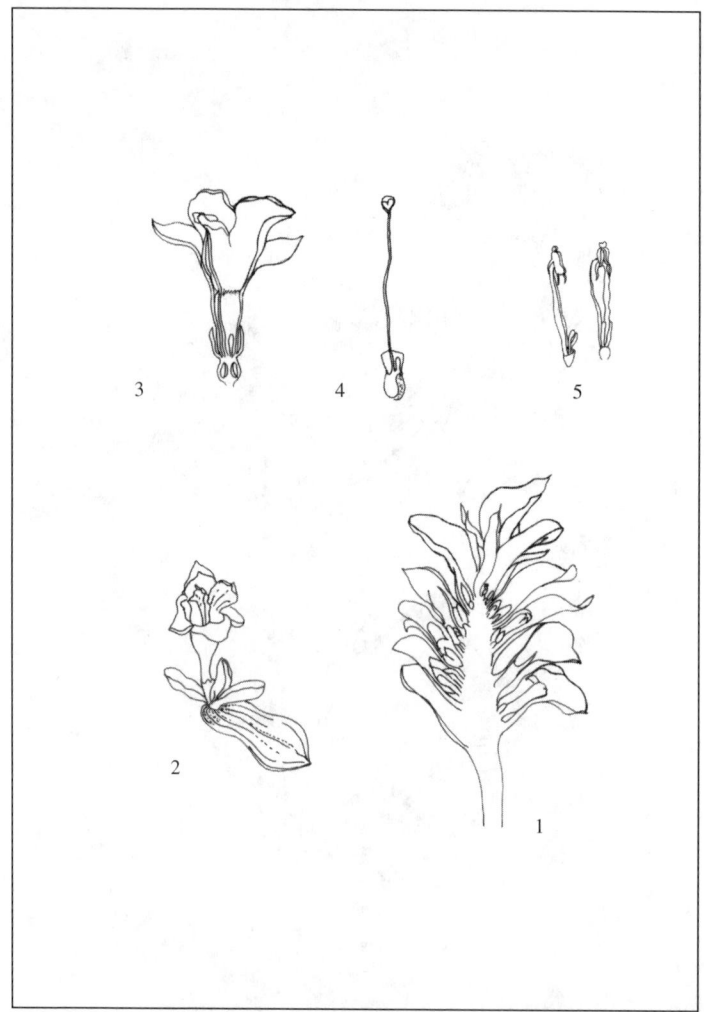

FIGURE 2.2 Line diagram depicting floral parts of turmeric: 1. LS of inflorescence. 2. A flower with bract and bracteoles. 3. LS of a flower. 4. Pistil. 5. Stamen.

The rhizome of this plant is widely used as a vegetable, in pickles, and as a spice for adding a ginger-cum-mango flavor. It is used in traditional medicine as well as in local and tribal systems for various ailments such as prurigo and rheumatism, as an expectorant, and as an astringent; useful in diarrhea.

2.2.3.2 *C. aromatica* Salisb.

C. aromatica Salisb. Parad. London t. 96, 1805; Wight, Icon. Pl. Indiae Orient. 6: t. 2005, 1853; Baker in Hooker f. Fl. Brit. India 6:210, 1890; Trimen, Hand. Fl. Ceylon 4:241, 1898; Schuman in Engler, Pflanzen. 4(46):111, 1904; Cooke, Fl. Pre. Bombay 2:730, 1907; Fische. Rec. Bot. Sur. India 9:177, 1921; in Gamble, Fl. Pre. Madras, 8:1483, 1928; Burtt and Smith, in Dassanayaka, Rev. Hand. Fl. Ceylon, 4:503, 1983.

The specific epithet of this species is derived from the camphoraceous aroma of rhizome. Primary rhizome large, 3 to 5 × 3 to 4 cm; yellow within, aromatic; sessile tubers many, yellow to pale yellow. Leafy shoot 1 m or more, leaves distichous, five to seven; petiole long; lamina 40

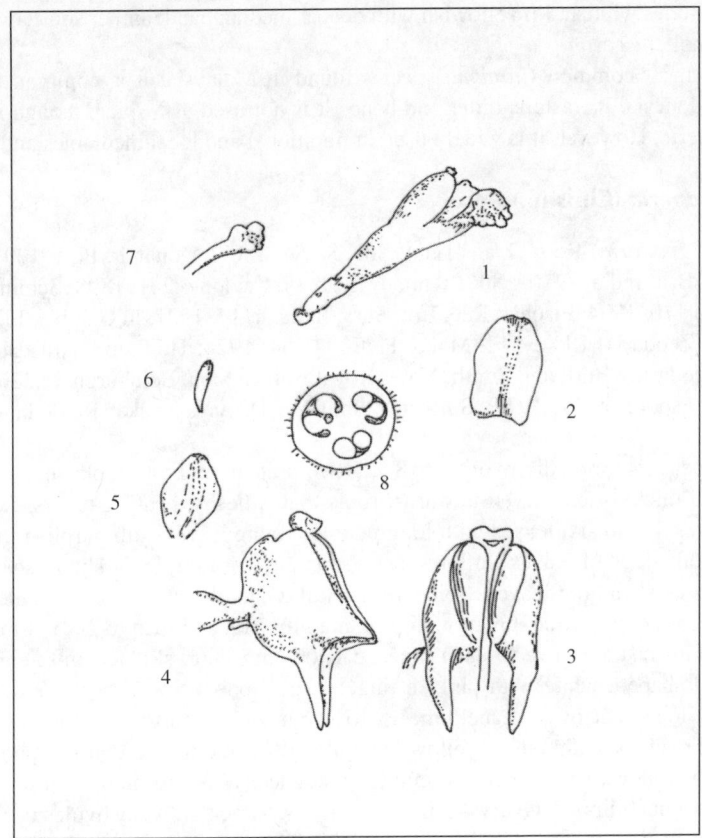

FIGURE 2.3 Line diagram depicting floral parts of turmeric: 1. Flower. 2. Lateral staminode. 3. Anther front view. 4. Anther lateral view. 5. Outer bracteole. 6. Inner bracteole. 7. Stigma. 8. TS of ovary.

to 70 × 10 to 14 cm; broadly lanceolate, acuminate, densely pubescent below. Inflorescence 15 to 30 × 9 cm, lateral, produced early in the season; peduncle 5 to 8 cm long, covered by sheaths. Coma bracts large, pink. Fertile bracts up to 6 cm long, sparsely pubescent. Calyx 2 cm long, tip three-lobed, split on one side, sparsely pubescent. Corolla tube just exceeding the calyx, funnel shaped, lobes unequal, pinkish white; dorsal side broadly ovate; arching over the anther, hooded; lateral lobes narrower. Labellum orbicular, deep yellow. Lateral staminodes oblong, obtuse; anther thecae parallel, each ending in a long sharp spur at the base. Style long, filiform; stigma two-lobed, with a perforation in the center.

2.2.3.3 *C. zanthorrhiza* Roxb.

C. zanthorrhiza Roxb. Fl. Ind. 1:25, 1820; Ridel; Fl. 4:254; Valet; Bull. Btzg. 2nd Ser. XXVII: 61, t. 28; t. 8f.1; Holttum, Gard. Bull. Singapore, 13(1), 72, 1950.

The primary tuber is large, often 10 cm long; rhizomes few and rather short, thick, with few branches, externally pale orange; internally deep orange or orange-red, root tubers large, 5 to 30 cm long. Leaf shoot up to 2 m tall, bearing up to eight leaves. Leaf blades c. 40 × 15 to 90 × 21 cm; with a dark purple feather-shaped stripe 10 mm wide on either side of the midrib, often not reaching the base; petiole up to 30 cm long; ligule small. Inflorescence separate from the leaf shoot (lateral), scape 15 to 20 cm long, spike 16 to 20 cm long, 8 to 10 cm wide. Coma bracts purple; flowering bracts light green, bracteoles up to 25 mm long. Flowers as long as bracts. Corolla lobes light red, staminodes whitish, lip yellowish with deeper median band; anther short and broad, spurs as long as the pollen sacs.

This is the most common *Curcuma* species found in Malaysia; it is common in North East India and in Indonesia. Its taste is bitter and hence it is not used as a spice, though it is as deeply colored as turmeric. However, it is widely used in traditional and local medicines and in cosmetics.

2.2.3.4 *C. zedoaria* (Christm.) Rosc.

C. zedoaria (*C. zeodaria*) Rosc. Trans-Linn. Soc. 8:354, 1807; Monandr. Pl. t. 109, 1828; Baker in Hooker f. Fl. Brit. India 6:210, 1890; Triman, Hanb. Fl. Ceylon 4:241, 1898; Schuman in Engler; Pflanzenr. 4(46):110, 1904; Fischer. Rec. Bot. Surv. India 9:117, 1921; in Gamble, Fl. Pre. Madras, 8:1482, 1928; (Zeodaria) Ridley, Fl. Malay Penin. 4:254, 1924; Holttum, Gard. Bull. Singapore 13:71, 1950 (Zeodaria); Burtt and Smith. Notes Royal Bot. Garden, Edinburgh 31:226, 1972; Burtt, Gard. Bull. Singapore 30:59, 1977; Burtt and Smith in Dassavanayaka, Rev. Hand. Fl. Ceylon 4:501, 1983.

Primary tuber large, broadly ovoid, ca. 8 × 5 cm; branches, short, thick; secondary branches many, short and thick, often curved upwards; roots many, fleshy, bearing tubers. Leaf shoots up to 100 cm; leaves five to six, leaves with long petiole. Young leaves with purplish flush, about 15 mm wide on both sides of leaves. Inflorescence separate from the leafy shoot, scape lateral, 22 cm tall, with three sheathing leaves. Spike 16 cm tall. Coma bracts purple or dark pink. Fertile bracts 20 to 25, 5 × 4 cm, ovate, green with pink margin. Bracteole 1.5 to 2 × 0.5 to 1 cm, white, inner smaller. Flower about five to each bract; calyx 8 mm long, slightly pink; corolla tube 2.5 to 3 cm, funnel shaped, white with pinkish tinge; dorsal lobe 1.5 × 1.2 cm; broadly triangular, hooded; lateral lobes narrower. Labellum 1.5 to 2 cm wide, shortly three-lobed, middle lobe emarginate, pale yellow with a deep yellow band along the middle. Lateral staminodes 1.4 × 0.3 cm, oblong, pale yellow. Anther thecae 6 mm long, connective not forming a crest; spurred. Style long, filiform, stigma bilipped. Ovary 4 mm, tricarpillary, trilocular, many ovules in axile placenta, pubescent. Fruit ovoid, smooth, dehiscing regularly (Holttum, 1950; Sabu, 1991). This species is believed to be native to northeast India but occurs in the wild in Kerala and Karnataka forest and low-lying wastelands.

The rhizome is pale sulfur yellow to bright yellow inside, turning brownish when old; taste strong and bitter. The rhizomes are used in traditional and tribal medicines. It is used as a stomachic, and is applied to bruises and sprains. It is used in alleviating cold, and is mixed with lime and used in cases of scorpion bite.

There is even a view that the turmeric of ancient India was either this species or *C. xanthorrhiza* and not *C. longa*. The latter was introduced to India probably much later by the Budhist monks from Southeast Asia, and this plant, being much less bitter than *zedoaria* and *xanthorrhiza* and having a high color intensity, became the favorite of Indians, and its cultivation spread all over India in course of time.

2.3 GENERAL MORPHOLOGY AND GROWTH OF RHIZOME

The rhizome of turmeric consists of two parts, the central, pear-shaped "mother rhizome" or bulb and its lateral, axillary branches known as the "fingers." Usually there is only one main axis. The planting unit (seed rhizome) consists of either a bulb or a complete finger. The main axis develops into the aerial leafy shoot, and the seed rhizome produces usually only one main axis. The base of the main axis enlarges and becomes the first formed unit of the rhizome, the mother rhizome or the bulb.

The bulb soon undergoes branching, due to the development of axillary buds from the lower nodes of the main axis. These axillary buds on development give rise to the first-order branches, often called the primary fingers. The number of primary fingers usually varies from 2 to 5. The

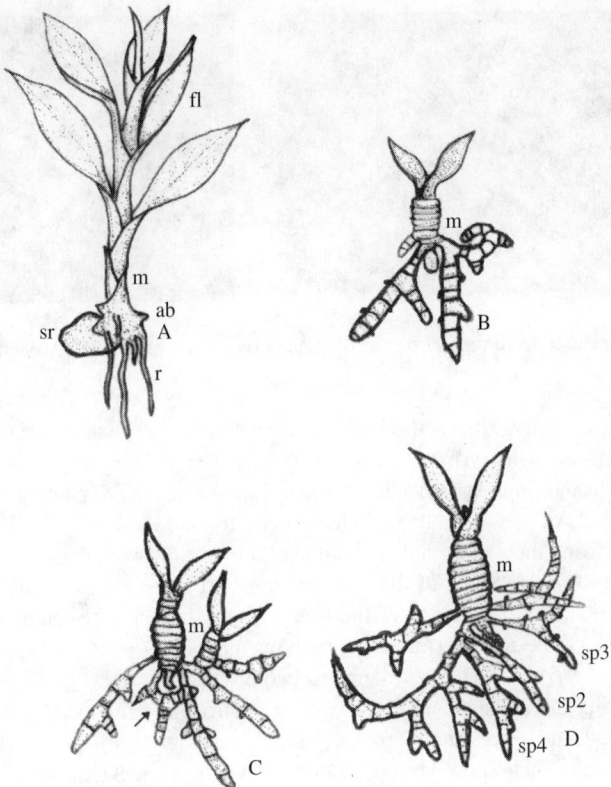

FIGURE 2.4 A–D, line diagrams depicting mode of rhizome development in turmeric: ab, axillary bud; fl, foliage leaf; m, main axis; r, root; sr, seed rhizome; sp2, sp3, sp4, secondary branches.

primary branches grow to some length and then either develop into an aerial shoot or stop growing further. The primary fingers grow in different directions, and in certain cases they grow up to the ground level with one or two or even without any foliage leaves, and the secondary branches developing at higher nodes of the primary branches are diageotropic (Shah and Raju, 1975). Certain primary branches after reaching the ground level do not form any aerial shoot. Instead they show positive geotropic growth. Some such branches arising from the bulb may be diageotropic, ortho-geotropic, or plagiotropic (Figure 2.4).

The primary fingers undergo further branching, producing secondary and tertiary branches, and these branches do not produce aerial shoots. The majority of them show positive geotropic growth or obliquely downward growth. The *longa* types have more sideward growth while the *aromatica* types have more downward growth (Figure 2.5).

2.3.1 NODE

At maturity, the mother rhizomes (bulbs) may have 7 to 12 nodes, and the internodal length varies from 0.3 to 0.6 cm. However the first few internodes at the proximal end are elongated due to which the bulb reaches the ground level (Shah and Raju, 1975). The internodes of the primary and secondary fingers are longer (about 2 cm). Except the first one or two nodes produced, all other nodes in the bulb and the finger show axillary buds.

The bulb has scale leaves only at the first two to four nodes; the rest of them have sheath leaves and foliage leaves. The secondary and tertiary branches have only scale leaves. The branches with negative geotropic growth have pointed scale leaves or sheath leaves.

FIGURE 2.5 Rhizome architecture of turmeric: (a) Longa type cv. Pratibha. (b) Aromatica type cv. Kasturi Tanuku.

The foliage leaves emerge from the buds on the axils of the nodes of the underground bulb and sometimes from the primary finger also. The lamina of the foliage leaf is large and flattened, having a long petiole and thick leaf sheath. The long leaf sheaths overlap and give rise to the aerial shoot. Phyllotaxy is distichous in both scale leaves and foliage leaves.

Roots emerge from the bulbs and sometimes from the primary fingers, but not from the secondary or tertiary fingers. Some of the roots enlarge and become fleshy due to storage of food material (see Figure 2.8a,b,c). They serve the functions of absorption, anchorage, and storage. In certain species some of the roots terminate in bulbous tubers.

Shah and Raju (1975) observed that there are two physiologic states in the development of the rhizome. In one state, the apices that show negative geotropism produce sheath leaves or foliage leaves. In the second, the apices that form only scale leaves grow in various directions. Further, the primary branches, which show various geotropic responses, develop from the axillary buds, and only the rhizome apex can change directly into a leafy shoot. The active axillary branches are mainly in the abaxial side or on the lower side of a horizontally developing branch. This is probably due to the higher concentration of auxins toward the lower side, which in turn might trigger the development of branches in this region.

2.3.2 SHOOT APEX

The shoot apical meristem has the tunica-corpus type of configuration. The tunica is two-layered, with cells dividing anticlinally, while in the corpus, which is the region proximal to the tunica, the cells divide in all directions. The central region underlying the corpus layer is the rib meristem that gives rise to files of cells, which later become the ground meristem. The central region is surrounded by the flank meristem, which produces the procambium, cortical region, and leaf primordium.

2.3.3 ROOT APICAL ORGANIZATION

In cross section, a very young rhizome shows a zone of narrow cells separating the inner and outer ground tissues. This zone has been named as diffuse meristem, which is the extension of the primary elongating meristem, and is noticeable below the second or third node. The diffuse meristem gives rise to the root meristem.

The root apex of turmeric shows three sets of initials developing from the diffuse meristem; one each for the root cap and plerome, and a common zone for dermatogen and plerome. The root cap has two regions: columella in the middle and calyptra at the periphery. The columella consists of five to seven layers of vertical files of cells, which divide mostly peridermally. Surrounding the columella is the peripheral region of the calyptra, the cells of which undergo the kappa-type divisions, followed by cell enlargement resulting in broadening of this region toward the distal end (Raju and Shah, 1977).

There is a single tier of common initials for the epidermis and the cortex — the protoderm–periblem complex. The initials in this complex are one to seven cells in the horizontal row. From this row, the epidermis and cortex are established by the Korper-type division (T-divisions). These divisions followed by cell enlargement enable the tissue to widen toward the proximal end. The outermost layer differentiates into the epidermis. The stellar and pith cells are formed from a group of cells located above the epidermis (the cortical initial) (Raju and Shah, 1977).

2.3.3.1 Dermatogen

This separates out as a distinct layer from the products of division of the protoderm–periblem complex at some distance on the flanks. This tier then divides only anticlinally and develops into the epidermis (Pillai et al., 1961).

2.3.3.2 Periblem and Its Derivatives

The term periblem is applied to the initial of the cortex extending from the hypodermis to the endodermis. Toward the tip are the cells of the calyptra. The cell rows toward the inside are arranged in the opposite direction, with the T-head toward the plerome dome (Pillai et al., 1961). After the periblem is wide enough, the cells divide mostly anticlinally like a rib meristem.

2.3.3.3 Hypodermis

The hypodermis arises from the inner of the two daughter cells of the T-divisions of the protoderm–periblem initials. The endodermis differentiates from the innermost periblem cells, derived from many previous T-divisions. These cells subsequently exhibit anticlinal division, forming the endodermal layer.

2.3.3.4 Plerome

Plerome gives rise to the pericycle, vascular elements, and pith. At its tip there is a group of more or less isodiametric cells. On the side of the plerome dome is the uniseriate pericycle. The metaxylem vessel elements get differentiated near the plerome dome, and they get vacuolated early. The isodiametic cells at the very center of the plerome divide like a ribmeristem to give rise to the pith (Pillai et al., 1961).

Pillai et al. (1961) through cyto-physiological and histochemical studies distinguished two zones at the root tip (excluding the root cap): the quiescent zone (center) and the meristematic zone. The quiescent center is found at the tip of the root body and is characterized by:

cells having the cytoplasm lightly stained with pyronin-methyl green and hematoxylin
cells having smaller nuclei and nucleoli
cell divisions being less frequent
vacuolation being noticeable in most

In medium-longisection this group of cells is in the shape of a cup, and the cells are in a state of relative inactivity.

The meristematic zone is shaped like an arch, surrounding the quiescent center on the side of the root body in longisections. The cells of this zone:

have cytoplasm more deeply stained with pyronin-methyl green and hematoxylin
show divisions more frequently
show larger nuclei and nucleoli
do not express prominent vacuolation

This zone includes cells of the structural histogens of the root body as mentioned earlier. Quiescent center is a general phenomenon in roots, and it is regarded as a reservoir of cells relatively resistant to damage, because of their inactivity and function as a permanent source of active initials (Clowes, 1967). However Pillai et al. (1961) pointed out that since the nearest mature phloem cells occur above 200 to 500 μm, the cells in the region of quiescent center may not be receiving enough nutrients and that they may go into a state of comparative inactivity.

2.3.4 FOLIAR MORPHOLOGY AND ANATOMY

Das et al. (2004) investigated the foliar anatomy of *C. longa, C. caesia*, and *C. amada*. Scanning electron microscopy of the turmeric leaf shows dense, uneven waxy, cuticle depositions, uniformly spread over epidermal cell boundaries. The stomatal aperture is elliptic with a somewhat incomplete cuticle rim around it. In *C. caesia* the stomatal aperture is elliptic with distinct rib-like cuticle depositions girdling the stomata. In *C. amada* the leaf shows profuse trichomes throughout the surface. Trichomes are unicellular with a broad base and a tapering apex. The epidermal surface shows somewhat parallelly undulated, dense cuticular striations giving characteristic rectangular shapes to the cells. The stomatal aperture is elliptic and elongated, and is guarded on either side by unevenly thickened cuticular ribs. The turmeric leaf TS across the mid-vein region shows the following features (Das et al., 2004).

Epidermis — Uniseriate, thin-walled, barrel-shaped parenchymatous cells. Trichomes are less frequent at the adaxial surface; at the abaxial surface, these are small, unicellular, hook-like with a slightly bulbous base.

Hypodermis — Multiseriate, mostly one- or two-layered, composed of irregularly polygonal colorless cells, present interior to both upper and lower epidermis.

Mesophyll — Not differentiated into palisade and spongy tissue. Mesophyll tissue is traversed by a single layer of abaxial air canals alternating with vascular bundles, which are embedded in a distinct abaxial band of chlorenchyma. Air canals are traversed by thin-walled trabeculae, which form a loose mesh within.

Vascular bundles — These are arranged in three rows, developing unequally at different levels. The main vascular bundles form a single conspicuous abaxial arc, alternating with air canals and embedded in chlorenchyma. The abaxial conducting system consists of an arc of vascular bundles of different sizes that are circular in outline. The adaxial conducting system consists of vascular bundles that are similar in appearance to the main vascular bundles but are smaller in size. The main vascular bundles are furnished with a massive fibrous or sclerenchymatous sheath above the xylem and below the phloem, extruded protoxylem, small mass of metaxylems, and phloem tissue. Vascular bundles of accessory arcs have reduced vascular tissues and contracted protoxylem. Abaxial bundles are enveloped within almost a complete fibrous sheath. Both *C. caesia* and *C. amada* have similar anatomical features. In *C. amada* the main vascular bundles are diporate. In *C. caesia* the vascular bundles are without distinct fibrous sheaths.

2.3.5 LEAF SHEATH

The transverse section of the sheathing petiole is horseshoe shaped in outline; the marginal parts are inflexed adaxially. Vascular bundles are arranged in three systems, forming arcs. The abaxial main bundles alternate with large air canals. Air canals are traversed internally by trabeculae. However toward the margin, the vascular bundles seem to arrange themselves in a single row. Both the upper and lower epidermis is uniseriate, consisting of rectangular cells (Das et al., 2004).

2.3.6 EPIDERMIS AND STOMATA

Raju and Shah (1975) studied the stomata in turmeric and its allied species *C. amada* (mango ginger). The following discussion is based mainly on their studies. The upper epidermis consists of polygonal cells that are predominantly elongated at right angles to the long axis of the leaf. Irregular polygonal cells are present on the lower epidermis except at the vein region, where they are vertically elongated and thick-walled. The epidermal cells in the scale and sheath leaves (the first two to five leaves above ground without the leaf blade) are elongated parallel to the long axis of the leaf. Oil cells present in the epidermis are rectangular, thick-walled, and suberized. They are frequent in the lower epidermis.

The leaves are amphistomatic. A distinct substomatal cavity is present. Stomata may be diperigenous or tetraperigenous. Occasionally anisocytic stomata are also observed. Mostly two lateral subsidiary cells align completely with the guard cells. The polar subsidiary cells in most of the cases have oil bodies. The lateral subsidiary cells may divide to form anisocytic stomata. Occasionally three to five lateral subsidiary cells are formed due to further divisions. The guard cells are 40.6 µm long while those on the sheath and scale leaves are 28.9 µm long. The nuclei in the guard cells are smaller than those in the subsidiary cells. Rarely, the guard cells divide, resulting in the formation of three guard cells.

Stomatal development starts with the differentiation of a guard mother cell or meristemoid through an asymmetrical division of a protodermal cell. The smaller cell accumulates dense cytoplasm, deeply stainable and having little vacuolation. The epidermal cells abutting on either side of the meristemoid divide to form subsidiary cells. The meristemoid later divides to form a pair of guard cells. Each epidermal cell lying at the polar region may completely abut the stomatal complex and thereby become a subsidiary cell, or it may remain similar to other epidermal cells in stainablity and size (Raju and Shah, 1975).

2.4 DEVELOPMENTAL ANATOMY

Rhizome anatomy and development were studied by Ravindran et al. (1998) and Sherilija et al. (1998, 1999, 2001), and the following discussions is based on these studies.

Transections of rhizome show an outer zone and an inner zone, separated by intermediate layers. Vascular bundles are present in both zones, but more in the inner zone. The vessels are with spiral and scalariform thickening and scalariform perforation plates. The phloem contains sieve tubes and two or three companion cells.

In the early stage of the rhizome growth (4 to 7 mm diameter), the outer zone is 1.2 to 2.5 mm, inner zone 2.5 to 3.5 mm, and the intermediate layer about 0.50 mm in thickness. At maturity the mother rhizome or bulb measures about 2 to 3 cm across, having an outer zone of about 6 to 10 mm, inner zone about 10 to 12 mm, and the intermediate layer about 1 to 1.5 mm. At this stage the primary finger has about 1 to 2 cm diameter, having an outer zone of 4 to 5 mm, inner zone about 9 to 10 mm, and an intermediate layer of about 1 mm.

The initial rhizome enlargement takes place by the activity of meristematic cells present below the young primordia of the developing rhizome. These meristamatic cells develop into the primary thickening meristem (PTM). These cells are responsible for the initial thickening in the width of the developing cortex by producing primary vascular bundles that are collateral. At the lower levels of the developing rhizome, the PTM becomes primarily a root-producing meristem. Soon after the development of the primary vascular cylinder is completed, some of the pericyclic cells at different places undergo one or more periclinal divisions, forming secondary thickening meristems (STMs), which vary from two to six layers. This meristem produces secondary vascular bundles and parenchyma cells on its inner side. These parenchyma cells become packed with starch grains on maturity. The crowded arrangement of the secondary vascular bundles, which are amphicribal, and their distribution clearly distinguish them from the primary bundles that are collateral and scattered. The cambium-like zones (PTM and STM) constitute the ray initials and the fusiform initials, which are visible at certain loci. In addition to this cambial activity, increase in the size of the rhizome is also the result of the activity of a ground meristem that divides at many loci, followed by cell enlargement.

Oil canals are present in ground parenchyma in actively growing regions along with phloem and xylem. Oil canals are formed lysigenously by the disintegration of entire cells. The process of formation of oil canals seems to be similar to that in ginger rhizomes described by Remashree et al. (1999) and Ravindran et al. (2005).

Chakravarty (1939) reported the occurrence of a fugacious cambium just below the endodermoid layer. It is a short-living cambium (meristematic cells) and may vary from two to six layers. It soon disappears almost completely, leaving behind a ring of crowded bundles. Though a true cambial ring does not form in monocots, however, the presence of ray and fusiform initials points to a true cambial nature. This cambial zone produces secondary vascular bundles and parenchyma cells centripetally to form a distinct zone of secondary thickening.

Remashree et al. (2003) reported the comparative rhizome anatomy of *C. longa, C. aromatica, C. amada,* and *C. zedoaria*. They identified the key characters for distinguishing the four species (Table 2.2). All the species are similar in their basic anatomical features; however, variations exist in the number and arrangement of primary and secondary bundles, orientation of tissues, number and shapes of curcumin cells, starch grains, and oil cells. The primary vascular bundles are abundant and scattered in both outer and inner zones in *C. longa* and *C. amada,* less in *C. zedoaria,* and least in *C. aromatica*. The numbers of oil and curcumin cells vary in the four species, and they are higher in the apical and nodal region than in the internodal region. Although *Curcuma* is a monocot, Remashree et al. (2003) reported the presence of cambial elements in the rhizomes of the four *Curcuma* spp.

Oil cell initials are present in the meristematic region. They are more or less spherical and densely stained with Sudan Black. Small irregular bodies, clusters of globular drop-like structures, and granular bodies are seen inside the oil cell. Large oil cells are present in *C. zedoaria*; the size decreases in *C. aromatica, C. amada,* and *C. longa* in that order. Oil cells are round-to-ovoid and more or less globular; the oil cell index decreases in the order: *C. aromatica* → *C. zedoaria* → *C. amada* → *C. longa.*

Curcumin cells appear yellowish-red following boric acid treatment. The cell wall is thick and filled with curcumin, which appears orange with boric acid treatment. The longisection of rhizome shows a long canal containing curcumin deposit. The number of curcumin cells is very few in *C. amada* and *C. aromatica* and numerous in *C. longa* and *C. zedoaria* (Remashree et al., 2003).

2.4.1 ONTOGENY OF OIL CELLS AND DUCTS

Oil cells and ducts develop in the meristematic region early in the development of rhizomes. The development of oil cells, curcumin cells, secretary ducts, and stages of secretion were reported by Remashree (2003) on which the following discussion is based. Both schizogenous and lysigenous types of development are found in turmeric.

2.4.1.1 Schizogenous Type

This type of secretary duct originates in the intercalary meristem of the developing region. The ducts are initiated by the separation of a group of densely stained meristematic cells through the dissolution of middle lamella. Concurrent separation of the cells leads to the formation of an intercellular space bordered by parenchymatous cells possessing dense protoplasmic content. These ducts anastamose and appear branched. The widening of the duct is affected by the separation of the bordering cells along the radial walls.

2.4.1.2 Lysigenous Type

This type of secretary duct formation is met with both in the meristematic regions and in the mature part of the rhizomes. There are four stages in its development — initiation, differentiation, secretion, and quiescence. These steps take place acropetally and are a gradual process (Figure 2.6).

TABLE 2.2
Comparative Rhizome Anatomy of *C. amada*, *C. aromatica*, *C. longa*, and *C. zedoaria*

Characters	C. amada	C. aromatica	C. longa	C. zedoaria
Rhizome nature	Highly branched and bulbous	Less branched and fingers present	Highly branched finger up to 4th order	Less branched bulbous rhizome
Rhizome color	Creamy-white	Creamy	Yellow	Orange yellow
Trichome	Numerous, uniseriate	Absent	Absent	Rare, uniseriate
Epidermis	Single layered	Single layered	Single layered	Single layered
Periderm	6–8 layers	8–10 layers	2–4 layers	14–15 layers
Outer zone	Both zones are almost equal sized and contain primary vascular bundles. Starch and oil cells evenly distributed	Small outer zone contains lot of oil cells, few curcumin cells, and primary vascular bundles	Small outer zone contains primary vascular bundles, starch, oil cells, and curcumin cells	Both zones are almost of equal size, contain primary vascular bundles, starch, oil cells and curcumin cells
Primary vascular bundles	60–70 vascular bundles in inner core of outer zone	50–60 vascular bundles in inner core of the outer zone	70–80 vascular bundles evenly distributed	90–120 bundles distributed evenly in the outer zone
Endodermoidal layer	Discontinuous	Discontinuous	Continuous	Continuous
Cambium	2-layered	2-layered	2-layered	2–3 layered
Inner zone	Both zones are equal sized. Secondary vascular bundles and starch deposition are higher, just below the endodermoidal layer than inner core	Large inner zone; secondary vascular bundles are more, just below the endodermoidal layer than inner core	Large inner zone; secondary vascular bundles and starch deposition are higher, just below the endodermoidal layer than inner core	Both zones are equal sized; secondary vascular bundles and starch deposition are higher, just below the endodermoidal layer than inner core
Secondary vascular bundles	In groups below the endodermoidal layers and also scattered in the inner core	In groups below the endodermoidal layers and also scattered in the inner core	In groups below the endodermoidal layers and also scattered in the inner core	In groups below the endodermoidal layers and few in the inner core
Xylem tracheids and vessels	Helical and spiral thickening	Helical and spiral thickening	Helical and spiral thickening	Helical and spiral thickening
Fibers	Present (rare)	Absent	Absent	Absent
Phloem	Sieve tube, 1–2 companion cells and phloem parenchyma	Sieve tube, 1–2 companion cells and phloem parenchyma	Sieve tube, 1–2 companion cells and phloem parenchyma	Sieve tube, 1–2 companion cells and phloem parenchyma
Bundle sheath	Present in the upper side of the bundle	Present in the lower side of the bundle	Absent	Absent
Oil cell	Plenty, small sized compared to the other three	Plenty and more than the other three species	Plenty but less than *C. aromatica* and *C. zedoaria*	Plenty and large sized
Curcumin cells	Rare with smaller size compare to other three	Few	Plenty	Plenty with large size
Starch grains	Numerous in inner and outer core, rod shaped; number varies from 8 to 16/cell	Numerous in inner and outer core, spindle shaped, eccentric; 5–20/cell.	Numerous in inner and outer core, triangular shaped. Number varies from 12 to 20/cell	Numerous in inner and outer core, large and rod shaped. Number varies from 5 to 20/cell

Source: Remashree, A.B., Balachandran, I., and Ravindran, P.N. (2003).

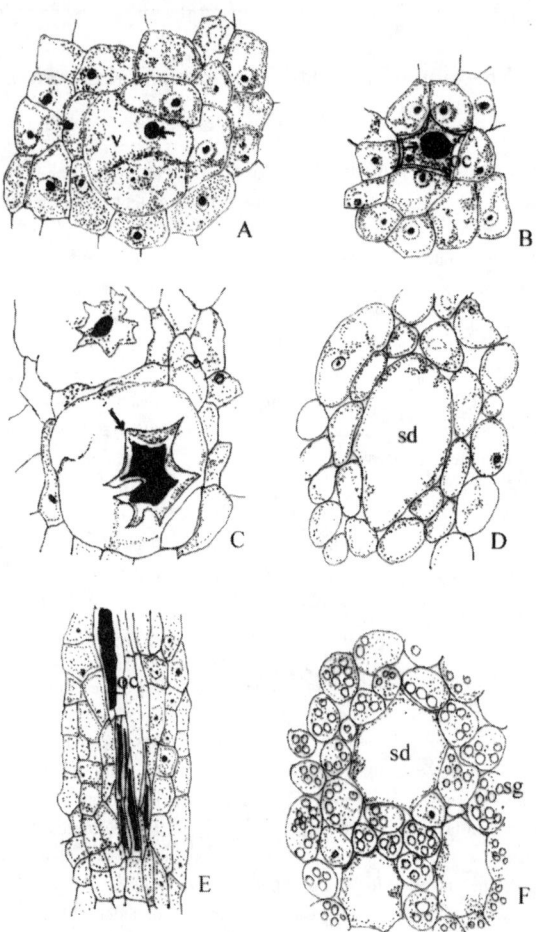

FIGURE 2.6 Ontogeny of oil cells and ducts in turmeric: (A) Lysigenous development of duct in the intercalary meristem. (B) Darkening of cell content and vacuole formation. (C,D) Complete disintegration of central cell. (E) LS of oil duct filled with secretory substances. (F) Mature oil duct and neighboring cell containing starch grains.

2.4.1.2.1 Initiation and Differentiation

Some of the cells in the meristematic shoot apex differentiate into oil cell mother cells, which are distinguishable by their larger size, dense cytoplasm, and prominent nucleus. A group of oil initials arise from the mother cells through anticlinal and periclinal division. These initials undergo cytoplasmic vacuolation at a distance of 410 to 130 µm from the shoot apex. Subsequently the surrounding cells also enlarge, undergoing cytoplasmic and nuclear disintegration. In due course the central cell enlarges and then gets lysed, thereby leading to the formation of a duct, which can be either an articulated or a nonarticulated type. The ducts get filled up with the contents of the lysed cells. Often the wall dissolves completely and the entire process leads to the formation of long ducts. Subsequently the cells surrounding the duct also get lysed, thereby leading to the widening of the duct lumen. These stages can be seen about 3500 µm behind the apex.

2.4.1.2.2 Secretion

The oil cells that differentiate from the initials start a holocrine type of secretion and expel content into the duct. Then the adjacent cell undergoes differentiation and the process continues. Simultaneously, cells in the primary tissue continue to become differentiated into new oil cells, and they reach the secretion stage. The secretion fills the secretary ducts in the young stage, but the quantity gets reduced gradually and finally the duct seems to be empty. Thus once the secretary duct is filled with oil content, further basipetal development leads to the radial spreading of the contents. Oil is spread radially through the intercellular space, and the contents of the duct gradually get dissipated. Such secretary stages are noticed about 3250 to 3750 μm behind the shoot apex (i.e., about 120 d old rhizome).

2.4.1.2.3   Maturing Stage

In the mature rhizome the ground parenchyma do not undergo further cell division and differentiation into a duct. In this stage the cells adjacent to the duct become storage cells, and they contain a large number of starch grains and possess large vacuoles.

The process of development of oil ducts and cells is more or less similar to that in ginger (Remashree et al., 1999; Ravindran et al., 2005).

2.4.2   Microscopical Characters of Cured Turmeric Rhizome

Cured (processed) and dried turmeric and its powder are the raw drugs used in the production of many ayurvedic and other traditional medicines. Raghunathan and Mitra (1982) described the anatomical features of dry, processed turmeric. The TS of cured and uncured turmeric are similar to a large extent, differing only in the nature of starch. Independent starch grains are present in the uncured samples; in the cured rhizome, gelatinized lumps of starch are present. In transection, the rhizome consists of an outer zone of cork, followed by a wide zone of cortex, and an endodermoid ring closely covering a discontinuous ring of vascular bundles. A large number of vascular bundles are scattered throughout the section. The parenchyma cortex and the pith are full of starch grains, the yellow pigment being present at some places only.

In the uncured sample, the cork is intact throughout, while in the cured sample, it is broken at several places. The rhizome cortex is demarcated into two zones. The outer cortex contains a few layers of irregular-shaped to rounded parenchyma with a few starch grains, while the inner cortex, which is large, has parenchyma full of starch grains. In the cured rhizome, these starch grains get gelatinized due to the boiling process and form a compact mass in each of the cells.

The vascular bundles present in the cortex have been called cortical vascular bundles, leaf trace bundles, or cortical meristeles. These bundles are closed types, consisting of phloem and xylem only. These vascular bundles measure 80 to 115 to 124 μm along their major axis. The vascular bundles found toward the outside are smaller, having three to seven xylem elements, while those present toward the inside are bigger, having up to 12 elements.

Vascular bundles are arranged as a circle below the endodermoid ring. The ring encloses a central medulla in which are scattered several larger vascular bundles; these resemble the cortical vascular bundles in all respects. In unstained sections, a large number of cells having curcumin deposit are found throughout.

2.4.3   Microscopic Characters of Turmeric Powder

2.4.3.1   C. longa

The diagnostic characters of turmeric powder are:

abundant groups of parenchymatic cells which are filled with gelatinized starch and permeated with a bright yellow coloring matter
fairly abundant fragments of pale brown cork composed of thin-walled cells, which in surface view appear large and polygonal

The epidermis is composed of a layer of straight-walled tubular cells, polygonal to elongated in surface view; the walls are sometime slightly thickened and pitted.

The covering trichomes are unicellular, elongated, conical, and bluntly pointed with moderately thickened walls.

The vessels are fairly abundant, mostly large, and reticulately thickened with regularly arranged rectangular pits, with rare spiral or annular thickening.

Occasionally, nongelatinized starch grains are present, simple, flattened, oblong to oval or irregular in outline, with a small pointed hilum at the narrower end (Jackson and Snowden, 1990).

2.4.3.2 *C. amada*

The TS of the section shows the periderm, which consists of eight to ten layers of thin-walled cork cells, tangentially elongated into an outer and inner zone of parenchyma. Scattered in the cortex are numerous oil cells with suberized walls enclosing yellowish brown oleoresin contents. The inner cortical zone consists of three rings of collateral, closed vascular bundles. The larger bundles are enclosed in a sheath of fibers. Each vascular bundle contains phloem showing well-marked sieve tubes, and a xylem composed of vessels with annular, spiral, or reticulate thickening. The inner limit of the cortex is marked by a single layer of endodermis. The vascular bundles of stele are scattered as in a typical monocot stem. The ground mass of the stele is composed of parenchyma containing prismatic crystals of calcium oxalate, abundant starch grains, and numerous oil cells. The starch grains are flattened and ovoid, oblong, and have concentric striations. There are also cells filled with a yellow mass of oleoresin, having cuticulized walls.

The vascular bundles are of a closed type, consisting of xylem and phloem only. There are no fibrous zones above the vascular bundles. The vascular bundles toward the outside are smaller, having three to seven xylem elements, while those present toward the inside are comparatively bigger, having up to 12 elements.

2.5 CYTOLOGY AND CYTOGENETICS

Cytological studies on *Curcuma* have been limited to chromosome number reports. It was Suguira who first reported the chromosome number of turmeric as $2n = 64$ (Suguira, 1931, 1936). Raghavan and Venkatasubban (1943) reported the number of *C. amada* ($2n = 42$), *C. aromatica* ($2n = 42$), and *C. longa* ($2n = 62$). Venkatasubban (1946) further reported the chromosome numbers of both *C. zedoaria* and *C. petiolata* as $2n = 64$. An unusual number of $2n = 34$ was reported by Sato (1948). Chakravorti (1948) studied 48 taxa of Zingiberaceae cytologically and reported the somatic chromosome numbers of *C. amada* (42), *C. aromatica* (42), *C. angustifolia* (42), *C. longa* (62, 63, 64), and *C. zedoaria* (63, 64). He found that the chromosome numbers of seven cultivars studied exhibited variation from 62 to 64. Sharma and Bhattacharya (1959) reported $2n = 42$ for *C. amada* and $2n = 62$ for *C. longa*.

Ramachandran (1961, 1969) carried out detailed study of the chromosome numbers in Zingiberaceae, which included six species of *Curcuma*. He also studied meiosis in *C. decipiens* ($2n = 42$) and *C. longa* ($2n = 63$) and reported regular formation of bivalents in the former and a high percentage of trivalent association in the latter. The sterility of *C. longa* has been attributed to the triploidy. Various reports of chromosome numbers in *Curcuma* are given in Table 2.3.

Ramachandran's study confirmed that *C. longa* has a somatic chromosome number of $2n = 63$ (Figure 2.7). He suggested that turmeric is a triploid and might have evolved as a hybrid between tetraploid *C. aromatica* and an ancestral diploid *C. longa* (having $2n = 42$) types or one of these is evolved from the other by mutational steps, represented by the intermediate type that is known to occur. He has also suggested that the basic chromosome number of 21 might have been derived either by dibasic amphidiploidy (by combination of lower basic numbers of 9 and 12 found in some genera in the family) or by secondary polyploidy. The herbaceous perennial habit of this

TABLE 2.3
Chromosome Numbers in the Genus *Curcuma*

Name	Chromosome Number (2n)	Ref.
Curcuma decipiens, Dalz	42	Ramachandran (1961)
C. neilgherrensis Wight	42	Ramachandran (1961)
C. amada Roxb.	42,42,42	Chakravorti (1948), Sharma and Bhattacharya (1959), Ramachandran (1961)
C. zeodaria Rosc	64	Venkatasubban (1946)
	63,64	Chakravorti (1948)
	63	Ramachandran (1961)
C. anguistifolia Roxb	42,42	Chakravorti (1948), Sharma and Bhattacharya (1959)
C. petiolata Roxb	64	Venkatasubban (1946)
C. aerugenosa Roxb.	63	Joseph (1999)
C. caesia Roxb.	63	Joseph (1999)
C. comosa Roxb.	42	Joseph (1999)
C. haritha Mangaly & Sabu	42	Joseph (1999)
C. malabarica Velayudhan	42	Joseph (1999)
C. raktacanta Mangaly & Sabu	63	Joseph (1999)
C. aromatica Salisb	42	Raghavan and Venkata Subban (1943)
	63	Ramachandran (1961)
G.L. Puram type	63	Ramachandran (1961)
Kasturi Amalapuram	86	Ramachandran (1961)
Polavaram	86	Ramachandran (1961)
C. longa Linn.	64	Sugiura (1931)
	62	Raghavan and Venkata Subban (1943)
	32	Sato (1948)
	62,63,64	Chakravorti (1948)
	63	Ramachandran (1961)
Duggirala type	63	Ramachandran (1961)
Kovvur Desavali	63	Ramachandran (1961)
Tekurpeta	63	Ramachandran (1961)
G.L. Puram 2	63	Ramachandran (1961)
Kasturi Duggirala	63	Ramachandran (1961)
Nallakatla pasupu	63	Ramachandran (1961)

species, its vegetative mode of propagation, and the small size of chromosomes favor the perpetuation of polyploidy.

Sato (1960) carried out karyotype analyses in Zingiberales in which he reported the chromosome number of *C. longa* as $2n = 32$ and suggested that the species could be an allotetraploid with a basic number of $x = 8$. Karyologically, there is one pair of A chromosomes with medium constriction, five pairs of A chromosomes with submedian constriction, six pairs of B chromosomes with submedian and subterminal constrictions, and four pairs of C chromosomes with submedian and subterminal constrictions. One pair each of A and C has satellites.

Nambiar (1979) studied the cytology of five *C. longa* types, six *aromatica* types, and one *C. amada* collection. He found the somatic chromosome numbers $2n = 63$ in all *C. longa* types consistently. The *C. aromatica* types had $2n = 84$ and *C. amada* had $2n = 42$. In *C. longa* cvs. in Mydakur and Coll. No. 24, $2n = 62$ occurred in 7.0 and 4.0% of cells respectively. Among *C. aromatica* cv. Kasturi had 19.5% cells having $2n = 82$ and 86; in Kasturi Tanuka 14.0% had $2n = 86$; in cv. Dahgi 26.0% cells gave a count of $2n = 86$. Joseph et al. (1999) studied the cytology of six species of *Curcuma* (*C. aeruginosa, C. caesia, C. comosa, C. haritha, C. malabarica,* and *C.*

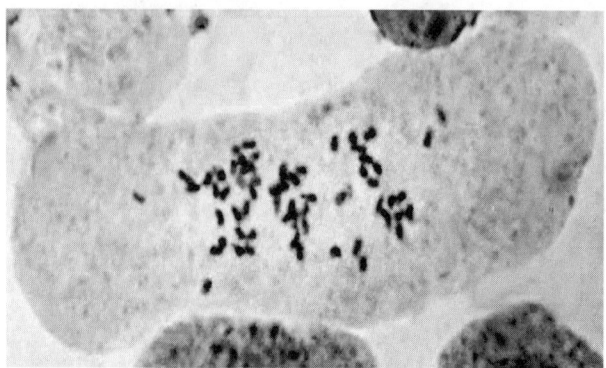

FIGURE 2.7 Somatic chromosomes of turmeric variety Prabha (2n: 63).

raktacanta). Three of them (*aeruginosa, caesia,* and *raktacanta*) possess triploid somatic chromosome number of 2n = 63, while the others are diploids with 2n = 42. The karyotypes are reported
to be symmetrical in all the above species. The chromosomes fall into three categories:

Type A: Comparatively short chromosomes with primary and secondary constrictions; one
  median and the other subterminal (size: 0.99 to 0.7 μm)
Type B: Short chromosomes with median and submedian primary constriction (size: 0.95
  to 0.5 μm)
Type C: Very small chromosomes with median to submedian primary constriction (size:
  0.49 to 0.24 μm)

They felt that *Curcuma* is karyologically more advanced than other members of the Zingiberaceae. Meiosis exhibited varying degrees of chromosome abnormalities and chromosome associations. Quadrivalents, trivalents, bivalents, and univalents were recorded, but their relative frequency
varied among cultivars. In cv. Kuchipudi, 28 bivalents, one quadrivalent, and three univalents were
noted in half of the cells analyzed. In cv. Nandyal 1 III + 29 II + 1 I was recorded in 25% of cells,
while in 75% of the cells, the association frequency was 30 II and 3 I. In cv. No. 24, 1 IV, 1 III, 27
II, and 2 I were present in 16.7% cells, while in the remaining cells 1 III, 29 II, and 2 I were present.
  In *C. aromatica* the meiosis was more normal. In cv. Kasturi (2n = 84) 44.4% cells had 42
trivalents. The maximum association recorded was four quadrivalents, 32 bivalents, and four
univalents. The separation was normal in 85.7% cells. In cv. Uadayagiri (2n = 84), one hexavalent
occurred in 40% of mother cells at diakinesis, while 42 II were recorded in 30% of cells. Normal
separation observed in 46.7% cells. In cv. Kasturi Tanuku (2n = 84), 41.7% cells had 42 trivalents
in diakinesis, while in the remaining one hexavalent, one quadrivalent and 37 bivalents were
observed. The anaphase separation was normal in 66.7% of cells. In cv. Jobedi, 4 quadrivalents
and 34 bivalents were recorded. In cv. Dahgi, 1 hexavalent and 39 bivalents were present at
diakinesis in 33.3% of cells, in the remaining 42 bivalents were observed. The abnormalities in
meiosis led to lower pollen fertility in *C. longa* types than in *C. aromatica* types (Table 2.4).
  Based on his studies, Nambiar (1979) concluded that turmeric is a natural hybrid between two
species having 2n = 42 and 2n = 84 chromosomes. Though turmeric is relatively sterile, the
chromosome pairing in meiosis is nearly normal, with few trivalents and univalents. In a triploid
plant, meiosis is very irregular, with high frequency of trivalents, univalents, bridge, and fragments.
The presence of near-normal meiosis was explained by assuming that the putative parents (with
2n = 42 and 2n = 84) were evolved from parents having secondary basic number of x = 21, which
in turn forms the primary basic number of x = 9 and x = 12. Nambiar (1979) has proposed the
following scheme.

TABLE 2.4
Pollen Fertility in *C. longa*
and *C. aromatica* Types

Acc. No.	Pollen Fertility (%)
C. longa	
Kuchipudi	45.7
Nandyal type	46.4
No. 24	48.5
C. aromatica	
Kasturi	74.5
Udayagiri	70.0
Katergia	73.6
Dahgi	68.6
Amalapuram	70.0

Source: Nambiar, M.C. (1979).

Possible triploid origin of turmeric:

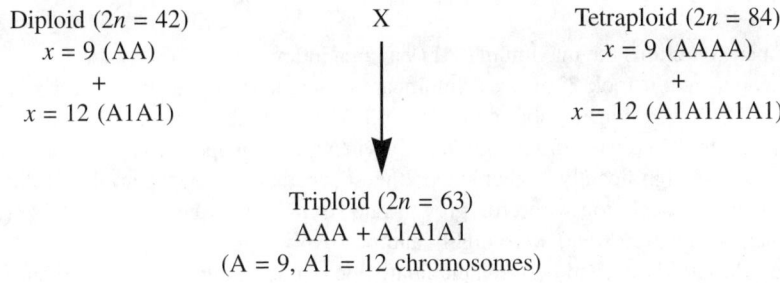

Diploid (2*n* = 42)   X   Tetraploid (2*n* = 84)
x = 9 (AA)   *x* = 9 (AAAA)
+   +
x = 12 (A1A1)   *x* = 12 (A1A1A1A1)

Triploid (2*n* = 63)
AAA + A1A1A1
(A = 9, A1 = 12 chromosomes)

A certain extent of homology existed between A and A₁ genomes, and this is reflected in the presence of a few tetravalents in *C. longa* and hexavalents in *C. aromatica*.

Microsporogenesis and megasporogenesis in *C. aurantiaca* and *C. lorgengii* were reported by Sastrapradja and Aminali (1970). There is seed set in *C. aurantiaca*, while the seed set is absent in *C. lorgengii*, possibly due to pollen abortion.

2.6 MOLECULAR TAXONOMY AND INTERRELATIONSHIPS

With the advancement of molecular biology, molecular data is being increasingly used to understand the phylogeny and interrelationships in Zingiberaceae. Chase (2004) attempted to have an overview of the phylogeny and relationships in monocots based on analysis of DNA sequence data of seven genes representing all three genomes and reported to have high bootstrap support to the clades comprising Acorales followed successively by Alismatales (including Araceae and Tofieldiaceae), Petrosaviales, Dioscoreales/Pandanales, Liliales, Asparagales, and finally a polytomy of Arecales, Commelinales/Zingiberales, Dasypogonaceae, and Poales.

A phylogenetic analysis of the tribe Zingibereae (Zingiberaceae) was performed by Ngamri-absakul et al. (2003) using nuclear ribosomal DNA (ITS1, 5.8S and ITS2) and chloroplast DNA (*trn*L [UAA] 5[prime prime or minute] exon to *trn*F [GAA]). The study indicated that the tribe Zingibereae is monophyletic with two major clades, the *Curcuma* clade and the *Hedychium* clade. The genera *Boesenbergia* and *Curcuma* are apparently not monophyletic.

Kress et al. (2002) studied the phylogeny of the gingers (Zingiberaceae) based on DNA sequences of the nuclear internal transcribed spacer (ITS) and plastid *matK* regions and suggested a new classification. Their studies suggest that at least some of the morphological traits based on which the gingers are classified are homoplasious and three of the tribes are paraphyletic. The African genus *Siphonochilus* and Bornean genus *Tamijia* are basal clades. The former Alpinieae and Hedychieae for the most part are monophyletic taxa with the Globbeae and Zingiberaceae included within the latter. They proposed a new classification of the Zingiberaceae that recognizes four subfamilies and four tribes: Siphonochiloideae (Siphonochileae), Tamijioideae (Tamijieae), Alpinioideae (Alpinieae, Riedelieae), and Zingiberoideae (Zingibereae, Globbeae).

2.7 PHYSIOLOGY

Very few studies have gone into the physiology of turmeric. Satheesan (1984) and Satheesan and Ramadasan (1987, 1988a) studied the physiology, growth, and productivity of turmeric in monoculture and under the shade in a coconut garden. The turmeric plant showed three distinct stages in growth and development:

1. the phase representing growth up to 8 weeks after planting
2. the period from 8 to 10 weeks, characterized by initiation of finger and maximization of shoot growth
3. the period dominated by rhizome growth

Under open conditions the maximum LAI (leaf area index) was gained by the 18th week; under shade of coconut tree, it took 22 weeks. Significant differences existed in the response to shade; cv. Cls. 24 (C1) showed considerable reduction in LAI under the coconut shade; in cv. CLL 328 Sugadham (C2), the LAI did not differ significantly when grown in open or in shade. In cv. Duggirala (C3), the LAI was significantly higher under the shade than in the open. The LAI showed a significant positive association with crop growth rate (CGR) up to the 3 months of growth. CGR and NAR (net assimilation state) were highly and positively correlated.

A linear relationship existed between cumulative leaf area development (LAD) and dry matter production. This relationship was not influenced by cultivars or changes in cropping system. In cv. CLL-24, a significant positive correlation of rhizome yield on LAD existed. In this cultivar, the higher dry matter production is preceded by the attainment of higher relative leaf growth rate (RLGR) under the open system than under the intercropping system. In other cultivars such differences did not exist.

The highest NAR and LAR (leaf area ratio) were recorded during Phase I, thereby resulting in the highest relative growth rate (RGR) during this period. Cultivar CLL-24 showed a further remarkable increase in NAR in Phase III under the monoculture system, which had resulted in a secondary peak in CGR and a leveling of RGR during this period. The increase in NAR is interpreted as a response of the photosynthetic apparatus to an increased demand for assimilates caused by the rapid bulking of the rhizomes. The cultivar differences in final rhizome yield can be well explained by the differences in CGR during Phase III, which in turn is influenced by the NAR.

The RGR of the cultivars were reduced significantly under shade, as the NAR rate was reduced markedly. The effect of shade is more drastic in CLL-24. The cultivars showed a higher solar energy conversion efficiency under shade than under open condition, and there existed differences among cultivars. Significant variation in harvest index (HI) existed among cultivars. The HI was significantly higher in CLL-24, indicating higher partitioning efficiency of this cultivar. The efficiency was higher under open than under shade.

Li et al. (1997) found that the growth of turmeric can be divided into three stages. About 13 leaves are produced during the life span of a turmeric plant. Leaf formation takes place rapidly in the seedling stage. The leaf area and increase of dry matter per day in roots and leaves are smaller and lower in the seedling stage. In the daughter rhizome formation stage, net assimilation rate (NAR) reaches the maximum (3.54 m^2/d), and the leaf area also reaches the maximum (3302.9 $cm^2/plant$, leaf area index [LAI] 4.95). Before the daughter rhizome formation stage, more than 75% of dry matter is distributed in leaves, and in the daughter rhizome formation stage, 50 to 75% of dry matter is distributed in leaves. In the late growing period, more than 40% of dry matter is distributed in the daughter rhizome. The growth center of turmeric changes twice during the life span of a turmeric plant.

Panja and De (2001) studied the stomatal frequency and its relationship with leaf biomass and rhizome yield. The stomatal frequency is positively correlated with rhizome yield than leaf dry weight in genotypes resistant to leaf blotch disease, while in those that are susceptible, it exhibited significant and negative correlation with leaf dry weight and rhizome yield. Stomatal frequency of 12 genotypes varied from 426 to 1,773/cm^2 in the upper leaf surface and 7,131 to 12,270/cm^2 in the lower surface. The highest frequency was in cv. TCP-9.

Dixit et al. (2000a)) studied the physiological efficiency in terms of photosynthetic capacity and total $^{14}CO_2$ incorporation in relation to essential oil and curcumin accumulation. There were significant differences among the genotypes in yield components such as leaf area, leaf area ratio, and photosynthetic characters, namely CO_2 exchange rate, initial transpiration rate, stomatal conductance, and chlorophyll content. Total $^{14}CO_2$ incorporation was highest in the genotype Krishna and lowest in CL-16. The highest percentage of incorporation of ^{14}C in oil was in genotype CL-315 and in curcumin in the cv. Rashmi.

Dixit and Srivastava (2000d)) also investigated the distributions of photosynthetically fixed $^{14}CO_2$ into curcumin and essential oil in relation to primary metabolites in developing turmeric leaves. They determined the distribution of photosynthates and found that of the total $^{14}CO_2$ assimilated by plants, the first, second, third, and fourth leaves fixed 31, 23, 21, and 9%; roots 4%, rhizome 6%, oil 0.01%, and curcumin 4.6% of the fresh weight of rhizome. Leaf area, its fresh and dry weight, and $^{14}CO_2$ exchange rate increased up to the third leaf. The incorporation of $^{14}CO_2$ into sugar was maximal, followed by incorporation into organic acid, amino acid, and essential oil at all stages of leaf development. Assimilates translocated to roots, and rhizome showed similar trends of incorporation in fractions as in leaves. The youngest developing leaf assimilated maximum $^{14}CO_2$ into metabolites and essential oil. In the rhizome, curcumin constituted the major metabolite. The incorporation of $^{14}CO_2$ into metabolites and oil declined as the leaves matured. A major portion of $^{14}CO_2$ assimilated was translocated to roots and to rhizome for curcumin formation.

Dixit et al. (1999, 2001, 2002) investigated boron deficiency–induced changes in translocation of $^{14}CO_2$ photosynthate into primary metabolites in relation to essential oil and curcumin accumulation by growing turmeric plants in B-deficient and B-containing (varying concentrations) media. B-deficiency resulted in decrease in leaf area, fresh and dry mass, chlorophyll (Chl) content, and photosynthetic rate. Total $^{14}CO_2$ incorporated was highest in young growing leaves. The incorporation of $^{14}CO_2$ declined, with leaf positions being maximal in the youngest leaf. B-deficiency resulted in reduced accumulation of sugars, aminoacids, and organic acids at all leaf positions. Translocation of metabolites toward rhizome and roots decreased. In rhizome, the amount of aminoacids increased but organic acids remain unchanged. Photoassimilate partitioning to essential oil in leaf and to curcumin in rhizome decreased. Curcumin showed a slight increase, but the overall yield per unit area decreased.

Dixit and Srivastava (2000b,c)) studied the physiological effect of iron deficiency on turmeric genotypes (Rashmi, Krishna, Roma, CL-315, CEL-6, and CEL-70). All the genotypes exhibited a decrease in plant growth, fresh weight, rhizome size, photosynthetic rate, and chlorophyll content, whereas curcumin content increased significantly in all the genotypes under Fe deficiency. The oil content of rhizome increased in all genotypes except CL-315 and CEL-6, in which it remained unaffected. Among the biochemical changes, the leaf and root nitrate reductase activity was highest in the genotype CL-70 and minimum in Rashmi. Peroxidase activity increased in genotypes Rashmi and Krishna, and protein content in leaves decreased in all the genotypes except Krishna.

TABLE 2.5
Effect of Stage of Harvesting on Curcumin Content of
Improved Varieties

Variety	June–November	June–December	June–January	June–February
Alleppy	9.5	8.0	7.4	7.0
Suvarna	6.0	3.8	3.5	3.5
Suguna	8.5	7.0	6.0	5.8
Sudarsana	8.5	6.7	6.0	5.5

Source: Zachariah et al. (1999).

Turmeric is cultivated for its underground rhizomes. During growth, the photosynthates are translocated to the rhizomes and thus contribute to the economic yield (economic sink). For yield, one of the major contributing factors is the accumulation of dry matter in the parts of the plant, which is more useful than the total dry matter produced (Asana, 1968), and thus, HI would be indicative of yield potential. In general, increased productivity is associated with a higher HI in the root crops. Hence, for obtaining higher economic yields, cultivars that are efficient in mobilizing dry matter content to economic sink are desired, registering a higher HI. In a trial with five cultivars of turmeric, the cultivars differed significantly in their HI, reflecting the physiological variations in their yield potentials. Among the cultivars, Duggirala (70.50%) recorded higher values of HI, followed by Tekurpet (68.70%), which indicates their higher physiological efficiency compared to other cultivars (Sreenivasulu and Rao, 2002).

Mehta et al. (1980) estimated the curcumin content in leaves and rhizomes of three cultivars of *C. longa* and one type of *C. amada* during various stages of growth, starting from the 100th day of planting up to final harvest. With increasing maturity, curcumin content of leaf decreased while that of rhizome increased, which indicates that the site of biosynthesis of curcumin is the leaves, from where it gets translocated to rhizome.

It has already been shown by many researchers that the secondary metabolites are synthesized much too early. In turmeric the synthesis of curcumin starts as early as 120 d after planting and reaches an optimum at 180 to 190 d after flowering (Indian Institute of Spices Research [IISR], Calicut). A study carried out at IISR, Calicut, Kerala showed that if turmeric is planted in June and harvested in November, it yields 30% more curcumin/kg compared to regular harvest at full maturity (Table 2.5). The relative reduction in curcumin content in the latter stage of the plant can be attributed to accumulation of starch and fiber (Zachariah et al., 1999).

Curcumin content of turmeric grown at Coimbatore, Tamil Nadu, is about 40% less compared to that grown in Calicut and Moovattupuzha in Kerala (Table 2.6). This indicated the fact that curcumin is highly dependent on location and other agro climatic conditions and is influenced by genotypes. In Kerala, the peak growing season of turmeric is flanked by heavy monsoons during June to August while in Coimbatore, the rainfall is hardly 100 to 150 mm during the period. It has also been reported that the dry recovery of turmeric is relatively high in Coimbatore compared to locations in Kerala.

Balashanmugam et al. (1993) obtained early sprouting and enhanced vigor with finger rhizomes of turmeric cv. BSR-1 in 0.2% potassium nitrate than that of mother rhizomes and CO-1.

Jayakumar et al. (2001) reported the changes in proteins and RNA during storage of turmeric rhizomes. They found that protein and RNA content increased gradually in the initial stages and rapidly in the final stages of the storage period. SDS-PAGE analysis of the storage proteins revealed syntheses of 56, 52, and 47 KDa proteins during the later stages of storage, with concomitant disintegration of 23 and 16 KDa proteins.

TABLE 2.6
Effect of Location on Curcumin Content of Improved Varieties

	Curcumin (%)		
Variety	Coimbatore, Tamil Nadu	Calicut, Kerala	Moovattupuzha, Kerala
Acc. 363	4.6	6.5	6.0
Prabha	5.5	7.0	7.3
Suguna	4.4	6.5	4.8
Alleppey	5.5	6.4	6.0
Prathiba	4.5	5.8	6.8
Sudarsana	4.5	5.8	6.0

Source: Zachariah et al. (1999).

2.7.1 EFFECT OF GROWTH REGULATORS

Satheesan and Ramadasan (1988b)) reported the effect of CCC on growth and productivity of turmeric under monocropping and under the shade of coconut. Ravindran et al. (1998) tested the effect of triacontanol, paclobutrazol, and GA_3 application on turmeric. Triacontanol application did not produce any significant change in growth of turmeric plants. Paclobutrazol produced a significant dwarfening effect, while GA_3 increased plant height significantly.

Paclobutrazol treatment (0.2 and 0.4%) led to 53 and 60% reduction in growth respectively. However the root length increased significantly. The number of primary, secondary, and tertiary rhizomes increased significantly, while the length decreased proportionately. The leaves became darker and larger and rhizomes thicker with shorter internodes. In the treated plants the girth of mother rhizome, outer zone, inner zone, and intermediate layers are significantly higher compared to the other treatments and control. Paclobutrazol stimulated vascular tissue formation, which is evident from the expanded inner region. As a result the rhizome growth increased 23 and 38%, respectively, in the two treatments. Due to higher procambial activity the number and size of xylem elements and phloem tissue increased significantly. Both oil cells and curcumin cells increased as a result of the treatment. Starch grains increased significantly, indicating increased photosynthate transport and higher carbohydrate accumulation.

2.8 FLORAL BIOLOGY

Pathak et al. (1960) studied the flowering behavior and anthesis of *C. longa*. Flowering in turmeric is reported to vary depending on the cultivars and climatic conditions. Flowering takes place between 109 and 155 days after planting, depending upon the variety and the environment. Turmeric inflorescence takes 7 to 11 days for blossoming after the emergence of the inflorescence. The duration of flower opening within an inflorescence lasts for 7 to 11 days. The anthesis starts from 7 A.M. and continues up to 9 A.M., with the maximum occurring around 8 A.M. Anther dehiscence takes place between 7.15 and 7.45 A.M. (Nazeem et al., 1993; Rao et al., 2006).

Nambiar et al. (1982) were the first to report seed propagation in turmeric especially in *aromatica* types. They studied the flowering behavior, fruit set, and germination in *C. longa* and *C. aromatica*. In *C. aromatica*, the flowering period was July to September, whereas in *C. longa*, it was September to December. Opening of the flowers took place between 6 and 6.30 A.M. at Kasargod, Kerala. Mature capsules were observed during October to November. The number of days taken for flowering in *C. longa* varied from 118 to 143 days, whereas in *C. aromatica*, it varied from 95 to 104 days. The time taken for maturation of seeds from flowering ranged from 23 to 29 days in aromatica. Two distinct types of seeds, dark heavy and light brown, were extracted from mature capsules of *C. aromatica*. Seeds had a white aril, smooth surface and an apical micropylar ring with a wavy outline. The percentage of germination varied from cultivar to cultivar and even plant to plant, and 70 to 90% was recorded in 8 to 20 days after sowing. No germination was observed after 20 days. All the progenies had the somatic chromosome number $2n = 84$ like the parental clones, and seeds of individual plants had remarkable morphological similarities to the parental clones. The seedling progenies produced mainly roots; root tubers and rhizomes were very small during the first year. Normal development occurred during the second year. According to them *C. aromatica* being a tetraploid semi wild species, it produces viable seeds and the production seems to be mainly controlled by environmental factors. In *C. longa* two distinct types, flowering and nonflowering, was noticed. Seed production was not observed in any flowering group. They were completely sterile and had greater vegetative vigor and ability to spread by rhizome compared to *C. aromatica*. Hybridization of *C. aromatica* ($2n = 4\times = 84$) with *C. longa* ($2n = 3\times = 63$) could be attempted through polyploid and haploid breeding techniques to produce plants with differential chromosome types for selecting better types with short duration as in *C. aromatica* and the high rhizome yield and curcumin content as in *C. longa*. Lad (1993) reported a case of seed-setting in cultivated turmeric types (*Curcuma longa*).

Nazeem et al. (1993) studied the floral biology and the nature of seed set in a few turmeric cultivars. The mean number of flowers per inflorescence ranged from 26 to 35.2. The first flower opening commenced 7.6 to 11.6 days after the emergence of the inflorescence. The duration of flower opening in an inflorescence lasted for 8 to 11 days. The anthesis started from 7 A.M. and continued up to 9 A.M. The maximum number of flowers (72%) opened before 8 A.M. The pollen grains of turmeric are ovoid-to-spherical, light yellow, and slightly sticky. Pollen grains show heterogeneity in size among cultivars. Pollen stainability ranges from 71% (cv. Kodur) to 84.46% (cv. Kuchipudi) in the cultivars studied. Among the various media, for pollen germination *in vitro*, modified ME_3 medium at pH 6 gave the best result. The mean pollen length ranged from 7.0 μm (cv. Amalapuram) to 7.2 μm (cv. Dindrigam), while the breadth varied between 4.2 μm (cv. Kodur) and 4.95 μm (cv. VK5). Pollen fertility as well as viability vary with the position of flowers in the inflorescence. It is high in the flowers in the lower portion and low at middle and upper portions (KAU, 2000).

2.8.1 SEED SET

The fruit is a thick-walled trilocular capsule with numerous arillate seeds. The arillate seeds have two seed coats, the outer thick and the inner thin. Seeds are filled with massive endosperm and the embryo is seen toward the upper side of the ovule. Seed set was observed in 8 of the 11 crosses tried in the type Nandyal on open pollination. Setting recorded in crosses involving the aromatica types was obviously high. The type VK5 of the longa group produced seeds when crossed with Kasthuri Thanuku and Dindigam, which are aromatica types. But it failed to yield seeds in combination with Amalapuram. However, seed set in the above crosses cannot be completely ruled out since the type Amalapuram indicated a pollen fertility of 74.75%. No set was observed in the case of Kasthuri Thanuku × Kodur, while its reciprocal cross produced seeds. The seeds were later found to be nonviable. Kodur × Amalapuram also failed to produce seed set. Results obtained on seed set among the various crosses revealed the possibility of combining the high curcumin content of selected *longa* types with the high curing percentage of selected aromatica types. Hybridization among *aromatica* types to evolve superior ones was also found to be feasible. Furthermore, turmeric being vegetatively propagated, any variability obtained through hybridization is fixed immediately and true to types could be multiplied. The failure of all the types to set seeds on selfing indicated a self-incompatibility mechanism. The type Nandyal was found to set seeds when open pollinated. Thus, evaluation of the open pollinated progeny of the aromatica types could be taken up as a breeding method to select superior types (Nazeem et al., 1993).

Renjith et al. (2001) reported *in vitro* pollination and hybridization between two short duration types VK-70 and VK-76 and reported seed set and seed development. This reduces the breeding time and helps in recombination breeding, which has not been attempted in turmeric earlier.

2.8.2 SEED GERMINATION AND SEEDLING GROWTH

Turmeric seed germination commences 17 to 26 days after sowing, and its duration ranged from 10 to 44 days in the various crosses. Crosses VK5 × Dindgam and VK5 × Kasthuri Thanuku gave 100% germination, while Amalapuram × Amruthapani Kothapetta gave only 17.22%. The seeds of the crosses, Amalapuram × Nandyal and Kodur × Kasthuri Thanuku failed to germinate. Nambiar et al. (1982) reported that the percentage of seed germination varied with cultivars and 90% of the seeds germinated within 20 d of sowing. The maximum mean height (70 cm) was observed in the progenies of VK5 × Kasthuri Thanuk and minimum (44 cm) with Kuchipudi × Amalapuram. The mean number of tillers varied between one (VK5 × Kasthuri Thanuku and Nadyal × Amalapuram) and two (Kuchipudi × Amalapuram and VK5 × Dindigam). The mean number of leaves was also maximum in the above crosses (6.5), while minimum (2.18) was recorded in the progenies of Nadyal × Amalapuram. Flowering was observed in Amrithapani Kothapetta × Amalapuram and Nandyal–open pollinated progenies. The seedlings produced only one mother rhizome with root tubers during the first year of growth and the weight ranged from 14.18 to 49.4 g (Nazeem et al., 1993). The size of the mother rhizomes progressively increases over the years and full growth is observed during the third year (Figure 2.8). As the size of the rhizome increases, the number of root tubers declines, especially under cultivation.

George (1981) and Menon et al. (1992) reported variability in the open pollinated progenies of turmeric.

2.8.3 PROPAGATION

The seed set is rare and hence turmeric is propagated vegetatively through rhizome bits. Both the mother and finger rhizomes are used for commercial cultivation. The higher seed weight of the mother rhizomes undoubtedly promotes seedling vigor and better early growth of the plant that is responsible for increased yields but requires a higher seed rate. Although research results favor planting of whole mother rhizomes, planting of primary fingers is the common practice in the major turmeric growing belts. As seed source, fingers have been observed to remain better in storage, remain more tolerant to wet soil conditions at planting, and involve a 33% lower seed rate and lesser cost of seed material (Rao et al., 2006).

2.9 GENETIC RESOURCES

Curcuma is widespread in the tropics of Asia, Africa, and Australia from sea level to an altitude of 2000 m msl (Purseglove, 1974; Purseglove et al., 1981). *Curcuma* consists of about 117 species; from India around 40 species are reported (Velayudhan et al., 1999a,b; Sasikumar, 2005; Ravindran et al., 2006) (Table 2.7). Apart from *C. longa* there are several other species that are used mainly as coloring agents, for production of arrowroot, cosmetics, and medicinal purposes. Species such as *C. aromatica, C. caesia* (black turmeric), *C. amada* (mango ginger), *C. zedoaria, C. purpurescens, C. mangga, C. heyneana, C. xanthorrhiza, C. aeruginosa, C. phaeocaulis* and *C. petiolata* are also cultivated in different places and regions. The genetic resources of turmeric have recently been reviewed by Sasikumar (2005).

India is the world's largest producer, consumer, and exporter of turmeric with an annual production of about 658,400 t from an area of 142,900 ha. Andhra Pradesh is the foremost state in turmeric production with an area of 64,100 ha and a production of 346,400 t, followed by Tamil Nadu, Orissa, Karnataka, and West Bengal (Rao et al., 2006). India has good diversity in turmeric

FIGURE 2.8 Various stages of rhizome development in turmeric. Formation of rhizome in the (a) first, (b,c) second, (d) third year of turmeric seedling and development of plantlets from the rhizome bits — (e) mother rhizome and (f) primaries.

cultivars. Collection and conservation of genetic resources of turmeric have been given great importance in India. In addition to the IISR, which maintains the National Conservatory for Turmeric Germplasm, good regional collections of turmeric germplasm are also maintained at various research centers (Table 2.8). The germplasm is usually maintained in field gene banks, being replanted each year. But planting in the same field year after year may lead to mixing and loss of purity. At IISR the nucleus germplasm is planted in tubs to maintain purity (Figure 2.9). To augment the conservation efforts an *in vitro* gene bank of important genotypes is also operational at IISR and National Bureau of Plant Genetic Resources, New Delhi (Ravindran and Peter, 1996; Ravindran, 1999; Geetha, 2002; Ravindran et al., 2006).

2.9.1 CULTIVAR DIVERSITY

Curcuma collections and species differ in floral characters, aerial morphology, rhizome morphology, and chemical constituents (Valeton, 1918; Velayudhan et al., 1999a,b)). The distinguishing features of *Curcuma* species useful in characterization are given by Velayudhan et al. (1999b). Character-

TABLE 2.7
Curcuma **Species Occurring in India and Their Distribution**

S. No.	Species	Distribution
1	C. aeruginosa	West Bengal
2	C. albiflora	Kerala
3	C. amada	All over India
4	C. amarissima	West Bengal
5	C. angustifolia	Uttar Pradesh, Madhya Pradesh, Himachal Pradesh, North East
6	C. aromatica	South India, Orissa, Bihar
7	C. caesia	West Bengal
8	C. caullina	Maharashtra
9	C. comosa	West Bengal
10	C. petiolata	West Bengal
11	C. decipiens	Kerala, Karnataka
12	C. rubescens	West Bengal
13	C. ferruginea	West Bengal
14	C. longa	All over India
15	C. montana	South India
16	C. neilgherransis	South India
17	C. oligantha	Kerala
18	C. pseudomontana	South India
19	C. reclinata	Madhya Pradesh
20	C. xanthorrhiza	West Bengal
21	C. zedoara	All over India
22	C. sylvatica	Kerala
23	C. aurantiaca	South India
24	C. sulcata	Maharashtra
25	C. inodora	Gujarat, Maharashtra, Karnataka
26	C. ecalcarata	Kerala
27	C. soloensis	West Bengal
28	C. brog	West Bengal
29	C. harita	Kerala
30	C. raktakanta	Kerala
31	C. kudagensis	Karnataka
32	C. thalakaveriensis	Karnataka
33	C. malabarica	Kerala, Karnataka
34	C. karnatakensis	Karnataka
35	C. cannanorensis	Kerala
36	C. vamana	Kerala
37	C. lutea	Kerala, Karnataka
38	C. coriacea	Kerala
39	C. nilamburensis	Kerala
40	C. leucorhiza	West Bengal
41	C. mangga	Andamans[a]

Source: Velayudhan et al. (1999); Shiva et al. (2003).

TABLE 2.8
Germplasm Holdings of Turmeric

Institution	Number of Accessions
Indian Institute of Spice Research, Calicut, Kerala	936
National Bureau of Plant Genetic Resources, Regional Station, Trichur, Kerala	954
Orissa University of Agriculture and Technology, High Altitude Research Station, Pottangi, Orissa	193
NGR Agricultural University, Regional Agriculture Research Station, Jagtial, Andhra Pradesh	352
Y.S. Parmar University of Horticulture and Forestry, Solan, Himachal Pradesh	145
Rajendra Agricultural University, Dholi, Bihar	85
Tamil Nadu Agricultural University, Horticultural Research Station, Bhavanisagar, Tamil Nadu	124
Indira Gandhi Agricultural University, Regional Station, Raigarh, Chattisgarh	42
N.D. University of Agriculture and Technolgy, Faizabad, Uttar Pradesh	114
Uttar Banga Krishi Vishwa Vidyalaya, Pundibari, West Bengal	145
Tamil Nadu Agricultural University, Coimbatore, Tamil Nadu	261

ization is done based on vegetative floral and underground rhizome characterss. Milian et al. (2002) suggested a list of descriptors and characterization of germplasm of spice and medicinal species.

More than 70 turmeric types belonging to *C. longa* and a few cultivars that belong to *C. aromatica* are known to be under cultivation in India. Most of these cultivars go by local names, derived mostly from the place of occurrence (Nair et al., 1980). Popular cultivars of turmeric grown in different places in India and their quality parameters are given in Table 2.9 and Table 2.10.

These germplasm collections are evaluated, characterized, and a database on *Curcuma* species is prepared using the database in MS access and visual basic. The database gives the details of the important species present in India and discriminative features of each species. The data can be retrieved by using serial number and species name (IISR, 2003a).

2.9.2 VARIABILITY

Existence of wide variability among the cultivars with respect of growth parameters, yield attributes, resistance to biotic and abiotic stresses, and quality characters was reported by various workers (Dash et al., 2002; Gopalam and Ratnambal, 1987; Govindarajan, 1980; Hazra et al., 2000; Hegde et al., 1997; Indiresh et al., 1992; Jalgaonkar and Jamdagni, 1989; Jana and Bhattacharya, 2001; Jana et al., 2001; Kumar and Jain, 1998; Kumar et al., 2004; Kumar, 1999; Lynrah and Chakraborty, 2000; Mohanty, 1979; Nambiar et al., 1998; Natarajan, 1975; Nirmal and Yamgar, 1998; Panja and De, 2000; Panja et al., 2001; Palarpawar and Ghurde, 1989; Pathania et al., 1988, 1990; Peter and Kandiannan, 1999; Philip and Nair, 1981, 1986; Pino et al., 2003; Poduval et al., 2001; Prabhakaran, 1991; Pujari et al., 1987; Rakhunde et al., 1998; Rao et al., 1994, 2006; Ratnambal et al., 1986; Raveendra et al., 2001; Ravindran et al., 2006; Shahi et al., 1992, 1994a,b; Shanmugasundaram and Thangaraj, 2002; Shanmugasundaram et al., 2000, 2001; Shridhar et al., 2002; Singh and Tiwari, 1995; Singh et al., 1995, 2001, 2002, 2003; Velayudhan and Liji, 2003; Velayudhan et al., 1999a,b; Venkatesha et al., 2000; Yadav and Singh, 1989, 1996).

Ratnambal et al. (1986) studied variability in quality parameters in over 100 collections of turmeric collected from all over India, belonging to both *longa* and *aromatica* types as well as a few exotic and wild collections. She reported wide variability in curcumin, oleoresin, essential oil contents, and dry recovery (Table 2.10).

Rao et al. (1975) collected more than 100 accessions from within the country and studied their performance. They were broadly classified into short duration *Kasthuri* types belonging to

FIGURE 2.9 Maintenance of turmeric germplasm at National Conservatory, IISR conserving nucleus germplasm in tubs (to maintain purity) and in field.

C. aromatica, maturing in 7 months, and referred to as Ca types, medium duration *Kesari* types, referred to as intermediate or CLi types, and long duration types belonging to *C. longa,* maturing in 9 months and are referred to as CLL types. Clonal selections were made among the more popular cultivars in these categories (Rao et al., 1975; Subbarayudu et al., 1976), which resulted in identification of eight promising types in short duration, six types in medium duration, and eight in long duration group.

Turmeric is affected by foliar as well as rhizome diseases. Among the foliar diseases, leaf spot caused by *Colletotrichum capsici* and leaf blotch caused by *Taphrina maculans* are serious. Rhizome rot caused by *Pythium graminicolum* is the most serious malady of the crop. Identifying disease-resistant varieties is an important breeding objective. The various collections of turmeric germplasm

TABLE 2.9
Popular Cultivars of Turmeric Grown in India

S. No.	Popular Cultivar	Growing Area/Suitability	Yield (t/ha)[a]	Reaction to Pest and Diseases
Andhra Pradesh — short duration Kasthuri (aromatica) types				
1	Kasturi Kothapeta	East and West Godavari districts	15–20	Susceptible to leaf spot
2	Kasthuri Tanuku	West Godavari	12–15	Susceptible to leaf spot
3	Kasthuri Amalapuram	Central delta in East Godavari	10–12	Susceptible to leaf spot
4	Chaya Pasupu	Agency area of Godavari, Vizag, and Srikakulam	—	—
Andhra Pradesh — medium duration Kesari types (C. longa)				
1	Kesari Duvvur	Cuddapah	—	Susceptible to leaf blotch
2	Amruthapani Kothapeta	East Godavari	25	
Andhra Pradesh — long duration types				
1	Duggirala	Krishna and Guntur districts	32	Susceptible to leaf blotch
2	Tekurpeta	Rayalaseema	—	Tolerant to leaf spot
3	Mydukur	Cuddapah	32	Susceptible to rot and leaf spot
4	Armoor	Northern districts of Telangana	25	Susceptible to leaf spot, rot and fly
5	Sugandham	Cddapah	20–25	Susceptible to blotch and rot
6	Vontimitta	Cuddapah	20	—
7	Nandyal	Kurnool	—	—
8	Avanigadda	Krishna	15–18	—
Tamil Nadu				
1	Erode	Tamil Nadu	30–32	
2	Salem	Tamil Nadu	—	—
Kerala				
1	Alleppey	—	25	—
2	Manuthy Local	—	24	—
Assam				
1	Shillong		40	Tolerant to leaf blotch and rot
2	Tall Karbi	Karbi region of Assam	30–40	Tolerant to leaf spot and rot
Maharastra				
1	Rajapuri	Maharashtra and Gujarat	20	Resistant to leaf spot and susceptible to blotch and rot
2	Eavaigon	—	45	—

TABLE 2.9 *(Continued)*
Popular Cultivars of Turmeric Grown in India

S. No.	Popular Cultivar	Growing Area/Suitability	Yield (t/ha)[a]	Reaction to Pest and Diseases
		Meghalaya		
1	Lakadong	Meghalaya & Northeast states	20	—
		Orissa		
1	Duhgi	—	10	—
2	Jobedi	—	—	—
3	Katergia	—	8	—
		Uttar Pradesh		
1	Gorakhpur	—	15	—

[a] Fresh weight.

Source: Rao and Rao (1994).

also exhibited high variability for resistance to various pests and diseases. A list of identified sources for disease resistance especially to nematode (*Meloidogyne incognita*), rhizome rot, leaf blotch, and leaf spot are given in Table 2.11.

Philip and Nair (1986) recorded heritability estimates of 0.52, 0.99, and 0.70 and mean genetic advances (GAs) of 62.6, 45, and 29.7% for the characters fresh rhizome yield, curing percentage, and plant height, indicating high scope for improvement of these characters through selections. The mother rhizome diameter had the highest percentage of heritability and GA (Subramanyam, 1986). In Andhra Pradesh, Reddy (1987) found high heritability with medium GA for the characters rhizome yield, crop duration, number of leaves, number of primary fingers, and height of pseudostem. Characters such as yield of cured turmeric, number of primary fingers, and yield of secondary fingers showed high heritability with appreciable GA (Jalgaonkar et al., 1990). Indiresh et al. (1992) recorded higher heritability estimates with high GA for the characters rhizome yield, internodal distance of primary and secondary fingers, and number of secondary fingers per plant, Yield per plant had more GA, whereas length and width of rhizome showed high heritability (Yadav and Singh, 1996). However, superior genotypes may be obtained through selection based on the number and weight of mother, primary, and secondary rhizomes (Singh et al., 2003).

Association of various characters among themselves and with yield is essential, as this information can provide the criteria for indirect selection. Studies on correlation indicated the magnitude of association between any pair of characters, hence aid in selection. Leaf length exerted the highest influence on yield followed by leaf width, plant height, number of leaves, and primary and secondary fingers (Natarajan, 1975). Plant height, number of leaves, and leaf area are positively correlated to the yield of rhizomes but number of tillers is negatively correlated (Mohanty, 1979). The yield per plant was highly associated with length of primary fingers, and length and girth of secondary fingers. The correlation coefficients of these yield components were positive and significant (George, 1981). Mother rhizome diameter has significant and positive association with yield (Subramaniyam, 1986). Number of leaves, number of primary fingers, and crop duration had shown positive association with rhizome yield at both genotypic and phenotypic levels (Reddy, 1987; Panja et al., 2002). Panja et al. (2000) reported that plant height, leaf number, number of primary fingers, and number and weight of secondary fingers had positive correlation with yield. Path analysis indicated the importance of plant height, leaf number, number of primary fingers, and weight of secondary

TABLE 2.10
Cultivar Diversity of Turmeric and Their Quality Parameters

S. No.	Cultivar/Accessions	Dry Recovery (%)	Oleoresin (%)	Oil (%)	Curcumin (%)
		C. longa			
1	Maran	26.0	13.5	7.0	8.7
2	Jorhat	21.7	10.8	7.5	6.9
3	Dadra, Gauhati	23.2	16.6	7.0	7.7
4	Kaziranga, Jorhat	24.5	18.2	6.0	10.2
5	Anogiri, Garohills	26.9	13.6	6.0	5.2
6	Nowgong, Assam	20.0	10.0	5.0	4.0
7	Mekhozer	20.0	12.0	5.0	4.0
8	Hajo,Gauhati	21.0	13.0	7.5	5.5
9	Rajasagar	16.6	10.3	6.0	5.0
10	Teliamura, Agarthala	23.5	13.0	7.0	5.5
11	Barhola, Jorhat	25.0	13.3	7.0	5.3
12	Kahikuchi	21.2	12.1	9.5	3.1
13	Along	21.7	13.0	5.0	6.6
14	Besar, along	20.6	11.6	5.0	6.1
15	Gaspani, Nagaland	24.4	11.0	8.0	4.5
16	Singhat, Manipur	19.7	15.0	7.0	7.9
17	Kongpopkri	21.4	13.2	7.5	5.6
18	Aigal	20.0	14.0	5.0	9.0
19	Amampuri, jumpoi Hills	25.7	11.5	5.0	4.0
20	Amkara, Tripura	22.7	12.7	8.0	5.6
21	Torku	19.5	10.9	6.0	6.0
22	Barpather, Galoghat	23.3	12.0	3.0	4.3
23	Rorathong, E. Sikkim	22..8	16.8	4.0	4.7
24	C11 316, Gorakpur	18.7	15.0	6.0	6.0
25	Pusa	13.5	14.5	7.5	7.2
26	PTS-5	20.0	12.7	5.0	6.0
27	PTS-10	22.5	15.0	5.0	7.7
28	PTS-24	23.0	14.1	5.0	7.9
29	PTS-68	22.6	12.9	5.0	5.1
30	Amalapuram	20.2	16.0	8.0	6.0
31	Cls No. 34	23.8	14.0	5.0	5.0
32	Amalapuram II	20.0	13.3	4.0	4.3
33	Cls No. 15	21.8	16.0	5.5	4.8
34	Cls. No. 3	22.3	12.5	6.6	5.0
35	Amalapuram Sel. III	30.0	16.5	6.0	5.0
36	Cll 390 Amalapuram	19.4	13.7	7.5	7.0
37	Amrithapani	23.2	19.0	7.0	7.0
38	Amrithapani, Kothapetta	32.4	15.0	4.0	7.0
39	Nandyal Type	22.0	13.5	8.0	4.7
40	Cls No. 13	23.4	16.8	5.0	5.4
41	Vontimitta	18.0	10.5	6.5	5.4
42	Cls No. 11A	21.2	13.0	5.0	4.0
43	Cll 322 Vontimitta	24.5	11.4	6.0	7.4
44	GL Puram II	19.6	13.1	6.0	6.2
45	Cls No. 5A	30.1	13.5	6.0	4.9
46	GL Puram III	21.8	12.9	6.0	5.1

TABLE 2.10 *(Continued)*
Cultivar Diversity of Turmeric and Their Quality Parameters

S. No.	Cultivar/Accessions	Dry Recovery (%)	Oleoresin (%)	Oil (%)	Curcumin (%)
47	Cll 324 Armoor	18.0	15.6	6.5	7.0
48	Cls No. 1	25.0	10.8	5.0	6.0
49	Cls No. 1A	24.0	15.8	5.0	6.4
50	Cls No. 1C	20.0	11.1	5.0	6.3
51	Ethamukula	22.5	15.0	5.0	5.5
52	Cls No. 26	24.6	12.0	5.0	5.7
53	Cll 321 Ethamukula	26.2	11.3	6.5	6.0
54	Cls No. 27B	19.9	10.2	5.0	4.0
55	Duggirala	17.6	14.6	5.0	7.5
56	Cls No. 22	18.1	15.4	6.6	5.2
57	CII 325 Duggirala	21.0	11.7	8.0	5.0
58	Kuchipudi	18.6	14.0	7.0	7.5
59	Cls No. 8B	14.7	12.9	5.0	6.0
60	Cls No. 8C	19.4	14.3	6.0	7.9
61	T Sundar	20.0	16.0	6.0	6.9
62	Sugandham	22.6	12.0	8.5	9.1
63	Cls No. 19	19.0	13.8	6.0	5.4
64	Dindigam	17.0	10.6	6.0	6.4
65	CII 327, Takkurpet	21.0	14.0	6.5	6.1
66	CII 326, Mydukkur	19.4	11.8	6.0	2.8
67	Karhadi Local	18.3	13.0	6.0	4.9
68	Cls No. 7	26.9	12.7	7.0	5.0
69	CII 323 Avanigadda	19.4	14.5	7.5	7.0
70	Cls No. 30	22.0	13.0	7.0	6.5
71	CIIs 328 Sugandam	20.0	12.8	6.0	9.0
72	Cls No. 9A	14.5	11.6	4.0	7.9
73	No. 24	23.0	16.0	5.0	5.0
74	Cls No. 24	22.0	14.3	6.0	4.5
75	Cls No. 6	21.0	13.9	4.0	6.7
76	Cls No. 6A	24.0	15.8	5.0	6.4
77	Palani	21.0	16.5	5.0	7.6
78	Kayyam, Gudalur	23.0	16.5	5.0	8.4
79	Pathavayal, Gudalur	25.0	14.0	4.0	3.6
80	Upper Dinamala	24.0	12.0	5.5	7.8
81	Rajpuri Local	16.3	13.8	7.0	7.8
82	Cls No. 14B	18.5	13.5	5.0	6.5
83	CII 390 Rajpuri	18.3	12.2	8.0	6.0
84	Moovattupuzha	20.1	11.5	5.0	7.0
85	Varapetty, Kothamangalam	18.0	10.9	8.0	5.4
86	Pathanapuram	17.0	14.0	8.0	6.7
87	Karimala, Mannarghat	27.0	12.5	8.0	5.5
88	Ochira	21.8	12.0	5.0	5.6
89	Cls No. 29	23.5	11.7	4.0	4.0
90	Alleppey	17.2	13.0	8.0	6.0
91	Cls No. 21	25.5	12.1	8.0	6.2
92	Valra falls, Adimali	20.0	14.0	6.5	6.0
93	Mundakkayam	23.4	10.5	8.5	3.2

Continued

TABLE 2.10 *(Continued)*
Cultivar Diversity of Turmeric and Their Quality Parameters

S. No.	Cultivar/Accessions	Dry Recovery (%)	Oleoresin (%)	Oil (%)	Curcumin (%)
94	Mananthody	22.5	16.5	8.5	9.1
95	Cls No. 16	25.0	18.3	9.0	7.0
96	Vandoor, Nilambur	22.6	10.5	4.0	5.4
97	Manjapally, Perumbavoor	16.4	10.6	6.5	5.6
98	Murangathapally	20.0	13.0	8.5	7.8
99	Puthuppadi, Meenangadi	24.4	11.3	5.9	5.4
100	Edapalayam	22.3	14.5	6.0	10.9
101	Erathupetta	20.0	11.2	5.0	6.0
102	Erathukunnam	21.0	12.0	6.0	10.3
103	Idukki No. 1	21.6	10.8	4.0	8.5
104	Idukki No. 2	28.5	13.7	4.0	9.0
105	Thodupuzha	21.2	14.8	6.9	9.5
106	Cls No. 28	24.0	13.0	5.0	5.7
107	Palapally, Trichur	21.0	15.3	6.0	10.7
108	Kolathuvayal	20.0	13.5	7.0	4.2
109	Elanji, Idukki	20.0	12.6	7.0	2.7
110	Karuvilangad	21.0	13.9	4.0	6.2
111	Ayur	21.5	12.4	4.6	5.1
112	Kothamangalam	23.5	10.7	5.0	4.2
113	Kakkayam Local	20.5	14.5	9.0	7.5
114	Chamakuchi	26.5	16.9	4.0	7.0
115	Anchal	28.0	12.4	5.0	5.4
116	Muringakalla	23.1	11.8	5.0	7.0
117	Mongam, Malappuram	26.2	12.0	7.0	5.7
118	Maramboor	18.0	13.0	4.0	5.6
119	Ernad	30.5	12.9	9.0	5.2
120	Wynad local	20.0	15.3	7.0	9.4
	C. aromatica				
1	Silapather, N. Lekhimpur	21.7	12.5	5.0	3.2
2	Burahazer, Dibrugarh	28.0	11.6	4.0	3.1
3.	Tura, Garohills	25.0	13.9	4.0	4.3
4	Dibrugarh	20.3	12.5	6.5	8.0
5	Hahim	27.2	13.0	6.0	2.3
6	Aseemgiri, Garohills	17.8	10.5	7.0	3.5
7	Bahumura, Agarthala	18.0	13.5	4.0	5.0
8		18.5	10.5	5.0	2.5
9	Nagsar, Titasar, Jorhat	18.8	14.4	5.0	5.0
10	Besar, Along	24.1	12.1	5.5	4.1
11	Kanchanpur, Tripura	20.0	9.6	8.0	3.5
12	Namachi	18.7	12.4	6.0	3.7
13	Pakyong	22.4	12.5	5.5	3.9
14.	Nayabunglow, Meghalaya	26.0	14.0	5.0	4.0
15	Shillong	23.2	14.0	5.5	4.7
16.	Phu, E. Sikkim	20.6	14.6	5.0	4.7
17	Pedong, Kalimpong	21.0	13.0	5.5	4.0
18	Ca 72 Udayagiri	22.5	12.9	9.0	7.4

TABLE 2.10 *(Continued)*
Cultivar Diversity of Turmeric and Their Quality Parameters

S. No.	Cultivar/Accessions	Dry Recovery (%)	Oleoresin (%)	Oil (%)	Curcumin (%)
19.	Cas No.57	21.0	10.9	8.0	4.1
20.	G L Puram I	23.0	11.5	8.5	4.0
21.	Ca 66 GL Puram	17.8	12.8	9.0	3.5
22.	Armoor	20.6	14.5	8.0	4.0
23.	Kodúr	20.3	14.2	8.0	2.5
24.	Chayapasupu	17.2	16.0	5.0	3.5
25.	Ca I Chayapasupu	18.9	12.0	6.0	2.8
26	Ca 69 Dindigam	20.2	12.4	8.0	3.1
27.	Ca 68 Dhagi	20.6	13.5	7.0	2.8
28.	Katirgia	18.8	13.0	8.5	4.1
29.	Jobedi	25.7	12.0	9.0	3.2
30	Kasturi	18.8	10.5	6.5	2.9
31.	KasturiTanuku	19.2	14.0	6.0	3.0
32	Ca 73 Amalapuram	22.0	13.3	8.0	3.8
33	Cas No.58	14.0	11.7	8.0	6.0
34.	Cas No. 58B	20.9	12.0	7.0	3.1
35	Erode	25.5	11.5	9.0	4.7
36	Nadavayal	19.8	10.3	5.0	5.0
37	Keeranthode	21.3	12.5	5.0	4.0
38	Makkapuzha, Ranni	26.1	19.2	5.0	5.0
39	Konni	24.2	10.7	8.5	6.6
40	Thachanatukara, Mannarghat	23.8	11.7	4.0	5.7
41	Mampad, Nilambur	18.5	16.0	5.0	3.0
42	Chamakuchi	18.0	12.0	5.0	2.3
43	Adimali	20.0	10.3	8.5	4.0
44	Amnicad	22.0	10.3	8.5	4.0
	Exotic types (Solomon Islands)				
1	Mamarei	20.0	13.2	6.0	3.0
2	Vatuloro	18.0	10.0	6.0	3.4
3	Vanagobulu	17.5	11.0	5.0	3.1
4	Cokuma	—	12.0	6.0	4.1
5	Tsavana	—	10.5	6.0	3.9
6	Tuva Vitalio	—	12.0	6.0	2.8
	Other *Curcuma* sp.				
1	*C. angustifolia* Roxb.	30.0	7.6	3.0	0.2
2	*C. xanthorrhiza* Roxb.	25.8	10.0	2.0	1.5
3	*C. zedoaria* (Berg.) Rosc.	20.5	6.0	2.0	2.0
4	Wild unidentified (1)	22.7	6.4	2.0	0.3
5	Wild unidentified (2)	30.0	7.9	2.0	2.2
6	Wild unidentified (3) from Uttar Pradesh	23.6	7.2	5.0	1.3
7	Wild unidentified (4)	26.5	8.6	3.0	0.8
8	Wild unidentified (5)	30.8	4.5	4.0	0.2
9	Wild unidentified (6) from Nagsar, Titsar, Jorhat	24.0	6.2	1.0	0.5
10	Wild unidentified (7) (Dergroni, Jorhat)	21.4	5.8	4.0	1.6

Continued

TABLE 2.10 *(Continued)*
Cultivar Diversity of Turmeric and Their Quality Parameters

S. No.	Cultivar/Accessions	Dry Recovery (%)	Oleoresin (%)	Oil (%)	Curcumin (%)
11	Wild unidentified (8) (Kattapana, Idukki)	31.4	8.6	2.0	0.02
12	Wild unidentified (9) (Taranagar, Agarthala)	30.7	4.0	1.5	2.6
13	Wild unidentified (10) (Sibsagar)	25.0	9.6	3.0	0.05
14	*C. amada* Roxb.	30.0	5.0	1.5	—

Source: Ratnambal, M.J. (1986) *Qualities Plantarum,* 36, 243–252.

fingers as criteria for selection. Correlation and path analysis studies were also reported by Nandi et al. (1994), Singh and Tiwari (1995), Hazra et al. (2000), etc., Tomar et al. (2005).

The yield of cured turmeric correlated significantly with the secondary fingers (Jalgaonkar et al., 1990). Cholke (1993) observed significant and positive correlation between primary fingers, plant height, and yield. Shashidhar et al. (1997) noted positive correlations of fresh rhizome yield with all growth and yield parameters investigated, including dry matter accumulation and nutrient uptake. In the character association studies by Rao (2000), all growth and yield characters studied showed positive correlation with cured rhizome yield. The quality characters (curcumin, essential oils, and oleoresins) had shown negative correlation. The genotypic coefficient of variation was high for curcumin content followed by length of secondary fingers and number of tillers. Cured rhizome yield per hectare was positively and significantly correlated with plant height, number of leaves, leaf area, LAI, and fresh yield per plant. For crop improvement in turmeric, plant height, and number of leaves determines the yield potential of the genotype (Narayanpur and Hanamashetti, 2003).

Number of tillers, plant height, and number of fingers had significant correlation with yield. The number of tillers, number of leaves, plant height, and leaf length × breadth were found to have highly significant intercorrelation among themselves. Path coefficient analyses indicated that whenever significant positive correlation between yield and morphological characters was established, it was mainly due to the substantial positive contributions by plant height and number of fingers either directly or indirectly. It is therefore concluded that plant height is the single important morphological character on which selection for yield could be made.

Tomer et al. (2005) studied character association and path analysis for yield components in turmeric and found plant height, leaf length, thickness of primary and secondary rhizomes, and number of secondary rhizomes had significant positive association with rhizome yield. Path analysis showed positive direct effect of plant height, leaf length, and thickness of primary and secondary rhizomes on rhizome yield. Hence these traits may be given due emphasis while making selections for higher rhizome yield. Jirali et al. (2003) reported that cultivars Amalapuram, Bidar-4 and Cuddapah are physiologically more efficient and hence can be used as genetic sources for yield improvement.

Yield being a complex and polygenically controlled character, direct selection for yield may not be a reliable approach because it is highly influenced by environmental factors. Therefore, it is essential to identify the direct effects and indirect effects of the component characters through which yield improvement could be obtained. Nambiar (1979) indicated that wherever significant positive correlations between yield and morphological characters were established, they were mainly due to substantial positive contribution by plant height and number of fingers either directly or indirectly. Based on this, it was concluded that plant (pseudostem) height in turmeric is the single important morphological character for which selection for yield could be made.

Subramaniyam (1986) revealed that the diameter of mother rhizome is the main determinant of yield in turmeric. Geetha and Prabhakaran (1987) observed plant height and length of secondary

TABLE 2.11
Sources for Pest and Diseases Resistance in Turmeric[a]

Disease Trait	Pest Reaction	Promising line/cultivar/variety	Location[a]	Reference
Nematode (*Meloidogyne incognita*)	Resistant	Acc. 43, 56, and. 57	Calicut, Kerala	IISR (2005)
		Acc Nos. 31, 82, 84, 142, 178, 182, 198, and 200	Calicut, Kerala	Sarma et al. (2001)
Rhizome rot (*Pythium graminicolum*)	Resistant	PCT-13, PCT-14, and Shillong	Calicut, Kerala	Sarma and Anandaraj (2000)
	Resistant	GS, JTS-303, CLI-370, JTS-604, JTS-308, PTS-9, PTS-10, CLI-325, CLI-330, PCT-10, CLI-320, PCT-7, and PCT-14; RH-5, NDH-18, TCP-2, JTS-12, and JTS-15 (long duration), JTS-314, 320, 321, 325, and 319 (medium duration) and PCT-13, JTS-612, and 607 (short duration)	Jagtial, A.P.	AICRPS (2003, 2005)
	Resistant/tolerant	PCT-13 and PCT-14	Jagtial, A.P.	Rao et al. (1992)
Leaf blotch (*Taphoina maculans*)	Resistant/tolerant	Long-duration types	Jagtial, A.P.	Rao et al. (1992)
	Resistant	CLL-324 Ethamukala, CLL-316 Gorakhpur, CLL-326 Mydukur, Alleppey, PCT-12 and PCT-13	Calicut, Kerala	Sarma and Anandaraj (2000)
	Resistant	ACC-360, ACC-361, ACC-585, ACC-126, and T4-11	Raigarh, Chattisgarh	AICRPS (2003)
	Resistant	PTS-11, 15, 52, 55, and 59 and JTS-319 (medium duration)	Jagtial, A.P.	AICRPS (2005)
	Highly resistant	Kohinoor, G.L. Puram, Rajendra Sonia, RH-5, and RH-24	Dholi, Bihar	AICRPS (2003)
	Immune	PTS-3, PTS-4, PTS-11, Roma and Rashmi	Pantnagar, U.P.	Verma and Tiwari (2002)
	Resistant	PTS-10 and PTS-36	Pantnagar, U.P.	Verma and Tiwari (2002)
Leaf spot and leaf blotch	Resistant	CL-32, CL-34, CL-54, and CL-55	Coimbatore, T.N.	Subramanian et al. (2004)
	Tolerant	Clone No. 326 and 327	Guntur, A.P.	Sri Ramarao (1982)
	Resistant	CL-32, CL-34, CL-54, and CL-55	Coimbatore, T.N.	AICRPS (2003, 2005)
Leaf spot (*Colletotrichum capsici*)	Resistant	Bhendi, Gadhait, and Krishna	Calicut, Kerala	Sarma and Anandaraj (2000)

Continued

TABLE 2.11 *(Continued)*
Sources for Pest and Diseases Resistance in Turmeric[a]

Disease Trait	Pest Reaction	Promising line/cultivar/variety	Location[a]	Reference
	Resistant	JTS-606; PTS-11, JTS-10 to 15, and Duggirala Red (long duration), JTS-314 to 326, and CLI-317 (medium duration), JTS-607 to 612 and PCT-13 (short duration)	Jagtial, A.P.	AICRPS (2003, 2005)
	Resistant	Sudarshan and RRTS-1	Raigarh, Chattisgarh	AICRPS (2003)
	Resistant	PTS-39 and PTS-27	Pantnagar, U.P.	Verma and Tiwari (2002)
	Resistant/tolerant	Short-duration PCT cultures	Jagtial, A.P.	Rao et al. (1992)

[a] A.P.: Andra Pradesh; T.N.: Tamil Nadu; U.P.: Uttar Pradesh.

fingers to be the major contributing characters toward rhizome yield. Nandi et al. (1992) observed the highest direct effect by girth of finger followed by weight of finger. The path analysis studies of Rao (2000) indicated that selection for greater weight of fresh finger and mother rhizome and curing percentage would be more effective as they have a positive direct effect as well as significant positive association with cured turmeric yield.

In the genetic divergence studies of Nambiar (1979), cultivars of *C. longa* and *C. aromatica* were grouped into four clusters based on generalized D^2 statistics. The cultivars belonging to *C. aromatica* were grouped separately from those belonging to *C. longa*. Estimation of oil and curcumin contents in different cultivars of *C. longa* and *C. aromatica* indicated that the variability was high in *C. longa* as compared to *C. aromatica*.

Chandra et al. (1997) reported that the cvs. PCT-13 and Lakadong formed solitary groups and were genetically most distant. The land races of the northeast region were almost clustered in low-to-moderate yield groups, while genotypes from the southern region were scattered among different complexes ranging from moderate to high yielders (Chandra et al., 1998, 1999).

Genetic divergence studies (D^2 analysis) with 54 genotypes showed wider diversity among genotypes studied and were grouped in as many as six clusters (Rao, 2000). The genotypes PTS-38 and Duggirala in cluster I for the highest cured yield, PCT-5 and 8 in cluster III for more curcumin, essential oils, and oleoresins, PCT-14, PCT-13, and PCT-10 in cluster IV for shorter duration, medium yield with good curcumin content were identified as potential parents for future breeding programs.

Although a detailed morphological characterization of turmeric was done, its molecular characterization is still in a nascent stage except for some genetic fidelity studies of micropropagated plants and isozyme-based characterization (Sasikumar, 2005). Isozyme markers were used to characterize turmeric germplasm by Shamina et al. (1998). Allozyme profiles, molecular markers, and sequence analysis were used to study genetic diversity in populations of *C. alismatifolia* (Paisooksantivatana et al., 2001). PCR-based detection of adulteration in market samples of turmeric powders was attempted (Remya et al., 2004; Syamkumar et al., 2005; Sasikumar et al., 2005). Identification of somaclonal variation (Salvi et al., 2001) and PCR-based techniques have been made use of to differentiate within and among *C. wenyujin, C. sichuanensis,* and *C. aromatica* (Chen et al., 1999). Pimchai et al. used molecular markers for the identification of early flowering *Curcumas*. Kress et al. (2002) employed molecular techniques to work out the phylogeny of

Zingiberaceae. Cao et al. (2001), Sasaki et al. (2002), and Xia et al. (2005) employed such techniques for the authentication of Chinese and Japanese curcuma drugs, while Chase (2004) utilized it to have an overview of the phylogeny and relationships in monocots. PCR-based techniques are being used for germplasm characterization at IISR (IISR, 2003b,c, 2004).

2.10 CROP IMPROVEMENT

Crop improvement programs in turmeric were limited to selection and mutation breeding for many years, mainly to identify turmeric types with high yield potential, high curing percentage, and high curcumin content. Studies carried out on blossom biology and viable seed set obtained through successful hybridization have opened the way for recombination breeding programs (Nambiar et al., 1982; Nazeem et al., 1993; Sasikumar et al., 1994, 1996b).

Presently, crop improvement work in turmeric in India is under progress at the IISR, and under the aegis of All India Coordinated Research Project on Spices (ICAR) in centers such as Pottangi (Orissa), Coimbatore (Tamil Nadu), Regional Agricultural Research Station (RARS), Jagtial (Andhra Pradesh), Pundibari (West Bengal), and Solan (Himachal Pradesh) besides a few agricultural universities in the country.

2.10.1 SELECTION

The varietal improvement work so far attempted is only through clonal selection because hybridization could not be practiced due to sterility and rare seed set, and to a certain extent incompatibility. Most of the selections have been made from the landraces collected from various parts of the country. Improved selections were developed mainly in *C. longa* and to a lesser extent in aromatica types and *C. amada*.

Pujari et al. (1986) released a high-yielding variety Krishna (7.2 t/ha) for general cultivation in Maharashtra. This variety also performed well (53.86 t/ha of fresh rhizomes) in the Konkan region (Jalgaonkar et al., 1988).

After many years of research work at the High Altitude Research Station in Pottangi (Orissa), PTS-10 and PTS-24 were identified for high rhizome yield and high dry recovery (31 and 26% respectively). Curcumin content, essential oil, and oleoresin contents were 9.3, 4.2, and 13.5% for PTS-10 and 9.3, 4.4, and 13.1% for PTS-24. They are orange-yellow and comparatively less susceptible to leaf spot (*Colletotrichum* sp), rhizome rot, and scale infection and were released as Roma and Suroma and recommended for large-scale cultivation to replace local varieties (Anon., 1986).

At the IISR, evaluation of a large number of accessions were carried out, and 19 high-yielding accessions designated as PCT-1 to PCT-19 were selected (Ratnambal and Nair, 1986) and given for multilocation testing. Among them, the best selections PCT-8, PCT-13, and PCT-14 with high yield potential and high curcumin content were released as Survarna, Suguna, and Sudarshana. Suguna and Sudarshana are short-duration varieties and are also field tolerant to rhizome rot (Figure 2.10). Subsequently two more high-yielding high-quality lines, IISR Alleppy Supreme and IISR Kedaram, were released (Ratnambal et al., 1992; Sasikumar et al., 2005).

Maurya (1990) recommended RH-10 Rejendra sonia for Bihar. It has a yield potential of 42.0 t/ha fresh rhizomes, 7.5 t/ha dry rhizomes, and 8.4% curcumin content.

In Northern Telangana Zone of Andhra Pradesh, fresh and cured rhizome yields were higher in Armoor than in other cultivars. In short-duration cultivars, Reddy et al. (1989) recorded the highest rhizome yields in PCT-13 and PCT-14 (28.04 and 26.71 t/ha fresh and 5.69 and 5.44 t/ha cured yield, respectively).

Nandi (1990) recorded the highest rhizome yield in PTS-25 (27.0 t/ha), followed by CLS-9 (24.6 t/ha). Indiresh et al. (1992) found the highest fresh yield in PCT-8 (32.3 t/ha) followed by c.v. Waigon (31.6 t/ha). The highest cured yield was recorded in PCT-8 (6.5 t/ha). Ratnambal et al. (1992) reported a fresh rhizome yield of 29 and 28.82 t/ha in PCT-13 and PCT-14, respectively.

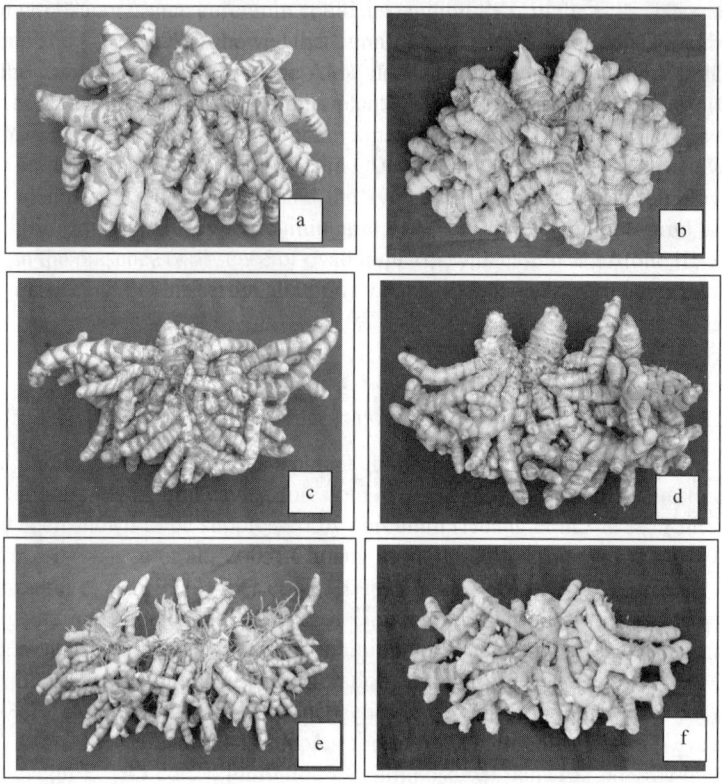

FIGURE 2.10 Rhizomes of improved varieties and important species: (a) Sudarshana (short duration longa type). (b) Pratibha (longa type seedling selection). (c) Suvarna (longa type). (d) Amalapuram (aromatica type). (e) Black turmeric (*C. caesia*). (f) Mango ginger (*C. amada*).

Hegde (1992) found higher fresh rhizome yield in the cultivar Cuddapah (22.19 t/ha) followed by Bidar (21.69 t/ha). Cholke (1993) recorded the highest fresh rhizome yield in Cuddapah (33.37 t/ha), followed by Amalapuram (31.44 t/ha). However the cured yield was higher in the case of Amalapuram followed by Cuddapah.

Under rainfed conditions, BSR-1 was found by Sheshagiri and Uthaiah (1994) to be the best in fresh rhizome yield (16.57 t/ha) followed by Waigon (15.45 t/ha). Radhakrishna et al. (1995) recorded the highest green turmeric yield in CO-1 (16.54 t/ha), followed by BSR-1 (14.74 t/ha). Kurian and Valsala (1996) reported that VK-5 was superior in fresh rhizome yield (5.85 kg/m^2) followed by VK-155 (5 kg/m^2). The cultivar Chayapasupu recorded the highest yield (43.73 t/ha) of fresh rhizome followed by VK-31 (42.42 t/ha).

The yield of cured rhizome was highest in VK-116 (5.63 t/ha) under 25 to 30% shaded conditions. But in open conditions, VK-31 (35.06 t/ha) and VK-55 (34.66 t/ha) were better yielders in terms of fresh rhizome yield (Latha et al., 1995). Ramakrishna et al. (1995) found that PCT-13 was superior in yield (19.15 t/ha). Kurian and Nair (1996) reported higher fresh rhizome yield in VK-121 (43.02 t/ha) and cured yield in VK-116 (8.43 t/ha). Kumar et al. (1992) reported that the cultivars RH-10 produced the highest yields of fresh rhizomes (30.1 t/ha) and cured product (6.1 t/ha). Curing percentage was highest in G.L.Puram (22.3%).

At present there are about 24 improved varieties of turmeric in India and one in *C. amada*, evolved through selection from germplasm, clonal selection, mutant selection, and selection from open-pollinated progeny (seedlings selection). Important attributes of improved varieties are given in Table 2.12 (Johny and Ravindran, 2006).

TABLE 2.12
Important Characters of Improved Varieties of Turmeric

Sn. No.	Variety	Pedigree	Crop Duration (days)	Mean Yield (Fresh) (t/ha)	Dry Recovery (%)	Curcumin (%)	Oleoresin (%)	Essential Oil (%)	Important Characters
1	CO-1	Mutant (X-ray) selection from Erode local	270	30.5	19.5	3.2	6.7	3.7	Bold, bright orange rhizomes, suitable for drought prone areas
2	BSR-1	Mutant (X-ray) selection from Erode local	285	30.7	20.5	4.2	4.0	3.7	Suitable for drought prone areas
3	BSR-2	Mutant selection from Erode local	245	32.7	—	—	—	—	Bold rhizomes resistant to scale insects
4	Krishna	Clonal selection from Tekurpeta	240	9.2	16.4	2.8	3.8	2.0	Moderately tolerant to pests and diseases
5	Sugandham	Germplasm selection	210	15.0	23.3	3.1	11.0	2.7	Moderately tolerant to pests and diseases
6	Roma	Clonal selection from Tsundur	250	20.7	31.0	6.1	13.2	4.2	Suitable for hilly areas
7	Suroma	Mutant (X-ray) selection from Tsundur	253	20.0	26.0	6.1	13.1	4.4	Field tolerant to leaf blotch, leaf spot and rhizome scale
8	Ranga	Clonal selection from Rajpuri local	250	29.0	24.8	6.3	13.5	4.4	Bold rhizomes, moderately resistant to leaf blotch and rhizome scales
9	Rasmi	Clonal selection from Rajpuri local	240	31.3	23.0	6.4	13.4	4.4	Bold rhizomes
10	Rajendra Sonia	Local germplasm selection	225	23.0	18.0	8.4	—	5.0	Bold and plumpy rhizomes
11	Megha turmeric	Selection from Lakadong types	300–315	20.0	16.37	6.8	—	—	Bold rhizomes
12	Pant Peetabh	Selection from germplasm	—	29.0 (potential)	18.5	7.5	—	1.0	Resistant to rhizome rot

Continued

TABLE 2.12 (Continued)
Important Characters of Improved Varieties of Turmeric

Sn. No.	Variety	Pedigree	Crop Duration (days)	Mean Yield (Fresh) (t/ha)	Dry Recovery (%)	Curcumin (%)	Oleoresin (%)	Essential Oil (%)	Important Characters
13	Suranjana	Local germplasm selection	235	—	21.2	5.7	10.9	4.1	Tolerant to rhizome rot and leaf blotch; resistant to rhizome scales and moderately resistant to shoot borer
14	Suvarana	Germplasm selection	200	17.4	20.0	4.3	13.5	7.0	Bright orange colored rhizome with slender fingers
15	Suguna	Germplasm selection	190	29.3	20.4	7.3	13.5	6.0	Short duration type, field tolerant to rhizome rot
16	Sudarshana	Germplasm selection	190	28.8	20.6	5.3	15.0	7.0	Short duration type, field tolerant to rhizome rot
17	IISR Prabha	Selection from open pollinated seedlings	205	37.47	19.5	6.5	15.0	6.5	—
18	IISR Pratibha	Selection from open pollinated seedlings	225	39.12	18.5	6.21	16.2	6.2	—
19	IISR Alleppey Supreme	Selection from Alleppey Finger Turmerics	210	35.4	19.0	5.55	16.0	—	Tolerant to leaf blotch
20	IISR Kedaram	Germplasm selection	210	35.5	18.9	5.9	13.6	—	Tolerant to leaf blotch
21	Kanthi	Clonal selection from Mydukur	240–270	37.65	20.15	7.18	8.25	5.15	Big mother rhizomes and bold fingers with short internodes
22	Sobha	Germplasm selection	240–270	35.88	19.38	7.39	9.65	4.24	Big mother rhizomes and bold fingers with short internodes
23	Sona	Germplasm selection	240–270	4.02 t dry	18.88	7.12	10.25	4.4	Field tolerant to leaf blotch
24	Varna	Germplasm selection	240–270	4.16 t dry	19.05	7.87	10.8	4.56	Bold rhizome with short internodes, field tolerant to leaf blotch

Source: Edison et al. (1991); Johny and Ravindran (2006a,b).

2.10.2 Mutation Breeding

Induced mutation was attempted in turmeric, which led to the release of two high-yielding mutants, CO-1 and BSR-1, both from Erode local (Muthuswamy and Shah, 1982). Chezhiyan and Shanmuga Sundaram (2000) reported a new high-yielding mutant BSR-2 from Erode local through ionizing radiations (X-rays). BSR-2 gives 690 to 890 g of fresh rhizome yield per plant, with a curcumin content of 3.75% and a curing of 20%. The effect of gamma rays in inducing mutations in turmeric was reported by Rao (1999). Preliminary trials conducted at Anantharajupeta using chemical mutagens viz., Colchicine, EMS, and MNG each at 250, 500, 1000 ppm on cv. Mydukar resulted in taller and higher yielding colchiploids (Anon., 1986).

2.10.3 Somaclonal Variation

Somaclonal variation is an important source of variation in predominantly vegetatively propagated crops such as turmeric. Variations in rhizome characters of somaclones were reported by Nirmal Babu et al. (2003) and Praveen (2005). Variants with high curcumin content were isolated from tissue-cultured plantlets (Nadgauda et al., 1982). Root rot disease–tolerant clones of turmeric cv. Suguna were isolated using continuous *in vitro* selection technique against pure culture filtrate of *P. graminicolum* (Gayatri et al., 2005).

2.10.4 Hybridization

Sasikumar et al. (1996a) were the first to report the development of high yielding varieties from open pollinated seedling progenies of turmeric. They released two varieties, Prabha and Prathibha, which are also high in quality.

Nazeem et al. (1993) reported VK-5 of *C. longa* ($2n = 63$) produced seeds when it was crossed with Kasthuri and Dindigam, which belong to *C. aromatica* ($2n = 84$). They indicated the possibility of crossing high yielding *longa* types with selected *aromatica* types having high curing percentage. However, such a strategy may lead to the deterioration of the quality of turmeric, as *C. aromatica* cannot be used for any spice purpose because of its bitterness. Breeders have been cautioned against the use of aromatica types in turmeric breeding programs. A viable seed set in turmeric reported by Lad (1993) has given hope for a recombination breeding program.

2.11 FUTURE STRATEGY

Turmeric being the source of curcuminoids (curcumin) has numerous pharmaceutical, nutraceutical, and phytoceutical properties in addition to being a nontoxic constituent of the diet and a natural colorant; the most important priority is to develop varieties with high yield of curcumin. This coupled with good organic fertilizer–responsive varieties resistant to biotic and abiotic stresses will be the objective of turmeric breeders. Since most of the available variability in turmeric was exploited in selection, use of recombination breeding will open up hitherto nonavailable variability for developing new, improved varieties. Increasing the spectrum of variation through somaclonal variation and *in vitro* mutagenesis is another important approach that needs to be intensified. Identifying varieties with newer pharmaceutically important compounds also must make a beginning in turmeric because turmeric is considered as a wonder plant.

REFERENCES

AICRPS (2003) Annual Report: 2002-03 of All India Coordinated Research Project on Spices, Indian Institute of Spices Research, Calicut, Kerala, p.9-12.

AICRPS (2005) Annual Report: 2003-04 of All India Coordinated Research Project on Spices, Indian Institute of Spices Research, Calicut, Kerala, p.12-16.

Aiyadurai, S.G. (1966) *A Review of Research on Spices and Cashewnut in India.* Indian Council of Agricultural Research, New Delhi.

Anonymous (1986) Annual Report of the Regional Fruit Research Station, Anantharajupeta, Andhra Pradesh.

Baker, J.G. (1882) Scitaminae. In : Hooker , J.D. *The Flora of British India,* Vol. VI, Bishen Singh Mahendrapal Singh, Dehradun, Rep. 1978, pp. 198-264.

Balashanmugam, P.V., Chezhiyan, N. and Ahmad Shah, H. (1986). BSR-1 Turmeric : *South Indian Horticulture,* 34: 60-61.

Balashanmugam, P.V., Khader, M.A., Shanmugasundaram, V.S. and Manavalan, R.S. (1993) Influence of chemicals on sprouting of turmeric (*Curcuma domestica* L.) rhizomes. *South Indian Horticulture,* 41(3): 152-154.

Burt, B. L. (1977) *Curcuma zedoaria. Gard. Bull.* Singapore, 30, 59-62.

Burt, B.L. and Smith , R.M. (1983) Zingiberaceae. In: Dasanayake, M.D. (ed) *A Revised Hand Book to the Flora of Ceylon,* Vol. IV, pp 488- 532, Amerind Publishers, New Delhi.

Cao, H., Komatsu, K., Yao Xue and Xue Bao. (2003) Molecular identification of six medicinal *Curcuma* plants produced in Sichuan: evidence from plastid trnK gene sequences. *Biol. Pharm. Bull.* 22, 871-875.

Cao, H., Sasaki, Y., Fushimi, H., Komatsu, K. and Cao, H. (2001) Molecular analysis of medicinally used Chinese and Japanese *Curcuma* based on 18S rRNA and trnK gene sequences. *Biol. and Pharm. Bull.,* 24, 1389-1394.

Chakravarty, D.N. (1939) The occurrence of fugacious cambium in the rhizome of *Curcuma longa* L. *Philippine J. Sci.,* 69, 191-196.

Chakravorti, A.K. (1948) Multiplication of chromosome numbers in relation to speciation of Zingiberaceae. *Sci. & Cul.,* 14, 137-140.

Chandra, R., Desai, A.R., Govind, S. and Gupta, P.N. (1997) Metroglyph analysis in turmeric (*Curcuma longa* L.) germplasm in India. *Scientia Hort.,* 70, 211-222.

Chandra, R., Govind, S. and Desai, A.R. (1999) Growth, yield and quality performance of turmeric (*Curcuma longa* L.) genotypes in mid altitudes of Meghalaya. *J. Applied Hort.,* Lucknow, 1, 142-144.

Chandra, R., Singh, A.K. and Desai, A.R. (1998) Crop improvement in turmeric. *In*: Mathew, N.M., Kuruvilla Jacob, C., Licy, J., Joseph, T., Meenattoor, J.R. and Thomas, K.K. (eds.). *Developments in Plantation Crops Research.* Allied Publishers Ltd., New Delhi, pp.58-62.

Chase M.W. (2004) Monocot relationships: an overview, *American J. Bot.* 91,1645-1655.

Chaurasia, L.D., Kulakarni, B.M., Nair, K.N.G. and Mathew, T.V. (1974) Curcumin content of Indian turmeric. In *Proc. Symposium on Development and Prospects of Spices Industry in India,* pp.55-56.

Chempakam B., Leela, N.K. and John, S. P. (2000) Distribution of curcuminoids is turmeric during rhizome development. In: Ramana, K.V., Eapen, S.J., Nirmal Babu, K, Krishnamurthy, K. S. and Kumar, A.(Eds.) *Spices And Aromatic Plants – Challenges and Opportunities in the New Century,* Calicut, Indian Society for Spices, Calicut, Kerala, pp.293-296

Chen, Y., Bai, S., Cheng, K., Zhang, S. and Nian, L.Z. (1999) Random amplified polymorphic DNA analysis on *Curcuma wenuujin* and *C. sichuanensis. Zhongguo ZhongYao Za Zhi.,* 24, 131-133, 189.

Chezhiyan, N. and Shanmugasundaram, K.A. (2000) BSR-2 - A promising turmeric variety from Tamil Nadu. *Indian J. Arecanut, Spices and Medicinal Plants,* 2, 24-26.

Chezhiyan, N., Thangaraj, T., Vijayakumar, M., Mohanalakshmi, M. and Ramar, A. (1999) Evaluation and selection for yield in turmeric. In: Sasikumar, B., Krishnamurthy, B., Rema, J., Ravindran, P. N. and Peter, K. V. (Eds). *Biodiversity, Conservation and Utilization of Spices, Medicinal and Aromatic Plants.* Indian Institute of Spices Research, Calicut, Kerala, pp 114-115.

Cholke, S.M. (1993) Performance of turmeric (*Curcuma longa* L.) cultivars. M.Sc.(Ag.), thesis, University of Agricultural Sciences, Dharwad, Karnataka.

Choudhuri, P. and Hore, J.K. (2004) Studies on growth, bulking rate and yield of some turmeric cultivars. *J. Plantation crops,* 32(1): 47-50.

Clowes, F.A.L. (1967) The quiescent centre. *Phytomorphology,* 17, 132-140.

Das, D., Bhattacharjee, A., Biswas, I. and Mukherjee, A. (2004) Foliar characteristics of some medicinal plants of Zingiberaceae. *Phytomorphology,* 54, 291-302.

Dash, S.K and Jana, J.C. (2004) Genetic variability in a collection of turmeric (*Curcuma longa* L.) genotypes. In: Korikanthimath, V.S., John Zachariah, T., Nirmal Babu, K., Suseela Bhai, R. and Kandiannan, K. (eds.) New Perspectives in Spices, Medicinal and Aromatic Plants, ICAR Research Complex for Goa, pp. 56-61.

Dash, S.K., De, D.K., Bhattacharya, P.M. and Jana, J. (2002) Genetic variability, heritability of rhizome yield and its component traits in turmeric. In: National seminar on strategies for increasing production and export of spices, Indian Institute of Spices Research, Calicut, Kerala, p. 7 (Abstract).

Dhandar, D.G. and Varde, N.P.S. (1980) Performance of selected clones of turmeric under Goa conditions. *Indian Cocoa, Arecanut and Spices J.*, 3, 83-84.

Dikshit, N., Dabas, B.S. and Gautam, P.L (1999) Genetic variation in turmeric germplasm of Kandhamal District of Orissa. In: Sasikumar, B., Krishnamurthy, B., Rema, J., Ravindran, P. N. & Peter, K.V. (eds.) Biodiversity, Conservation and Utilization of Spices, Medicinal and Aromatic Plants. Indian Institute of Spices Research, Calicut, Kerala, pp 110-113.

Dixit, D. and Srivastava, N.K. (2000a) Distribution of photosynthetically fixed $^{14}CO_2$ into curcumin and essential oil in relation to primary metabolites in developing turmeric (*Curcuma longa*) leaves. *Plant Sci.* (Limerick), 152, 165-171.

Dixit, D. and Srivastava, N.K. (2000b) Effect of iron deficiency stress on physiological and biochemical changes in turmeric. (*Curcuma longa*) genotypes. *J. Med. Aromatic Plant Sci.*, 22, 652-658.

Dixit, D. and Srivastava, N.K. (2000c) Partitioning of photosynthetically fixed $^{14}CO_2$ into oil and curcumin accumulation in *Curcuma longa* grown under iron deficiency. *Photosynthetica*, 38, 193-197.

Dixit, D. and Srivastava, N.K. (2000d) Partitioning of 14C-photosynthate of leaves in roots, rhizome, and in essential oil and curcumin in turmeric(*Curcuma longa* L.). *Photosynthetica*, 38, 275-280.

Dixit, D., Srivastava, N.K. and Kumar, R. (2001) Intraspecific variation in yield capacity of turmeric, *Curcuma longa* with respect to metabolic translocation and partitioning of $^{14}CO_2$ photoassimilate into essential oil and curcumin. *J. Medicinal and Aromatic Plant Sci.*, 22-23, 4A-1A, 269-274.

Dixit, D., Srivastava, N.K. and Sharma, S. (1999) Effect of Fe deficiency on growth, physiology, yield and enzymatic activity in selected genotypes of turmeric (*Curcuma longa* L.). *J. Plant Biology*, 26: 237-241.

Dixit, D., Srivastava, N.K. and Sharma, S. (2002) Boron deficiency induced changes in translocation of $^{14}CO_2$ -photosynthate into primary metabolites in relation to essential oil and curcumin accumulation in turmeric (*Curcuma longa* L.). *Photosynthetica*, 40: 109-113.

Dixit, D., Srivatava, N.K., Kumar, R. and Sharma, S. (2002) Cultivar variation in yield, metabolite translocation and partitioning of $^{14}CO_2$ assimilated photosynthate into essential oil and curcumin of turmeric (*Curcuma longa* L.). *J. Plant Biology*, 29: 65-70.

Edison, S., Johny, A.K., Nirmal Babu, K.and Ramadasan, A. (1991) Spice varieties. A compendium of morphological and agronomic characters of improved varieties of spices in India. National Research Centre for Spices, Calicut, Kerala, 68 p.

Gayatri, M.C., Roopa, D.V. and Kavyashree, R. (2005) Selection of turmeric callus for tolerant to culture filtrate of *Pythium graminicolum* and regeneration of plants. *Plant Cell, Tissue and Organ Culture*, 83: 33-40.

Geetha, S.P. (2002) *In vitro Technology for Genetic Conservation of Some Genera of Zingiberaceae*. Unpublished Ph.D Thesis. Calicut University

Geetha, V. and Prabhakaran, P.V. (1987) Genotypic variability, correlation and path coefficient analysis in turmeric. *Agric. Res. J. Kerala*, 25: 249-254.

George, H. (1981) Variability in the open pollinated progenies of turmeric (*Curcuma longa*). M.Sc. (Horti.) thesis, Faculty of Agriculture, Department of Plantation Crops, Vellanikarra, Trichur.

Ghosh, S.P. and Govind, S. (1982) Yield and quality of turmeric in North Eastern hills. *Indian Horticulture*, 39: 230-232.

Gogoi, R., Bokadia, D. and Das, D.S. (2002) Leaf epidermal morphology of some species of Zingiberaceae. *Plant Archives*, 2, 257-262.

Gopalam, A. and Ratnambal, M. J. (1987) Gas chromatographic evaluation of turmeric essential oils. *Indian Perfumer*, 31, 245-296.

Govindarajan, V.S. (1980) Turmeric - chemistry, technology and quality. *CRC Critical Reviews in Food Science and Nutrition*, 12, 199-310.

Hazra, P., Roy, A. and Bandopadhyay, A. (2000) Growth characters as rhizome yield components of turmeric (*Curcuma longa* L.). *Crop Research*, Hisar, 19, 235-240.

Hegde, G.S. (1992) Studies on the performance of turmeric (*Curcuma longa* L.) cultivars. M.Sc.(Ag) thesis, University of Agricultural Sciences, Dharwad.

Hegde, S., Venkatesha, J. and Chandrappa, H. (1997) Performance of certain promising cultivars of turmeric (*Curcuma domestica* Val.) under Southern dry region of Karnataka. *Indian-Cocoa, Arecanut and Spices J.*, 21, 11-13.

Holttum, R.E. (1950) The Zingiberaceae of the Malay Peninsula. *Garden's Bull.* (Singapore), 13, 1-249.

IISR (2003) Proceedings of the workshop on Agricultural Bioinformatics, 29-30 October 2003, Calicut, Kerala, P. 15.

IISR (2003). Annual report: 2002-03, Indian Institute of Spices Research, Calicut, Kerala.

IISR (2003). Research Highlights: 2002-03, Indian Institute of Spices Research, Calicut, Kerala.

IISR (2004) Annual Report 2003-2004, Indian Institute of Spices Research, Calicut, Kerala.

IISR (2005). Research Highlights: 2004-05, Indian Institute of Spices Research, Calicut, Kerala.

Indiresh, K.M., Uthaiah, B.C. Reddy, M.J. and Rao, K.B. (1992) Genetic variability and heritability studies in turmeric. *Indian Cocoa, Arecanut & Spices J.*, 16, 52-54.

Indiresh, K.M., Uthaiah, B.C., Herle, P.S., and Rao, K.B. (1990) Morphological, rhizome and yield characters of different turmeric varieties in coastal Karnataka. *Mysore J. Agricultural Sci.*, 24, 484-490.

Jadhao, B.J., Jogdande, N.D., Gonge, V.S., Dalal, S.R. and Panchbhai, D.M.. (2004) Integrated nutrient management studies in different turmeric varieties. In: National seminar on New Perspectives in Spices, Medicinal and Aromatic Plants, ICAR Research Complex for Goa, Goa, p. 161-162 (Abstract)

Jadhao, B.L., Mahorkar, V.K., Dalal, S.R., Mohariya, A., Warade, A.D., Panchabhai, M., Jogdande, N.D., Ingle, V.G., Gonge, V.S. and Hussain, I.R. (2004) Influence of nutrient levels on yield and quality of turmeric varieties. In: Commercialization of Spices, Medicinal Plants and Aromatic Crops, Indian Institute of Spices Research, Calicut, Kerala, p.9. (Abstract)

Jalgaonkar, J., Patil, M.M and Rajput, J.C. (1988) Performance of different varieties of turmeric under Konkan conditions of Maharastra. In : Proceedings of the National Seminar on Chillies, Ginger and Turmeric, *Hyderabad.*, pp.102-105.

Jalgaonkar, R. and Jamdagni, B.M. (1989) Evaluation of turmeric genotypes for yield and yield determining characters. *Annals of Plant Physiology*, 3, 222-228.

Jalgaonkar, R., Jamadagni, B.M. and Selvi, M.J. (1990) Genetic variability and correlation studies in turmeric. *Indian Cocoa Arecanut and Spices J.*, 14: 20-22.

Jana, J.C. and Bhattacharya, B. (2001) Performance of different promising cultivars of turmeric (*Curcuma domestica* Val.) under terai agro-climatic region of West Bengal. *Environment and Ecology,* 19, 463-465.

Jana, J.C., Dutta, S. and Chatterjee, R. (2001) Genetic variability, heritability and correlation studies in turmeric (*Curcuma longa* L.). *Research on Crops*, 2, 220-225.

Jayakumar, M., Eyini, M., Lingakumar, M. and Kulandaivelu, G. (2001) Changes in proteins and RNA during storage of *Curcuma longa* L. rhizome. *Biologia Plantarum*, 44, 297-299.

Jirali, D.I., Hiremath, S.M., Chetti, M.B. and Patil, B.C. (2003) Association of yield and yield components with growth, biochemical and quality parameter for enhancing the productivity in turmeric (*Curcuma longa* L.) genotypes. In: National seminar on New Perspectives in Spices, Medicinal and Aromatic Plants, ICAR Research Complex for Goa, Goa, p. 5-6 (Abstract).

Johny A.K and Ravindran P.N (2006) Over 225 High yielding spice varieties in India. Part -2. *Spice India*.18, 40- 44.

Joseph, R., Joseph, T. and Jose, J. (1999) Karyomorphological studies in the genus *Curcuma* Linn. *Cytologia*, 64, 313-317.

KAU (2000) Three decades of spices research at KAU. Directorate of Extension, Kerala Agricultural University, Thrissur, Kerala.

Kress, W.J., Liu, A.Z., Newman, M. and Li, Q.J (2005) The molecular phylogeny of *Alpinia* (Zingiberaceae): a complex and polyphyletic genus of gingers. *American J.Botany* 92,167-178.

Kress, W.J., Prince, L.M. and Williams, K.J. (2002) The phylogeny and a new classification of the gingers (Zingiberaceae): evidence from molecular data. *American J. Botany,* 89,1682-1696.

Kumar, G.V., Reddy, K.S. Rao, M.S. and Ramavatharam, M. (1992) Soil and plant characters influencing curcumin content of turmeric. *Indian Cocoa Arecanut and Spices J.* 15, 102-104.

Kumar, R. and Jain, B.P. (1996) Growth and rhizome characters of some turmeric (*Curcuma longa* L.) cultivars. *J. Research,* Birsa Agricultural University, 8,131-133.

Kumar, R. and Jain, B.P. (1998) Evaluation of growth and rhizome characters of some turmeric (*C. longa* L.) cultivars under plateau region of Bihar. *Souvenir, National Seminar on Recent Development in Spices Production Technology,* Bihar Agricultural College, Sabour. pp. 9-10.

Kumar, R., Pandey, V.P. and Dwevedi, A. (2004) Genetic variability in turmeric germplasm. In: *Commercialization of Spices, Medicinal Plants and Aromatic Crops,* Indian Institute of Spices Research, Calicut, Kerala, p.4. (Abstract)

Kumar, S. (1999) A note on conservation of economically important Zingiberaceae of Sikkim Himalaya. In:. Sasikumar, B., Krishnamurthy, B., Rema, J., Ravindran, P. N. and Peter, K. V. (Eds) *Biodiversity, Conservation and Utilization of spices Medicinal and Aromatic Plants.* Indian Institute of Spices Research, Calicut, Kerala, pp 201-207.

Kumar, T.V., Reddy, M.S. and Krishna, V.G. (1997) Nutrient status of turmeric growing soils in Northern Telangana Zone of Andhra Pradesh. *J. Plantation Crops,* 25, 93-97.

Kurian, A. and Nair, G.S. (1996) Evaluation of turmeric germplasm for yield and quality. *Indian J. Pl. Genetic Resources,* 9, 327-329.

Kurian, A. and Valsala, P.A. (1996) Evaluation of turmeric types for yield and quality. *J. Tropical Agriculture,* 33, 75-76.

Lad, S.K. (1993) A case of seed-setting in cultivated turmeric types (*Curcuma longa* Linn.). *J. Soils and Crops,* 3, 78-79.

Latha, P., Giridharan, M.P. and Naik, B.J. (1995). Performance of turmeric (*Curcuma longa* L.) cultivars in open and partially shaded conditions. *J. Spices and Aromatic Crops,* 4, 139-144.

Lawande, K.E., Raijadhav, S.B., Yamgar, V.T. and Kale, P.N. (1991) Turmeric research in Maharashtra. *J. Plantation crops,* 18(Suppl.), 404-408.

Lewis, Y.S. (1973) The importance of selecting the proper variety of a spice for oil and oleoresins extractions. In: *Proc. Conference on Spices,* T.P.I., London. pp.183-185.

Li, L., Zhang, Y., Oin, S. and Liao, G. (1997) Ontogeny of *Curcuma longa* L. *Zhongguo Zhong Yao Za Zhi,* 10, 587-590.

Linnaeus, C. (1753) *Specius Plantarum.* London

Lynrah, P.G. and Chakraborty, B.K. (2000) Performance of some turmeric and its close relatives/genotypes. *J. Agricultural Sci. Soc. North East India,* 13, 32-37.

Lynrah, P.G., Chakraborty, B.K. and Chandra, K. (2002) Effect of CCC, kinetin, NAA and KNO_3 on yield and curcumin content of turmeric. *Indian J. Plant Physiology,* 7, 94-95.

Manglay Jose, K. and Sabu, M. (1993) A taxonomic revision of the South Indian species of *Curcuma*. (Zingiberae). *Rheedea,* 3,139-171.

Mathai, C.K. (1974) Quality studies in cashew and spices. Annual Report, CPCRI, Kasargod. pp.166-167.

Maurya, K.R. (1990) RH 10, A promising variety of turmeric to boost farmer's economy. *Indian Cocoa Arecanut and Spices J.,* 13: 100-101.

Mehta, K.G. and Patel, R.H. (1980) Phenotypic stability for yield in turmeric. In: Nair M.K, Premkumar T., Ravindran P.N. and Sarma Y.R. (Eds.) *Proc. National Seminar on Ginger and Turmeric,* Central Plantation Crops Research Institute, Kasaragod. pp.34-38.

Mehta, K.G., Raghava Rao, D.V. and Patel, S.H. (1980) Relative curcumin content during various growth stages in the leaves and rhizomes of three cultivars of *Curcuma longa* and *C. amada*. In: Nair M.K, Premkumar T., Ravindran P.N. and Sarma Y.R. (Eds.) Proc. *National Seminar on Ginger and Turmeric,* Central Plantation Crops Research Institute, Kasaragod. pp.76-78.

Menon, R., Valsala, P.A. and Nair, G.S. (1992) Evaluation of open pollinated progenies of turmeric. *South Indian Horticulture,* 40, 90-92.

Milian. J. M. and Sanchez, R. I. (2002) List of descriptors and characterization of germplasm of spice and/or medicinal species. *Centro Agricola,* 29, 37-41.

Mohanty, D.C. (1979) Genetic variability and inter relationship among rhizome yield and yield components in turmeric. *Andhra Agricultural J.,* 26, 77-80.

Muthuswamy, S. and Shah, A.H. (1982) Comparative quality evaluation of Selam and Erode turmeric types. *Indian Cocoa Arecanut and Spices J.* 5, 77.

Nadgauda, R.S., Khuspe, S.S. and Mascarenhas, A.F. (1982) Isolation of high curcumin varieties of turmeric from tissue culture In R. D. Iyer (Ed.) *Proceedings V Annual Symposium on Plantation Crops,* pp: 143-144, CPCRI Kasargod.

Naidu, M.M., Murty, P.S.S., Kumari, P. and Murthy, G.N. (2003) Evaluation of turmeric selections for high altitude and tribal areas of Visakhapatnam (Dist.), Andhra Pradesh. In: National Seminar on New Perspectives in Spices, Medicinal and Aromatic Plants, ICAR Research Complex for Goa, p. 29 (Abstract).

Nair, M.K., Nambiar, M.C. and Ratnambal, M.J. (1980) Cytogenetics and crop improvement of ginger and turmeric. In: Nair M.K, Premkumar T., Ravindran P.N. and Sarma Y.R.(eds.) Proc. National Seminar on Ginger and Turmeric, Central Plantation Crops Research Institute, Kasaragod. pp.15-23.

Nambiar K.K.N, Sarma Y.R. and Brahma R.N. (1998) Field reaction of turmeric types to leaf blotch. *Journal plantation Crops*, 5, 124-125.

Nambiar, M.C. (1979) Morphological and cytological investigations in the genus *Curcuma* L. Unpublished *Ph.D. thesis,* University of Bombay, pp: 95.

Nambiar, M.C., Thankamma Pillai, P.K. and Sarma, Y.N. (1982) Seedling propagation in turmeric (*Curcuma aromatica* Salisb). *J. Plantation Crops*, 10, 81-85.

Nandi, A. (1990) Evaluation of turmeric (*Curcuma longa*) varieties for North Eastern plateau zone of Orissa under rainfed conditions. *Indian J. Agril. Sci.,* 60, 760-761.

Nandi, A., Lenka, D. and Singh, D.N. (1992) Path analysis in turmeric. *Indian Cocoa Arecanut and Spices J.,* 27, 54-55.

Narayanpur, V.B. and Hanamashetti, S.I. (2003) Genetic variability and correlation studies in turmeric (*Curcuma longa* L.). *J. Plantation Crops*, 31 (2), 48-51.

Natarajan, S.T. (1975) Studies on the yield components and gamma ray induced variability in turmeric (*Curcuma longa* L.). M.Sc.(Ag.) Thesis, Tamil Nadu Agricultural University, Coimbatore.

Nazeem, P.A., Menon, R. and Valsala, P.A. (1994) Blossom, biological and hybridization studies in turmeric (*Curcuma* spp.). *Indian Cocoa Arecanut and Spices J.,* 16, 106-109.

Ngamriabsakul, C., Newman, M.F. and Cronk, Q.C.B. (2003) The phylogeny of tribe *Zingibereae* (*Zingiberaceae*) based on its (nrDNA) and *trnl*–f (cpDNA) sequences. *Edinburgh J. Botany*, 60, 483-507.

Nirmal Babu K, Sasikumar B, Ratnambal MJ, George JK and Ravindran PN (1993) Genetic variability in turmeric (*Curcuma longa* L.). *Indian J. Genetics and Plant Breeding*, 53, 91-93.

Nirmal Babu, K., Ravindran, P.N. and Sasikumar, B. (2003) Field evaluation of tissue cultured plants of spices and assessment of their genetic stability using molecular markers. Final Report submitted to Department of Biotechnology, Government of India. pp. 94

Nirmal, S.V. and Yamgar, V.T. (1998) Variability in morphological and yield characters of turmeric (*Curcuma longa* L.) cultivars. *Advances in Plant Sciences*, 11, 161-164.

Paisooksantivatana, Y., Kako, S. and Seko , H. (2001) Isozyme polymorphism in *Curcuma alismatifolia* Gagnep. (Zingiberaceae) populations from Thailand. *Scientia Horticulturae*, 88, 299-307.

Palarpawar, M.Y. and Ghurde, V.R. (1989) Sources of resistance in turmeric against leaf spots incited by *Colletotrichum capsici* and *C. curcumae. Indian Phytopathology,* 42, 171-173.

Pandey, G., Sharma, B.D. and Hore, D.K. (1990) Metroglyph and index score analysis of turmeric germplasm in North Eastern region of India. *Indian J. Pl .Genet. Resources*, 3, 59-66.

Pandey, V.P., Dixit, J. Saxena, R.P. and Gupta, R.K. (2004) Comparative performance of turmeric genotypes in Eastern Uttar Pradesh. In: Commercialization of Spices, Medicinal Plants and Aromatic Crops, Indian Institute of Spices Research, Calicut, Kerala, p.4. (Abstract)

Pandey, V.S., Pandey, V.P. and Singh, P.K. (2004) Path analysis in turmeric. In: Korikanthimath, V.S., John Zachariah, T., Nirmal Babu, K., Suseela Bhai, R. and Kandiannan, K. (eds.) New Perspectives in Spices, Medicinal and Aromatic Plants, ICAR Research Complex for Goa, pp. 32-36.

Pandey, V.S., Pandey, V.P., Pandey, S. and Dixit, J. (2004) Divergence analysis of turmeric. In: Korikanthimath, V.S., John Zachariah, T., Nirmal Babu, K., Suseela Bhai, R. and Kandiannan, K. (eds) New perspectives in Spices, Medicinal and Aromatic Plants, ICAR Research Complex for Goa, pp. 62-65.

Pandey, V.S., Pandey, V.P., Singh, T. and Srivastava, A.K. (2003) Genetic variability and correlation studies in turmeric. In: Proc. National Seminar on strategies for increasing production & export of spices, IISR, Calicut, Kerala, pp. 195-200.

Panja, B.N. and De, D.K. (2000) Characterization of blue turmeric: a new hill collection. *J. Interacademicia,* 4, 550-557.

Panja, B.N. and De, D.K. (2001) Studies on stomatal frequency and its relationship with leaf biomass and rhizome yield of turmeric (*Curcuma longa* L.) genotypes. *J. Spices and Aromatic Crops,* 10, 127-134.

Panja, B.N., De, D.K. and Majumdar, D. (2001) Evaluation of turmeric (*Curcuma longa* L.) genotypes for yield and leaf blotch disease (*Taphrina maculans* Butl.) for tarai region of West Bengal. *Environment and Ecology,* 19, 125-129.

Panja, B.N., De. D.K., Basak, S. and Chattopadhyay, S.B. (2002) Correlation and path analysis in turmeric (*Curcuma longa* L.). *J. Spices and Aromatic Crops,* 11, 70-73.

Pathak, K.A., Kishore, K., Singh, A.K. and Bharali, R. (2003). Varietal evaluation of turmeric germplasm under Mizoram conditions. In: National seminar on New Perspectives in Spices, Medicinal and Aromatic plants, ICAR Research Complex for Goa, Goa, p. 50 (Abstract).

Pathak, S., Patra, B.C. and Mahapatra, K.C. (1960) Flowering behaviour and anthesis of *Curcuma longa. Curr. Sci.,* 29, 402.

Pathania, N.K., Arya, P.S. and Singh, M. (1988) Variability studies in turmeric (*Curcuma longa* L.). *Indian J. Agricultural Research,* 22, 176-178.

Pathania, N.K., Singh, M. and Arya, P.S. (1990) Variation for volatile oil content in turmeric cultivars. *Indian Cocoa, Arecanut & Spices J.,* 14(1), 23-24.

Patil, D.V., Kuruvilla, K.M. and Madhusoodanan, K.N. (1995) Performance of turmeric varieties in lower pulney hills of Tamil Nadu. *Indian Cocoa Arecanut and Spices J.,* 4: 156-158.

Peter, K.V. and Kandiannan, K. (1999) Turmeric. In: T.K. Bose , S.K. Mitra, A.A. Faroogi and M.K. Sadhu (eds.) *Tropical Horticulture* Vol .1, pp. 683-691.

Philip, J. (1983) Studies on growth, yield and quality component in different turmeric types. *Indian Cocoa Arecanut and Spices J.,* 6, 93-97.

Philip, J. and Nair, P.C.S. (1981) Field reaction of turmeric types to important pests and diseases. *Indian Cocoa Arecanut and Spices J.,* 4(4), 107-109.

Philip, J. and Nair, P.C.S. (1986) Studies on variability, heritability and genetic advance in turmeric. *Indian Cocoa Arecanut and Spices J.,* 10, 29-30.

Philip, J., Sivaraman Nair, P.C.S., Nybe, E.V. and Mohan, N.K. (1982) Variation of yield and quality of turmeric. In: Nair M.K, Premkumar T., Ravindran P.N. and Sarma Y.R. (Eds.) *Proc. National Seminar on Ginger and Turmeric,* Central Plantation Crops Research Institute, Kasaragod. pp.42-46.

Pillai, P.K.T. and Nambiar, M.C. (1975) Germplasm collection and cataloguing of turmeric, Annual Report, CPCRI, Kasargod: 138-140.

Pillai, S.K., Pillai, A. and Sachdeva, S. (1961) Root apical organization in monocotyledons-Zingiberaceae. *Proc. Indian Acad. Sci.,* (B), 53, 240-256.

Pillai,.P.K.T., Nambiar, M.C. and Ratnambal, M.J. (1976) Germplasm collections and cataloguing of turmeric. Annual Report, CPCRI, Kasargod. pp.256-258.

Pimchai, A., Somboon, A.I., Puangpen, S. and Chiara (1999) Molecular markers in the Identifcation of some early flowering *Curcuma* L. (Zingiberaceae) Species. *Annals of Botany,* 84, 529-534.

Pino, J., Marbot, R., Palau, E. and Roncal, E. (2003) Essential oil constituents from Cuban turmeric rhizomes. *Revista-Latinoamericana-de-Quimica,* 31, 16-19.

Poduval, M., Mathew, B., Hasan, M.A. and Chattopadhyay, P.K. (2001) Yield and curcumin content of different turmeric varieties and species. *Environment and Ecology,* 19, 744-746.

Prabhakaran, P.V. (1991) Factor analysis in turmeric. *Annals of Agricultural Research,* 12, 151-155.

Praveen, K. (2005) *Variability in Somaclones of Turmeric (Curcuma longa* L.). Unpublished Ph.D Thesis. University of Calicut, Kerala, India.

Pujari, P.P., Patil, R.B. and Sakpal, B.T. (1987). Studies on growth yield quality components in different varieties. *Indian Cocoa Arecanut and Spices J.,* 11, 15-17.

Pujari, P.P., Patil, R.B. and Sakpal, R.T. (1986) Krishna, a high yielding variety of turmeric. *Indian Cocoa Arecanut and Spices J.,* 9, 65-66.

Purseglove, J.W. (1974) *Tropical Crops - Monocotyledons.* Longman Group Ltd., London.

Purseglove, J.W., Brown E.G., Green C.L. and Robin, S.R.J. (1981) *Turmeric.* In: *Spices,* Vol 2 Long man, New York, pp.532-580.

Radhakrishna, V.V., Madhusoodanan, K.J. and Kuruvilla, K.M. (1995) Performance of different varieties of turmeric (*Curcuma longa* L.) in the high ranges of Idukki district of Kerala. *Indian Cocoa, Arecanut and Spices J.,* 29, 8-10.

Raghavan, T.S. and Venkattasubban, K R. (1943) Cytological studies in the family Zingiberaceae with special reference to chromosome number and cytotaxonomy. *Proc. Ind. Acade. Sci.,* Ser. B, pp.118-132.

Raju, E.C. and Shah, J.J. (1975) Studies in stomata of ginger, turmeric and mango ginger. *Flora., Bd.,* 164, 19-25.

Raju, E.C. and Shah, J.J. (1977) Root apical organization in some rhizomatous spices: ginger, turmeric and mango ginger. *Flora Bd.,* 166 , 105-110.

Rakhunde, S.D., Munjal, S.V. and Patil, S.R. (1998) Curcumin and essential oil contents of some commonly grown turmeric (*Curcuma longa* L.) cultivars in Maharashtra. *J. Food Science and Technology Mysore,* 35, 352-354.

Ramachandran, K. (1961) Chromosome numbers in the genus, *Curcuma* Linn. *Current Science,* 30, 194-196.

Ramachandran, K. (1969) Chromosome numbers in Zingiberaceae. *Cytologia,* 34, 213-221.

Ramakrishna, M., Reddy, R.S. and Padmanabham, V. (1995) Studies on the performance of short duration varieties / cultures of turmeric in Southern Zone of Andhra Pradesh. *J. Plantation Crops,* 23, 126-127.

Raghunathan, K and Mitra, R. (1982) *Pharmacognosy of Indigenous Drugs,* CCRAS, New Delhi.

Rao, A. M., Jagdeeshwar, R. and Sivaraman K. (2006) Turmeric In : Ravindran, P.N., Nirmal Babu, K., Shiva. K.N. and Johny, A.K. (eds) *Advances in Spices Research,* Agribios, Jodhpur. pp. 433-492

Rao, A.M. (2000) Genetic variability, yield and quality studies in turmeric (*Curcuma longa* L.). Unpublished Ph.D. thesis, ANGRAU, Rajendranagar, Hyderabad.

Rao, A.M., Rao, P.V. and Reddy, Y.N. (2004) Evaluation of turmeric cultivars for growth, yield and quality characters. *J. Plantation Crops,* 32, 20-25.

Rao, D.V.R. (1999) Effect of gamma irradiation on growth, yield and quality of turmeric. *Advances in Horticulture and Forestry,* 6, 107-110.

Rao, M.R. and Rao, D.V.R. (1992) Genetic resources of turmeric in India. *J. Plantation crops,* 20 (Suppl.), 212-217.

Rao, M.R. and Rao, D.V.R. (1994) Genetic resources of turmeric in India. In: *In*: Chadha, K.L. and Rethinam, P. (eds.) *Advances in Horticulture. Vol.2- Plantation and Spices Crops.* 9, 131-148.

Rao, M.R., Reddy, K.R.C. and Subbarayudu, M. (1975) Promising turmeric types of Andhra Pradesh. *Indian Spices,* 12, 2-5.

Rao, P.S., Ramakrishna, M., Srinivas, C., Meena Kumari, K. and Rao, A.M. (1994) Short duration, disease resistant turmerics for Northern Telangana. *Indian Horticulture,* 39: pp.55.

Rao, P.S., Reddy, M.L.N., Rao, T.G.N., Krishna, M.R. and Rao, A.M. (1992) Reaction of turmeric cultivars to *Colletotrichum* leaf spot, *Taphrina* leaf blotch and rhizome rot. *J. Plantation Crops,* 20, 131-134.

Rao, T.G.N., Reddy, B.N and Sasikumar, B. (1993) Turmeric production constraints in Andhra Predhesh. *Spices India,* 6, 8-11.

Rao, V.P., Rao, A.M., Ramakrishna, M. and Rao, P.S. (1994) Leaf area estimation by linear measurements in turmeric. *Annals of Agricultural Research,* 15, 231-233.

Ratnambal, M.J. (1986) Evaluation of turmeric accession for quality. *Qualities Plantarum,* 36, 243-252.

Ratnambal, M.J. and Nair, M.K. (1986) High yielding turmeric selection PCT - 8. *J. Plantation Crops,* 14, 94-98.

Ratnambal, M.J., Nirmal Babu, K.N., Nair, M.K. and Edison, S. (1992) PCT 13 and PCT 14 - two high yielding varieties of turmeric. *J. Plantation Crops,* 20, 79-84.

Raveendra, B.H., Hanamashetti, S.I., and Hegde, L.N. (2001) Correlation studies with respect to growth and yield of sixteen cultivars of turmeric (*Curcuma longa* L.). *J. Plantation Crops,* 29, 61-63.

Raveendra, B.H., Hanamashetti, S.I., Kotikal, Y.K. and Shashidhar, T.R. (2004) Evaluation of sixteen turmeric (*Curcuma longa* L.) cultivars gown under Ghataprabha left bank command area of Northern Karnataka with respect to growth, yield and rhizome characters. In: Korikanthimath, V.S., John Zachariah, T., Nirmal Babu, K., Suseela Bhai, R. and Kandiannan, K. (Eds.) New Perspectives in Spices, Medicinal and Aromatic Plants, ICAR Research Complex for Goa, Goa, pp. 72-76.

Ravindran, P.N. and Peter, K.V. (1996) Biodiversity of major spices and their conservation in India. *In*: Arora, R.K., and Rao, V.R. (eds). Proceedings of the South Asia national coordinators meeting on plant genetic resources, Dhaka, Bangladesh, 10-12 January 1995. IPGRI, Office for South and Southeast Asia (IPGRI/RECSEA), New Delhi, pp. 123-134.

Ravindran, P.N. (1999) Genetic diversity of major spices and their conservation in India. In: Sasikumar B, Krishnamurthy B, Rema, J, Ravindran, P. N. & Peter, K. V. (eds). *Biodiversity, Conservation and Utilization of Spices, Medicinal and Aromatic Plants*. Indian Institute of Spices Research, Calicut, Kerala, pp 16-44.

Ravindran, P.N., Nirmal Babu, K. and Shiva, K.N. (2006) Genetic resources of spices and their conservation. In: P.N. Ravindran, K. Nirmal Babu, K.N. Shiva and A.K. Johny (eds) *Advances in Spices Research*, Agrobios, Jodhpur. pp. 63-91.

Ravindran, P.N., Nirmal Babu, K., Peter, K.V., Abraham, Z. and Tyagi, R.K. (2005) Spices. In: B.S. Dhillon, R.K. Tyagi, S. Saxena and G.J.Randhawa (eds) *Plant Genetic Resources: Horticultural Crops*, Narosa Publishing House, New Delhi . pp. 190 - 227

Ravindran, P.N., Remashree, A.B. and Sherlija, K.K. (1998) *Developmental morphology of rhizomes of ginger and turmeric*. Final Report, ICAR Ad-hoc Project, Indian Institute of Spices Research, Calicut.

Reddy, G.V., Reddy, M.L. and Naidu, G.S. (1989) Estimation of leaf area in turmeric (*Curcuma longa* Linn.) by non-destructive method. *J. Research*, APAU, 17, 43-44.

Reddy, M.L.N. (1987) Genetic variability and association in turmeric (*Curcuma longa* L.). *Progressive Horticulture*, 19, 83-86.

Reddy, M.L.N., Rao, A.M., Rao, D.V.R. and Reddy, S.A. (1989) Screening of short duration turmeric varieties / Cultures suitable for Andhra Pradesh. *Indian Cocoa Arecanut and Spices J.*, 12, 87-89.

Remashree, A.B., Balachandran, I. and Ravindran, P.N. (2003) Pharmacognostic studies in four species of *Curcuma*. Proc. 12th Swadeshi Sci. Cong., Nov. 6-9.

Remashree, A.B., Ravindran, P.N. and Balachandran, I. (2006) Anatomical and histochemical studies on four species of *Curcuma*. *Phytomorphology* (In Press).

Remashree, A.B., Unnikrishnan, K. and Ravindran, P.N. (1999) Development of oil cells and ducts in ginger (*Zingiber officinale* Rosc.). *J. Spices and Aromatic Crops*, 8, 163 – 170.

Remya, R., Syamkumar, S. and Sasikumar, B. (2004) Isolation and amplification of DNA from turmeric powder . *British Food J.*, 106, 673-678.

Renjith, D., Valsala, P.A. and Nybe, E.V. (2001) Response of turmeric (*Curcuma domestica* Val.) to *in vivo* and *in vitro* pollination. *J. Spices and Aromatic Crops*, 10, 135-139.

Rheede, H. von (1685) Hortus Indicus Malabaricus, Amsterdomi, Holland.

Roxburgh W.(1832) *Flora Indica or Description of Indian plants*, W. Carey (ed) Serampur.

Sabu, M. (1991*) A Taxonomic and Phylogenetic study of South Indian Zingiberaceae*, Ph. D Thesis, University of Calicut, Kerala.

Salvi, N. D., Eapen, S. and George, L. (2003) Biotechnological studies of turmeric (*Curcuma longa* L.) and ginger (*Zingiber officinale* Roscoe). *Advances in Agricultural Biotechnology*. 1, 11-32 .

Salvi, N.D., George, L. and Eapen, S. (2000) Plant regeneration from leaf base callus of turmeric and random amplified polymorphic DNA analysis of regenerated plants. *Plant Cell Tissue and Organ Culture*, 66, 113-119.

Sasaki, Y., Fushimi, H. and Komatsu, K. (2004) Application of single-nucleotide polymorphism analysis of the trnK gene to the identification of *Curcuma* plants. *Biological and Pharmaceutical Bulletin*, 27, 144-146.

Sasaki, Y., Fushimi, H., Cao, H., Cai, S.Q. and Komatsu, K. (2002) Sequence analysis of Chinese and Japanese Curcuma drugs on the 18S rRNA gene and trnK gene and the application of amplification-refractory mutation system analysis for their authentication. *Biol. Pharm. Bull.*, 25,1593-9.

Sasikumar, B.(2005) Genetic resources of *Curcuma*: diversity, characterization and utilization. *Plant Genetic Resources: Characterization and Utilization*, 3, 230-251.

Sasikumar, B., George, J.K., Saji K.V. and Zacharaiah, T.J. (2005) Two new high yielding, high CURCUMIN turmeric (*Curcuma longa* L.) varieties- IISR Kedaram and IISR Alleppey Supreme . *J. Spices and Aromatic crops*, 14, 71-74.

Sasikumar, B., George, J.K., Zacharaiah, T.J., Ratnambal, M.J., Nirmal Babu K. and Ravindran, P.N. (1996) IISR Prabha and IISR Prathibha two new high yielding and high quality turmeric (*Curcuma longa* L.) varieties. *J. Spices and Aromatic crops*, 5, 41-48.

Sasikumar, B., Krishnamoorthy, B., Saji, K.V., George, J.K., Peter, K.V. and Ravindran, P.N. (1999) Spice diversity and conservation of plants that yield major spices in India. *Plant Genetic Resources Newsletter*, 118, 19-26.

Sasikumar, B., Ravindran, P.N. and George, J.K. (1994) Breeding ginger and turmeric. *Indian Cocoa Arecanut and Spices J.* 18 (1), 10-12.

Sasikumar, B., Ravindran, P.N., George, J.K. and Peter K.V. (1996) Ginger and turmeric breeding in Kerala. In P.I. Kuriachan (ed) Proc. Sem. Crop Breeding in Kerala, University of Kerala. pp. 65-72.

Sasikumar, B., Syamkumar, S., Remya and John Zacharia, T. (2004) PCR based detection of adulteration in the market samples of turmeric powder. Food Biotechnology, 18, 299-306.

Sastrapradja, S. and Aminali, S.H. (1970) Factors affecting fruit production in *Curcuma* species. *Ann. Bogor.* 5, 99-107.

Satheesan, K. V. and Ramadasan, A. (1987) Curcumin and essential oil contents of three turmeric (*Curcuma domestica* Val.) cultivars grown in monoculture and as inter crop in cocunut gardens. *Journal of plantation Crops*, 15, 31-37.

Satheesan, K.V. (1984) *Physiology of growth and productivity of turmeric (Curcuma domestica Val.) in monoculture and as an intercrop in coconut garden.* Ph.D Thesis, University of Calicut.

Satheesan, K.V. and Ramadasan, A. (1988) Changes in carbohydrate levels and starch/sugar ratio in three turmeric (*Curcuma domestica* Val.) cultivars grown in monoculture and as an intercrop in coconut garden. *J. Plantation Crops*, 16, 45-51.

Satheesan, K.V. and Ramadasan, A. (1988) Effect of growth retardant CCC on growth and productivity of turmeric under monoculture and in association with coconut. J. *Plantation Crops*, 16, 140-143.

Sato, D. (1948) The karyotype and phylogeny of Zingiberaceae. *Jap. J. Genet.,* 23, 44.

Sato, D. (1960) The karyotype analysis in Zingiberales with special reference to the protokaryotype and stable karyotype. *Sci. papers of the College of General Education*, Univ. Tokyo, 10, 225-243.

Schuman K.(1904) Zingeberaceae. In: *Engler's Pflanazenreich*, 4,1-428.

Shah, J.J. and Raju, E.C. (1975) General morphology, growth, and branching behaviour of the rhizome of ginger, turmeric and mango ginger. *New Bot.*, 11(2), 59-69.

Shahi, R.P., Shahi, B.G. and Yadava, H.S. (1994) Stability analysis for quality characters in turmeric (*Curcuma longa* L.). *Crop Research,* Hisar, 8, 112-116.

Shahi, R.P., Yadava, H.S. and Sahi, B.G. (1994) Stability analysis for rhizome yield and its determining characters in turmeric. *Crop Research,* Hisar, 7, 72-78.

Shamina A, Zachariah T J, Sasikumar B and George J K (1998) Biochemical variation in turmeric based on isozyme polymorphism. *Journal of Horticulture Science and Biotechnology*, 73, 477-483.

Shankaracharya, N.B. and Natarajan, C.P. (1974) Technology of spices. *Arecanut and Spices Bulletin.* 7, 27-43.

Shanmugasundaram, K.A. and Thangaraj, T. (2002) Variation in inner core diameter of rhizome and its significance in curing percentage of turmeric (*Curcuma longa* L.). In: National seminar on strategies for increasing production & export of spices, IISR, Calicut, Kerala, p. 15 (Abstract).

Shanmugasundaram, K.A., Thangaraj, T. and Chezhiyan, N. (2000) Variability, heritability and genetic advance studies in turmeric. *South Indian Horticulture*, 48, 88-92.

Shanmugasundaram, K.A., Thangaraj, T., Azhakiamanavalan, R.S. and Ganga, M. (2001) Evaluation and selection of turmeric (*Curcuma longa* L.) genotypes. *J. Spices and Aromatic Crops*, 10, 33-36.

Sharma, A.K. and Bhattacharya, N.K. (1959) Cytology of several members of Zingiberaceae and study of the inconsistency of their chromosome complement. *La Cellule*, 59, 279-349.

Shashidhar, T.R. and Sulikeri, G.S. (1997) Correlation studies in turmeric (*Curcuma longa* L.). *Karnataka J. Agricultural Sciences*, 10, 595-597.

Shashidhar, T.R., Sulikeri, G.S. and Gasti, V.D. (1997) Correlation studies in turmeric (*Curcuma longa* L.). *Mysore J. Agricultural Sci.,* 31, 217-220.

Sherilija, K.K., Unnikrishnan, K. and Ravindran, P.N. (1999) Bud and root development of turmeric (*Curcuma longa* L.) rhizome. *J. Spices and Aromatic Crops,* 8, 49-55.

Sherilija, K.K., Unnikrishnan, K. and Ravindran, P.N. (2001) Anatomy of rhizome enlargement in turmeric (*Curcuma longa* L.). In Recent Research in Plant Anatomy and Morphology, *J. Eco. Tax. Bot.,* Addl. Series 19, 229-235.

Sherlija, K.K., Remashree, A.B., Unnikrishnan, K. and Ravindran, P.N. (1998) Comparative rhizome anatomy of four species of *Curcuma*. *J. Spices and Aromatic Crops*, 7, 103-109.

Sheshagiri, K.S. and Uthaiah, B.C. (1994) Performance of turmeric (*Curcuma longa* L.) varieites in the hill zone of Karnataka. *Indian Cocoa Arecanut Spices J.,* 3: 161-163.

Shiva, K.N., Suryanarayana, M.A. and Medhi, R.P. (2003) Genetic resources of spices and their conservation in Bay Islands. *Indian J. Plant Genet. Resour.,* 16, 91-95.

Shridhar, S.V., Prasad, S. and Gupta, H.S. (2002) Stability analysis in turmeric. In: National Seminar on Strategies for Increasing Production & Export of Spices, IISR, Calicut, Kerala, p. 13 (Abstract).

Singh, B., Yadav, J.R. and Srivastava, J.P. (2003) Azad Haldi-1 A disease resistant variety turmeric (*Curcuma longa*). *Plant Archives,* 3, 151-152.

Singh, D.P. and Tiwari, R.S. (1995) Path analysis in turmeric (*Curcuma longa* L.). *Recent Horticulture,* 2, 113-116.

Singh, J.P., Singh, M.K., Singh, P.K. and Singh, R.D. (1995) Phenotypic stability in turmeric (*Curcuma longa* L.). *Indian Cocoa, Arecanut and Spices J.,* 19, 40-42.

Singh, P.K., Srivastava. R., Sharma, A., Hore, D.K. and Panwar, B.S. (2001) Genetic variability and correlation in turmeric (*Curcuma longa* L.). *Indian J. Hill Farming,* 14, 24-28.

Singh, S. and Singh, S. (1987) Evaluation of some turmeric types for the Andamans. *J. Andaman Sci. Assoc.* 3, 38-39.

Singh, Y., Mittal, P. and Katoch, V. (2003) Genetic variability and heritability in turmeric (*Curcuma longa* L.). *Himachal J. Agricultural Research,* 29, 31-34.

Sirirugsa, P. (1999) Thai Zingiberaceae.: Species diversity and their uses. International conference on biodiversity and bioresources: Conservation, utilization. Phuket, Thailand.

Sreenivasulu, B. and Rao, D.V.R. (2002) Physiological studies in certain turmeric cultivars of Andhra Pradesh. In: Rethinam, P., Khan, H.H., Reddy, V.M., Mandal, P.K., Suresh, K. (eds) *Plantation crops research and development in the new millenium.* Coconut Development Board, Kochi, pp: 442-443.

Sriramarao, T. (1982) Commercial varieties of turmeric in Andhra Pradesh. In: Nair M.K, remkumar T., Ravindran P.N. and Sarma Y.R. Proceedings of the National Seminar on Ginger and Turmeric, Central Plantation Crops Research Institute, Kasaragod. pp.213-215.

Subbarayudu, M., Reddy, K.R.C. and Rao, M.R. (1976) Studies on varietal performance of turmeric. *Andhra Agricultural J.,* 23, 195-198.

Subramaniyam, S. (1986) Studies on growth and development turmeric (*Curcuma longa* L.). M.Sc (Hort.), thesis, Tamil Nadu Agricultural University, Coimbatore.

Sugiura, T. (1928) (Cited from Sugiura, 1936)

Sugiura, T. (1931) *Bot. Maj.* Tokyo, 45, 353(Cited from Sugiura, 1936)

Sugiura, T. (1936) Studies on the chromosome number of higher plants. *Cytologia,* 7, 544-595.

Syamkumar, S., Lowarence, B. and Sasikumar, B. (2003) Isolation and amplification of DNA from rhizomes of turmeric and ginger. *Plant Molecular Biology Reporter,* 21, 171 ñ171.

Tomar, N.S. and Singh, P. (2002). Collection, evaluation and characterization of turmeric (*Curcuma longa* L.) germplasm. In: National Seminar on Strategies for Increasing Production & Export of Spices, IISR, Calicut, Kerala, p. 14-15 (Abst.).

Tomer, N.S., Nair, S.K. and Gupta, C.R.(2005) Character association and path analysis for yield components in turmeric (*Curcuma longa* L.). *J. Spices and Aromatic Crops,* 14, 75-77.

Valeton, T.H. (1918) New notes on Zingiberaceae of Java and Malaya. *Bull.Jard. Buitenzorg ser.,* II, 27, 1-8.

Velayudhan, K. C., Asha, K. I., Mithal, S. K. and Gautam, P. L. (1999). Genetic Resources of turmeric and its relatives in India. In: Sasikumar, B., Krishnamurthy, B., Rema, J., Ravindran, P.N. & Peter, K.V. (eds.) *Biodiversity, conservation and utilization of spices medicinal and aromatic plants.* Calicut. Indian Institute of Spices Research, Calicut, Kerala, India, pp: 101-109.

Velayudhan, K. C., Muralidharan, V. K., Amalraj, V. A., Gautam, P. L., Mandal, S. and Dinesh Kumar (1999) *Curcuma Genetic Resources.* Scientific monograph No. 4. National Bureau of Plant Genetic Resources, New Delhi. p.149.

Velayudhan, K.C. and Liji, R.S. (2003) Preliminary screening of indigenous collections of turmeric against shoot borer (*Conogethes punctiferalis* Guen.) and scale insect (*Aspidiella hartii* Sign.). *J. Spices and Aromatic Crops,* 12, 72-76.

Venkatasubban, K.R. (1946) A preliminary survey of chromosome numbers in Scitamineae of Bentham and Hooker. *Proc. Indian Acad. Sci.,* 23B, 281-300.

Venkatesha, J. (1994) *Studies on the evaluation of promising cultivars and nutrient requirement of turmeric (Curcuma domestica* Val.), Ph.D. Thesis, Division of Horticulture, UAS, Bangalore.

Venkatesha, J., Chandrappa, H. and Goud, B.M. (2000) Performance of certain turmeric cultivars (*Curcuma domestica* VAL.) in Southern dry region of Karnataka. In: Muraleedharan, N. and Rajkumar, R. (eds) *Recent Advances in Plantation Crops Research*. Allied Publishers Ltd., Chennai, pp. 94-96.

Venkatesha, J., Jagadesh, S.K. and Umesha, K. (2002) Evaluation of promising turmeric cultivars for rainfed conditions under hill zone of Karnataka. In: (Rethinam, P. Khan, H.H. Reddy, V M., Mandal, P K., Suresh K. (eds.) *Plantation Crops Research and Development in the New Millenium*. Coconut Development Board, Kochi, pp: 249-251.

Venkatesha, J., Khan, M.M. and Chandrappa, H. (1998) Studies on character association in turmeric. In: Mathew, N.M., Kuruvilla Jacob, C., Licy, J., Joseph, T., Meenattoor, J.R. and Thomas, K.K. (eds.). *Developments in Plantation Crops Research*. Allied Publishers Ltd., New Delhi, pp. 54-57.

Verma, A. and Tiwari, R.S. (2002) Genetic variability and character association studies in turmeric (*Curcuma longa* L.). In: National seminar on strategies for increasing production & export of spices, IISR, Calicut, Kerala, p. 8 (Abstract).

Verma, A. and Tiwari, R.S. (2002) Path coefficient analysis in turmeric (*Curcuma longa* L.). In: National seminar on strategies for increasing production & export of spices, IISR, Calicut, Kerala, p. 14 (Abstract).

Xia, Q., Zhao, K.J., Huang, Z.G., Zang, P., Dong, T.T., Li S.P.,Tsim, K.W. (2005) Molecular genetic and chemical assessment of Rhizoma Curcumae in China. *J. Agric Food Chem*.27;53, 6019-26

Yadav, D.S. and Singh, R. (1996) Studies on genetic variability in turmeric (*Curcuma longa*). *Journal of Hill Research*, 9, 33-36.

Yadav, D.S. and Singh, S.P. (1989) Phenotypic stability and genotype X environment interaction in tumeric (*Curcuma longa* L.). *Indian J. Hill Farming*, 2, 35-37.

Zachariah, T.J., Sasikumar, B. and Nirmal Babu, K. (1999) Variation for quality components in ginger and turmeric and their interaction with environments. In: Sasikumar B, Krishnamurthy B, Rema, J, Ravindran, P. N. & Peter, K. V. (eds.) *Biodiversity, Conservation and Utilization of Spices Medicinal and Aromatic Plants*. Indian Institute of Spices Research, Calicut, Kerala, pp 116-120.

3 Phytochemistry of the Genus *Curcuma*

Lutfun Nahar and Satyajit D. Sarker

CONTENTS

3.1 INTRODUCTION

The genus *Curcuma* L. of the family Zingiberaceae is well known as the turmeric genus, because of *Curcuma longa* (turmeric, commonly known as "Haldi" or "Holud"). *C. longa* is the most investigated species of this genus, although there are over 100 others in this genus (Table 3.1) (GRIN Database, 2005; *Curcuma* list, 2005). The name of the genus *Curcuma* came from the Arabic word "kurkum," which originally meant "saffron," but is now used for turmeric only. Most of the species of this genus are perennials and grow well in tropics and subtropics where it requires a hot and moist climate and a fairly light soil. *Curcuma* species are native to the countries of the

TABLE 3.1
Species of the Genus *Curcuma*, Their Trivial Names and Distribution

Species	Trivial Names	Distribution (Native to)
C. aeruginosa Roxb.[a]	Gajutsu, pink and blue ginger	Cambodia, Myanmar, Thailand, Vietnam, Indonesia, Malaysia, Japan
C. alismatifolia Gagnep.	Siam tulip	Southeast Asian countries
C. amada Roxb.	Amada, mango-ginger, mangoingwer	India, Bangladesh
C. angustifolia Roxb.	East indian arrowroot, tikur, tikor	India, Laos, Myanmar, Nepal, Pakistan
C. aromatica Salisb.	Wild turmeric, yellow zedoary, safran des indes	India, Nepal, Sri Lanka
C. caesia Roxb.	—	India
C. caulina J. Graham	Chavar	India
C. chuanyujin Roxb	—	China, Nepal, India
C. comosa Roxb.	Na-nwin-khar	Thailand, Myanmar
C.cochinchinensis Gagnep.	—	China, Nepal, India, Thailand
C. heyneana Val. and Zijp	—	India
C. harmandii Gagnep.	—	Thailand, China, India
C. kwangsiensis S. K. Lee and C. F. Liang	Guang xi e zhu	Temperate Asian countries
C. longa L.	Indian-saffron, turmeric, yü chiu, curcuma, safran des indes, gelbwurzel, kurkuma, ukon, açafrão-da-Índia, azafrán de la India, holud, haldi	India–Pakistan–Bangladesh subcontinent
C. parviflora Wall	—	Thailand
C. petiolata Roxb.	—	Myanmar, Indonesia
C. phaeocaulis Valeton	E zhu, peng e zhu	China, Indonesia, Vietnam
C. rotunda L.	Chinese-keys	China, Indonesia, Thailand, Malaysia
C. pierreana Gagnep.		
C. wenyujin Y. H. Chen and C. Ling	Wen yu jin	Temperate Asian countries, China
C. xanthorrhiza Roxb.	Temu lawak	Indonesia, Malaysia
C. zedoaria (Christm.) Roscoe[a]	Kua, zedoary, zedoaire, zitwer, temu putih, temu kuning, zedoária, cedoaria, cetoal, er-chu	India–Pakistan–Bangladesh subcontinent, China

Southeastern Asia and extensively cultivated in Bengal (Bangladesh and India), China, Taiwan, Sri Lanka, Indonesia, Peru, Australia, and the West Indies. However, India is the world's primary producer of the commercially most important species, *C. longa.* Table 3.1 presents a list of *Curcuma* species and their geographical distribution.

Previous phytochemical investigations on several *Curcuma* species revealed the presence of various types of plant-secondary metabolites, including diphenylheptanoids, monoterpenes, sesquiterpenes, etc. (ISI Database, 2005; DNP, 2001). However, curcumin and its structural analogues, predominantly found in *C. longa,* are the medicinally most valuable compounds of this genus. The aim of this chapter is to present an overview of the phytochemistry of the genus *Curcuma* on the basis of published literature of the last five decades.

Only about 20 *Curcuma* species have been studied phytochemically, to date (Table 3.2). The most investigated species are *C. longa, C. xanthorrhiza,* and *C. zedoaria.* Various compounds isolated from *Curcuma* species belong to three major classes: diphenylalkanoids (Figure 3.1 and Figure 3.2), phenylpropene derivatives of cinnamic acid type (Figure 3.3 and Figure 3.4), and terpenoids (Figure 3.5 to Figure 3.17).

TABLE 3.2
Compounds Isolated from Various Species of the Genus *Curcuma*

Species	Isolated Compounds	Ref.
C. aeruginosa	Aerugidiol (**152**)	(Masuda et al., 1991)
	Camphor (**40**)	(Jirovetz et al., 2000)
	1,8-Cineole (**43**)	
	Curcumenol (**155**)	
	Curdione (**128**)	
	Curzerenone (**114**)	
	Dehydrocurdione (**129**)	(DNP, 2001)
	Difurocumenone (**188**)	
	β-Elemene (**116**)	(DNP, 2001; Jirovetz et al., 2000;
	Isocurcumenol (**158**)	Bats et al., 1999)
	β-Pinene (**48**)	
	Zedoalactone A (**175**)	(Takano et al., 1995)
	Zedoalactone B (**176**)	
	Zedoarondiol (**173**)	
C. alismatifolia	Malvidin 3-rutinoside (**211**)	(Nakayama et al., 2000)
C. amada	Bis-demethoxycurcumin (**3**)	(Gupta et al., 1999)
	Curcumin (**1**)	
	Demethoxycurcumin (**2**)	
	Calarene (**189**)	(Singh et al., 2002)
	β-Caryophyllene (**197**)	
	1,8-Cineole (**43**)	
	α-Copaene (**190**)	
	β-Curcumene (**88**)	
	Curzerenone (**114**)	
	α-Humulene (**198**)	
	Limonene (**49**)	
	Myrcene (**61**)	
	(Z)-β-Ocimene (**64**)	
	(E)-β-Ocimene (**63**)	
	Perillene (**58**)	
	α-Pinene (**47**)	
	β–Pinene (**48**)	
	Terpinen-4-ol (**56**)	
	ar-Turmerone (**92**)	(Jain et al., 1964; DNP, 2001)
C. angustifolia	Curzerene (**113**)	(DNP, 2001)
C. aromatica	Bis-demethoxycurcumin (**3**)	(DNP, 2001)
	Curcumin (**1**)	
	Demethoxycurcumin (**2**)	
	Acetoxyneocurdione (**127**)	(Kuroyanagi et al., 1990; Shiobara et al.,
	13-Acetoxydehydrocurdione (**131**)	1986)
	β-Bisabolene (**67**)	(Singh et al., 2002)
	Bisabola-3,10-diene-2-one (**80**)	(Ho and Hall, 1983; Kim et al., 2002)
	1,3,5,10-Bisabolatetraene (**78**)	
	1,3,5,11-Bisabolatetraene (**79**)	
	Bisacumol (**73**)	(He et al., 1998)
	Borneol (**36**)	(DNP, 2001)
	β-3-Carene (**42**)	(Singh et al., 2002)
	Carvacrol (**46**)	
	1,8-Cineole (**43**)	

Continued

TABLE 3.2 *(Continued)*
Compounds Isolated from Various Species of the Genus *Curcuma*

Species	Isolated Compounds	Ref.
	Curcumadione (**110**)	(Kuroyanagi et al., 1990; Shiobara et al.,
	Curmadione (**109**)	1986)
	Curcumol (**154**)	(Yang and Chen, 1980)
	α-Curcumene (**87**)	(DNP, 2001; Kojima et al., 1998)
	β-Curcumene (**88**)	
	Curdione (**128**)	
	Curzerene (**113**)	
	Curzerenone (**114**)	
	p-Cymene (**44**)	
	p-Cymene-8-ol (**45**)	(Singh et al., 2002)
	β-Elemene (**116**)	(Kojima et al., 1998)
	4,5-Epoxy-12-hydroxy-1(10),7(11)-germacradien-8-one (**144**)	(Kuroyanagi et al., 1990; Shiobara et al., 1986)
	4,5-Epoxy-12-acetoxy-1(10),7(11)-germacradien-8-one (**145**)	
	4,5-Epoxy-12-acetoxy-7α,11-dihydrogermacradien-8-one (**146**)	
	β-Farnesene (**195**)	(Singh et al., 2002)
	Germacrone (**138**)	(Kojima et al., 1998)
	Germacrone-4,5-epoxide (**141**)	
	13-Hydroxydehydrocurdione (**130**)	(Kuroyanagi et al., 1990; Shiobara et al.,
	13-Hydroxygermacrone (**147**)	1986)
	Isocurcumadione (**112**)	
	Isoprocurcumenol (**165**)	
	Isozedoarondiol (**174**)	(Kuroyanagi et al., 1987)
	Linalool (**60**)	(Kojima et al., 1998)
	p-Methoxycinnamic acid (**28**)	(DNP, 2001)
	Methylzedoarondiol (**178**)	(Kuroyanagi et al., 1987)
	Neoprocurcumenol (**167**)	(Kuroyanagi et al., 1990; Shiobara et al., 1986)
	9-Oxo-neoprocurcumenol (**166**)	(Etoh et al., 2003)
	α-Pinene (**47**)	(Singh et al., 2002)
	β-Pinene (**48**)	
	α-Terpineol (**57**)	
	ar-Turmerone (**92**)	(He et al., 1998)
	Xanthorrhizol (**97**)	(Kojima et al., 1998)
	Zederone (**151**)	(Pant et al., 2001)
	Zedoarondiol (**173**)	(Kuroyanagi et al., 1987)
	(1β,4β,5β,10β)-Zedoarondiol (**177**)	
	Zingiberene (**98**)	(Singh et al., 2002)
C. caesia	Borneol (**36**)	(Pandey and Chowdhury, 2003)
	Bornyl acetate (**37**)	
	1,8-Cineole (**43**)	(Behura and Srivastava, 2004)
	Camphor (**40**)	
	α-Curcumene (**87**)	(Pandey and Chowdhury, 2003)
	γ-Curcumene (**89**)	
	β-Elemene (**116**)	
	(*E*)-β-Ocimene (**63**)	
	ar-Turmerone (**92**)	

TABLE 3.2 *(Continued)*
Compounds Isolated from Various Species of the Genus *Curcuma*

Species	Isolated Compounds	Ref.
C. caulina	Curzerene (**113**)	(DNP, 2001)
C. chuanyujin	Bis-demethoxycurcumin (**3**)	(Huang et al., 2000)
	Curcumin (**1**)	
	Demethoxycurcumin (**2**)	
	1-*p*-Coumaroyl-cinnamic acid (**32**)	
	1-Feruloyloxy-2-methoxycinnamic acid (**33**)	
	1-Feruloyloxy-cinnamic acid (**34**)	
C. cochinchinensis	1,8-Cineole (**43**)	(Grayson, 1998)
	4-*Epi*-curcumenol (**157**)	
	Isocurcumenol (**158**)	(DNP, 2001)
C. comosa	(*E*)-5-Acetoxy-1,7-diphenyl-1-heptene (**14**)	(Suksamrarn et al., 1997)
	(*E*)-1,7-Diphenyl-1-hepten-5-one (**17**)	
	E)-1,7-Diphenyl-1-hepten-5-ol (**13**)	
	(*E*)-1,7-Diphenyl-3-hydroxy-1-hepten-5-one (**18**)	
	(*E*)-7-(3,4-Dihydroxyphenyl)-5-hydroxy-1-phenyl-1-heptene (**15**)	
	(1*E*,3*E*)-1,7-Diphenyl-1,3-heptadien-5-ol (**19**)	
	(*E*)-1,7-Diphenyl-1-hepten-5-one (**17**)	(Jurgens et al., 1994; Claeson et al., 1993)
	(*E*)-5-Hydroxy-7-(4-hydroxyphenyl)-1-phenyl-1-heptene (**16**)	(Suksamrarn et al., 1997)
	Myrciaphenone A (**207**)	(DNP, 2001; Yoshikawa et al., 1998a)
	Phloracetophenone (**208**)	(Piyachaturawat et al., 2002)
C. heyneana	Bis-demethoxycurcumin (**3**)	(Jitoe et al., 1992)
	Curcumin (**1**)	
	Demethoxycurcumin (**2**)	
	Curcumanolide A (**193**)	(DNP, 2001)
	Curcumanolide B (**194**)	
	Isocurcumenol (**158**)	(Firman et al., 1988; Bats et al., 1999)
	8(17),12-Labdadiene-15,16-dial (**35**)	(Firman et al., 1988)
	Oxycurcumenol (**159**)	(Firman et al., 1988; Bats et al., 1999)
C. harmandii	Isocurcumenol (**158**)	(DNP, 2001; Bats et al., 1999)
C. kwangsiensis	Gweicurculactone (**168**)	(DNP, 2001; Bats et al., 1999)
	Isocurcumenol (**158**)	
C. longa (syn. *C. domestica*)	Bis-demethoxycurcumin (**3**)	(Park and Kim, 2002; Kiuchi et al., 1993; Nakayama et al., 1993; Sastry, 1970; Ravindranath and Satyanarayana, 1980)
	(*E*)-1,7-Bis-(4-hydroxy-3-methoxyphenyl)-1-heptene-3,5-dione (**6**)	
	(1*E*, 4*E*, 6*E*)-1,7-Bis-(4-hydroxy-3-methoxyphenyl)-1,4,6-heptatrien-3-one (**22**)	
	(1*E*, 4*E*, 6*E*)-1,7-Bis-(4-hydroxyphenyl)-1,4,6-heptatrien-3-one (**21**)	
	(*E*)-1,7-Bis-(4-hydroxyphenyl)-1-hepten-3,5-dione (**7**)	
	Curcumin (**1**)	
	Cyclocurcumin (**9**)	
	Demethoxycurcumin (**2**)	
	Dihydrocurcumin (**10**)	
	(*E*)-7-Hydroxy-1,7-bis-(4-hydroxy-3-methoxyphenyl)-1-heptene-3,5-dione (**8**)	(Nakayama et al., 1993; Park and Kim, 2002)

Continued

TABLE 3.2 *(Continued)*
Compounds Isolated from Various Species of the Genus *Curcuma*

Species	Isolated Compounds	Ref.
	(1*E*, 6*E*)-1-(4-Hydroxy-3-methoxyphenyl)-7-(3,4-dihydroxyphenyl)-1,6-hepta diene-3,5-dione (**4**)	
	(1*E*, 4*E*)-1,5-Bis-(4-hihydrox-3-methoxyphenyl)-1,4-pentadien-3-one (**23**)	(Masuda et al., 1993; Park and Kim, 2002)
	(1*E*, 4*E*)-1-(4-Hydroxy-3-methoxyphenyl)-5-(4-hydroxyphenyl)-1,4-pentadien –3-one (**24**)	
	α-Atlantone (**65**)	(Roth et al., 1998)
	β-Atlantone (**66**)	
	β-Bisabolene (**67**)	(Singh et al., 2002)
	1,10-Bisaboladiene-3,4-diol (**75**)	(DNP, 2001; Ohshiro et al., 1990; Kuroyanagi
	2,10-Bisaboladiene-1,4-diol (**76**)	et al., 1990; Ohshiro et al., 1990)
	Bisabola-3,10-diene-2-one (**80**)	
	Bisacumol (**73**)	
	Borneol (**36**)	(Roth et al., 1998)
	Calebin A (**31**)	(Park and Kim, 2002)
	Caffeic acid (**25**)	(Roth et al., 1998)
	Camphene (**41**)	
	Camphor (**40**)	
	Cinnamic acid (**26**)	
	p-Coumaric acid (**27**)	
	β-Caryophyllene (**197**)	(Sacchetti et al., 2005; Bansal et al., 2002;
	α-Curcumene (**87**)	Singh et al., 2002)
	1,8-Cineole (**43**)	
	β-Curcumene (**88**)	
	Curcumenol (**155**)	(DNP, 2001; He et al., 1998; Hikino et al.,
	Curcumenone (**111**)	1968a; Ohshiro et al., 1990)
	Curdione (**128**)	(Roth et al., 1998)
	Curlone (**90**)	(He et al., 1998; Ohshiro et al., 1990; Kiso et al., 1983)
	Epi-curcumenol (**156**)	(DNP, 2001)
	Curzerene (**113**)	
	Curzerenone (**114**)	(Roth et al., 1998)
	p-Cymene (**44**)	(Grayson, 1998)
	Dehydrocurdione (**129**)	(He et al., 1998; Ohshiro et al., 1990)
	2,5-Dihydroxybisabola-3,10-diene (**81**)	
	4,5-Dihydroxybisabola-2,10-diene (**82**)	
	Eugenol (**205**)	(Roth et al., 1998)
	β-Farnesene (**195**)	(Singh et al., 2002)
	Germacrone-13-al (**139**)	(Ohshiro et al., 1990; He et al., 1998)
	Germacrone-4,5-epoxide (**141**)	
	4-Hydroxybisabola -2,10-dien-9-one (**85**)	(Ohshiro et al., 1990)
	4-Hydroxy-3-methoxy-2,10-bisaboladien-9-one (**83**)	(DNP, 2001; Ohshiro et al., 1990; Kuroyanagi et al., 1990; Ohshiro et al., 1990)
	3-Hydroxy-1,10-bisaboladien-9-one (**84**)	
	Procurcumadiol (**161**)	
	Procurcumenol (**163**)	
	1-*Epi*-procurcumenol (**164**)	
	Isoprocurcumenol (**165**)	

TABLE 3.2 *(Continued)*
Compounds Isolated from Various Species of the Genus *Curcuma*

Species	Isolated Compounds	Ref.
	Isoborneol (**38**)	(Roth et al., 1998)
	8(17),12-Labdadiene-15,16 dial (**35**)	
	Limonene (**49**)	
	4-Methoxy-5-hydroxy-bisabola-2,10-diene-9-one (**86**)	
	1-(4-Methylphenyl)ethanol (**206**)	
	Myrcene (**61**)	(Behura and Srivastava, 2004)
	α-Phellandrene (**50**)	(Sacchetti et al., 2005; Bansal et al., 2002)
	α-Pinene (**47**)	(Roth et al., 1998)
	β-Pinene (**48**)	(Grayson, 1998)
	D-Sabinene (**52**)	(Rothe et al., 1998)
	Syringic acid (**210**)	
	Terpenolene (**55**)	(Sacchetti et al., 2005; Behura and Srivastava, 2004; Bansal et al., 2002)
	Terpinen-4-ol (**56**)	(Raina et al., 2002)
	ar-Turmerone (dehydroturmerone, **92**)	(Singh et al., 2002; Su et al., 1982)
	α-Turmerone (**93**)	(Golding et al., 1982)
	Turmerone (**91**)	(DNP, 2001; Su et al., 1982)
	β–Turmerone (**94**)	(He et al., 1998)
	Turmeronol A (**95**)	(Imai et al., 1990)
	Turmeronol B (**96**)	
	Turmerin (noncyclic polypeptide)	(Srinivas et al., 1992)
	Vanillic acid (**209**)	(Rothe et al., 1998)
	Zingiberene (**98**)	(Singh et al., 2002; Ohshiro et al., 1990)
	Zedoarondiol (**173**)	
	Ukonans **A-D** (neutral polysaccharides)	(DNP, 2001)
C. parviflora	Cadalenequinone (**99**)	(Toume et al., 2004)
	8-Hydroxycadalene (**101**)	
	Parviflorene A (**182**)	
	Parviflorene B (**183**)	
	Parviflorene C (**184**)	
	Parviflorene D (**185**)	
	Parviflorene E (**186**)	
	Parviflorene F (**187**)	
C. petiolata	Curzerene (**113**)	(DNP, 2001)
C. phaeocaulis	Curdione (**128**)	(Chen et al., 2001)
	Isocurcumenol (**158**)	(DNP, 2001; Bats et al., 1999)
	Camphor (**40**)	
	Isoborneol (**38**)	
	Isobornyl acetate (**39**)	
C. rotunda	Curzerene (**113**)	(DNP, 2001)
C. wenyujin	Curcumalactone A (**192**)	(Harimaya et al., 1991)
	Curcumol (**154**)	(DNP, 2001; Hikino et al., 1965; Hariyama et al., 1991; Gao et al., 1989)
	Curdione (**128**)	
	(1α,4β,5α,10β) 1,10:4,5-Diepoxy-7(11)-germacren-8-one (**132**)	
	(1*R*, 10*R*)-Epoxy-(–)-1,10-dihydrocurdione (**134**)	
	3,4-Epoxy-6,9-germacranedione (**133**)	(Gao et al., 1991)

Continued

TABLE 3.2 *(Continued)*
Compounds Isolated from Various Species of the Genus *Curcuma*

Species	Isolated Compounds	Ref.
	β–Elemene (**116**)	(Guo, 1983)
	Germacrone (**138**)	(Yan et al., 2005)
	(1*S*,10*S*),(4*S*,5*S*)-Germacrone-1(10),4-diepoxide (**143**)	(DNP, 2001; Hikino et al., 1965; Hariyama et al., 1991; Gao et al., 1989)
	Neocurdione (**149**)	
	Wenjine (**150**)	(Gao et al., 1991)
C. xanthorrhiza	Bis-demethoxycurcumin (**3**)	(Uehara et al., 1992; Yasni et al., 1993)
	Curcumin (**1**)	
	Demethoxycurcumin (**2**)	
	(1*E, 3E*)-1,7-Diphenyl-1,3-heptadien-5-ol (Alnustone, **19**)	(Claeson et al., 1993; Claeson et al., 1996)
	(1*E*,3*E*)-1,7-Diphenyl-1,3-heptadien-5-one (**20**)	
	(1*E*,3*E*)-1,7-Diphenyl-1,3-heptadien-5-ol (**19**)	
	(*E*)-1,7-Diphenyl-1-hepten-5-ol (**13**)	
	(*E*)-7-(3,4-Dihydroxyphenyl)-5-hydroxy-1-phenyl-1-heptene (**15**)	(Suksamrarn et al., 1994)
	(*E*)-5-Hydroxy-7-(4-hydroxyphenyl)-1-phenyl-1-heptene (**16**)	
	7-Hydroxy-1,7-bis-(4-hydroxy-3-methoxyphenyl)-1-heptene-3,5-dione (**8**)	(Masuda et al., 1992)
	5′-Methoxycurcumin (**5**)	
	β-Atlantone (**66**)	(Itoawa et al., 1985)
	1,3,5,10-Bisabolapentaen-9-ol (**77**)	(Uehara et al., 1989)
	Bisacumol (**73**)	
	Bisacurol (**72**)	
	Bisacurone A (**69**)	(Ohshiro et al., 1990; Uehara et al., 1990, 1989)
	Bisacurone B (**70**)	
	Bisacurone C (**71**)	
	Bisacurone (**68**)	
	Bisacurone epoxide (**74**)	
	Borneol (**36**)	Phytochemical and Ethnobotanical Database (2005)
	Cinnamaldehyde (**30**)	(Claeson et al., 1993; Claeson et al., 1996)
	Camphor (**40**)	(Uehara et al., 1992; Yasni et al., 1993)
	α-Curcumene (**87**)	
	β-Curcumene (**88**)	(DNP, 2001; Yasni et al., 1994)
	Curzerene (isofuranogermacrene, **113**)	(DNP, 2001)
	Curzerenone (**114**)	(Uehara et al., 1992; Yasni et al., 1993)
	Germacrone (**138**)	
	α-Phellandrene (**50**)	(Ohshiro et al., 1990; Uehara et al., 1990, 1989)
	β-Phellandrene (**51**)	
	ar-Turmerone (**92**)	
	α-Turmerone (**93**)	
	Turmerone (**91**)	
	Xanthorrhizol (**97**)	(Rimpler et al., 1970; Hwang et al., 2000)
	Zingiberene (**98**)	Phytochemical and Ethnobotanical Database (2005)
C. zedoaria	Bis-demethoxycurcumin (**3**)	(DNP, 2001; Matsuda et al., 2004)
	Curcumin (**1**)	

TABLE 3.2 *(Continued)*
Compounds Isolated from Various Species of the Genus *Curcuma*

Species	Isolated Compounds	Ref.
	Demethoxycurcumin (**2**)	
	Dihydrocurcumin (**10**)	
	Tetrahydrodemethoxycurcumin (**11**)	
	Tetrahydro-bis-demethoxycurcumin (**12**)	
	(1*E*, 4*E*, 6*E*)-1,7-Bis-(4-hydroxyphenyl)-1,4,6-heptatrien-3-one (**21**)	(Jang et al., 2004)
	Aerugidiol (**152**)	(DNP, 2001; Mau et al., 2003; Matsuda et al., 2001)
	Alismoxide (**153**)	(Jang et al., 2004; Matsuda et al., 2001)
	β-Bisabolene (**67**)	(DNP, 2001; Mau et al., 2003; Matsuda et al., 2001)
	Bisacumol (**73**)	(Jang et al., 2004; Matsuda et al., 2001)
	Bisacurone (**68**)	
	Borneol (**36**)	(Yoshihara et al., 1984; Shiobara et al., 1985)
	1,10-Bisaboladiene-3,4-diol (**75**)	(DNP, 2001; Matsuda et al., 2004)
	α-Cadinol (**102**)	(Mau et al., 2003; Matsuda et al., 2001)
	Calarene (**189**)	(DNP, 2001; Mau et al., 2003; Matsuda et al., 2001)
	α-Calacorene (**100**)	
	Camphene (**41**)	
	Camphor (**40**)	
	β-Caryophyllene (**197**)	(Singh et al., 2002; Singh et al., 2003)
	β-3-Carene (**42**)	
	1,8-Cineole (**43**)	(Mau et al., (2003); Matsud*a* et al., 2001)
	Citronellol (**59**)	(Singh et al., 2002; Singh et al., 2003)
	Curcolonol (**120**)	(DNP, 2001; Syu et al., 1998)
	Curcolone (**121**)	(Hikino et al., 1967; DNP, 2001)
	Curcumadiol (**162**)	(Yoshihara et al., 1984; Shiobara et al., 1985)
	Curcumadione (**110**)	Jang et al., 2004; Matsuda et al., 2001)
	Curcumalactone A (**199**)	(DNP, 2001; Mau et al., 2003; Matsuda et al., 2001)
	Curcumalactone B (**200**)	
	Curcumalactone C (**201**)	
	4-*Epi*-curcumenol (**157**)	
	Curcumanolide A (**193**)	(Shiobara et al., 1985)
	Curcumanolide B (**194**)	
	α-Curcumene (**87**)	(Mau et al., 2003; Matsuda et al., 2001)
	Curcumenol (**155**)	(DNP, 2001; Hikino et al., 1968a; Bats et al., 1999; Mau et al., 2003)
	Curcumenone (**107**)	(Yoshikawa et al., 1998; Shiobara, 1985; Mau et al., 2003; Matsuda et al., 2001)
	Curcurabranol A (**105**)	
	Curcurabranol B (**106**)	
	Curcumol (**154**)	(Hikino et al., 1965; Hariyama et al., 1991)
	Curdione (**128**)	(DNP, 2001; Mau et al., 2003; Matsuda et al., 2001)
	Curzerene (**113**)	(Mau et al., 2003)
	Curzeone (**103**)	(Shiobara et al., 1986)
	Curzerenone (Zedoarone, **114**)	(Fukushima et al., 1970; Matsuda et al., 2001)
	5-*Epi*-curzerenone (**115**)	(DNP 2001; Mau et al., 2003)
	p-Cymene (**44**)	(Singh et al., 2002; Singh et al., 2003)
	p-Cymene-8-ol (**45**)	

Continued

TABLE 3.2 *(Continued)*
Compounds Isolated from Various Species of the Genus *Curcuma*

Species	Isolated Compounds	Ref.
	Dehydrocurdione (**129**)	(Yoshikawa et al., 1998; Shiobara 1985; Mau
	4*S*-Dihydrocurcumenone (**108**)	et al., 2003; Matsuda et al., 2001)
	1,4-Dihydroxyfuranoeremophilan-6-one (**126**)	(DNP, 2001; Syu et al., 1998)
	3,7-Dimethyl-5-indanecarboxylic acid (**204**)	
	β-Dictyopetrol (**122**)	(Mau et al., 2003; Matsuda et al., 2001)
	Ethyl-*p*-methoxycinnamate (**29**)	(Phytochemical and Ethnobotanical database 2005; Gupta et al., 1976)
	β-Elemene (**116**)	(DNP 2001; Mau et al., 2003; Matsuda et al.,
	γ-Elemene (**117**)	2001; Jang et al., 2004; Matsuda et al., 2001)
	β-Elemenone (**119**)	
	Elemol (**118**)	
	7α,11α,-Epoxy-5β-hydroxy-9-guaiaen-8-one (**172**)	
	β-Eudesmol (**123**)	
	β-Farnesene (**195**)	
	Farnesol (**196**)	
	Furanodiene (**135**)	
	Furanodienone (**136**)	(DNP 2001; Pandji et al., 1993)
	Furanogermenone (**137**)	(DNP, 2001; Shibuya et al., 1986)
	Gajutsulactone A (**202**)	(DNP, 2001; Mau et al., 2003; Matsuda et al.,
	Gajutsulactone B (**203**)	2001)
	Germacrene B (**140**)	
	Germacrone 4,5-epoxide (**141**)	(Yoshihara et al., 1984; Shiobara et al., 1985)
	*G*lechomanolide (**142**)	(DNP, 2001; Mau et al., 2003; Matsuda et al.,
	α-Humulene (**198**)	2001)
	β-Himachalene (**191**)	
	13-Hydroxygermacrone (**147**)	(Matsuda et al., 2001; Shiobara et al., 1986)
	4-Hydroxy-7(11),10(14)-guaiadien-8-one (**171**)	(Kuroyanagi et al., 1990)
	Isoborneol (**38**)	(Kouno et al., 1985; Mau et al., 2003)
	Isocurcumenol (**158**)	(DNP, 2001; Hikino et al., 1968a; Bats et al.,
	Isofuranodienone (**148**)	1999; Mau et al., 2003)
	Isoprocurcumenol (**165**)	(Jang et al., 2004; Matsuda et al., 2001)
	Isospathulenol (**170**)	(DNP, 2001; Mau et al., 2003; Matsuda et al., 2001)
	Isozedoarondiol (**174**)	(Jang et al., 2004; Matsuda et al., 2001)
	Linalool (**60**)	
	Myrcene (**61**)	(Singh et al., 2002); Singh et al., 2003)
	Neocurcumenol (**160**)	(DNP, 2001; Mau et al., 2003; Matsuda et al.,
	Neocurdione (**149**)	2001)
	Nerol (**62**)	(Singh et al., 2002; Singh et al., 2003)
	(*E*)-β-Ocimene (**63**)	
	(*Z*)-β-Ocimene (**64**)	
	9-Oxo-neoprocurcumenol (**166**)	(Etoh et al., 2003)
	α-Phellandrene (**50**)	(Singh et al., 2002; Singh et al., 2003)
	β-Phellandrene (**51**)	
	α-Pinene (**47**)	(DNP, 2001; Mau et al., 2003; Matsuda et al.,
	β-Pinene (**48**)	2001)
	Procurcumenol (**163**)	(Jang et al., 2004; Matsuda et al., 2001)
	1-*Epi*-procurcumenol (**164**)	

TABLE 3.2 *(Continued)*
Compounds Isolated from Various Species of the Genus *Curcuma*

Species	Isolated Compounds	Ref.
	Pyrocurzerenone (**104**)	(DNP, 2001)
	α-Selinene (**124**)	(DNP, 2001; Mau et al., 2003; Matsuda et al.,
	β-Selinene (**125**)	2001)
	Spathulenol (**169**)	
	Terpinen-4-ol (**56**)	(DNP, 2001; Singh et al., 2002; Singh et al.,
	α-Terpineol (**57**)	2003)
	Terpinolene (**55**)	
	α-Terpinene (**53**)	
	γ-Terpinene (**54**)	
	ar-Turmerone (**92**)	
	β-Turmerone (**94**)	
	Zederone (**151**)	(Matsuda et al., 2001; Hikino et al., 1968)
	Zedoarol (**179**)	(Matsuda et al., 2001; Shiobara et al., 1986)
	Zedoarondiol (**173**)	(Jang et al., 2004; Matsuda et al., 2001)
	Zedoalactone B (**176**)	
	Zedoarolide A (**180**)	(DNP, 2001; Mau et al., 2003; Matsuda et al.,
	Zedoarolide B (**181**)	2001)
	Zingiberene (**98**)	(DNP, 2001; Matsuda et al., 2004)

3.2 DIPHENYLALKANOIDS

There are two classes of diphenylalkanoids in the genus *Curcuma*: diphenylheptanoids (Figure 3.1) and diphenylpentanoids (Figure 3.2). However, diphenylheptanoids are the most abundant group of compounds in this genus.

3.2.1 DIPHENYLHEPTANOIDS

Diphenylheptanoids (Figure 3.1), sometimes refered to as curcuminoids, are a broad class of naturally occurring compounds. So far, over 100 diarylheptanoid compounds have been discovered, isolated, and identified in nature. The diarylheptanoids having an aryl–C7–aryl skeleton are common in the genus *Curcuma* and also in the family Zingiberaceae. These compounds are present in many, but not all, *Curcuma* species. Structurally, diphenylheptanoids are composed of two phenyl (or substituted phenyl) groups linked through a seven-carbon unit alkane (or modified alkane) chain. Diphenylheptanoids found in the *Curcuma* differ in the substitution patterns in the phenyl rings as well as in the heptane chain (e.g., number and position of the double bonds, carbonyl functionalities, or hydroxyl moeities). Many of these compounds have phenolic hydroxyl groups in their structures (**1** to **12, 21, 22**) and thus exhibit significant antioxidant properties. Oxygenation on the phenyl rings are limited to C-3 and C-4 positions, with the only exception in 5'methoxycurcumin (**5**) isolated from *C. xanthorrhiza* (Masuda et al., 1992), where an additional oxygenation was found at C-5. Cyclization of the seven-carbon unit, leading to the formation of a pyrone ring as in cyclocurcumin (**9**), is not common, but was only found in *C. longa* (Kiuchi et al., 1993; Nakayama et al., 1993; DNP, 2001). Curcumin (**1**), demthoxycurcumin (**2**), and *bis*-demethoxycurcumin (**3**), also known as curcumin I, II, and III, respectively, are the most widely distributed diphenylheptanoids in the genus *Curcuma*. To date, over 20 different diphenylheptanoids have been reported from at least eight different species of this genus (Figure 3.1, Table 3.2).

The biosynthesis of cucuminoids has been studied extensively by means of ^{13}C labeling and nuclear magnetic resonance (NMR) spectroscopy in various plant species (Hölscher and Schneider,

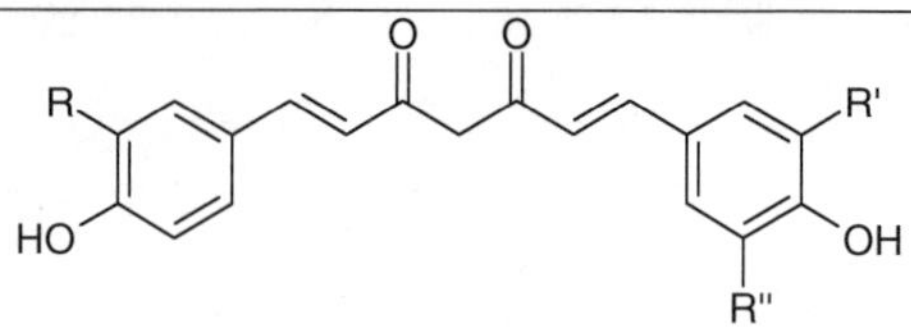

Curcumin (**1**)	R' = R'' = H
Demethoxycurcumin (**2**)	R = OMe; R' = H; R'' = H
Bis-demethoxycurcumin (**3**)	R = R' = R'' = H
(1*E*, 6*E*)-1-(4-Hydroxy-3-methoxyphenyl)-7-(3,4- dihydroxy phenyl) -1,6-heptadiene-3,5-dione (**4**)	R = OMe; R' = OH; R'' = H
5'-Methoxycurcumin (**5**)	R = R' = R'' = OMe

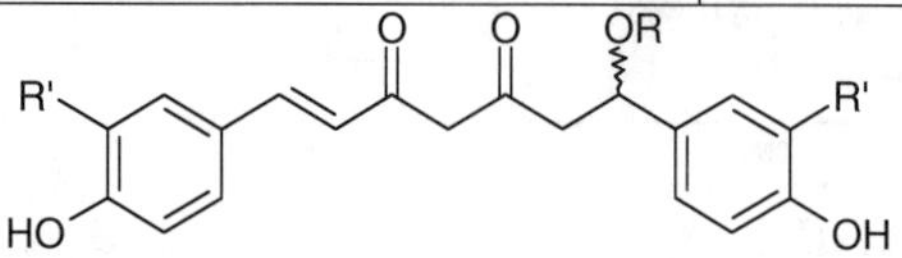

(*E*)-1,7-*Bis*-(4-hydroxy- 3-methoxyphenyl)- 1-heptene-3,5-dione (**6**)	R = H
	R' = OMe
(*E*)-1,7-*Bis*-(4-hydroxyphenyl)-1-hepten-3,5-dione (**7**)	R = R' = H
(*E*)-7-Hydroxy-1,7-*bis*-(4-hydroxy-3-methoxyphenyl)-1-heptene-3,5- dione (**8**)	R = OH
	R' = OMe

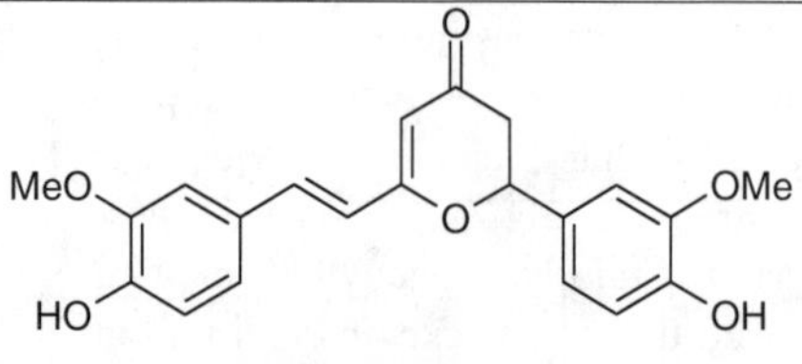

Cyclocurcumin (**9**)

FIGURE 3.1 Diphenylheptanoids from the genus *Curcuma*.   *Continued*

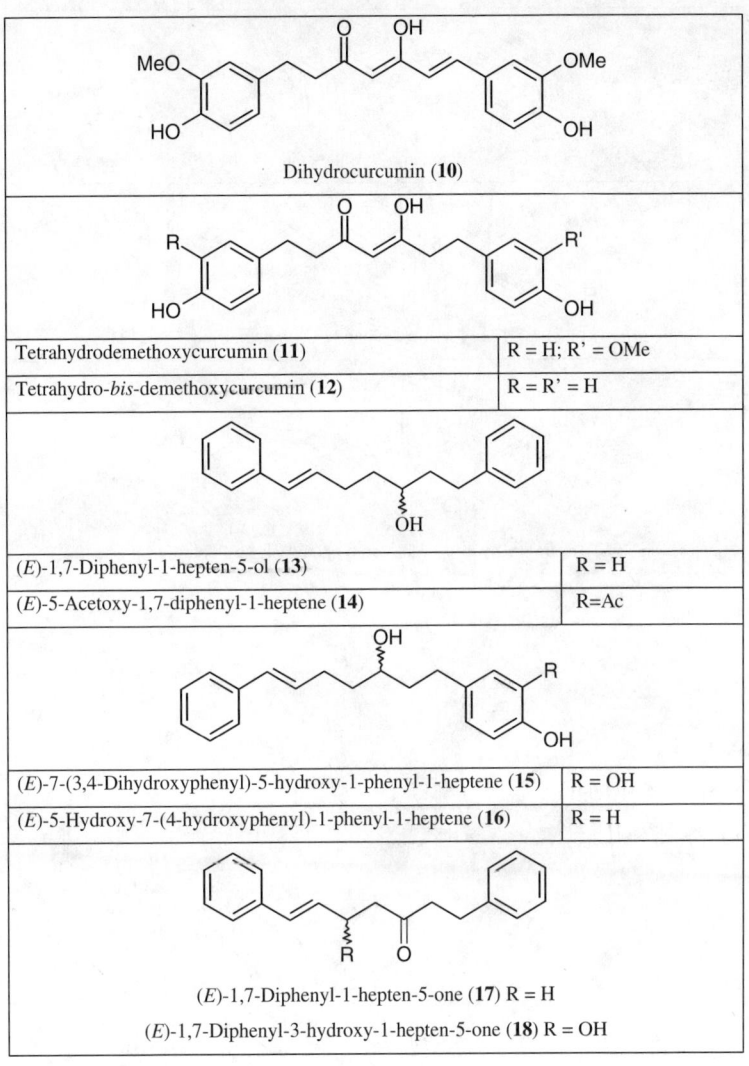

Dihydrocurcumin (**10**)

| Tetrahydrodemethoxycurcumin (**11**) | R = H; R' = OMe |
| Tetrahydro-*bis*-demethoxycurcumin (**12**) | R = R' = H |

| (*E*)-1,7-Diphenyl-1-hepten-5-ol (**13**) | R = H |
| (*E*)-5-Acetoxy-1,7-diphenyl-1-heptene (**14**) | R=Ac |

| (*E*)-7-(3,4-Dihydroxyphenyl)-5-hydroxy-1-phenyl-1-heptene (**15**) | R = OH |
| (*E*)-5-Hydroxy-7-(4-hydroxyphenyl)-1-phenyl-1-heptene (**16**) | R = H |

(*E*)-1,7-Diphenyl-1-hepten-5-one (**17**) R = H

(*E*)-1,7-Diphenyl-3-hydroxy-1-hepten-5-one (**18**) R = OH

FIGURE 3.1 *Continued*

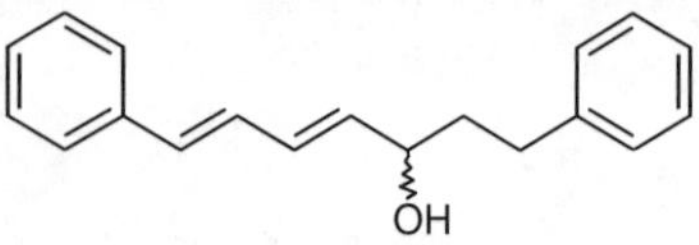

(1*E*,3*E*)-1,7-Diphenyl-1,3-heptadien-5-ol (**19**)

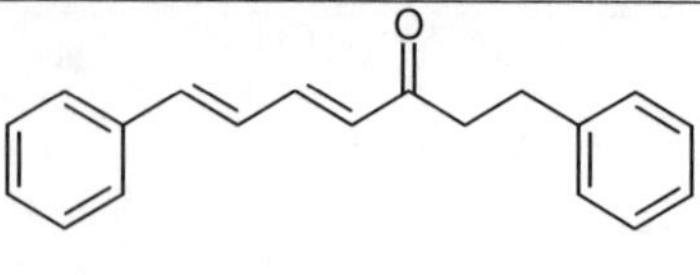

(1*E*, 3*E*)-1,7-Diphenyl-1,3-heptadien-5-one (**20**)

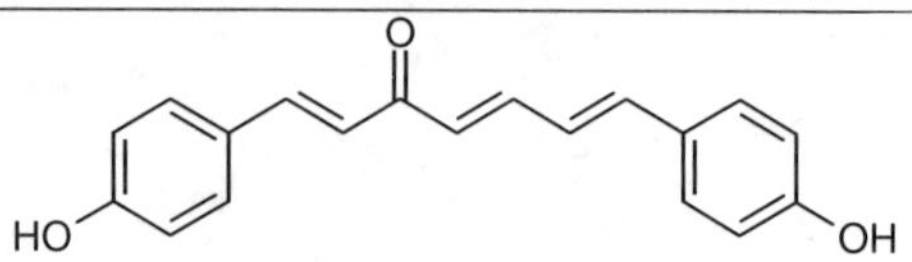

(1*E*, 4*E*, 6*E*)-1,7-*Bis*-(4-hydroxyphenyl)-1,4,6-heptatrien-3-one (**21**)

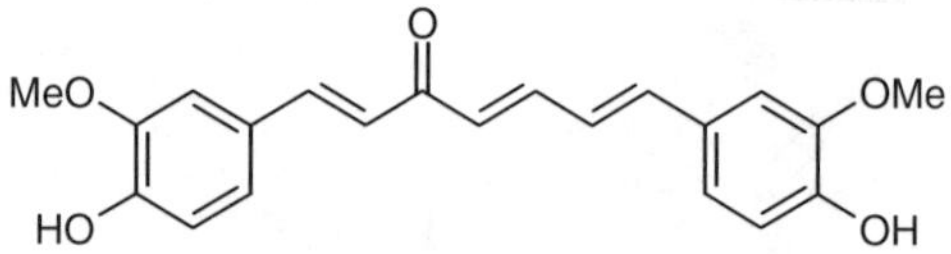

(1*E*, 4*E*, 6*E*)-1,7-*Bis*-(4-hydroxy-3-methoxyphenyl)-1,4,6-heptatrien-3-one (**22**)

FIGURE 3.1 *Continued*

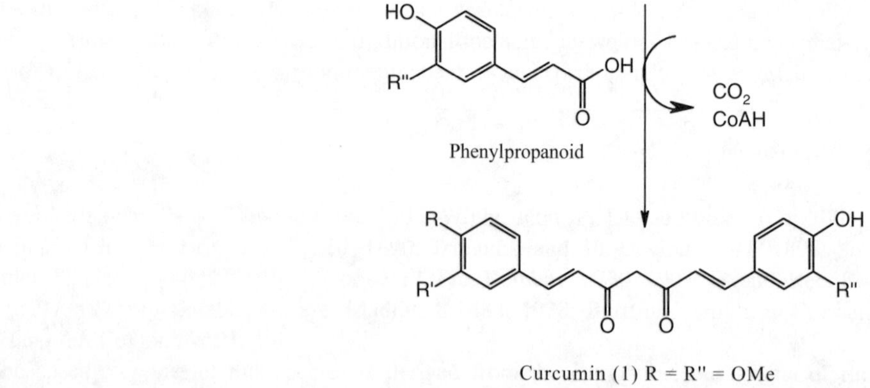

Curcumin (1) R = R'' = OMe

SCHEME 3.1 Biosynthesis of curcuminoids.

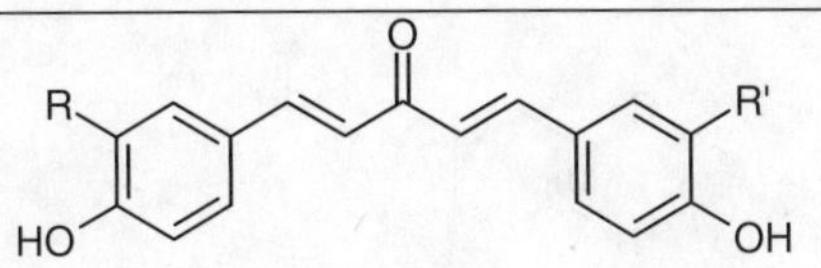

| (1*E*, 4*E*)-1,5-*Bis*-(4-hihydroxy-3-methoxyphenyl)-1,4-pentadien-3-one (**23**) | R = R' = OMe |
| (1*E*, 4*E*)-1-(4-Hydroxy-3-methoxyphenyl)-5-(4-hydroxyphenyl)-1,4-pentadien-3-one (**24**) | R = H, R' = OMe |

FIGURE 3.2 Diphenylpentanoids from the genus *Curcuma*.

1995a,b; Schmitt et al., 2000; Schmitt and Schneider, 1999). It has been demonstrated that cucuminoids are biosynthesized via diketide intermediates (Scheme 3.1). Phenylpropanoid CoA is believed to be the starter. The biosynthesis of these compounds involves one condensation reaction by a chalcone synthase (CHS)-related type III plant polyketide synthase (PKS), followed by a number of reactions, leading to the formation of various cucuminoids (Brand et al., 2006). However, not many studies have been carried out on the biosynthesis of these compounds in relation to any species of the genus *Curcuma*.

3.2.2 DIPHENYLPENTANOIDS

Instead of a seven-carbon alkane chain, there is a five-carbon chain between two phenyl groups in diphenylpentanoids (Figure 3.2). There are only two such compounds, (1*E*, 4*E*)-1,5-bis-(4-hydroxy-3-methoxyphenyl)-1,4-pentadien-3-one (**23**) and (1*E*, 4*E*)-1-(4-hydroxy-3-methoxyphenyl)-5-(4-hydroxyphenyl)-1,4-pentadien-3-one (**24**), isolated from *C. longa* (Park and Kim, 2002; Masuda et al., 1993).

3.3 PHENYLPROPENE DERIVATIVES (CINNAMIC ACID-TYPE)

Phenylpropene derivatives (cinnamic acid-type), both monomers and dimers, are quite common in nature. A number of species of the genus *Curcuma* produce mainly monomeric phenylpropene derivatives (Figure 3.3). However, dimeric phenylpropene derivatives (lignans) are also reported (Figure 3.4).

3.3.1 MONOMERIC PHENYLPROPENE DERIVATIVES

There are at least seven monomeric phenylpropene derivatives, caffeic acid (**25**), cinnamic acid (**26**), *p*-coumaric acid (**27**), *p*-methoxycinnamic acid (**28**), ethyl-*p*-methoxycinnamic acid (**29**), cinnamaldehyde (**30**) and calebin A (**31**) in this genus (Table 3.2). However, calebin A (**31**), isolated from *C. longa* (Park and Kim, 2002), is in fact an ester formed between ferulic acid and a phenyl butane derivative.

3.3.2 DIMERIC PHENYLPROPENE DERIVATIVES (LIGNANS)

Three dimeric phenylpropene derivatives of biphenyl ether-type (lignans), 1-*p*-coumaroyl-cinnamic acid (**32**), 1-feruloyl-2-methoxy-cinnamic acid (**33**) and 1-feruloyl-cinnamic acid (**34**), have been isolated only from *Curcuma chuanyujin* (Huang et al., 2000). Within the genus *Curcuma*, the distribution of these compounds (**32** to **34**) seems to be restricted to this particular species, and to our knowledge, there is no report on the occurrence of any other lignans in *Curcuma*.

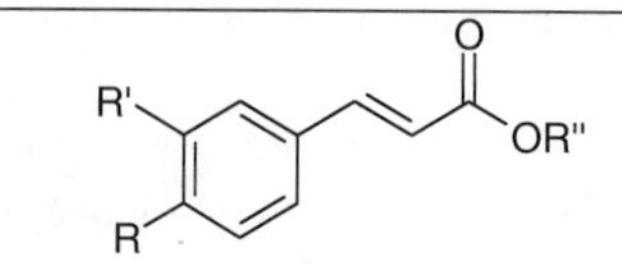

Caffeic acid (**25**)	R = R' = OH; R'' = H
Cinnamic acid (**26**)	R = R' = R'' = H
p-Coumaric acid (**27**)	R = OH; R' = R' = H
p-Methoxycinnamic acid (**28**)	R = OMe; R' = R'' = H
Ethyl-*p*-methoxycinnamate (**29**)	R = OMe;R' = H; R'' = CH₃CH₂

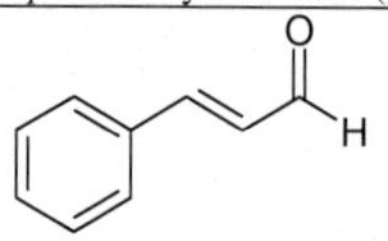

Cinnamaldehyde (**30**)

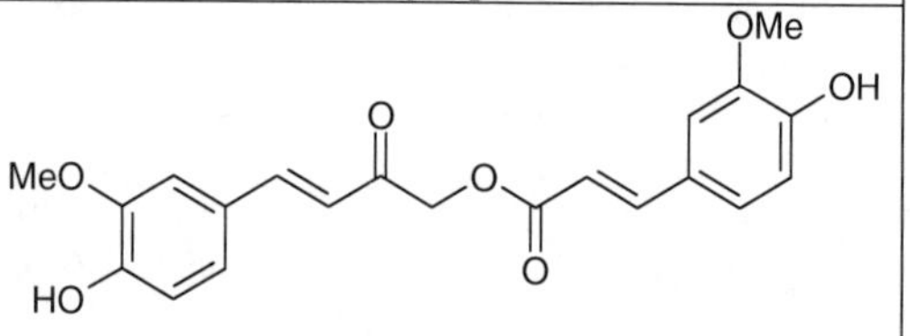

Calebin A (**31**)

FIGURE 3.3 Monomeric phenylpropene derivatives from the genus *Curcuma*.

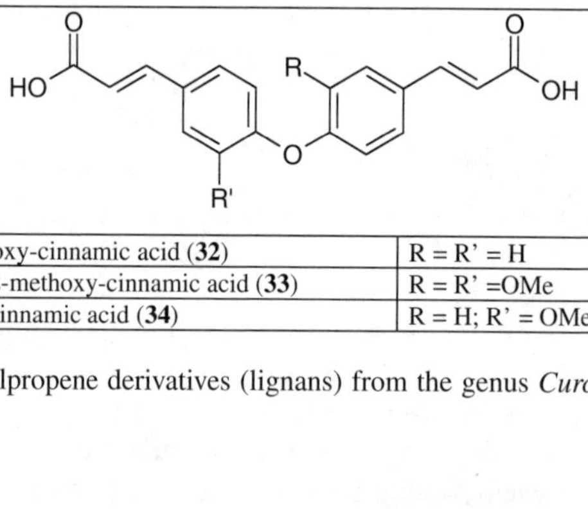

1-*p*-Coumaroyloxy-cinnamic acid (**32**)	R = R' = H
1-Feruloyloxy-2-methoxy-cinnamic acid (**33**)	R = R' =OMe
1-Feruloyloxy-cinnamic acid (**34**)	R = H; R' = OMe

FIGURE 3.4 Dimeric phenylpropene derivatives (lignans) from the genus *Curcuma*.

3.4 TERPENOIDS

Terpenoids (Figure 3.5 to Figure 3.17) are another major group of plant secondary metabolites found in the genus *Curcuma*. While there is no report on the occurrence of any tri- or tetraterpenoids to date, a great variety of mono- and sesquiterpenes have been isolated from various species of the *Curcuma* (Figure 3.6 to Figure 3.17).

3.4.1 DITERPENES

The sole diterpene 8(17),12-labdadiene-15,16-dial (**35**) of this genus was reported from *C. heyneana* (Firman et al., 1988) and *C. longa* (Roth et al., 1998) (Figure 3.5).

3.4.2 MONOTERPENES

Monoterpenoids have been isolated from the fragrant oils of many plants and are important in the perfumery and flavor industries. *Curcuma* essential oils, obtained from various species, are rich in monoterpenes (Figure 3.6, Table 3.2). About 30 different monoterpenoids have been reported from this genus. The *Curcuma* species that produce monoterpenes in high amounts include *C. aeruginosa*,

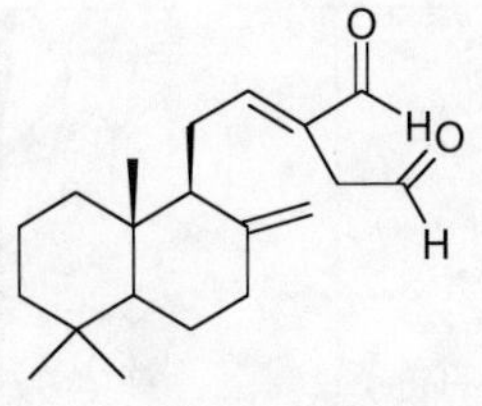

8(17),12-Labdadiene-15,16-dial (**35**)

FIGURE 3.5 Labdane-type diterpenoid from the genus *Curcuma*.

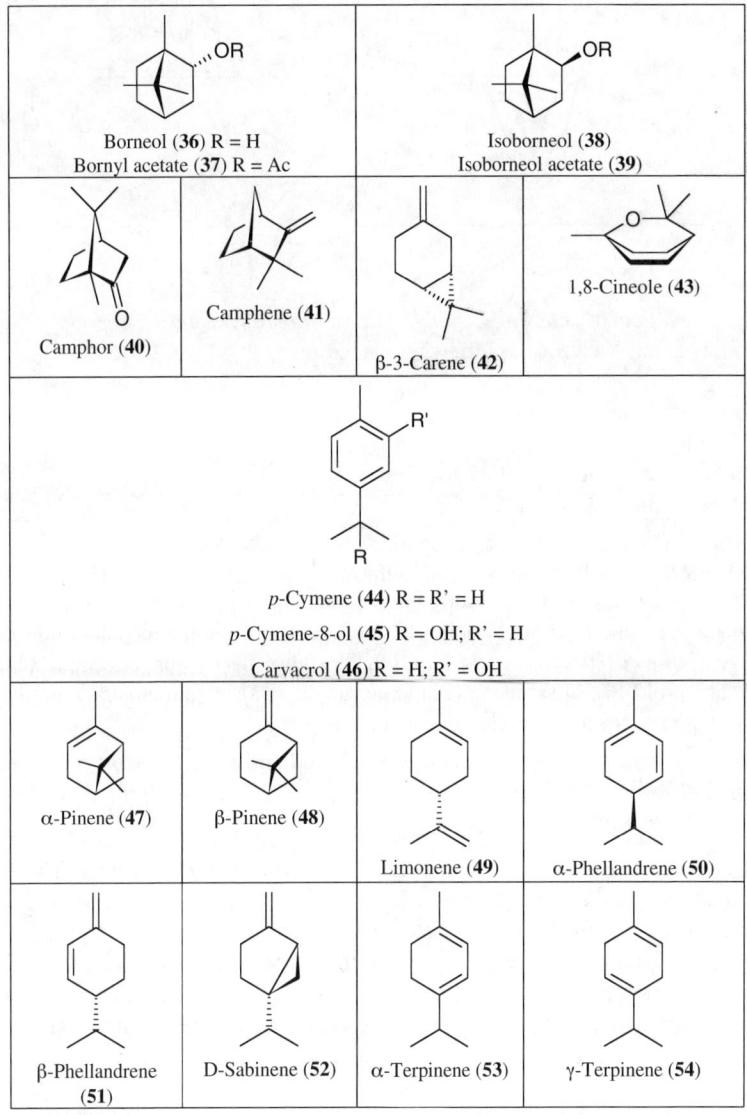

FIGURE 3.6 Monoterpenoids from the genus *Curcuma*.   *Continued*

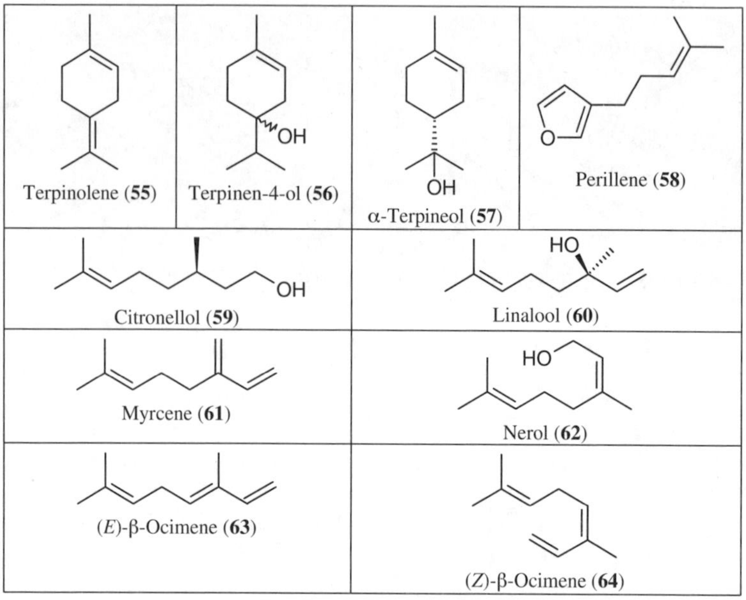

Terpinolene (**55**)	Terpinen-4-ol (**56**)	α-Terpineol (**57**) OH	Perillene (**58**)
Citronellol (**59**) OH		Linalool (**60**) HO	
Myrcene (**61**)		Nerol (**62**) HO	
(*E*)-β-Ocimene (**63**)		(*Z*)-β-Ocimene (**64**)	

FIGURE 3.6 *Continued*

C. amada, C. aromatica, C. caesia, C. longa, C. xanthorrhiza, and *C. zedoaria.* Monoterpenes isolated form this genus were of the following categories:

1. Cineole derivatives (**42** and **43**)
2. 2,6-Dimethyloctane derivatives (**59** to **64**)
3. Menthane-type (**44** to **46** and **49** to **57**)
4. Pinane-type (**36** to **41,47, 48**)

Perillene (**58**), isolated from *C. amada* (Singh et al., 2002), is a modified 2,6-dimethyloctane derivative with a furan ring system. Among these monoterpenes, borneol (**36**), camphor (**40**), 1,8-cineole (**43**), and β-pinene (**48**) are widely distributed within this genus, and were reported from more than five *Curcuma* species (Table 3.2). On the other hand, bornyl acetate (**37**), isobornyl acetate (**39**), carvacrol (**46**), D-sabinene (**52**), α-terpinene (**53**), γ-tepinene (**54**), perillene (**58**) and nerol (**62**) are the least common monoterpenes of this genus.

3.4.3 SESQUITERPENES

The sesquiterpenoids are C_{15} compounds formed by the assembly of three isoprenoid units (DNP, 2001). They are distributed in many living systems including higher plants. There are a large number of sesquiterpenoid carbon skeletons, all of which arise from the common precursor, farnesyl pyrophosphate, by various modes of cyclizations followed, in many cases, by skeletal rearrangement. About 140 different sesquiterpenes have been isolated from the genus *Curcuma,* and they can be classified into ten distinctly different structural types (Figure 3.7 to Figure 3.17). However, most of these compounds fall into one of the three major categories, bisabolane, germacrane, or guaiane types.

3.4.3.1 Bisabolane-Type Sesquiterpenes

The bisabolanes are one of the three fairly large groups of sesquiterpenes found in the genus *Curcuma* (Figure 3.7). At least 34 different bisabolane-type sesquiterpenes (**65** to **98**) have been

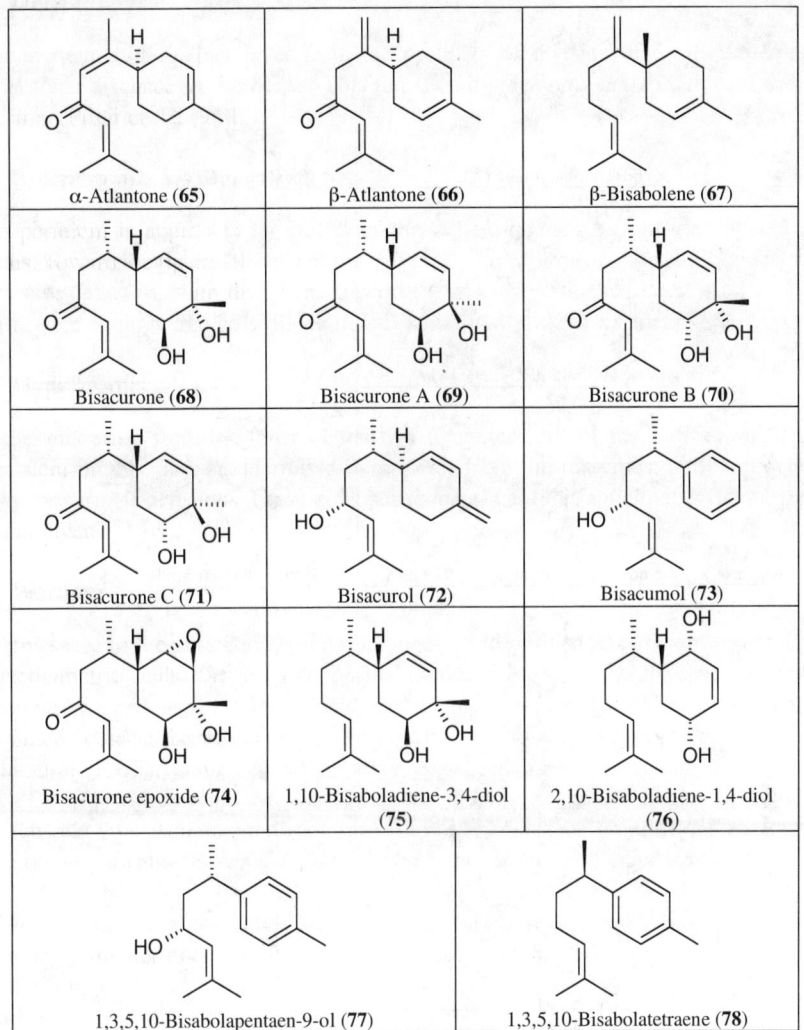

FIGURE 3.7 Bisabolane-type sesquiterpenes from the genus *Curcuma*. *Continued*

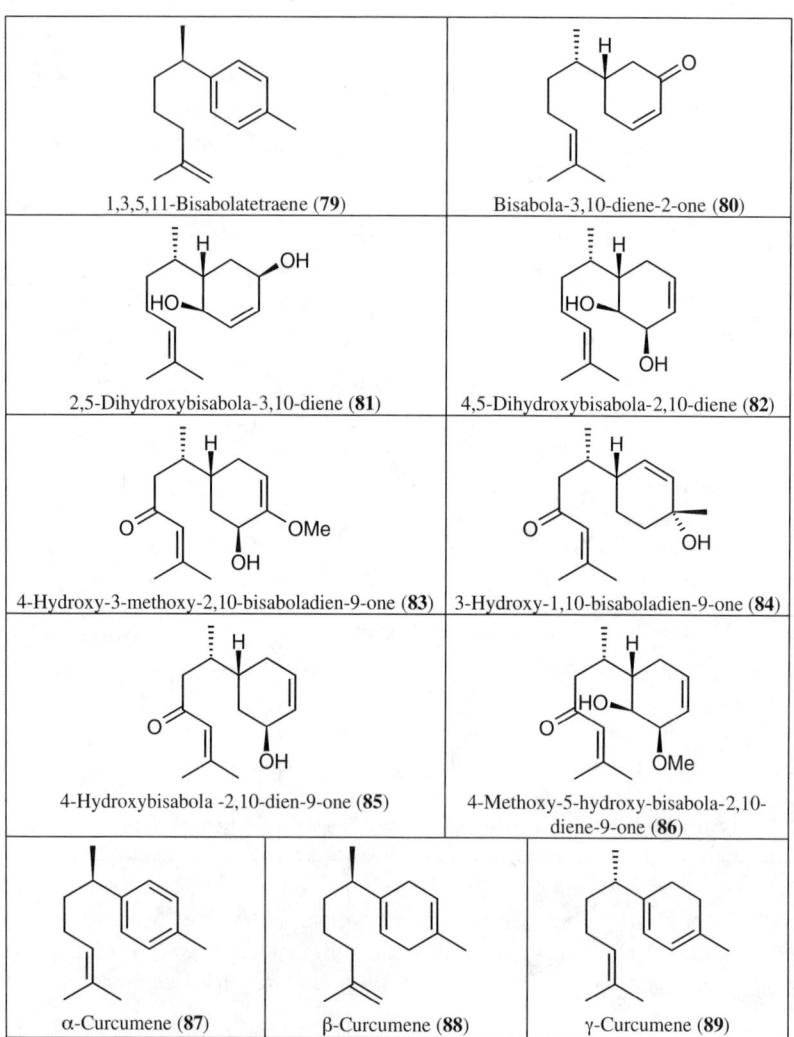

FIGURE 3.7 *Continued*

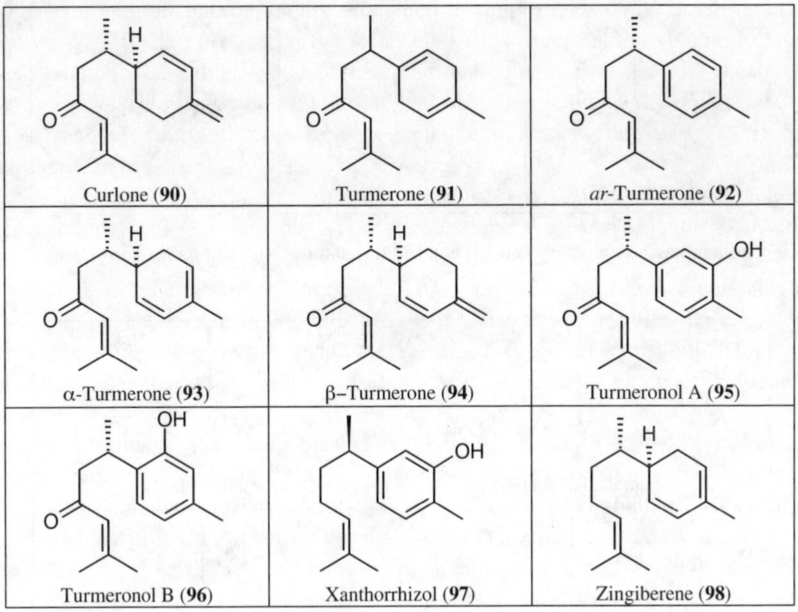

Curlone (**90**)	Turmerone (**91**)	*ar*-Turmerone (**92**)
α-Turmerone (**93**)	β–Turmerone (**94**)	Turmeronol A (**95**)
Turmeronol B (**96**)	Xanthorrhizol (**97**)	Zingiberene (**98**)

FIGURE 3.7 *Continued*

reported from nine species of the *Curcuma* (Table 3.2). *C. aromatica, C. longa. C. xanthorrhiza,* and *C. zedoaria* are the four major sources of these compounds. *ar*-Turmerone (**92**) is the most widely distributed bisabolane-type sesquiterpene within this genus. Bisacumol (**73**), α-curcumene (**87**), β-curcumene (**88**), and zingiberene (**98**) were reported from four or more species. While compounds **87** and **88** are fairly common, γ-curcumene was only isolated from *C. caesia* (Pandey and Chowdhury, 2003). Xanthorrhizol (**97**), isolated from *C. aromatica* (Kojima et al., 1998) and *C. xanthorrhiza* (Rimpler et al., 1970; Hwang et al., 2000), is one of the most important biologically active components of this genus.

3.4.3.2 Cadalene-Type Sesquiterpenes

Cadalene group of sesquiterpenes are structurally related to cadinane-type (DNP, 2001). Only three sesquiterpenes (Figure 3.8) of this type have ever been reported from the genus *Curcuma*. While α-calacorene (**100**) was isolated from *C. zedoaria* (Mau et al., 2003; Matsuda et al., 2001), cadalenequinone (**99**) and 8-hydroxycadalene (**101**) were found in *C. parviflora* (Toume et al., 2004).

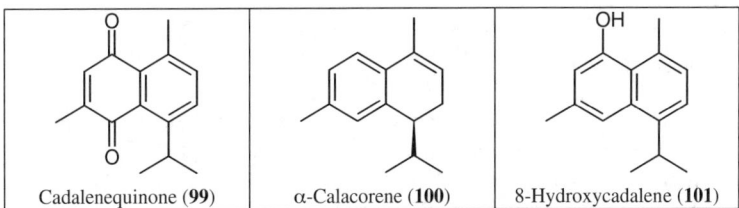

Cadalenequinone (**99**)	α-Calacorene (**100**)	8-Hydroxycadalene (**101**)

FIGURE 3.8 Cadalene-type sesquiterpenoids from the genus *Curcuma*.

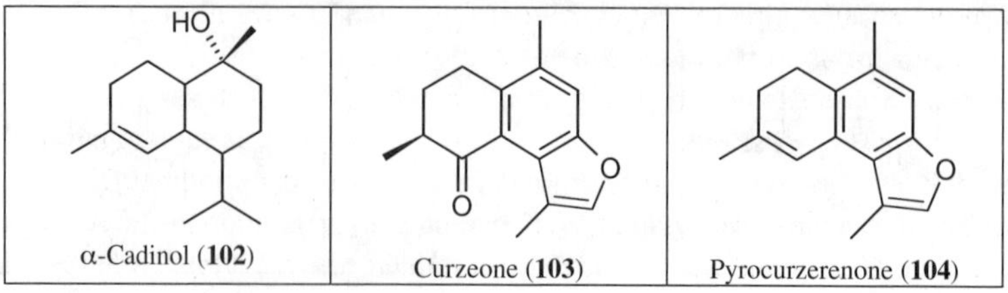

FIGURE 3.9 Cadinane and furanocadinane-type sesquiterpenoids from the genus *Curcuma*.

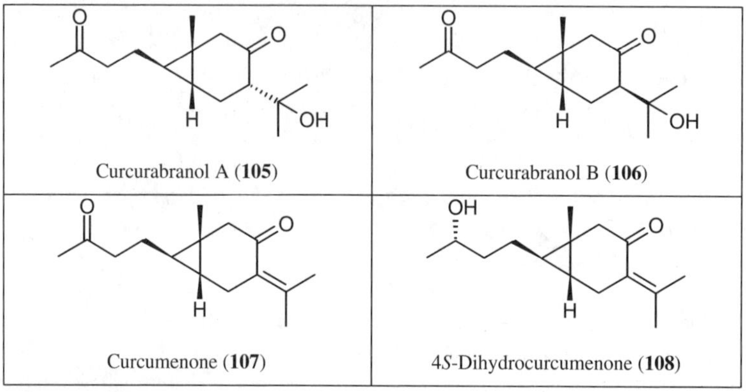

FIGURE 3.10 Carabrane-type sesquiterpenoids from the genus *Curcuma*.

3.4.3.3 Cadinane and Furanocadinane-Type Sesquiterpenes

One cadinane, α-cadinol (**102**), and two furanocadinane-type sesquiterpenes, curzeone (**103**) and pyrocurzerenone (**104**), were reported from this genus (Figure 3.9, Table 3.2). The distribution of these compounds is restricted only to *C. zedoaria* (Matsuda et al., 2004; Mau et al., 2003).

3.4.3.4 Carabrane-Type Sesquiterpenes

The carabranes are a small group of 5,10-cycloxanthanes. These sesquiterpenes contain a cyclo-hexane- and cyclopropane-fused ring system (Figure 3.10). Four sesquiterpenes of this type, cur-curabranol A (**105**), curcurabranol B (**106**), curcumenone (**107**), and 4S-dihydrocurcumenone (**108**), were reported from *C. zedoaria* (Mau et al., 2003; Matsuda et al., 2001; Yoshikawa et al., 1998; Shiobara, 1985). Three other compounds, curcumenolactones A, B, and C (**199 to 201**), isolated from the same plant, are the lactones derived from basic carabrane skeleton (Matsuda et al., 2001) (Figure 3.17).

3.4.3.5 Curcumane-Type Sesquiterpenes

Curcumane-type sesquiterpenes are based on a cycloheptane ring system (Figure 3.11). Four cur-cumane-type sesquiterpenes were found in the genus *Curcuma* (Table 3.2). Curmadione (**109**), curcumadione (**110**), and isocurcumadione (**112**) were isolated from *C. aromatica* (Kuroyanagi et al., 1990; Shiobara et al., 1986). Curcumenone (**111**) was found in *C. longa* (He et al., 1998; Ohshiro et al., 1990; Hikino et al., 1968a) and *C. zedoaria* (Mau et al., 2003; Matsuda et al., 2001; Yoshikawa et al., 1998; Shiobara, 1985).

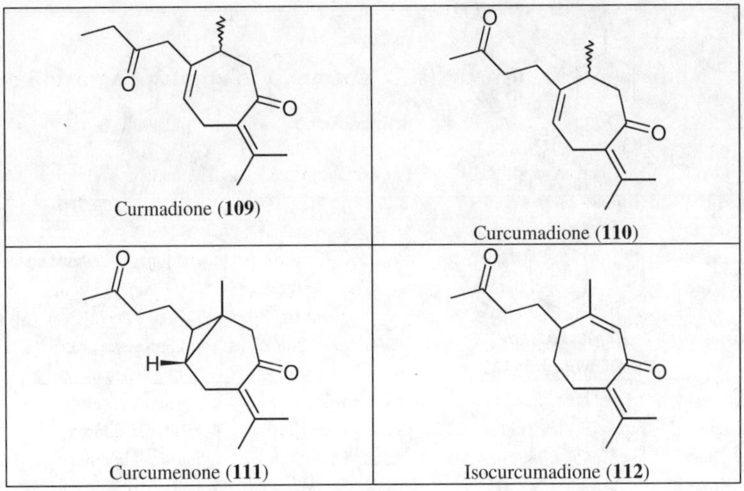

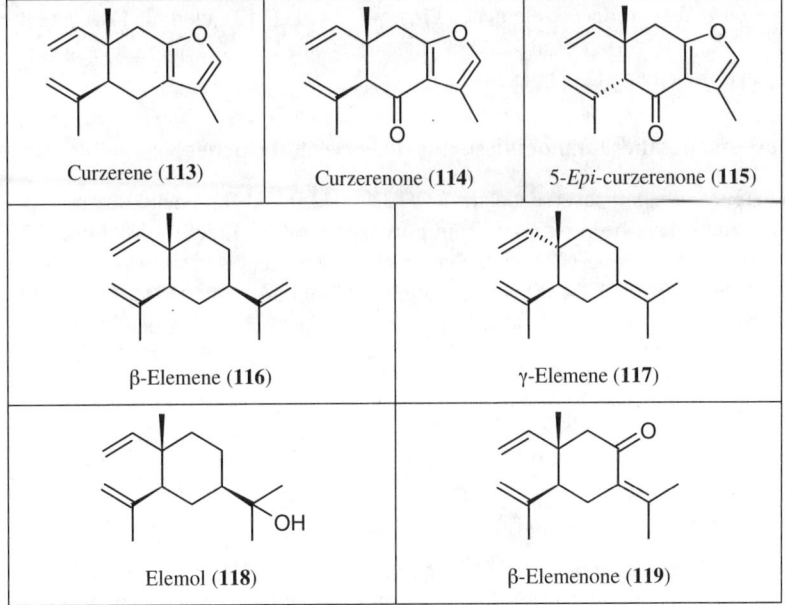

FIGURE 3.11 Curcumane-type sesquiterpenoids from the genus *Curcuma*.

3.4.3.6 Elemane-Type Sesquiterpenes

Elemane-type sesquiterpenes are rapidly formed *in vitro* by Cope rearrangement of the corresponding 1(10),4-germacradienes and it is believed that they are artifacts produced during the isolation procedures (DNP, 2001). However, some elemanes are oxidatively modified, which presumably is not formed by a Cope rearrangement during isolation. Seven sesquiterpenes (**113 to 119**) of this class were reported from the genus *Curcuma* (Figure 3.12, Table 3.2). Among these compounds, curzerenone (**113**) and curzerenone (**114**) are widely distributed in this genus and were reported

FIGURE 3.12 Elemane-type sesquiterpenoids from the genus *Curcuma*.

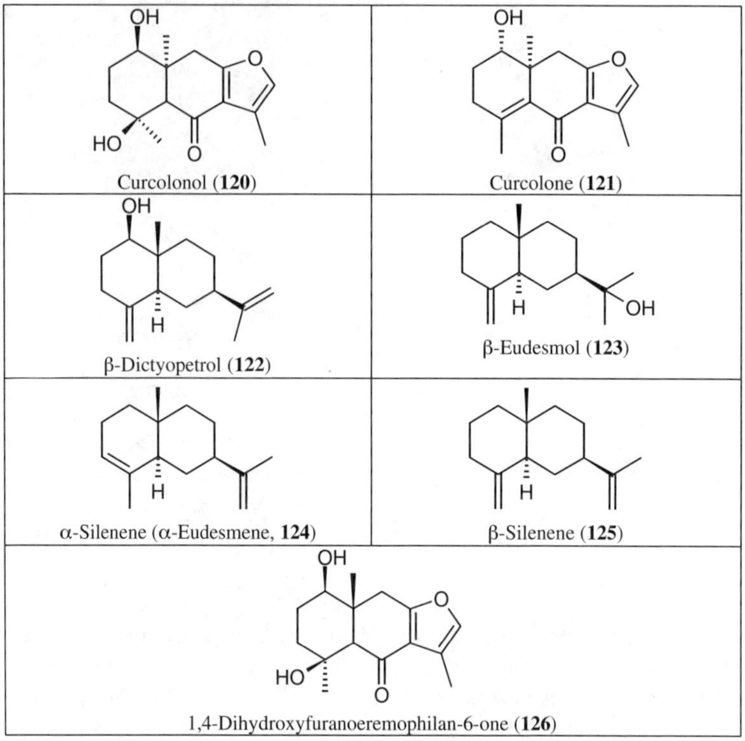

FIGURE 3.13 Eudesmane and furanoeudesmane-type sesquiterpenoids from the genus *Curcuma*.

from at least six different species. An epimer of 114 was isolated from *C. zedoaria* (Mau et al., 2003). *C. zedoaria* also produced -elemene (**116**), γ-elemene (**117**), elemol (**118**), and β-elemenone (**119**) (Jang et al., 2004; Mau et al., 2003; Matsuda et al., 2001). Three other species were also reported to have β-elemene (**116**) (Table 3.2).

3.4.3.7 Eudesmane and Furanoeudesmane-Type Sesquiterpenes

Within the genus *Curcuma*, seven eudesmane- (**122 to 125**) and furanoeudesmane-type (**120, 121, 126**) sesquiterpenes have been reported from only *C. zedoaria* (Figure 3.13, Table 3.2). Hikino et al., (1967) isolated curcolone (**121**) and Syu et al., (1998) reported the isolation of β-eudesmol (**123**), curcolonol (**120**), and 1,4-dihydroxyfuranoeremophilan-6-one (**126**). α-Silenene (α-eudesmene, **124**), β-silenene (**123**), and β-dictyopetrol (**122**) have recently been reported (Mau et al., 2003; Matsuda et al., 2001).

3.4.3.8 Germacrane-Type Sesquiterpenes

Germacrane-type sesquiterpenes form another fairly large group of sesquiterpenes found in several species of the genus *Curcuma* (Figure 3.14, Table 3.2). At least 25 different germacrane derivatives (**127 to 151**), curdione (**128**) being the most abundant one, have been found in this genus. *C. aeruginosa* was found to produce **128** and dehydrocurdione (**129**) (DNP, 2001; Jirovetz et al., 2000). *C. aromatica, C. longa, C. wenyujin,* and *C. zedoaria* are the four major sources of these sesquiterpenes. In addition to **128**, *C. aromatica* afforded acetoxyneocurdione (**127**), 13-acetoxydehydrocurdione (**131**), 4,5-epoxy-12-hydroxy-1(10),7(11)-germacradien-8-one (**144**), 4,5-epoxy-12-acetoxy-1(10),7(11)-germacradien-8-one (**145**), 4,5-epoxy-12-acetoxy-7α,11α-dihydrogermacra-

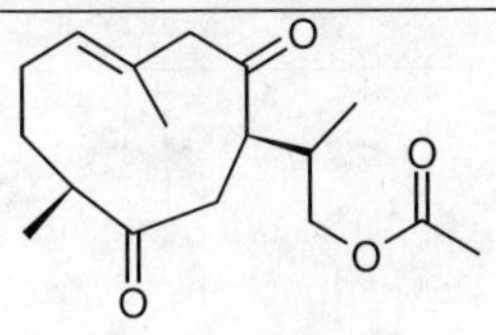

FIGURE 3.14 Germacrane-type sesquiterpenoids from the genus *Curcuma*.

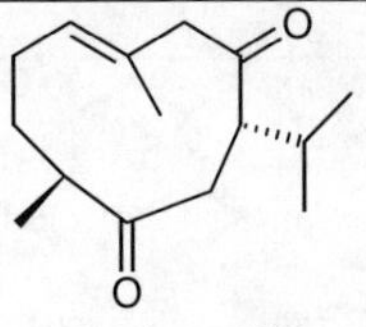

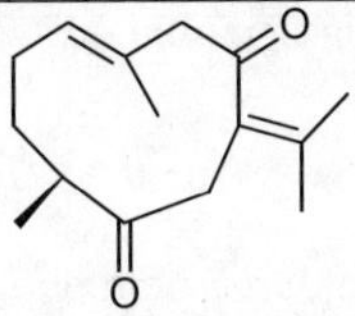

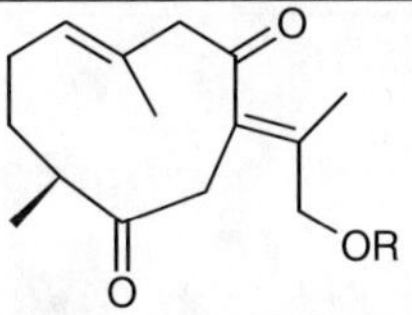

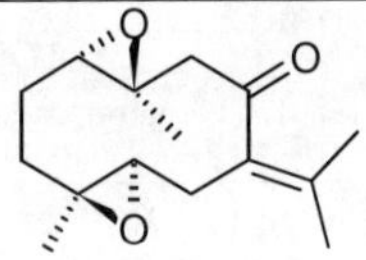

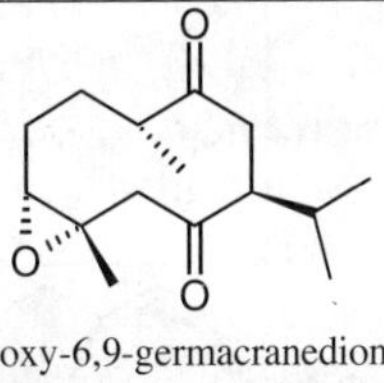

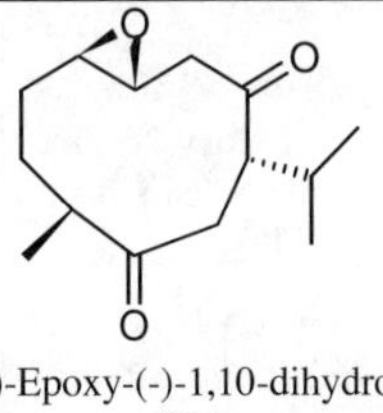

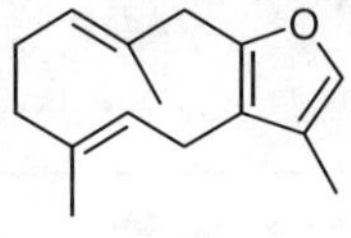

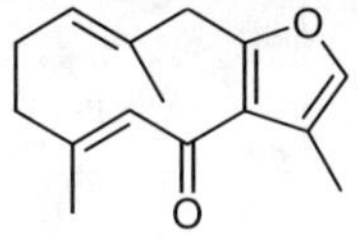

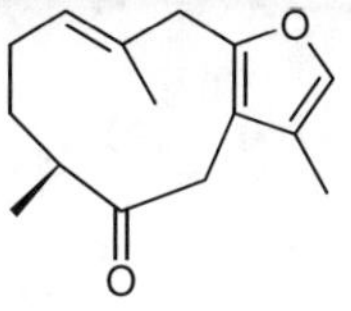

Continued

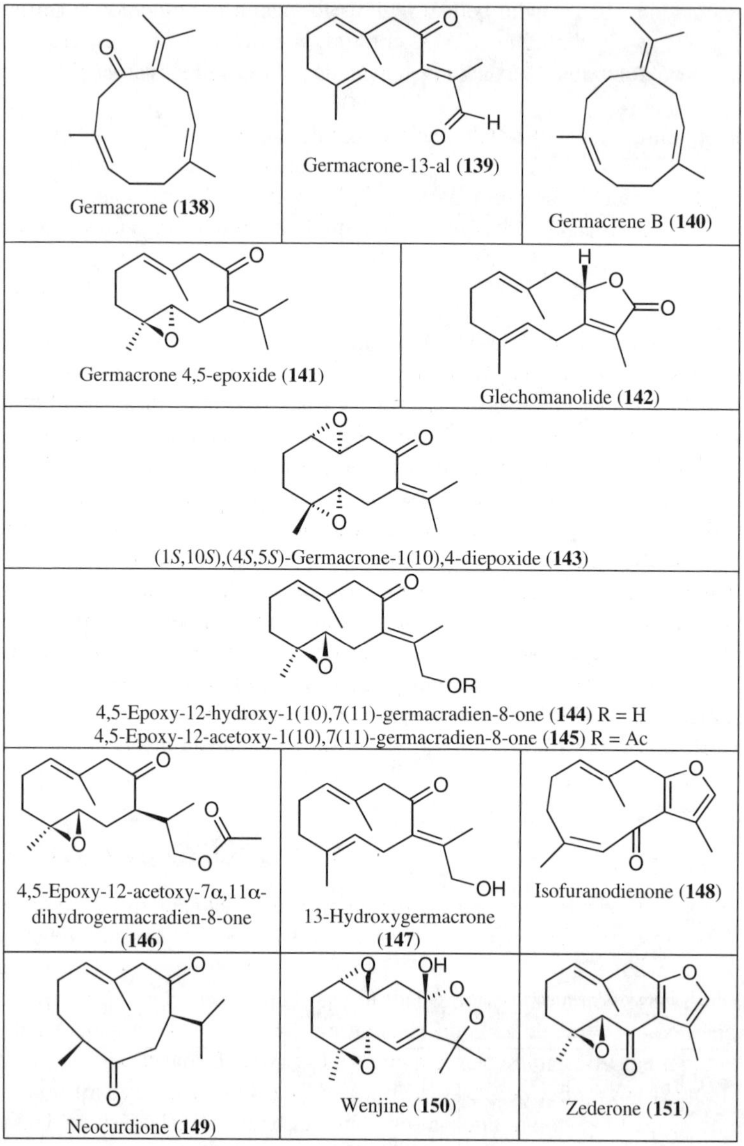

Germacrone-13-al (**139**)

Germacrone (**138**)

Germacrene B (**140**)

Germacrone 4,5-epoxide (**141**)

Glechomanolide (**142**)

(1*S*,10*S*),(4*S*,5*S*)-Germacrone-1(10),4-diepoxide (**143**)

4,5-Epoxy-12-hydroxy-1(10),7(11)-germacradien-8-one (**144**) R = H
4,5-Epoxy-12-acetoxy-1(10),7(11)-germacradien-8-one (**145**) R = Ac

4,5-Epoxy-12-acetoxy-7α,11α-
dihydrogermacradien-8-one
(**146**)

13-Hydroxygermacrone
(**147**)

Isofuranodienone (**148**)

Neocurdione (**149**)

Wenjine (**150**)

Zederone (**151**)

FIGURE 3.14 *Continued*

dien-8-one (**146**), germacrone (**138**), gemacrone-4,5-epoxide (**141**), 13-hydroxydehydrocurdione
(**130**), 13-hydroxygermacrone (**147**), and zederone (**151**) (Pant et al., 2001; Kojima et al., 1998;
Kuroyanagi et al., 1990; Shiobara, 1986). Together with **128**, **129**, and **141**, a sesquiterpene
aldehyde, germacrone-13-al (**139**), was isolated from *C. longa* (He et al., 1998; Ohshiro et al.,
1990). A series of germacrane-type sesquiterpenes, including **128**, **138**, (1α,4β,5α,10β) 1,10:4,5-
diepoxy-7(11)-germacren-8-one (**132**), 3,4-epoxy-6,9-germacranedione (**133**), (1*R*, 10*R*)-epoxy-
(–)-1,10-dihydrocurdione (**134**), (1*S*,10*S*),(4*S*,5*S*)-germacrone-1(10),4-diepoxide (**143**), neocurdione (**149**), and wenjine (**150**), were isolated from *C. wenyujin* (Hikino et al., 1965; Harimaya et
al., 1991; Gao et al., 1989, 1991). Pandji et al., (1993) isolated furanodienone (**136**) and Shibuya
et al., (1986) reported furanogermenone (**137**) from *C. zedoaria*. A number of other germacrane-
type sesquiterpenes including **128**, furanodiene (**135**), germacrene B (**140**), **141**, glechomanolide

(142), 147, isofuranodienone (148), 149, and zederone (151) were also isolated from this species (Mau et al., 2003; Matsuda et al., 2001; Shiobara et al., 1985; Yoshihara et al., 1984). Among these sesquiterpenes, 135 to 137, 142, 148, and 151 belong to the structural subgroup, furanogermacranes.

3.4.3.9 Guaiane-Type Sesquiterpenes

About 30 different guaiane-type sesquiterpenes (152 to 181), isocurcumenol (158) being the most widely distributed compound, were found in various species of the genus *Curcuma* (Figure 3.15, Table 3.2). *C. zedoaria* alone produces at least 22 different sesquiterpenes of this class. Among these sesquiterpenes, distinct structural differences do exist. For example, curcumol (154), curcumenol (155), *epi*-curcumenol (156), 4-*epi*-curcumenol (157), isocurcumenol (158), oxycurcumenol (159), and neocurcumenol (160) contain a 5,8-cyclic ether, and zedoalactones A and B (175 and 176) and zedoarolides A and B (180 and 181) possess a five-membered lactone ring. Zedoarol (179), isolated from *C. zedoaria* (Matsuda et al., 2001; Shiobara et al., 1986), is the only guaiane-type sesquiterpene that contains a furan ring system. Spathulenol (169) and isospathulenol (170), isolated from *C. zedoaria*, are modified guaiane-type, also known as aromadendranes, and have a fused cyclopropane ring (Mau et al., 2003; Matsuda et al., 2001).

3.4.3.10 Sesquiterpene Dimers

Sesquiterpene dimers (Figure 3.16), parviflorene A-F (182 to 187) were isolated from the underground parts of *C. parviflora* (Toume et al., 2004). These sesquiterpene dimers belong to three structural classes:

1. biscardinanes (182, 183, 185, and 187)
2. a cardinane-isocardinane adduct (184)
3. a biscardinane with an alternate bond connection (186)

Another sesquiterpene dimer, difurocumenone (188), was found in *C. aeruginosa* (DNP, 2001).

3.4.3.11 Other Sesquiterpenes

There are at least 12 other sesquiterpenes of various types, isolated from the genus *Curcuma* (Figure 3.17, Table 3.2). Calarene (189) and α-copaene (190), isolated from *C. amada* (Singh et al., 2002), are aristolane-type and copaane-type sesquiterpenes, respectively. There are two farnesane-type sesquiterpenes, β-farnesene (195) and farnesol (196), reported from this genus. Curcumalactones A, B, and C (199 to 201), isolated from *C. zedoaria* (Mau et al., 2003; Matsuda et al., 2001), are a rather unique type of sesquiterpenes, and only found in this genus. There are two caryophyllane-type sesquiterpenes, α- and β-caryophyllene (197 and 198), distributed in the *Curcuma* species. Two other unique sesquiterpene lactones, gajutsulactones A and B (202 and 203), were isolated from *C. zedoaria* (Mau et al., 2003; Matsuda et al., 2001).

3.5 MISCELLANEOUS COMPOUNDS

Apart from the secondary metabolites already discussed above, there are a few other types of compounds reported from the *Curcuma* (Figure 3.18, Table 3.2). The only flavanol compound, malvidin 3-rutinoside (211), was isolated from *C. alismatifolia* (Nakayama et al., 2000). Acetophenone derivatives, phloracetophenone (208) and myrciaphenone (207), isolated from *C. comosa* (Piyachaturawat et al., 2002), are two pharmacologically active compounds. A number of other simple aromatic compounds including 3,7-dimethyl-5-indanecarboxylic acid (204), eugenol (205), 1-(4-methylphenyl)ethanol (206), vanillic acid (209), and syringic acid (210) have also been found in this genus.

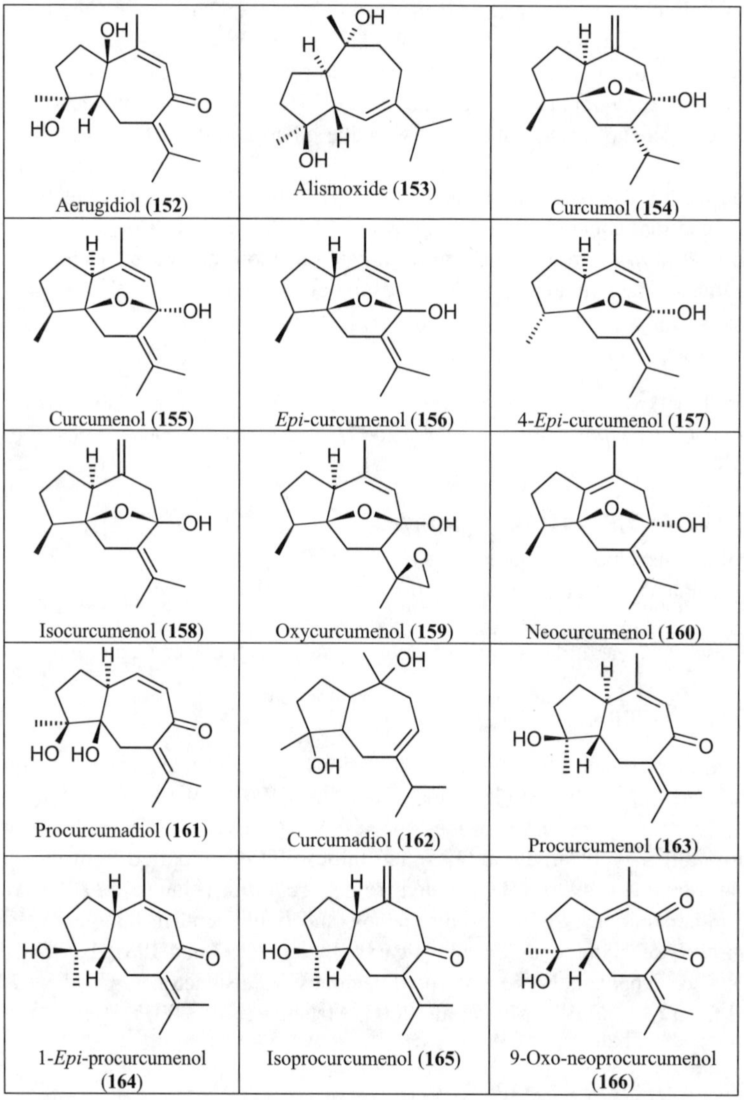

FIGURE 3.15 Guaiane-type sesquiterpenoids from the genus *Curcuma*.

Continued

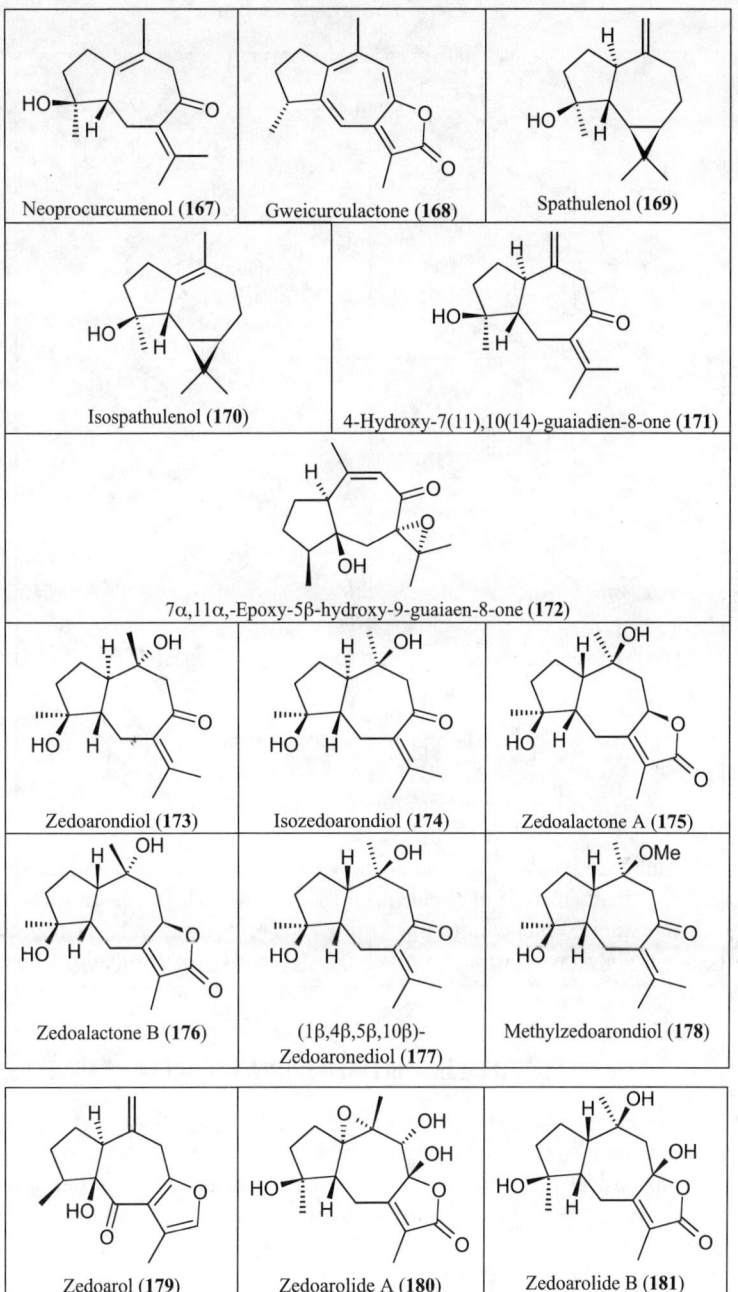

Neoprocurcumenol (**167**)

Gweicurculactone (**168**)

Spathulenol (**169**)

Isospathulenol (**170**)

4-Hydroxy-7(11),10(14)-guaiadien-8-one (**171**)

7α,11α,-Epoxy-5β-hydroxy-9-guaiaen-8-one (**172**)

Zedoarondiol (**173**)

Isozedoarondiol (**174**)

Zedoalactone A (**175**)

Zedoalactone B (**176**)

(1β,4β,5β,10β)-
Zedoaronediol (**177**)

Methylzedoarondiol (**178**)

Zedoarol (**179**)

Zedoarolide A (**180**)

Zedoarolide B (**181**)

FIGURE 3.15 *Continued*

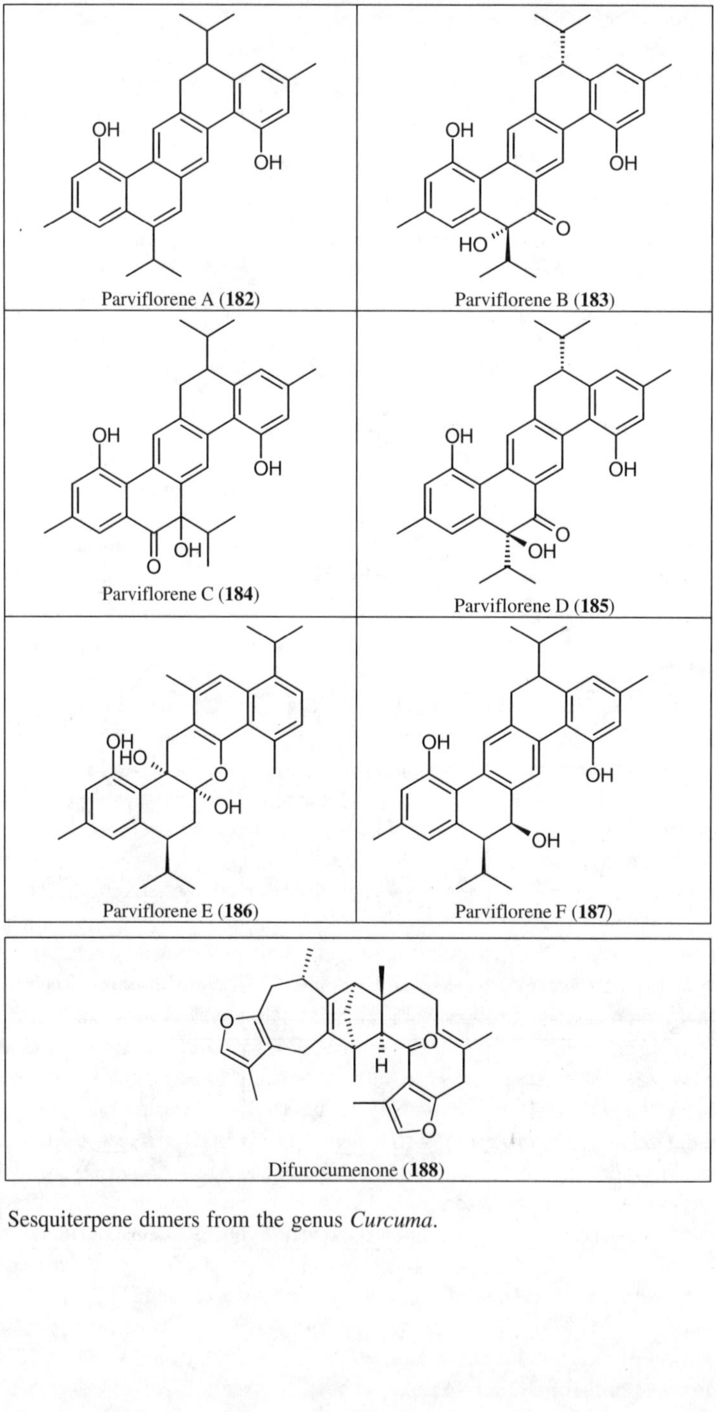

FIGURE 3.16 Sesquiterpene dimers from the genus *Curcuma*.

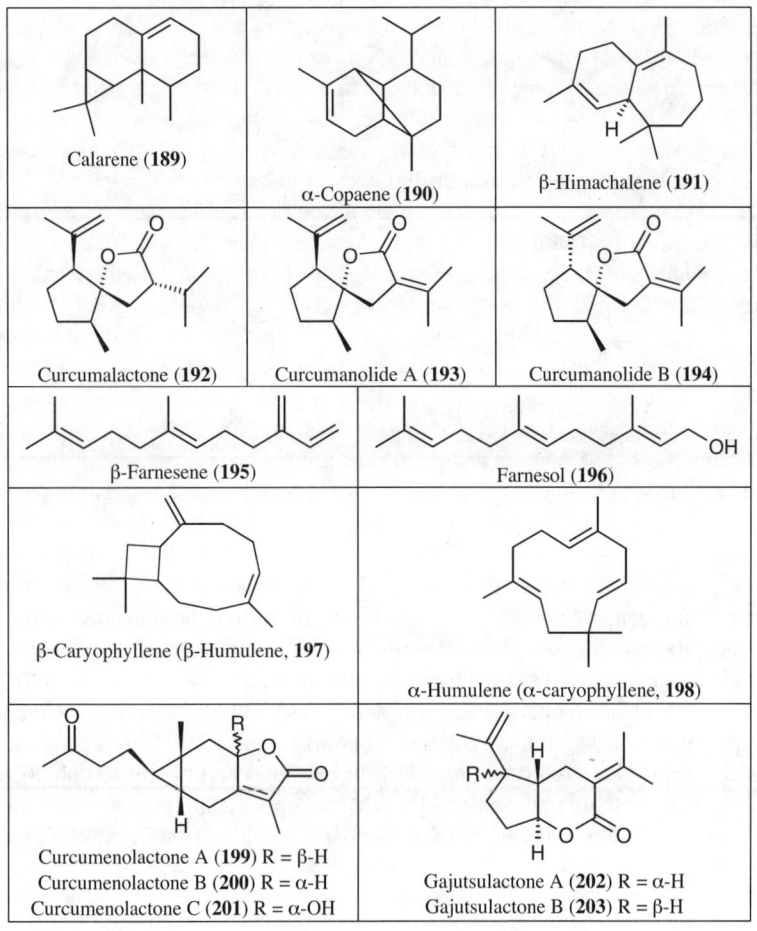

FIGURE 3.17 Other sesquiterpenoids from the genus *Curcuma*.

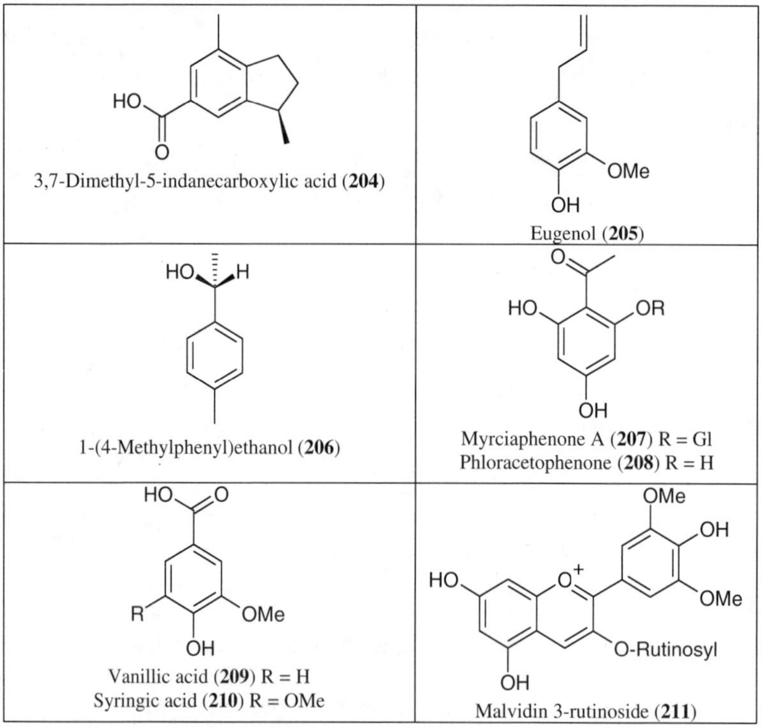

FIGURE 3.18 Miscellaneous compounds from the genus *Curcuma*.

3.6 CONCLUSION

It is evident that the genus *Curcuma* is a rich source of various biologically active compounds including diphenylheptanoids, sesquiterpenes, monoterpenes, etc. The pharmacological properties of curcumin (**1**) and its derivatives, *ar*-turmerone (**92**) and phloracetophenone (**208**), have extensively been researched (see chapter 9 by Sarker and Nahar) and they clearly have the potential for being developed as new drugs for the treatment of various ailments. Only a small proportion of the available species of the large genus *Curcuma* have ever been investigated phytochemically. It is reasonable to believe that investigations on other untapped *Curcuma* species will lead to the isolation of a number of novel and interesting plant-secondary metabolites in the years to come.

REFERENCES

Bats, J.W., and Ohlinger, S.H. (1999) Absolute configuration of isocurcumenol. *Acta Crystallographica Section C-Crystal Structure Communications* 55, 1595-1598.

Bansal, R.P., Bahl, J.R., and Garg, S.N., Naqvi A.A., and Kumar, S. (2002) Differential chemical compositions of the essential oils of the shoot organs, rhizomes and rhizoids in the turmeric *Curcuma longa* grown in indo-grangetic plains. *Pharm Biol* 40, 384-389.

Behura, S. and Srivastava, V.K. (2004) Essential oils of leaves of *Curcuma* species. *J Essential Oil Res* 16, 109-110.

Brand, S., Hölscher, D., Schierhorn, A., Svatoš, A., Schroder, J., and Schneide,r B. (2006) A type III polyketide synthase from *Wachendorfia thyrsiflora* and its role in diarylheptanoid and phenylphenalenone biosynthesis. *Planta* (in press).

Chen, S.L., You, J., Wang, G.J. (2001) Extraction of active compound in *Curcuma phaeocaulis* valeton by supercritical fluid extraction and trapping with silica gel column. *Chinese J Anal Chem* 29, 664-666.

Claeson, P., Pongprayoon, U., Sematong, T., Tuchinda, P., Reutrakul, V., Soontornsaratune, P., and Taylor, W.C. (1996) Non-phenolic linear diarylheptanoids from Curcuma xanthorrhiza: A novel type of topical anti-inflammatory agents: Structure-activity relationship. *Planta Medica* 62, 236-240.

Claeson, P., Panthong, A., Tuchinda, P., Reutrakul, V., Kanjanapothi, D., Taylor, W.C., and Santisuk, T. (1993) 3 Nonphenolic diarylheptanoids with antiinflammatory activity from *Curcuma xanthorrhiza*. *Planta Medica* 59, 451-454.

Curcuma List (2005). Available on-line at: http://www.golatofski.de/Pflanzenreich/gattung/c/curcuma.htm

DNP (2001) Dictionary of Natural Products, CD ROM, Chapman and Hall-CRC, USA.

Etoh, H., Kondoh, T., Yoshioka, N., Sugiyama, K., Ishikawa, H., and Tanaka, H. (2003) 9-Oxo-neoprocurcumenol from *Curcuma aromatica* (Zingiberaccae) as an attachment inhibitor against the Blue Mussel, *Mytilus edulis galloprovincialis. Biosci Biotech Biochem* 67, 911-913.

Firman, K., Kinoshita, T., Itai, A., and Sankawa, U. (1988) Terpenoids from *Curcuma heyneana. Phytochemistry* 27, 3887-3891.

Fukushima, S., Kuroyanagi, M., Ueno, A., Akahori, Y., and Saiki, Y. (1970) Structure of curzerenone, a new sesquiterpene from *Curcuma zedoaria. Yakugaku Zasshi* 90, 863-869.

Gao, J.F, Xie, J.H., Harimaya, K., Kawamata, T., Iitaka, Y., and Inayama, S. (1991) The absolute structure and synthesis of wenjine isolated from *Curcuma wenyujin. Chem Pharm Bull* 39, 854-856.

Gao, J.F, Xie, J.H., Iitaka, Y., Inayama, S. (1989) The stereostructure of wenjine and related (1S,10S),(4S,5S)-germacrone-1(10),4-diepoxide isolated from *Curcuma wenyujin. Chem Pharm Bull* 37, 233-236.

Garg, S.N., Naquvi, A.A., Bansal, R.P., Bahl, J.R., and Kumar, S. (2005) Chemical composition of the essential oil from the leaves of Curcuma zedoaria Rosc. of Indian origin. *J Essential Oil Res* 17, 29-31.

Golding, B.T., Pombo, E., and Samuel, C.J. (1982) Turmerones - isolation from turmeric and their structure determination. *J Chem Soc-Chem Commun* 6, 363-364.

Grayson, D.H. (1998) Monoterpenoids. *Nat Prod Rep* 15, 439-475.

GRIN Database (2005) USDA, ARS, National Genetic Resources Program. Germplasm Resources Information Network - (GRIN), National Germplasm Resources Laboratory, Beltsville, Maryland. Available online at: http://www.ars-grin.gov/cgi-bin/npgs/html/taxgenform.pl

Gupta, .AP., Gupta, M.M., and Kuma,r S. (1999) Simultaneous determination of curcuminoids in *Curcuma* samples using high performance thin layer chromatography. *J Liq Chromatog Related Technol* 22, 1561-1569.

Gupta, S.K., Banerjee, A.B., Achari, B. (1976) Isolation of Ethyl *p*-methoxycinnamate, the major antifungal principle of *Curcumba zedoaria. Lloydia* 39, 218-222.

Guo, Y.T. (1983) Isolation and identification of elemene from the essential oil of *Curcuma wenyujin. Zhong Yao Tong Bao* 8, 31.

Harimaya, K., Gao, J.F., Ohkura, T., Kawamata, T., Iitaka, Y., Guo, Y.T., and Inayama, S. (1991) A series of sesquiterpenes with a 7-alpha-isopropyl side-chain and related-compounds isolated from *Curcuma wenyujin. Chem Pharm Bull* 39, 843-853.

He, X.-G., Lin, L.-Z., Lian, L.-Z., and Lindenmaier, M. (1998) Liquid chromatography–electrospray mass spectrometric analysis of curcuminoids and sesquiterpenoids in turmeric (*Curcuma longa*). *Journal of Chromatography A* 818, 127-132.

Hikino, H., Takahashi, S., Sakurai, Y., Takemoto, T., and Bhacca, N.S. (1968) Structure of zederone. *Chem Pharm Bull* 16, 1081-1087.

Hikino, H., Sakurai, Y., Numabe, S., and Takemoto, T. (1968a) Structure of curcumenol. *Chem Pharm Bull* 16, 39-42.

Hikino, H., Sakurai, Y., and Takemoto, T. (1967) Structure of curcolone. *Chem Pharm Bull* 15, 1065-1066.

Hikino, H., Meguro, K., Sakurai, Y., and Takemoto, T. (1965) Structure of curcumol. *Chem Pharm Bull* 13, 1484-1485.

Ho, T.L., Hall TW (1983) A short synthesis of iso-a--curcumene. *Chemistry & Industry* 22, 862-862.

Hölscher, D. and Schneider, B. (1995a) A diarylheptanoid intermediate in the biosynthesis of phenylphenalenones in *Anigozanthos preissii. J. Chem. Soc. Chem. Comm.* 525-526.

Hölscher, D. and Schneider, B. (1995b) The origin of the central one carbon unit in the biosynthesis of phenylphenalenones in *Anigozanthos preissii. Natural Products Letters* 7, 177-182.

Huang, J., Ogihara, Y., Gonda, R., and Takeda, T. (2000) Novel biphenyl ether lignans from the rhizomes of *Curcuma chuanyujin. Chem Pharm Bull* 48, 1228-1229.

Hwang, J.K., Shim, J.S., Baek, N.I., and Pyun, Y.R. (2000) Xanthorrhizol: A potential antibacterial agent from *Curcuma xanthorrhiza* against *Streptococcus* mutans. *Planta Medica* 66, 196-197.

Imai, S., Morikiyo, M., Furihata, K., Hayakawa, Y., and Seto, H. (1990) Turmeronol A and Turmeronol-B, new inhibitors of soybean lipoxygenase. *Agric Biol Chem* 54, 2367-2371.

ISI Database (2005) ISI Web of Knowledge, Thompson-ISI, London. Available on-line at: http://portalt.wok.mimas.ac.uk/portal.cgi?DestApp=WOS&Func=Frame

Itokawa, H., Hirayama, F., Funakoshi, K., and Takeya, K. (1985) Studies on the antitumor bisabolane sesquiterpenoids isolated from Ccurcuma xanthorrhiza. *Chem Pharm Bull* 33, 3488-3492.

Jain, M.K., Mishra, R.K. (1964) Chemical examination of *Curcuma amada* Roxb. *Ind J Chem* 2, 39-41.

Jang, M.K, Lee, H.J., Kim, J.S., Ryu, J.H. (2004) A curcuminoid and two sesquiterpenoids from *Curcuma zedoaria* as inhibitors of nitric oxide synthesis in activated macrophages. *Arch Pharm Res*. 27, 1220-1225.

Jirovetz, L., Buchbauer, G., Puschmann, C., Shafi, M.P., Nambiar, M.K.G. (2000) Essential oil analysis of *Curcuma aeruginosa* Roxb. leaves from South India. *J Essential Oil Res* 12, 47-49.

Jitoe, A., Masuda, T., Tenga,h I.G.P., Suprapta, D.N., Gara, I.W., Nakatani, N. (1992) Antioxidant activity of tropical ginger extracts and analysis of the contained curcuminoids. *J Agric Food Chem* 40, 1337-1340.

Jurgens, T.M., Frazier, E.G., Schaeffer, J.M., Jones, T.E., Zink, D.L., Borris, R.P., Nanakorn, W., Beck, H.T., Balick, M.J. (1994) Novel nematocidal agents from *Curcuma comosa*. *J Nat Prod* 57, 230-235.

Kim, J.H., Shim, J.S., Lee, S.K., Kim, K.W., Rha, S.Y., Chung, H.C., Kwon, H.J. (2002) Microarray-based analysis of anti-angiogenic activity of demethoxycurcumin on human umbilical vein endothelial cells: Crucial involvement of the down-regulation of matrix metalloproteinase. *Japanese J Cancer Res* 93, 1378-1385.

Kiso, Y., Suzuki, Y., Watanabe, N., Oshima, Y., Hikino, H. (1983) Stereostructure of curlone, a sesquiterpenoid of *Curcuma longa* rhizomes. *Phytochemistry* 22, 596-597.

Kojima, H., Yanai, T., and Toyota, A. (1998) Essential oil constituents from Japanese and Indian *Curcuma aromatica* rhizomes. *Planta Medica* 64, 380-381.

Kiuchi. F., Goto, Y., Sugimoto, N., Akao, N., Kondo, K., Tsuda, Y. (1993) Nematocidal activity of turmeric: synergistic action of curcuminoids. *Chem Pharm Bull* 41, 1640-1641.

Kuroyanagi, M., Ueno, A., Koyama, K., Natori, S. (1990) Structures of sesquiterpenes of *Curcuma aromatica* Salisb .2. studies on minor sesquiterpenes. *Chem Pharm Bull* 38, 55-58.

Kuroyanag, M., Ueno, A., Ujiie, K., Sato, S. (1987) Structures of sesquiterpenes from *Curcuma aromatica* Salisb. *Chem Pharm Bull* 35, 53-59.

Masuda, T., Jitoe, A., Isobe, J., Nakatani, N., Yonemori, S. (1993) Antioxidative and antiinflammatory curcumin-related phenolics from rhizomes of *Curcuma domestica*. *Phytochemistry* 32, 1557-1560.

Masuda. T., Isobe. J., Jitoe, A., Nakatani, N. (1992) Antioxidative curcuminoids from rhizomes of *Curcuma xanthorrhiza*. *Phytochemistry* 31, 3645-3647.

Masuda, T., Jitoe, A., Nakatani, N. (1991) Structure of aerugidiol, a new bridge-head oxygenated guaiane sesquiterpene. *Chem Letts* 9, 1625-1628.

Matsuda, H., Tewtrakul, S., Morikawa, T., Nakamura, A., Yoshikawa, M. (2004) Anti-allergic principles from Thai zedoary: structural requirements of curcuminoids for inhibition of degranulation and effect on the release of TNF-a and IL-4 in RBL-2H3 cells. *Bioorganic & Medicinal Chemistry* 12, 5891-5898.

Matsuda, H., Morikawa, T., Ninomiya, K., Yoshikawa, M. (2001) Absolute stereostructure of carabrane-type sesquiterpene and vasorelaxant-active sesquiterpenes from Zedoariae Rhizoma. *Tetrahedron* 57, 8443-8453.

Mau, J.-L., La,i E.Y.C., Wang N.-P., Chen, C.-C., Chang, C.-H., Chyau, C.-C. (2003) Composition and antioxidant activity of the essential oil from *Curcuma zedoaria*. *Food Chem* 82, 583-591.

Nakayama, R., Tamura, Y., Yamanaka, H., Kikuzaki, H., Nakatani, N. (1993) 2 Curcuminoid pigments from *Curcuma domestica*. *Phytochemistry* 33, 501-502.

Nakayama, M., Roh, M.S., Uchida, K., Yamaguchi., Y., Takano, K., Koshioka, M. (2000) Malvidin 3-rutinoside as the pigment responsible for bract color in *Curcuma alismatifolia*. *Biosci Biotech Biochem* 64, 1093-1095.

Ohshiro, M., Kuroyanagi. M., Ueno, A. (1990) Structures of sesquiterpenes from *Curcuma longa*. *Phytochemistry* 29, 2201-2205.

Pandey, A.K., Chowdhury, A.R. (2003) Volatile constituents of the rhizome oil of *Curcuma caesia* Roxb. from central India. *Flavour and Fragrance Journal* 18, 463-465.

Pandji, C., Grimm, C., Wray, V., Witte, L, Proksch, P. (1993) Insecticidal constituents from 4 species of the Zingiberaceae. *Phytochemistry* 34, 415-419.

Pant, N., Jain, D.C., Bhakuni, R.S., Prajapati, V., Tripathi, A.K., Kuma,r S. (2001) Zederone: A sesquiterpenic keto-dioxide from *Curcuma aromatica*. *Ind J Chem Sec B- Org Chem Including Med Chem* 40, 87-88.

Park, S.Y, Kim, D.S.H.L. (2002) Discovery of natural products from *Curcuma longa* that protect cells from beta-amyloid insult: A drug discovery effort against Alzheimer's disease. *J Nat Prod* 65, 1227-1231.

Phytochemical and Ethnobotanical Database (2005) USDA-ARS-NGRL, Beltsville, Agricultural Research Center, Beltsville, Maeyland. Available on-line at: http://www.ars-grin.gov/duke/plants.html

Piyachaturawat, P., Srivoraphan, P,, Chuncharunee, A., Komaratat, P., Suksamrarn, A. (2002) Cholesterol lowering effects of a choleretic phloracetophenone in hypercholesterolemic hamsters. *Eur J Pharmacol* 439, 141-147.

Raina, V.K., Srivastava, S.K., Jain, N., Ahmad, A., Syamasundar, K.V., Aggarwal, K.K. (2002) Essential oil composition of *Curcuma longa* L. cv. Roma from the plains of northern India. *Flavour and Fragrance Journal* 17, 99-102.

Ravindranath, V., Satyanarayana, M.N. (1980) An unsymmetrical diarylheptanoid from *Curcuma longa*. *Phytochemistry* 19, 2031-2032.

Rimpler, H., Hansel, R., Kochendoerfer, L. (1970) Xanthorrhizol, a new sesquiterpene from *Curcuma xanthorrhiza*. *Z Naturforsch B*. 25, 995-998.

Roth, G.N., Chandra, A., Nair, M.G. (1998) Novel bioactivities of Curcuma longa constituents. *J Nat Prod* 61, 542-545.

Sacchetti, G., Maietti, S., Muzzoli, M., Scaglianti, M., Manfredini, S., Radice, M., Bruni, R. (2005) Comparative evaluation of 11 essential oils of different origin as functional antioxidants, antiradicals and antimicrobials in foods. *Food Chem* 91, 621-632.

Sastry, B.S. (1970) Curcumin content of turmeric. *Res. Ind.* 15, 258–260.

Schmitt, B., Schneider, B. (1999) Dihydrocinnamic acids are involved in the biosynthesis of phenylphenalenones in *Anigozanthos preissii*. *Phytochemistry* 52, 45-53.

Schmitt B, Hölscher D, Schneider B (2000) Variability of phenylpropanoid precursors in the biosynthesis of phenylphenalenones in *Anigozanthos preissii*. *Phytochemistry* 53, 331-337.

Shibuya, H., Yoshihara, M., Kitano, E., Nagasawa, M., Kitagawa, I. (1986) Qualitative and quantitative analysis of essential oil constituents in various *Zedoariae rhizoma* (Gajutsu) by means of gas-liquid chromatography-mass spectrometry. *Yakugaku Zasshi-J Pharm Soc Jap* 106, 212-216.

Shiobara, Y., Asakawa, Y., Kodama, M., Takemoto, T. (1986) Zedoarol, 13-hydroxygermacrone and curzeone, 3 sesquiterpenoids from *Curcuma zedoaria*. *Phytochemistry* 25, 1351-1353.

Shiobara, Y., Asakawa, Y., Kodama, M., Yasuda, K., Takemoto, T. (1985) Curcumenone, curcumanolide-A and curcumanolide-B, 3 sesquiterpenoids from *Curcuma zedoaria*. *Phytochemistry* 24, 2629-2633.

Singh, G., Singh, O.P., Prasad, Y.R., Lampasona, M.P., Catalan, C. (2003) Chemical and biocidal investigations on rhizome volatile oil of Curcuma zedoaria Rosc - Part 32. *Ind J Chem Technol* 10, 462-465.

Singh, G., Singh, O.P., Maurya, S. (2002) Chemical and biocidal investigations on essential oils of some Indian *Curcuma* species. *Progress in Crystal Growth and Characterization of Materials* 45, 75-81.

Srinivas, L., Shalini, V.K., Shylaja, M. (1992) Turmerin - a water-soluble antioxidant peptide from turmeric [*Curcuma longa*]*Arch Biochem Biophys* 292, 617-623.

Su, H.C.F., Horvat, R., Jilani, G. (1982) Isolation, purification, and characterization of insect repellents from *Curcuma longa* L. *J Agric Food Chem* 30, 290-292.

Suksamrarn, A., Eiamong, S., Piyachaturawat, P., Byrne, L.T. (1997) A phloracetophenone glucoside with choleretic activity from *Curcuma comosa*. *Phytochemistry* 45, 103-105.

Suksamrarn, A., Eiamong, S., Piyachaturawat, P., Charoenpiboonsin, J. (1994) Phenolic diarylheptanoids from *Curcuma xanthorrhiza*. *Phytochemistry* 36, 1505-1508.

Syu, W.J., Shen, C.C., Don, M.J., Ou, J.C., Lee, G.H., Sun, C.M. (1998) Cytotoxicity of curcuminoids and some novel compounds from *Curcuma zedoaria*. *J Nat Prod* 61, 1531-1534.

Takano, I., Yasuda, I., Takeya, K., Itokawa, H. (1995) Guaiane sesquiterpene lactones from *Curcuma aeruginosa*. *Phytochemistry* 40, 1197-1200.

Toume, K., Takahashi, M., Yamaguchi, K., Koyano, T., Kowithayakorn, T., Hayashi, M., Komiyama, K., Ishibashi, M. (2004) Parviflorenes B–F, novel cytotoxic unsymmetrical sesquiterpene-dimers with three backbone skeletons from *Curcuma parviflora*. *Tetrahedron* 60, 10817-10824.

Uehara, S., Yasuda, I., Takeya, K., Itokawa, H. (1989) New bisabolane sesquiterpenoids from the rhizomes of *Curcuma xanthorrhiza* (zingiberaceae). *Chem Pharm Bull* 37, 237-240.

Uehara, S., Yasuda, I., Takeya, K., Itokawa, H., Iitaka, Y. (1990) New bisabolane sesquiterpenoids from the rhizomes of *Curcuma xanthorrhiza* (Zingiberaceae) ii. *Chem Pharm Bull* 38, 261-263.

Uehara, S., Yasuda, I., Takeya, K, Itokawa, H. (1992) Terpenoids and curcuminoids of the rhizome of *Curcuma xanthorrhiza* Roxb. *Yakugaku Zasshi* 112, 817-823.

Yan, J., Chen, G., Tong, S., Feng, Y., Sheng, L., Lou, J. (2005) Preparative isolation and purification of germacrone and curdione from the essential oil of the rhizomes of *Curcuma wenyujin* by high-speed counter-current chromatography. *Journal of Chromatography A* 1070, 207-210.

Yang, S.D., Chen, Y.H (1980) The determination of curcumol in the volatile oil of *Curcuma aromatica* by phloroglucinol spectrophotometry. *Yao Xue Xue Bao* 15, 228-233.

Yasni, S., Imaizumi, K., Sin, K., Sugano, M., Nonaka, G., Sidik, S. (1993) Identification of an active principle in essential oils and hexane-soluble fractions of *Curcuma xanthorrhiza* Roxb. Showing triglyceride-lowering action in rats. *Food and Chemical Toxicology* 32, 273-278.

Yoshikawa, M., Shimada, H., Nishida, N., Li, Y.H., Toguchida, I., Yamahara, J., Matsuda, H. (1998a) Antidiabetic principles of natural medicines. II. Aldose reductase and alpha-glucosidase inhibitors from Brazilian natural medicine, the leaves of *Myrcia multiflora* DC. (Myrtaceae): Structures of myrciacitrins I and II and myrciaphenones A and B. *Chem Pharm Bull* 46, 113-119.

Yoshikawa, M., Murakami, T., Morikawa, T., Matsuda, H. (1998) Absolute stereostructures of carabrane-type sesquiterpenes, curcumenone, 4S-dihydrocurcumenone, and curcarabranols A and B: Vasorelaxant activity of zedoary sesquiterpenes. *Chem Pharm Bull* 46, 1186-1188.

Yoshihara, M., Shibuya, H., Kitano, E., Yanagi, K., Kitagawa, I. (1984) The absolute stereostructure of (4*S*,5*S*)-(+)-germacrone 4,5-epoxide from Zedoariae rhizoma cultivated in Yakushima island. *Chem Pharm Bull* 32, 2059-2062.

4 Biotechnology of Turmeric and Related Species

K. Nirmal Babu, D. Minoo, S.P. Geetha, V. Sumathi, and K. Praveen

CONTENTS

4.1 INTRODUCTION

Turmeric of commerce is the dried rhizome of *Curcuma longa* L (syn *C. domestica* Val.). *C. aromatica* Salisb (*yellow zedoary*), *C. amada* Roxb (mango ginger), *C. caesia* Roxb (black turmeric), *C. angustifolia* Roxb (Indian arrow root), and *C. zedoaria* Rosc (*zedoary*) are other important related species of turmeric. Turmeric is conventionally propagated vegetatively through rhizome bits carrying one or two buds. The species is a sterile triploid. Rich morphological variability is observed among the cultivated types of turmeric, probably due to vegetative mutations accumulated over a period of time. The rarity of seed set hampers recombination breeding. In such circumstances, biotechnological tools gain relevance in solving many crop-specific problems and for crop improvement.

The past few years have witnessed a quantum jump in the utilization of biotechnological tools in commercial propagation and in the development of novel varieties and new breeding lines. Somaclonal variation, anther culture, protoplast fusion, recombinant DNA technology, etc. have been made use of for the improvement, conservation, and utilization of the diversity and in increasing the utility of this spice. This chapter summarizes the work so far carried out in the area of turmeric tissue culture and biotechnology.

4.2 MICROPROPAGATION

4.2.1 CLONAL MULTIPLICATION *IN VITRO*

Micropropagation is the true-to-type propagation of selected genotypes using *in vitro* culture techniques (Debergh and Read, 1991). Many workers reported successful micropropagation of turmeric. (Nadgauda et al., 1978; Shetty et al., 1982; Yasuda et al., 1988; Keshvachandran and Khader 1989; Nirmal Babu et al., 1997; Meenakshi et al., 2001; Sunitibala et al., 2001; Nirmal Babu and Minoo, 2003; Mogov et al., 2003; Ali et al., 2004; Rahman et al., 2004 and Praveen 2005). The *in vitro* responses of various *Curcuma* species are given in Table 4.1. Such microprop-agation techniques (Figure 4.1) could be used for the production of disease-free planting material of elite genotypes.

Nadgauda et al. (1978) were the first to report micropropagation of turmeric. They cultured young vegetative buds excised from cultivars (cvs) Duggirala and Tekkurpeta on MS (Murashige and Skoog, 1962) medium supplemented with coconut milk, kinetin, and BA (6-benzyl adenine) or on Smith's medium supplemented with coconut milk, kinetin, BA, and inositol. The elongated shoots, on transfer to White's liquid medium with 2% sucrose, developed a healthy root system and could then be transferred to pots and grown in the field. They again subcultured explants from the shoot region of the newly formed sterile plantlets on the same medium to regenerate multiple shoots. The capacity for shoot formation increased after the first subculture and became consistent after three to four passages. The number of plantlets obtained was also higher if the cultures were incubated for 4 weeks instead of a shorter period. Plantlet formation by subculture was observed to take place throughout the year without showing the normal dormancy period of field-grown plants.

Shetty et al. (1982) reported 10 to 12 multiple shoots per explant (in clone 15B) when cultured on modified MS medium (pH 5.6) containing sucrose (40 gl⁻¹) and kinetin (0.5 mgl⁻¹). Keshvachan-dran and Khader (1989) reported an average of 2.5 and 2.1 shoots in turmeric cv. Co-1 and BSR-1, respectively on MS medium supplemented with 1 mgl⁻¹ kinetin, 1 mgl⁻¹ BA, and 40 g sucrose/l. Balachandran et al. (1990) reported that rhizome explants cultured on MS medium supplemented with 3 mgl⁻¹ BA gave the highest multiplication rate in all the three species — *C. longa*, *C. aeruginosa*, and *C. caesia*. Within 4 weeks, the regenerated plants produced prolific root system in all the treatments irrespective of the concentration of BA. Nirmal Babu et al. (1997) reported micropropagation of turmeric using young vegetative bud explants in MS medium supplemented with BA 1.0 mgl⁻¹ and α-naphthalene acetic acid (NAA) 0.5 mgl⁻¹. These explants responded readily to culture conditions, producing 8 to 10 shoots in 40 days of culture (Figure 4.1). Meenakshi et al. (2001) reported that the explants of 2.5 and 3.0 mm size on MS + 1.0 mgl⁻¹ BA + 0.1 mgl⁻¹ GA (Giberellic acid) + 0.1 mgl⁻¹ NAA were the best for initiation of cultures of turmeric cv. Cuddapah, which on transfer to 0.3 mgl⁻¹ NAA gave healthy roots. Sunitibala et al. (2001) reported multiple shoots on 1 mgl⁻¹ NAA + 1 mgl⁻¹ kinetin or 1 mgl⁻¹ NAA + 2 mgl⁻¹ BA. Salvi et al. (2002) reported higher shoot production in turmeric cv. *Elite* on 10 μM BA and 1 μM NAA with agar at 0.4% and 0.6%, respectively. Further, they reported that carbohydrate sources like xylose, rhamnose, lactose, and soluble starch were inhibitory, and no variation was observed in Random amplified polymor-phine DNA (RAPD) analysis. Rahman et al. (2004) developed multiple shoots (14.5) on MS medium supplemented with 2.0 mgl⁻¹ BA and rooting was obtained on half strength of MS with 0.2 mgl⁻¹ indole-3 butyric-acid (IBA), which gave 70% survival rate on hardening. Prathanturarug et al. (2003, 2005) were able to induce 11 to 18 multiple shoots in 8 to 12 weeks when the bud explants were supplemented with 18.17 to 72.64 μM Thiaduzuron (TDZ) for 1 week prior to culture on MS

TABLE 4.1
In Vitro Responses of *Curcuma* and Related Species

Explant Used	*In Vitro* Response	Media Composition	Refs.
		Turmeric	
Vegetative buds, rhizome bits with axillary buds	MS + 1 mgl^{-1} NAA	Multiple shoots and *in vitro* rooting	(Nirmal Babu et al., 1997)
	MS + 10% coconut milk, 2–5 mgl^{-1} BA	Plantlet formation	(Nadgauda et al., 1978)
Vegetative buds	MS + 2 mgl^{-1} 2,4-D	Callus	(Nirmal Babu et al., 1997)
	MS + 3 mgl^{-1} BA	Multiple shoots and *in vitro* rooting	(Balachandran et al., 1990)
Callus derived from vegetative buds	MS + 10% CM, 2–5 mgl^{-1} BA	Shoot regeneration	(Shetty et al., 1982)
	MS + 1 mgl^{-1} BA, 0.5 mgl^{-1} NAA	Organogenesis and plant regeneration	(Nirmal Babu et al.,1997)
Callus derived from leaf	MS + 10 µM TDZ and 22.2 µM BA	Plant regeneration	(Praveen, 2005)
In vitro plantlets	MS + 0.3 mgl^{-1} BA + 0.1 mgl^{-1} NAA + 0.5 mgl^{-1} ancymidol + 10% sucrose	Microrhizomes	(Raghu Rajan, 1997)
	MS + 0.3 mgl^{-1} BA + 7–9% sucrose	Microrhizomes	(Peter et al., 2002)
	MS + 1 mgl^{-1} Kin + 0.1 mgl^{-1} NAA + 8% sucrose	Microrhizomes	(Sunitibala et al., 2001)
		C. aromatica	
Vegetative buds and rhizome bits	MS + 1 mgl^{-1} BA, 0.5 mgl^{-1} NAA/IBA	Multiple shoots and *in vitro* rooting	(Nirmal Babu et al., 1997)
	MS + 5 mgl^{-1} BA	*In vitro* multiplication	(Sanghamitra, 2000)
	MS + 5 mgl^{-1} BA, 60 gl^{-1} sucrose + 8 h photoperiod	Microrhizome induction	(Sanghamitra, 2000)
		C. amada	
Vegetative buds and rhizome bits	MS + 1 mgl^{-1} BA, 0.5 mgl^{-1} NAA/IBA	Multiple shoots and *in vitro* rooting	(Nirmal Babu et al., 1997)
Vegetative buds and callus	MS + 8.88 µM BA and 2.7 µM NAA	Multiple shoots and plant regeneration from callus	(Prakash et al., 2004)
		C. aeruginosa	
Rhizome buds	MS + 3 mgl^{-1} BA	Multiple shoots and *in vitro* rooting	(Balachandran et al., 1990)
		C. caesia	
Rhizome buds	MS + 3 mgl^{-1} BA	Multiple shoots and *in vitro* rooting	(Balachandran et al., 1990; Raju et al., 2005)
		C. zedoaria	
Rhizome buds	MS + 2 mgl^{-1} BA, and 2 mgl^{-1} IBA	Multiple shoots and *in vitro* rooting; multiple shoots	(Chan and Thong, 2004; Raju et al., 2005; Nguyen et al., 2005)
	MS + 3 mgl^{-1} BA, 0.5 mgl^{-1} IBA, 20 (v/v) CW		

Continued

TABLE 4.1 *(Continued)*
In Vitro **Responses of** *Curcuma* **and Related Species**

Explant Used	*In Vitro* Response	Media Composition	Refs.
	Amomum subulatum		
Vegetative buds and rhizome bits	MS + 1 mgl^{-1} BA, 0.5 mgl^{-1} NAA/IBA	Multiple shoots and *in vitro* rooting	(Sajina et al., 1997b; Nirmal Babu et al., 1997)
	Alpinia conchigera		
Rhizome buds	MS + 2 mgl^{-1} BA, and 2 mgl^{-1} IBA.	Multiple shoots and *in vitro* rooting	(Chan and Thong, 2004)
	A. galanga		
Rhizome buds	MS + 2 mgl^{-1} BA, and 2 mgl^{-1} IBA	Multiple shoots and *in vitro* rooting	(Chan and Thong, 2004)
Hedychium coronarium	VW + 5 mgl^{-1} 2,4-D, 0.5 mgl^{-1} BA, 200 ml^{-1} coconut water	Somatic embryogenesis.	(Huang and Tsai, 2002)
	Kaempferia galanga		
Vegetative buds and rhizome bits	MS + 1 mgl^{-1} BA, 0.5 mgl^{-1} NAA/IBA	Multiple shoots and *in vitro* rooting	(Geetha et al., 1997; Nirmal Babu et al., 1997)
Callus	MS + 0.5 mg/l 2,4-D, 0.2 mg/l BAP and 16.336 mg/l tryptophan (80 µM)	Somatic embryogenesis plant regeneration	(Lakshmi and Mythili, 2003; Rahman et al., 2004)
Rhizome	MS + 12 µM BA, 3 µM NAA	Multiple shoots and plant lets	(Fatima et al., 2001)
Shoot buds	MS + 4 mgl^{-1} BA	High frequency multiple shoots and plant lets	(Jose et al., 2002)
	K. rotunda		
Vegetative buds and rhizome bits	MS + 1 mgl^{-1} BA, 0.5 mgl^{-1} NAA/IBA	Multiple shoots and *in vitro* rooting	(Geetha et al., 1997; Nirmal Babu et al., 1997)

Abbreviation: MS, Murashige and Skoog (1962) medium; SH, Schenk and Hildebrandt (1972) medium; WPM, Woody Plant Medium (Mc Cown, 1978) medium; CW, coconut water.

medium. The regenerated plants spontaneously rooted. These results were substantiated by Praveen (2005) who also got higher rate of multiplication when TDZ was used in culture medium.

Adelberg and Cousins (2006) compared both solid and liquid media and large as well as small culture vessels and reported that use of liquid media in large culture vessel with gentle tilting agitation gave bigger plantlets, while liquid culture in small vessels on a shaker gave the most plants. Increased biomass was observed in liquid cultures compred with solid (agar) cultures.

Most of the earlier workers used MS basal medium for *in vitro* culture of turmeric and related species. But Mukhri and Yamaguchi (1986) used Ringe and Nitsch media containing 1 mgl^{-1} BA for shoot and root development in turmeric. On increasing the concentration of BA to 10 mgl^{-1} and 1 mgl^{-1} 2, 4-dichloro phenoxy acetic acid (2,4-D) or 10 mgl^{-1} BA and 10 mgl^{-1} NAA produced callusing.

Salvi et al. (2002) reported that MS medium solidified with agar (0.4 and 0.6%) was superior to liquid media for production of multiple shoots in turmeric. Prathanturarug et al. (2005) also preferred solid medium for a better response in turmeric.

FIGURE 4.1 Micropropagation of turmeric (*C. longa* L.).

4.2.2 INDUCTION OF MICRORHIZOMES

In turmeric, micropropagation by *in vitro* microrhizomes is an ideal method for the production of disease-free planting material and also for conservation and exchange of germplasm. Since minimal level of growth regulators are used and the number of subculture cycles are reduced in microrhizome production, the pathway may be better suited for the production of genetically stable planting material. Microrhizomes can be produced *in vitro*, independent of seasonal fluctuations. *In vitro* induction of microrhizomes (Figure 4.2) was reported in turmeric (Raghu Rajan 1997; Nirmal Babu et al., 1997, 2003; Nayak, 2000; Sunitibala et al., 2001; Shirgurkar et al., 2001, Peter et al., 2002). Culture of plantlets on MS medium with high sucrose level was the most common method used.

Raghu Rajan (1997) induced microrhizome production on MS medium supplemented with 0.3 mgl⁻¹ BA + 0.1 mgl⁻¹ NAA and 0.5 mgl⁻¹ ancymidol, and 10% sucrose. Sanghamitra and Nayak

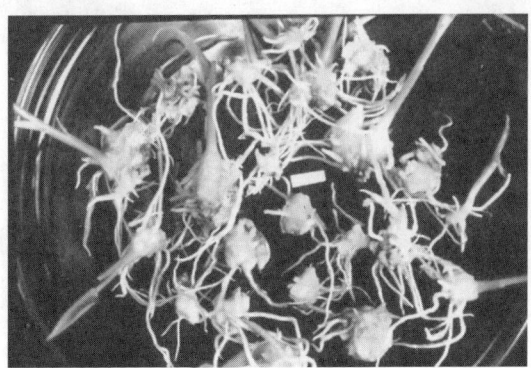

FIGURE 4.2 Microrhizomes of turmeric.

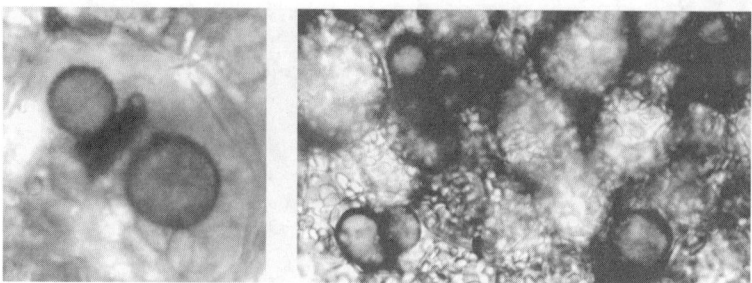

FIGURE 4.3 A section of microrhizomes showing curcumin cells and starch cells.

(2000) reported *in vitro* microrhizome production in four turmeric varieties (Ranga, Rashmi, Roma, and Suroma) after 30 days of culturing on MS liquid medium, supplemented with 3 mgl⁻¹ BA and 60 gl⁻¹ sucrose and 4 h photoperiod. One to four buds were produced per microrhizome weighing 40 to 700 mg each. Microrhizomes harvested after 8 weeks from the induction medium without an intervening phase of hardening gave the highest germination under field conditions. Sunitibala et al. (2001) reported that *in vitro* rhizome formation was obtained with MS + 0.1 mgl⁻¹ NAA + 1.0 mgl⁻¹ kinetin and 8% sucrose, whereas Shirgurkar et al. (2001) reported *in vitro* rhizome production on half strength MS + 80 gl⁻¹ sucrose. BA had an inhibitory effect on microrhizome production. At 35.2 μM concentration of BA, microrhizome production was totally inhibited.

Microrhizomes were successfully induced in 80% of the turmeric cultures on MS basal medium with 9 to 12% of sucrose. These microrhizomes have a good number of cells storing starch and curcumin (Figure 4.3). Microrhizome production depended on the size of the multiple shoot used. Microrhizomes of 0.1 to 2.0 g fresh weight per explant were produced in 1 to 6 months of culture and could be planted directly in the field without hardening, with 80% success. The microrhizomes need more care, especially ensuring moisture and shade until germination. Although under *in vitro* conditions, small (0.1 to 0.4 g), medium (0.41 to 0.8 g), and large (>0.81 g) microrhizomes regenerated, plantlets developed from large microrhizomes grew faster (Geetha 2002). Field evaluation of microrhizhome-derived plants of turmeric cv. Prabha was conducted by Nirmal Babu et al. (2003). The microrhizome-derived plants are smaller but gave higher number of mother rhizomes per plant and gave reasonably large (Figure 4.4) rhizomes. These microrhizomes gave a fresh rhizome yield of 100 to 900 g per plant. The fresh rhizome yield of control plants was 400 to 1300 g per plant, when 20 to 30 g seed rhizomes were used. Thus microrhizomes in turmeric, though

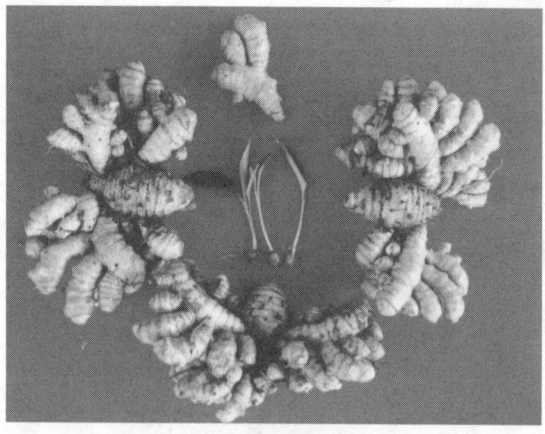

FIGURE 4.4 Rhizomes harvested using microrhizomes (seen at center) as planting material.

gave lesser yield per bed, gave higher recovery *vis-a-vis* the weight of seed material used. This, coupled with its disease-free nature will make microrhizomes an ideal source of planting material when new areas are brought under turmeric cultivation, thus preventing disease incidence and spread. Field data and RAPD profiling also indicated that microrhizomes are more genetically stable than micropropagated plants.

Adelberg and Cousins (2006) reported *in vitro* induction of functional storage organs (rhizomes) in turmeric using liquid cultures. These rhizomes had normal morphology, characteristic pigments, and fragrance. They also reported that the antioxidant activity of fresh rhizome tissue extracts is 30 times higher than that of standard food preservative butylated hydroxy toluene (BHT) showing high potency of *in vitro* produced rhizomes.

Sanghamitra and Nayak (2000) reported induction of microrhizome in *C. aromatica*. MS basal medium with 5 mgl^{-1} BA, 60 gl^{-1} sucrose, and an 8 h photoperiod was optimum for induction of microrhizomes within 30 d of culture. Harvested microrhizomes were stored in moist sand kept in poly-bags. They sprouted after 2 months of storage at room temperature.

4.2.3 INFLORESCENCE CULTURE AND *IN VITRO* POLLINATION

Salvi et al. (2000) reported direct shoot regeneration from cultured immature inflorescences of turmeric cv. Elite on MS basal medium, supplemented with BA (10 mgl^{-1}) in combination with 2,4-D (0.2 mgl^{-1}) or NAA (0.1 mgl^{-1}) and TDZ (1 or 2 mgl^{-1}) in combination with indole-3-acetic acid (IAA) (0.1 mgl^{-1}). Regenerated shoots were grown on MS medium for further development and later transferred to medium supplemented with 0.1 mgl^{-1} NAA for induction of roots. Salvi et al. (2003) reported that immature inflorescence of turmeric showed direct regeneration of shoots when cultured on different concentrations of BA in combination with 2,4-D and TDZ in combination with IAA. An average of five to six shoots per inflorescence segment was obtained, which later differentiated into well-developed roots.

4.2.4 *IN VITRO* POLLINATION AND HYBRIDIZATION

Absence of seed set hinders improvement in turmeric. Nazeem and Menon (1994) reported seed set in controlled crosses of short duration turmeric types. Renjith et al. (2001) reported *in vitro* pollination and hybridization between two short duration (*aromatica*) types VK-70 and VK-76 to develop a high yielding variety with high curcumin and curing percentage. Seed set as well as seed development was obtained by *in vivo* stigmatic pollination. Hybrid seed germinated under *in vitro* condition and the seedlings were micropropagated to establish a population of six plantlets/seedling within a period of 10 months. The multiple shoots were obtained in 1/2 MS + 3% sucrose + BA (2.5 mgl^{-1}) + NAA (0.5 mgl^{-1}). This reduces the breeding time and helps in recombination breeding which was so far not attempted in turmeric.

4.2.5 PLANT REGENERATION FROM CALLUS CULTURES AND SOMACLONAL VARIATION

Efficient plant regeneration protocol is essential for genetic manipulation of any crop species. Organogenesis and plantlet formation were achieved from the callus cultures of turmeric (Shetty et al., 1982; Nirmal Babu et al., 1997; Sunitibala et al., 2001; Salvi et al., 2001, 2002; Praveen 2005). Shetty et al. (1982) reported 10 to 12 multiple shoots per explant in clone 15B, when cultured on modified MS medium (pH 5.6) containing sucrose (40 gl^{-1}) and kinetin (0.5 mgl^{-1}). On transfer to a similar medium, callus was produced. When subcultured and exposed to light, the callus produced several buds that turned green and later developed into plantlets.

Salvi et al. (2000, 2001) also reported plant regeneration from leaf base callus of turmeric and random amplified polymorphic DNA analysis of regenerated plants showed variation at DNA level. Variants with high curcumin content were isolated from tissue-cultured plantlets (Nadgauda et al.,

1982). Root rot disease tolerant clones from the cv. Suguna were isolated using continuous *in vitro* selection technique against pure culture filtrate of *Pythium graminicolum* (Gayatri et al., 2005).

Sunitibala et al. (2001) reported callus induction and plant regeneration by organogenesis on MS + 3 mgl^{-1} 2,4-D and 1 mgl^{-1} kinetin whereas Salvi et al. (2001) induced callusing on MS medium supplemented with 2.0 mgl^{-1} dicamba or picloram or 5 mgl^{-1} NAA + 0.5 mgl^{-1} BA and plant regeneration on MS with 5 mgl^{-1} BA. RAPD analysis of eight regenerated plants showed variation at DNA level indicating somaclonal variation.

Praveen (2005) obtained efficient micropropagation and plant regeneration from callus using TDZ in two turmeric cvs Prabha and Prathibha and obtained both organogenesis and embryogenesis in MS medium supplemented with 0.1 to 0.5 µM TDZ (Figure 4.5 and Figure 4.6). When the concentration of TDZ was raised to 10 µM, direct plant regeneration was observed from leaf tissues without intervening callus. This study revealed the occurrence of cytotypes and morphological variants among callus regenerated progenies. RAPD profiling also indicated the occurrence of variants.

FIGURE 4.5 Embryogenic calli from turmeric on MS medium containing TDZ.

FIGURE 4.6 Plant regeneration from turmeric embryogenic callus.

4.2.6 Micropropagation and Plant Regeneration of Related Taxa

Protocols for micropropagation of many economically and medicinally important zingiberaceous species like *Amomum subulatum* (large cardamom), *C. aromatica* (kasturi turmeric), *C. amada* (mango ginger), *Kaempferia galanga, K. rotunda, Alpinia* spp. were developed (Table 4.1) (Borthakur and Bordoli, 1992; Thomas et al., 1996; Vincent et al., 1992; Geetha et al., 1997; Sajina et al., 1997; Fatima et al., 2001; Jose et al., 2002; Prakash et al., 2004; Chithra et al., 2005). Most of the workers used MS basal media supplements with BA or kinetin 1 to 5 mgl^{-1} for multiplication and around 1 mgl^{-1} NAA for rooting. Balachandran et al. (1990) reported that rhizome explants cultured on MS medium supplemented with 3 mgl^{-1} BA gave the highest multiplication rate in all the three species studied (*C. longa, C. aeruginosa* and *C. caesia*). Successful plant regeneration and variations among regenerated plants were reported in *K. galanga* (Ajith Kumar and Seeni, 1995). Chan and Thong (2004) reported *in vitro* propagation of *Alpinia conchigera, A. galanga, C. longa, C. zedoaria,* and *K. galanga. In vitro* clonal propagation of *C. caesia* and *C. zedoaria* from rhizome bud explants was reported by Raju et al. (2005). Nguyen et al. (2005) reported micropropagation of zedoary (*C. zedoaria*).

Yasuda (1988) reported successful callus induction from rhizomes of *C. zedoaria, C. longa,* and *C. aromatica* on MS medium supplemented with NAA (1 ppm) and kinetin (0.1 ppm). Prakash et al. (2004) reported multiple shoots from rhizome explants of *C. amada* on MS + 4.44 μM BA and 1.08 μM NAA. MS + 9.0 μM 2,4-D resulted in semi friable callus from leaf sheath, which on transfer to 8.88 μM BA and 2.7 μM NAA resulted in shoot initiation and development. Lakshmi and Mythili (2003) induced somatic embryogenesis from callus cultures of *K. galanga* on MS medium containing 0.5 mgl^{-1} 2,4-D, 0.2 mgl^{-1} BAP, and 16.336 mgl^{-1} tryptophan (80 μM). Rahman et al. (2004) also reported efficient plant regeneration through somatic embryogenesis from leaf base–derived callus of *K. galanga* L.

Miachir et al. (2004) reported micropropagation and callogenesis of *C. zedoaria*. Shoot apices were micropropagated on MS medium supplemented with 2.0 mgl^{-1} BAP and 30 gl^{-1} sucrose. Treatment with endomycorrhiza during the *ex vitro* transferring was beneficial for hardening, plant growth, and development. Callus induction and growth were obtained by inoculating root segments on MS medium supplemented with 1.0 mgl^{-1} NAA and incubation in the dark at 25 ± 2°C. Cell suspension cultures were established on liquid medium of same chemical composition and same culture conditions and a growth curve was obtained.

Marcia et al. (2000) developed a formula for quantifying the micropropagation of *C. zedoaria* to estimate the number of plants that can be produced at *n* number of transfers after 120 days from inoculation based on the following equation: $y = 8.74 \, x^2 - 21.19 \, x + 23.47$, where *y* is the number of plants produced and *x* is the number of transfers to be performed from each explant. He estimated that the cost of *C. zedoaria* plants produced *in vitro* was US $0.38/plant when a production of 200,000 plants/year is considered.

4.2.7 Cell Suspension

The establishment of callus or cell suspension culture is the first step toward *de novo* organogenesis. Plant cell suspension cultures provide a relatively homogeneous population of cells, readily accessible to exogenously applied chemical and growing under defined, aseptic conditions. Cell suspension culture is a model system for studying pathways of secondary metabolism, enzyme induction, gene expression, and degradation of xenobiotics and as a source material for enzyme purification. A major advantage of the use of plant suspension cultures in mutant selections is the capacity to apply nearly standard microbiological techniques. The practical utility of induced mutations for improvement of quantitatively inherited characters in zingiberaceous crops is well recognized.

Efficient cell suspension systems were developed by Praveen (2005) that were suitable for *in vitro* mutant selection. A good cell suspension culture of small aggregates of cells of turmeric was established by subculturing at 200 rpm at weekly intervals. The cell suspension has to be aseptically diluted to 1:3 in fresh medium for maintenance. Two ml of the diluted suspension was plated in petriplate (90 mm) containing solidified MS + BA (4.4 μM) for further growth and development. The plating efficiency was high with 40 to 60 well-developed microcalli. These microcalli were further proliferated on callus proliferating medium. The cell suspension can form an ideal system for *in vitro* mutagenesis (Pierik, 1987).

4.3 FIELD EVALUATION

Tissue-cultured plants of turmeric, kasturi turmeric (*C. aromatica*), mango ginger, *K. galanga* etc. behave similar to those of seedlings and hence require at least two crop seasons to develop rhizomes of normal size that can be used as seed rhizomes for commercial cultivation. The various stages of rhizome development in micropropagated plants of turmeric are represented in Figure 4.7 and Figure 4.8.

Salvi et al. (2002) reported 95% of the micropropagated plants survived in sterilized soil in paper cups and all of them survived in the field.

FIGURE 4.7 Different stages of rhizome development from micropropagated plants of turmeric — after 6, 12, 18, and 24 months.

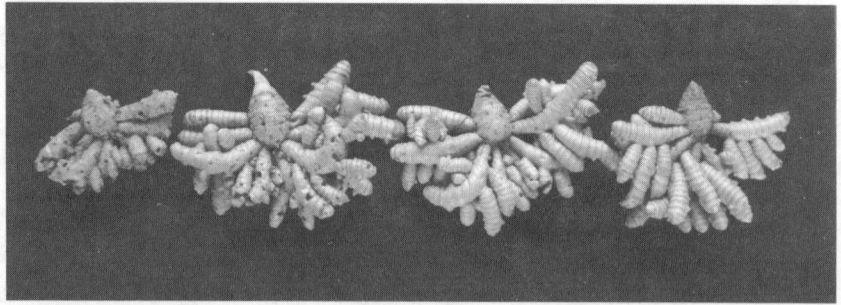

FIGURE 4.8 Rhizomes derived from tissue-cultured plants of turmeric in second year.

4.4 INDUCED VARIATIONS IN TURMERIC AND RAPD CHARACTERIZATION OF SOMACLONES

Improvement in turmeric is hindered by lack of seed set. No source of resistance either to rhizome rot or leaf spot disease is available in the germplasm (Nair et al., 1980). Improvement in turmeric is also sought for high curcumin content and yield. Somaclonal variation is a tool that can be used by plant breeders for improving crop plants and it is the more useful in crops like turmeric with narrow genetic base, sterility, limited scope for hybridization and vegetative propagation (Karp, 1995). Considering these factors, somaclonal variations find a prime place in the breeding program of turmeric.

Salvi et al. (2002) reported that of the 48 plants, two showed variegated leaves in the tillers. The micropropagated plants showed a significant increase in shoot length, number of tillers, number and length of leaves, number of fingers, and total fresh rhizome weight per plant when compared with conventionally propagated plants. RAPD analysis of 11 regenerated plants using sixteen decamer primers did not show any polymorphism.

Nirmal Babu et al. (2003) studied morphological and molecular variations among microprop-agated and callus regenerated plants and found variations in both. They reported that somaclones obtained through callus pathway exhibited greater variation. *In vitro* plants developed through microrhizome pathway has exhibited least variation. They inferred that this is due to the accumulated vegetative mutations (mosaic) in turmeric rhizomes, which is the major source of variation in turmeric germplasm in the absence of sexual recombination. This must be the most probable reason for the existence of large numbers of cultivars with reasonably good amount of variability.

Salvi et al. (2003) reported that micropropagated plants from shoot tips showed significant increase in plant height, number of tillers, number and length of leaves, number of fingers, and total rhizome weight per plant in the first generation when compared with conventionally propagated (control) plants. Second and third generation of regenerated plants tested under the same field conditions confirmed the stability of character observed in the first generation. RAPD analysis of 10 turmeric plants, each propagated conventionally, regenerated from shoot tips, leaf base callus and immature inflorescence was carried out using 15 primers. Plants regenerated from shoot tips showed uniform banding pattern, whereas, callus-derived and inflorescence-derived plants showed polymorphism in banding pattern when compared to conventionally propagated plants.

Praveen (2005) also observed variations in turmeric somaclones with respect to both morpho-logical as well as RAPD profiles but could not detect any variations among micropropagatd plants. He reported that *in vitro* mutagenesis increased the spectrum of somaclonal variation. He found variability among somaclones and isolated somaclones having short nodes.

4.5 ISOLATION OF PROTOPLASTS

The protoplast is a naked cell without cell wall, the absence of which makes it suitable for a variety of manipulations that are not normally possible with intact cells and hence protoplast is an important tool for parasexual modification of genetic content of cells (Vasil and Vasil, 1980). Protoplasts could be isolated successfully from leaf tissues as well as from cell suspension cultures of turmeric (Nirmal Babu et al., 2005).

4.6 SYNTHETIC SEEDS

Artificial or synthetic seeds can be an ideal system for low cost plant movement, propagation, conservation, and exchange of germplasm (Redenbaugh et al., 1986). Sajina et al. (1997c) reported development of synthetic seeds in turmeric by encapsulating the somatic embryos and shoot buds in calcium alginate. These synthetic seeds are viable up to 7 months. Gayatri et al. (2005) reported encapsulation and regeneration of aseptic shoot buds of turmeric. Lakshmi and

Mythili (2003) reported development of synthetic seeds by encapsulating somatic embryos in calcium alginate beads.

4.7 CONSERVATION OF GENETIC RESOURCES

In many crop species, conventional seed storage can satisfy most of the conservation requirements. But in crops with recalcitrant seeds, the conservation needs cannot be satisfied by seed storage, they have to be stored *in vitro*. The genetic resources of most of the spices are conserved either in seed gene banks and or in field repositories. In case of vegetatively propagated perennials, the germplasm is conserved in clonal field repositories. Storage of germplasm in seed banks is not practical in some crops as they are sterile hence propagated vegetatively or their seeds are recalcitrant and heterozygous. India is the center of diversity for major spices like black pepper, cardamom, large cardamom, etc. However, disturbances in their natural habitats have resulted in the loss of these valuable materials. Conservation of the germplasm in *in vitro* gene bank and cryo bank is a viable method and a safe alternative to augment the conventional conservation strategies (Ashmore, 1997; Rekha et al., 2004).

Plant regeneration and successful cloning of genetically stable plantlets in tissue culture is an important prerequisite in any *in vitro* conservation effort. These techniques form the base for establishing tissue cultures and developing *in vitro* technology for conservation. The basal media used is MS (Murshige and Skoog, 1962) for crops like cardamom, ginger, turmeric, kasturi turmeric (*C. aromatica*), mango ginger, large cardamom, kaempferia, etc. Though micropropagation protocols were standardized using growth regulators, *in vitro* storage were carried out using growth regulator-free media in order to reduce the rate of multiplication, which in turn will reduce the extent of variation.

4.7.1 *IN VITRO* CONSERVATION

In vitro conservation involves maintenance of explants in a sterile, pathogen-free environment and is widely used for the conservation of species that produce recalcitrant seeds, or that which do not produce seeds (Engelmann, 1997). Various *in vitro* conservation methods are being used. For short- and medium-term storage, the aim is to increase the intervals between subcultures by reducing growth. This is achieved by modifying the environmental conditions, including the culture medium to realize slow-growth conservation. The principle of slow growth storage is that the safety of *in vitro* culture be ensured without the disadvantage of frequent subculturing. Thus the risk of contamination at each transfer interval, inputs in terms of labor and consumables, are reduced. Several methods, such as temperature reduction, medium modification, use of osmoticums, etc., have been found to reduce the rate of growth of tissue cultures, so that they can be kept unattended for moderate length of time (Withers, 1980,1985,1991; Withers and Williams, 1986; Chandel and Pandey, 1991).

Normal culture vessel allows comparatively better gaseous exchange, but there is a faster rate of moisture depletion in culture media and drying up of cultures between 120 and 180 days depending upon the species. However, use of screw caps, polypropylene caps, or aluminum foil as vessel closures minimizes the moisture loss, and helps to increase longevity of cultures; the subculture period can be prolonged up to 360 days or more (Nirmal Babu et al., 1994; Geetha et al., 1995; Geetha, 2002). Sealing the culture tube with parafilm helped in reducing chance contamination, moisture loss, and increased the longevity of culture in ginger and turmeric (Balachandran et al., 1990). Normal culture room temperature of $22 \pm 2°C$ is suitable for the storage of spices germplasm (Nirmal Babu et al., 1994,; Geetha et al., 1995; Dekkers et al., 1991).

Conservation of germplasm in zingiberaceous crops like turmeric, kasthuri turmeric, mango ginger, *Kaempferia* spp. (*K. galanga* and *K. rotunda*), *Alpinia* spp. in *in vitro* gene banks by slow growth was reported (Balachandran et al., 1990; Dekkers et al., 1991; Geetha et al., 1995; Tyagi et al., 1996; Nirmal Babu et al., 1997, 1998, 1999, 2000; Peter et al., 2002; Geetha, 2002; Ravindran

TABLE 4.2
In Vitro **Storage of** *Curcuma* **and Related Genera — Media Constituents and Survival of Cultures**

Species	Basal Medium	Carbon Source (g/l)	Growth	Survival Period (d)	Survival (%)	Establishment in Nursery
Elettaria cardamomum	MS	30S	Fast	200	55	85
	1/2 MS	20S	Normal	250	45	
	1/2 MS	10S + 10M	Slow	360	85	
		15S + 15M	Slow	360	90	
Zingiber offiicinale	1/2 MS	10S + 10M	Slow	360	85	85
C. longa	1/2 MS	10S + 10M	Slow	360	90	90
C. aromatica	1/2 MS	10S + 10M	Slow	360	85	90
C. amada	1/2 MS	10S + 10M	Slow	360	80	85
K. galanga	1/2 MS	10S + 10M	Slow	360	90	100
				450	85	
K. rotunda	1/2 MS	10S + 10M	Slow	360	80	90
A. purpurata	1/2 MS	10S + 10M	Slow	240	85	90
A. subulatum	1/2 MS	10S + 10M	Slow	180	90	90

Abbreviation: MS, Murashige and Skoog basal medium; S, sucrose; M, mannitol.

Source: Nirmal Babu et al., 1996.

et al., 2004). All these taxa could be stored up to 1 year without subculture in half strength MS medium with 10 g1^{-1} each of sucrose and mannitol in sealed culture tubes (Geetha et al., 1995, Geetha, 2002). *Alpinia purpurata* and *A. subulatum* cultures could be stored up to 8 and 6 months, respectively, in the same medium. Refining the techniques in such cases are in progress to increase the subculture intervals for over 360 days. Tyagi et al. (2004) reported genotype conservation of eight wild species of *Curcuma*. The conserved materials showed normal rate of multiplication when transferred to multiplication medium after storage. The normal sized plantlets when transferred to soil established with over 80% success. They developed into normal plants without any visible morphological variations and were similar to mother plants. RAPD profiling of these conserved plants also showed their genetic integrity (Nirmal Babu et al., 1997, 1999; Peter et al., 2002; Geetha, 2002; Ravindran et al., 2004). The highlights of the present status of information on *in vitro* and cryo conservation of important spices are given in Table 4.1 and Table 4.2. About 100 accessions of *Curcuma* germplasm are currently kept in *in vitro* repository of Indian Institute of Spices Research and at the National Facility for Plant Tissue Culture Repository at the National Bureau of Plant Genetic Resources (NBPGR) in India.

4.7.2 GERMPLASM EXCHANGE

Germplasm exchange using *in vitro* culture is to a certain extent practiced in some horticulture crops. In view of the necessity of reducing the possibility of introduction of new pathogens and pests along with the introduction of new plants, it is imperative that we use *in vitro* technology for plant introduction wherever possible, especially in spices. Utilization of microrhizomes and synthetic seeds in ginger, *Kaempferia*, and turmeric can be utilized for disease-free transport and germplasm exchange.

4.7.3 CRYOPRESERVATION

For long-term conservation of plant species, cryopreservation is the only method currently available. Plant germplasm stored in liquid nitrogen (–196°C) does not undergo cellular divisions. In addition, all metabolic and physical processes are stopped at this temperature. As such, plants can be stored for very long time periods, avoiding both the problem of genetic instability and the risk of losing accessions due to contamination or human errors during subculturing. Most cryopreservation endeavors deal with recalcitrant seeds and *in vitro* tissues from vegetatively propagated crops.

Cryopreservation of ginger and turmeric shoot tips was successfully done with 80% of recovery using vitrification method. But the rate of recovery was only 40% when encapsulated shoot tips were dehydrated in progressive increase of sucrose concentration together with 4 to 8 h of desiccation (Peter et al., 2002; Ravindran et al., 2004).

4.7.4 DNA STORAGE

Concurrent with the advancements in gene cloning and transfer, technology for the removal and analysis of DNA has also been progressing. DNA from the nuclei, mitochondria, and chloroplast are now routinely extracted and immobilized onto nitrocellulose sheets where the DNA can be probed with numerous cloned genes. In addition, the rapid development of polymerase chain reaction (PCR) means that one can routinely amplify specific oligonucleotides or genes from the entire mixture of genomic DNA. These advances, coupled with the prospect of the loss of significant plant genetic resources throughout the world, have led to the establishment of DNA bank for the storage of genomic DNA. The vast resources of dried specimens in the world's herbaria may hold considerable DNA that would be suitable for PCR. The advantage of storing DNA is that it is efficient and simple and overcomes many physical limitations and constraints of other forms of storage (Adams, 1988, 1990, 1997; Adams and Adams, 1991).

A simple and rapid method for isolating good quality DNA from mature rhizome tissues of turmeric and ginger, which are rich in polysaccharides, polyphenols and alkaloids, was standardized. The isolated DNA proved amenable to PCR amplification and restriction dissection. DNA was isolated from leaf tissues of 200 turmeric accessions for characterization by RAPD (IISR, 2002, 2003, 2004).

4.8 MOLECULAR CHARACTERIZATION

In recent times there is an increased emphasis in molecular markers for characterization of the genotypes, genetic fingerprinting in identification and cloning of important genes, marker-assisted selection, and in understanding of interrelationships at the molecular level.

Although work on morphological characterization of *Curcuma* species has been attempted, its molecular characterization is still in a nascent stage except for some genetic fidelity studies of micropropagated plants and isozyme-based characterization (Sasikumar, 2005).

Genetic diversity of *Curcuma alismatifolia* Gagnep. populations from both cultivated and wild habitats in Thailand was studied using allozyme polymorphism at seven loci (TPI, G6PD-2, IDH-1, SKD-2, MDH-1, GOT-1, and GOT-2) (Paisooksantivatana et al., 2001). High diversity was observed in all populations with relatively lower values in cultivated populations. Percentage of polymorphic loci (P) varied from 85.7 to 100% in cultivated populations compared with 100% in all natural populations. Allele number per locus (A_L) was 3.14 in cultivated populations and from 2.86 to 4 in natural populations. Allele number per polymorphic locus (A_P) of cultivated and natural populations ranged from 3.14 to 3.5 and 2.86 to 4, respectively. Genetic diversity within populations (H_S) varied from 0.586 to 0.611 in cultivated and from 0.621 to 0.653 in natural populations. The genetic identity (I_{SP}) for the species was 0.833. The cultivated populations yielded higher value of genetic identity with highland populations ($I_{C/H} = 0.776$) than with the lowland ones ($I_{C/L} = 0.754$). The analysis of genetic similarities with the Neighbor-Joining algorithm resulted in the separation of cultivated populations from all wild populations. One highland population from the tourist spot, H_2, was placed in a separate cluster between the cultivated and other wild populations, and hence is considered as the possible origin of the cultivated populations. Cultivated populations from Japan (cJ) and Thailand (cT) had the lowest percentage of polymorphic loci ($P = 40$ to 60%), alleles per locus ($A_1 = 1.8$), alleles per polymorphic locus ($A_p = 2.33$ to 3.00), and gene diversity ($H_s = 0.216$ to 0.304) compared with two lowland populations (L_1, L_2) with $P = 100\%$, $A_1 = 3.0$ to 3.2, $A_p = 3.0$ to 3.2, $H_s = 0.465$ to 0.496; and six highland populations (H_1 to H_6) with $P = 80$ to 100%, $A_1 = 2.4$ to 3.8, $A_p = 2.4$ to 3.8, $H_s = 0.342$ to 0.659. Within population, H_1 (the highest elevation sampled in this study) had the greatest genetic diversity ($H_s = 0.659$). Mean genetic diversity over all loci across all populations was 0.444. Mean genetic identity between cultivated populations (I_C), lowland populations (I_L), among highland populations (I_H), and across all populations (I_{SP}) were 0.950, 0.947, 0.944, and 0.922, respectively. Using cluster analysis, H1 and cJ were separated first from the rest into distinct groups. Two lowland populations were placed together with H6, whereas cT was grouped in the same cluster of H2 to H5.

An efficient protocol for the isolation of high molecular weight DNA from dry powdered samples of turmeric including market samples is described by Remya et al. (2004); Sasikumar et al. (2004); and Syamkumar et al. (2005). This will help in PCR-based detection of adulteration in marketed turmeric powders. The method involves a modified CTAB (Cetyl Trimethyl Ammonium Chloride-3%) procedure with 2 M NaCl, 0.3% β-mercaptoethanol coupled with purification of DNA in 30% polyethylene glycol (8000). The yield of the DNA obtained from the samples varied from 2 to 4 μg/g tissue. The DNA obtained from the five different samples were consistently amplifiable (RAPD primers).

Salvi et al. (2001) analyzed turmeric plants regenerated from leaf base callus using RAPD. This study of eight regenerated plants using 14 primers when separated on nondenaturing poly-acrylamide gels showed 38 novel bands. About 51 bands present in the control were absent in the regenerants. The result indicates that variation at DNA level has occurred during *in vitro* culture.

Chen et al. (1999) used RAPD polymorphism to differentiate within and among *Curcuma wenyujin*, *C. sichuanensis*, and *C. aromatica*. It is hard to differentiate between *C. wenyujin* and *C. sichuanensis* from the DNA level. The relationship between *C. wenyujin* and *C. aromatica* was also analyzed and based on the morphological and chemical data, it is suggested that these two species should be combined into one and that classification based on inflorescence central or lateral may not be right.

Kress et al. (2002) studied the phylogeny of the gingers (*Zingiberaceae*) based on DNA sequences of the nuclear internal transcribed spacer (ITS) and plastid *matK* regions and suggested a new classification. Their studies suggest that at least some of the morphological traits based on which the gingers are classified are homoplasious and three of the tribes are paraphyletic.

Cao et al. (2001) and Sasaki et al. (2002) used sequence analysis of Chinese and Japanese curcuma drugs on the 18S rRNA gene and trnK gene and the application of amplification-refractory mutation system (ARMS) analysis for their authentication. The botanical origins of Chinese and Japanese *Curcuma* drugs were determined to be *C. longa*, *C. phaeocaulis*, the Japanese population of *C. zedoaria*, *C. kwangsiensis*, *C. wenyujin*, and *C. aromatica* based on a comparison of their 18S rRNA gene and *trnK* gene sequences with those of six *Curcuma* species reported previously. Moreover, to develop a more convenient identification method, ARMS analysis of both gene regions was performed on plants. The ARMS method for the 18S rRNA gene was established using two types of forward primers designed based on the nucleotide difference at position 234. When DNAs of four *Curcuma* species were used as templates, PCR amplification with either of the two primers only generated a fragment of 912 base pairs (bp). However, when DNA of the purple-cloud type of *C. kwangsiensis* and *C. wenyujin* were used, PCR amplifications with both primers generated the fragment, suggesting that these two were heterozygotes. The ARMS method for the *trnK* gene was also established using a mixture of four types of specific reverse primers designed on the basis of base substitutions and *indels* (insertion deletions) among six species, and common reverse and forward primers. *C. phae-ocaulis* or the Chinese population of *C. zedoaria*, the Japanese population of *C. zedoaria* or the purple-cloud type of *C. kwangsiensis*, the pubescent type of *C. kwangsiensis* or *C. wenyujin*, and *C. aromatica* were found to show specific fragments of 730, 185, 527 or 528, and 641 or 642 bp, respectively. All species, including *C. longa*, showed a common fragment of 897 to 904 bp. Using both ARMS methods, together with information on producing areas, the *Curcuma* plants were identified. This ARMS method for the trnK gene was also useful for authentication of *Curcuma* drugs.

Cao et al. (2003) used a molecular approach — *trnK* nucleotide sequencing — for identification of six medicinal curcuma — *C. longa*, *C. phaeocaulis*, *C. sichuanensis*, *C. chuanyujin*, *C. chuan-huangjiang,* and *C. chuanezhu* found in Sichuan, China. The *matK* gene (an intron embodied in *trnK* gene) sequence and the intron spacer region of the *trnK* gene have great diversity within these six medicinal *Curcuma* species. There were six single bases substitutions between *trnK* coding region and *matK* region, the 9-bp deletion and 4-bp or 14-bp insertion repeat at some sites of *matK* region in each taxon. These relatively variable sequences were potentially informative in the identification of these six *Curcuma* species at the DNA level.

Chase (2004) presented an overview on the phylogeny and relationships in monocots based on analysis of DNA sequence data of seven genes representing all three genomes. Monocots have been shown in molecular clock studies to be at least 140 million years old, and all major clades and most families date to well before the end of the Cretaceous.

A phylogenetic analysis of the tribe Zingibereae (Zingiberaceae) was performed by Ngamri-absakul et al. (2003) using nuclear ribosomal DNA (ITS1, 5.8S, and ITS2) and chloroplast DNA (trnL [UAA] 5′ exon to trnF [GAA]). The study indicated that tribe Zingibereae is monophyletic with two major clades, the *Curcuma* clade, and the *Hedychium* clade. The genera *Boesenbergia* and *Curcuma* are apparently not monophyletic.

Sasaki et al. (2004) applied single-nucleotide polymorphism (SNP) analysis of the *trnK* gene to the identification of *Curcuma* plants. *Curcuma* plants and drugs derived from *C. longa*, *C. phaeocaulis*, *C. zedoaria*, and *C. aromatica* could be identified by the nucleotide differences at two sites and the existence of a 4-base *indel* on *trnK* gene. Thus, SNP analysis was developed to identify four *Curcuma* plants.

Sasikumar et al. (unpublished) studied over 96 Indian cultivars and related species of turmeric using RAPD profiling for establishing their interrelationships. Sixty random decamer primers were screened among which 34 gave good, clear consistent bands. RAPD analyses of DNA from 96 accessions of turmeric with 15 random decamer primers showed good polymorphism among the accessions studied. Five species of *Curcuma* were characterized using 12 primers. The intra species polymorphism in curcuma was high as compared to the interspecies polymorphism (IISR, 2003, 2004).

A PCR-based method for detection of extraneous *Curcuma* species contamination in the pow-dered market samples of turmeric was developed by Sasikumar et al. (2004). The study revealed the presence of *C. zedoaria* samples mixed with true turmeric (*C. longa*) samples.

Xia et al. (2005) undertook molecular genetic and chemical assessment of rhizoma curcumae in China. Rhizoma curcumae (*Ezhu*) is a tradition Chinese medicine that has been used in removing blood stasis and alleviating pain for over 1000 years. Three species of *Curcuma* rhizomes are being used, which include *C. wenyujin*, *C. phaeocaulis*, and *C. kwangsiensis*. Chemical fingerprints were generated from different species of *Curcuma*, which could serve as identification markers. For molecular identification, the 5S-rRNA spacer domains of five *Curcuma* species, including the common adulterants of this herb, were amplified, and their nucleotide sequences were determined. Diversity in DNA sequences among various species was found in their 5S-rRNA spacer domains. Thus, the chemical fingerprint together with the genetic distinction could serve as markers for quality control of *Curcuma* species.

Pimchai et al. (1999) reported association of a few isozyme markers in the identifcation of some early flowering *Curcuma* species.

4.9 CONCLUSION

Viable protocols for micropropagation, plant regeneration, microrhizome induction, synthetic seed production, and *in vitro* conservation are available for cloning disease-free planting materials of elite genotypes, conservation, movement, and exchange of germplasm. In turmeric somaclonal variation and *in vitro* mutagenesis were not exploited fully to isolate useful variants. Molecular characterization of germplasm has made reasonable progress. Identifying markers like important agronomic characters will help in marker-assisted selection. Application of recombinant DNA technology for the production of resistant types to biotic and abiotic stress has to go a long way before they can be effectively used in spices improvement.

REFERENCES

Adams, R.P. and Adams, J.E. (1991) *Conservation of Plant Genes: DNA Banking and In Vitro Biotechnology.* New York: Academic Press.

Adams, R.P. (1988) The preservation of genomic DNA: DNA Bank Net. *Am J Bot* 75:156.

Adams, R.P. (1990) The preservation of Chihuahuan plant genomes through in vitro biotechnology: DNA Bank–Net, a genetic insurance policy. In: Powell, A.M., Hollander, R.R., Barlow, J.C., McGillivray, W.B. and Schmidly, D.J., eds. Third Symposium on Resources of the Chihuahuan Desert Region. Lubbock, TX: Printech Press, pp. 1–9.

Adams, R.P. (1997) DNA Banking. In: Callow, J.A. Ford–Loyd, B.V. and Newbury, H.J. eds Biotechnology and Plant Genetic Resources: Conservation and Use. Biotechnology in Agriculture Series, No. 19. CAB International, pp 163–174.

Adelberg, J. and Cousins, M. (2006) Thin film of liquid media for heteromorphic growth and storage organ development: Turmeric (*Curcuma longa*) as plant. *Hort Sci* 41, 539–542.

Ajith Kumar, P. and Seeni, S. (1995) Isolation of somaclonal variants through rhizome explant cultures of *Kaempferia galanga* L. In: All India Symposium on Recent Advances in Biotechnological Applications of Plant Tissue and Cell Culture. CFTRI Mysore: pp. 43.

Ali, A., Munawar, A. and Siddiqui, F.A. (2004) In vitro propagation of turmeric, *Curcuma longa* L. *Int J Biol Biotechnol* 1:511–518.

Ashmore, S.E. (1997) Status report on the development and application of in vitro techniques for the conservation and use of plant genetic resources. International Plant Genetic Resources Institute, Rome, Italy, pp. 67.

Balachandran, S.M., Bhat, S.R. and Chandel, K.P.S. (1990) In vitro clonal multiplication of turmeric (*Curcuma longa*) and ginger (*Zingiber officinale* Rosc.). *Plant Cell Rep* 3:521–524.

Borthakur, M.P. and Bordoloi, D.N. (1992) Micropropagation of *Curcuma amada* (Roxb.). *J Spices Aromatic Crops* 1:154–156.

Cao, H., Komatsu, K., Yao Xue and Xue Bao. (2003) Molecular identification of six medicinal Curcuma plants produced in Sichuan: evidence from plastid trnK gene sequences. *Biol Pharm Bull* 11:871–875.

Cao, H., Sasaki, Y., Fushimi, H., Komatsu, K. and Cao, H. (2001) Molecular analysis of medicinally used Chinese and Japanese Curcuma based on 18S rRNA and trnK gene sequences. *Biol Pharm Bull* 24:1389–1394.

Chan, L.K. and Thong, W.H. (2004) In vitro propagation of *Zingiberaceae* species with medicinal properties. *J Plant Biotechnol* 6:181–188.

Chandel, K.P.S. and Pandey, R. (1991) Plant Genetic Resources Conservation — Recent Approaches. In: Paroda, R.S. and Arora, A.K., eds. Plant Genetic Resources Conservation and Management — Concepts and Approaches. New Delhi: IBPGR, Regional Office for South and Southeast Asia (ROS-SEA), 248–272.

Chase M.W. (2004) Monocot relationships: an overview. Am J Bot 91:645–1655.

Chen, Y., Bai, S., Cheng, K., Zhang, S. and Nian, L.Z. (1999) Random amplified polymorphic DNA analysis on *Curcuma wenuujin* and *C. sichuanensis*; Zhongguo ZhongYao Za Zhi 24:131–133, 189.

Chithra, M., Martin, K.P., Sunandakumari, C. and Madhusoodanan, P.V. (2005) Protocol for rapid propagation, to overcome delayed rhizome formation in field established in vitro derived plantlets of *Kaempferia galanga* L *Scientia-Horticulturae* 104:113–120.

Debergh, P.C. and Read, P.E. (1991) *Micropropagation*. In: Debergh P.C. and Zimmerman R.H., eds. The Netherlands: Kluwer Academic Publication, 1–13.

Dekkers, A.J., Rao, A.N. and Ghosh, C.J. (1991) In vitro storage of multiple shoot culture of gingers at ambient temperatures of 24–29°C. *Sci Hortic* (Amsterdam) 47, 157-167.

Engelmann, F.(1997) *In vitro* conservation methods. In: Callow, J.A., Ford–Loyd, B.V. and Newbury, H.J,
    eds. Biotechnology and Plant Genetic Resources: Conservation and Use. Biotechnology in Agriculture
    Series, No. 19. Oxfordshire, U.K., CAB International, pp. 119–161.

Fatima, S., Sandeep, K. and Mishra, Y. (2001) In vitro plantlet production system for *Kaempferia galanga*, a
    rare Indian medicinal herb. *Plant Cell Tissue Organ Cult* 63:193–197.

Gayatri, M.C., Roopa, D.V. and Kavyashree, R. (2005) Selection of turmeric callus for tolerant to culture
    filtrate of *Pythium graminicolum* and regeneration of plants. *Plant Cell Tissue Organ Cult* 83:33–40.

Gayatri, M.C., Roopadarshini, V., Kavyashree, R. and Kumar, C.S. (2005) Encapsulation and regeneration of
    aseptic shoot buds of turmeric (*Curcuma longa* L.) *Plant Cell Biotechnol Mol Biol* 6:89–94.

Geetha, S.P. (2002) In vitro technology for genetic conservation of some genera of *Zingiberaceae*. Ph.D Thesis.
    Calicut University, Calicut, Kerala: India.

Geetha, S.P., Manjula, C. and Sajina, A. (1995) *In vitro* conservation of genetic resources of spices. In
    Proceedings Seventh Kerala Science Congress Palakkad, Kerala, India, January 27–29, 12–16.

Geetha, S.P., Manjula, C., John, C.Z., Minoo, D., Nirmal Babu, K. and Ravindran, P.N. (1997) Micro-
    propagation of *Kaempferia* spp. (*K. galanga* L. and *K. rotunda* L.) *J Spices Aromatic Crops*,
    6:129–135.

IISR (2002) Annual Report 2001–2002, Indian Institute of Spices Research, Calicut, Kerala, India.

IISR (2003) Annual Report 2002–2003, Indian Institute of Spices Research, Calicut, Kerala, India.

IISR (2004) Annual Report 2003–2004, Indian Institute of Spices Research, Calicut, Kerala, India.

Jose, A.S.R, Thomas R. and Nair G.M. (2002) Micropropagation of Kaempferia galanga Linn. through high
    frequency *in vitro* shoot multiplication. *J Plant Biol* 29:97–100.

Karp, A. (1995) Somaclonal variation as tool for crop improvement. *Euphytica* 85, 295–302.

Keshvachandran, R, and Khader, M.D.A. (1989) Tissue culture propagation of turmeric. *South Indian Horti-
    culture* 37:101–102.

Kress, W.J., Liu, A.Z., Newman, M. and Li, Q.J (2005) The molecular phylogeny of *Alpinia* (Zingiberaceae):
    a complex and polyphyletic genus of gingers. *Am J Bot* 92:167–178.

Kress, W.J., Prince, L.M. and Williams, K.J. (2002) The phylogeny and a new classification of the gingers
    (*Zingiberaceae*): evidence from molecular data. *Am J Bot* 89:1682–1696.

Lakshmi, M. and Mythili, S. (2003) Somatic embryogenesis and plant regeneration from callus cultures of
    Kaempferia galanga — a medicinal plant. *J Med Aromatic Plant Sci* 25:947–951.

Marcia, O.M., Antônio, F.C.A. and Murilo M. (2000) Quantifying the micro propagation of *Curcuma zedoaria*
    Roscoe. *Sci Agricola* 57:703–707.

Meenakshi, N., Suliker G.S., Krishnamoorthy, V., and Hegde, R.V. (2001) Standardization of chemical envi-
    ronment for multiple shoot induction of turmeric (*Curcuma longa* L) for in vitro clonal propagation.
    Crop Research Hissar 22:449–453.

Miachir, Jeanette Inamine, Romani, Vera Lúcia Moretti, Amaral, Antônio Francisco De Campos (2004)
    Micropropagation and callogenesis of *Curcuma zedoaria* Roscoe. *Sci agric* (Piracicaba, Braz.)
    61:427–432.

Mogor, A.F., Mogor, G., Ono, E.O. and Rodrigues, J.D. (2003) Effect of benzylaminopurine (BAP) on clonal
    propagation rate of *Curcuma longa* L. *Acta Hortic* 597:321–323.

Mukhri, Z. and Yamaguchi, H. (1986) In vitro plant multiplication from rhizomes of turmeric (*Curcuma
    domestica* Val.) and Temeoe Lawak (*C. xanthorhiza* Robx.) *Plant Tissue Culture Letters* 3:28–30.

Murashige, T. and Skoog, F. (1962) A revised medium for rapid growth and bioassays with tobacco tissue
    cultures. *Physiol Plant* 15:473–493.

Nadgauda, R.S., Khuspe, S.S. and Mascarenhas, A.F. (1982) Isolation of high curcumin varieties of turmeric
    from tissue culture In: Iyer RD, ed. Proceedings V Annual Symposium on Plantation Crops, CPCRI
    Kasargod, Kerala, India, pp. 143–144.

Nadgauda, R.S., Mascarenhas, A.F., Hendre, R.R. and Jagannathan, V. (1978) Rapid clonal multiplication of
    turmeric *Curcuma longa* L. plants by tissue culture. *Ind J Exp Biol* 16:120–122.

Nair, M.K., Nambiar, M.C. and Ratnambal, M.J. (1980) Cytogenetics and crop improvement in ginger and
    turmeric. In: Nair, M.K., Premkumar,T., Ravindran P.N. and Sarma Y.R., eds. Proc. National Seminar
    on Ginger and Turmeric, CPCRI, Calicut, Kerala, India, pp. 15–23.

Nayak, S (2000) *In vitro* microrhizome production in four cultivars of turmeric (*Curcuma longa* L.) as regulated
    by different factors. *Spices and Aromatic Plants,* Challenges and opportunities in the new century.
    20–23 September.

Nazeem, P.A. and Menon, R. (1994) Blossom biology and hybridization studies in turmeric (*Curcuma longa* L.) *South Indian Hort,* 42:161–167.

Ngamriabsakul, C., Newman, M.F. and Cronk, Q.C.B. (2003) The phylogeny of tribe Zingibereae (*Zingiberaceae*) based on its (nrDNA) and *trn*l–f (cpDNA) sequences. *Edinb J Bot* 60:483–507.

Nguyen, H.L., Doan, T.D., Tae, H.K. and Moon, S.Y. (2005) Micropropagation of zedoary (*Curcuma zedoaria* Roscoe) — a valuable medicinal plant. *Plant Cell Tiss Org Cult* 81:119–122.

Nirmal Babu, K., and Minoo, D. (2003) Commercial Micropropagation of Spices. In: Ramesh Chandra and Maneesh Misra, eds. *Micropropagation of Horticultural Crops.* Lucknow: International Book Distributing Company, pp. 345.

Nirmal Babu, K., Geetha, S.P., Minoo, D., Ravindran, P.N. and Peter, K.V. (1999) *In vitro conservation of germplasm.* In: S.P. Ghosh, ed. Biotechnology and its application in Horticulture. New Delhi: Narosa Publishing House, pp. 106–129.

Nirmal Babu, K., Minoo, D., Geetha, S.P., Ravindran, P.N. and Peter, K.V. (2005) Advances in Biotechnology of Spices and Herbs. *Ind J Bot Res* 1:155–214.

Nirmal Babu, K., Minoo, D., Geetha, S.P., Samsudeen, K., Rema, J., Ravindran, P.N. and Peter, K.V. (1998) Plant biotechnology — its role in improvement of spices. *Ind J Agril Sci* 68:533–547.

Nirmal Babu, K., Ravindran, P.N. and Peter, K.V. (2000) Biotechnology of spices. In: K.L. Chadha, P.N. Ravindran and Leela Sahijram, eds. *Biotechnology of Horticulture and Plantation Crops.* New Delhi: Malhotra Publishing House.

Nirmal Babu, K., Ravindran, P.N. and Peter, K.V., eds (1997). Protocols for micropropagation of spices and aromatic crops. Indian Institute of Spices Research, Calicut, Kerala, India, pp. 35.

Nirmal Babu, K., Ravindran, P.N., and Sasikumar, B. (2003) Field evaluation of tissue cultured plants of spices and assessment of their genetic stability using molecular markers. Final Report submitted to Department of Biotechnology, Government of India. pp.. 94.

Paisooksantivatana, Y., Kako, S. and Seko, H. (2001) Isozyme polymorphism in *Curcuma alismatifolia* Gagnep. (*Zingiberaceae*) populations from Thailand. *Sci Hortic* (Amsterdam) 88:299–307.

Peter, K.V., Ravindran, P.N., Nirmal Babu, K., Sasikumar, B., Minoo, D., Geetha, S.P. and Rajalakshmi, K. (2002) Establishing *in vitro* Conservatory of Spices Germplasm. ICAR Project report. Indian Institute of Spices Research, Calicut, Kerala, India, pp. 131.

Pimchai, A., Somboon, A.I., Puangpen, S. and Chiara, S. (1999) Molecular Markers in the Identifcation of Some Early Flowering *Curcuma* L. (*Zingiberaceae*) Species. *Ann Bot* (Lond) 84:529–534.

Pierik, R.L.M. (1987) *In vitro culture of higher plants.* Dordrecht: Martinus Nijhott publisher, pp. 101–105.

Prakash, S., Elangomathavan, R., Seshadri, S., Kathiravan and Ignacimuthu, S. (2004) Efficient regeneration of *Curcuma amada* Roxb. plantlets from rhizome and leaf sheath explants. *Plant Cell Tiss Org Cult* 78:159–165.

Prathanturarug, S., Soonthornchareonnon, N., Chuakul, W., Phaidee, Y. and Saralamp, P. (2003) High frequency shoot multiplication in *Curcuma longa* L. using thidiazuron. *Plant Cell Rep* 21:1054–1059.

Prathanturarug, S., Soonthornchareonnon, N., Chuakul, W., Phaidee, Y. and Sarakamp, P. (2005) Rapid micropropagation of *Curcuma longa* using bud explants pre-cultured in thidiazuron-supplemented liquid medium. *Plant Cell Tiss Org Cult* 80:347–351.

Praveen, K. (2005) Variability in Somaclones of Turmeric (*Curcuma longa* L.). Ph.D Thesis. Calicut University, Kerala, India.

Raghu Rajan, V. (1997) Micropropagation of turmeric (*Curcuma longa* L.) by *in vitro* microrhizomes. In: Edison, S., Ramana, K.V., Sasikumar, B., Nirmal Babu, K. and Santhosh, J.E. eds. *Biotechnology of Spices,* Medicinal and Aromatic Crops, Indian Society for Spices, pp. 25–28.

Rahman, M.M., Amin, M.N., Ahamed, T., Ali, M.R. and Habib, A. (2004) Efficient plant regeneration through somatic embryogenesis from leaf base-derived callus of *Kaempferia galanga* L. *Asian J Plant Sci* 3:675–678.

Rahman, M.M., Amin, M.N., Jahan, H.S. and Ahmed, R. (2004) *In vitro* regeneration of plantlets of *Curcuma longa* L. a valuable spice plant in Bangladesh. *Asian J Plant Sci* 3:306–309.

Raju, B., Anita, D. and Kalita, M.C. (2005) In vitro clonal propagation of *Curcuma caesia* Roxb, and *Curcuma zedoaria* Rosc. from rhizome bud explants. *J Plant Biochem Biotechnol* 14:61–63.

Ravindran, P.N., Nirmal Babu, K., Saji, K.V., Geetha, S.P., Praveen, K. and Yamuna, G. (2004) Conservation of Spices genetic resources in *in vitro* gene banks. ICAR Project report. Indian Institute of Spices Research, Calicut, Kerala, India, pp. 81.

Redenbaugh, K., Brian, D.P., James, W., Mary, E., Peter, R. and Keith, A.W. (1986) Somatic seeds: encapsulation of asexual plant embryos. *Biotechnol* 4:9–83.

Rekha, C. and Malik, S.K. (2004) Genetic conservation of plantation crops and spices using cryopreservation. *Indian J Biotechnol* 3:348–358.

Remya, R., Syamkumar, S. and Sasikumar, B. (2004) Isolation and amplification of DNA from turmeric powder. *Br Food J* 106:673–678.

Renjith, D., Valsala, P.A. and Nybe, E.V. (2001) Response of turmeric (*Curcuma domestica* Val.) to *in vivo* and *in vitro* pollination. *J Spices Aromatic Crops* 10:135–139.

Sajina, A., Minoo, D., Geetha, P., Samsudeen, K., Rema, J., Nirmal Babu, K., Ravindran, P.N. and Peter, K.V. (1997) Production of synthetic seeds in few spice crops. In: Edison, Ramana, S., Sasikumar, K.V., Nirmal Babu, B Santhosh K and Eapen J., eds. Biotechnology of Spices, Medicinal and Aromatic Plants, Calicut, India: Indian Society for Spices, pp. 65–69.

Salvi, N.D., Eapen, S. and George, L. (2003) Biotechnological studies of turmeric (*Curcuma longa* L.) and ginger (*Zingiber officinale* Roscoe). *Adv Agricul Biotechnol* 1:11–32.

Salvi, N.D., George, L., and Eapen S. (2001) Plant regeneration from leaf base callus of turmeric and random amplification polymorhic DNA analysis of regenerated plants. *Plant Cell Tiss Org Cult* 66:113–119.

Salvi, N.D., George, L., and Eapen, S. (2000a) Direct regeneration of shoots from immature inflorescence cultures of turmeric. *Plant Cell Tiss Org Cult* 62, 235–238.

Salvi, N.D., George, L., and Eapen, S. (2000b) Plant regeneration from leaf base callus of turmeric and random amplified polymorphic DNA analysis of regenerated plants. *Plant Cell Tiss Org Cult* 66:113–119.

Salvi, N.D., George, L., and Eapen, S. (2002) Micropropagation and field evaluation of micropropagated plants of turmeric. *Plant Cell Tiss Org Cult* 68:143–151.

Sanghamitra, N. and Nayak, S. (2000) *In vitro* microrhizome production in four cultivars of turmeric (*Curcuma longa* L.) as regulated different factors. In: Spices and Aromatic plants — Challenges and opportunities in the new century. Calicut, Kerala, India: Indian Society for Spices.

Sanghamitra, N., and Nayak, S. (2000) *In vitro* multiplication and microrhizome induction in *Curcuma aromatica* Salisb. *Plant Growth Regul* 32:41–47.

Sasaki, Y., Fushimi, H., and Komatsu, K. (2004) Application of single-nucleotide polymorphism analysis of the trnK gene to the identification of Curcuma plants. *Biol Pharm Bull* 27:144–146.

Sasaki, Y., Fushimi, H., Cao, H., Cai, S.Q. and Komatsu, K. (2002) Sequence analysis of Chinese and Japanese Curcuma drugs on the 18S rRNA gene and trnK gene and the application of amplification-refractory mutation system analysis for their authentication. *Biol Pharm Bull* 25:1593–1599.

Sasikumar, B. (2005) Genetic resources of Curcuma: diversity, characterization and utilization. Pant Genetic Resources: Conservation and Utilization, 3:230–251.

Sasikumar, B., Syamkumar, S., Remya and John Zacharia, T. (2004) PCR based detection of adulteration in the market samples of turmeric powder. *Food Biotechnol* 18:299–306.

Shetty, M.S.K., Hariharan, P. and Iyer, R.D. (1982) Tissue culture studies in turmeric. In: Nair, M.K., Premkumar, T., Ravindran, P.N. and Sarma, Y.R., eds. Proceedings of National Seminar on Ginger and Turmeric, Calicut: CPCRI, Kasaragod, pp. 39–41.

Shirgurkar, M.V., John, C.K. and Nadgauda, R.S. (2001) Factors affecting in vitro microrhizome production in turmeric. *Plant Cell Tissue Organ Cult* 64:5–11.

Sunitibala, H., Damayanti, M. and Sharma, G.J. (2001) *In vitro* propagation and rhizome formation in *Curcuma longa* Linn. *Cytobios* 105:71–82.

Syamkumar, S., Lowarence, B. and Sasikumar, B. (2003) Isolation and amplification of DNA from rhizomes of turmeric and ginger. *Plant Mol Biol Rep* 21:171–171.

Thomas, K.A., Martin, K.P. and Molly, H. (1996) *In vitro* multiplication of *Alpinia calcarata* Rosc. *Phytomorphology* 46:133–138.

Tyagi, R.K., Bhat, S.R. and Chandel, K.P.S. (1996) *In vitro* conservation strategies for spices crop germplasm-*Zingiber*, *Curcuma* and *Piper* species. In: Proc. 12th Symp. Plantation Crops (PLACROSYM-XII) Kottayam, India: pp. 77–82, 27–29 Nov.

Tyagi, R.K., Yusuf, A., Dua, P. and Agrawal, A. (2004) *In vitro* plant regeneration and genotype conservation of eight wild species of *Curcuma*. *Biol Plant* 48:129–132.

Vasil, I.K., and Vasil, V. (1980) Isolation and culture of protoplasts. In Vasil I K, ed. Perspectives in Plant Cell and Tissue Culture. International Review of Cytology (Supplement) IIB 10–19.

Vincent, K.A., Mary, M. and Molly, H. (1992) Micropropagation of *Kaempferia galanga* L. — a medicinal plant. *Plant Cell Tissue Org Cult* 28:229–230.

Withers, L.A. (1980.) Tissue culture storage for genetic conservation. IBPGR Technical Report. International Board for Plant Genetic Resources, Rome: pp. 91.

Withers, L.A. (1985) *Cryopreservation of cultured cells and meristems.* In: Vasil, I.K, ed. Cell culture and somatic cell genetics of plants, Vol.2. Cell growth, nutrition, cytodifferentiation and cryopreservation. Orlando, Florida: Academic Press, p. 253–316.

Withers, L.A. (1991) Biotechnology and plant genetic resources Conservation. In: Paroda, R.S and Arora, R.K., eds. Plant Genetic Resources Conservation and Management — Concepts and Approaches. New Delhi: IBPGR, pp. 273–297.

Withers, L.A. and Williams, J.T. (1986) In vitro conservation; International Board of Plant Genetic Resources — Research Highlights. 1984–8, Rome: IBPGR, pp. 21.

Xia, Q., Zhao, K.J., Huang, Z.G.,,Zang, P., Dong, T.T., Li S.P., and Tsim, K.W. (2005) Molecular genetic and chemical assessment of Rhizoma Curcumae in China. *J Agric Food Chem* 27;53:6019–6026.

Yasuda, K., Tsuda, T., Shimiju, H. and Sugaya, A. (1988) Multiplication of Curcuma sp. by tissue culture. *Planta Med* 54:75–79.

5 Agronomy of Turmeric

K. Sivaraman

CONTENTS

5.1 INTRODUCTION

Agronomy is the art and underlying science of handling the crops and soils to produce the highest possible quantity and quality of the desired crop products from each unit of land and soil, water and light (Ball, 1925), with a minimum use of resources in raising productivity associated with profitability, lower costs, and sustained competitiveness. Crop management and its scientific study — agronomy — encompasses the physical elements of the climate, soil and land, the biological constituents of the vegetation and soil, the economic opportunities and constraints of markets, sales and profit, and the social circumstances and preferences of those who manage crop production. Such a management acts directly on a part of a plant, a whole plant, or a small group of plants in a stand, or else an amount of soil that can be worked by a person, animal, or machine. Each act of management influences the physiological processes of the plants, which, in turn, modify or regulate the flow of environmental resources — sunlight, water, and nutrients — to economic or useful products. However, management and its effects influence and are influenced by processes and events at much smaller and larger scales. Any human influence on a plant organ affects the metabolic pathways that give the organ its nutritional and biochemical character, and any disturbance to the soil ultimately alters its minute architecture and the organisms that inhabit it. Conversely, the climate, the topography of the land, and the economic and social factors such as markets and roads determine the type of management that is possible and feasible. Many individual acts of management together influence the pattern of land usage, the fertility of the land and its local microclimate, and the wealth and health of people and nations (Gregory, 1988; Loomis and Amthor, 1999; Azam-Ali and Squire, 2002).

The tropics have the potential to be the most productive cropping environments in the world. Plants need heat, light, and moisture to grow, and all of these are available in abundance in the tropics. Where rainfall is sufficient, crops can be grown year round, rather than only in the warm seasons, as in temperate regions. And yet, despite these natural advantages, yields in tropical cropping systems, in general, are often abysmally low. The unpredictability of the climate — in particular, the timing of the rains — and the lack of nutrients for plant growth in many soils combine to limit crop production in the tropics.

Agricultural production systems in the tropics depend heavily on the character of production, i.e., whether crops are produced in a subsistence or a market economy. One of the main features of subsistence farming is that the farmer has to produce in order to live. Consequently farmers often resist changing production methods since, when the changes turn out to be unproductive, their livelihood and survival are threatened. The way farming is organized further depends on the level of technology and the land area available. At high level of technology and land abundance, generally, a high level of mechanization, uniformity of land, soil fertility, and genotypes are needed. On the other hand, when land is scarce, cropping systems tend to be more intensive and less mechanized (Beets, 1990).

Turmeric, the dried underground rhizomes of *Curcuma longa* L (Syn. *Curcuma domestica* Val.,) is one of the horticultural and commercial crops produced in India as well as in other tropical regions of the world. Though turmeric is a tropical spice, it is also grown under subtropical conditions in India, Pakistan, Nepal, China, and Japan. The major importers of Indian turmeric are the Middle East countries, Japan, and U.S. The International Trade Centre (ITC), Geneva, has estimated an annual growth rate of 10% in the world demand for turmeric. Turmeric is also grown in Bangladesh, Pakistan, China, Myanmar, Sri Lanka, Indonesia, Malaysia, Thailand, Taiwan, Iran, Vietnam, West Indies, Argentina, Brazil, Cuba, Peru, Venezuela, Haiti, Liberia, Nigeria, Sudan, and Maldives, to a smaller extent, mostly in home gardens usually for home consumption. Cultivation of turmeric is governed by the prevalence of suitable climatic conditions occurring within a given region. India is the single largest producer, consumer, and exporter of turmeric in the world, with an annual production of about 650,000 t from an area of about 1,61,000 ha. The production has moved up with an annual growth rate of 7.6% and area with 2.8% from 1970.

There is a lack of published information on the agronomical aspects of turmeric in general and weather relationships with turmeric in particular. Results of most of the systematic work on agronomical aspects of turmeric are available from India. The status of scientific study on agronomical aspects of turmeric in other growing countries is meager, except a few publications on cultivation of turmeric from Bangladesh (Ahmed and Rahman, 1987; Choudhury et al., 2000; Islam et al., 2002; Alam et al., 2003), Brazil (Pereira and Stringheta, 1998; Bambirra et al., 2002; Silva et al., 2004), China (Chiu et al., 1993; Hu et al., 1996; Li and Oin, 1996; Li et al., 1996, 1997, 1999a,b), and Japan (Ishimine et al., 2003, 2004; Hossain, 2005; Hossain et al., 2005a, 2005b).

5.2 CROP ENVIRONMENT

Growth and development of turmeric rhizome and leaves are dependent on several factors, such as nutrition, cultivation practices, genotype, and environmental factors. The crop endures an annual average rainfall of 640 to 4200 mm and optimum annual mean temperatures of 18.2 to 27.4°C.

The seed rhizomes planted in the field take about a month to produce new shoots. The weather during this period had no significant effect on yield, and it is probably significant only after emergence of the crop (Kandiannan et al., 2002a). However a temperature range of 25 to 35°C is optimum for the sprouting of turmeric rhizome buds, and sprouting does not occur below 10 or above 40°C. Seedlings elongate well in the temperature range between 25 and 30°C, but do not survive above 40°C (Ishimine et al., 2004).

Turmeric grows luxuriantly in shades, but it produces larger and better rhizomes in the open ground exposed to sun (Ridley, 1912). Turmeric comes up well under partial shaded conditions, but thick shade affects yield adversely (Sundararaj and Thulasidas, 1976; Singh and Edison, 2003). Growth parameters showed a positive beneficial effect up to 25 and 50% shade, respectively. The yields at 25, 50, and 70% shade levels expressed as percentage of that in the open were 74, 55, and 30% on fresh weight basis, respectively. The general trend indicated the superiority of full light in most of the varieties tested (KAU, 1991; Satheesan and Ramadasan, 1988) since most of the photosynthates of shade grown plants were utilized for shoot growth affecting rhizome development significantly and limiting the productivity of turmeric grown under shade (Sivaraman, 1992).

With access to long sequences of historical weather data, crop models could be an excellent tool for assessing the production variability associated with weather for various strategies (Thornton and Wilkens, 1998; Matthews and Stephens, 2002). Crop–weather models using mathematical or statistical techniques provide a simplified representation of the complex relationships between weather/climate and crop performance (growth, yield, and yield components) (Baier, 1979). Crop–weather relations, in general, and for other crops are discussed in detail (Baier, 1977; Frere and Popov, 1979; Monteith, 1981; WMO, 1982; Venkatraman and Krishnan, 1992; Jain and Ranjana, 2000). Management of spice crops, including turmeric, is governed by the prevalence of requisite climatic conditions occurring within a region, and information on crop–weather relationships on these crops is meagre (Venkatraman and Krishnan, 1992). Prediction of yield of turmeric is useful for the policy framers and traders in future markets. Based on the data for 20 years on area and production of turmeric in Coimbatore District, Tamil Nadu, India, Kandiannan et al. (2001, 2002a) developed a yield prediction model ($Y = -11675.5119 - 591.0167 EVPN_3 + 810.3569 TMIN_5 + 12.1481 RAINF_2 + 91.7499 RHMIN_9$) ($R^2 = 0.8181$) and reported that calculated absolute values did not differ much from the actual yield, and similar models could be developed for other agroclimatic zones for the prediction of turmeric yield/production. Among the variables tested in this model, second month (July) rainfall, third month (August) evaporation, fifth month (October) minimum air temperature (TMIN), and ninth month (February) minimum relative humidity (RHMIN) account for maximum yield variability. Turmeric takes one month to germinate after planting and the crop is in the establishment stage during the second month. Rainfall (RAINF) during this time had a significant correlation ($r = 0.6024$) with yield. Probably, the rainfall during this period helps in better establishment. In the third month, low rainfall coupled with high wind speed enhances the evaporation rate and induces water stress, resulting in a negative relation ($r = -0.4889$) of evaporation (EVPN) with yield. Low temperature might favor the rhizome formation as indicated from the study that TMIN of the fifth month had significant association ($r = -0.4538$) with yield. Similarly, RHMIN of ninth month contributes toward yield ($= 0.6575$). Thus, the environmental effects on the yield of turmeric could be studied by developing simulation models (Penning de Vries, 1983; Ritchie, 1993).

Li and Oin (1996) and Li et al. (1996, 1997, 1999a, 1999b) developed a simulation model using five main parameters that affect fresh rhizome yield. They made use of the second-order rotative regression design for the purpose of developing a simulation model. They reported that sowing time, plant population, and potassium play an important role in increasing the fresh rhizome yield. Sowing time is interrelated with plant population, and plant population is clearly interrelated with the fertilizer used. Both N and K are important in raising the yield. The relative effect of agronomic measures on the tuber yield is plant density > phosphorous > sowing date > nitrogen > potassium.

5.3 SOILS

Turmeric thrives well on loose and friable, well-drained, loamy or alluvial soils. Coarse or heavy soils hinder rhizome development. The crop is raised without irrigation, where rainfall is bimodal and ample and with irrigation in plains where rainfall is low and unimodal. The crop is sensitive to saline soils as well as saline water irrigation. It grows in soils with a pH range of 4.3 to 7.5. The crop is grown up to 1200 m above the mean sea level. Turmeric is cultivated as a component crop in rotation with rice, sugar cane, cotton, and banana in southern states of India. It is raised as a sole crop or intercropped with cereals, pulses, vegetables, etc. It is also raised with tree-based cropping/alley cropping/agroforestry systems, and in hill slopes along with grass hedges in jhum cultivation in northeastern states of India. It is intensively grown in Bangladesh in the highland with sandy loam soil (Sivaraman and Palaniappan, 1994; Datta and Dhiman, 2001; Islam et al., 2002; Kandiannan et al., 2002b, 2002c; Nybe et al., 2003; Sarma et al., 2003, Pradhan et al., 2003; Bandyopadhyay et al., 2003; Singh and Rai, 2003; Chenchaiah et al., 2002a, 2002b). Li et al. (1999a, 1999b) employed correlation analysis, path analysis, and cluster analysis to study the type of soil that influence favorably the yield and quality of turmeric. The rhizome yield and quality varied greatly with the types of soil texture. They got clear correlation among the soil factors as well as between the soil factors, rhizome yield, and quality. They also developed a regression equation of the rhizome yield and soil factors.

5.4 TIME OF PLANTING

Though turmeric is a perennial rhizomatous crop, it is domesticated and grown as a rainfed or irrigated annual crop. Adjusting the time of planting to coincide with the rainfall pattern has been a major consideration as it determines the rhizome yield to a large extent. The optimum time for planting turmeric in India varies with varieties, planting materials, and climatic conditions (Randhawa and Mahey, 2002). Planting season for turmeric in India begins in the second fortnight of April and continues up to the first week of July. It has been suggested that turmeric should be planted by the end of April or beginning of May under the conditions prevailing in Punjab, because later plantings, particularly in June, do not allow the crop to attain sufficient growth to withstand the monsoon rains (Randhawa and Nandpuri, 1974; Randhawa and Mishra, 1974). Yield of fresh and dry rhizomes decreased with any delay in planting after the end of April. Fresh and dry yields were 254 and 227% higher, respectively, for plants planted on April 30th than for those planted on June 9th (Randhawa et al., 1984). Plantings of turmeric extend from April to August in various Indian states, being primarily from May to mid-June in Orissa and Assam, in the first three weeks of June in Madhya Pradesh and Maharashtra, and in May, June through August in Andhra Pradesh and Tamil Nadu (Ilyas, 1978; Balashanmugam et al., 1989). Planting long-duration varieties of turmeric during the first fortnight of June and planting short- and medium-duration varieties during the second fortnight of June and the first fortnight of July are recommended in Andhra Pradesh (Philip et al., 1981). In Andhra Pradesh, delayed plantings reduced the vegetative growth and yield and increased the incidence of leaf spot diseases. In Kerala, turmeric is planted during April and May, beginning with the onset of premonsoon showers, whereas in Brazil, the time of planting is from December to January (Silva et al., 2004) and March to May in Japan (Ishimine et al., 2004). The time of planting in various regions are adjusted in such a way to suit the prevailing environmental conditions in a region. Working under the conditions existing in China, Li et al. (1999a, 1999b) reported that planting time is a major parameter influencing fresh rhizome yield.

5.5 PLANTING MATERIAL

Turmeric is propagated vegetatively through rhizomes, and both the mother and the finger rhizomes are used for planting. Plants raised from mother rhizomes may yield seedlings 50% larger than those raised from fingers, but this difference is not consistent (Desai, 1939). The higher seed weight of the mother rhizomes promotes seedling vigor and early growth of the plant that is responsible for increased yields (Ibrahim and Krishnamurthy, 1955; Philip et al., 1981; Singh et al., 2000; Datta et al., 2001; Meenakshi et al., 2001), but using the mother rhizomes requires a higher seed rate (more turmeric per unit area of land). The possibility of decreasing the higher seed rate needed with mother rhizomes has been tested by slicing whole rhizomes in half longitudinally. Studies at the turmeric research station, Padapalam (Andhra Pradesh), indicated that mother rhizomes planted whole or split halves produce more vigorous sprouts than fingers. The initial superiority of the whole-mother rhizomes is sustained throughout the growing season, but differences in vigor and growth with split-mother rhizomes and fingers eventually disappear. However, a reduction in yield of rhizomes to the extent of 20 and 30%, respectively, for planting of split-mother rhizomes and finger rhizomes, have also been reported in comparison to using whole-mother rhizomes (Patil and Borse, 1980). Sarma and Ramesam (1958), using finger and mother rhizomes (whole and sliced halves) as a planting material, recorded a significantly higher production of rhizomes where fingers were used, as opposed to whole-mother rhizomes. Sarma and Murty (1961) used whole, one-half, one-third, and one-quarter size mother rhizomes (sliced transversely into equal-sized pieces) as planting material and reported that a reduction in seed size to thirds without economic loss was possible. Given the choice, the use of mother rhizomes for seed material is generally preferred to the finger rhizomes, although a higher seed rate is required as the mother rhizomes are larger. Economically, the extra costs associated with planting whole-mother rhizomes may offer no special advantage over split-mother rhizomes or fingers. The use of rhizome halves is recommended to solve any problem of seed scarcity.

Aiyadurai (1966) states that mother rhizomes were found to be better than finger rhizomes in Orissa. In Andhra Pradesh, mother rhizomes when planted as whole or even as split halves, gave rise to more vigorous sprouts than the fingers. But fingers keep better in storage and remained more tolerant to wet soil conditions, and seed rate can be lower. Though the research results favor planting of whole-mother rhizomes (Umrani et al., 1980; Philip, 1984), planting primary fingers is the common practice with the major turmeric-growing belts. It might be due to lower requirement of planting material. Primary fingers were also reported to be efficacious, and the cost of seed material was cheaper in Andhra Pradesh (Anjaneyulu and Krishnamurthy, 1979). As a seed source, fingers have been observed to keep better in storage, remain more tolerant to wet soil conditions at planting, and involve a 33% lesser seed rate. Thus, by planting first-order (primary) finger rhizomes, instead of mother rhizomes, the grower can save nearly one-third the seed cost, with essentially no reduction in yield. The market value of cured mother rhizomes is generally less than that of fingers, but the higher (about 4%) curing value for rounds over fingers offsets this loss to some extent (Randhawa and Mahey, 2002). However, whole- or split-mother rhizomes are used for planting, and well-developed healthy and disease-free rhizomes are to be selected as seed material as recommended by the Indian Institute of Spices Research at Calicut, India (IISR, 2001). However, studies carried out in Japan indicated that seed rhizomes should weigh 30 to 40 g with a larger diameter, and seed-mother rhizomes should be free from daughter rhizomes (finger rhizomes) (Hossain et al., 2005a).

5.6 DEPTH OF PLANTING

Turmeric rhizomes planted at the depths of 8, 12, and 16 cm emerged earlier and more evenly than those planted at a shallower depth (4 cm) in both glasshouse and field experiments conducted in dark red soils of southern Japan. The overall results suggested that rhizomes of turmeric should be planted at a depth of 8 to 12 cm in dark red soil for higher yield and lower weed competition (Ishimine et al., 2003). In India, germination, growth, crop yield, yield attributes, and fresh rhizome yield were positively affected by increasing planting depth (Mishra, 2000). The recommended depth of planting in Tamil Nadu (India) is to dibble in the sides of ridges, 45 cm apart at 15 cm spacing at a depth of 4 cm (TNAU, 2004). In the state of Kerala, turmeric is planted in small pits made with a hand hoe in the raised beds of about 15 cm high in rows with a spacing of 25 × 30 cm and covered with soil or dry-powdered cattle manure. The optimum spacing in ridges and furrows system of planting are between 45 and 60 cm between the rows and 25 cm between the plants. A seed rate of 2500 kg of rhizomes is required for planting 1 ha of turmeric with single seed bit weighing about 15 g (IISR, 2001).

5.7 SEED RATE

The seed rate for turmeric generally ranges between 1000 and 1200 kg/ha, differing slightly for differences in the size and types of rhizomes and plant spacing (Agnihotri, 1949). Rao (1957) observed that usually 1800 kg of mother rhizomes or 1200 kg of fingers are used for planting 1 ha. Adhate (1958) recommended 2000 to 2300 kg/ha of mother rhizome for planting. Rao et al. (1975) observed that about 1000 kg of finger rhizomes are required per ha. Rao (1978) suggested that with a spacing of 30 × 20 cm, 1500 kg of rhizomes are needed per ha. The recommended seed rate for planting turmeric under Tamil Nadu conditions is 1500 to 2000 kg/ha, and Indian Institute of Spices Research recommends about 2500 kg/ha under Kerala conditions (TNAU, 2004; IISR, 2001).

5.8 PLANTING METHODS

Turmeric is planted by the flat bed and the ridges and furrow methods. For the flat bed method, the field is leveled. In the ridges and furrow method, the field is harrowed to produce a series of raised ridges. In Andhra Pradesh, the flat bed system of planting has produced yields 80% higher than the ridges and furrow method. The higher yield is probably due to higher plant densities and more uniform stands of the crop. The flat bed system of planting is also better adaptable to poor drainage situations, and hence could be recommended for growing turmeric in low-lying and ill-drained soils. However, the spacing of turmeric depends on the method of planting. Aiyadurai (1966) observed that flat beds were preferable in Orissa, while in Andhra Pradesh, the broad-ridge method of planting was found to be superior and more profitable than the ridge and furrow method, as the elevated beds provided better drainage. According to Aiyadurai, in Orissa, the spacing of 22.5 × 22.5 cm gave the best yield, whereas in Andhra Pradesh, using the ridge and furrow method of planting, the optimum spacing was found to be 45 and 60 cm between the rows and 22.5 cm between the plants. In heavy black soils of Andhra Pradesh, Rao et al. (1975) recommended a wider spacing of 46 × 23 cm due to excessive vegetative growth. In Tamil Nadu, Ponnuswamy and Muthuswamy (1981) recommended 45 × 20 cm for cultivar Erode local, whereas in Maharashtra, Rajput et al. (1980) advocated 45 × 30 cm for local cultivars. A closer spacing of 22 to 25 cm with high plant densities was suggested in Kerala by Philip (1985) and in Punjab (Randhawa and Misra, 1974). However, Philip (1985) found that further increase in population would decrease the growth, yield, and yield components due to severe competition for nutrients. In Kerala, the land is prepared soon after the early monsoon showers. The soil is brought to a fine tilth by giving about four deep ploughings. Hydrated lime at the rate of 400 kg/ha has to be applied for laterite soils and thoroughly ploughed. Immediately with the receipt of premonsoon showers, beds of 1.0 to 1.5 m width, 15 cm height, and of convenient length are prepared with spacing of 50 cm between beds (IISR, 2001) (Figure 5.1).

An evaluation on the planting of turmeric rhizomes and transplanted seedlings of different ages by Patil and Borse (1980) indicated that *in situ* planting of whole-mother rhizomes produced significantly superior yields over transplants of 30, 37, 44, and 51 day-old seedlings raised from cut-mother rhizomes. The increase in yield by planting rhizomes as compared to transplants may be due to a lack of space during early growth of transplant or a disturbance of root system of the transplant, which induces a setback in the initial growth phase (transplant shock). Randhawa et al. (1984) in a comparison of direct-planted rhizomes and transplanted seedlings at ten-day intervals

FIGURE 5.1 Various stages of land preparation and planting of turmeric a) preparation of beds and planting of turmeric rhizome bits, b and c) mulching of beds after planting, d) growth of turmeric plants after 45 d of planting, and e) second-stage earthing up of beds after 3 months of growth.

from April to June observed that the direct-planted crop yielded an average of 49.7% more than the transplanted seedlings. The possibility of raising turmeric from transplants is of importance, where the crop, due to its long-growing period requirement, cannot be fitted in to existing rotations. The decreased yield, usually associated with delayed planting, could be reduced through transplants (Randhawa et al., 1984). Turmeric planted on June 24th, using transplants, yielded 83% higher than a crop directly planted on June 9th. The cost of production should be reduced if the crop is to be in the field for a shorter period due to transplanting. Studies in Japan indicated that for reducing weed interference and obtaining higher yield, turmeric should be planted in a 30 cm triangular pattern on two-row ridge in a 75 to 100 cm width (Hossain et al., 2005b). A spacing of 45×10 cm was suitable, but economically 45×20 cm² plant spacing is viable for turmeric production in Bangladesh (Islam et al., 2002).

5.9 MULCHING

Mulch is a cover to the soil surface. It may be comprising plant residues from the previous crop or imported for the purpose, e.g., straw, wood dust, saw dust, gravel, or plastic sheeting. The effect of mulch is complex. Reduction in soil water loss occurs not only because the mulch acts as a barrier preventing loss, but also because the soil radiation balance and its thermal regime are usually altered, thus influencing the evaporation rate at the surface. The most usual mulch material is plant residues. They may be ineffective at reducing evaporation rates if present only as a thin layer. Usually very rapid evaporation from wet soil is prevented, but slow drying may continue thereafter. The effect of the mulch may therefore be beneficial only where frequent wetting occurs. The advantages of mulching for preserving soil water have to be weighed against the disadvantages. The surface of the plant residue mulch is usually more reflective than the soil surface, and therefore the soil remains cooler than in the absence of the mulch. Mulches of plant residues may harbor pests and weed seeds, which will cause problems later and which should be weighed against the advantages of mulching.

Due to slow germination (2 to 4 weeks) and slow initial growth, different types of mulches and short-duration fast-growing intercrops have been tried to protect the sprouts and also to conserve moisture, besides reducing the weed population before turmeric makes good growth. In Ananthara-jupeta (India), mulching with dried mango leaves was found beneficial. Mulching reduced weed growth and enhanced germination. In Orissa, mulching with daincha (*Sesbania aculeata*) and sunhemp (*Crotalaria juncea*) leaves was found beneficial (Aiyadurai, 1966). In Bihar, increased yields were recorded in turmeric with the application of *Dalbergia* leaves as mulch (Jha et al., 1983).

In the Indian state of Orissa, application of 12 t of sal (*Shorea robusta*) (green leaves with small branches) and 10 t of ashes are applied as mulch per acre (Nair, 1946). It has been suggested that after planting turmeric, the field may be covered with palas (*Butea* spp.) leaves until the shoots emerge (the leaves by this time will have decomposed) (Choudhri, 1952). Mulch should be applied soon after turmeric is planted to encourage early sprouting and to control weed growth (Rao et al., 1975). Dusting with wood ash is advocated periodically to supply potassium to the crop (Choudhri, 1952). The material used for mulching should be cheap and available in adequate quantities. Mulching with daincha and sun hemp is as good as mulching with sal leaves (Philip et al., 1981). Application of a mulch of paddy husk or wheat straw increased the rhizome yield by 59.5 and 21.8%, respectively, compared with unmulched plots (Mahey et al., 1986).

A long-term study was conducted since 1995, in Orissa, to compare the relative efficacy of different types of mulch materials (paddy straw, grasses, and gliricidia) applied at 6, 8, and 10 t/ha on the performance of rainfed turmeric. Application of mulches at 10 t/ha conserved more moisture and increased the yield of turmeric by 12%. Application of paddy straw mulch resulted in 18% increase in yield over gliricidia mulch. The quality of mulch was more effective in conserving soil moisture and increasing the growth and yield of turmeric (Kumar and Savithri, 2003). There was no significant difference in rhizome yield and growth due to application of fresh sal leaf or paddy straw as mulch (Mishra et al., 2000). Application of wheat straw mulch significantly improved growth and yield of turmeric by 46 and 44% in Punjab during 1995 and 1996, respectively (Mohanty et al., 1991; Gill et al., 1999).

On a previously deforested area in Pottangi, Orissa, turmeric as a sole crop without mulching (control), as an intercrop under a canopy of guava (*Psidium guajava*), pummelo (*Citrus grandis*), rose apple (*Syzygium jambos*), silver oaks (*Grevillea robusta*), mango (*Mangifera indica*), and coffee (*Coffea arabica*), with the natural leaf fall as a mulch or as a sole crop with dry leaf mulching, soil erosion was significantly lower in the intercropping (0.235 t/ha soil washed away) and mulching (0.318 t/ha) treatments than in the control (0.936 t/ha). Both treatments also substantially increased the organic carbon, and the available P and K in the soil after harvest, and significantly increased plant height, number of tillers/plant and leaves/tiller, and rhizome yield, compared to the control (Mohanty et al., 1998). Similar results have been reported by Grewal et al. (1995) on the turmeric growth, in agroforestry systems near Chandigarh, India.

5.10 EMERGENCE

The rate of emergence in turmeric depends upon soil moisture, variety, and size and type of the seed rhizomes. Sprouting starts about 2 weeks after planting and may continue over 1 month. An aerial shoot generally emerges from a seedbed within 15 days of planting under irrigated conditions and within 30 days under rainfed conditions. Sprouting generally increases with an increase in the size of the rhizomes, while a reduction in the size of the seed rhizome tends to be associated with a higher mortality of seedling due to infection through the cut rhizomes surfaces. The infection is lower in lengthwise division than in cross section of tissues. The dormant bud of turmeric rhizomes requires about 15 days to sprout, and the growth of the root starts later. Treatment of rhizomes with naphthalene acetic acid (NAA) for 30 min in 0.5 or 1 ppm significantly lowered germination rate in turmeric rhizomes, and the emergence was reduced by over 25%. Treatment of rhizomes with thiourea, Agallol, auxin, and gibberellic acid did not affect germination (Randhawa and Mahey, 1984).

5.11 NUTRITION

Supply and absorption of chemical compounds needed for growth and metabolism may be defined as nutrition, and the chemical compounds required by an organism are termed as nutrients. The mechanism by which nutrients are converted into cellular constituents or used for energetic purposes is the metabolic process. The term "metabolism" encompasses the various reactions occurring in a living cell in order to maintain life and growth. Nutrition and metabolism are thus very closely interrelated. For growth and crop production, one may distinguish between organs in which primary organic molecules are produced and organs in which organic molecules are stored or consumed. In crop physiology, the former are called sources and the latter physiological sinks (Mengel et al., 2001). Plants suffer nutrient deficiency stress when the availability of soil nutrients (and/or the amount of nutrients taken up) is lower than that required for sustaining metabolic processes in a particular growth stage. Deficiency may occur as a result of (1) an inherently low amount of nutrients in the soil, (2) low mobility of nutrients in the soil, or (3) poor solubility of given chemical forms of the nutrients. Plant genotypes differ in the capacity to convert nonavailable forms of nutrients to available forms and to take them up. Factors underlying the differential capacities of plant genotypes to access soil nutrients include differences in the surface area of contact between roots and soil, the composition and amount of root exudates, and rhizosphere microflora, all of them together resulting in differences in the chemistry and biology of the rhizosphere (Rengel and Marschner, 2005). For agronomic purposes, the mineral macronutrients required from soil are separated arbitrarily into two groups: (1) nitrogen, phosphorus, and potassium are taken up by plants in moderate-to-large amounts, deficiencies are common, and they are the major constituents of commercial fertilizers and (ii) calcium, magnesium, and sulfur are taken up by plants in moderate amounts, and, although deficiencies are less common, they can be very important regionally. The micronutrients, iron, manganese, copper, zinc, boron, molybdenum, and chlorine, also known as trace elements, are believed to be required by all higher plants, whereas requirement for cobalt has only been established for biological nitrogen-fixing systems, as in legume nodules. Requirement of sodium and silicon for crop plants is generally considered to be beneficial rather than essential.

Turmeric generally responds to increased soil fertility by producing higher yields, but the quantity of fertilizers (inorganic or organic) required to produce a crop varies with variety, soil, and weather conditions prevailing during crop growth. Crop growth period of turmeric could be divided into four phases, viz., a phase of moderate vegetative growth, a phase of active vegetative growth, a period of slow vegetative growth, and a phase-approaching senescence. Generally, as dry matter increases, the uptake also increases, and the phase of active vegetative growth is also the period during which maximum uptake of nutrients takes place. It has been observed that the uptake of nutrients was higher up to the third month for potassium, up to the fourth month for nitrogen, and up to the fifth month for phosphorus with subsequent decrease. The crop attains maximum vegetative growth during the fourth and fifth month, suggesting the need for earlier application of N, P, and K for increasing the plant growth (Rao and Rao, 1988). It has been reported that turmeric removes 16.5 kg of N, 3.1 kg of P_2O_5, and 44.5 kg of K_2O per t of produce, and the ratio of K_2O and P_2O_5 removal relative to N works out to 100, 19, and 270, respectively (PPIC, 2001). It was

TABLE 5.1
Uptake of Nutrients by Turmeric at Harvest (kg/ha)

Location/Soil Type	Nutrients					Refs.
	N	P$_2$O$_5$	K$_2$O	Ca	Mg	
Kasaragod-Laterite	124	30	236	73	84	Nagarajan and Pillai (1979)
Vellanikkara-Laterite	72–115	14–17	141–233	—	—	KAU (1991)
Bhavanisagar-Sandy loam	166	37	285	—	—	Sivaraman (1992)
Coimbatore-clay loam	187	39	327	—	—	Sivaraman (1992)

Source: Rethinam et al., 1994.

noted that the rhizome yield and nutrient uptake by the crop were significantly higher following inoculation of either single or combined inoculants of *Azotobacter* and *Azospirillum*, irrespective of fertilizer application. This increase was greatest with 60 kg N/ha. The percentage increase in yield due to inoculation integrated with fertilizer N ranged from 15.2 to 30.5%. The N-use efficiency was enhanced due to inoculation and showed higher values with 30 kg N than with 60 kg N/ha. The soil was left with a positive N-balance in the integrated treatments only, indicating a buildup of soil fertility (Jena et al., 1999). Fertilizer N balance calculations clearly indicated that the recovery of N in turmeric was high in favor of four splits at 180 days growth stage (19.46%) as well as at harvest (30.76%) (Jagadeeswaran et al., 2004). However, turmeric is a heavy feeder of nutrients as seen from the data on nutrient uptake in Table 5.1.

As turmeric is also raised as a component crop, the intensity of shade by the other component crop also influences uptake of nutrients. However, a 2-year study in Kerala did not show a conclusive trend. Increasing shade levels (0, 25, 50, and 75% shade) resulted in a steady decrease in nutrients uptake, whereas during the subsequent year, the trend was reversed. The range of uptake of N varied from 66 to 137 kg/ha, P from 10 to 20 kg/ha, and K from 158 to 314 kg/ha, depending upon the level (KAU, 1991; Meerabai et al., 2000).

Location-specific fertilizer recommendations have been made in different states based on the prevailing agroecological conditions, variety, and the crop management. It varies widely from 30 kg in Kerala to 375 kg N/ha in Andhra Pradesh, 30 kg in Kerala to 200 kg P$_2$O$_5$/ha in Andhra Pradesh, and 50 kg in Bihar to 200 kg K$_2$O/ha in West Bengal (Rao, 1973; Kundu and Chatterjee, 1981; Randhawa et al., 1984; Rao et al., 1975; Rao and Reddy, 1977; Shankaraiah and Reddy, 1987; Panigrahi et al., 1987; Umate et al., 1984; TNAU, 2004; KAU, 1993; IISR, 2001; Committee on Spices, 1989; Thamburaj, 1991; Yamgar and Pawar, 1991; Medda and Hore, 2003; Venkatesh et al., 2003; Majumdar et al., 2002; Attarde et al., 2003; Dubey and Yadav, 2001; Upadhyay and Misra, 1999). Spraying 1.4% MgSO$_4$ ameliorated the deficiency of magnesium in Meghalaya, India (Chandra et al., 1997) (Table 5.2).

5.12 MICRONUTRIENTS

Micronutrient requirements are steadily increasing worldwide, in response to intensive cultivation practices. The use of high-yielding cultivars that remove high amounts of micronutrients from the soil leads to micronutrient deficiencies in many countries. Among the micronutrients reported as becoming deficient worldwide, zinc (Zn) and iron (Fe) are the most important elements.

Iron is an important micronutrient involved in various photosynthetic processes, namely chlorophyll biosynthesis, photosynthetic electron transport system, formation of chloroplast ultrastructure, photosynthetic enzyme activity, and carotenoid content. Structure and function of the whole photosynthetic apparatus is affected by iron deficiency. Iron deficiency is usually observed in turmeric grown in calcareous or alkaline soils. The presence of high amount of phosphate may also

TABLE 5.2
Fertilizer Recommendations for Turmeric in Different States of India

State	Recommendations
Andhra Pradesh	200–300 kg N, 125–150 kg P_2O_5, 100–150 kg K_2O in three Splits viz. 60 N, 60 P, and 60 K kg/ha as basal dose, 60 N kg/ha and 65 K kg/ha at 60 DAS, 60 kg N/ha at 120 DAS
Tamil Nadu	200:200:200 kg NPK/ha as basal and top dressing with 25 kg N/ha + 18 kg K /ha at 30, 60, 90th, and 120th DAS
Orissa	60:30:90 kg NPK/ha in two split dose as basal, whole FYM and half K on furrows before planting (7.2 kg/m^2 and green mulch 2.5 kg/m^2), half N at 45 DAS, half N and half K at 90 DAS
Kerala	30:30:60 kg NPK/ha, full and half applied as basal dose, two third after 30 d and rest N and 100:50:150 kg NPK/ha at 60 DAS
Bihar	100:50:150 kg NPK/ha. Application of 15-kg/ha iron sulphate and 20 kg/ha of zinc sulphate is also recommended
Chattisgarh	150:125:125 kg NPK/ha, in three split doses 30, 60, and 90 d after planting

induce this condition in acid soil. Iron deficiency in six varieties (Rashmi, Krishna, Roma, CL-315, CEL-6, and CL-70) studied under culture in glasshouse conditions exhibited decrease in plant growth, fresh weight, rhizome size, photosynthetic rate, and chlorophyll content, whereas curcumin content increased significantly in all the genotypes under Fe deficiency. Fe deficiency resulted in higher accumulation of sugars, amino acids, and organic acids in leaves. This is due to poor translocation of metabolites. Roots and rhizomes of Fe-deficient plants had lower concentrations of total photosynthates, sugars, and amino acids, whereas organic acid concentration was higher in rhizomes. $^{14}CO_2$ incorporation in essential oil was lower in the youngest leaf as well as incorporation in curcumin content in rhizome (Dixit and Srivastava 2000a, 2000b). Application of $FeSO_4$ at 30 kg/ha recorded the highest yield of rhizomes (24% more than control) followed by foliar spray of 0.5% $FeSO_4$ during third, fourth, and fifth month after planting (Balashanmugam et al., 1990).

Zinc deficiency is a common micronutrient deficiency in plants growing in different climatic regions of the world, particularly in arid and semiarid regions where alkaline soils predominate. Several physiological processes are impaired in plants suffering from Zn deficiency. Zinc deficiency causes rapid inhibition of plant growth and development, and thus of final yield. Zinc plays a fundamental role in several critical cellular functions, such as protein metabolism, gene expression, structural and functional integrity of biomembranes, photosynthetic-C metabolism, and Indole 3 acetic-acid (IAA) metabolism (Marschner, 1995). Compared with other micronutrients, Zn exists in biological systems in high concentrations, particularly in biomembranes.

Zinc deficiency in turmeric can be identified by the occurrence of light green, yellow, or white areas between the veins of leaves, particularly the older ones. Other symptoms include small, narrow, thickened leaves, early loss of foliage, and stunted growth (Rethinam et al., 1994; Tiwari et al., 1995). Application of $ZnSO_4$ at 15 kg/ha increased the rhizome yield by 15% over control (Velu, 1988; Balashanmugam et al., 1990; Thamburaj, 1991).

Although the curcumin content of rhizome increased due to boron deficiency, the overall rhizome yield and curcumin yield decreased. The influence of boron deficiency on leaf area, fresh and dry masses, CO_2 exchange rate, oil content, and rhizome and curcumin yields can be ascribed to reduced photosynthate formation and translocation (Dixit et al., 1999; 2000; 2001; 2002a; 2002b). For correcting deficiency of micronutrients especially boron, iron, and zinc at rhizome development stage, application of 375 g $FeSO_4$, 375 g $ZnSO_4$, 375 g borax, 375 g of urea in 250 l of water/ha spraying twice at a 25-day interval is recommended. The above micronutrients are dissolved in super phosphate slurry (15 kg super phosphate is dissolved in 25 l of water stored overnight, and the supernatant solution is made up to 250 l). In this solution, the micronutrients are added (TNAU, 2004).

5.13 ORGANIC MANURING

Soil organic matter (SOM) refers to the sum total of all organic carbon-containing substances in the soil. It consists of a mixture of plant and animal residues in various stages of decomposition, substances synthesized microbiologically and/or chemically from the breakdown products, and the bodies of live and dead microorganisms and their decomposing remains. SOM represents a key indicator for soil quality, both for agricultural and for environmental functions. SOM is the main determinant of biological activity. The amount, diversity, and activity of soil fauna and microorganisms are directly related to the organic matter. Organic matter and the biological activity that it generates have a major influence on the physical and chemical properties of soils. Aggregation and stability of soil structure increase with organic matter content. These, in turn, increase infiltration rate and available water capacity of the soil, as well as resistance against erosion by water and wind. SOM also improves the dynamics and bioavailability of main plant nutrient elements (Sanchez and Logan, 1992; Greenland et al., 1992).

Turmeric responds to heavy dressings of organic matter, and many experimental evidences are available on the beneficial effects of organic manures either alone or in combination with inorganic fertilizers on the growth and productivity of turmeric. Under mid-hill conditions of Mizoram, the highest yield and profit were obtained for the application of 50 t/ha of cow dung manure alone, followed by the application of 90 kg N, 60 kg P_2O_5, and 90 kg K_2O/ha in the form of urea, diammonium phosphate, and the muriate of potash, respectively (Saha, 1988). An increase of over 37% in fresh rhizome yield was recorded over control no farm yard manure (FYM) by the application of 25 t FYM/ha (Balashanmugam et al, 1989). Usually large quantities of organic manures in the form of FYM, oilcakes, and green leaves (as mulch) are applied in different turmeric-growing regions. Recommendations differ according to the variety, soil conditions, initial soil test values, and the climatic conditions. Application of tank silt and sheep penning is also practiced in some areas of Tamil Nadu and Andhra Pradesh for improving soil fertility (Rao et al., 1975; KAU, 1993; Nair, 1982; Inamdar and Diskalkar, 1987; Panigrahi et al., 1987; Saha, 1988; Rethinam et al., 1994; Vadiraj et al., 1998; Majumdar et al., 2002; Venkatesh et al., 2003). Sadanandan and Hamza (1998) found highest rhizome yield (4884 kg/ha, averaged over the four cultivars) in NPK fertilizer plots followed by neem cake, groundnut cake, and cotton cake treatments (4818, 4809, and 4623 kg/ha, respectively), or curcumin yield (average of four cultivars) was highest in plots supplied with neem cake (287 kg/ha), followed by cotton cake, groundnut cake, and NPK fertilizer (284, 277, and 268 kg/ha, respectively). Application of organic fertilizers increased nutrient availability and improved soil physical conditions. Economic analysis showed that the application of NPK fertilizer gave the highest benefit:cost ratio, followed by groundnut cake. In the context of growing market for organically produced commodities and sustaining productivity, detailed work on organic farming and integrated management of nutrient, pests including weeds, and diseases need to be carried out. Li and Oin (1996) used regression method of orthogonal substitution with factors to build up a model to study the influence of various organics on rhizome yield.

5.14 FACTORS INFLUENCING NUTRIENT MANAGEMENT

Nutrient management in a region is influenced by the factors such as planting material (Umrani et al., 1982; Tayde and Deshmukh, 1986; Valsala et al., 1988; Rao and Rao, 1988; Singh et al., 1988; Meenakshi et al., 2001; Medda and Hore, 2003; Attarde et al., 2003), spacing, and variety (Shankaraiah and Reddy, 1987; Shashidhar and Sulikeri, 1996; Shashidhar et al., 1997), prevailing weather conditions during crop growth (Kandiannan et al., 2002a, 2002b, 2002c), cropping systems (Balashanmugham et al., 1987; Rao and Reddy, 1990; Jayaraj, 1990; Sivaraman, 1992; Sivaraman and Palaniappan, 1994), soil fertility (Thankamani et al., 1998; Singh et al., 1988,2000; Kumar and Savithri, 2003), and other management practices. It is necessary to apply extra amounts of nutrients to meet the requirements of component crops in the system (Balashanmugham et al.,

1987; Rao and Reddy, 1990; Sivaraman, 1992). Rao and Reddy (1990) reported that sowing maize in every second and third inter-row space of turmeric and maintaining 100% population of both turmeric and maize together with the application of 188 kg N/ha, 70 kg P_2O_5/ha, and 125 kg K_2O/ha for turmeric and 120 kg N/ha, 60 kg P_2O_5/ha, and 40 kg K_2O/ha for maize resulted in higher rhizome yield and net returns compared to the farmers' practice of growing turmeric and maize in every inter-row space without fertilizer application. Similarly, when onion is raised as an intercrop in turmeric, application of 50% higher levels of NPK than that recommended for the monocrop of turmeric (120:60:60 kg NPK/ha) has been suggested (Balashanmugham et al., 1987).

Turmeric performed well under 25% shade with 90 kg K_2O/ha when the entire quantity of K was applied in a single basal dose. The curcumin content was also found to be higher at 25% shade and showed an increase with increase in K levels.

5.15 NUTRIENTS ON QUALITY

Dryage or curing percentage is the proportion of boiled and dried product to the fresh rhizomes, which is influenced by the nutrients application. Though curing percentage is a varietal character, it is also influenced by other factors such as soil moisture, soil fertility, and bright sunshine (NRCS, 1993). Crops raised under higher fertility levels of N, P, and K recorded a lower percentage of curing (Rao and Reddy, 1977). High moisture content and less lignification of the rhizomes are considered to be the probable reasons for the low curing percentage. Higher levels of N, P, and K fertilization have also been reported to decrease the curcumin content of rhizomes (Rao et al., 1975). However, graded doses of potassium as well as nitrogen and phosphorus significantly improved the curcumin content of rhizomes at harvest (Mohanbabu and Muthuswamy, 1984; Ahmed Shah et al., 1988; Meenakshi et al., 2001). It has also been reported that iron and boron deficiencies in some genotypes under glasshouse conditions recorded relatively higher curcumin content (Dixit et al., 1999, 2002).

5.16 CROPPING SYSTEMS

A cropping system usually refers to a combination of crops in time and space. Combination in time occurs when crops occupy different growing periods, and combination in space occurs when crops are interplanted. When annual crops are considered, a cropping system usually means the combination of crops within a year. Cropping system approach seeks to increase the benefits derived from crop production by efficient utilization of both natural (soil, light, and moisture) and socio-economic (labor, credit, and market demand) resources. Turmeric is usually rotated in sequential cropping systems with rice, cotton, sugar cane, onion, garlic, and some cereals in assured irrigated areas (Douglas, 1973).

5.17 RESPONSE TO SOLAR RADIATION

Solar radiation is an essential determinant of crop yield in many ways — via photosynthesis, sunlight fuels crop growth and development. Solar radiation is also a major determinant of the energy balance of the soil and the plants, and drives water and nutrient transport. In addition, light perceived by specific photoreceptors (nonphotosynthetic pigments), including phytochromes, cryptochromes, and phototropin, induces photomorphogenic responses, influencing both the pattern of investment of captured resources and the ability of the plant to capture further resources. The effects of light signals perceived by these photoreceptors can be different for the crop and the accompanying weeds. Consequently, photomorphogenic responses of crop and weed plants are predicted to alter the interactions between these two components of the system.

Ridley (1912) observed that turmeric grew luxuriantly in shade, but it produced larger and better rhizomes in the open ground exposed to sun. Turmeric grows well under partial shade conditions, but thick shade affects the yield adversely (Sundararaj and Thulasidas, 1976). Because turmeric is a long-term crop, fields are usually intercropped with crops of very short duration to ensure some early returns for the farmers even though the intercrop may reduce the yield of turmeric by 5 to 20%. It is customary to grow a number of subsidiary or mixed crops along with turmeric in the same field, generally along the borders and near the irrigation channels (Aiyer, 1947). At least one shade crop is included as turmeric is believed to be a shade-loving crop by the farmers. Castor (*Ricinus communis*) is the one usually selected, but *agathi* (*Sesbania grandiflora*) is also grown in some places (Aiyer, 1949). According to the experience of various farmers and growers in Tamil Nadu, chillies, castor, maize, finger millet, pearl millet, and vegetables such as radish, beetroots, sweet potato, brinjal, and pulses are grown as mixed crops with turmeric, with various population levels and without any row arrangement.

A variety of annual crops have been observed to be intercropped with turmeric viz., cereals (rice, maize, and finger millet), pulses (Pigeon pea, green gram, and black gram), vegetables (bhendi, brinjal, radish, beetroots, and chillies), spices (fenugreek and coriander) (Nair, 1946; Ibrahim and Krishnamurthy, 1955; Wani and Bhandare, 1957; Chaugule and Mohite, 1962; Kundu and Chatterjee, 1981; Singh and Randhawa, 1985).

Turmeric is a component of perennial crop-based cropping systems such as coconut (*Cocos nucifera* L) and arecanut (*Areca catechu* L)-based intercropping and high-density multispecies-cropping systems (Balasimha, 2004; Nybe et al., 2003; Chenchaiah et al., 2002a, 2002b; Bandyo-padhyay et al., 2003; Nath, 2002; Hegde and Sulikeri, 2001; Ray and Reddy, 2001; Anilkumar and Reddy, 2000; Ray et al., 2000; Hegde et al., 1998; Narayanpur and Sulikeri, 1996; Sairam et al., 1997; Sarma et al., 1996; Shanmugasundaram and Subramanian, 1993; Satheesan and Ramadasan, 1988). Turmeric can be successfully grown in forest plantations aged 2 yr or more, providing both substantial income and improved growth of turmeric as compared with unshaded locations (Dagar and Singh, 2001; Chaturvedi and Pandey, 2001; Mishra and Pandey, 1998; Narain et al., 1997a, 1997b; Narain et al., 1998; Grewal, 1996; Singhal and Panwar, 1991; Grewal et al., 1992; Lahiri, 1989; Shrivastava, 1988) and alley cropping and agrisilvi-horticultural system (Pradhan et al., 2003; Koppad et al., 2001; Bisht et al., 2000 & 2004; Taylor, 1993; Jasural et al., 1993). Turmeric may also be seen growing as an intercrop in mango (*M. indica* L) litchi (*Litchi chinensis* Sonn.) and jack (*Artocarpus heterophyllus* Lam.) orchards or plantations, and other fruit orchards (Douglas, 1973; Singh and Ram, 1994 &1995; Singh and Rai, 2003).

5.18 IRRIGATION

For vegetative growth and development, plants require, within reach of their roots, water in adequate quality, in appropriate quantity, and at the right time. Most of the water a plant absorbs, performs the function of raising dissolved nutrients from the soil to the aerial organs; from there it is released to the atmosphere by transpiration; agricultural water use is intrinsically consumptive. Crops have specific water requirements, and these vary depending on local climatic conditions.

In irrigated agriculture, water taken up by crops is partly or totally provided through human intervention. Irrigation water is withdrawn from a water source (river, lake, or aquifer) and led to the field through an appropriate conveyance infrastructure. To satisfy their water requirements, irrigated crops benefit both from more or less unreliable natural rainfall and from irrigation water. Irrigation provides a powerful management tool against the vagaries of rainfall, and makes it economically attractive to grow high-yielding crop varieties and to apply adequate plant nutrition as well as pest control and other inputs, thus giving room for a boost in yields.

Highest yields of turmeric are obtained in well-drained loamy soils under irrigation with quality water. The actual number of irrigations required to produce a good crop depends on the type of soil and prevailing weather conditions during crop growth and rainfall. In Maharashtra, first irrigation is applied at the time of planting, followed by a second irrigation 2 to 3 days later, and subsequent irrigations every 8 to 10 days, depending upon the arrival of monsoon rains, the dryness of the soil, and the temperature, until the last month of crop growth when irrigation is totally stopped (Wani and Bhandare, 1957). Hence, about 17 to 20 irrigations are given in a season, depending upon the rainfall received. A total of 150 to 165 cm of water (irrigation and rainfall) are required to produce a good crop of turmeric (Adhate, 1958).

Most of the water is required during the early part of the crop growth when the rhizomes are vigorously developing (Lakshmanan, 1949). Red, loamy soils are usually irrigated every 5 days, whereas, black, loamy soils are irrigated every 7 to 9 days for optimum growth (Philip et al, 1981). The crop apparently requires 15 to 20 irrigations in clay soils and about 40 irrigations in light sandy soils for growth.

Measurements of the production of turmeric under combinations of three cumulative evaporation rates (U.S. open pan evaporimeter rate) of 40, 60, and 80 mm and three mulches (no mulch, paddy husk, and wheat straw mulch at 6 t/ha) have indicated that irrigation significantly increased the rhizome yield when application of water is scheduled at 40 mm evaporation rate (Mahey et al., 1986). Yields on fields irrigated at 40 mm evaporation rate averaged 36 and 105% above fields irrigated at 60 and 80 mm evaporation rates, respectively. On an average, water expenditure was 193 cm for the 40-mm evaporation rate, 151 cm for the 60-mm evaporation rate, and 140 cm for the 80-mm evaporation rate. Application of mulch saved irrigation water without any adverse effect on yield. Without mulch, 22 irrigations produced 2.9 t/ha of rhizomes with irrigation at 40 mm evaporation rate; with mulch, 14 irrigations produced 6.1 t/ha of rhizomes with irrigation at the 60 mm evaporation rate (Randhawa and Mahey, 2002). Drip irrigation was found to reduce the irrigation water requirement of turmeric 20 to 60% in Tamil Nadu and Maharashtra (Selvaraj et al., 1997; Singte et al., 1997). However, the capital investment as well as the fact that turmeric is an annual crop limits the adoption of rotational cropping.

5.19 WEED MANAGEMENT

Weeds are certainly as old as agriculture, and from the very beginning, farmers realized that the presence of those unsown species interfered with the growth of the crop they were intending to produce. Competition between the undesired plants and the crop was to be avoided, if reasonable yields were to be achieved. Weeds were removed first by hand and then mechanically when new farming tools were developed. During the 1950s, the first hormonal herbicides were developed; from that moment onward, the spectrum of molecules known to have deleterious physiological activity on weeds has increased dramatically and so has the effectiveness of new herbicides.

Many different types of weeds occur in turmeric fields and cause considerable losses to the farmer. This is because weeds compete with the turmeric crop for nutrients, water, sunlight, and space. Weeds may harbor pests and diseases or physically injure turmeric plants and rhizomes. Weeds may also harbor natural enemies that control pests; so certain weeds can be left on turmeric fields, provided they are not many enough to compete with the crop.

The traditional method of weed management, mulching, which is a common practice in turmeric, is being replaced by chemical methods. Three preemergence herbicides, alachlor, dichlormate, and nitrofen at 1.5 and 2 kg/ha each were individually compared with that of hand-weeded and unweeded control. The common weeds were *Trianthema portulacastrum*, *Eclipta alba*, *Euphorbia hirta*, *Phyllanthus amarus*, *Cyperus* spp., *Echinochloa colonum*, and *Echinochloa grusgalli*. The effect of herbicides lasted up to 35 days with alachlor and 25 days with nitrofen and dichlormate. The rhizome weight of plant was more in alachlor, followed by dichlormate at 2 kg/ha. Alachlor at 2 kg/ha gave the highest yield (16.83 t/ha) followed by dichlormate at 2 kg/ha (15.2 t/ha), hand-weeded control (13.52 t/ha), and unweeded check (5.34 t/ha). The cost of chemical weed control was found to be economical as compared to hand weeding (Rethinam et al., 1977). Simazine or atrazine at 2 kg/ha effectively controlled weed growth in turmeric-based cropping systems with pulses; however, preemergence application of alachlor at 2 kg/ha was found to be more effective and economical (Mishra and Mishra, 1982). Higher rhizomes yield 37.9 t/ha with the preemergence application of oxyflourfen (0.15 kg/ha). Unweeded control recorded the lowest yield of 17.4 t/ha. The herbicide had no residual effect on the succeeding groundnut crop. Integrated weed management of preemergence application of herbicide followed by one manual weeding was more economical than farmers' practice of two manual weedings (Balashanmugam et al., 1985).

Mulching with green leaves and straw applied just after planting controlled weeds and enhanced earliness in sprouting of turmeric (Mohanty et al., 1991). The highest gross returns were obtained with fluchloralin plus hand weeding, followed by atrazine plus hand weeding in turmeric plus maize intercropping system (Reddy et al., 1992).

The dominant weed species recorded in turmeric in Punjab were *Digitaria ischaemum, Cynodon dactylon, Cyperus rotundus, Eleusine aegypticum (Dactyloctenium aegyptium), E. hirta, Commelina benghalensis,* and *Eragrostis pilosa.* The herbicide treatments alone did not provide season-long weed control, but the integrated treatments achieved similar levels of control to the hand weeding plus hoeing treatment. The highest percentage increase in yield over the control was achieved by 0.70 kg metribuzin/ha (mean of 17.09 t/ha, compared to 11.62 t/ha in the control), followed by 1.0 kg diuron/ha and the integrated control treatments of diuron, metribuzin, and atrazine (Gill et al., 2000).

Among the six weed control treatments (weedy check, six hand weedings, and preemergence application of atrazine at 1.0 kg/ha, pendimethalin at 1.0 kg/ha, directed spray application of glyphosate at 0.5 kg/ha at 15 days after planting, and preplant incorporation of fluchloralin at 1.5 kg/ha with four hand weedings in all the herbicides treated plots) tried in both turmeric plus maize-intercropping system and sole turmeric, application of pendimethalin at 1.0 kg/ha resulted significantly in lower weed dry matter as compared to atrazine at 1.0 kg/ha, fluchloralin 1.0 kg/ha, and glyphosate 0.5 kg/ha. At 30 days after planting and at maize harvest, the weed dry matter was significantly lower in intercropping system than that of sole crop of turmeric. Application of atrazine at 1.0 kg/ha and pendimethalin at 1.0 kg/ha resulted in maize grain yield on par with hand weeding. The turmeric rhizome yield and equivalent yield were comparable in all the weed control treatments and significantly superior over unweeded control. There was significant decrease in turmeric rhizome and equivalent yield due to maize intercropping (Anilkumar and Reddy, 2000).

In Uttar Pradesh, application of pendimethalin and oxyfluorfen, each followed by hand weeding, resulted in 45 and 39% more fresh rhizome yield, respectively, compared to the weedy control (Singh et al., 2002).

5.20 HARVESTING AND SEED PRESERVATION

Turmeric crop is usually ready for harvest after about 6 to 9 months of growth, depending upon the variety. At maturity, the leaves turn yellow, fade, and, subsequently wither and dry. Maximum rhizome yield and dry rhizomes are obtained at this stage. At harvest, the leaves and stems are cut close to the ground to maximize the removal of vegetative material. Because the rhizomes are dug by manual labor, the fields are irrigated prior to digging to ease and speed the harvesting process. Rhizomes are subsequently cleaned, and fingers are separated from mother rhizomes.

Although the harvest period varies for different areas, in most of the regions of India, harvests begin in December or January and continue through March. The main harvesting period is February to April in Andhra Pradesh, the end of February in Maharashtra, from January to March in Kerala and Tamil Nadu, and January to February in Madhya Pradesh.

Harvesting of turmeric rhizome is labor intensive, requiring skilled labor to dig out the crop. The nonavailability of such skilled labor, the high wages demanded by them to harvest the crop, and the higher field losses and damage to the crop by manual harvesting, necessitate the need to develop a suitable mechanical harvester for turmeric. A prototype harvester with optimized design parameters is being evaluated for its performance and the further line of work is in process (Kathirvel and Manian, 2002).

About 15 to 20% of the harvested rhizomes is retained by the farmers as seed material. Preservation of seed rhizomes is one of the most important aspects in the cultivation of turmeric. At the time of harvest, a requisite quantity of sound, undamaged, well-developed turmeric fingers must be carefully selected and cleaned for preserving as seed. From the time of harvesting of rhizomes (January to February) and till subsequent planting of the crop (May to June), the seed rhizomes are to be stored (90 to 105 d) in healthy and viable conditions. Generally, a pit of 60 cm deep and large enough to hold the seed rhizomes is dug in a cool, shady, dry location. The seed rhizomes are placed in the pit and covered lightly with loose, dry soil and turmeric, banana, or other leaves. Ilyas (1978) observed that the seed rhizomes stored under shade and covered with turmeric leaves plastered together with mud and cow dung. In this manner, the seed rhizomes are preserved in good condition until needed for planting. Gorabal et al. (2002) concluded on the basis of their experiment that the turmeric seed rhizomes could be best stored in a zero energy cool chamber with minimum storage losses due to physiological loss in weight, rotting, and insect damage. Under ambient conditions, rhizomes could be stored in polyethylene bags with 0.5% ventilation in a dry and cool room, and under field conditions, they could be stored in pits lined with wheat straw.

5.21 CONCLUSION

Turmeric is a tropical rhizomatous crop having its importance as a spice, flavoring agent, colorant, and its use in most of the systems of medicine in the treatment of various illnesses. Tropical regions especially India, where more than 90% of the world's turmeric is produced, processed, consumed, and exported, enjoy a competitive advantage in terms of variability, in the availability of germplasm, production and processing technologies with an effective marketing linkage with the farmers, and the traders with the government support in the field of research, development, quality assurance, and export.

Turmeric has been used for centuries as a food additive, spice and food colorant. The medicinal properties of this spice were recognized in Indian folklore medicine and in Ayurveda, which is an ancient Indian traditional system of medicine. Though large number of studies unequivocally identified the numerous pharmaceutical actions of curcuminoids (Khanna, 1999; Chattopadhyay et al., 2004; Joe et al., 2004, Aggarwal et al., 2004), its acceptance as a "wonder compound" is slowly forthcoming. Because curcumin is a constituent of the diet, it is nontoxic in nature. Curcumin has a plethora of beneficial effects and certainly qualifies for serious consideration as a cosmetic, pharmaceutical/nutraceutical/phytoceutical agent.

Rapid and dramatic technological changes have occurred in the food and agricultural sector throughout the world during the last 15 years. These include increased global competition, the advent of biotechnology and precision production, changes in intellectual property rights, increased product differentiation/value addition, greater demand for ecosystem services from agriculture, and changes in farm and market structure. The ITC, Geneva has estimated an annual growth rate of 10% in the world demand for turmeric. To meet increasing world demand for turmeric, it is of utmost importance to increase the productivity with the desired quality parameters at an affordable cost. Many of the key issues related to the agronomical aspects of turmeric have been highlighted throughout this chapter mainly based on the work done in India. It is expected that this review will be a basis for the other countries to pick up the thread in pursuing the purpose of improving the productivity in their respective regions.

REFERENCES

Adhate, S. (1958) Turmeric. *Farmer*, 9, 21-27.

Aggarwal, B.B., Kumar, A., Aggarwal, M.S. and Shishodia, S. (2004) Curcumin derived from turmeric (*Curcuma longa* L): a spice for all seasons. pp. 349-387. *In*: Bagchi, D., and Preuss, H.G. (Ed.). *Phytopharmaceuticals in Cancer Chemoprevention*. pp. 688. CRC Press, LLC. ISBN: 0849315603.

Agnihotri, B.N. (1949) Turmeric. *Indian J. Hort.*, 6,28-31.

Ahmed Shah, H., Vedamuthu, P.G.B., Abdul Khader, Md. and Prakasam, V. (1988) Influence of different levels on potassium on yield and curcumin content of turmeric. In: Satyanarayana et al, (Eds.), *Proc. Nat. Seminar on Chillies, Ginger and Turmeric*. pp. 108-113.

Ahmed, N.U. and Rahman, M.M. (1987) Effect of seed size and spacing on the yield of turmeric, *Bangladesh J. Agric. Res.*, 12, 50-54.

Aiyadurai, S.G. (1966) A Review of Research on Spices and Cashewnut in India. Regional Office (Spices and Cashew). Indian Council of Agricultural Research, Ernakulam, Kerala, India.

Aiyer, A.K.Y.N. (1947) *Condiments and spices*. In: Field Crops of India with special reference to Mysore, pp.311-384. Government Press, Bangalore, India.

Aiyer, A.K.Y.N. (1949) Mixed cropping in India. *Indian J. Agric. Sci.*, 19, 439-543.

Alam, M.K., Islam, Z., Rouf, M.A., Alam, M.S. and Mondal, H.P. (2003) Response of turmeric to planting material and mulching in the hilly region of Bangladesh. *Pakistan J. Biol. Sci.,* 6, 7-9.

Anilkumar, K. and Reddy, M.D. (2000) Integrated weed management in maize + turmeric intercropping system. *Indian J. Weed Sci.*, 32, 59-62.

Anjaneyulu, V.S.R. and Krishnamurthy, D. (1979) The efficiency of broad ridge method of planting turmeric. *Indian Cocoa Arecanut and Spices J.*,2, 115-116.

Attarde, S.K., Jadhao, B.J., Adpawar, R.M. and Warade, A.D. (2003) Effect of nitrogen levels on growth and yield of turmeric. *J. Spices and Aromatic Crops*, 12, 77-79.

Azam-Ali, S.N. and Squire, G.R. (2002) *Principles of Tropical Agronomy*. CABI Publishing, CAB International, Wallingford, Oxon, UK. pp.236,

Baier, W. (1977) *Crop–weather models and their use in yield assessments*, WMO Technical Note no. 151. World Meteorological Organization, Geneva, p. 48.

Baier, W. (1979) Note on the terminology of crop–weather models. *Agric. Meteorol.*, 20, 137–145.

Balashanmugam, P.V., Mohamed Ali, A. and Chamy, A. (1985) Annual grass and broad leaved weed control in turmeric. Annual conference Indian Society of Weed Science, Anand, Gujarat, p.25.

Balashanmugam, P.V., Vanangamudi, K. and Chamy, A. (1989) Studies on the influence of farm yard manure on the rhizome yield of turmeric. *Indian Cocoa Arecanut and Spices J.,* 12, 126.

Balashanmugam, P.V., Vijayakumar, R.M. and Subramanian, K.S. (1990) Effect of zinc and iron on turmeric yield. *South Indian Hort.,* 38,284-285.

Balashanmugham, P.V., Vanangamudi, K., Thirumoorthy, S. and Chamy, A. (1987) Additional application of fertilizer and yield of onion as intercrop. *Seeds and Farms.*, 13 (9),21-22.

Balasimha, D. (2004) *Cropping systems*. In: Balasimha, D. and Rajagopal, V. (Eds.). *Arecanut*, pp. 103-130. Central Plantation Crops Research Institute, Kasaragod – 671 124, Kerala, India.

Ball, C.R. (1925) Why agronomy needs research in plant physiology. *J. Amer. Soc. Agron.*, 17, 661-675.

Bambirra, M.L.A., Junqueira, R.G. and Gloria, M.B.A. (2002) Influence of post harvest processing conditions on yield and quality of ground turmeric. *Brazilian Archives of Biology and Technology*, 45, 423-429.

Bandyopadhyay, A., Chattopadhyay, N., Ghosh, D. and Hore- J.K. (2003) Performance of different turmeric germplasms as intercrop in young arecanut plantation. *Orissa J. Horticulture.*, 31, 22-25.

Beets, W.C. (1990) *Raising and sustaining productivity of smallholder farming systems in the tropics*. AgBe Publishing, P.O. Box 9125, 1800 GC, Alkmaar, Holland. Pp.xvi.+ 738.

Bisht, J.K., Chandra, S. and Singh, R.D. (2004) Performance of taro (*Colocasia esculenta*) and turmeric (*Curcuma longa*) under fodder trees in a agri-silvi-horti system of western Himalaya. *Indian J. Agric. Sci.,* 74, 291-294.

Bisht, J.K., Chandra, S., Chauhan, V.S. and Singh, R.D. (2000) Performance of ginger (*Zingiber officinale*) and turmeric (*Curcuma longa*) with fodder tree based silvi-horti system in hills. *Indian J. Agric. Sci.,* 70: 431-433.

Chandra, R., Desai, A.R. and Govind, S. (1997) Effect of foliar spray of magnesium and planting materials of growth and yield of turmeric. *J. Hill Research*, 10, 1-4.

Chattopadhyay, I., Biswas, K., Bandyopadhyay, U. and Banerjee, R.K. (2004) Turmeric and curcumin: Biological actions and medicinal applications. *Curr. Sci.*, 87, 44-53.

Chaturvedi, O.P. and Pandey, I.B. (2001) Yield and economics of *Populus deltoides* G3 marsh based inter-cropping system in Eastern India. *Forests, Trees and Livelihoods*, 11, 207-216.

Chaugule, B.A. and Mohite, B.V. (1962) Yield of turmeric in the broad ridge and ridge and furrow layouts with and without maize as an intercrop. *Poona Agric. Coll. Mag.*, 52 (3 & 4),14-17.

Chenchaiah, K.C., Sit, A.K. and Biswas, C.R. (2002a) Evaluation of some annual and perennial intercrops in areca garden under sub-Himalayan Terai region of West Bengal., *J. Plantation Crops,* 30 (3), 41-43.

Chenchaiah., K.C., Biddappa, C.C. and Acharya, G.C. (2002b) Arecanut based high density multispecies cropping system for North Bengal. *Orissa J. Hort.,* 30: 114-120.

Chiu, S.M., Liu, H.I. and Chu, C.L. (1993) Growth and development of turmeric (*Curcuma aromatica* Salisb.) plants: I. The formation and chemical composition of rhizome finger sets of different orders. *J. Agric. Res. China,* 42, 154-161.

Choudhri, B.L. (1952) Don't buy turmeric, grow it. *Indian Farming,* 7 (5),10-12.

Choudhury, A.K., Hoque, A.F.M.E., Firoz, Z.A. and Quayyum, M.A. (2000) Performance of turmeric-legume inter cropping system. *Bangladesh J. Agric.. Res.,* 25, 325-332.

Committee on Spices (1989) *Report of working group-II-Spices research,* Dept. of Agriculture and Coopera-tion, Govt of India.

Dagar, J.C. and Singh, G. (2001) Evaluation of crops in agroforestry with *Casuarina equisetifolia* (Linn.) plantations. *Indian J. Agroforestry,* 3, 47-50.

Datta, M. and Dhiman, K.R. (2001) Effect of some multipurpose trees on soil properties and crop productivity in Tripura area. *J. Indian Soc. Soil Sci.,* 49, 511-515.

Datta, S., Chatterjee, R. and Chattopadhyay, P.K. (2001) Turmeric propagated from mother rhizome -- its morphological and yield characters. *Environment and Ecology.* 19: 114-117.

Desai, H.M. (1939) Scope of improvement in the technique of cultivation of some of the important garden crops under the Poona conditions. *Department of Agriculture. Bombay Bull. No., 183.*

Dixit, D. and Srivastava, N.K. (2000a) Partitioning of photosynthetically fixed $^{14}CO_2$ into oil and curcumin accumulation in *Curcuma longa* grown under iron deficiency. *Photosynthetica,* 38, 193-197.

Dixit, D. and Srivastava, N.K. (2000) Partitioning of ^{14}C-photosynthate of leaves in roots, rhizome, and in essential oil and curcumin in turmeric (*Curcuma longa* L.). *Photosynthetica,* 38, 275-280.

Dixit, D. and Srivastava, N.K. (2000b) Effect of iron deficiency stress on physiological and biochemical changes in turmeric (*Curcuma longa*) genotypes. *J. Medicinal and Aromatic Plant Sci.,* 22 1B, 652-658.

Dixit, D., Srivastava, N.K. and Kumar, R. (2001) Intraspecific variation in yield capacity of turmeric *Curcuma longa* with respect to metabolic translocation and partitioning of $^{14}CO_2$ photoassimilate into essential oil and curcumin. *J. Medicinal and Aromatic Plant Sci.,* 22-23, 4A-1A, 269-274.

Dixit, D., Srivastava, N.K. and Sharma, S. (1999) Effect of Fe-deficiency on growth, physiology, yield and enzymatic activity in selected genotypes of turmeric (*Curcuma longa* L.). *J. Plant Biol.,* 26, 237-241.

Dixit, D., Srivastava, N.K. and Sharma, S. (2002a) Boron deficiency induced changes in translocation of $^{14}CO_2$ -photosynthate into primary metabolites in relation to essential oil and curcumin accumulation in turmeric (*Curcuma longa* L.). *Photosynthetica,* 40, 109-113.

Dixit, D., Srivatava, N.K., Kumar, R. and Sharma, S. (2002b) Cultivar variation in yield, metabolite translo-cation and partitioning of $^{14}CO_2$ assimilated photosynthate into essential oil and curcumin of turmeric (*Curcuma longa* L.). *J. Plant Biol.,* 29, 65-70.

Douglas, J.S. (1973) Commercial *Scitamineae.* III. Profitable turmeric cultivation with special reference to production in India. *Flavour India,* 4 (9), 387-388.

Dubey, A.K. and Yadav, D.S. (2001) Response of turmeric (*Curcuma longa* L.) to NPK under foothill conditions of Arunachal Pradesh. *Indian J. Hill Farming,* 14, 144 -146.

Frere, M. and Popov, G.F. (1979) *Agro meteorological crop monitoring and forecasting.* Plant Production and Protection Division, no. 17. Food and Agriculture Organization of the United Nations, Rome, p. 64.

Gill, B.S., Randhawa, G.S. and Saini, S.S. (2000) Integrated weed management studies in turmeric (*Curcuma longa* L.). *Indian J. Weed Sci.,* 32, 114-115.

Gill, B.S., Randhawa, R.S., Randhawa, G.S. and Singh, J. (1999) Response of turmeric (*Curcuma longa* L.) to nitrogen in relation to application of farm yard manure and straw mulch. *J. Spices and Aromatic Crops,* 8, 211-214.

Gorabal, K., Rokhade, A.K., and Hanamashetti, S.I. (2002) Effect of methods of storage on post-harvest losses and viability of seed rhizomes in turmeric (*Curcuma longa* L.). *J. Plantation Crops,* 30(2), 68-70.

Greenland, D.J., Wild, A. and Adams, D. (1992) Organic matter dynamics in soils of the tropics — from myth to complex reality, In: Lal, R and Sanchez, P.A. (eds.), *Myths and Science of Soils in the Tropics.* Soil Science Society of America Special Publication No. 29. Madison, WI, U.S.A. pp. 28-29.

Gregory, P.J. (1988) Crop growth and development. In: Wild, A (Ed.). *Russell's Soil Conditions and Plant Growth* (Eleventh edition). pp. 31-68. ELBS, English Language Book Society and Longman Group, UK Ltd. ISBN. 0582 006181.

Grewal, S.S. (1996) Competitive and synergistic effects of some agroforestry systems evaluated for optimising production and resource conservation. In: Kohli, R.K., Arya, K.S., and Atul (ed.). Proc. IUFRO-DNAES international meeting: Resource Inventory Techniques to Support Agroforestry and Environment, pp. 275-285. HKT Publications; Chandigarh, India.

Grewal, S.S., Mittal, S.P., Dyal, S. and Agnihotri, Y. (1992) Agroforestry systems for soil and water conservation and sustainable production from foothill areas of north India. *Agroforestry Systems*, 17, 183-191.

Grewal, S.S., Singh, K. and Juneja, M.L. (1995) Conservation and production potential of an agro-forestry system integrating grey gum (*Eucalyptus tereticornis*), white popinac (*Leucaena latisiliqua*) and turmeric (*Curcuma longa*). *Indian J. Agric. Sci.*, 65, 191-195.

Hegde, N.K. and Sulikeri, G.S. (2001) Land equivalent ratio and economics of intercropping systems with tuber and rhizomatous crops in arecanut plantation. *Karnataka J. Agric. Sci.,* 14, 853-857.

Hegde, S., Venkatesha, J. and Chandrappa, H. (1998) Performance of promising turmeric cultivars (*Curcuma domestica* Val.) under coconut cropping systems. In: Mathew, N.M., Kuruvilla Jacob, C., Licy, J., Joseph, T., Meenattoor, J.R. and Thomas, K.K. (ed.). *Developments in Plantation Crops Research.* pp. 220-222. Proc. Allied Publishers Ltd; New Delhi.

Hossain, M.A., Ishimine, Y., Akamine, H. and Motomura, K. (2005a) Effects of seed rhizome size on growth and yield of turmeric (*Curcuma longa* L.). *Plant Prod. Sci.*, 8, 86-94.

Hossain, M.A., Ishimine, Y., Motomura, K. and Akamine, H. (2005b) Effects of planting pattern and planting distance on growth and yield of turmeric (*Curcuma longa* L.). *Plant Prod. Sci.*, 8, 95-105.

Hu, M.F., Chiu, S.M., Liu, H., Lai, M.H. and Liu, S.Y. (1996) Effects of planting date and density on the rhizome yield and curcumin content of turmeric plant (*Curcuma aromatica* Salisb.). *J. Agric. Res. China*, 45,164-173.

Ibrahim, S. and Krishnamurthy, N.H.V. (1955) Preliminary studies in turmeric in Godavari delta. *Andhra Agric. J.*, 2, 241-246.

IISR (2001) Turmeric. Extension Phamplet. Agricultural Technology Information Centre, Indian Institute of Spices Research (ICAR), Calicut, Kerala, India.

Ilyas, M. (1978) Turmeric. The spice of India. II. *Economic botany.*, 32, 239-263.

Inamdar, K.S. and Diskalkar, P.D. (1987) Turmeric — its cultivation and marketing in Tasgoan taluka (Sangli District). *J. Maharashtra Agric. Univ.*, 12, 241-242.

Ishimine, Y., Hossain, M.A., Ishimine, Y. and Murayama, S. (2003) Optimal planting depth for turmeric (*Curcuma longa* L.) cultivation in dark red soil in Okinawa Island, Southern Japan. *Plant Production Sci.*, 6 (1), 83-89.

Ishimine, Y., Hossain, M.A., Motomura, K., Akamine, H. and Hirayama, T. (2004) Effects of planting date on emergence, growth and yield of turmeric (*Curcuma longa* L.) in Okinawa prefecture, Southern Japan. *Japanese J. Trop. Agric.*, 48, 10-6, 20.

Islam, F., Karim, M.R., Shahjahan, M., Hoque, M.O., Alam, M.R. and Hossain, A. (2002) Study on the effect of plant spacing on the production of turmeric at farmer's field. *Asian J. Plant Sci.*, 1, 616-617.

Jagadeeswaran, R., Arulmozhiselvan, K., Govindaswamy, M. and Murugappan, V. (2004) Studies on nitrogen use efficiency in turmeric using ^{15}N tagged urea. *J. Nuclear Agric. Biol.*, 33, 69-76.

Jain, R.C. and Ranjana, A. (2000) Statistical models for crop yield forecasting. *In*: S.V. Singh, L.S. Rathore, S.A. Saseendran, K.K. Singh (Eds.), *Dynamic Crop Simulation Modeling for Agrometeorological Advisory*. pp. 295–304, Proc. National Centre for Medium Range Weather Forecasting, New Delhi.

Jasural, S.C., Mishra, V.K. and Verma, K.S. (1993) Intercropping ginger and turmeric with poplar (*Populus deltoides* 'G-3' Marsh.). *Agroforestry Systems*, 22, 111-117

Jayaraj, P. (1990) *Effect of potassium in mitigating the effects of shade in intercrops.* M.Sc (Agriculture) Thesis. Kerala Agric. Univ., Vellanikkara, Trichur, Kerala, India.

Jena, M.K., Das, P.K. and Pattanaik, A.K. (1999) Integrated effect of microbial inoculants and fertilizer nitrogen on N-use efficiency and rhizome yield of turmeric (*Curcuma longa* L.). *Orissa J. Hort.*, 27(2), 10-16.

Jha, R.C., Sharma, N.N. and Maurya, K.R. (1983) Effect of sowing dates and mulching on the yield and profitability of turmeric. *Bangladesh Horticulture,* 11,1-4.

Joe, B., Vijayakumar, M. and Lokesh, B.R. (2004) Biological properties of curcumin — cellular and molecular mechanisms of action. *Critical Reviews in Food Science and Nutrition,* 44, 97-111.

Kandiannan, K., Chandaragiri K.K., Sankaran, N., Balasubramanian, T.N. and Kailasam, C. (2002 a) Crop-weather model for turmeric yield forecasting for Coimbatore district, Tamil Nadu, India. *Agric. Forest Meteorol.,* 112, 133-137.

Kandiannan, K., Chandragiri, K.K., Govindaswamy, M., Subbian, P. and Sankaran, N. (2002b) Analysis of spatial variability of turmeric (*Curcuma longa* L. syn. *C. domestica* Val.) yield in India. *J. Spices and Aromatic Crops*, 11, 155-159.

Kandiannan, K., Chandragiri, K.K., Sankaran, N., Balasubramanian, T.N. and Kailasam, C. (2002c) Weather during turmeric production and its relation to turmeric yield under tropical irrigated condition. *J. Plantation Crops*, 30, 45-49.

Kandiannan. K., Chandragiri K.K., Sankaran, N., Balasubramanian, T.N. and Kailasam, C. (2001) Empirical model for turmeric (*Curcuma domestica* Val.) yield prediction. *J. Spices and Aromatic Crops,* 10: 59-61.

Kathirvel, K. and Manian, R. (2002) Effect of tool geometry on harvesting efficiency of turmeric harvester. AMA, Agricultural Mechanization in Asia, Africa and Latin America, 33, 39-42.

KAU (1991) ICAR-ad-hoc scheme on shade studies on cocnut based intercropping systems. Final research report. Kerala Agricultural University. Vellanikkara, Trichur (mimeographed).

KAU (1993) *Package of practices*. Directorate of Extension. Kerala Agricultural University, Vellanikarra, Trichur. Kerala, India.

Khanna, N.M. (1999) Turmeric — nature's precious gift. *Curr. Sci.*, 76, 1351-1356.

Koppad, A.G., Manjappa, K. and Chandrasekharaiah, A.M. (2001) Impact of different conservation measures on rehabilitation of degraded lands and agricultural productivity. *Indian J. Soil Conservation*, 29, 148-151.

Kumar, P.S.S. and Savithri, P. (2003) Characteristics of turmeric growing areas of the Coimbatore district in Tamil Nadu. *Crop Res. Hisar*, 26, 365-369.

Kundu, A.L. and Chatterjee, B.N. (1981) Effect of major nutrients on turmeric production as a sole crop or in mixture with other turmeric crops. *Indian J. Agric. Sci.*, 51, 504-508.

Lahiri, A.K. (1989) Turmeric based agroforestry trials in West Bengal. *Indian Forester,* 115, 127

Lakshmanan, S.S. (1949) Turmeric survey. *Madras Agric. J.*, 36, 367-377

Li, L. and Oin, S. (1996) Effect of organic fertilizer and mineral on the tuber yield of *Curcuma longa* L. *Zhongguo Zhong Yao Za Zhi*, 24, 651-652.

Li, L. Oin, S., Song, H., Zhang, Y. and Liao, G. (1996) *Curcuma longa* L. tuber yield simulation model and its application under combined agronomic measures for good quality, high yield and obvious economic results. *Zhongguo Zhong Yao Za Zhi*, 21, 527-529, 574.

Li, L., Oin, S. and Liao, G., Fang, O. and Yang, S. (1997) Optimal high-yield agronomic measures for *Curcuma longa* L. *Zhongguo Zhong Yao Za Zhi*, 22, 145-147.

Li, L., Song, H., Zhang, Y and Fu, S. (1999a) A study on fresh rhizome simulation model and its application to comprehensive agronomic measures for good quality and high yield of *Curcuma longa* L. *Zhongguo Zhong Yao Za Zhi*, 24, 654-657, 701.

Li, L., Zhang, Y. and Song, H. (1999b) A study on soil suitability for growth of rhizomes of *Curcuma longa* L. *Zhongguo Zhong Yao Za Zhi*, 24, 718-721.

Loomis, R.S. and Amthor, J.S. (1999) Yield potential, plant assimilatory capacity, and metaboloic efficiencies. *Crop Sci.*, 39, 1584-1596.

Mahey, R.K., Randhawa, G.S. and Gill, S.R.S. (1986) Effect of irrigation and mulching on water conservation, growth and yield of turmeric. *Indian J. Agron.*, 31, 79-82.

Majumdar, B., Venkatesh, M. S. and Kumar, K. (2002) Effect of nitrogen and farmyard manure on yield and nutrient uptake of turmeric (*Curcuma longa* L.) and different forms of inorganic N build-up in an acidic Alfisol of Meghalaya. *Indian J.Agric. Sci.,* 72, 528-531.

Marschner, H. (1995) *Mineral Nutrition of Higher Plants*, 2nd edn. London, UK., Academic Press.

Matthews, R.B. and Stephens, W. (2002) *Crop-soil Simulation Models: Applications in Developing Countries*, CAB International, Wallingford, Oxon, UK. Pp. 304.

Medda, P.S. and Hore, J.K. (2003) Effects of N and K on the growth and yield of turmeric in alluvial plains of West Bengal. *Indian J. Hort.*, 60, 84-88.

Meenakshi, N., Sulikeri, G.S. and Hegde, R.V. (2001) Effect of planting material and P & K nutrition on yield and quality of turmeric. *Karnataka J. Agric. Sci.*, 14, 197-198.

Meerabai, M., Jayachandran, B.K., Asha, K.R. and Geetha, V. (2000) Boosting spice production under coconut gardens of Kerala: maximizing yield of turmeric with balanced fertilization. *Better Crops International.*, 14, 10-12.

Mengel, K., Kirkby, E.A., Kosengarten, H. and Appel, T. (2001) *Principles of Plant Nutrition* (5th Edition). Kluwer Academic Publishers, Dordrecht, The Netherlands.

Mishra, M. (2000) Effect of no-mulch production technology and depth of planting on turmeric (*Curcuma longa*). *Indian J. Agric. Sci.*, 70, 613-615.

Mishra, R.K. and Pandey, V.K. (1998) Intercropping of turmeric under different tree species and their planting pattern in agroforestry system. *Range Management and Agroforestry*, 19, 199-202.

Mishra, S. and Mishra, S.S. (1982) Effect of mulching and weedicides on growth and fresh rhizome yield of ginger. Annual conference, Indian Society of Weed Science (Abstract), p. 33.

Mishra., M., Mishra., S.N. and Patra. G.J. (2000) Effect of depth of planting and mulching on rainfed turmeric (*Curcuma longa*). *Indian J. Agron.*, 45, 210-213.

Mohanbabu, N. and Muthuswamy, S. (1984) Influence of potassium on the quality of turmeric. *South Indian Hort.*, 32, 343-46.

Mohanty, D.C., Pattnaik, A.K., Senapati, A.K. and Dash, D.K. (1998) Ecological implication of turmeric cultivation in the Eastern Ghats of Orissa. *Environment and Ecology*, 16, 468-470.

Mohanty, D.C., Sarma, Y.N. and Panda, B.S. (1991) Effect of mulch materials and intercrops on the yield of turmeric cv. Suroma under rainfed conditions. *Indian Cocoa, Arecanut and Spices J.,* 15, 8-11.

Monteith, J.L. (1981) Climatic variation and the growth of crops. *Quart. J. R. Meteorol. Soc.,* 107, 749–774.

Nair, C.P.K. (1946) Turmeric (*Curcuma longa* L), its cultivation and uses. *Allahabad Farmer.*, 20 (5), 146-151.

Nair, P.C.S. (1982) *Agronomy of ginger and turmeric.* pp. 63-68. In: Nair, M.K., Premkumar, T., Ravindran, P.N and Sarma, Y.R. (Eds.). *Ginger and Turmeric.* Central Plantation Crops Research Institute, Kasaragod, Kerala, India.

Narain, P., Singh, R.K., Sindhwal, N.S. and Joshie, P. (1997a) Agroforestry for soil and water conservation in the western Himalayan valley region of India. 1. Runoff, soil and nutrient losses. *Agroforestry Systems*, 39, 175-189.

Narain, P., Singh, R.K., Sindhwal, N.S. and Joshie, P. (1997b) Agroforestry for soil and water conservation in the western Himalayan valley region of India. 2. Crop and tree production. *Agroforestry Systems*, 39, 191-203.

Narain, P., Singh, R.K., Sindhwal, N.S. and Joshie, P. (1998) Water balance and water use efficiency of different land uses in western Himalayan valley region. *Agricultural Water Management*, 37, 225-240.

Narayanpur, M.N. and Sulikeri, G.S. (1996) Economics of companion cropping system in turmeric (*Curcuma longa* L.). *Indian Cocoa, Arecanut and Spices J.*, 20, 77-79.

Nath, J.C. (2002) Prospects of coconut based high density multistoreyed cropping in Assam. *Indian Coconut J.*, 33 (3), 10-11.

NRCS (1993) Annual Report for 1992 – 93. National Research Centre for Spices, Calicut, Kerala, India (1993).

Nybe, E.V., Peter, K.V. and Raj, N. M. (2003) Integrated cropping in coconut involving spice crops. *Indian Coconut J.*, 34, 3-9.

Panigrahi, U.C., Patra, G.K. and Mohanty, G.C. (1987) Package of practices for turmeric cultivation in Orissa. *Indian Farming.*, 37 (4), 4-6.

Patil, R.B. and Borse, C.D. (1980) Effect of planting material and transplanting of seedlings of different age on the yield of turmeric (*Curcuma longa* L). *Indian Cocoa, Arecanut and Spices J.*, 4,1-5.

Penning de Vries, F.W.T. (1983) Modelling of growth and production. In: Lange, O.L., Nobel, P.S., Osmond, C.B., Ziegler, H. (Eds.), *Encyclopedia of Plant Physiology*, New series, Vol. 12D, Physiological Plant Ecology IV. Springer, Berlin, Pp. 117–150.

Pereira, A.S. and Stringheta, P.C. (1998) Considerations on turmeric culture and processing (Consideracoes sobre a cultura e processamento do acafrao). *Horticultura Brasileira*, 16, 102-105.

Philip, J. (1984) Effect of different planting materials on growth, yield and quality of turmeric. *Indian Cocoa Arecanut and Spices J.,* 7, 8-11.

Philip, J. (1985) Effect of plant density on yield and yield components of turmeric. *Indian Cocoa Arecanut and Spices J.,* 8, 93-96.

Philip, J., Sethumadhavan, P. and Vidhyadharan, K.K. (1981) Turmeric cultivation – an appraisal of agronomic practices. *Indian Farmers Digest*, 14 (3),19-21.

PPIC (2001) *Nutrient removal by crops* — Fertilizer knowledge No.1. Potash & phosphate institute of Canada–India programme, Gurgaon, p. 4.

Pradhan, U.B., Maiti, S. and Pal, S. (2003) Effect of frequency of pruning and tree spacing of *Leucaena* on the growth and productivity of turmeric when grown under alley cropping system with *Leucaena leucocephala. J. Interacademicia*, 7, 11-20.

Rajput, S.G., Patil, V.K., Warke, D.C., Ballal, A.L. and Gunjkar, (1980) Effect of Nitrogen and spacings on the yield of turmeric rhizomes. In: *Proceedings of the National Seminar on Ginger and Turmeric*, pp. 83-85, CPCRI, Kasaragod.

Randhawa, G.S. and Mahey, R.K. (1984) Effect of chemicals on emergence of turmeric. *J. Res. Punjab Agric.Univ.*, 21, 470-471.

Randhawa, G.S. and Mahey, R.K. (2002) Advances in agronomy and production of turmeric in India. *In*: Cracker, L.E. and Simon, J.E. (Editors-in-chief), *Herbs, Spices and Medicinal plants — Recent Advances in Botany, Horticulture and Pharmacology*, volume – 3 (Indian reprint) pp : 71-101, CBS Publishers and Distributers, Darya Ganj, New Delhi.

Randhawa, G.S., Mahey, R.K., Saini, S.S. and Sidhu, B.S. (1984) Nitrogen requirements of turmeric under Punjab conditions. *J. Res. Punjab Agric. Univ.*, 21, 308-310.

Randhawa, K.S. and Mishra, K.A. (1974) Effect of sowing dates, seed size and spacing on the growth and yield of turmeric. *Punjab Hort. J.*, 14, 53-55.

Randhawa, K.S. and Nandpuri, K.S. (1974) Cultivation of turmeric. Punjab Agricultural University Publication. 12 p.

Rao, A.M. and Reddy, M.L.N. (1990) Population and fertilizer requirement of maize in turmeric + maize intercropping systems. *J. Plantation Crops*, 18, 44-49.

Rao, C.H. (1957) Profitable intercrops in coconut plantations of East Godavari district. *Andhra Agric. J.*, 4 (3),73-75.

Rao, M.R. and Reddy, V.R. (1977) Effect of different levels of Nitrogen, Phosphorus and Potassium on yield of turmeric. *J. Plantation Crops*, 5, 61-62.

Rao, M.R., Reddy, K.R.C. and Subbarayudu, M. (1975) Promising turmeric types of Andhra Pradesh. *Indian Spices*, 12 (2), 2-13.

Rao, P.S. and Rao, T.G.N. (1988) Diseases of Turmeric in Andhra Pradesh. In: *Proceedings of National Seminar on Chillies, Ginger and Turmeric*, Pp. 162-167.

Rao, R.D.V. (1973) *Studies on the nutrition of turmeric*. Unpublished M.Sc (Ag) thesis, Andhra Pradesh Agric. Univ., Hyderabad.

Rao, S.V. and Reddy, P.S. (1990) The rhizome fly *Calobata albimana* Macq. A major pest of turmeric. *Indian Cocoa, Arecanut and Spices J.*, 14, 67-69.

Rao, T.S. (1978) Turmeric cultivation in Andhra Pradesh. *Indian Arecanut, Spices and Cocoa J.*, 2 (2), 31-32.

Ray, A.K. and Reddy, D.V.S. (2001) Performance of areca-based high-density multispecies cropping system under different levels of fertilizer. *Trop. Agric.*, 78, 152-155.

Ray, A.K., Reddy, D.V.S., Sairam, C.V. and Gopalasundaram, P. (2000) Performance of areca based high density multispecies cropping system under different levels of fertilizers. *J. Plantation Crops*, 28 (2), 110-116.

Reddy, C.N., Reddy, N.V., Padmavathi Devi, P.C. and Kondap, S.M. (1992) Weed management systems in turmeric + maize intercropping system. Annual conference, Indian Society of Weed Science (abst.) p. 99-100.

Rengel, Z. and Marschner, P. (2005) Nutrient availability and management in the rhizosphere: exploiting genotypic differences. *New Phytologist*, 168, 305-312.

Rethinam, P., Sankaran, N. and Sankaran, S. (1977) Chemical weed control in turmeric. Proceedings of Weed Science Conference, Indian Society of Weed Science (ISWS), 218-219.

Rethinam, P., Sivaraman, K. and Sushma, P.K. (1994) *Nutrition of Turmeric*. In : Advances in Horticulture, Vol-9, Plantation and Spices Crops, Part 1: pp.477-488, Malhotra Publishing House, New Delhi.

Ridley, R.N. (1912) *Spices*. Macmillan and Co., London.

Ritchie, J. (1993) Simulation and agroforestry. *Agroforestry Today*, 10(3), 10–13.

Sadanandan, A.K. and Hamza, S. (1998) Effect of organic manures on nutrient uptake, yield and quality of turmeric (*Curcuma longa* L.). *In*: Mathew, N.M., Kuruvilla Jacob, C., Licy, J., Joseph, T., Meenattoor, J.R. and Thomas, K.K. (ed.). *Developments in plantation crops research*. pp.175-181. Allied Publishers Ltd; New Delhi.

Saha, A.K. (1988) Note on response of turmeric to manure and source of N and P under terrace conditions of mid altitude of Mizoram. *Indian J. Hort.*, 45, 139-140.

Sairam, C.V., Gopalasundaram, P. and Umamaheswari, L. (1997) Capital requirements for adoption of coconut-based inter cropping systems in Kerala. *Indian Coconut J.,* 27 (10), 2-4.

Sanchez, P.A. and Logan, T.J. (1992) Myths and science about the chemistry and fertility of soils in the tropics. In: Lal, R., Sanchez, P.A. (Eds.), *Myths and Science of Soils of the Tropics*. SSSA Special Publication, vol. 29. pp. 35–46, Soil Science Society of America, Madison, WI, USA.

Sarma, N.N., Paul, S.R., Dey, J.K., Sarma, D., Baruah, G.K.S., Sarma, R.K., Maibangsa, M.M. and Dey, A.K. (2003) Improved practices for management of shifting cultivation in Assam. *Indian J. Agric. Sci.*, 73, 315-321.

Sarma, R., Prasad, S., Mohan, N.K. and Medhi, G. (1996) Economic feasibility of growing some root and tuber crops under intercropping system in coconut garden. *Hort. J..* 9, 167-170.

Sarma, S.S. and Murty, D.K. (1961) Preliminary note on the influence of the size and seed rhizomes (fingers) on yield of turmeric. *Andhra Agric J.*, 8, 231-233.

Sarma, S.S. and Ramesam, M. (1958) Root tubers in turmeric – factors influencing their formation. *Andhra Agric. J.,* 5,216-219.

Satheesan, K.V. and Ramadasan, A. (1988) Effect of growth retardant CCC on growth and productivity of turmeric under monoculture and in association with coconut. *J. Plantation Crops*, 16, 140-143.

Selvaraj, P.K., Krishnamurthi, V.V., Manickasundaram, P., Martin, G.J. and Ayyaswamy, M. (1997) Effect of irrigation schedules and nitrogen levels on the yield of turmeric through drip irrigation. *Madras Agric. J.*, 84, 347-348.

Shankaraiah, V. and Reddy, I.P. (1987) Studies on the effect of different manurial and spacing levels with local and high yielding turmeric varieties. *Indian Cocoa, Arecanut and Spices J.*, 11,94-95.

Shanmugasundaram, V.S. and Subramanian, K.S. (1993) Coconut based integrated farming system. *Indian Coconut J.*, 24(1), 3-4.

Shashidhar, T.R. and Sulikeri, G.S. (1996) Effect of spacing and nitrogen levels on nutrient uptake and yield of turmeric (*Curcuma longa* L.) cv. Amalapuram. *Karnataka J. Agric. Sci.*, 9, 649-656.

Shashidhar, T.R., Sulkeri, G.S. and Gasti, V.D. (1997) Correlation studies in turmeric (*Curcuma longa* L.). *Mysore J. Agric.Sc.,* 31, 217-220.

Shrivastava, M.B. (1988) Economics of agro-forestry in Indo-Gangetic alluvium of Uttar Pradesh. *Indian J. Forestry*, 11, 265-272

Silva, N.F.da., Sonnenberg, P.E. and Borges, J.D. (2004) Growth and production of turmeric as a result of mineral fertilizer and planting density. *Horticultura Brasileira*, 22, 61-65.

Singh, A., Singh, B. and Vaishya, R.D. (2002) Integrated weed management in turmeric (*Curcuma longa*) planted under poplar plantation. *Indian J. Weed Sci.*, 34, 329-330.

Singh, A.K. and Edison, S. (2003) Eco-friendly management of leaf spot of turmeric under partial shade. *Indian-Phytopathology,* 56, 479-480.

Singh, I.P. and Ram, S. (1994) Effect of inter-crops on growth, yield and quality of Dashehari mango (*Mangifera indica* L). *Recent Hort.*, 1, 23-29.

Singh, I.P. and Ram, S. (1995) Nutritional status of Dashehari mango (*Mangifera indica*) as influenced by various intercrops. *Recent Hort.*, 2 (2), 8-15.

Singh, J. and Randhawa, G.S. (1985) Effect of intercropping and mulch on yield and quality of turmeric (*Curcuma longa* L) . *Proceedings of 5th International Society of Horticulture Science Symposium on Medicinal, Aromatic and Spice Plants*, Darjeeling, India. pp. 183-185.

Singh, J., Malik, Y.S., Nehra, B.K. and Partap, S. (2000) Effect of size of seed rhizomes and plant spacing on growth and yield of turmeric (*Curcuma longa* L.). *Haryana J. Hort. Sci.*, 29, 258-260.

Singh, J.P., Prasad, U.K. and Singh, H. (2000) Effect of irrigation and drainage requirement on water-use efficiency and yield of pure [crops] and maize (*Zea mays*)-based intercrops. *Indian J. Agric. Sci.*, 70, 65-68.

Singh, R.V. and Rai, M. (2003) Standardization of mango based cropping system for sustainable production. *J. Res. Birsa Agric. Univ.*, 15, 61-63.

Singh, S., Omprakash. and Singh, S. (1988) Performance of turmeric clones under South Andaman conditions. *J. Andaman Science Association*, 4, 81-82.

Singhal, R.M. and Panwar, B.P.S. (1991) A study of cropping pattern of poplar (*Populus deltoides* Marsh)-based agro-forestry system in North Western Uttar Pradesh. *Van Vigyan.*, 29, 187-191.

Singte, M.B., Yamger, V.T., Kathmale, D.K. and Gaikwad, D.T. (1997) Growth, productivity and water use of turmeric (*Curcuma longa*) under drip irrigation. *Indian J. Agronomy.*, 42, 547-549.

Sivaraman, K. (1992) *Studies on productivity of turmeric — maize and onion intercropping systems under varied population and nitrogen levels*. PhD. thesis. Tamil Nadu Agricultural University, Coimbatore, India.

Sivaraman, K., and Palaniappan, S.P. (1994) Turmeric – maize and onion intercropping systems. I. Yield and land use efficiency. *J. Spices and Aromatic Crops*, 3, 19-27.

Sundararaj, D.D. and Thulasidas, G. (1976) *Botany of Field Crops*. The Macmillan company of India Ltd. Delhi.

Tayde, G.S. and Deshmukh, V.D. (1986) Yield of turmeric as influenced by planting material and nitrogen levels. *PKV Res. J.*, 10, 63-66.

Taylor, S. (1993) Sustainable aqua-agro-silvo-touro-pastoralism on a few hectares in Goa state, India. *Sylva.*, 56, 14-16.

Thamburaj, S. (1991) Research on spice crops at TNGDNAU. *Spice India*, 4 (3),17-21.

Thankamani, C.K., Kandiannan, K. and Sivaraman, K. (1998) Production potential of four turmeric (*Curcuma longa* L.) cultivars in laterite soil. In: Mathew, N.M., Kuruvilla Jacob, C., Licy, J., Joseph, T., Meenattoor, J.R. and Thomas, K.K. (ed.). *Developments in Plantation Crops Research*. pp. 239-243, Allied Publishers Ltd; New Delhi.

Thornton, P.K. and Wilkens, P.W. (1998) Risk assessment and food security. *In*: Tsuji, G.Y., Hoogenboom, G. and Thornton, P.K. (eds) *Understanding Options for Agricultural Production Systems Approaches for Sustainable Agricultural Development*. Kluwer, Dordrecht, The Netherlands, 329–345.

Tiwari, R.J., Banafar, R.N.S. and Jain, R.C. (1995) Response of turmeric to zinc sulphate application in medium black soils of Madhya Pradesh. *Crop Res. Hisar.*, 10, 107-108.

TNAU (2004) Tamil Nadu Agricultural University. *Crop Production Techniques of Horticultural Crops*, Directorate of Horticulture and Plantation Crops, Chepauk, Chennai, India, 289

Umate, M.G., Latchanna, A. and Bidigir, U.S. (1984) Growth and yield of turmeric varieties as influenced by varying levels of nitrogen. *Indian Cocoa Arecanut and Spices J.*, 8, 23-57.

Umrani, N.K., Patel, R.B. and Pawar, H.K. (1980) Effect of planting materials and transplanting on yield of turmeric cv. Tekurpeta. In: Nair, M.K., Premkumar, T., Ravindran, P.N., Sarma, Y.R. (eds). *Ginger and turmeric*, p.79-82, CPCRI, Kasaragod, Kerala, India.

Umrani, N.K., Patil, R.B. and Pawar, H.K. (1982) Studies on time of harvesting turmeric, pp.90-92. *In*: Nair, M.K., Premkumar, T., Ravindran, P.N and Sarma, Y.R. (Eds.). *Ginger and Turmeric*. Central Plantation Crops Research Institute, Kasaragod, Kerala, India.

Upadhyay, D.C. and Misra, R.S. (1999) Nutritional study of turmeric (*Curcuma longa* Linn.) cv. Roma under agroclimatic conditions of Eastern Uttar Pradesh. *Progressive Hort.*, 31, 214-218.

Vadiraj, B.A., Siddagangaiah. and Poti, N. (1998) Effect of vermicompost on the growth and yield of turmeric. *South Indian Hort.*, 46, 176-179.

Valsala, P.A., Abraham, K. and Nair, G.S. (1988) Nutritional and agronomical aspects of ginger and turmeric with reference to Kerala State. A Review. pp. 114-120. In: Sathyanarayana et, al (Eds.). Proceedings of National Seminar on Chillies, ginger and Turmeric. January, 11-12, 1988. Spices Board and Andhra Pradesh Agricultural University, Hyderabad.

Velu, G. (1988) Response of turmeric to iron and zinc nutrition. *South Indian Hort.*, 36, 152-153.

Venkatesh, M.S., Majumdar, B. and Kumar, K. (2003) Effect of rock phosphate, single super phosphate and their mixtures with FYM on yield, quality and nutrient uptake by turmeric (*Curcuma longa* L.) in acid alfisol of Meghalaya. *J. Spices and Aromatic Crops,* 12, 47-51.

Venkatraman, S. and Krishnan, A. (1992) *Crops and Weather*. Indian Council of Agricultural Research, New Delhi.

Wani, B.V. and Bhandare, K.S. (1957) It pays to grow turmeric in North Satara. *Farmer*, 8 (9):23-27.

WMO (1982) The effect of meteorological factors on crop yields and methods of forecasting the yield, Technical Note no. 174. World Meteorological Organization, Geneva, p. 54.

Yamgar, V.T. and Pawar, H.K. (1991) Studies on the fertilizer sources on yield of turmeric. *J. Plantation Crops,* 19, 61-62.

6 Diseases of Turmeric

N.P. Dohroo

CONTENTS

6.1 INTRODUCTION

Turmeric, being a Zingiberaceous species, has disease problems similar to that of ginger (Dohroo, 2005). There are diseases that affect rhizomes, such as the rhizome rot, which kills the plant, and those affecting the aerial shoot, such as leaf blotch and leaf spots, which lead to severe reduction in yield. These diseases occur in all turmeric-growing regions, but their occurrence and severity vary much, depending upon the growing region and growing conditions. Among the diseases affecting turmeric, rhizome rot and foliar diseases are the most serious. A brief review of these diseases affecting turmeric is presented here.

FIGURE 6.1 Rhizomes of turmeric affected by rhizome rot and wilt (*Pythium* sp.).

6.2 RHIZOME ROT

Rhizome rot of turmeric was first reported from Ceylon (Park, 1934). In India, it was first reported from the Krishna district of Andhra Pradesh, Tiruchirapally, Coimbatore (Ramakrishnan and Soumini, 1954) and losses to the tune of 50% were reported from Telangana areas of Andhra Pradesh (Rao and Rao, 1988). This disease has also been reported from the Kasaragod area of Kerala (Anon, 1975) and the Assam region (Rathaiah, 1982b). Shankiaraiah et al. (1991) reported high disease incidence of rhizome rot in Andhra Pradesh, where turmeric is a very important crop, growing year after year in the same soil.

6.2.1 Symptoms and Epidemiology

The diseased plant shows progressive yellowing of leaves, which proceeds first along the margins and later covers the entire leaf, causing it to dry up. The base of the aerial shoots shows water-soaked soft lesions. The root system is adversely affected, with only a few decaying brown roots. As the disease progresses, infection gradually passes to the rhizomes, which begin to rot and become soft. The rhizome color changes to different shades of brown. The disease may be confined to a few isolated plants, may occur in patches, or may spread in a whole field if flow irrigation is practiced. The yield is considerably reduced in case of severe attack (Joshi and Sharma, 1980). The rotten rhizomes emit foul smell during advanced stages (Figure 6.1).

6.2.2 Causal Organism

Park (1934) reported that the causal organism of the rhizome rot occurring in Ceylon as *Pythium aphanidermatum* (Edson). Middleton (1943) reported that turmeric is susceptible to *Pythium graminicolum* Subram, which has also been reported from Madras (Ramakrishnan and Soumini, 1954). In addition to the above-mentioned species of *Pythium*, *Fusarium solani* also causes rhizome rot (Figure 6.2) in certain areas of Northern India (Dohroo, 1988).

   Pythium belongs to the group Oomycetes (class Phycomycetes) and is a widely distributed facultative pathogen. It grows over a wide range of pH (3 to 9). The best growth of the fungus is

FIGURE 6.2 Foliage symptoms of turmeric affected by Rhizome rot caused by (*Fusarium* sp.).

obtained between pH 7 and 8, while oospore production is higher between pH 6 and 9 (Ramakrishnan, 1954). The pathogen is both soil and rhizome borne (Chattopadhyay, 1967; Rathaiah, 1982b). This fungus is also pathogenic to seedlings of sorghum, wheat, maize, barley, oats, arrowroots, and cotton. Rhizome rot incidence was noticed in all types of soils. *Mimegralla coerulifrons* was reported to be associated with rhizome rot in the state of Maharashtra (Ajiri et al., 1982). However, it was not found to play any role in disease development in Kerala, as confirmed by the detailed studies of Premkumar et al., (1982). The association of *Colobata albimana* with rhizome rot has been reported (Nair, 1978). Root knot infestation (*Meloidogyne incognita*) in turmeric was noticed in Telangana where the incidence of rhizome rot is severe.

6.2.3 DISEASE MANAGEMENT

6.2.3.1 Use of Healthy Rhizome

Selection of healthy rhizomes from disease-free areas is imperative in disease management. It can also be multiplied through tissue culture and through the microrhizome technology (Geetha, 2002; Nirmal Babu et al., 2006) to avoid rhizome-borne inoculum.

6.2.3.2 Phytosanitation

The infected clumps together with the soil should be removed from the field and burnt in order to check the further spread of inoculum. It is followed by drenching the soil with chestnut compound or ceresan wet (0.1%, 5.5 l/m2), as reported by Ramakrishnan and Soumini (1954).

6.2.3.3 Crop Rotation

Crop rotation may be followed to check the buildup of soil-borne inoculum. A restricted crop rotation is followed in Telangana, where turmeric is intercropped with maize and chillies.

6.2.3.4 Host Resistance

Elite turmeric cultivars, PCT-18 (Suvarna), PCT-13 (Suguna), and PCT-14 (Sudarsana), are found to be field resistant to rhizome rot (Rao and Rao 1992). Cv. Shillong, a traditional cultivar from Assam, was reported to be resistant to *Pythium myriotylum* (Rathaiah, 1982a, 1982b). Aromatic types such as Ca-69, Ca-17/1, and Ca-146/4 are found to be tolerant. The popular cultivars from Andhra Pradesh, Mydukur, Tekurpet, and Duggirala were reported to be highly susceptible (Rathaiah, 1982a, 1982b). Anandan et al. (1996) also reported that CA-17/1, CA-146/4, and Suvarna are tolerant.

6.2.3.5 Chemical Control

Dipping seed rhizomes in Ridomil (0.25%) for 40 min, followed by a soil drench at the first appearance of symptoms gave considerable control of rhizome rot and increased the yield (Rathaiah, 1982a, 1982b). Shankiaraiah et al. (1991) reported that Carbendazim as well as Bordeaux mixture controlled the disease when applied at the first appearance of symptoms. The combination of *Trichoderma harzianum* and Ridomil is effective against rhizome rot caused by *F. solani* (Reddy et al., 2003). Anandan et al. (1996) reported that Ridomil was most effective in controlling the disease, followed by mancozeb.

6.3 LEAF BLOTCH

Leaf blotch was first described by Butler (1911) from Gujarat and Saharanpur (India) and Rangpur (Bangladesh). Later, it was observed in all turmeric-growing regions of India (Upadhyay and Pavgi, 1967a). Severe outbreak of this disease was also reported from the Rayalaseema area of Andhra Pradesh (Sarma and Dakshinamurthy, 1962).

6.3.1 SYMPTOMS AND ETIOLOGY

The disease is characterized by the appearance of several spots on both the surfaces of leaves, being generally more numerous on the upper surface. The leaf spots first appear as pale yellow discoloration not sharply defined from the rest of the tissue, become dirty yellow, and then deepen to a bay shade. The individual spots are small — 1 to 2 mm in diameter — and coalesce freely. The infected leaves are distorted, with reddish brown appearance in comparison with the normal ones. In severe cases, hundreds of spots appear on both sides of the leaves (Figure 6.3). The spots are discrete, brown to black, and mostly confined to lower leaves (Rao, 1995).

FIGURE 6.3 Leaves of turmeric affected by leaf blotch caused by *Taphrina maculans* and leaf spot (*Colletotrichum capsici*).

The disease is caused by *Taphrina maculans*. Pavgi and Upadhyay (1964), for the first time, obtained culture of *T. maculans* from turmeric. *Taphrina* belongs to the class Ascomycetes. The fungus produces brown leaf spots and the mycelium grows in the subcuticular interspaces of the epidermis (Upadhyay and Pavgi, 1979) and produces cuboid ascogenous cells. The fungus persists during the summer by means of ascogenous cells on leaf debris and as desiccated ascospores or blastospores in the soil and on fallen leaves (Upadhyay and Pavgi, 1967b). Secondary infection was demonstrated under controlled conditions. The ascospores discharged from the successively maturing asci grow into eight-spored microcolonies and infect fresh leaves. Secondary infection causes profuse spotting all over the leaves (Upadhyay and Pavgi, 1966).

The early appearance and severity of *T. maculans* depends on the concentration of inoculum in the soil (Upadhyay and Pavgi, 1967c). Primary infection occurs on the lower leaves in October to November at 80% relative humidity and 21 to 23°C. Secondary infection is related to the availability of large inoculum potential periodically produced under cool and humid conditions. The plant debris, rhizomes, etc. of the previously infected crop or soil from infested fields did not serve as a primary source of infection (Ahmad and Kulkarni, 1968).

Latent period (5 to 6 d) and infectious period (24 to 25 d) are not found to vary in different cultivars (Prasadji et al., 2004a). Agarwal et al. (1982) have shown that peroxidase, amylase, invertase indoleacetic acid (IAA) oxidase and polyphenol oxidase activities decreased. The alternate hosts of *T. maculans* include *Curcuma aromatica* (Nambiar et al., 1977), *Curcuma amada* (Upadhyay and Pavgi, 1967a), *Cassia angustifolia, Zingiber cassumunar, Z. zesumbet*, and *Hedychium* sp. (Butler, 1918).

6.3.2 DISEASE MANAGEMENT

6.3.2.1 Host Resistance

Upadhyay and Pavgi, 1967a reported that a variant clone of *Curcuma longa* selected from a local susceptible variety was immune to the disease. The intermediate-duration *C. longa* types such as CLL were more susceptible to leaf blotch. Sarma and Dakshinamurthy (1962) reported that the varieties that are susceptible to *Colletotrichum* leaf spot are practically unaffected by *T. maculans*. Nambiar et al. (1977) reported that the intensity of attack by the pathogen was more in *C. longa* types than in *C. aromatica* types. The cultivars CLL 324, Amalapuram, Mydukur, Karhadi local, CLL 326, Ochira 24, and Alleppey among *C. longa* group and Ca68, Ca 67, Dahgi, and Kasthuri among *C. aromatica* types were free from infection. Panja et al. (2001) showed that PTS 62, Acc.360, Acc.361, Roma, BSR1, and Kasturi were highly resistant to leaf blotch disease. Singh et al. (2002) showed PTS 62, Acc.360, and JTS-1 as disease resistant, while Rajendra Sonia and RH-5 as highly susceptible. Singh et al. (2003) reported that turmeric cv. Azad Haldi-1 developed through clonal selection was resistant. Reaction of turmeric accessions to leaf blotch disease is given in Table 6.1.

6.3.2.2 Chemical Control

In field trials, spraying with Dithane Z-78 (0.2%) gave best control of the disease, followed by spraying with Dithane M-45, Blitox-50, Bavistin, and Cuman L. (Srivastava and Gupta, 1977). The disease can be controlled to a considerable extent by Bordeaux mixture and Aureofungin (2.5g/l) (Thirumalachar et al., 1969) and Zineb (Nirwan et al., 1972). Ridomil, thiophanate methyl, carbendazim, blitox, and antracol managed the disease (Singh et al., 2000). Propiconazole (0.1%), bitertenol (0.1%) and chlorothalonil (0.2%) were effective and at par in managing the disease (Prasadji et al., 2004b). Lowest disease severity and highest yield are obtained in the three-spray treatment (Singh et al., 2003). Mancozeb and carbendazim (ST and spray) gave the best control of both the diseases, i.e., leaf blotch and leaf spot in Rajendra Sonia and RH-5.

TABLE 6.1
Reaction of Turmeric Accessions to Leaf Blotch Disease

Disease Incidence (%)	Rating	Accessions
0–20	Resistant	PCT14, JTS604, TC2
20–40	Moderately resistant	PCT10
40–60	Moderately susceptible	JTS602.JTS601, JTS306, PTS38, PTS9, CLI330, CLI325, TS361, JTS7
More than 60	Susceptible	GS, JTS308, PTS19, CLI320

Source: Annual Report (2001–2002), AICRPS, IISR, Calicut, India.

TABLE 6.2
Effect of Fungicides and Bioagent on Incidence of Leaf Blotch and Leaf Spot Diseases as well as Yield of Turmeric cvs. Rajendra Sonia (RS) and RH-5

Treatment	Leaf Blotch (%)		Leaf Spot (%)		Yield (qha⁻¹)	
	RS	RH-5	RS	RH-5	RS	RH-5
Mancozeb + carbendazim (ST + spray)	42.78	40.29	41.04	128.15	355.99	349.71
Carbendazim + *Trichoderma harzianum* (seed treatment + soil application)	42.12	41.02	44.47	146.88	367.47	353.68
Control	60.01	61.98	62.66	189.71	253.30	244.35

Source: Annual Report (2002–2003), AICRPS, IISR, Calicut, India.

6.4 LEAF SPOT

The disease was first reported from Coimbatore district in Tamil Nadu by McRae (1917, 1924). The disease is prevalent in all the turmeric-growing tracts in India, but it has been reported to be severe only in Madras and Andhra Pradesh (Ramakrishnan, 1954). Nair and Ramakrishnan (1973b) observed that infections of *Colletotrichum capsici* reduced dry rhizome yield by 62.7%, although the percentage of oil and curcumin yield were slightly increased. The disease is also severe in regions like Cuddapah, Kurnool, Guntur, Krishna, and Godavari districts of Andhra Pradesh.

6.4.1 SYMPTOMS AND ETIOLOGY

The leaf spot disease appears usually during August and September when there is high and continuous humidity in the atmosphere, but incidence was also observed during October and November. The pathogen attacks mostly leaves, and the infection is confined to leaf blades, but occasionally seen on leaf sheaths also. The disease manifests in the form of elliptic or oblong spots of variable size. In the initial stage, the spots are small and measure up to 40 mm in size, but very soon many of them increase in size. Two or more such spots coalesce, developing into irregular patches often involving a major portion of the leaf, which eventually dries up. Each individual spot has a characteristic appearance. The center is grayish white and thin, with numerous black dot-like acervuli on both surfaces (Figure 6.3 and Figure 6.4). These are arranged in concentric rings. Beyond the grayish white portion is a brown margin all around the spot. There is an indefinite yellowish region outside, forming a halo around the spot. The spots, although visible on both surfaces, are more marked on the upper surface in new leaves. The field presents a scorched

FIGURE 6.4 Leaves of turmeric affected by *Colletotrichum* leaf spot (*Colletotrichum capsici*).

appearance during severe infection. In such cases, the loss may be more than 50% (Ramakrishnan, 1954). The stromatoid bodies were also found on the scales of rhizomes. The infection is also noticed on turmeric flowers, as spots (Palarpawar and Ghurde, 1989b).

The disease is caused by *C. capsici* (Syd.) Butler and Bisby, a member of the class Deutero-mycetes. Palarpawar and Ghurde (1995) found three different strains of the fungus. Strains 1 and 3 were more virulent than strain 2. The fungus grows best at 27°C on Czapek agar, with optimum pH of 7 and relative humidity of 80.5%. Continuous red light provided conditions for maximum growth, while abundant sporulation occurred under continuous blue light + 10 min ultraviolet (UV) light (Ghogre and Kodmelwar, 1986).

The pathogen may be carried over in the rhizomes as dormant stromata between the scales of the rhizomes (Sundararaman, 1925a and b). But Ramakrishnan (1954) pointed out that the method of preservation of seed rhizomes during the summer months may not allow the dormant stromata to survive. As the causal organism is of common occurrence on a large number of hosts, infection is always possible in the season. Dissemination is possible only in wet weather, as the conidia are held firmly around the base of the setae in dry weather. The pathogen survives on infected leaf debris for 1 yr and also infects Pigeon pea, *Cyamopsis tetragonoloba*, sorghum, ginger, and pawpaws (Palarpawar and Ghurde, 1989c). In pot trials, the pathogen remained viable for 3 months in leaf debris buried at 0, 5, 10, and 15 cm depths, although viability decreased gradually from 3 months (Palarpawar and Ghurde, 1992).

6.4.2 DISEASE MANAGEMENT

6.4.2.1 Sanitation

Ali et al. (2002) reported that the incidence and severity of the disease can be reduced by sanitation.

6.4.2.2 Host Resistance

The cvs. Nallakatla, Sugandham, Duvvur, and Gandikota are resistant to leaf spots (Reddy et al., 1963). Sarma and Dakshinamurthy (1962) reported that long-duration *C. longa* types were highly susceptible to *Colletotrichum* sp. while the early-duration P73 types were found to have some resistance. The source of resistance to three isolates of *C. curcuma* has also been located in the varieties Bhendi, Gadhavi and Krishna in Vidarbha region of Maharashtra (Palarpawar and Ghurde,

1989a). Of the 150 germplasm screened against leaf spot at Coimbatore, Sugantham was highly resistant (Thamburaj, 1991). The varieties PCT8, PCT10, Suguna, and Sudarshana performed well in disease-prone areas of Andhra Pradesh (Rao et al., 1994). TC-17 was found to be highly resistant (Chandra et al., 1998). Some of varieties showed tolerant reaction (11%) to the disease, which include CL 520, Ranaga, Roma, BSRI, Armoor, Kasturi, and PTS 43 (Tarafdar and Chatterjee, 2003).

6.4.2.3 Shade Effect

The turmeric plants under the shade of trees were found free of the disease (Subha Rao, 1985). It was found that the stomata of shaded plants were either closed or partially open. Excess shade may be avoided (Rangaswami, 1972). The severity of the disease was found to be reduced (2%) under heavy shade of Pigeon pea and partial shade (4.5%) of maize, in comparison to open (24%) cultivation (Singh and Edison, 2003).

6.4.2.4 Chemical Control

The disease can be checked by spraying 1% Bordeaux mixture once in early August, before disease appearance (Sundararaman, 1925b; Govinda Rao, 1951; Ramakrishnan, 1954). Dakshinamurthy et al. (1966) found Captan and Dithane Z-78 applied at monthly intervals during September to December controlled the disease adequately. The growth of the fungus was inhibited by carbendazim and benomyl when applied to the medium at 100, 250, and 1000 ppm. Carbendazim inhibited protein, DNA, and RNA synthesis during germination (Raja, 1989). Rao and Rao (1987) recommended six sprays of Dithane M-45 at 0.25% at 15-d interval for controlling leaf spot and increasing the yield. Carbendazim and Edifenphos sprays gave very good control of leaf spot of turmeric (Palarpawar and Ghurde, 1989b). Carbendazim gave the most effective control, followed by thiophanate methyl, edifenphos, and dinocap. Two sprays of carbendazim at disease initiation and 15 days later were found to control the disease effectively (Thamburaj, 1991). Narasimhudu and Balasubramanian (2002) report that Topsin M was the most effective chemical for the disease management, followed by Indofil M-45 and Bavistin.

6.5 LEAF BLAST

The disease attacks plants when they are about 5 months old, causing dark spots on the older leaves, leaf sheaths, and petals. The disease is caused by *Pyricularia curcumae*. Turmeric cvs. Mikkir and Kasturi were highly susceptible while cvs. Sugandham and Rajpuri were highly resistant to the disease (Rathaiah, 1980).

6.6 STORAGE ROTS

The extent of damage due to storage rots of turmeric is sometimes more than 60% (Reddy and Rao, 1973). Diverse group of pathogens are associated with turmeric in storage (Rathaiah, 1982a). Sharma and Roy (1984) reported the association of *Aspergillus niger* in more than 70% of the samples of turmeric. *Macrophomina phaseolina* and *Cladosporium cladosporioides* were found, causing storage rots in turmeric (Sharma and Roy, 1981). *Aspergillus, Fusarium* (Figure 6.5), *Rhizoctonia,* and *Sclerotium* are also associated with storage rots of turmeric (Kumar and Roy, 1990). Ten fungi associated with turmeric rhizomes were *Aspergillus flavus, A. niger, C. cladosporioides, Drechslera rostrata, Fusarium moniliforme, Fusarium oxysporum, Macrophomina phaseolina, Pythium aphanidermatum, Rhizoctonia solani,* and *Sclerotium (Corticium) rolfsii* (Kumar and Roy, 1990). Favorable incubation temperature and relative humidity (60%) led to

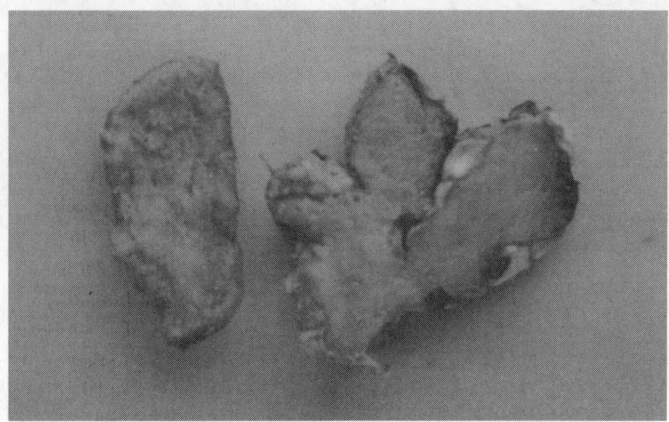

FIGURE 6.5 Turmeric rhizomes affected by storage rot caused by *Fusarium* sp.

high rate of spoilage, but no rotting occurred at 15°C even when the relative humidity varied from 30 to 90%.

6.6.1 DISEASE MANAGEMENT

A. flavus was effectively controlled with pre- and postinoculation treatments of Bavistin (carbendazim) and benomyl at 10 ppm (Sharma and Roy, 1984). Storage rot caused by *S. (Corticium) rolfsii* was controlled by treating the seed rhizomes with ceresan wet (Reddy and Rao, 1973). Hot water treatment (50°C) for 30 min eradicated storage fungi without affecting germination (Sharma et al., 1987).

6.7 NEMATODE INFESTATION

Yield losses due to nematode infestation are reported in turmeric-growing areas, especially in Andhra Pradesh. Seven genera of plant parasitic nematodes, namely, *Meliodogyne* sp., *Rhadopholus similis*, *Rotylenchulus* sp., *Hoplolamus* sp., *Criconemoides* sp., *Longidoros* sp., and *Pratylenchus* sp., were associated with this crop. Root knot nematode was the most predominant, causing considerable crop loss in the states of Kerala, Tamil Nadu, and Andhra Pradesh. *R. similis* is found to cause considerable crop loss of turmeric in the state of Kerala. High populations of *Pratylenchus* sp. were also found to be associated with turmeric in Andhra Pradesh. (Eapen et al., 1997; Haidar et al.,1995, 1998a,b; Mani and Prakash, 1992; Poornima and Vadivelu, 1999; Rajendran et al., 2003; Ramana and Eapen, 1995, 1999; Ray et al., 1995; Sukumaran et al., 1989).

Nematode infestation resulted in significant reduction in growth parameters and yield. Efforts to identify resistant genotypes to root knot nematode, at the Indian Institute of Spices Research (IISR), resulted in identifying eight accessions as resistant. Studies at IISR also identified five biocontrol agents, namely *Verticillium chlamydosporium*, *Paecilomyces lilacinus*, *Fusarium* sp., *Aspergillus nidulans*, and *Scopulariopsis* sp., that suppressed root knot nematode populations significantly (Sarma et al., 2001).

6.7.1 MANAGEMENT

Healthy nematode-free planting material soil application of aldicarb at 1 kg/ha twice (third and fifth month after planting) should be used, followed by irrigation. Application of 10 to 15 kg/ha at the time of field preparation is also recommended.

REFERENCES

Agarwal, M.L., Kumar, S., Goel, A.K., and Tayal, M.S. (1982) Biochemical analysis in leaf spot disease of turmeric: some hydrolyzing and oxidative enzymes and related chemical metabolites. *Indian Phytopath* 35:438–441.

Ahmad, L., and Kulkarni, N.B. (1968a) Studies on *Taphrina maculans* Butler inciting leaf spot of turmeric (*Curcuma longa* L.) I. Isolation of the pathogen. *Mycopath Mycol Appl* 34:40–46.

Ahmad, L. and Kulkarni, N.B. (1968b) Studies on *Taphrina maculans* Butler inciting leaf spot of turmeric (*Curcuma longa* L.) III. Perpetuation and transmission of the disease. *Mycopath Phycol Appl* 35:325–328.

Ajiri, D.A., Ghorpade, S.A., and Jadhav, S.S. (1982) Research on rhizome fly *Mimegrella coerulifrons* on turmeric and ginger in Maharashtra. ICAR Ad-hoc scheme (Unpublished) Report.

Ali, M.A., Fakir, G.A., and Sarker, A.K. (2002) Research on seed-borne fungal diseases of spices in Bangladesh Agricultural University. *Bangladesh J Train Dev* 15:245–250.

Anandan, R.J., Rao, A.S., and Babu, K.V. (1996) Studies on rhizome rot of turmeric (*Curcuma longa* L.). *Indian Cocoa Arecanut Spices J* 20:17–20.

Anonymous (1953) Detailed report of the subordinate officers of Dep. Agric., Madras for 1943–44, 1944–45, 1948–49, 1949–50, 1950–51, pp. 84.

Anonymous (1972) Annual report of the research branch Dep. Agric. Sarawak for 1971. pp. 191.

Anonymous (1974) Central Plantation Crops Research Institute, Annual report 1973. Kasaragod, India.

Anonymous (1975) Central Plantation Crops Research Institute, Annual report 1974. Kasaragod, India.

Anonymous (1976) Annual report 1975–76. Queensland Dep. Primary Indus. Brisbane, Australia. pp. 91.

Anonymous (1992) Agricultural Situation in India. 47, 296.

Butler, E.J. (1907) An account of genus Pythium and some Chytridiaceae. Mem Dep Agric India (Bot Ser) 1(5):70.

Butler, E.J. (1911) Leaf spot of turmeric (*Taphrina maculans* sp. nov.). Ann Mycol 36–39.

Butler, E.J. (1918) Fungi and diseases in plants. Thacker and Spink, Calcutta, pp. 598.

Chandra, R., Singh, A.K., Desai, A.R., Chandra, R., Mathew, N.M., Kurivilla, J.C., Licy, J., Joseph, T., Meenattoor, J.R., and Thomas, K.K. (1998) Crop Improvement in turmeric. Developments in Plantation crops research-Proceedings of the 12th Symposium on Plantation Crops, PLACROSYM XII, Kottayam, India, 27–29 November, 1996. pp 58–62.

Chattopadhyay, S.B. (1967) Diseases of plants yielding Drugs, Dyes and Spices. ICAR, New Delhi.

Chattopadhyay, S.B., and Maiti, S. (1990) Diseases of betel vine and spices, Publication and Information Division, ICAR, New Delhi.

Dakshinamurthy, V., Reddy, G.S., Rao, D.K., and Rao, P.G. (1966) Fungicidal control of turmeric leaf spot caused by *Colletotrichum capsici. Andhra Agric J* 13:69–72.

Dohroo, N.P. (1988) *Fusarium solani* on *curcuma longa. Indian Phytopath* 41, pp. 504.

Dohroo, N.P. (2005) Diseases of ginger. In: Ravidram, P.N. and Nirmal Babu, K. (eds.) *Ginger — the genus Zingiber.* Boca Raton, Florida, U.S.A.: CRC Press. pp. 305–340.

Dohroo, N.P. (2006) Diseases of ginger and tumeric. In: Ravindran, P.N., Nirmal Babu, K., Shiva, K.N. and Kallupurackal, J.A., eds. *Advances in Spices Research: History and Achievements of Spices Research in India since Independence.* India: Agrobios Pub. pp. 493–513.

Eapen, S.J., Ramana, K.V. and Sarma, Y.R. (1997) Biological control of plant parasitic nematodes of major spices. In: Symposium on Economically important Diseases of Crop Plants. Indian Phytopathological Society, Bangalore.

Geeta, G.S., and Reddy, T.K.R. (1990) *Aspergillus flavus* Link and its occurrence in relation to other mycoflora on stored spices. *J Stored Prod Res* 26:211–213.

Geetha, S.P. (2002) *In vitro* technology for genetic conservation of some genera of *Zingiberaceae.* Ph.D. thesis, Calicut University, Calicut, India.

Ghogre, S.B., and Kodmelwar, R.V. (1986) Influence of temperature, humidity, pH and light on mycelial growth and sporulation of *Colletotrichum curcumae* (Syd.) and Bisby. *PKV Research Journal* 10:77–79.

Govinda Rao, P. (1951) Control of plant diseases in the northern region of the Madras State in the year 1949–50. *Plant Prot Bull* 3:21–23.

Haidar, M.G., Jha, R.N., and Nath, R.P. (1995) Studies on the nematodes of spices. 1. Nemic association of turmeric in Bihar and reaction of certain lines to some of the dominating nematodes. *Indian J Nematology* 25:212–213.

Haidar, M.G., Jha, R.N., and Nath, R.P. (1998a) Efficacy of some nematicides and organic soil amendments on nematode population and growth characters of turmeric (*Curcuma longa* L.). *J Research Birsa Agricultural University*, 10:211–213.

Haider, M.G., Jha, R.N., and Nath, R.P. (1998b) Studies on the nematodes of spices. Pathogenic effect of root-knot (*Meloidogyne incognita*) and reniform nematodes (*Rotylenchulus reniformis*) alone and in combination on turmeric (*Curcuma longa* L.).

Jamil, M. (1964) Annual Report of the Department of Agriculture, Federation of Malaya, 1961. pp. 85.

Joshi, L.K., and Sharma, N.D. (1980) Disease of ginger and turmeric. In: Nair MK, Premkumar T, Ravindran, PN and Sharma YR, eds. Proc. Nat. Sem. Ginger and Turmeric. Calicut. April 8–9, CPCRI, Kasaragod, pp.104–119.

Kumar, H., and Roy, A.N. (1990) Occurrence of fungal rot of turmeric (*Curcuma longa* L.) rhizome in Delhi market. *Indian J Agric Sci* 60:189–191.

Kuruvina Shetty, M.S., Handasan, P., Padmaja, H., and Iyer, R.D. (1982) Tissue culture studies in ginger and turmeric. CPCRI, Kasaragod, Kerala, 39–41.

Mani, A., and Prakash, K.S. (1992) Plant parasitic nematodes associated with turmeric in Andhra Pradesh. *Current Nematology* 3:103–104.

McRae, W. (1917) Notes on South Indian Fungi. Year Book, Madras Agric Dep pp.110.

McRae, W. (1924) Economic Botany. II (Mycology) *Rep Bio Sci Adv* India 1922–23, pp.31–35.

Middleton, J.T. (1943) The taxonomy, host range and geographic distribution of the genus *Pythium*. Mem Torrey Bot Cl 20:1–140.

Nair, M.C., and Ramakrishnan, K. (1973) Effect of *Colletotrichum* leaf spot disease of Curcuma longa L. on the yield and quality of rhizomes. *Curr Sci* 42:549–550.

Nair, M.R.G.K. (1978) A Monograph on Crop Pests of Kerala and their Control. COA, KAU, Trivandrum, Kerala, pp. 162.

Nambiar, K.K.N. Sarma, Y.R. and Brahma, R.N. (1977) Field reaction of turmeric types to leaf blotch disease. *J Plant Crops,* 5, 121–125.

Narasimhudu, Y., and Balasubramanian, K.A. (2002) Fungicidal management of leaf spot of turmeric incited by *Colletotrichum capsici*. *Indian Phytopath* 55:527–528.

Nirmal Babu, K., Minoo, D., Geetha, S.P., and Jayakumar, V.N. (2006) Progress of spices biotechnology in India. In: Ravindran, P.N., Nirmal, Babu K., Shiva, K.N. and Kallupurackal, J.A. (eds.) *Advances in Spices Research.* Agrobios, India, pp. 169–194.

Nirwan, P.S., Ram, G., and Upadhyaya, J. (1972) Chemical control of turmeric leaf spot incited by *Tapherina maculans* Butler. *Horticultural Advance* 9:47–48.

Orian, G. (1953) Botanical Division. Rep Dep Agric Mauritius 1952:37–40.

Palarpawar, M.Y., and Ghurde, V.R. (1989a) Sources or resistance in turmeric against leaf spots incited by *Collectotrichum capsici* and *C. curcumae*. *Indian Phytopath* 42:171–173.

Palarpawar, M.Y., and Ghurde, V.R. (1989b) Perpetuation and host range of *Colletotrichum curcumae* causing leaf spot disease of turmeric. *Indian Phytopath* 42:576–578.

Palarpawar, M.Y., and Ghurde, V.R. (1989c) Fungicidal control of leaf spot of turmeric incited by *Colletot-richum curcumae*. *Indian Phytopath* 42:578–579.

Palarpawar, M.Y., and Ghurde, V.R. (1992) Survival of *Colletotrichum curcumae* causing leaf spot of turmeric. *Indian Phytopath* 45:255–256.

Palarpawar, M.Y., and Ghurde, V.R. (1995) Variability in *Colletotrichum curcumae,* the incitant of leaf spot of turmeric (*Curcuma longa* L.) in Vidarbha, Maharashtra, India *J Spices Arom Crops* 4:74–77.

Panja, B.N., and De, D.K. (2001) Economic control of leaf blotch disease of turmeric with fungicide in Tarai region of West Bengal. *Indian Agriculturist* 45:115–120.

Panja, B.N., De, D.K., and Majumdar, D. (2001) Evaluation of turmeric (*Curcuma longa* L.) genotypes for yield and leaf blotch disease (c.o. *Taphrina maculans* Butl.) from Tarai region of West Bengal *Environment Ecology* 19:125–129.

Park, M. (1934) Report on the work of mycological Division Ceylon Admn. Rep. Rep Dir Agric 1933:D126.

Park, M. (1935) Report on the work of mycological Division Ceylon Admn. Rep Rep Dir Agric 1934:D124–131.

Pavgi, M.S., and Upadhyay, R. (1964) Artificial culture and pathogenicity of *Taphrina maculans* Butler. *Sci Cult* 30:558–559.

Peter, K.V. (1997) Fifty years of research on major spices in India. Employment News 25–31 January, 1997:2–4.

Poornima, K., and Vadivelu, S. (1999) Occurrence and seasonal population behaviour of phytonematodes in turmeric (*Curcuma longa* L.). *Pest Management in Horticultural Ecosystems* 5: 42–45.

Prasadji, J.K., Murthy, K.V.M.K., Rama, P.S., and Muralidharan, K. (2004b) Management of *Taphrina maculans* incited leaf blotch of turmeric. *J Mycol Pl Pathol* 34:446–449.

Prasadji, J.K., Ramapandu, S., and Muralidhran, K. (2004) Pathogen and disease development in *Curcuma longa — Taphrina maculans* pathosystem. *J Mycol Pl Pathol* 34:371–375.

Premkumar, T., Sarma, Y.R., and Gautam, S.S.S. (1982) Association of dipteran maggots in rhizome rot of ginger. In: Nair, M.K., Premkumar, T., Ravindran, P.N., and Sarma, Y.R., eds. Proc. National Seminar on Ginger and Turmeric. 8–9 April, 1980. Calicut pp. 128–130. CPCRI, Kasargod, Kerala.

Quinon, V.L., Aragaki, M., and Ishii, M. (1964) Pathogenicity and serological relationships of three strains of *Pseudomonas solanacearum* in Hawaii. *Phytopath* 41:1096–1099.

Raja, K.T.S. (1989) In vitro effect of fungicides on mycelial growth and macromolecular synthesis during spore germination of *Collectotrichum capsici* causing leaf spot disease of turmeric. *Pestology* 13:28–31.

Rajendran. G., Shanthi, A., and Senthamizh, K. (2003) Effect of potensized nematode induced cell extract against root-knot nematode, *Meloidogyne incognita* in tomato and reniform nematode, *Rotylenchulus reniformis* in turmeric. *Indian J Nematology* 33:67–69.

Ramakrishnan, T.S. (1954) Leaf spot disease of turmeric (Curcuma longa L.) caused by *Colletotrichum capsici* (Syd.) Buil and Bisby. *Indian Phytopath* 7:111–117.

Ramakrishnan, T.S. and Soumini, C.R. (1954) Rhizome rot and root rot of turmeric caused by *Pythium graminicolum* subr. *Indian Phylopath,* 7:152–153.

Ramana, K.V., and Eapen, S.J. (1995) Parasitic nematodes and their management in major spices. *J Spices Aromatic Crops* 4:1–16.

Ramana, K.V., and Eapen, S.J. (1999) Nematode pests of spices and their management. *Indian J Arecanut Spices Medicinal Plants* 1:4.

Rangaswami, G. (1961) Pythiaceous fungi — Monograph, ICAR, New Delhi.

Rao, P.S., and Rao, T.G.N. (1987) Leaf spot of turmeric caused by *Colletotrichum capsici,* yield loss and fungicidal control. *J Plant Crops* 15:47–51.

Rao, P.S., and Rao, T.G.N. (1988) Disease of turmeric in Andhra Pradesh. Proc. Nat. Sem. on Chillies, Ginger and Turmeric. Spices Board, Cochin, Kerala, India, pp. 162–167.

Rao, P.S., Krishna, M.R., Srinivas, C., Meenakumari, K., and Rao, A.M. (1994) Short duration, disease resistant turmerics for northern Telangana. *Ind Hort* 39:3, 55–56.

Rao, T.G.N. (1995) Disease of turmeric (*Curcuma longa* L.) and their management. *J Spices Aromatic Crops* 4:49–56.

Rathaiah, Y. (1980) Leaf blast of turmeric. Pl Dis 64, 104–105.

Rathaiah, Y. (1982a) Ridomil for control of rhizome rot of turmeric. *Indian Phytopath* 35:297–299.

Rathaiah, Y. (1982b) Rhizome rot of turmeric. *Indian Phytopath* 35:415–417.

Ray, S., Mohanty, K.C., Mohapatra, S.N., Patnaik, P.R., and Ray, P. (1995) Yield losses in ginger (*Zingiber officinale* Rosc.) and turmeric (*Curcuma longa* L.) due to root knot nematode (*Meloidogyne incognita*). *J Spices Aromatic Crops* 4:67–69.

Reddy, G.S, Dakshinamurthy, V., and Sarma, S.S. (1963) Notes on varietal resistance against leaf spot diseases in turmeric. *Andhra Agric J* 10:146–148.

Reddy, G.S., and Rao, P.G. (1973) Storage rot in seed rhizomes of turmeric in Andhra Pradesh. *Indian Phytopath* 26:24–27.

Reddy, M.N., Devi, M.C., and Sreedevi, N.V. (2003) Biological control of rhizome rot of turmeric (*Curcuma longa* L.) caused by *Fusarium solani. J Biol Control* 17:193–195.

Sarma, S.S., and Dakshinamurthy, D. (1962) Varietal resistance against leaf spot diseases of turmeric. *Andhra Agric J* 9:61–64.

Sarma, Y.R., Ramana, K.V., Devasahayam, S., and Rema, J. (2001) The Saga of Spices Research. Indian Institute of Spices Research, Calicut, India. pp. 95–96.

Shankaraiah, V., Zaheeruddin, S.M., Reddy, L.K., and Vijaya (1991) Rhizome rot complex on turmeric crop in Nizamabad District in Andhra Pradesh. *Indian Cocoa Arecanut Spices J* 14:104–106.

Sharma, M.P., and Roy, A.N. (1981) Rhizome rot of turmeric (Curcuma longa L.) caused by Aspergillus flavus. *Indian Phytopath* 34:12–18.

Sharma, M.P., and Roy, A.N. (1984) Storage rot in seed rhizome of turmeric (*Curcuma longa* L.) and its control. *Pesticides* 18:26–28.

Singh, A., Basandrai, A.K., and Sharma, B.K. (2003) Fungicidal treatment of *Taphrina* leaf spot of turmeric. *Indian Phytopath* 56:119–120.

Singh, A.K., and Edison, S. (2003) Eco-friendly management of leaf spot of turmeric under partial shade. *Indian Phytopath* 56:479–480.

Singh, A.K., Edison, S., and Singh, S. (2000) Evaluation of fungicides for the management of Taphrina leaf blotch of turmeric (Curcuma longa L.) *J Spices Arom Crops* 9:69–71.

Singh, A.K., Gupta, C.R., and Edison, S. (2002) Yield performance of different cultivars of turmeric in relation to leaf blotch disease under field conditions of Chattisgarh. *Progressive Horticulture* 34:236–239.

Singh, B., Yadav, J.R., and Srivastava, J.P. (2003) Azad Haldi-1, a disease resistant variety turmeric (*Curcuma longa*) *Plant Archives* 3:151–152.

Sivaprasad, P., Arthur, J., Nair, S.K., and Balan, G. (1990) Influence of VA-mycorrhizal Mycorrhizal Research. In: Jalali, B.L., and Chand A., eds. Proc. National Conference on Mycorrhizae 14–16 February, 1990, HAU, Hisar, India.

Srivastava, V.P., and Gupta, J.H. (1977) Fungicidal control of turmeric leaf spot incited by *Taphrina maculans. Ind J Mycol Pl Pathol* 7:76–77.

Subha Rao, I.V. (1985) Stomatal opening and incidence of *Colletotrichum* leaf spot disease inturmeric cultivars. *Indian Cocoa Arecanut spices J* 8:97–99.

Sukumaran, S., Koshy, P.K., and Sundararaju, P. (1989) Effect of root-knot nematode, *Meloidogyne incognita* on the growth of turmeric. *J Plantation Crops* 16 (Suppl.):293–295.

Sundararaman, S. (1925a) *Vermicularia curcumae* Syd. on *Curcuma longa.* Year Book, Madras Agric Dept pp. 18–19.

Sundararaman, S. (1925b) Report Subor Offrs. Dep Agric Madras 1924–25, pp. 290.

Tarafdar, J., and Chatterjee, R. (2003) Differential reactions to *Colletotrichum* leaf spot and genetic variability in some cultivars of turmeric. *Ann Pl Prot Sci* 11:300–303.

Thamburaj, S. (1991) Research in spice crops at TNGDNAU. Spice India, 4:17–18.

Thirumalachar, M.J., Pavgi, M.S., and Singh, R.A. (1969) In vitro activity of three antifungal antibiotics against *Protomyces macrosporus, Taphrina maculans* and *Corticium sasakii. Hindustan Antibiot Bull* 11:89–90.

Upadhyay, R., and Pavgi, M.S. (1966) Secondary host infection by *Taphrina maculans* Butler. *Phytopath Z* 56:151–154.

Upadhyay, R., and Pavgi, M.S. (1967a) Varietal resistance in turmeric to leaf spot disease. *Indian phytopath* 20:29–31.

Upadhyay, R., and Pavgi, M.S. (1967b) Perpetuation of *Taphrina maculans* Butler, the incitant of turmeric leaf spot disease. *Phytopath Z* 59:136–140.

Upadhyay, R., and Pavgi, M.S. (1967c) Some factors affecting incidence of leaf spot of turmeric by *Taphrina maculans* Butler. *Ann Phytopath Soc Japan* 33:176–180.

Upadhyay, R., and Pavgi, M.S. (1979) Biphasic diurnal cycle of ascus development of *Taphrina maculans* Butler. *Mycopathologica* 69:33–41.

7 Insect Pests of Turmeric

S. Devasahayam and K.M. Abdulla Koya

CONTENTS

7.1 INTRODUCTION

The turmeric (*Curcuma longa* L.) crop is infested by over 70 species of insects among which the shoot borer (*Conogethes punctiferalis* Guen.) and rhizome scale (*Aspidiella hartii* Sign.) are major ones in the field and storage, respectively. In addition to these two major insect pests, various sap feeders, leaf feeders, and rhizome borers also occasionally infest the crop causing serious damage in localized areas. The cigarette beetle (*Lasioderma serricorne* Fab.), drugstore beetle (*Stegobium paniceum* L.), and coffee bean weevil (*Araecerus fasciculatus* DeG.) are serious insect pests of dry turmeric rhizomes. Information on insect pests of turmeric is available mainly from India and has been previously reviewed by Jacob (1980), Butani (1985), Koya et al. (1991), and Premkumar et al. (1994). The distribution, damage, life history, seasonal incidence, host plants, resistant sources, natural enemies, and management of insect pests of turmeric have been consolidated in this chapter. A list of insects recorded on turmeric in the world has also been tabulated (Table 7.1).

7.2 MAJOR INSECT PESTS

7.2.1 SHOOT BORER (*CONOGETHES PUNCTIFERALIS* GUEN.)

7.2.1.1 Distribution

The shoot borer is the most serious insect pest of turmeric in India that has been recorded on the crop as early as 1914 (Fletcher, 1914). Although the pest is known to occur in most of the turmeric growing areas in the country, authentic information on the distribution of the pest is limited. Kotikal and Kulkarni (2000a) conducted surveys in the northern districts of Karnataka and reported that the incidence of the pest in the field was higher at Raibag, Chikodi (Belgaum District), Jamakhandi, Indi (Bijapur District), Basavakalyan, and Humanabad (Bidar District) taluks. Although the shoot borer is widely distributed throughout Asia, Africa, America, and Australia, authentic records of the pest on turmeric are lacking. The shoot borer is known by many other common names generally indicative of the crop and plant part infested. The shoot borer has been suggested to be a complex of more than one species, especially in Australia and South East Asia (Honda, 1986a; 1986b; Honda et al., 1986; Robinson et al., 1994; Boo, 1998).

7.2.1.2 Damage

The newly hatched larvae of shoot borer scrape and feed on the margins of unopened leaf or newly opened leaf of turmeric plants. Later, the larvae bore into the pseudostem and feed on the growing shoot, resulting in yellowing and drying of the central shoot. The presence of boreholes in the pseudostem through which frass is extruded and the withered central shoot are a characteristic symptoms of pest infestation (Figure 7.1). Sometimes, the larvae also bore into the rhizome near the base of the pseudostem.

7.2.1.3 Life History

The adults are medium-sized moths with a wingspan of 18 to 24 mm; the wings and body are pale straw yellow with minute black spots (Figure 7.2). The adults have been described by Hampson (1856). Freshly laid eggs are elliptical, pitted on the surface, and creamy white. There are five larval instars; fully grown larvae are light brown with sparse hairs and measure 16 to 26 mm in length (Figure 7.3). The dimensions of adults and larvae may vary depending on the host in which they are raised. The morphometrics of various stages when reared on turmeric has been provided by Jacob (1981). Thyagaraj et al. (2001) suggested a method for determination of sex in the shoot borer based on the size and morphology of pupae.

Studies on life history of the pest on turmeric conducted at Kasaragod (Kerala, India) under laboratory conditions (temperature range: 30 to 33°C; relative humidity range: 60 to 90%) indicated that the preoviposition and egg periods lasted for 4 to 7 and 3 to 4 d, respectively. The five larval instars lasted for 3 to 4, 5, 3 to 7, 3 to 8, and 7 to 14 d, respectively. The prepupal and pupal periods lasted for 3 to 4 and 9 to 10 d, respectively. Adult females laid 30 to 60 eggs during its life span, and six to seven generations were completed during a crop season in the field. Variations were also

TABLE 7.1
List of Insects Recorded on Turmeric

Genus/Species	Plant Part Affected	Distribution
Order: Orthoptera		
Family: Tettigonidae		
Letana inflata Bru.	Leaf	India
Phenoroptera gracillus Burm.	Leaf	India
Family: Acrididae		
Orthacris simulans B.	Leaf	India
Cyrtacanthacris ranacea Stoll.	Leaf	India
Order: Hemiptera		
Family: Membracidae		
Oxyrachis tarandus F.	Leaf, psuedostem	India
Tricentrus bicolor Dist.	Leaf	India
Family: Cicadellidae		
Tettigoniella ferruginea Wlk.	Leaf	India
Family: Aphididae		
Pentalonia nigronervosa Coq.	Leaf	China, India
Uroleucon compositae (Theobald)	Leaf	India
Family: Pseudococcidae		
Pseudococcus sp.	Rhizome	Fiji
Planococcus sp.	Rhizome	India
Unidentified	Leaf	India
Family: Coccidae		
Aspidiella hartii Ckll.	Pseudostem, rhizome, leaf	India, West Africa, West Indies
Ccoccus hesperidium (Ckll.)	—	Fiji, Papua New Guinea
Aspidotus curcumae Gr.	Pseudostem, rhizome	India
Family: Diaspididae		
Hemiberlesia palmae (L.)	—	Fiji, Papua New Guinea
Howardia biclavis (Com.)	Rhizome	—
Family: Tingidae		
Stephanitis typica (Dist.)	Leaf	India
Family: Coreidae		
Cletus rubidiventris Westd.	Leaf	India
C. bipunctatus Westd.	Leaf	India
Riptortus pedestris F.	Leaf	India
Family: Pentatomidae		
Coptosoma cribraria Fab.	Leaf	India
Nezarsda viridula (L.)	Leaf	India
Order: Thysanoptera		
Family: Thripidae		
Anaphothrips sudanensis (Trybom)	Leaf	India

Continued

TABLE 7.1 *(Continued)*
List of Insects Recorded on Turmeric

Genus/Species	Plant Part Affected	Distribution
Asprothrips indicus Bagn.	Leaf	India
Panchaetothrips indicus Bagn.	Leaf	India
Family: Phlaeothripidae	Leaf	India
Haplothrips sp.	Leaf	India

Order: Coleoptera

Genus/Species	Plant Part Affected	Distribution
Family: Scarabaeidae		
Holotrichia sp.	Rhizome, root	India
H. serrota (F.)	Rhizome	India
Family: Anobiidae		
Lasioderma serricorne (Fab.)	Dry rhizome	Bangladesh, India
Stegobium paniceum L.	Dry rhizome	India
Family: Bostrychidae		
Tribolium castaneum (Hbst.)	Dry rhizome	India
Family: Sylvanidae		
Oryzaephilus surinamensis (L.)	Dry rhizome	India
Rhizopertha dominica (F.)	Dry rhizome	—
Family: Coccinellidae		
Epilachna sparsa (Hbst.)	Leaf	India
Family: Tenebrionidae		
Tenebroides mauritanicus (L.)	Dry rhizome	India
Family: Chrysomelidae		
Chirida bipunctata F.	Leaf	India
Ceratobasis nair Loc.	Leaf	India
Colasposoma splendidum F.	Leaf	India
Cryptocephalus rajah Jac.	Leaf	India
C. schestedti F.	Leaf	India
Lema fulvicornis Jac.	Leaf	Sri Lanka
L. lacordairei Baly.	Leaf	India
L. praeusta F.	Leaf	India
L. semiregularis Jac.	Leaf	India
L. signatipennis Jac.	Leaf	India
Psuedocophora sp.	Leaf	India
Monolepta signata (Olivier)	Leaf	India
Raphidopalpa abdominalis (Fab.)	Leaf	India
Family: Anthribiidae		
Araecerus fasciculatus (Deg.)	Dry rhizome	India
Family: Curculionidae		
Hedychorus rufomaculatus M.	Leaf	India
Myllocerus discolor Boheman	Leaf	India

Continued

TABLE 7.1 *(Continued)*
List of Insects Recorded on Turmeric

Genus/Species	Plant Part Affected	Distribution
M. undecimpustulatus Faust.	Leaf	India
M. viridanus Fab.	Leaf	India
Order: Diptera		
Family: Tipulidae		
Libnotes punctipennis Meij	Rhizome	India
Family: Syrphidae		
Eumerus albifrons Wlk.	Rhizome	India
E. pulcherrimus Bru.	Rhizome	India
Family: Micropezidae		
Calobata sp.	Rhizome	India
C. albimana Macq.	Rhizome	India
Mimegralla coeruleifrons Macq.	Rhizome	India
M. albimina (Doleschall)	Rhizome	India
Order: Lepidoptera		
Family: Tineidae		
Setomorpha rutella Zell.	Dry rhizome	India
Family: Oecophoridae		
Blastobasis nair	Leaf	India
Family: Pyralidae		
Conogethes punctiferalis Guen.	Shoot	India, Sri Lanka
Ephestia sp.	Dry rhizome	India
Family: Hesperiidae		
Notocrypta curvifascia (C and R Felder)	Leaf	India
Udaspes folus Cram.	Leaf	India
Family: Pieridae		
Catopsila pomona F.	Leaf	India
Family: Arctiidae		
Spilarctia obliqua (Wlk.)	Leaf	India
Creatonotus gangis (L.)	Leaf	India
Amata passalis Fab.	Leaf	India
Family: Noctuidae		
Bombotelia nugatrix Gr.	Leaf	India
Spodoptera litura (F.)	Leaf	India
Family: Lymantriidae		
Euproctis lutifolia Hamp.	Leaf	India

FIGURE 7.1 Turmeric pseudostem infested by shoot borer.

FIGURE 7.2 Shoot borer — adult.

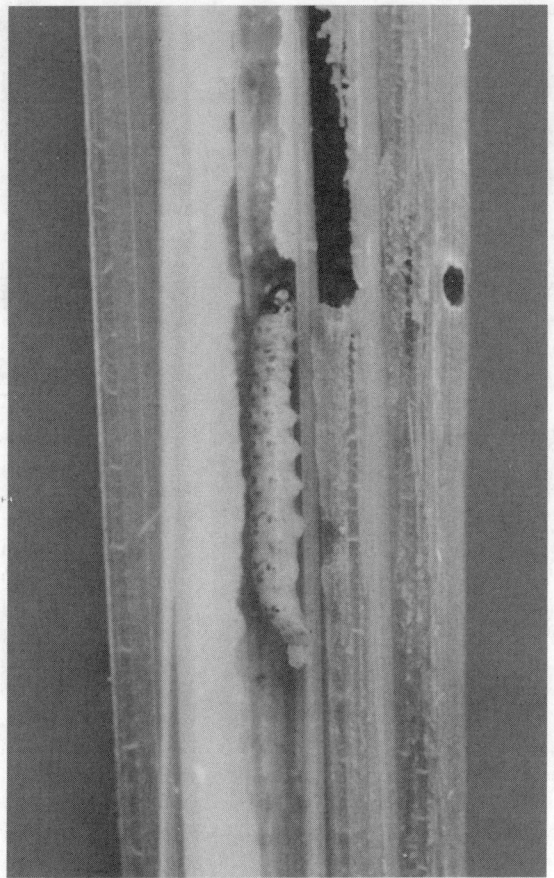

FIGURE 7.3 Shoot borer — larva.

observed in the duration of the life cycle (up to 30 d during August to October and up to 38 d during November to December) during various seasons (Jacob, 1981).

7.2.1.4 Seasonal Incidence

No information is available on the seasonal population dynamics of the shoot borer on turmeric, though the pest is known to occur throughout the crop period during June to December in Kerala.

7.2.1.5 Host Plants

The shoot borer is highly polyphagous and has been recorded on 65 host plants belonging to 30 families (Table 7.2). Many of the hosts of shoot borer are economically important plants and the pest infests various parts of these plants, such as buds, flowers, shoots, and fruits.

7.2.1.6 Resistance

The reaction of various turmeric types to shoot borer in the field has been studied. Sheila et al. (1980) reported that among the 13 types of turmeric screened at Vellanikkara (Thrissur District, Kerala), Dindigam Ca-69 (an aromatica type) was the least susceptible and Amruthapani Kothapeta Cll-317 (a longa type), the most susceptible. Philip and Nair (1981) reported that among the 19 turmeric types screened, Manuthy Local was the most tolerant. Velayudhan and Liji (2003) recorded

TABLE 7.2
Host Plants of *Conogethes punctiferalis*

Common Name	Scientific Name	Family	Distribution
Custard apple	*Annona* sp.	Annonaceae	Australia
Cherimoya	*Annona cherimola* Mill.	Annonaceae	India
Hollyhocks	*Alcea rosea* L.	Malvaceae	India
Cotton	*Gossypium* sp.	Malvaceae	Australia, India
Silk cotton tree	*Ceiba pentandra* (L.) Gaertn	Bombacaceae	India, Indonesia
Cocoa	*Theobroma cacao* L.	Sterculiaceae	India, Sri Lanka
Carambola	*Averrhoa carambola* L.	Oxalidaceae	—
Orange	*Citrus* sp.	Rutaceae	Australia, China, Japan
Tangor	*Citrus nobilis* Lour.	Rutaceae	—
—	*Fortunella* sp.	Rutaceae	China
Grape	*Vitis vinifera* L.	Vitaceae	India
Longan	*Dimocarpus longan* Lour.	Sapindaceae	China
Rambutan	*Nephelium lappaceum* L.	Sapindaceae	Malaysia
Soapnut	*Sapindus emarginatus* Vahl.	Sapindaceae	India
Soapnut	*Sapindus laurifolius* Vahl.	Sapindaceae	India
Mango	*Mangifera indica* L.	Anacardiaceae	India
Sumac	*Rhus chinensis* Mill.	Anacardiaceae	Japan
Bean	*Canavalia indica*	Fabaceae	Australia
—	*Cassia* sp.	Fabaceae	Australia
Fever nut	*Caesalpinia bonducella* Flem.	Fabaceae	India
Soybean	*Glycine max* (L.) Merr.	Fabaceae	Australia
Tamarind	*Tamarindus indica* L.	Fabaceae	India
Hawthorn	*Crataegus pinnatifida* Bunge	Rosaceae	China
Loquat	*Eriobotrya japonica* (Thunb.) Lindl.	Rosaceae	China
Apple	*Malus domestica* Borkh.	Rosaceae	Japan
Cherry	*Prunus japonica* Thunb.	Rosaceae	China, Japan
Peach	*Prunus persica* (L.) Batsch	Rosaceae	Australia, India, Thailand
Pear	*Pyrus communis* L.	Rosaceae	China, India
Granadilla	*Passiflora* sp.	Passifloraceae	Australia
Papaya	*Carica papaya* L.	Caricaceae	Australia, Philippines
Garuga	*Garuga pinnata* Roxb.	Rubiaceae	India
Dahlia	*Dahlia* sp.	Compositae	Australia
Sunflower	*Helianthus annuus* L.	Compositae	Sri Lanka
Perssimon	*Diospyros kaki* Thunb.	Ebenaceae	Japan, Korea
Teak	*Tectona grandis* L.	Verbenaceae	Burma, Indonesia
Amaranth	*Amaranthus* sp.	Amaranthaceae	India
Black pepper	*Piper nigrum* L.	Piperaceae	—
Guava	*Psidium guajava* L.	Myrtaceae	Australia, India
Pomegranate	*Punica granatum* L.	Lythraceae	India
Avocado	*Persia americana* Mill.	Lauraceae	India
Queensland nut	*Macadamia integrifolia* Maiden & Betche	Proteaceae	Australia
Castor	*Ricinus communis* L.	Euphorbiaceae	Australia, Bangladesh, India, Indonesia, Papua New Guinea
Chestnut	*Castanea mollissima* Blume	Fagaceae	China
Oak	*Quercus* spp.	Fagaceae	Korea
Oak	*Quercus acutissima* Carrutt.	Fagaceae	Japan
Jack	*Artocarpus heterophyllus* Lam.	Moraceae	India
Mulberry	*Morus* sp.	Moraceae	India
Fig	*Ficus carica* L.	Moraceae	India

Continued

TABLE 7.2 *(Continued)*
Host Plants of *Conogethes punctiferalis*

Common Name	Scientific Name	Family	Distribution
Alligator pepper	*Aframomum melegueta* Schum.	Zingiberaceae	India
—	*Alpinia* sp.	Zingiberaceae	India
Galangal	*Alpinia galanga* (L.) Sw.	Zingiberaceae	India
—	*Amomum* sp.	Zingiberaceae	India
—	*Alpinia microstephanum* Baker	Zingiberaceae	India
Greater cardamom	*Alpinia subulatum* Roxb.	Zingiberaceae	India
Turmeric	*Curcurma longa* L.	Zingiberaceae	India, Sri Lanka
Yellow zedoary	*Curcuma aromatica* Salisb.	Zingiberaceae	India
Mango ginger	*Curcuma amada* Roxb.	Zingiberaceae	India
Cardamom	*Elettaria cardamomum* Maton	Zingiberaceae	India, Sri Lanka
Ginger lily	*Hedychium coronarium* J. Konig	Zingiberaceae	India
Yellow ginger lily	*Hedychium flavescens* Carey ex Rosc.	Zingiberaceae	India
Ginger	*Zingiber officinale* Rosc.	Zingiberaceae	India, Sri Lanka
Banana	*Musa* sp.	Musaceae	Australia
Sugarcane	*Saccharum officinarum* L.	Poacea	Australia
Sorghum	*Sorghum bicolor* (L.) Moench	Poacea	Australia, India
Maize	*Zea mays* L.	Poacea	Australia, China
Cedar	*Cryptomeria japonica* (L. f.) D. Don	Taxodiaceae	—

Source: *Review of Applied Entomology-Series A/Review of Agricultural Entomology*, CAB International, Wallingford; *Crop Protection Compendium* (2002), CAB International, Wallingford; *CABPESTCD*, CAB International, Wallingford; Koya, K.M.A., Devasahayam, S., and Premkumar, T. *J Plantation Crops*, 19, 1–13, 1991.

the incidence of the shoot borer on 489 accessions belonging to 21 morphotypes and the lowest incidence was observed in morphotype II with a mean score of 2 on a 0 to 9 scale; 22 accessions were tolerant with a score of less than 3. Kotikal and Kulkarni (2001) screened eight genotypes of turmeric in the field at Belgaum (Karnataka) and reported that all of them were susceptible with more than 10% damage.

7.2.1.7 Natural Enemies

Various natural enemies have been recorded on the shoot borer, especially from Sri Lanka, China, Japan, and India. The only record of a virus infecting shoot borer is *Dichocrocis punctiferalis* NPV (Baculoviridae) (Murphy et al., 1995). *Dolichurus* sp. (Sphegidae), *Xanthopimpla* sp. (Ichneumonidae), and *Phanerotoma hendecasisella* Cam. (Braconidae) were recorded as parasitoids of shoot borer from Sri Lanka (Rodrigo, 1941). *Apanteles* sp. (Braconidae), *Brachymeria lasus* West. (Chalcidae), and *Temelucha* sp. (Ichneumonidae) were recorded as parasitoids of shoot borer infesting longan (*Dimocarpus longan* Lour.) in China (Huang et al., 2000). *Trathala flavoorbitalis* (Cam.) (Icheumonidae) and *Brachymeria obscurata* Wlk. from China, *Apechthis scapulifera*, *Scambus persimilis* (Ichneumonidae), and *B. obscurata* from Japan, have also been documented as natural enemies of the shoot borer (CABI, 2002).

A number of natural enemies have been documented on the shoot borer in India. The entomopathogenic nematode *Steinernema glaseri* Steiner (Steinernematidae) has been recorded on larvae (CABI, 2002). *Angitia trochanterata* Morl. (Ichneumonidae), *Theromia inareolata* (Braconidae), *Bracon brevicornis* West., *Apanteles* sp. (Braconidae), *Brachymeria euploeae* West. (Chalcidae) (David et al., 1964), and *Microbracon hebetor* Say. (Braconidae) (Patel and Gangrade, 1971) were documented as natural enemies of the pest infesting castor. *Brachymeria nosatoi* Habu and *B. lasus* West. were recorded as parasitoids of the pest by Joseph et al. (1973). Various parasitoids have been recorded on the shoot borer infesting cardamom and they include *Palexorista parachrysops* (Tachinidae), *Agrypon* sp., *Apechthis copulifera, Eriborus trochanteratus* (Morl.), *Friona* sp., *Gotra* sp., *Nythobia* sp., *S. persimilis, Temeluca* sp., *Theronia inareolata, Xanthopimpla australis* Kr., *Xanthopimpla kandiensis* Cram. (Ichneumonidae), *B. brevicornis* West., *Microbracon hebator, Apanteles* sp., *Phanerotoma hendecasisella* Cram. (Braconidae), *Synopiensis* sp., *Brachymeria* sp. nr. *australis* Kr. and *B. obscurata* (Chalcidae) (CPCRI, 1985; Varadarasan, 1995). *Hexamermis* sp. (Mermithidae) and *Apanteles taragamme* (Braconidae) have been documented on the shoot borer infesting ginger in Kerala (Devasahayam, unpublished).

Mermithid nematode (Mermithidae), *Myosoma* sp. (Braconidae), *X. australis* Kr. and general predators such as dermapteran [*Euborellia stali* Dohrn (Carcinophoridae)], asilid flies (*Philodicus* sp. and *Heligmoneura* sp.) (Asilidae), and spiders (*Araneus* sp., *Micaria* sp., and *Thyene* sp.) have been recorded on the shoot borer infesting turmeric in Kerala (Jacob, 1981).

7.2.1.8 Management

7.2.1.8.1 Chemical Control

In spite of the serious nature of damage caused by the shoot borer, no field trials have been conducted for the control of the pest on turmeric. The insecticides generally recommended for the management of shoot borer on ginger have also been recommended against the pest on turmeric (IISR, 2001).

7.2.1.8.2 Biopesticides

Two commercial products of *Bacillus thuringiensis*, namely, Bioasp and Dipel, were evaluated along with malathion for the management of the shoot borer in the field at Peruvannamuzhi (Kerala, India). The trials indicated that all the treatments were effective in reducing the damage caused by the pest compared to control when sprayed at 21-d intervals during July to October. Spraying of Dipel 0.3% was the most effective treatment resulting in significantly lower percentage of infested pseudostems (Devasahayam, 2002).

Choo et al. (1995) evaluated the pathogenecity of entomopathogenic nematodes against the shoot borer. *Steinernema* sp. and *Heterorhabditis* sp. caused 90% and 100% mortality of test insects, respectively, in the laboratory, when 20 nematodes per larva were inoculated. Choo et al. (2001) later reported that the LC_{50} for *S. carpocapsae* Pocheon strain and *H. bacteriophora* Hamyang strain were 5.6 and 5.8, while the mortalities were 96.9% and 96.5%, respectively.

7.2.1.8.3 Plant Products

Evaluation of neem oil 1% and commercial neem products (1%) in the field for the management of the shoot borer indicated that consistent results could not be obtained (IISR 2002; 2003; 2004).

7.2.1.8.4 Sex Pheromones

Many workers have demonstrated the presence of sex pheromones in the shoot borer (Konno et al., 1980; 1982; Liu et al., 1994; Kimura and Honda, 1999). Trials on the efficacy of sex pheromones in the field have also been reported on various crops (other than turmeric) in China, Japan, Korea, and India (Cai and Mu, 1993; Liu et al., 1994; Chakravarthy and Thygaraj, 1997; 1998; Jung et al., 2000).

7.2.2 Rhizome Scale (*Aspidiella hartii* Ckll.)

7.2.2.1 Distribution

The rhizome scale is distributed mainly in the tropical regions in Asia, Africa, Central America, and Caribbean Islands, but authentic records of the pest infestation on turmeric in various parts of the world including India are limited. Apart from India, the pest has also been recorded to infest turmeric rhizomes and leaves in West Africa and West Indies (Hill, 1983).

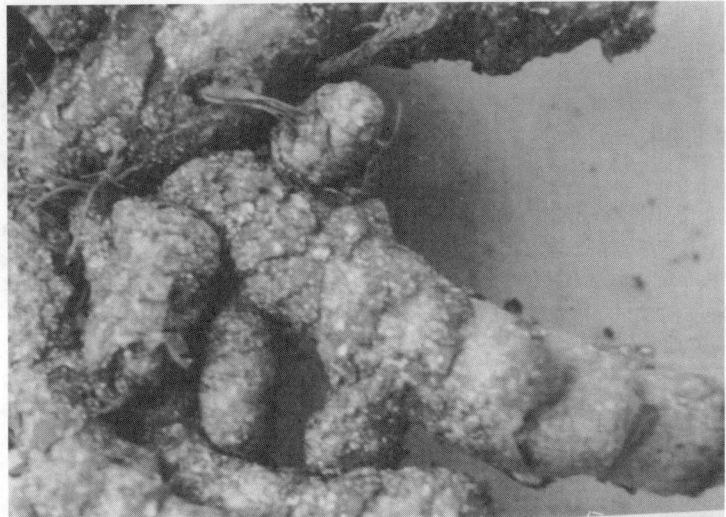

FIGURE 7.4 Turmeric rhizome infested by rhizome scale.

7.2.2.2 Damage

The rhizome scale infests rhizomes of turmeric both in the field and in storage. In the field, the pest is generally seen during the later stages of the crop, and in severe cases of infestation, the plants wither and dry. In storage, the pest infestation results in shriveling of buds and rhizomes; when the infestation is severe, it adversely affects the sprouting of rhizomes (Figure 7.4).

7.2.2.3 Life History

The adult females of rhizome scale are minute, circular, and light brown to gray measuring about 1.5 mm in diameter (Figure 7.5). Females are ovo-viviparous and also reproduce parthenogeneti-

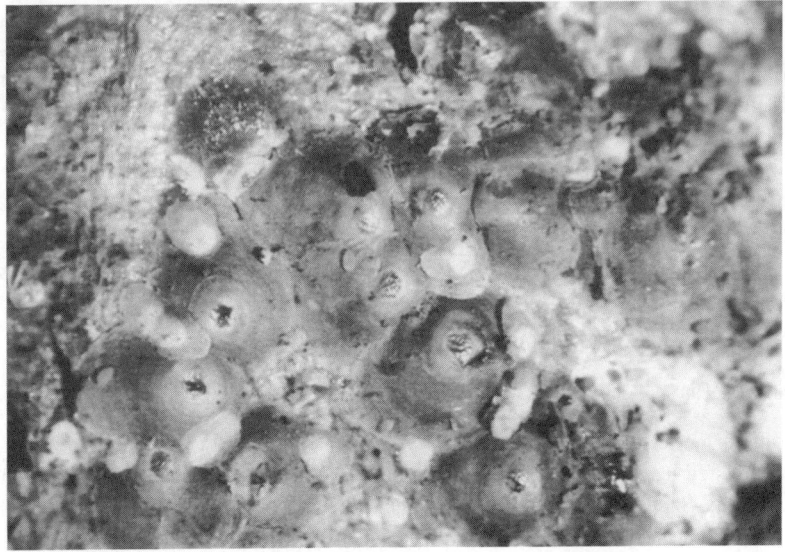

FIGURE 7.5 Rhizome scale — adults.

cally. Little information is available on the life history of the pest on turmeric. A single female lays about 100 eggs and the life cycle from egg to adult is completed in about 30 d (Jacob, 1982; 1986). The pest completes its life cycle in 11 to 20 d on yams (*Dioscorea* spp.) (Palaniswami, 1994).

7.2.2.4 Host Plants

In India, the rhizome scale has been recorded to infest ginger (*Zingiber officinale* Rosc.) (Ayyar, 1940), elephant foot yam [*Amorphophallus paeoniifolius* (Dennst.) Nicolson (=*A. companulatus*) (Regupathy et al., 1976)], yams [*Dioscorea alata* L., *Dioscorea esculenta* (Lour.) Burkill, and *Dioscorea rotundata* Poir (Palaniswami et al., 1979)], tannia [*Xanthosoma sagittifolium* (L.) Schott] (Jacob, 1986), and taro [*Colocasia esculenta* (L.) Schott.] (Pillai and Rajamma, 1984).

   In other countries, the rhizome scale has been recorded to infest yams and tannia in West Indies (Ballou, 1916; Catoni, 1921), yams in Panama Canal Zone in Central America (Fisher, 1920), Ivory Coast (Sauphanor and Ratnadass, 1985), and Nigeria (Onazi, 1969); and sweet potato in Africa (Sasscer, 1920).

7.2.2.5 Resistance

Regupathy et al. (1976) studied the reaction of 191 turmeric types to rhizome scale at Coimbatore (Tamil Nadu, India) and found that 87 accessions were free from infestation. Velayudhan and Liji (2003) recorded the incidence of the rhizome scale on 489 accessions belonging to 21 morphotypes at Vellanikkara (Kerala). Eighty accessions were free of infestation and the lowest scale incidence was observed in morphotype VI with a mean score of 0.6 on a 0 to 9 scale.

7.2.2.6 Natural Enemies

The natural enemies recorded on rhizome scale at Kasaragod (Kerala, India) include *Physcus* (*Cocobius*) *comperei* Hayat (Aphelinidae), *Adelencyrtus moderatus* Howard (Encyrtidae) and two species of mites. Parasitization by *P. comperei* brought down the population of rhizome scale by about 80% in three months (Jacob, 1986). At Peruvannamuzhi (Kerala, India), apart from *Cocobius* sp., a predatory beetle and ant were observed to predate on the rhizome scale (Devasahayam, 1996).

7.2.2.7 Management

Very few trials have been conducted for the control of rhizome scale on turmeric, and the insecticides generally recommended for the management of the pest on ginger have also been recommended against the pest on turmeric.

   Dipping the seed rhizomes of turmeric in quinalphos 0.1% for 5 min after harvest and before planting was found effective in controlling rhizome scale infestation (CPCRI, 1985). Discarding of severely infested and dipping the seed rhizomes in quinalphos 0.075% after harvest and before planting has also been recommended for the management of rhizome scale infestation (IISR, 2001).

7.3 MINOR INSECT PESTS

7.3.1 Sap Feeders

Infestation of turmeric leaves by *Stephanitis typica* Dist. in India was first reported by Fletcher (1914). The infested leaves turned pale and dried up. In Maharashtra, the pest population appeared in the field during September and reached its peak during November (Patil et al., 1988). Kotikal and Kulkarni (2000a) reported that the incidence of *S. typicus* in the field was higher at Hukkeri (Belgaum District), Mudhol (Bijapur District), Chincholi (Gulbarga District), and Basavakalyan (Bidar District) taluks in northern Karnataka and caused more than 10% damage to the crop.

Spraying of dimethoate or phosphamidon (0.05%) was more effective in controlling the pest (Thangavelu et al., 1977).

Ayyar (1920) included *Panchaetothrips indicus* Bagn. infesting turmeric in his list of economically important Thysanoptera from India. The infested leaves roll up, turn pale, and gradually dry up; the development of rhizomes is reduced when the pest infestation is severe. Ananthakrishnan (1973) mentions that this species is always associated with *Asprothrips indicus* (Bagn.). The incidence of *P. indicus* in the field was higher at Gokak (Belgaum District), Indi (Bijapur District), Aland (Gulbarga District), and Basavakalyan (Bidar District) taluks in the northern districts of Karnataka causing more than 10% damage to the crop (Kotikal and Kulkarni, 2000a). The other species of thrips recorded by these authors in this region include *A. indicus* (Bagn.), *Anaphothrips sudanensis* (Trybom), and *Haplothrips* sp. Spraying of dimethoate 0.06%, fenpropathrin 0.02%, bendiocarb 0.08%, and methyl demeton 0.05% were more effective in controlling the pest (Balasubramanian, 1982).

Apart from *A. hartii*, *Aspidiotus curcumae* Gr. has been recorded on turmeric infesting pseudostems and rhizomes (Nair, 1975). The other scale insects recorded on turmeric include *Howardia biclavis* (Com.) (Chua and Wood, 1990), *Coccus hesperidium* (Ckll.), and *Hemiberlesia palmae* L. (Anonymous, 2005). *Planococcus* sp. was observed to infest rhizomes in the field at Wayanad District in Kerala, India. *Pseudococcus* sp. has been recorded to infest rhizomes in the field at Fiji (Anonymous, 2005). Unidentified mealybugs were also recorded to infest turmeric leaves in northern Karnataka (Kotikal and Kulkarni, 2000b).

The aphid species recorded to infest turmeric include *Pentalonia nigronervosa* Coq. at Fujian in China (Zhou et al. 1995) and India (Kotikal and Kulkarni, 2000b), and *Uroleucon compositae* (Theobald) from India (Kotikal and Kulkarni, 2000b). The other hemipterans recorded to feed on sap from leaves and pseudostems in northern Karnataka in India include *Nezara viridula* (L.), *Coptosoma cribraria* Fab., *Tettigoniella ferruginea* Wlk., *Oxyrhachis tarandus* F., *Tricentrus bicolor* Dist, *Cletus rubidiventris* West., *Cletus bipunctatus* West., and *Riptortus pedestris* F. (Kotikal and Kulkarni, 2000b).

7.3.2 LEAF FEEDERS

Two species of grasshoppers, namely, *Orthacris simulans* B. and *Cyrtacanthacris ranacea* Stoll, caused more than 10% damage to the crop by feeding on the leaves in the northern districts of Karnataka, whereas *Letana inflata* Bru. and *Phenoroptera gracillus* Burm. caused minor damage to the crop in the region (Kotikal and Kulkarni, 2000b).

Among the leaf-feeding caterpillars recorded on turmeric, the turmeric skipper, *Udaspes folus* Cram. is the most serious, especially in India where it has been recorded as early as 1909 (Maxwell-Lefroy and Howlett, 1909). The larvae cut and fold leaves, remain within, and feed on them. Studies on the biology of the pest conducted at Godavari Delta (Andhra Pradesh) indicated that egg, larva, prepupa, and pupal stages lasted for 5, 25 to 30, 2, and 7 to 8 d, respectively (Sujatha et al. 1992). At Kasaragod (Kerala, India), egg, larval, and pupal periods lasted for 4 to 5, 13 to 25, and 6 to 7 d, respectively, and the pest was abundant in the field during August to October (Abraham et al., 1975). At Maharashtra, the pest population reached its peak in the field during October (Patil et al., 1988). The natural enemies and alternate hosts of the pest have been reviewed by Koya et al. (1991).

Ramakrishna and Raghunath (1982) reported a heavy incidence of *Creatonotus gangis* (L.) on turmeric in Andhra Pradesh causing severe defoliation. The biology of the pest on turmeric was also studied by them. The hesperid leaf roller *Notocrypta curvifascia* C and R Felder was recorded in Andaman Islands feeding on turmeric leaves especially during November (Veenakumari et al., 1994). The other leaf feeding caterpillars recorded on turmeric in India include *Bombotelia nugatrix* Gr., *Catopsila pomona* F., *Spilarctia obliqua* Walk. (Nair, 1975), *Amata passalis* Fab., *Euproctis lutifolia* Hamp., and *Spodoptera litura* (F.) (Kotikal and Kulkarni, 2000b).

Leaf-feeding beetles such as *Lema praeusta* Fabr., *L. signatipennis* Jac., and *L. semiregularis* Jac. have been recorded on turmeric in Orissa in India (Sengupta and Behura, 1955), whereas *L. fulvicornis* Jac. has been recorded from Sri Lanka (Hutson, 1936). Life histories of *L. praeusta* and *L. semiregularis* have been studied. The egg, larval, and pupal periods of the former species lasted for 8 to 10, 10 to 12, and 15 to 25 d, respectively, whereas that of the latter species lasted for 4 to 5, 15, and 19 d, respectively (Sengupta and Behura, 1956; 1957).

The leaf beetle *Pseudocophora* sp. congregate and feed on tender and older leaves of turmeric. The larvae feed on the roots. The incubation, larval, and pupal periods lasted for 4 to 6, 15 to 20, and 6 to 20 d, respectively (CPCRI, 1979). The other leaf feeding beetles recorded on turmeric include *Colasposoma splendidum* (F.), *Cryptocephalus rajah* Jac., *Epilachna sparsa* (Hbst), *Ceratobasis nair* (Loc.), *Myllocerus viridanus* Fab., *Lema lacordairei* Baly. (CPCRI 1977), *Hedychorus rufomaculatus* M. (Nair, 1975), *Chirida bipunctata* F., *Cryptocephalus schestedti* F., *Monolepta signata* (Olivier), *Raphidopalpa abdominalis* (Fab.), *Myllocerus discolor* Boheman, and *Myllocerus undecimpustulatus* Faust. (Kotikal and Kulkarni, 2000b).

7.3.3 RHIZOME BORERS

Various species of dipteran maggots bore into rhizomes and roots and are generally seen in plants affected by rhizome rot disease. The maggots recorded on turmeric rhizomes include *Calobata* sp. (Fletcher, 1914), *C. albimana* Macq. (Rao and Reddy, 1990), *Mimegralla coeruleifrons* Macq. (CPCRI, 1977), *M. albimina* (Doleschall) (Kotikal and Kulkarni, 2000b), *Eumerus pulcherrimus* Bru. (Ghorpade et al., 1988), *Libnotes punctipennis* (Shylesha et al., 1998), and *Tipula* sp. (Kotikal and Kulkarni, 2000b).

Ghorpade et al. (1983) conducted surveys in Maharashtra (India) and reported that *M. coeruleifrons* was endemic in Sangli and Satara districts of Maharashtra (India) and resulted in reduction in 25.4% yield of turmeric. The population of adult *M. coeruleifrons* in the field was higher at Raibag, Athani (Belgaum District), Mudhol, Indi (Bijapur District), Chincholi (Gulbarga District), and Basavakalyan (Bidar District) taluks in the northern districts of Karnataka (Kotikal and Kulkarni, 2000b). Ghorpade et al. (1988) studied various aspects of bioecology of *M. coeruleifrons* on turmeric at Kholapur (Maharashtra). Females laid 76 to 150 eggs in soil, and the egg, larval, and pupal stages lasted for 2 to 5, 13 to 25, and 5 to 15 d, respectively. The pest infestation was at its peak in the field during mid-August to mid-October. The biology of rhizome fly during different seasons was studied in the laboratory and field at Dharwad and the total life cycle lasted from 28.3 to 38.0 d in the laboratory and 28.3 ± 3.6 d in the field. The adults were preyed upon by spiders (Araneae) and dragonflies (Odonata); eggs and larvae by earwigs (*Forficula auricularia* L.) and the pupae were parasitized by *Trichopria* sp. (Ichneumonidae) and *Spalangia* sp. (Pteromalidae) (Kotikal and Kulkarni, 2000c).

In trials involving systemic granular insecticides, application of phorate 0.75 kg ai/ha thrice at 30-d intervals was more effective for the management of the rhizome fly (Dhoble et al., 1978). Later, trials conducted at Sangli District (Maharashtra) indicated that spraying parathion 0.03% (six sprays) along with soil application of diazinon, chlordane, or aldrin (0.75 kg/acre) (twice) was more effective in controlling the pest infestation (Dhoble et al., 1981). Kotikal and Kulkarni (1999) evaluated the efficacy of six insecticides as seed rhizome treatment for the management of the pest and among the various insecticides, soaking of seed rhizomes in dimethoate (1.5 ml/l) for 4 h and 8 h resulted in least adult fly emergence, and maximum germination was obtained when the seed rhizomes were soaked in dimethoate (1.5 ml/l) which was on par with monocrotophos (1 ml/l). Reddy and Reddy (2000) mention that in Andhra Pradesh, fields that received neem cake as dressing were free of rhizome fly.

Jadhav et al. (1982) screened several varieties of turmeric against the rhizome fly at Maharashtra and found that Sugandham and Duggirala were more resistant. Kotikal and Kulkarni (2001) screened eight genotypes of turmeric in the field at Belgaum (Karnataka) and reported that all of them were susceptible, recording more than 10% damage.

Nair (1978) reported that maggots of *Calobata* sp. were observed in rotting turmeric rhizomes. Rao and Reddy (1990) mention that *C. albimana* Macq. was a major pest on turmeric in the coastal regions of Andhra Pradesh. The pest damage was independent and rarely associated with rhizome rot caused by fungi. The pest infestation was more common in the crop raised in ill-drained soils and reached its peak during October to February causing 6.7 to 13.8% yield loss. Studies on the biology of the pest indicated that the egg, larval, and pupal periods lasted for 3 to 4, 23 to 25, and 10 to 15 d, respectively.

The larvae of the crane fly, *L. punctipennis* Meij, were observed to bore into turmeric rhizomes at Meghalaya. The life cycle was completed in 40 to 55 d under laboratory conditions. Rhizome damage was very low in local cultivars such as Lakadong, VK-77, RCT-1, and PTS-11 (Shylesha et al., 1998).

Although the rhizome fly has been categorized as a major insect pest of turmeric in many areas in India, critical studies have not been conducted on the role of the insect in rhizome rot disease. Surveys conducted by Ghorpade et al. (1983) in Maharashtra indicated that the pest infestation was less in light- and well-drained soils (where the incidence of disease is lower). Rao and Reddy (1990) also mention that the incidence of the pest was higher in ill-drained soils. Cultivars such as PCT-8, PCT-10, Suguna, and Sudarshana that were free from rhizome rot disease in northern Telengana were also free of rhizome fly infestation (Rao et al., 1994). In the northern districts of Karnataka, the insect was generally observed in rhizomes affected by rhizome rot disease (Kotikal and Kulkarni, 2000b). In Kerala, where the incidence of rhizome rot on turmeric is negligible, the rhizome fly rarely infests turmeric rhizomes, whereas ginger rhizomes affected by rhizome rot are seriously infested indicating that the rhizome fly is a secondary pest on the crop. The association of dipteran maggots with diseased rhizomes of ginger has been investigated by various workers (Iyer et al., 1981; Premkumar et al. 1982; Radke and Borle 1982; Koya 1988; 1990), and these studies indicate that the maggots could infest only diseased ginger rhizomes and hence cannot be considered as a primary pest of the crop. A similar situation may exist in turmeric also.

The white grub *Holotrichia serrota* F. was reported to infest rhizomes and roots of turmeric at western districts of Maharashtra (Patil et al., 1988). An unidentified species of *Holotrichia* was observed to feed on roots in northern Karnataka (Kotikal and Kulkarni, 2000b).

7.4 MAJOR INSECT PESTS OF STORED TURMERIC

Various insects have been recorded on dry turmeric used as a spice, especially from India. These insects mainly belong to the orders Coleoptera and Lepidoptera, among which the cigarette beetle (*L. serricorne* Fab.), drugstore beetle (*S. paniceum* L.), and coffee bean weevil (*A. fasciculatus* DeG.) are serious.

7.4.1 DISTRIBUTION

The insect pests of dry turmeric are widely distributed in the warmer parts of the world occurring mainly in Asia and Africa. In temperate regions they are common in heated stores. Abraham (1975) reported that the cigarette beetle and coffee bean beetle were the most common insect pests of dry turmeric in Kerala, and 30 to 60% of the samples were infested by these pests. Srinath and Prasad (1975) collected market samples of stored turmeric across India, and 88 of 115 infested samples were infested by the cigarette beetle. In Udaipur (Rajasthan), 67.7% of the market samples were infested by the cigarette beetle (Kavadia et al., 1978). In Bangladesh, the cigarette beetle was the most serious pest of turmeric based on various market samples of spices collected from Dhaka (Rezaur et al., 1982).

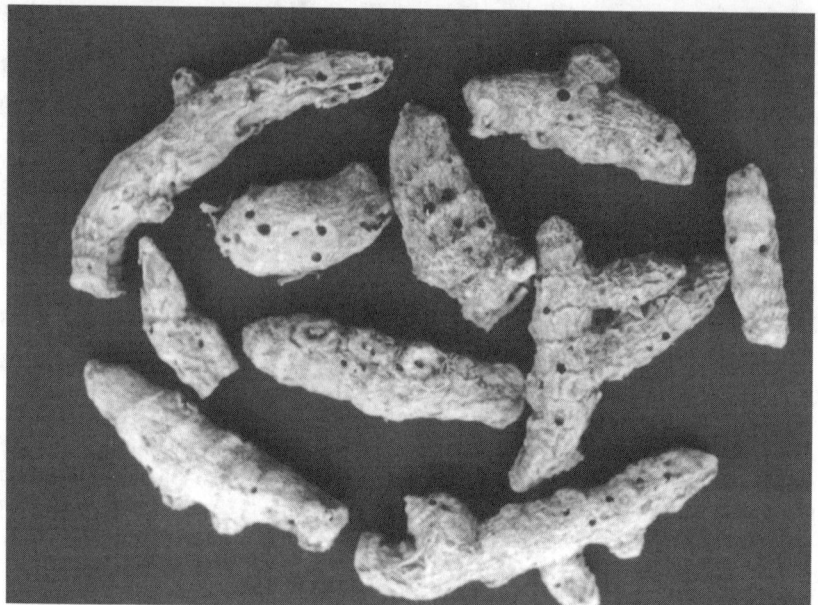

FIGURE 7.6 Dry turmeric rhizomes damaged by cigarette beetle.

7.4.2 DAMAGE

The larvae of cigarette beetle and drugstore beetle tunnel into dry turmeric and also contaminate the produce with abundant production of frass (Figure 7.6). The larvae and adults also make extensive holes in the produce. The adults of cigarette beetle do not feed but tunnel through the produce to leave the cocoon, making extensive holes. Both adults and larvae of coffee bean weevil are injurious to dry turmeric rhizomes and completely feed on it leaving only the outer covering intact. The cigarette beetle was reported to cause 39.8% weight loss at Udaipur (Kavadia et al., 1978). Studies on the effect of infestation by cigarette beetle and drugstore beetle on proximate composition and uric acid levels of turmeric rhizomes indicated that after 3 and 6 months of infestation, there were significant reduction in protein, fat, and ash contents and an increase in the uric acid level (Gunasekaran et al., 2003).

7.4.3 LIFE HISTORY

The cigarette beetle is a small (3 to 4 mm) brown beetle with smooth elytra with fine hairs. The head is strongly deflected under the pronotum, especially when alarmed, and the antennae are serrated. The eggs are creamy white; the larvae are gray-white with dense hairs and are very active when young but become sluggish as they age. There are four to six larval instars and the later instars are scarabaeiform. Pupation occurs within a silken cocoon, and the pupa is brown. The incubation period lasts for 9 to 14 d and the larval and pupal periods are 17 to 29 and 2 to 8 d, respectively, in Kerala (Abraham, 1975). The biology of the pest on turmeric has also been studied in Bangladesh (Rezaur et al., 1982).

The drugstore beetle resembles the cigarette beetle superficially but is smaller with striated elytra, and the distal segments of the antenna are clubbed. The larvae are pale white with the abdomen terminating in two dark horny points in fully grown specimens. The eggs are cigar shaped and hatch in 6 d. The larval and pupal periods last for 10 to 20 and 8 to 12 d, respectively (Abraham, 1975). In Rajasthan, Srivastava (1959) mentioned that the life cycle of the pest was completed in about 6 weeks.

The coffee bean weevil is a small (3- to 5-mm long), gray, stout beetle with pale marks on the elytra and with a long clubbed antennae. The eggs are oval and are laid in small pits dug on the rhizomes by the female beetles. Pupation takes place within the infested rhizomes. The entire life cycle lasts for 21–28 d (Abraham, 1975).

7.4.4 HOSTS

All the storage pests infest a wide range of produce including cocoa and coffee beans, cereals, spices, dried fruits, oil seeds, confectionery products, processed foodstuffs, and even animal products. Studies on olfactory response of adults of *L. serricorne* and *S. paniceum* to various spices including turmeric indicated that in *L. serricorne*, the highest attraction value of 42.6% was observed in turmeric; however, in case of *S. paniceum*, the attraction value was only 1.1% (Jha and Yadhav, 1991a). Laboratory studies on growth and food intake by the pest on various spices indicated the order of preference as cumin > anise > ginger > turmeric powder > turmeric (Jacob 1992).

7.4.5 NATURAL ENEMIES

Several natural enemies including predatory mites, hemipterans and coleopterans, and hymenopterous parasitoids have been recorded on major storage pests. Predatory mites such as *Acaropsis docta* (Berl.), *Acaropsellina solers* (Kuzin), *Cheyletus* spp., *Tydeus* sp. (Cheyletidae), *Tyrophagus putrescentiae* (Sch.), *Pyemotus tritici* (Lagreze-Fossat and Montane) (Pyemotidae), *Chortoglyphus gracilipes* (Chortoglyphidae), *Blattisocius tarsalis* (Berl.), and *B. keegan* Fox (Ascidae) are commonly associated with storage pests (Al-Badry et al., 1980; Stusak et al., 1986; Rizk et al., 1980; Kumar, 1997; CABI, 2002; Rao et al., 2002; Riudavets et al., 2002).

The predatory bugs recorded on major stored pests include *Xylocoris flavipes* (Reuter) (Anthocoridae), *Peregrinator biannulipes* (Montr. and Sign.), *Alloeocranum biannulipes* (Montr. and Sign.) (Reduviidae), and *Termatophyllum insigne* (Miridae) (Yao et al., 1982; Tawfik et al., 1984–1985). The predatory beetles include *Tribolium castaneum* (Hbst.) (Bostrychidae), *Alphitobius diaperinus* (Panzer), *Tenebroides mauritanicus* (L.) (Tenebrionidae), *Thaneroclerus buqueti* (Lefevre), and *Tilloidea notata* (Klug) (Cleridae) (Jacob and Mohan, 1973; Gautam, 1989; Iwata, 1989; CABI, 2002).

The hymenopterous parasitoids recorded include *Cephalonomica gallicola* (Ashmead), *Israelius carthami*, *Perisierola gestroi* (Bethylidae), *Anisopteromalus calandrae* (Howard), *Lariophagus distinguendus* (Forst), and *Pteromalus cerealellae* Ashmead (Pteromalidae) (Kohno et al., 1987; Brower, 1991; CABI, 2002).

7.4.6 MANAGEMENT

Various strategies have been suggested for the management of storage pests including storage in suitable containers, fumigation, heat treatment, radiation, and application of insecticides.

Polypropylene (2 μm) was resistant to biting by *L. serricorne* and *S. paniceum*. *S. paniceum* did not bite low-density black polyethylene (250 μm), aluminium foil laminated with low-density polyethylene (50 μm), or printed polypropylene (50 μm) indicating the potential of using these materials for packaging spice products including turmeric (Jha and Yadav, 1991b).

Srivastava (1959) mentioned that in fumigation tests, a 1:19 mixture of ethylene dibromide and carbon tetrachloride applied at 8 lb per 1000 cubic ft. for 24 h was most effective for controlling storage pest infestations in turmeric. Abraham (1975) suggests impregnation of jute bags lined with alkathene (500 guage) with malathion 0.2% or fumigation with methyl bromide for 6 h for preventing the pest infestation. However, regulations on pesticide residues in many importing countries render these fumigants unsafe for application. Kavadia et al. (1978) suggested fumigation of godowns with phosphine gas from celphos tablets used at a rate of 140 tablets/100 m³ which resulted in mortality of adults and larvae. Jacob (1986) suggests fumigation with aluminium phosphide tablets in an airtight store for 2 to 3 d for controlling the pest infestation. Muthu and Majumdar (1974) have furnished the concentration, time of exposure, and residual effects of various fumigants recommended for controlling insect infestations in various spices including turmeric. Fumigation of dry turmeric with fumes of burning sulphur in a chamber is widely practiced by many traders in India.

Padwal-Desai et al. (1987) suggested treating whole and ground spices including turmeric with Co^{60} gamma radiation at 1 kGy and stored at 28 to 30°C. Rezaur et al. (1982) also mention that the *L. serricorne* could be effectively controlled by gamma radiation. Gamma radiation treatment of eggs and larvae of *L. serricorne* and *S. paniceum* indicated that a dose of 25 Gy prevented eggs from developing into the adult stages and a dose of 50 Gy was required to prevent older larvae from developing. Doses above 30 Gy produced a sterilizing effect on both the species (Harwalkar et al., 1995).

Sex pheromones have been identified in *L. serricorne* and *S. paniceum* (Barratt, 1974; 1977; Kuwahara et al., 1975; Chuman, 1984; Chuman et al., 1985) and have been used for monitoring of population of these species in stores. Aggregation pheromones have also been identified in *A. fasciculatus* (Singh, 1993; Novo, 1998).

7.5 MINOR INSECT PESTS OF STORED TURMERIC

The other insects infesting stored dry turmeric rhizomes include *Tenebriodes mauritanicus* (L.), *Ephestia* sp. (Abraham, 1975), *Oryzaephilus surinamensis* L. (Srinath and Prasad, 1977), *Setomorpha rutella* Zell. (Jacob, 1986), *Rhizopertha dominica* (F.) (CABI, 2002), and *Tribolium castaneum* (Hbst.) (Kotikal and Kulkarni, 2000b).

ACKNOWLEDGMENT

We are thankful to Ms. Tresa Thomas, Indian Institute of Spices Research, Calicut, who assisted us in various ways during the preparation of the manuscript.

REFERENCES

(CABI) CAB International (2002) *Crop Protection Compendium*. CAB International, Wallingford.
(CPCRI) Central Plantation Crops Research Institute (1977) *Annual Report for 1976*, Central Plantation Crops Research Institute, Kasaragod.
(CPCRI) Central Plantation Crops Research Institute (1979) *Annual Report for 1977*, Central Plantation Crops Research Institute, Kasaragod.
(CPCRI) Central Plantation Crops Research Institute (1985) *Annual Report for 1983*, Central Plantation Crops Research Institute, Kasaragod.
(IISR) Indian Institute of Spices Research (2001) (Turmeric extension pamphlet), Indian Institute of Spices Research, Calicut.
(IISR) Indian Institute of Spices Research (2002) *Annual Report 2001–02*, Indian Institute of Spices Research, Calicut.
(IISR) Indian Institute of Spices Research (2003) *Annual Report 2002–03*, Indian Institute of Spices Research, Calicut.
(IISR) Indian Institute of Spices Research (2004) *Annual Report 2003–04*, Indian Institute of Spices Research, Calicut.
Abraham, C.C. (1975) Insect pests of stored spices and their control. *Arecanut Spices Bull.*, 7, 4–6.
Abraham, V.A., Pillai, G.B. and Nair, C.P.R. (1975) Biology of *Udaspes folus* Cram. (Lepidoptera: Hesperidae), the leaf roller pest of turmeric and ginger. *J. Plantation. Crops*, 3, 83–85.
Al-Badry, E.A., Rizk, G.N. and Hafez, S.M. (1980) Biological studies of the predatory mite *Acaropsis docta* (Berlese) attacking stored product pests. *Mesopotamia J. Agric.*, 15, 179–202.

Ananthakrishnan, T.N. (1973) *Thrips: Biology and Control*. Macmillan India, Delhi.

Anonymous (2005) *Curcuma longa* Accessed on 30 March 2005 at http:// www.ecoport.org /perl/ecoport 15.

Ayyar, T.V.R. (1920) A note on the present knowledge of Indian Thysanoptera and their economic importance. *Report, Proceedings of the Third Entomological Meeting, Pusa, February 1919*. 2, 618-622.

Ayyar, T.V.R. (1940) *Handbook of Economic Entomology for South India*. Government Press, Madras.

Balasubramanian, M. (1982) Chemical control of turmeric thrips *Panchaetothrips indicus* Bagnall. *South Indian Hort.*, 30, 54-55.

Ballou, H.A. (1916) Report of the prevalence of some pests and diseases in the West Indies during 1915. Part I. Insect pests. *West Indian Bulletin, Barbados*, 16 (1), 1–30.

Barrat, B.I.P. (1977) Sex pheromone emission by female *Stegobium paniceum* (L.) (Coleoptera: Anobiidae) in relation to reproductive maturation and oviposition. *Bull. Entomol. Res.*, 67, 491–499.

Barratt, B.I.P. (1974) Timing of production of a sex pheromone by females of *Stegobium paniceum* (L.) (Coleoptera, Anobiidae) and factors affecting male response. *Bull. Entomol. Res.*, 64, 621–628.

Boo, K.S. (1998) Variation in sex pheromone composition of a few selected lepidopteran species. *J. Asia-Pacific Entomol.*, 1, 17–23.

Brower, J.H. (1991) Potential host range and performance of a reportedly monophagous parasitoid, *Pteromalus cerealellae* (Hymenoptera: Pteromalidae). *J. Entomol. News*, 102, 231–235.

Butani, D.K. (1985) Spices and pest problems: 10. Turmeric. *Pesticides*.19 (5), 22-25.

Cai, R.X. and Mu, Z.L. (1993) Trap trials for *Dichocrocis punctiferalis* Guenee with sex pheromone in citrus orchards. *China Citrus*, 22, 33.

Catoni L.A. (1921) Plant inspection and quarantine report (1919–20). *Bulletin, Puerto Rico Insular Experimental Station, Rio Piedras*, 27, 1–23.

Chakravarthy, A.K and Thyagaraj, N.E. (1997) Response of the cardamom (*Elettaria cardamomum* Maton) shoot and fruit borer (*Conogethes punctiferalis* Guenee Lepidoptera: Pyralidae) to different pheromone compounds. *Insect Environ.*, 2, 127–128.

Chakravarthy, A.K. and Thyagaraj, N.E. (1998) Evaluation of selected synthetic sex pheromones of the cardamom shoot and fruit borer, *Conogethes punctiferalis* Guenee (Lepidoptera: Pyralidae) in Karnataka. *Pest Manage. Trop. Ecosyst.*, 4, 78–82.

Choo, H.Y., Kim, H.H., Lee, S.M., Park, S.H., Choo, Y.M. and Kim, J.K. (2001) Practical utilization of entomopathogenic nematodes, *Steinernema carpocapsae* Pocheon strain and *Heterorhabditis bacteriophora* Hamyang strain for control of chestnut insect pests. *Korean J. Appl. Entomol.*, 40, 69–76.

Choo, H.Y., Lee, S.M., Chung, B.K., Park, Y.D. and Kim, H.H. (1995) Pathogenecity of Korean entomopathogenic nematodes (Steinernematidae and Heterorhabditidae) against local agricultural and forest insect pests. *Korean J. Appl. Entomol.*, 34, 314–320.

Chua, T.H. and Wood, B.J. (1990) Other fruit trees and shrubs. In Rosen, D. (ed.), *Armoured Scale Insects, Their Biology, Natural Enemies and Control. Volume B*. Elsevier Science Publishers B. V., Amsterdam, 543–552.

Chuman, T. (1984) Chemical study on the sex pheromone of the cigarette beetle (*Lasioderma serricorne* F.). *J. Agric. Chem. Soc. Japan*, 58, 1135–1146.

Chuman, T., Mochizuki, K., Mori, M., Kohno, M., Kato, K. and Noguchi, M. (1985) *Lasiderma* chemistry: sex pheromone of cigarette beetle (*Lasiderma serricorne* F.). *J. Agric. Chem. Ecol.*, 11, 417–434.

David, B.V., Narayanaswami, P.S. and Murugesan, M. (1964) Bionomics and control of the castor shoot and capsule borer *Dichocrocis punctiferalis* Guen. in Madras State. *Indian Oilseeds J.*, 8, 146–158.

Devasahayam, S. (1996) Biological control of insect pests of spices. In Anandaraj, M. and Peter, K.V. (eds.), *Biological Control in Spices*. Indian Institute of Spices Research, Calicut, 33–45.

Devasahayam, S. (2002) Evaluation of bioperticides for the management of shoot borer (*Conogethes punctiferalis* Guen.) on turmeric (*Curcuma longa* L.).In: Rethinam, P., Khan, H.H., Reddy, V.M., Mandal, P.K., and Suresh, K., eds. Plantation Crops Research and Development in the New Millenium, Coconut Development Board: Kochi: Kochi, 489-490.

Dhoble, S.Y., Kadam, M.V. and Dethe, M.D. (1978) Control of turmeric rhizome fly by granular systemic insecticides. *J. Maharashtra Agric. Univ.* 3, 209-210.

Dhoble, S.Y., Kadam, M.V. and Dethe, M.D. (1981) Chemical control of turmeric rhizome fly, *Mimegralla coeruleifrons* Macquart. *Indian J. Entomol.* 43, 207-210.

Fisher, H.C. (1920) *Report 1919, Health Department, Panama Canal, Mount Hope*, pp.1–134.

Fletcher, T.B. (1914) *Some South Indian Insects and Other Animals of Importance Considered Especially from an Economic Point of View*. Government Press, Madras.

Gautam, R.D. (1989) Exploration of the lesser meal worm for the control of storage insects together with its stages and effect on seed viability. *Agric. Situation India* 64, 487–489.

Ghorpade, S.A., Jadhav, S.S. and Ajri, D. (1983) Survey of rhizome fly on turmeric and ginger in Maharashtra. *J. Maharashtra Agric. Univ.*, 8, 292–293.

Ghorpade, S.A., Jadhav, S.S. and Ajri, D. (1988) Biology of rhizome fly, *Mimegralla coeruleifrons* Macquart (Micropezidae: Diptera) in India, a pest of turmeric and ginger crops. *Trop. Pest Manage.*, 34, 48–51.

Gunasekaran, N., Baskaran, V. and Rajendran, S. (2003) Effect of insect infestation on proximate composition of selected stored spice products. *J. Food Sci. Tech. Mysore*, 40, 239–242.

Hampson, G.F. (1856) *Fauna of British India Including Ceylon and Burma*. Moths. Taylor and Francis, London.

Harwalkar, M.R., Dongre, T.K. and Padwal-Desai, S.R. (1995) Radiation disinfestations of spices and spice products I. Radiation sensitivity of developmental stages of *Lasioderma serricorne* and *Stegobium paniceum*. *J. Food Sci. Tech.*, 32, 249– 251.

Hill, D.S. (1983) *Agricultural Insect Pests of the Tropics and their Control (2nd edition)*. Cambridge University Press, Cambridge.

Honda, H. (1986a) EAG responses of the fruit- and Pinaceae-feeding type of yellow peach moth, *Conogethes punctiferalis* (Guenee) (Lepidoptera: Pyralidae) to monoterpene compounds. *Appl. Entomol. Zool.*, 21, 399–404.

Honda, H. (1986b) Post-mating reproductive isolation between fruit- and Pinaceae- feeding types of the yellow peach moth, *Conogethes punctiferalis* (Guenee) (Lepidoptera: Pyralidae). *Appl. Entomol. Zool.*, 21, 489–491.

Honda, H., Maruyama, Y. and Matsumoto, Y. (1986) Comparisons in EAG response to n-alkyl compounds between the fruit- and Pinaceae-feeding type of yellow peach moth, *Conogethes punctiferalis* (Guenee) (Lepidoptera: Pyralidae). *Appl. Entomol. Zool.*, 21, 126–133.

Huang, Y,Q., Zhang, X.J., Wei, H., Hu, Q.Y. and Zhan, Z.X. (2000) Studies on the *Dichocrocis punctiferalis* Guenee and its enemies. *Acta Agric. Univ. Jiangxiensis*, 22, 523–525.

Hutson, J.C. (1936) Report on the Work of the Entomological Division. *Adminstrative Report, Division of Agriculture, Ceylon*. D22-D28.

Iwata, R. (1989) *Tilloidea notata* (Klug) (Coleoptera: Cleridae), as a predator of *Stegobium paniceum* (Linn.) (Coleoptera: Anobiidae). *Pan Pacific Entomol.*, 65, 449–450.

Iyer, R., Koya, K.M.A. and Banerjee, S.K. (1981) Relevance of soil and insects in the epidemiology of rhizome rot of ginger. In *Abstracts, Third International Symposium on Plant Pathology, 14–18 December 1981*, New Delhi.

Jacob, S. (1992) Host preference of the cigarette beetle, *Lasioderma serricorne* (F.) to few stored spices. *Plant Prot. Bull., Faridabad*, 44, 16–17.

Jacob, S.A. (1980) Pests of ginger and turmeric and their control. *Pesticides*, 14 (11), 36–40.

Jacob, S.A. (1981) Biology of *Dichocrocis punctiferalis* Guen. on turmeric. *J. Plantation. Crops*, 9, 119–123.

Jacob, S.A. (1982) Biology and bionomics of ginger and turmeric scale *Aspidiotus hartii* Green. In Nair, M.K., Premkumar, T., Ravindran, P.N. and Sarma, Y.R. *Proceedings, National Seminar on Ginger and Turmeric*. Central Plantation Crops Research Institute, Kasaragod, pp. 131–132.

Jacob, S.A. (1986) Important pests of ginger and turmeric and their control. *Indian Cocoa Arecanut Spices J.*, 9, 61–62.

Jacob, S.A. and Mohan, M.S. (1973) Predation on certain stored product insects by red flour beetle. *Indian J. Entomol.*, 35, 95–98.

Jadhav, S.S., Ghorpade, S.A. and Ajri, D.S. (1982) Field screening of some turmeric varieties against rhizome fly. *J. Maharashtra Agric. Univ.*, 7, 260.

Jha, A.N. and Yadav, T.D. (1991a) Olfactory response of L*asioderma serricorne* Fab. and *Stegobium paniceum* (Linn.) to different spices. *Indian J. Entomol.*, 53, 396–400.

Jha, A.N. and Yadav, T.D. (1991b) Insect proof packaging against *Lasioderma serricorne* Fab. and *Stegobium paniceum* (Linn.), pests of spices. *Indian J. Entomol.*, 53, 401–404.

Joseph, K.J., Narendran, T.C. and Joy, P.J. (1973) Taxonomic Studies of the Oriental Species of *Brachymeria* (Hymenoptera: Chalcididae). *Report, PL 480 Research Project*, University of Calicut, Calicut.

Jung, J.K., Han, K.S., Choi, K.S. and Boo, K.S. (2000) Sex pheromone composition for field trapping of *Dichocrocis punctiferalis* (Lepidoptera: Pyralidae) males. *Korean J. Appl. Entomol.*, 39, 105–110.

Kavadia, V.S., Pareek, B.L. and Sharma, K.P. (1978) Control of *Lasioderma serricorne* Fab. Infestation of turmeric by phosphine fumigation. *Entomon*, 3, 57–58.

Kimura, T. and Honda, H. (1999) Identification and possible functions of the hair pencil scent of the yellow peach moth *Conogethes punctiferalis* (Guenee) (Lepidoptera: Pyralidae). *Appl. Entomol. Zool.*, 34, 147–153.

Kohno, M., Hori, Y. and Iloh, H. (1987) Storage of flue-cured tobacco and the occurrence of *Cephalonomia gallicola* (Ashmead) (Hymenoptera: Bethylidae). *Japanese J. Appl. Entomol. Zool.*, 31, 260–261.

Konno, Y., Arai, K. and Matsumato, Y. (1982) (E)-10-Hexadecenal, a sex pheromone component of the yellow peach moth, *Conogethes punctiferalis* (Guenee) (Lepidoptera: Pyralidae). *Appl. Entomol. Zool.*, 17, 201–217.

Konno, Y., Honda, H. and Matsumato, Y. (1980) Observations on the mating behaviour and bioassay for the sex pheromone of the yellow peach moth, *Conogethes punctiferalis* (Guenee) (Lepidoptera: Pyralidae). *Appl. Entomol. Zool.*, 15, 321–327.

Kotikal, Y.K. and Kulkarni, K.A. (1999) Management of rhizome fly, *Mimegralla coeruleifrons* Macquart (Micropezidae: Diptera), a serious pest of turmeric in northern Karnataka. *Pest Manage. Hort. Ecosyst.*, 5, 62–66

Kotikal, Y.K. and Kulkarni, K.A. (2000a) Incidence of Insect Pests of Turmeric (*Curcuma longa*) in Northern Karnataka, India. *J. Spices Aromatic Crops*, 9, 51–54.

Kotikal, Y.K. and Kulkarni, K.A. (2000c) Studies on the biology of turmeric rhizome fly. *Karnataka J. Agric. Sci.*, 13, 593–596.

Kotikal, Y.K. and Kulkarni, K.A. (2001) Reaction of selected turmeric genotypes to rhizome fly and shoot borer. *Karnataka J. Agric. Sci.*, 14, 373–377.

Kotikal, Y.K. and Kulkarni, K.A. (2000b) Insect Pests Infesting Turmeric in Northern Karnataka. *Karnataka J. Agric. Sci.*, 13, 858–866.

Koya, K.M.A. (1988) Distribution of dipteran maggots associated with ginger (*Zingiber officinale* Rosc.) in Kerala. *J. Plantation Crops*, 16, 137–140.

Koya, K.M.A. (1990) Role of rhizome maggot *Mimegralla coeruleifrons* Macquart in rhizome rot of ginger. *Entomon*, 15, 75–77.

Koya, K.M.A., Devasahayam, S. and Premkumar, T. (1991) Insect pests of ginger (*Zingiber officinale* Rosc.) and turmeric (*Curcuma longa* Linn.) in India. *J. Plantation. Crops*, 19, 1–13.

Kumar, D. (1997) Mite infestation in stored grain pest culture. *Insect Environ.*, 3 (2), 42.

Kuwahara, Y., Fukami, H., Ishii, S., Matsumura, F. and Burkholderm W.E. (1975) Studies of the isolation and bioassay of the sex pheromone of the drugstore beetle, *Stegobium paniceum* (Coleoptera: Anobiidae). *J. Chem. Ecol.*, 1, 413–422.

Liu, M.Y., Tian, Y. and Li, Y.X. (1994) Identification of minor components of the sex pheromone of yellow peach moth, *Dichocrocis punctiferalis* Guenee, and field trials. *Entomol. Sinica*, 1, 150–155.

Maxwell-Lefroy, H. and Howlett, F.M. (1909) *Indian Insect Life-A Manual of the Insects of the Plains (Tropical India)*. Government Press, Calcutta.

Murphy, F.A., Fauquet, C.M., Mayo, M.A., Jarvis, A.W., Ghabrial, S.A., Summers, M.D., Martelli, G.P. and Bishop D.H.L. (eds.) (1995) *Sixth Report of the International Committee on Taxonomy of Viruses, Archives of Virology*, Springer Verlag, New York.

Muthu, M. and Majumdar, S.K. (1974) Insect control in spices. In *Proceedings, Symposium on Development and Prospects of Spice Industry in India*. Association of Food Science and Technology. p. 35.

Nair, M.R.G.K. (1975) *Insects and Mites of Crops in India*. Indian Council of Agricultural Research, New Delhi.

Nair, M.R.G.K. (1978) *A Monograph on Crop Pests of Kerala and Their Control*. Kerala Agricultural University, College of Agriculture, Vellayani, Trivandrum.

Nayar, K.K., Ananthakrishnan, T.N. and David B.V. (1976) *General and Applied Entomology*. Tata McGraw-Hill Publishing Company Limited, New Delhi.

Novo, J.P.S. (1998) Olfactory response of the coffee bean weevil *Araecerus fasciculatus* (Deg.) (Coleoptera: Anthribiidae) to pheromones. *J. Ann. Soc. Entomol. Brazil*, 27, 337–343.

Onazi, O.C. (1969) The infestation of stored potatoes (*Solanum tuberosum*) by *Planococcus citri* (Risso) (Homoptera: Pseudococcidae) on the Jos Plateau, Nigeria. *Nigerian Entomol. Mag.* 2, 17–1 8.

Padwal-Desai, S.R., Sharma, A. and Amonkar, S.V. (1987) Disinfestation of whole and ground spices by gamma radiation. *J. Food Sci. Tech.*, 24, 321–322.

Palaniswami, M.S. (1994) Scale insect (*Aspidiella hartii*) and its parasitoids. *J. Root Crops,* 17, 75–76

Palaniswami, M.S., Pillai, K.S. and Abraham, K. (1979) Note on the occurrence and nature of damage by the scale insect *Aspidiella hartii* Ckll. (Diaspididae: Homoptera) on different edible *Dioscorea. J. Root Crops,* 5, 65–66.

Patel, R.K. and Gangrade, G.A. (1971) Note on the biology of castor capsule borer, *Dichocrocis punctiferalis. Indian J. Agric. Sci.,* 41, 443–444.

Patil, A.P., Thakur, S.G. and Mohalkar, P.R. (1988) Incidence of pests of turmeric and ginger in western Maharashtra. *Indian Cocoa Arecanut Spices J.,* 12, 8–9.

Philip, J. and Nair, P.C.S. (1981) Field reaction of turmeric types to important pests and diseases. *Indian Cocoa, Arecanut Spices J.,* 4, 107–109.

Pillai, K.S. and Rajamma, P. (1984) New records on the incidence of spotted beetle *Oides affinis* Jacoby (Coleoptera: Chrysomelidae) on elephant foot yam and scale insect *Aspidiella hartii* Ckll. (Homoptera: Diaspididae) on taro. *Entomon,*12, 113–114.

Premkumar, T., Devasahayam, S. and Koya, K.M.A. (1994) Pests of spice crops. In Chadha, K.L. and Rethinam, P. (eds.), *Advances in Horticulture, Vol. 10, Plantation and Spices Crops, Part 2.* Malhotra Publishing House, New Delhi, 787–823.

Premkumar, T., Sarma, Y.R. and Gautam, S.S.S. (1982) Association of dipteran maggots in rhizome rot of ginger. In Nair, M.K., Premkumar, T., Ravindran, P.N. and Sarma, Y.R. *Proceedings, National Seminar on Ginger and Turmeric.* Central Plantation Crops Research Institute, Kasaragod, 128–130.

Radke, S.G. and Borle, H.N. (1982) Status of rhizome fly *Mimegralla coeruleifrons* Macquart on ginger. *Punjabrao Krishi Vidyapeeth Res. J.,* 6, 68–69.

Ramakrishna, T.A.V.S. and Raghunath, T.A.V.S. (1982) New record of *Creatonotus gangis* (Linneaus) on turmeric in India. *Indian Cocoa Arecanut Spices J.* 5, 63.

Rao, C.V.N., Rao, B.N. and Babu, T.R. (2002) New record of predation on the eggs of cigarette beetle, *Lasioderma serricorne* (Fabricius) (Coleoptera : Anobiidae), a stored tobacco pest. *J. Biol. Control ,* 16, 169–170.

Rao, P.S., Krishna, M.R., Srinivas, C., Meenakumari, K and Rao, A.M. (1994) Short duration, disease resistant turmerics for northern Telangana. *Indian Hort.,* 39, 55–56.

Rao, S.V. and Reddy, P.S. (1990) The rhizome fly, *Calobata albimana* Macq, a major pest of turmeric. *Indian Cocoa, Arecanut Spices J.,* 14, 67–69.

Reddy, M.R.S. and Reddy, P.V.R.M. (2000) Preliminary observations on neem cake against rhizome fly of turmeric (*Curcuma longa*). *Insect Environ.,* 6 (2), 62.

Regupathy, A., Santharam, G., Balasubramanian, M. and Arumugam, R. (1976) Occurrence of the scale *Aspidiotus hartii* C. (Diaspididae: Hemiptera) on different types of turmeric *Curcuma longa* Lin. *J. Plantation. Crops,* 4, 80.

Rezaur, R., Ahmed, M., Hossain, M. and Nazar, G. (1982) A preliminary report on the problems of dried spice pests and their control. *Bangladesh J. Zool.,* 10, 141– 144.

Riudavets, J., Maya, M., Monserrat, M., Adler, C., Navarro, S., Scholler, M. and Hansen, L.S. (eds.) (2002) Predation by *Blattisocium tarsalis* (Acari: Ascidae) on stored product pests. *Proceedings, IOBC-WRPS Working Group (Integrated Protection in Stored Products), Lisbon, Portugal, 3–5 September 2001. Bulletin OILB SROP,* 25, 121–126.

Rizk, G.N., Al Badry, E.A. and Hafez, S.M. (1980) Biological studies of the predatory mite *Acaropsis solers* (Kuzin) attacking stored product pests. *Mesopotamia J. Agric.,* 15, 203–222.

Robinson, G.S., Tuck, K.R. and Shaffer, M. (1994) *A Field Guide to the Smaller Moths of South-East Asia.* Malaysian Nature Society, Kuala Lumpur.

Rodrigo, E. (1941) *Administration Report 1940,* Director of Agriculture, Colombo, 1–18.

Sasser, E.R. (1920) Important foreign insect pests collected on imported nursery stock in 1919. *J. Eco. Entomol.* 13, 181–184.

Sauphanor, B. and Ratnadass, A. (1985) Entomological problems associated with storage of yams in the Ivory Coast. *Agron. Trop.,* 40, 261–270.

Sengupta, G.C. and Behura, B.K. (1955) Some new records of crop pests from India. *Indian J. Entomol.,* 17, 283–285.

Sengupta, G.C. and Behura, B.K. (1956) Note on the life history of *Lema semiregularis* Jac. (Coleoptera: Chrysomelidae). *J. Bombay Nat. Hist. Soc.* 53, 484–485.

Sengupta, G.C. and Behura, B.K. (1957) On the biology of *Lema praeusta* Fab. *J. Econ. Entomol.,* 50, 471–474.

Sheila, M.K., Abraham, C.C. and Nair, P.C.S. (1980) Incidence of shoot borer (*Dichocrosis punctiferalis*) Guen. (Lepidoptera: Pyraustidae) on different types of turmeric. *Indian Cocoa Arecanut Spices J.,* 3, 59–60.

Shylesha, A.N., Azad- Thakur, N.S. and Ramachandra, S. (1998) Occurence of *Libnotes punctipennis* Meij (Tipulidae: Diptera) as rhizome borer of turmeric in Meghalaya. *Developments in Plantation Crops Research, Proceedings of the 12th Symposium on Plantation Crops, PLACROSYM-XII, Kottayam, India, 27–29 November 1996.* pp. 276–278.

Singh, K. (1993) Evidence of male and female of coffee bean weevil, *Araecerus fasciculatus* (DeG.) (Coleoptera: Anthribidae) emitted aggregation and sex pheromones. *J. Crop Res. Hisar,* 6, 97–101.

Srinath, D. and Prasad, C. (1975) *Lasioderma serricorne* F. as a major pest of stored turmeric. *Bull. Grain Technol,.* 13, 170–171.

Srivastava, B.K. (1959) *Stegobium paniceum* as a pest of stored turmeric in Rajastan, India and its control by fumigation. *FAO Plant Protection Bull.,* 7, 113–114

Stusak, J.M., Verner, P.H and Tung, N.V. (1986) The stored pests of the winged bean seeds, *Psophocarpus tetragonolobus. Agric. Trop. Sub-trop.,* 19, 143–157.

Sujatha, A., Zaherudeen, S. M. and Reddy, R. V. S. K. (1992) Turmeric leaf roller, *Udaspes folus* Cram. And its parasitoids in Godavari Delta. *Indian Cocoa Arecanut Spices J.,* 15, 118–119

Tawfik, M.F.S., Awadallah, K.T., El-Husseini, M.M. and Afifi, A.L. (1984–1985) Survey on stored drug insect and mite pests and their associated natural enemies in Egypt. Bull. Soc. Entomol. Egypt, 65, 267–274.

Thangavelu, P., Sivakumar,C.V. and Kareem, A.A. (1977) Control of lace wing bug, (*Stephanitis typica* Distant) (Tingidae: Homoptera) in turmeric. South Indian Hort., 25, 170–172.

Thyagaraj, N.E., Singh, P.K. and Chakravarthy, A.K. (2001) Sex determination of cardamom shoot and fruit borer, *Conogethes punctiferalis* (Guenee) pupae. *Insect Environ.,* 7, 93.

Varadarasan, S. (1995) Biological control of insect pests of cardamom. In Ananthakrishnan, T.N. (ed.) *Biological Control of Social Forest and Plantation Crop Insects.* Oxford and IBH Publishing Company Private Limited, New Delhi. pp. 109–111.

Veenakumari, K., Mohanraj, P. and Ranganath, H.R. (1994) New records of insect and mite pests of spice crops in Andaman Islands, India. *J. Spices Aromatic Crops,* 3, 164–166.

Velayudhan, K.C. and Liji, R.S. (2003) Preliminary screening of indigenous collections of turmeric against shoot borer (*Conognethes punctiferalis* Guen.) and scale insect (*Aspidiella hartii* Sign.). *J. Spices Aromatic Crops,* 12, 72–76

Yao, K., Deng, W.X., Tao, J.P. and Hu, W.Z. (1982) Preliminary observations of the bionomics of *Peregrinator biannulipes* Montrouzier & Signoret. Kunchong Zhishi., 19, 24–27.

Zhou, Z.J., Lin, Q.Y., X, L.H., Zheng, G.Z. and Huang, Z.H. (1995) Studies on banana bunchy top. III. Occurrence of its vector Pentalonia nigronervosa. J. Fujian Agric. Univ., 24, 32–38.

8 Postharvest Technology and Processing of Turmeric

K.V. Balakrishnan

CONTENTS

8.1 INTRODUCTION

The spice turmeric comes from the underground rhizomes of *Curcuma longa*. The center of domestication of turmeric is the Indian subcontinent. It is now cultivated in India, China, Taiwan, Pakistan, Bangladesh, Myanmar, Thailand, Sri Lanka, and Indonesia. In India, turmeric is cultivated in almost all states; the main regions being the states of Andhra Pradesh, Maharashtra, Orissa, Tamilnadu, Karnataka, and Kerala. Turmeric is also grown in Africa, Australia, Japan, Caribbean and Latin America-Jamaica, Haiti, Costa Rica, Peru, and Brazil.

Turmeric requires a hot and moist climate, a liberal supply of water, and a well-drained soil. It grows both under rainfed and irrigated conditions. Turmeric grows in all elevations ranging from sea level to an altitude of 1200 m. It is usually planted on raised beds to avoid stagnant water. Beds of convenient length and width are prepared based on the topography of the land. In India, planting is done during May to June. Turmeric can be grown mixed with other crops, since it is not affected by partial shade. It could also be rotated with other short-term crops.

Turmeric is ready for harvest in 7 to 9 months after planting, depending on cultivar, soil, and growing conditions. In India, the harvest season begins from February. Harvesting at the right maturity is important for optimum color and aroma. When mature, the leaves turn yellow and the whole plant gradually dries out. Before harvest, the leafy tops are cut off to facilitate digging out the rhizomes. The soil is first loosened with a small digger, and clumps are manually lifted. Rhizomes are cleaned from adhering soil by soaking in water, and long roots are removed. Care needs to be taken to prevent the rhizomes from being cut or bruised. Harvested rhizomes must be dried as soon as possible to minimize contamination, mold growth, and fermentation.

Fresh turmeric rhizome has a brown skin and bright orange flesh. Each rhizome consists of a central bulb portion bearing a number of finger-like lateral offshoots, called the fingers. The fingers, sometimes called the daughter rhizomes, are separated from the mother rhizome. The fingers, about 2 to 3 in. long, break off easily from the bulb.

8.2 PROCESSING

Processing of raw rhizomes assumes importance from the point of view of the appearance and color of the end product. The processing consists of three stages — curing, drying, and polishing.

8.2.1 CURING

Turmeric rhizomes are cured before drying. Curing essentially involves boiling fresh rhizomes in water until soft before drying. Boiling destroys the vitality of fresh rhizomes, obviates the raw odor, reduces drying time, and yields a uniformly colored product.

Some traditional methods for curing turmeric are no longer in practice. In the contemporary curing process, the cleaned rhizomes are boiled in copper or galvanized iron or earthern vessels with water just enough to soak them. Boiling is stopped when froth comes out, with the release of white fumes having the typical turmeric aroma (Rao et al., 1975; Spices Board, 1995; Anandaraj et al., 2001). The cooking lasts for 45 to 60 min when the rhizomes become soft; but the duration could be longer depending on the batch size. The stage at which boiling is stopped largely influences the color and aroma of the final product. Over-cooking spoils the color of the final product while under-cooking renders the dried product brittle (Weiss, 2002). Rhizomes are tested by pressing with fingers. Optimum cooking is attained when the rhizome yields to finger pressure and can be perforated by a blunt piece of wood. The cooking should be thorough as otherwise the product is prone to insect attack.

In an improved method, the cleaned rhizomes are taken in a perforated trough made of galvanized iron or mild steel sheet with extended parallel handle. Perforated trough containing the raw turmeric is then immersed in a pan of boiling water, which can hold three to four troughs at a time. Boiling is continued till the material is soft. The cooked turmeric is taken out of the pan by lifting the trough and draining the water into the pan itself. The same hot water in the pan can be used for cooking several batches (Spices Board, 1995; Anandaraj et al., 2001; Weiss, 2002). Small batches of 50 to 75 kg can be conveniently cured by immersion in boiling water.

It is important to boil batches of rhizome that are equal in size since different size materials would require different cooking times. Fingers and bulbs are boiled separately. Bulbs require longer cooking time than the fingers. Bigger bulbs are split into halves or quarters. Curing is more uniform when done in small batches. Curing should be done within 2 or 3 days after harvest to avoid rhizome spoilage. Various improvements in the cooking vessels and heating furnaces have been introduced to handle large quantities.

Recommendations as to the acidity or alkalinity of the boiling water vary. If water is acidic, 0.05 to 0.1% sodium bicarbonate or carbonate is sometimes added to make it slightly alkaline. Boiling in alkaline water is said to improve the color (Pruthi, 1976; Govindarajan, 1980; Velappan et al., 1993; Weiss, 2002). Cooking of rhizomes with alkaline additives imparts turmeric tubers an orange yellow color (Krishnamurthy et al., 1975). Lead chromate-based chemicals were earlier used in the curing process to impart bright yellow color. However, such methods being potentially hazardous to health, are out of vogue now.

Cooking rhizomes prior to drying promotes gelatinization of starch, facilitates uniform drying, and increases dehydration rate. Other benefits include uniform distribution of pigments inside rhizome and a more attractive product (not wrinkled) that lends itself to easier polishing (Govindarajan, 1980; Sampathu et al., 1988; Bambirra et al., 2002). Cooking reduces the drying time from 30–35 to 10–15 days (Spices Board, 1995). Curing also destroys the vitality of the rhizomes and removes the raw odor. Curing by the boiling process has an additional advantage of significantly reducing the microbial load on the rhizomes, thereby imparting a sterilizing effect before drying. However, it is reported that while the total volatile oil and color remains unchanged, curcuminoid extractability may be lower (Buescher and Yang, 2000).

There are controversies on the effect of cooking by different methods on the curcuminoid pigments and the color of ground turmeric (Sampathu et al., 1988). In one study (Bambirra et al., 2002), the content of pigments in the product obtained by alkaline cooking (0.1% sodium bicarbonate solution; pH 8.6) was found to be only 91% of that obtained by plain water cooking. This reduction in pigment level may be attributed to the destruction of curcuminoid pigments in the presence of alkali at high temperatures (Tonnesen and Karlsen,1985a; Price and Buescher, 1992). The study also included the Hunter CIE color characteristics of ground turmeric obtained by different cooking processes, expressed as $L*a*b*$. A lower value of L (luminosity) was observed for the control sample which has not been subjected to any heat treatment, indicating that this was the darkest product compared to the processed ones. Products obtained from retort as well as plain and alkaline water cooking were lighter in that order. Heat treatment of the rhizome prior to dehydration is also reported to inactivate oxidative enzymes, which could cause browning of the product (Govindarajan, 1980). Another factor supporting the browning could be the occurrence of Maillard reaction (Bambirra et al., 2002). Immersion cooking can result in loss of soluble sugars in the cooking water, thereby limiting the amount of sugars available for the Maillard reaction. Turmeric cooked by immersion, therefore, should yield a higher quality powder compared to the autoclaved. The $a*$ value (intensity of red) of the control sample was significantly lower than the processed ones. The values for plain- and alkaline-water processed samples did not differ significantly. Cooking in alkaline media, thus, did not affect intensity of red in ground turmeric compared to regular cooking. Cooking prior to drying also improved the $b*$ value (intensity of yellow) of the material compared to the control. Cooking in alkaline medium provided a product with higher intensity of yellow compared to regular cooking. According to Tonnesen and Karlsen (1985a), the presence of alkali at high temperatures could favor the formation of ferulic acid and feruloylmethane. Part of the feruloylmethane formed during alkaline degradation can participate in condensation reactions originating compounds of yellow to yellow-brownish color, which could influence the spectrophotometric determination of curcuminoid pigments.

8.2.2 DRYING

The cooked rhizomes are allowed to cool gradually and spread out to dry in the open in 5 to 7 cm thick layers on uncoated plain bamboo mats or concrete drying floor. A thinner layer is not desirable as this may result in surface discoloration (Sankaracharya and Natarajan, 1975; Spices Board, 1995; Anandaraj et al., 2001). Turmeric should be dried on clean surfaces to ensure that the product does not get contaminated by any extraneous matter. Care should be taken to avoid mold growth on the rhizomes. The rhizomes are turned over intermittently to ensure uniformity in drying. During night time, they are heaped or covered with a material that allows adequate aeration. It may take 10 to 15 days for the rhizomes to become completely dry. In most growing areas, the cooked rhizomes are dried in the sun. But, where unfavorable seasonal conditions prevail, improved drying methods using mechanical dryers are also used. Drying using cross-flow hot air at a maximum temperature of 60°C is found to give a satisfactory product (Spices Board, 1995). Artificial drying gives a brighter product than sun drying. Solar driers can also be economically used for drying turmeric. However, the maximum temperature achieved by the drier depends on the outside climatic conditions. Satisfactory outputs cannot be achieved in regions where cloudiness and humidity are high.

When the dried finger breaks cleanly with a metallic sound, it is sufficiently dry. Moisture content at this level will be generally 5 to 10%. Improperly dried spice is susceptible to microbial growth and infestation by storage pests. Storage of dry turmeric for very long periods is not desirable. In such cases, fumigation with permitted chemicals is undertaken.

Slicing the rhizomes prior to drying reduces drying time and yields turmeric with lower moisture content and better curcuminoid extractability (Govindarajan, 1980; Buescher and Yang, 2000). Machines have also been developed to mechanize the slicing operation.

Yield of dry turmeric varies from 20 to 30% depending upon the variety and the region of cultivation. The dry rhizomes possess earthy, slightly unpleasant odor and a bitterish, mild acrid taste; they impart an exciting warmth in the mouth and color the saliva yellow.

8.2.3 Polishing

Dried turmeric has a rough appearance and dull surface color. The outer surface can be polished to give a better finish. Polishing removes the surface roughness by getting rid of the surface scales, the small rootlets, and any remaining soil particles. Polishing is done either by manual or mechanical means.

Manual polishing consists of rubbing the dried turmeric on a hard surface or trampling them under feet wrapped with gunny bags. Shaking the rhizomes with stones in a gunny bag or bamboo basket is also practiced.

Mechanical polishing is carried out in polishing drums. These are very simple drums rotated by hand or by power. The drum is made of expanded metal mesh fixed to solid, circular end plates and is mounted on a central axis. A door is provided for charging and discharging. The drum is covered with a tight wrapping of woven wire, the mesh of which is small enough to retain the turmeric, but large enough to allow dust, dirt, and rootlets to fall through. When the drum filled with turmeric is rotated, polishing is effected by abrasion of the surface against the mesh as well as by rubbing rhizomes against each other as they roll inside the drum (Charley, 1938; Pruthi, 1992; Natarajan and Lewis, 1980; Spices Board, 1995; Anandaraj et al., 2001). Sprinkling turmeric water during polishing is said to improve the color (CSIR, 1950) in manually operated drums. Fifty to eighty kg per batch can be polished at a time. Where large quantities are handled, power-driven drums are used. In this method, turmeric receives a higher degree of polishing and becomes smoother. The capacity of the drum is up to 800 kg of turmeric and polishing takes 4 to 5 hours. The polishing wastage is 2 to 8% of the weight of turmeric or higher, depending on cultivar and the extent of polishing.

Polishing of the rhizomes is undertaken by growers on a small scale, but largely by dealers or exporting companies in commercial quantities. Cured and polished turmeric is brittle; it has a smooth and shining yellow color. Polishing is not necessary for turmeric intended for solvent extraction and recovery of color matter.

8.2.4 Coloring

In order to improve the surface color, the dried rhizomes are sometimes coated with turmeric powder in the course of polishing. This is done to half-polished rhizomes in two ways, known as dry and wet coloring. In the dry process, turmeric powder is added to the polishing drum in the last 10 min of polishing, whereas in the wet process, turmeric powder suspended in water is sprinkled over the rhizomes at the final stage (Anandaraj et al., 2001; Govindarajan, 1980). Treatment with emulsions containing alum, turmeric powder, castor seed paste, sodium bisulfite and sulphuric acid, or hydrochloric acid in different combinations has also been recommended to impart attractive surface color (Rao et al., 1975; Pruthi, 1976; 1980). However, chemical treatments are largely discouraged. The use of lead chromate, once practiced to achieve the same result, has now been abandoned due to potential toxicity. The yield of polished turmeric from fresh rhizomes varies from 15 to 25%.

8.3 CLEANING, GRADING, PACKING, AND STORAGE

Apart from separating the fingers, bulbs, and splits, little grading of the spice is done at the growers' end. Different grades based on botanical identity, size, appearance, and color conforming to standard specifications are prepared by dealers for international trade. Being a natural produce, dried turmeric is bound to gather contaminants during various stages of processing. The spice is also cleaned at this stage to remove such foreign materials. A sifter, destoner, and an air screen separator will help remove materials such as stones, dead insects, excreta, and other extraneous matter (Pruthi, 1992).

Cleaned and graded material is packed generally in new double burlap gunny bags and stored over wooden pallets in a cool, dry place protected from light. The stores should be clean and free from infestation of pests and harborage of rodents. It is not recommended to apply pesticides on the dried/polished turmeric to prevent storage pests.

In India, Sangli district in Maharashtra, the country's key turmeric trading center, has been following a century-old practice of storing turmeric in pits. This is a unique indigenous agri-commodity storing system in the country (Rao et al., 1975). Pits of 4.5 to 5 m deep with 3 and 2 m sides are dug on raised ground. The pits are allowed to dry for a couple of days and the sides and bottom are padded with a thick layer of paddy straw or any such material. Over the layer, a date-mat is spread. Turmeric bags are then placed in the pit, covered with a layer of straw, and then with soil. Turmeric can be stored in these pits for 3 to 4 yr with no change in quality.

Temperature-controlled warehouses are now largely replacing the conventional underground pits.

8.4   ORGANIC TURMERIC

With the growing concern on the uncontrolled application of chemical pesticides and artificial crop improvement techniques, organic production and processing of agricultural produces are gaining importance.

In general, to be labeled "organic," a product must be grown following organic agricultural practices and certified by an accredited certification body. Postharvest handling and processing must be done in certified facilities, whether on the farm or in food packing or processing facilities. Only mechanical, thermal, or biological methods can be used in organic processing. The use of genetically modified organisms (GMO) (plants, animals or bacteria) and products of GMO are prohibited in organic production. Likewise, ionizing radiation and sewage sludge are prohibited. The identification of the certifying body must be stated on the labels of organic products. Currently, there are slight differences in standards between countries. IFOAM (International Federation of Organic Agriculture Movement) has established organic production, processing, and trading standards and is working to harmonize certification systems worldwide (Riddle and Ford, 2000).

8.5   CHEMICAL COMPOSITION OF TURMERIC

Rhizomes of *C. longa* contain pigments that contribute the color, along with the essential and fixed oils, flavonoids, bitter principles, carbohydrates, protein, minerals, and vitamins. Carbohydrates constitute the major fraction of the spice. A proximate chemical composition is given in Table 8.1 (CSIR, 1950; Natarajan and Lewis, 1980; Spices Board, 2002).

TABLE 8.1
Chemical Composition of Turmeric

Moisture	6–13%
Carbohydrates	60–70%
Protein	6–8%
Fiber	2–7%
Mineral matter	3–7%
Fat	5–10%
Volatile oil	3–7%
Curcuminoids	2–6%

TABLE 8.2
Nutritional Composition of Turmeric,
100 g

Moisture, g	6
Food energy, kcal	390
Protein, g	8.5
Fat, g	8.9
Carbohydrate, g	69.9
Ash, g	6.8
Calcium, g	0.2
Phosphorus, mg	260
Sodium, mg	10
Potassium, mg	2500
Iron, mg	47.5
Thiamine, mg	0.09
Riboflavin, mg	0.19
Niacin, mg	4.8
Ascorbic acid, mg	50

Phytosterols, tocopherols, and fatty acids have also been identified (Moon et al., 1976; USDA, 2004). The fatty acids are mainly saturated straight chain, saturated iso, monoenoic, and dienoic acids. Typical nutritional composition for turmeric is given in Table 8.2 (Spices Board, 2002).

Two active constituents in turmeric are the coloring matter and the volatile oil. The coloring matter, which represents 2 to 6% of turmeric, consists mainly of curcumin and small amounts of its analogues, mainly demethoxycurcumin and bisdemethoxycurcumin. The essential oil, which is present up to 5%, is composed mainly of sesquiterpenes, many of which are specific for the species. Most important for the aroma are α- and β-turmerones and *ar*-turmerone.

8.6 TURMERIC PRODUCTS

8.6.1 GREEN TURMERIC

This is the undried fresh turmeric. The fleshy rhizomes are washed clean to remove the adhering mud. In fresh state, the rhizome has an aromatic and spicy fragrance. Green turmeric has limited food application. Its main use is as an ingredient in ayurvedic and folk medicine. Fresh turmeric also finds application in traditional religious rituals.

Except for seed purpose, fresh rhizomes are not stored for a long time, since they are likely to get deteriorated. Seed rhizomes must be kept in well-ventilated rooms to minimize rot, covered with turmeric leaves to prevent dehydration (Govindarajan, 1980). They are also stored in pits covered with sawdust, sand, or leaves of the plant *Glycosmis pentaphylla*, which act as an insect repellent. Another effective method is storage in 100 gauge polythene bags with 3% ventilation (Venkatesha et al., 1997). By this method, a high percentage (98.88%) of healthy rhizomes was recovered, which exhibited 91.9% sprouting in the field. The Indian Institute of Spice Research recommends the prestorage dip treatment for rhizome seeds in a fungicide composition containing 0.075% quinalphos and 0.3% mancozeb (Anandaraj et al., 2001). Bulbs are preferred to fingers as seed stock.

8.6.2 DRIED WHOLE TURMERIC

This is the primary form of turmeric for trade. India is the world's leading producer of turmeric. Statistics on Indian production is displayed in Table 8.3 (Spices Board, 2004). Indian turmeric is

TABLE 8.3
Turmeric — Indian Production and Exports

	Production				
Year	1998–1999	1999–2000	2000–2001	2001–2002	2002–2003
Area (ha)	160,700	176,250	187,430	162,950	149,410
Production (Mt)	598,260	646,170	719,600	552,300	527,960
	Exports				
Year	1998–1999	1999–2000	2000–2001	2001–2002	2002–2003
Quantity (Mt)	37,297	37,776	44,627	37,778	32,402
Value (Rs., lakhs)	12,914.49	12,351.81	11,557.62	9,073.71	10,337.99

considered the best in the world. Some of the popular Indian varieties are: Alleppey, Erode, Rajapore, Sangli, and Nizamabad.

India consumes over 90% of its produce internally. Data on the export of turmeric from India during a 5-yr period are also given in Table 8.3. Though only a small fraction of its production is exported, India supplies most of the world's requirement for turmeric.

Turmeric of commerce is described in three ways:

Fingers: These are the lateral branches or secondary, "daughter" rhizomes, which are detached from the central rhizome before curing. Fingers usually range in size from about 2.5 to 7.5 cm in length and may be somewhat over 1 cm in diameter. Broken and very small fingers are combined and marketed separately from whole fingers.

Bulbs: These are the central "mother" rhizomes, which are ovate in shape and are of a shorter length but a greater diameter than fingers.

Splits: Splits are the bulbs that have been split into halves or quarters to facilitate curing and subsequent drying.

Polished fingers possess the best appearance and are generally regarded as superior in quality to other forms of turmeric. The splits and bulbs are more fibrous and harder to grind. All types are used for producing turmeric powder and for oleoresin extraction.

8.7 THE MAJOR TYPES OF TURMERIC

Around 30 turmeric varieties are grown in India, but only two designations are commercially significant in the world market: "Alleppey" and "Madras," both named after the places of export from India. The deep yellow to orange-yellow Alleppey turmeric, grown in the Thodupuzha and Muvattupuzha regions of Kerala State, is predominantly imported by the U.S. in unpolished form, where users prefer it as a spice and a food colorant (Spices Board, 1995; ASTA 2002). Alleppey turmeric contains about 3.5 to 5.5% volatile oil, and 4.0 to 7.0% curcumin (ASTA 2002; Buescher and Yang, 2000; Weiss, 2002). It has a peppery, earthy odor and a slightly aromatic, bitter taste. Its flavor also is said to have gingery and nut-like undertones. Alleppy variety is easier to grind than its Madras counterpart: In contrast, the Madras type contains only 2% of volatile oil and 2% of curcumin. Madras turmeric is comprised of as many as nine cultivars including Guntur, Salem, Rajamundry, Nizamabad, and Cuddappah. The Madras turmeric is preferred by the British and Middle Eastern markets for its more intense, brighter, and lighter yellow color and is better suited for the mustard paste and curry powder or paste used in oriental dishes (ASTA 2002, Govindarajan, 1980). India keeps most of its Madras turmeric for domestic use and exports most of the Alleppey type Rajpuri, which is a special grade of Madras turmeric, has thicker and stumpier fingers but about the same color content as Guntur (about 3.7%). It is sold unmixed with other cultivars at a premium and it is reputed to be popular because it is easier to hand-process in the household than other grades.

TABLE 8.4
Analysis of Popular Indian Turmeric Cultivars

Cultivar	Oil Content (%)	Curcumin Content (%)
Waigon	7.20	3.51
Kuchupudi	5.60	4.03
Sugandham	5.30	3.62
Erode	4.00	3.00
Rajpuri	4.50	3.45
Gadhvi	6.00	3.49
Tekurpeta	2.50	1.82
Kasturi	5.80	3.44
Miraj 26	6.30	2.87
Alleppy	4.00	5.44
Duggirala	3.80	2.22
Cuddappa	3.30	2.46

West Indian turmeric: This comes to the world market from Caribbean, Central and South American countries. It is dull yellowish brown in color, mostly small, and of poor appearance (Spices Board, 1995).

The important consideration for the grower to obtain good quality turmeric is the selection of varieties according to the following criteria:

1. Color
2. Aroma
3. Yield
4. Resistance to diseases

Several cultivars of turmeric have evolved through natural selection. Besides the cultivars grown since ancient times, about 24 improved ones are grown in India alone. The traditional cultivars in India are known by name of locality where they are grown. Typical analysis of some popular Indian cultivars are given in Table 8.4 (Krishnamurthy et al., 1976). Extensive research has been undertaken in India to develop cultivars of improved yield and quality characteristics.

8.8 QUALITY CRITERIA FOR TRADE

Quality of cured turmeric is assessed on the pigment (curcumin) content, organoleptic character, general appearance as well as the size and physical form of the rhizome. As in the case of other spices, trade in turmeric is also governed by numerous regional and national regulations.

8.8.1 CLEANLINESS

Cleanliness has always been a major concern for spices. Main contaminants encountered in spices are rodent, animal, and bird filth; field and storage insects; spiders, mites, and psocids; mold and bacteria; extraneous materials; chemicals, pesticides, and mycotoxins (Spices Board, 2002). Turmeric can pick up contamination during various stages from harvest to postharvest processing and storage. The most popular cleanliness specification for spices and herbs the world over is the American Spice Trade Association Cleanliness Specifications for Spices, Seeds and Herbs (ASTA, 1999). The unified ASTA, U.S. FDA Cleanliness Specifications for Spices, Seeds and Herbs was

TABLE 8.5
ASTA Cleanliness Specifications for Turmeric

Parameter	Upper Limit
Whole insects, dead (by count)	3
Excreta, mammalian (by mg/lb)	5
Excreta, other (by mg/lb)	5.0
Mold (% by weight)[a]	3.00
Insect defiled/infested (% by weight)[a]	2.50
Extraneous/foreign matter (% by weight)	0.50

[a] Moldy pieces and/or insect-infested pieces by weight.

made effective from 1 January 1990. Major producing countries have built up their facilities to meet the requirements as per ASTA Cleanliness Specification.

The specification laid down by ASTA for the cleanliness of turmeric imported into the U.S. is given in Table 8.5 (ASTA, 1999). Similar specifications have also been laid down by other importing countries like U.K., Germany, and The Netherlands.

In the case of defective spices, U.S. FDA has laid down some defect action levels (DAL). If the defects exceed the DAL, the spice will be detained and subjected to reconditioning (cleaning to remove the defect). If defects cannot be removed by reconditioning, the lot may be destroyed or returned to the supplier (FDA, 1995). However, for turmeric no DAL have so far been fixed.

8.8.2 MICROBIOLOGY

The presence of microorganisms in food products is critical from the point of view of human health. Strict regulations have been specified by the importing countries for the limits of microbial load on spices. Microbiological specifications for spices in Germany and The Netherlands are given in Table 8.6 (Spices Board, 2002). Inadequate and unhygienic drying and storage leads to accumulation

TABLE 8.6
Microbiological Specification for Spices

Parameter	Standard Value	Danger Value
Germany		
Total aerobic bacteria	10^5/g	10^6/g
Escherichia coli	Absent	Absent
Bacillus cereus	10^4/g	10^5/g
Staphylococcus aureus	100/g	1000/g
Salmonella	Absent in 25 g	Absent in 25 g
Sulfite reducing clostrides	10^4/g	10^5/g
The Netherlands		
Bacilus cereus	Absent in 20 g	Danger values as above.
E. coli	Absent in 20 g	
Clostridum perfringens	Absent in 20 g	
S. aureus	Absent in 20 g	
Salmonella	Absent in 20 g	
Total aerobic bacteria	1×10^6/g	
Yeast and mold	1×10^3/g	
Coliform	1×10^2/g	

TABLE 8.7
Tolerance Levels for Pesticide Residues in
Turmeric under U.S. Regulations

Pesticide	Tolerance Limit, ppm
Lindane	0.50
BHC	0.05
Heptachlor	0.01
Heptachlor epoxide	0.01
Trifluralin	0.05
Ethylene oxide	50
Propylene oxide	300
Diquat	0.02
Dichlorvos	0.50
Dalapon	0.20
Aluminium phosphide	0.10
2,4-D	0.10
Glyphosate	0.20
Methyl bromide	100

of microbial load on the spice. Standard methods are available to measure the microbial load in spices (AOAC, 1998).

8.8.3 PESTICIDE RESIDUES

Uncontrolled application of chemical pesticides at various stages of plant growth results in the accumulation of their residues in spices, sometimes to levels beyond the acceptable limits. Common pesticides can be grouped into organochlorine and organophosphorus. With the growing concern on the carcinogenic properties of various pesticide residues, the importing countries are tightening the tolerance limits. A number of pesticides have already been banned or restricted for use and many more are under vigilant scrutiny. Pesticide residue continues to be a serious problem in all the spices for export (Balakrishnan, 2005).

Individual countries have fixed maximum residue levels (MRLs) for common pesticides in spices. The tolerence levels for pesticide residues in turmeric under U.S., German, and Spanish regulations are given in Tables 8.7, 8.8, and 8.9, respectively (Spices Board, 2002). Similar regulations in The Netherlands and U.K. are shown in Table 8.10. Gas chromatography is generally employed for the determination of pesticide residues in the spice (AOAC, 2000a).

8.8.4 AFLATOXINS

Another major issue in the quality of the spice is the presence of aflatoxins. Aflatoxins are a group of secondary metabolites of the fungi, *Aspergillus flavus* and *Aspergillus parasticus* and are rated as potent carcinogens. Inadequate and unhygienic drying leads to the growth of these fungi on the spice. Aflatoxins in spices are generally classified into four categories — B_1, B_2, G_1, and G_2. B_1 and B_2 are produced by *A. flavus*, whereas G_1 and G_2 are produced by *A. parasticus*. Of these B_1 is the most virulent carcinogen and has received the most attention. Thin layer chromatography (TLC) and fluorescence measurements have been largely used for afltoxin determination till recently (Scott and Kennedy, 1975). However, with the increasing concern on these potent toxins, more precise determination techniques such as high performance liquid chromatography (HPLC) are recommended (ASTA, 1997a).

European Spice Association (ESA) comprising of the members of the European Union has prescribed limits for aflatoxin as 5 ppb for B_1 and l0 ppb for the group. Member countries and

TABLE 8.8
Tolerance Levels for Pesticide Residues in Spices under German Regulations

Pesticide	Tolerance Limit, ppm[a]
Aldrin and dieldrin	0.1
Chlordane	0.05
Sum of DDT isomers	1.0
Endrin	0.1
HCH without lindane	0.2
Heptachlor and epoxide	0.1
Hexachlor benzol	0.1
Lindane	0.01
HCN and cyanides	15.0
Bromides	400
Carbaryl	0.1
Carbofuran	0.2
Chlorpyrifos	0.05
Methyl chlorpyrifos	0.05
Cypermethrin (sum of all isomers)	0.05
Deltamethrin	0.05
Diazinol	0.02
Dichlorvos	0.1
Diclofop methyl	0.1
Dicofol (sum of isomers)	0.02
Dimethoate	0.5
Disulfoton	0.02
Dithiocarbamate	0.05
Endosulfan (sum of all isomers)	0.05
Ethion	0.05
Fenitrothion	0.05
Fenevalerate (sum of all isomers)	0.05
Copper based pesticides	40
Malathion	0.05
Methyl bromide	0.05
Mevinphos	0.05
Omethoate	0.05
Parathion and para oxon	0.1
Methyl parathion and methyl para oxon	0.1
Phorate	0.05
Phosalone	0.5
Phosphamidon	0.05
Pyrethrin	0.5
Quinalphos	0.01
Quintozen	0.01

[a] The limits mentioned against the first ten pesticides are specific for spices and the remaining are the general limits for all plant foods.

TABLE 8.9
Pesticide Residue Limits in Spices Prescribed by Spain

Pesticide	Tolerance Limit, ppm
Acephate	0.1
Atrazine	0.1
Bendiocarb	0.05
Carbaryl	0.1
Carbosulfan	0.1
Chlorpyrifos	0.05
Chlorpyrifos–methyl	0.05
Cypermethrin	0.05
Diazinon	0.05
Dicofol	0.02
Dimethoate	0.05
Ethun	0.1
Fentoato	0.05
Fenitrothion	0.05
Fenthron	0.05
Melathron	0.50
Metalaxyl	0.05
Methamidophos	0.01
Monocrotophos	0.02
Omethoate	0.10
Phosalone	0.10
Pirimicarb	0.05
Pirimiphos–methyl	0.01
Profenofos	0.02
Prothiofos	0.02
Pyrazophos	0.01
Terbuconazole	0.05
Tolclophos–methyl	0.01
Triazophos	0.01
Vinclozolin	0.05

TABLE 8.10
Pesticide Residue Limits in Spices in The Netherlands and U.K.

Pesticide	Tolerance Limit, ppm	
	The Netherlands	U.K.
HCH without lindane	0.02	0.02
Lindane	0.02	—
Hexachlorobenzene	—	0.01
Aldrin and dieldrin	0.03	0.01
Sum of DDT	0.15	0.05
Malathion	0.05	8.00
Dicofol	0.05	0.50
Chlorpyrifos	0.01	—
Ethion	0.01	—
Chlordan	0.01	0.02
Parathion	0.10	1.00
Parathion methyl	0.10	0.20
Mevinphos	0.05	—
Sum of endosulfan	0.02	0.10
Phosalon	1.00	0.10
Vinclozolin	—	0.10
Dimethoate	0.01	0.05
Quintozen	—	1.00
Metacriphos	—	—
Heptachlor and epoxide	0.21	0.01
Methidathion	—	—
Diazinon	0.05	0.05
Fenitrothion	0.05	0.05
Bromophos	—	—
Mecarbam	—	—
Methoxychlor	0.05	—
Omethoate	—	0.20
Dichlorvos	0.05	—
Phosmet	0.01	—
Methylbromide	—	0.10
Tetraditon	—	—

others have fixed individual limits ranging from 1 to 20 ppb. Table 8.11 displays the tolerance levels for aflatoxin in spices by the EU and major importing countries (Spices Board, 2002; Commission of the European Communities, 2002).

8.8.5 Heavy Metals

Level of heavy metals is also considered as a quality criterion for spices. Lead, cadmium, arsenic, and mercury are of major concern. Atomic absorption spectrophotometry is recommended as the standard method for the analysis (AOAC 2000b, 2000c). Limits specified for some trace metals in whole and ground turmeric under the Indian standards (BIS, 2002) can be found in Table 8.21.

TABLE 8.11
Tolerance Levels for Aflatoxins in Spices

Country	Aflatoxin	Limit, ppb max
Germany	B_1	2
	$B_1 + B_2 + G_1 + G_2$	4
EC	B_1	5
	$B_1 + B_2 + G_1 + G_2$	10
Austria	B1	1
The Netherlands	B1	5
Switzerland	B1	1
	B2 + G1 + G2	5
Spain	B1	5
	B1 + B2 + G1 + G2	10
Sweden	B1 + B2 + G1 + G2	5
Finland	B1 + B2 + G1 + G2	5
Italy and France	B1	10
U.S.		20

8.9 COMMERCIAL REQUIREMENTS

Turmeric produced in different regions exhibit quality variations due to the differences in soil, climatic conditions, and agricultural practices. The postharvest handling and processing of the commodity also add on to the quality differences. This necessitates the use of some standard specifications for trade.

To ensure the quality of spices exported from the country, Government of India introduced the scheme of compulsory Preshipment Inspection and Quality Control in 1963. Spices are graded based on the standards fixed for the purpose. These grades are popularly known as the Agmark Grades. Export Inspection Agency, under the Export Inspection Council of India, held the mandate for preshipment inspection and quality control certification. For turmeric, the grading takes into consideration the hardness of the rhizomes, percentage of small pieces, bulbs, foreign matter, as well as defectives. Indian Standards for turmeric follow the Agmark specifications to ensure quality and purity. Grade designations and respective specifications for Indian turmeric under Agmark are given in Tables 8.12 to 8.16 (Directorate of Marketing Inspection, 1964; Spices Board, 2001).

The European Spice Association (ESA) has come out with the "quality minima for herbs and spices." This serves as a guideline specifications for member countries in European Union. ESA specifications with respect to turmeric may be summarized as in Table 8.17 (Spices Board, 2002).

As per the Prevention of Food adulteration Act (PFA) of India (PFA, 2005), turmeric rhizomes shall:

- be free from lead chromate and other artificial coloring matter
- not contain extraneous matter exceeding 2% by weight
- not contain insect damaged rhizomes exceeding 5% by weight

Turmeric has traditionally been adulterated with related *Curcuma* species, specifically *Curcuma xanthorrhiza*, *Curcuma aromatica*, and *Curcuma zedoaria* (Weiss, 2002; Govindarajan, 1980; Dahal and Idris, 1999). Admixture of other varieties can adversely affect the quality of the spice. Adulterants found in ground turmeric are foreign starches (tapioca, arrowroot, cereal flours), husks, coaltar colors, lead chromate, etc.

TABLE 8.12
Turmeric Grading and Marking Rules[a]

Grade	Flexibility	Pieces % by wt, Max.[b]	Foreign Matter % by wt, Max.	Chura and Defective Bulbs, % by wt, Max.	Percentage of Bulbs by wt, Max.	General Characteristics
1	2	3	4	5	6	7
Special	Should be hard to touch and break with metallic twang	2	1.0	0.5	2.0	1. The turmeric fingers shall be secondary rhizomes of the plant *Curcuma longa* L.
Good	Should be hard to touch and break with metallic twang	3	1.5	1.0	3.0	2.a. They shall be well set and closely grained and be free from bulbs (primary rhizomes) and ill-developed porous fingers
Fair	Should be hard	5	2.0	1.5	5.0	b. Have the shape, length, color, and other characteristics of a variety
						c. Be perfectly dry and free from damage caused by weevils, moisture, overboiling or fungus attack except that 1.0% and 2.0% by weight of rhizomes damaged by moisture and over-boiling should be allowed in grades good and fair, respectively
Nonspecified	—	—	—	—	—	d. Have not been artificially colored with chemicals or dyes

Note: (1) Foreign matter includes chaff, dried leaves, clay particles, dust, dirt, and any other extraneous matter. (2) Length shall be reckoned from one tip of the finger to the other tip longitudinally. (3) Color of core and flexibility shall be reckoned from fingers freshly broken with hands. (4) Chura and defective bulbs include immature small fingers and/or bulbs, internally damaged, hollow, and porous bulbs, cut bulbs, and other types of damaged bulbs except weevilled bulbs. (5) Nonspecified — This is not a grade in its strict sense, but has been provided for the produce not covered by the other grades. Turmeric fingers under this grade shall be exported only against a firm order.

[a] Schedule II (see Rule 3 and 4). Grade designations and definitions of quality of Turmeric "fingers" produced in India (for varieties other than the Alleppey variety).
[b] Pieces are fingers, broken or whole, of 15 mm or less in length.

TABLE 8.13
Turmeric Grading and Marking Rules[a]

		Special Characteristics				
Grade	Flexibility	Pieces % by wt, Max.[b]	Foreign Matter % by wt, Max.	Chura and Defective Bulbs, % by wt, Max.	Percentag of Bulbs by wt, Max.	General Characteristics
1	2	3	4	5	6	7
Good	Should be hard to touch	5	1.0	3.0	4.0	1. The turmeric fingers shall be secondary rhizomes of the plant *Curcuma longa* L.
Fair	Should be hard to touch	7	1.5	5.0	5.0	2.a. They shall be well set and closely grained and be free from bulbs (primary rhizomes) and ill-developed porous fingers
						b. Have the shape, length, color, and other characteristics of a variety
						c. Be perfectly dry and free from damage caused by weevils, moisture, overboiling or fungus attack except that 1.0% and 2.0% by weight of rhizomes damaged by moisture and over-boiling should be allowed in grades good and fair
Nonspecified	—	—	—	—	—	d. Have not been artificially colored with chemicals or dyes

Note: (1) Foreign matter includes chaff, dried leaves, clay particles, dust, dirt, and any other extraneous matter. (2) Length shall be reckoned from one tip of the finger to the other tip longitudinally. (3) Color of core and flexibility shall be reckoned from fingers freshly broken with hands. (4) Chura and defective bulbs include immature small fingers and/or bulbs, internally damaged, hollow, and porous bulbs, cut bulbs, and other types of damaged bulbs except weevilled bulbs (5) Nonspecified — This is not a grade in its strict sense, but has been provided for the produce not covered by the other grades. Turmeric fingers under this grade shall be exported only against a firm order.

[a] Schedule II A (see Rule 3 and 4). Grade designations and definitions of quality of a variety of turmeric commercially known as Alleppey Finger Turmeric produced in India

[b] Pieces are fingers, broken or whole, of 15 mm or less in length.

TABLE 8.14
Turmeric Grading and Marking Rules[a]

				Special Characteristics			
Grade	Flexibility	Pieces % by wt, Max.[b]	Foreign Matter % by wt, Max.	Chura and Defective Bulbs, % by wt, Max.	Percentage of Bulbs by wt, Max.	Admixtures of Varieties of Turmeric, %	General Characteristics
1	2	3	4	5	6	7	8
Special	Should be hard to touch and break with metallic twang	3	1	3	2	2	1. The turmeric fingers shall be secondary rhizomes of the plant *Curcuma longa* L.
Good	Should be hard to touch and break with metallic twang	5	1.5	5	3	5	2.a. They shall be well set and closely grained and be free from bulbs (primary rhizomes) and ill-developed porous fingers
Fair	Should be hard to touch	7	2	7	5	10	b. Have the shape, length, color, and other characteristics of a variety
							c. Be perfectly dry and free from damage caused by weevils, moisture, overboiling or fungus attack except that 1.0% and 2.0% by weight of rhizomes damaged by moisture and over-boiling should be allowed in grades good and fair, respectively
Nonspecified	—	—	—	—	—	10	d. Have not been artificially colored with chemicals or dyes

Note: (1) Foreign matter includes chaff, dried leaves, clay particles, dust, dirt, and any other extraneous matter. (2) Length shall be reckoned from one tip of the finger to the other tip longitudinally. (3) Color of core and flexibility shall be reckoned from fingers freshly broken with hands. (4) Chura and defective bulbs include immature small fingers and/or bulbs, internally damaged, hollow, and porous bulbs, cut bulbs, and other types of damaged bulbs except weevilled bulbs. (5) Nonspecified —This is not a grade in its strict sense, but has been provided for the produce not covered by the other grades. Turmeric fingers under this grade shall be exported only against a firm order.

[a] Schedule II B (see Rule 3 and 4). Grade designations and definitions of quality of a variety of turmeric commercially known as "Rajapore" Finger Turmeric produced in India.

[b] Pieces are fingers, broken or whole, of 15 mm or less in length. Thumb fingers or Angatha gathes, i.e., Ungathas in Rajapore variety shall be taken as fingers for export only.

TABLE 8.15
Turmeric Grading and Marking Rules[a]

		Special Characteristics	
Grade	Foreign Matter % by wt, Max.	Chura and Defective Bulbs, % by wt, Max.	General Characteristics
1	2	3	4
Special	1.0	1.0	1. The turmeric fingers shall be primary rhizomes of the plant *Curcuma longa* L.
Good	1.5	3.0	2. a. They shall be well developed, smooth, sound, soft, and free from rootlets
			b. Have the shape, length (not below 15 mm), and color characteristic of the variety
			c. Be perfectly dry
Fair	2.0	5.0	d. Be free from damage caused by weevils, moisture, overboiling or fungus attack except that 0.1% and 0.2% by weight of rhizomes damaged by moisture and overboiling should be allowed in grades good and fair, respectively
Nonspecified	—	—	e. Have not been artificially colored with chemicals or dyes

Note: (1) Foreign matter includes chaff, dried leaves, powder, clay particles, dust, dirt, and any other extraneous matter. (2) Chura and defective bulbs include immature small fingers and/or bulbs, internally damaged, hollow bulbs except weevilled bulbs. (3) Length shall be reckoned at the points of greatest thickness of the bulbs. (4) Color of core shall be reckoned in bulbs freshly broken with hands. (5) Nonspecified: This is not a grade in its strict sense, but has been provided for the produce not covered by the other grades. Turmeric bulbs under this grade shall be exported only against a firm order.

[a] Schedule III (see Rule 3 and 4). Grade designations and definitions of quality of turmeric Bulbs (Round, Gathas, or Golas) produced in India

8.9.1 TURMERIC POWDER

This is the most common form of turmeric sold through the retail market and is used in home kitchen as well as by food processors. To make turmeric powder, the dried rhizomes are ground to the required particle size. The powder is designated as coarse if 98% passes through a 500-μm sieve, and fine if 98% passes through a 300-μm sieve (ISO, 1983). Since curcuminoids, the color constituents of turmeric, deteriorate with light, and to a lesser extent under heat and oxidative conditions, it is important that ground turmeric be packed in an ultraviolet (UV) protective packaging and appropriately stored (Buescher and Yang, 2000).

8.9.2 GRINDING PROCESS

Grinding is a simple process of disintegrating the rhizomes into small particles, then sifting through a series of screens. Depending on the type of mill, and the speed of crushing, the spice may heat up and volatiles may be lost. In the case of turmeric, heat and oxygen during the process may contribute to curcumin degradation. Cryogenic milling under liquid nitrogen prevents oxidation and volatile loss, but it is expensive and not widespread in the industry. Ground spices are size sorted through screens, and the larger particles can be further ground. Particle sizes are declared in mesh or micron. Specifications for turmeric powder under Indian Agmark rules are given in Tables 8.18 and 8.19 (Directorate of Marketing Inspection, 1964).

TABLE 8.16
Turmeric Grading and Marking Rules[a]

	Special Characteristics		
Grade	Foreign Matter % by wt, Max.	Chura and Defective Bulbs, Percentage by wt	General Characteristics
1	2	3	4
Special	1.0	3.0	1. The turmeric bulbs shall be primary rhizomes of the plant *Curcuma longa* L. (Syn: *C. domestica* Val.)
Good	1.5	5.0	2. They shall
			a. be well-developed, smooth, sound, soft, and free from rootlets
			b. have the shape, length (not below 15 mm) and color characteristic of the variety, (c) perfectly dry
Fair	2.0	7.0	d. Be free from damage caused by weevils, moisture, overboiling or fungus attack except that 0.1% and 0.2% by weight of rhizomes damaged by moisture and overboiling should be allowed in grades good and fair, respectively
Nonspecified			e. Have not been artificially colored with chemicals or dyes

Note: (1) Foreign matter includes chaff, dried leaves, powder, clay particles, dust, dirt, and any other extraneous matter. (2) Chura and defective bulbs include immature small fingers and/or bulbs, internally damaged, hollow bulbs except weevilled bulbs. (3) Length shall be reckoned at the points of greatest thickness of the bulbs. (4) Color of core shall be reckoned in bulbs freshly broken with hands. (5) Nonspecified: This is not a grade in its strict sense, but has been provided for the produce not covered by the other grades. Turmeric bulbs under this grade shall be exported only against a firm order.

[a] Schedule IIIA (see Rule 3 and 4). Grade designation and definitions of quality of a variety of turmeric bulbs (Round, Gathas, or Golas) commercially known as "Rajapore" turmeric bulbs produced in India.

Turmeric powder sold in the Indian domestic market should conform to the quality standards specified by the Prevention of Food Adulteration Rules (PFA, 2005), India. PFA specifications for turmeric powder are given in Table 8.20. Bureau of Indian Standards have also laid down specifications to ensure quality of turmeric powder. Indian Specifications (IS) for whole and ground turmeric (BIS, 2002) are displayed in Table 8.21. Requirements for ground turmeric in international trade as specified by ISO Specifications (ISO, 1983) are given in Table 8.22. Dried rhizomes are thoroughly cleaned prior to grinding. Table 8.23 lists the cleanliness specifications prescribed for ground turmeric in the U.K. (Spices Board, 2002).

In addition to alkaline pH, turmeric pigments are also sensitive to light and oxygen (Tonnesen and Karlsen, 1985b; Souza et al., 1997). Hence the powder must be protected from these factors. No change in the Hunter CIE color characteristic, expressed as $L*a*b*$, was observed in cured ground turmeric stored for 60 days at 25°C in polyethylene bags in the absence of excess air (Bambirra et al., 2002).

Moisture level above 12% can affect the free-flow characteristics of turmeric powder. Aluminium foil laminate packing offers maximum protection against loss of volatile oil and ingress of moisture. Polyethylene pouches alone are inadequate to give desired protection against loss of volatile oil. Printing on the polyethylene pouches may get disfigured and smudged and pouches can become sticky.

Turmeric is hard to grind, and this results in excessive wear of the contact parts inside the grinding machinery compared to other spices. This is revealed by the higher iron contamination in turmeric powder compared to other spice powders (Panduwawala et al., 1998). Turmeric is an important ingredient in curry powder, ranging from 10 to 30% (Govindarajan, 1980). Excessive turmeric imparts a bitter taste to the dish.

TABLE 8.17
European Spice Association Specifications of Quality Minima for Herbs and Spices

Spice	Turmeric
Extraneous matter	1%
Sampling	For routine sampling: square root of units/lots to a maximum of 10 samples
	For arbitration purposes: square root of all containers
Foreign matter	Maximum 2%
Ash, % w/w max.	Turmeric whole: 8 (BSI); ground: 9 (ISO)
Acid Insoluble Ash, % by w/w max	Turmeric whole: 2 (BSI); ground: 2.5 (ESA)
Moisture, % by w/w max	Turmeric whole: 12 (BSI); ground: 10 (ISO)
Heavy metals	Shall comply with national/EU legislation
Pesticides	Shall be utilized in accordance with manufacturers recommendations and good agricultural practice and comply with existing national and/or EU legislation
Treatments	Use of any EC approved fumigants in accordance with manufacturers' instructions, to be indicated on accompanying documents. (Irradiation should not be used unless agreed between buyer and seller)
Microbiology	Salmonella — absent in (at least) 25 g
	Yeast and mold — 10^5/g target, 10^6/g absolute maximum
	E. coli. — 10^2/g target, 10^3/g absolute maximum
	Other requirements to be agreed between buyer and seller
Off-odors	Shall be free from off odor or taste
Infestation	Should be free in practical terms from live and/or dead insects, insect fragments and rodent contamination visible to the naked eye (corrected in necessary for abnormal vision)
Aflatoxins	Should be grown, harvested, handled, and stored in such a manner as to prevent the occurrence of aflatoxins or minimize the risk of occurrence. If found, levels should comply with existing national and/or EU legislation
Volatile oil, % v/w min	Turmeric whole: 2.5 (BSI); ground: 1.5 (ESA)
Adulteration	Shall be free from
Bulk density	To be agreed between buyer and seller
Species	To be agreed between buyer and seller

Abbreviation: BSI, British Standards Institute; ESA, European Spice Association; ISO, International Organisation for Standardization.

8.9.3 CONTROLLING MICROORGANISMS

Like any other agricultural produce, herbs and spices are also prone to microbial contamination during harvest and varying stages of subsequent processing. Contaminated ingredients can be hazardous when used in processed foods, especially when such foods are consumed without further cooking. Therefore, it is important that the ingredients enter the food processing area with negligible bacterial and fungal count. In spite of the curing and drying process, turmeric can still carry a heavy bacterial load. Maintenance of high level of hygiene and sanitary conditions is important to reduce the risk of contamination by mold, spores, yeasts, aerobic/anaerobic bacteria and fungi during processing and storage of the spice.

The microbial load in herbs and spices can be controlled by three techniques, namely:

1. steam sterilization
2. fumigation
3. irradiation

TABLE 8.18
Grade Designation and Definition of Quality of Turmeric Powder[a]

Grade	Moisture % by wt. max.	Total ash % by wt max.	Acid insoluble ash % by wt. max.	Lead as Pb ppm, max.	Starch % by wt., max.	Chromate Test	General Characteristics
1	2	3	4	5	6	7	8
Standard	10	7	1.5	2.5	60	Negative	Turmeric powder shall be prepared by grinding clean, dry turmeric (*Curcuma longa* L.) rhizomes. It shall have its characteristic taste, flavor, and be free from musty odor. Shall be free from dirt, mold growth, and insect infestation. It shall be free from any coloring matter such as lead chromate, preservatives, and extraneous matter such as cereal or pulse, flour or any added starch. It shall be ground to such a fineness that all of it passes through a 300 μm sieve.

[a] Schedule IV (see Rule 3 and 4).

TABLE 8.19
Grade Designation and Definition of Quality of Turmeric Powder[a] (Coarse Ground)

Grade	Moisture % by wt. max.	Total ash % by wt max.	Acid insoluble ash % by wt. max.	Lead as Pb ppm, max.	Starch % by wt., max.	Chromate Test	General Characteristics
1	2	3	4	5	6	7	8
Standard	10	9	1.5	2.5	60	Negative	Turmeric powder shall be prepared by grinding clean, dry turmeric (*Curcuma longa* L.) rhizomes. It shall have its characteristic taste, flavor, and be free from musty odor. Shall be free from dirt, mold growth and insect infestation. It shall be free from any coloring matter such as lead chromate, preservatives, and extraneous matter such as cereal or pulse, flour or any added starch. It shall be ground to such a fineness that all of it passes through a 300 μm sieve.

[a] Schedule IV A (see Rule 3 and 4)

TABLE 8.20
Specifications for Turmeric Powder under PFA Regulations

Characteristics	PFA Limit (% by Wt., Max.)
Moisture	13.0
Total ash	9.0
Ash insoluble in dilute HCl	1.5
Test for lead chromate	Negative
Total starch	60.0

TABLE 8.21
Requirements of Turmeric, Whole and Ground, under Indian Standards

Characteristic	Requirement	
	Turmeric, Whole	Turmeric, Ground
Curcumin content (on dry basis) Percent by mass, min	2	3
Moisture content (on dry basis) Percent by mass, max	12	10
Total ash (on dry basis) Percent by mass, max	—	7
Acid insoluble ash (on dry basis) Percent by mass, max	—	1.5
Starch (on dry basis) Percent by mass, max	—	60
Lead (as Pb), mg/kg, max	10	10
Copper (as Cu), mg/kg, max	5	5
Arsenic (as As), mg/kg, max	0.1	0.1
Zinc (as Zn), mg/kg, max	25	25
Cadmium (as Cd), mg/kg, max	0.1	0.1
Tin (as Sn)	Nil	Nil

TABLE 8.22
Chemical Requirements for Turmeric, Ground (ISO)

Characteristic	Requirement
Moisture content % (m/m), max.	10
Total ash % (m/m) on dry basis, max.	9
Acid insoluble ash % (m/m) on dry basis, max.	1.5
Coloring power expressed as curcuminoids content % (m/m) on dry basis, min.	2

TABLE 8.23
Cleanliness Specification for Ground Turmeric in U.K.

Parameter	n	c	Counts m	M
Insect fragments	3	1	280/10 g	370/10 g
Heavy filth	3	1	1140/10 g	200/10 g
Rodent hairs	3	2	0/10 g	1/10 g

Note: n = the number of analytical units that must be examined from a lot to satisfy the requirements; c = the maximum allowable number of marginally acceptable analytical units in an individual sample; m = represents an acceptable level and values above m but below or equal to M are marginally acceptable; M = a criterion which separates marginally acceptable quality from defective quality. Values above M are unacceptable.

8.9.3.1 Steam Sterilization

Steam sterilization is one of the popular methods for bacterial reduction in herbs and spices. Steam is nonpollutant and nontoxic and complies with the strictest regulations governing food products worldwide. The process involves treating the spice to be sterilized with live steam for a short duration. In a commercial operation, the charge inside the sterilizer is preheated under vacuum to 50 to 55°C using hot water or jacket steam and, then treated with live steam at 110 to 140°C for 1 to 2 min. This treatment is able to achieve considerable reduction in microbial population. However, due to insulating effect of solid material, the exposure time and temperature required for effective killing of all the embedded microbes can lead to loss of quality. The steam condensing in the charge is removed under vacuum; excess residual moisture in the product can lead to spoilage during storage. Continuous steam sterilizers with on-line drying are commercially available.

8.9.3.2 Fumigation

Fumigation with suitable chemicals has developed into a quick, effective, and economic disinfestation procedure. Sulfur dioxide from burning sulfur in sealed warehouses has traditionally been used but has now largely been replaced by other gaseous fumigants, which are easier to use, more effective, and have minimum effect on the quality of spices. Main considerations in the use of fumigants are the extent of sterility required, permissible levels of fumigant residues, interaction products and their effect on the flavor quality of spice. The lethal doses of a number of fumigants for adult insects have been determined and a working dosage of four to five times the lethal dosage is recommended for treatment.

Methyl bromide is a widely used fumigant. This odorless gas is stable, has incredible penetration power, and is moderately priced. Normal aeration at the end of fumigation, processing, and cooking generally releases unreacted fumigant and brings down the residues to insignificant levels. The chemical, however, is a suspected carcinogen and mutagen and hence needs to be handled with care. With organic bromide fumigants, water-soluble inorganic bromides are formed as residues in the treated food. Their level is an indirect measure of bromine-containing fumigants originally used. Inorganic bromide residues up to 400 ppm is permitted by the U.S. Environmental Protection Agency (U.S. EPA) (CFR, 2005) in processed spices. The use of this chemical may be phased out, as methyl bromide comes under the Montreal Protocol for ozone depleting substances.

Ethylene oxide, first reported as a bactericidal fumigant in the late 1920s, is very effective for the microbial control of spices and curry powders. However, product treated with ethylene oxide may contain residues of unchanged ethylene oxide unless there is adequate holding time or aeration after treatment to allow the residue to volatilize. When treated commodities are freely ventilated, residual ethylene oxide usually drops to undetectable levels, but in sealed packages it is retained for longer periods. In addition, the inorganic halides naturally present in the substrate can react with ethylene oxide to produce ethylene chlorohydrin or bromohydrin, and the residues of these halohydrins do not disappear from the treated produce on storage so readily as the residual ethylene oxide (FAO, 1972). The formation of these halohydrins depends on the presence of ionic chlorine or bromine, and when both these are available, the bromohydrin is more readily formed than chlorohydrin. Ethylene chlorohydrin is more persistent, and substantial residue levels have been detected in sterilized materials having a high content of chloride. Ethylene bromohydrin is less persistent and decomposed slowly under storage conditions. Cooking results in the loss of residual halohydrins; the loss is almost complete when the conditions are alkaline (FAO, 1972). Ethylene glycol and diethylene glycol formed by hydrolysis of ethylene oxide or the halohydrins have also been detected. Residues may also include certain alkylated and hydroxyethylated reaction products. The composition and the amount of residues in the treated product depend upon the halide and moisture content of the starting material, the dosage of ethylene oxide used, the temperature during fumigation as well as the temperature and ventilation during subsequent storage.

Use of ethylene oxide has recently become more difficult as a result of environmental controls on its use. The instability and flammability of ethylene oxide requires it to be mixed with another gas. Ethylene oxide is usually applied in admixture with carbon dioxide or nitrogen to provide nonexplosive conditions. Chlorofluorohydrocarbons were formerly used, but being ozone-depleting substances, their use is no longer allowed. The most common stabilizer in commercial practice is carbon dioxide (usually 80 to 90%). Ethylene oxide can also be delivered with steam; however, the use of steam is discouraged as it can result in the loss of volatile aroma components, color changes and also increase the moisture content in the spice being treated. Treatments for sterilization are usually undertaken in specially designed vacuum fumigation installations. Spices destined for ethylene oxide treatment must be packed in materials that allow the gas to permeate. Bulk barrels must be opened for treatment. It is difficult for ethylene oxide to penetrate through large volumes of tightly packed spice.

Ethylene oxide is currently approved for use as a postharvest fumigant in agricultural commodities including whole spices in the U.S. The U.S. EPA has established a maximum tolerance level of 50 ppm for ethylene oxide in spices (CFR, 2005). An extension to the regulation also applies this tolerance to ground spices. The use of ethylene oxide is not permitted in the European Union and Japan as a sterilizing agent for spices (EC, 2002; Chatterjee et al., 2000). Ethylene oxide and ethylene chlorohydrin residues in spices are determined by gas chromatography (ASTA, 1997b; Brown, 1970; Heuser and Scudamore, 1967, 1968, 1969; Ragelis et al., 1968; Spitz and Weinberger, 1971; Weinberger, 1971; Woodrow et al., 1995). Microbiologically, ethylene oxide is far less effective than irradiation. Propylene oxide is an alternative to ethylene oxide, but is not as efficient (Tainter and Grenis, 2001).

Studies on the use of ethyl formate for the control of stored grain pests have shown that dosages of 300 to 400 g/m^3 in an exposure period of 48 to 72 h control insects in stored grain and their products (Muthu et al., 1984).

8.9.3.3 Irradiation

Irradiation is becoming an increasingly accepted technique to sterilize spices and other food products as the process does not affect the chemical or physical properties of the material (Lacroix et al., 2003). The process is also called "cold pasteurization" or "irradiation pasteurization," since the process is functionally similar to conventional pasteurization, except for the source of energy. In irradiation, the sterilization is effected by the energy of ionizing radiation. When ionizing radiation strikes bacteria and other microbes, its high energy breaks chemical bonds in molecules that are vital for cell growth and integrity. As a result, the microbes die, or can no longer multiply and cause illness or spoilage.

Irradiation process requires specially built and secure facilities. The common dose of 5 to 10 kGy applied to spices effectively kills bacteria, molds, and yeasts. Insects and other pests in all life stages are killed. Irradiation of turmeric rhizomes with 10 kGy at a dose rate of 19 Gy/min is reported not to modify the composition of volatile oils extracted after 1 week of storage at 5°C. The antioxidant activity was also reported to be unaffected by irradiation (Chatterjee et al., 1999). Storage after irradiation further enhances the sanitation effect because injured cells are unable to repair and die off over time. Spore-forming bacteria require higher dosage levels to kill because spores are less sensitive to radiation. The extent of reduction of microbial load on spices by irradiation is dependant on the radiation dose and the type and initial population of the microorganisms present. Most spice packaging materials are compatible with irradiation. Irradiation does not require the opening of packages, so the aroma characteristics of spices are better maintained. Color of spices like paprika and turmeric are stable to irradiation treatment.

By law, a spice irradiated once cannot be irradiated a second time; therefore, irradiated spices must be well labelled to avoid a second irradiation if it enters as an ingredient in a food product that will be further irradiated (Tainter and Grenis, 2001). The irradiation of spices is allowed in most countries. The laws allowing irradiation, however, vary from country to country, and most countries have set maximum dose regulatory limits. They require irradiated spices to be labeled with the international symbol for irradiation, and sometimes words to describe the process and/or the effect. Most countries do not require that the use of an irradiated spice be noted on the label of a processed food.

The American Society of Testing Materials recommends 3 to 30 kGy for the irradiation of dried spices, herbs, and vegetable seasonings to control pathogens and other microorganisms (ASTM, 1998). U.S. FDA regulations allow the irradiation of dehydrated herbs, spices, and vegetable seasonings for microbial disinfection up to 30 kGy (FDA, 2003). Cobalt-60 is the most commonly used radionuclide for food irradiation. Cobalt-60 emits ionizing radiation in the form of intense gamma rays. Cesium-137 is another gamma source used for irradiation. Cesium-137 has a less penetrating gamma beam, but a longer half-life, making it more suitable under certain circumstances. Other energy sources recommended by FDA for irradiation treatment of foods are high-energy electron beams and X-rays.

Methods are available to identify irradiated spices. One such method is based on the observation that irradiated spices exhibit thermoluminescence (Sharifzadeh and Sohrabpour, 1993). Inorganic dust present in spice powders are said to be responsible for this property. Light emission (chemiluminescence) from the reaction of irradiated foods in a luminol solution is also used an indicator of radiation treatment (Boegl and Heide, 1985; Heide and Boegl, 1985, 1987). An advanced method reported is the electron spin resonance spectroscopy (Lacroix et al., 2003; Tainter and Grenis, 2001).

Irradiation results in cleaner spices with fewer changes in the sensory characteristics than thermal or fumigation treatments. No detectable differences have been observed between the aroma impact compounds of the irradiated and nonirradiated samples (Chatterjee et al., 2000). Water content activity and characteristics of essential oil analyzed by gas chromatography are not changed by irradiation (Chosdu et al., 1985). Turmeric, often used for its color, does not undergo color change at the dosage level of 10 to 30 kGy usually required for microbial control. Moreover, irradiation results in a much lower microbial load than other treatments and is sometimes the only treatment effective enough to meet standards set by international regulations.

8.10 ESSENTIAL OIL

Turmeric owes its characteristic aroma to a volatile oil (essential oil) present in the rhizome. The oil, which can be recovered by hydrodistillation of turmeric powder, contains all the volatile aroma components of the spice. The yield of volatile oil from turmeric varies from 3 to 7% and contains primarily oxygenated sesquiterpenes. It is light orange-yellow in color with an odor reminiscent of turmeric powder. Turmeric oil is not sufficiently valuable or attractive to merit regular commercial distillation.

8.10.1 METHOD OF PRODUCTION

Essential oils are the volatile organic constituents of fragrant plant matter. They are generally composed of a number of compounds, possessing different chemical and physical properties. The aroma profile of the oil is a cumulative of the contribution from the individual compounds. The boiling points of most of these compounds range from 150 to 300°C at atmospheric pressure (Guenther, 1972). If heated to this temperature, labile substances would be destroyed and strong resinification would occur. Hydrodistillation permits the safe recovery of these heat sensitive compounds from the plant matter.

Hydrodistillation involves the use of water or steam or a combination of the two to recover volatile principles from plant materials. The fundamental feature of hydrodistillation is that it enables a compound or mixture of compounds to be distilled and subsequently recovered at a temperature substantially below that of the boiling point of the individual constituents.

8.10.1.1 Steam Distillation

Steam distillation is the most widely used industrial method for the isolation of essential oil from plant material. The spice is powdered and charged in a stainless steel still of optimum dimensions. The still is attached to a condenser and a separator. Direct steam is admitted from the bottom of the still. The steam, which rises through the charge, carries along with it the vapors of the volatile oil. The oil vapor-steam mixture is cooled in the condenser. The oil is separated from water in the separator and collected in glass or stainless steel bottles. The oil is thoroughly dried and stored airtight in full containers in a cool dry place protected from light (Balakrishnan, 2005). The release of oil from turmeric powder during steam distillation is comparatively slow because of the presence of high boiling sesquiterpene derivatives (Krishnamurthy et al., 1976). Hence, extended distillation time would be necessary for complete oil recovery. Volatile oil can also be recovered from the mother liquor left over after harvesting curcuminoids from the solvent extract of ground turmeric. The mother liquor is freed from the solvent and subsequently steam distilled to recover the volatile fraction. In another approach, the above extract is re-extracted with hexane to recover a composite of fixed and volatile oils, and this composite could be subjected to fractional distillation to get the volatile fraction (Jayaprakasha et al., 2001).

8.10.2 PROPERTIES OF TURMERIC OIL

Physical properties reported for essential oil obtained by steam distillation of dry rhizomes are given in Table 8.24 (Rupe et al., 1909; Schimmel & Co., 1911; Rustovskii and Leonov, 1924; Dieterle and Kaiser, 1932; Kelkar and Rao, 1933; Natarajan and Lewis, 1980).

8.10.3 COMPOSITION

Composition of turmeric oil has been studied extensively using gas chromatography (GC) and gas chromatography-mass spectrometry (GC-MS). The oil is dominated by oxygenated sesquiterpenes, along with small quantities of sesquiterpene hydrocarbons, monoterpene hydrocarbons and oxygenated monoterpenes. The main components are the turmerones, α- (30 to 32%) and β- (15 to 18%), and *ar*-turmerone (dehydroturmerone) (17 to 26%). Turmerone is unstable in the presence of air, yielding its dimmer or the most stable *ar*-turmerone (Su et al., 1982). The components identified in turmeric rhizome oil by different investigators are compiled in Table 8.25 (Kelkar and

TABLE 8.24
Physical Properties of Turmeric Oil

Color	Orange yellow
Odor	Peppery and aromatic
Taste	Sharp, burning
Specific gravity	0.9211–0.9430 (20°C–25°C)
Refractive index	1.4650–1.5130 (20°C–30°C)
Optical rotation	(–) 14°–(–) 24.76° (20°C–30°C)
Acid number	1.6–2.8
Ester value	3.2–12.1
Acetyl value	26.3–30

Rao, 1933; Carter et al., 1939; Khalique and Das, 1968; Purseglove et al., 1981; Golding et al., 1982; Kiso et al., 1983a; Gopalam and Ratnambal, 1987; Kapoor, 1990; Oshiro et al., 1990; Imai et al., 1990; Nigam and Ahmed, 1991; McCarron et al., 1995; Leung and Foster, 1996; Hu et al., 1997, 1998; Sharma et al., 1997; Richmond and Pombo-Villar, 1997; Kojima et al., 1998; Ibrahim et al., 1999; Chatterjee et al., 2000; Riaz et al., 2000; Raina et al., 2002; Singh et al., 2002; Chane Ming et al., 2002; Leela et al., 2002; Bansal et al., 2002; Tang and Chen, 2004). The relative composition varies considerably with the geographical origin.

The chemical structure of the key components, turmerone, and *ar*-turmerone are given in Figure 8.1. Figure 8.2 shows typical GC pattern of oil distilled from dried Indian turmeric (Alleppey finger).

The optimum time of harvest reported for maximum yield of turmeric oil is 7.5 to 8 months. The oil content in the bulbs is higher than that of finger rhizomes (Cooray et al., 1998). The composition of the oil is found to vary with maturity; sesquiterpenes (*ar*-turmerone and turmerones) increase while monoterpenes (1,8-cineole and α-phellandrene) decline in both bulb and finger rhizomes. Monoterpene content is lower in the mother sets during the early stages of growth.

8.11   TURMERIC LEAF OIL

Fresh turmeric leaves contain 0.2 to 0.3% essential oil. Monoterpenes form the major group of components in the leaf oil, dominated by α-phellandrene (18 to 32%). In addition to all components present in rhizome oil, the leaf oil also contained β-pinene up to 8.9%, which is rarely reported in rhizome oil (McCarron et al., 1995; Dung et al., 1995). Other main constituents are terpinolene, 1,8-cineole, and *p*-cymene (Oguntimein et al., 1990; McCaron et al., 1995; Dung et al., 1995; Sharma et al., 1997; Ramachandraiah et al., 1998; Raina et al., 2002; Leela et al., 2002; Chane-Ming et al., 2002; Behura et al., 2002; Garg et al., 2002; Behura and Srivastava, 2004). The relative amounts of these constituents vary widely with the geographical origin. Analysis of a typical sample of Indian turmeric leaf oil showed: specific gravity 0.8550; refractive index 1.4740 and optical rotation (+) 75°30′ at 25°C. A typical GC chromatogram of turmeric leaf oil is given in Figure 8.3.

8.12   TURMERIC FLOWER AND ROOT OILS

Flowers and roots of turmeric also contain essential oil. In a study (Leela et al., 2002), the flowers collected from 4-to-5-month-old plants yielded 0.3% oil composed of 60 constituents, the main ones being *p*-cymen-8-ol, followed by terpinolene and 1,8-cineole, when analyzed on GC-MS. The root oil revealed 43 components, with *ar*-turmerone, *ar*-curcumene, and dihydrocurcumene as the major components. The yield of oil from roots on dry basis was 4.3%. The same sesquiterpenoids were found in the rhizome, root, and flower oils, but in higher amounts in the rhizomes.

TABLE 8.25
Components Identified in Turmeric Rhizome Oil

- *alpha*-Turmerone
- ar-Turmerone (dehydroturmerone)
- Zingiberene
- Myrcene
- *1,8*-Cineole
- ar-Curcumene
- Borneol
- Nerolidol
- Curlone
- Germacrene
- Car-*3*-ene
- *gamma*-Terpinene
- Terpinen-4-ol
- *l*-Carveol
- Curcumenone
- Bisacumol
- Curcumanol
- *alpha*-Atlantone
- *beta*-Turmerone
- *a*-Curcumene
- *alpha*-Phellandrene
- p-Cymene
- *beta*-Caryophyllene
- *d*-Sabinene
- *beta*-Sesquiphellandrene
- Zerumbone
- Germacrone
- Dehydrozingerone
- *alpha*-Terpinene
- Terpinolene
- Linalool
- ar-Turmerol
- Dehydrocurdione
- Bisacurone
- Germacron-13-al
- *gamma*-Atlantone
- 1-(3-Cyclopentylformyl)-2,4-dimethylbenzene
- 6,6-Dimethy-1-bicyclo[3,1,1] hept-2-ene-ethanol
- 4-Cyano-2,2-dimethyl-1-methylene-cyclopentane
- 1-(3-Cyclopentylpropyl)-2,4-dimethylbenzene
- 4-Hydroxy bisabola-2,10-diene-9-one
- 4-Methoxy-5-hydroxy-bisabola-2,10-diene-9-one
- 2,5-Dihydroxybisabola-3,10-diene and procurcumadol
- (4S,5S)-Germacrone-4,5-epoxide
- Bisabola-3,10-dien-2-one
- 4,5-Dihydroxybisabola-2,10-diene
- Turmeronol A (2-methyl-6-[3-hydroxy-4-methylphenyl]-2-hepten-4-one)
- Turmeronol B (2-methyl-6-[2-hydroxy-4-methylphenyl]-2-hepten-4-one)

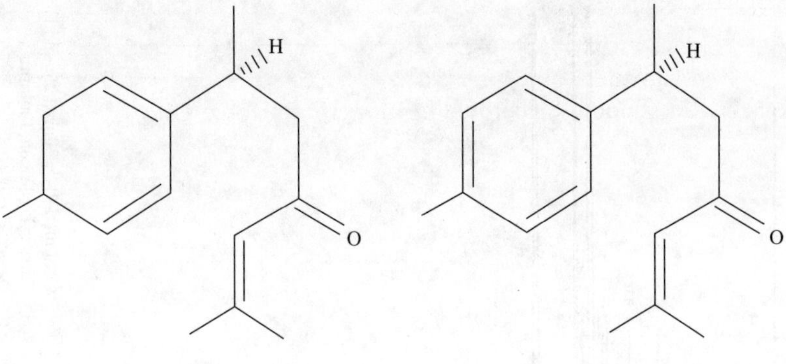

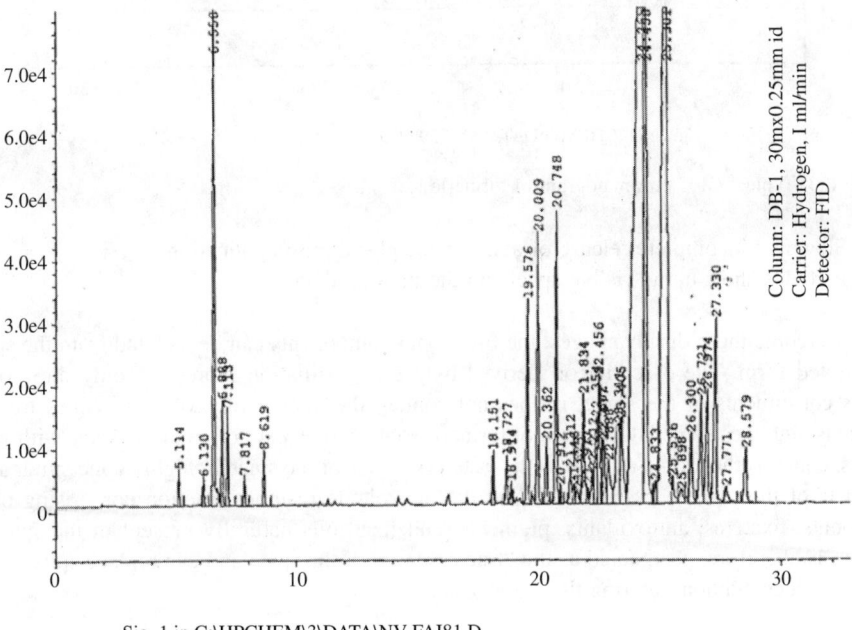

Turmerone   *ar*-Turmerone

FIGURE 8.1 Turmerone and *ar*-turmerone.

Sig. 1 in C:\HPCHEM\3\DATA\NV-FAI81.D

FIGURE 8.2 Typical gas-liquid chromatographic profile (GLC) chromatogram of turmeric rhizome oil.

8.13 OLEORESIN TURMERIC

Raw turmeric and turmeric powder suffer from certain drawbacks (Balakrishnan, 2005):

1. Large volume of inert matter associated with the color and flavor principles increases the transportation expenses and demands extensive storage facilities.
2. Lack of consistency in color and flavor due to variation in the agro-climatic conditions of growth, maturity at harvest, postharvest processing, and storage conditions.
3. Can pick up contamination from extraneous matter during harvesting, postharvest processing, storage, and transportation.
4. Low shelf life due to microbial contamination.

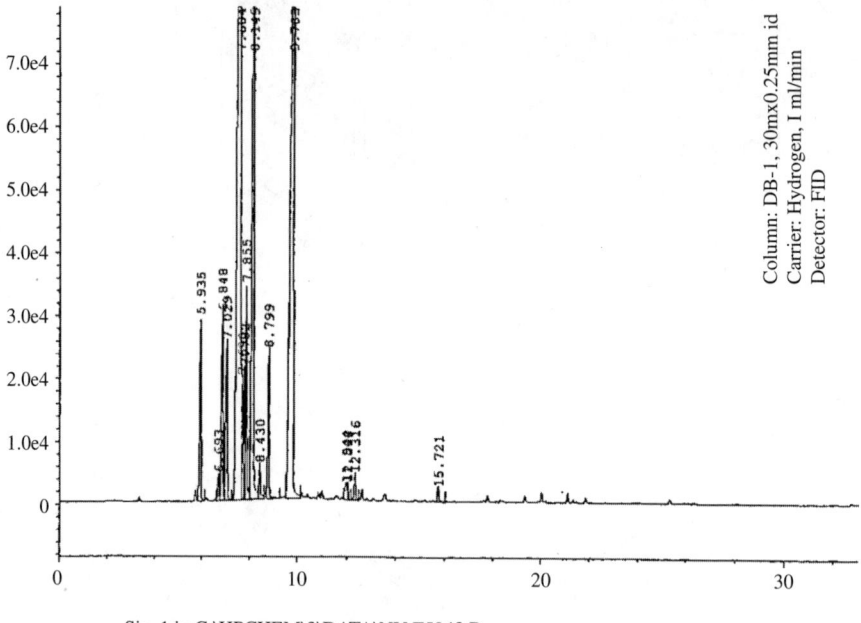

Sig. 1 in C:\HPCHEM\3\DATA\NV-F5942.D

FIGURE 8.3 Typical GLC chromatogram of turmeric leaf oil.

5. Slow and incomplete release of active principles during application.
6. Particles show up as visible specks in the final product.

To overcome these disadvantages, the functional components can be isolated from the spice in concentrated form. The essential oil derived by steam distillation represents only the aromatic, odorous constituents of the spice; it does not contain the nonvolatile color principles for which turmeric is highly esteemed. Oleoresin of turmeric, obtained by extraction of the spice with volatile solvents, contains the aroma as well as the taste principles of the spice in highly concentrated form. It consists of the volatile essential oil and the nonvolatile resinous fraction comprising of taste components, fixatives, antioxidants, pigments, and fixed oils naturally present in the spice. The oleoresin is, therefore, designated as the "true essence" of the spice and can replace spice powders in food products without altering the color or flavor profile.

8.13.1 ADVANTAGES OF OLEORESIN

1. Oleoresins can overcome the disadvantages associated with raw spices and spice powders.
2. They serve as standardized, convenient, and hygienic substitutes for spice powders and provide the stability and consistency required in product food formulations.
3. They can be customized to meet specific product needs for solubility/dispersibility, aroma, taste, and color, and are microbiologically stable.

8.13.2 MANUFACTURING PROCESS

Oleoresin of turmeric is obtained from the spice by solvent extraction. Solvent extraction technique involves the removal of a soluble fraction from a permeable solid phase with which it is associated by selective dissolution by a liquid solvent (Perry and Chilton, 1973). The process essentially comprises three steps: (1) contacting the spice powder and solvent in the extractor to effect the

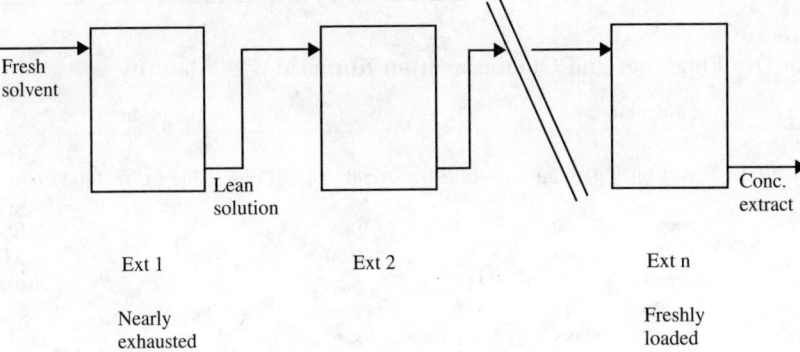

FIGURE 8.4 Extraction in a battery of extractors.

transfer of the functional components from the spice to the solvent, (2) separation of resulting solution from the powder, and (3) distillation of the solution to recover the product.

The spice is comminuted to predetermined particle size and loaded in the extractors. The raw material should be cleaned free from dirt, foreign matter, filth, insect, rodent, and microbial contamination. It should be dried to optimum moisture level since excessive moisture affects the percolation rate and product quality. Extractors are stainless steel cylindrical vessels with provisions for charging the feed from the top and removing the spent spice from the bottom. A perforated plate supported on a grid at the bottom of the extractor holds the charge. Extractor capacities range from 200 to 2000 kg based on the scale of operation. Solvent, admitted from the top, is sprayed on to the charge. As the solvent percolates down the charge, it dissolves the active principles from the spice. The concentrated solution obtained is filtered and the solvent distilled off. The distilled solvent is recycled. Bulk of the solvent from the extract is distilled out in an evaporator. The distilled solvent is recycled. Repeated washings would be necessary to exhaust the spice. The solute concentration in the extract progressively diminishes with each wash. Lean solutions are directed to the next freshly loaded extractor. Once the spice is fully exhausted, the solvent adhering to the spent spice is recovered by injecting live steam. Operation in a single extractor is rarely encountered in industrial practice. In industrial extraction, a number of batch contact units are arranged in a row called the extraction battery. Solids remain stationary in each extractor and are subjected to a multiple number of contacts with the extracts of progressively diminishing concentrations from the previous one. The final contact of the nearly exhausted solids is with fresh solvent, while the concentrated solution leaves in contact with the fresh solids in another extractor. The concentrated solution is distilled, while subsequent washings are directed to the next freshly loaded extractor. The extractors are discharged and reloaded one at a time. Depending on the size of the charge and the extraction parameters, each batch takes 6 to 24 h. Schematic layout of a simple extraction battery is shown in Figure 8.4 (Balakrishnan, 2005).

To make the oleoresin of turmeric, the residual solvent in the product is brought down below the limits specified by regulatory bodies. Bulk of the solvent in the extract is distilled off in an evaporator; the removal of the final traces is carried out in a desolventizer under controlled conditions of temperature and pressure to safeguard the active principles.

Oleoresin turmeric is a composite of the volatile aroma components and the nonvolatile taste and color principles of the spice. The fixed and essential oils in the oleoresin tend to dilute the color strength. These oils also contribute to the bitterness of the product. A variation suggested in the extraction process is consecutive extraction with two different solvents. The process involves leaching the disintegrated dried rhizomes with a solvent such as hexane to eliminate the oily components and recovering the pigment from the residue with a second more polar solvent such as methanol, isopropyl alcohol, acetone, or ethylene dichloride (Keil and Dobke, 1940; Sair and Klee, 1967; Verghese, 1993). However, this method is rarely practiced industrially.

TABLE 8.26
Yield of Dry Rhizomes and Oleoresins from Turmeric with Maturity

Rhizomes				Oleoresin		
1	2	3	4	5	6	7
Maturity Weeks	Wet (g)	Dry (g)	Yield (%)	Yield (%)	Color Value CU	Total Color (5 × 6)
19	390	30	7.69	10.50	4200	44100
21	450	42	9.33	10.00	4400	44000
23	680	75	11.02	9.55	4650	44407
25	1270	177	13.93	8.80	5200	45760
27	1230	185	15.04	8.40	5600	47040
29	1350	243	18.00	8.20	5800	47560
31	1120	215	19.19	8.20	5750	47150
33	1430	284	19.86	8.10	5800	46980
35	1000	210	21.00	8.15	5800	47270
37	1250	265	21.20	8.10	5750	46575
39	1620	337	20.80	8.00	5700	45600

Yield of oleoresin varies with the cultivar, solvent, and extraction conditions, but generally in the range of 10 to 12% and contains various proportions of the coloring matter, the volatile oils, and nonvolatile fatty and resinous materials. Large difference has been observed in the color strength of oleoresins from the fingers and the mother rhizomes. Generally the curcumin content of the oleoresin ranges from 30 to 40%.

Yield of dry matter and the color principles vary with the maturity of the rhizomes. Results of a typical study are displayed in Table 8.26 (Balakrishnan et al., 1992). It is found that the dry matter in the rhizomes steadily increases during the 19 to 37 weeks of growth. During this interval, the percentage yield of solvent extractives from dry spice gradually declined. This phenomenon is attributed to the accumulation of solvent insolubles like starch and fiber at a rate faster than the solvent solubles. Based on the recovery of total color, it was concluded that for producing oleoresin or dye, dry turmeric originating from rhizomes of about 29 weeks maturity is the best. For the maximum recovery of dry spice, it is beneficial to harvest the spice at about 37 weeks maturity.

Attempts have been made to derive extraction rate equations for spice extractions using a kinetic approach. The following equation is reported to fit concentration-time data reasonably well (Houser et al., 1975):

$$C/C_o = (1 + 2\ t^*)\ \exp\ [-k_1 t^*(1 + t^*)]$$

where t^* is the time after the start of the elution from the bed, C_o and C are initial concentration and concentration at t^*, respectively, and k_1 and are constants.

8.13.3 Factors Affecting the Extraction Efficiency

The major factors affecting the efficiency of solvent extraction are (Balakrishnan, 2005):

8.13.3.1 The Particle Size

To achieve adequate solute-solvent contact in solvent extraction, the raw material should be comminuted. Solute is usually surrounded by a matrix of insoluble matter. The solvent must, therefore, diffuse into the mass, and the resulting solution must diffuse out. Grinding of the raw material will greatly accelerate the leaching action, since more of the solute is exposed to the solvent. However, very fine powder tends to restrict the percolation of the solvent through the charge due to decrease

TABLE 8.27
Limits for Residual Solvents in
Spice Oleoresins

Solvent	Limit, ppm max
Acetone	30
Ethylene dichloride	30
Hexane	25
Isopropyl alcohol	50
Methanol	50
Methylene chloride	30

in the porosity of the bed. This will adversely affect the extraction rate. It is, therefore, necessary to select an optimum particle size for extraction.

Each plant material has characteristic shape, size, texture, and hardness. The choice of size reduction equipment depends on these factors. Turmeric is one of the hardest among spices. High content of starch in the spice tends to give excessive fine powder when comminuted.

8.13.3.2 Extraction Medium

The selection of solvent primarily focuses on optimum quantity of extractives of the desired quality and not necessarily maximum yield. A good extraction solvent should be:

- able to dissolve the active principles selectively and minimize the extraction of undesirable constituents.
- chemically pure, since residues of impurities can impart objectionable off-flavor to the product
- having a reasonably low boiling point to facilitate distillation; however, too low a boiling point can lead to excessive loss of solvent during processing
- chemically inert, i.e., should not react with the constituents of the product
- having low specific and latent heats
- nontoxic and should not pose health hazards
- nonflammable and nonexplosive
- readily available and reasonably priced
- acceptable under the food laws of the country where the product is to be used

Spice extraction involves the use of organic solvents. Permissible limits of the residues of various solvents in the oleoresin have been specified by regulatory bodies. Limits laid down by the Code of Federal Regulations of U.S. FDA (CFR, 1995) for common extraction solvents for spices are listed in Table 8.27.

Curcumin is soluble in polar solvents (acetone, ethyl acetate, methanol, ethanol), and quite insoluble in nonpolar solvents such as hexane. Acetone is a popular extraction medium for turmeric. Ethylene dichloride is an efficient extractant; however, its use is restricted due to alleged carcinogenicity. This is now being largely replaced by ethyl acetate (Rajaraman et al., 1981; Verghese and Joy, 1989). Hexane, isopropanol, and ethyl alcohol are also used at various stages in the retrieval and purification of curcuminoids from the oleoresin.

8.13.3.3 Temperature of Extraction

Generally, increase in temperature improves extraction efficiency. This, in turn, helps to reduce the solvent quantity and the process time. Extractors can be provided with steam jackets to heat the contents, or hot solvent may be sprayed on to the charge. However, high temperatures may lead to the extraction of excessive amounts of undesirable compounds from the spice, which can affect the quality of the product.

8.13.4 EXTRACTION USING SUPERCRITICAL FLUIDS

Conventional spice extraction involves the use of organic solvents, which can leave their residues in the final product. Moreover, at the distillation temperature, some deterioration or chemical modification of the labile components are also likely. Supercritical fluid extraction (SFE) is a novel isolation method that can overcome the above issues. An oleoresin, free from chemical alterations brought about by heat and water and without solvent residues and other artifacts, can be obtained by this method. Carbon dioxide (CO_2) is the popular medium for the supercritical extraction of spices.

8.13.4.1 Principle of Supercritical Fluid Extraction

Supercritical fluid extraction involves the use of a compound above its critical temperature and pressure as extraction medium. Figure 8.5 gives the phase diagram for a substance (Balakrishnan, 2005). The equilibrium curves for the three states — solid, liquid, and gas — meet at TP, the triple point of the substance. At the triple point, all the three phases co-exist. Moving along the gas-liquid equilibrium curve, we reach a point where a pure gaseous compound cannot be liquefied regardless of the pressure applied. This is the critical point for the compound. The temperature at critical point is the critical temperature and the vapor pressure of the gas at the critical temperature is the critical pressure. A supercritical fluid is any compound at a temperature and pressure above the critical values. In the figure, C is the critical point at the end of the gas-liquid equilibrium curve and the shaded area indicates the supercritical fluid region.

  In the supercritical environment only one phase exists. The fluid, as it is termed, is neither a gas nor a liquid and is best described as intermediate to the two extremes. This phase retains solvent power approximating liquids while the penetration power into the solid matrix is contributed by

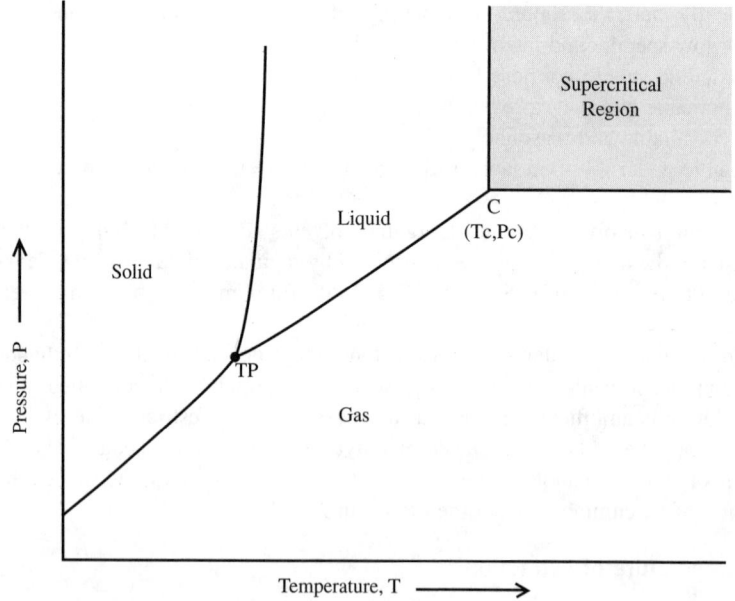

FIGURE 8.5 Phase diagram.

the transport properties common to gases. Therefore, the rates of extraction and phase separation are significantly faster than for conventional extraction processes. Furthermore, the pressure and temperature of the fluid in the supercritical region can be altered to effect selectivity in the extraction.

8.13.4.2 Carbon Dioxide as Extraction Medium

Carbon dioxide is the most commonly used supercritical fluid for spice extraction, primarily due to its low critical parameters (31.1°C, 73.8 bar), low cost, and nontoxicity. Spice extraction using supercritical carbon dioxide (SCO_2) has the following advantages over conventional solvent extraction:

- Carbon dioxide is nonflammable, noncorrosive, and nontoxic. Hence storage and handling do not pose any health hazard.
- It is a part of the environment and hence does not precipitate any environmental issues.
- CO_2 is relatively less expensive. It can be recycled and reused in the system.
- The operation can be carried out at lower temperature, thus preserving the heat-sensitive flavor components.
- CO_2 is chemically inert and does not react with the components of the solute. Moreover, CO_2 can act as a blanket and prevent oxidative degradation of these components.

Supercritical operation provides greater flexibility since the solvation power and selectivity for the solutes can be easily manipulated by altering the temperature–pressure conditions. This facilitates the recovery of products with predesigned physicochemical properties.

Conventional extraction media are organic solvents with varying levels of toxicity. Even with the most efficient solvent removal techniques, residual solvents in ppm (parts per million) levels are likely to be present in the end product. SCO_2 extraction eliminates such solvent residue problems.

However, the extractor and accessories for supercritical fluid extraction have to be designed to operate under high pressure.

Schematic layout of a supercritical extraction plant is displayed in Figure 8.6 (Balakrishnan, 2005). The pump draws liquid CO_2 from the collection/storage vessel, compresses it to the required pressure and transfers to the extraction vessel through a heat exchanger where it is heated to the extraction temperature. The CO_2 containing the dissolved product is directed to the separator. The pressure of the CO_2 is reduced in the separator causing the product to get separated. Several separators maintained at progressively decreasing pressure may be used so that fractions of the extract with different qualities can be collected separately. The gaseous CO_2 from the separator is liquefied in a refrigerated heat exchanger and collected in the collection/storage vessel for recycling.

Turmeric can be extracted using SCO_2. Optimum pressure, temperature, particle size, flow rate, and extraction time can yield product with quality characteristics superior to conventional solvent extracted oleoresin (Chassagnez et al., 1997; Ge et al., 1997; Gopalan et al., 2000; Baumann et al., 2000; Began et al., 2000; Marongiu et al., 2002; Maul et al., 2002; Braga et al., 2003; Su et al., 2004). An operating pressure of 250 bar, temperature 45 to 55°C and extraction time 2 to 4 h are recommended for turmeric. Use of co-solvents or modifiers enhance the extraction rate. Ethanol is the co-solvent of choice in spice extraction using CO_2. Ethanol up to 30% level is used for turmeric (Su et al., 2004). Methanol modified CO_2 is reported to give improved curcumin recovery (Sanagi et al., 1993).

8.13.5 PROPERTIES OF TURMERIC OLEORESIN

Oleoresin turmeric comprises of the flavor and color principles of the spice and devoid of the cellulose plant material of the rhizomes. Acetone, ethanol, and ethyl acetate are the common extraction solvents. Oleoresin turmeric contains 30 to 40% curcumin, 15 to 20% volatile oil, and 20 to 30% fixed oils, depending on the cultivar and extraction medium.

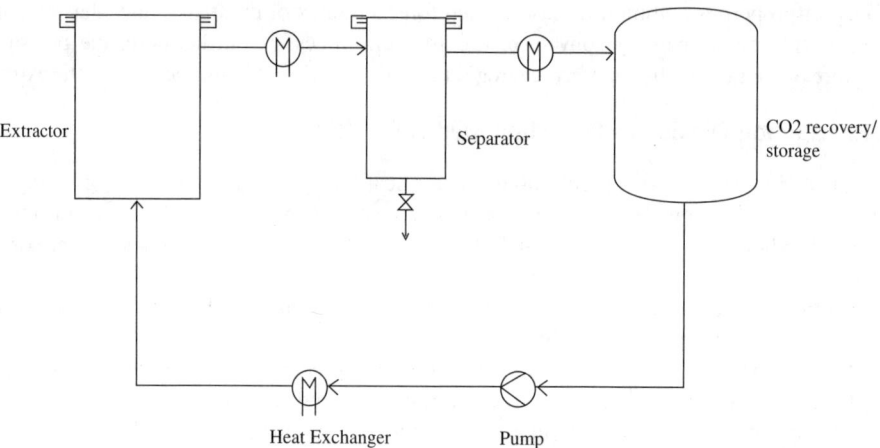

FIGURE 8.6 Schematic layout of a supercritical fluid extraction plant.

The U.S. Code of Federal Regulations (CFR, 1995) defines turmeric oleoresin as the combination of flavor and color principles obtained from turmeric by extraction using organic solvents. Oleoresin turmeric is a deep yellowish orange to red brown, highly viscous product difficult to handle at room temperature, but softens on warming. The prime quality determinant of the oleoresin is its color strength expressed in terms of curcumin content or color value. The product is normally standardized according to the label declaration. Food Chemicals Codex (FCC, 2004) specifies a curcumin content (or color value equivalent) between 1 and 45%. Oleoresin turmeric is generally regarded as safe (GRAS) as per U.S. FDA for application in food.

Manufacturers offer various grades in turmeric oleoresin, differing in color strength as well as the flow and solubility characteristics. A straight acetone or ethyl acetate extract is a highly viscous liquid, insoluble in water and oil. For convenience in application, it is usually extended with nonvolatile edible diluents or emulsifiers such as a propylene glycol or rendered water soluble with polysorbates. These modifications yield mobile homogeneous liquids with 8 to 20% curcumin. The oleoresin may be dispersed on a dry carrier or microencapsulated and spray dried to convert to powder form. The oleoresin concentration in these powders may be adjusted to give equivalence of a good-quality freshly ground spice.

The compounds of primary interest in turmeric oleoresin are the curcuminoids (30 to 40%) which contribute the characteristic yellow color. In addition to color matter, the oleoresin also contains the fixed and volatile oils that are responsible for the characteristic flavor and bitterness of turmeric. Turmeric flavor and bitterness are sometimes undesirable in the finished food product. The fixed and volatile oils can be removed from turmeric or its oleoresin by extraction with an aliphatic hydrocarbon, preferably hexane, having negligible solubility for curcumin (Sair and Klee, 1967). The resulting volatile oil-free product is known as debittered or defatted oleoresin. Removing the oils also serve to selectively enrich the color principles in the product.

The undesirable bitter principles in curcumin-containing turmeric or turmeric extracts may be reduced or eliminated by the addition of a glycine in sufficient dosages. The glycine suggested is at least one member selected from the group consisting of purified hydrolyzed vegetable protein, α-glycine, β-glycine, γ-glycine, glycine hydrochloride, a cationic salt of glycine and an anionic salt of glycine (Goldscher, 1979).

8.14 CURCUMINOIDS

The turmeric color is attributed primarily to a group of related compounds designated as curcuminoids with curcumin (I), diferuloyl methane [1,7-*bis*-(4-hydroxy-3-methoxy-phenyl) hepta-

1,6-diene-3,5-dione] as the principal component admixed with its two derivatives, demethoxy-curcumin (II) [4-hydroxycinnamoyl-(4-hydroxy-3-methoxycinnamoyl) methane], and *bis*-demethoxycurcumin (III) [*bis*-(4-hydroxy cinnamoyl) methane]. Curcumin with its two methoxy groups has a reddish orange color, whereas demothoxycurcumin with one methoxy group is orange-yellow, and *bis*-demothoxy curcumin without a methoxy group is yellow (Madsen et al., 2003). Other minor components are also present. The curcuminoids can be collectively isolated from the extract of the spice by crystallization. *C. longa* is the richest and cheapest source of these diferuloyl methane derivatives.

Though the rhizomes of *C. longa* have long been known to contain curcumin (Vogel and Pelletier, 1815), it was first obtained in crystalline state independently by Daube and Iwanof-Gajewsky in 1870 (Daube, 1870; Iwanof-Gajewsky, 1870). Since then, the isolation, characterization, and synthesis of curcumin have been attempted by a number of investigators (Milobedzka et al., 1910; Jackson and Clarke, 1911; Lampe and Milobedzka, 1913; Lampe, 1918; Ghosh, 1919; Pavolini et al., 1950; Lubis, 1968). Ciamician and Silber (1897) proposed a molecular formula $C_{21}H_{20}O_6$ for curcumin. The structure of curcumin as a diferuloylmethane was elucidated by Milobedzka et al. (1910), which was later confirmed by synthesis (Lampe and Milobedzka, 1913; Lampe, 1918). Srinivasan (1953), in his study on the constituents of the coloring matter in turmeric, subjected turmeric extract to column chromatography and the resolved fractions were analyzed through physical properties, absorption characteristics, and color-forming chemical reactions. He demonstrated that the pigment pool is composed of a mixture of analogues in which curcumin (I) was the main component. He proposed that the two main pigments accompanying curcumin were demethoxycurcumin (II) and *bis*-demethoxycurcumin (III) and all the three compounds exist in the *trans-trans* keto-enol form. Other minor pigments were also detected, which were possibly geometrical isomerides of the three main pigments. This pioneering work established the fact that the dye extracted is a mixture of I, II, and III, in which I dominated. The findings were backed by the observations of later researchers.

The structures of the three curcuminoids are given in Figure 8.7. The relative proportion of the three curcuminoids in turmeric dye vary with the cultivar; wide ranges have been reported (Perotti, 1975; Krishnamurthy et al., 1976; Govindarajan, 1980). Typical curcuminoid composition of popular Indian varieties are listed in Table 8.28 (Madsen et al., 2003) (C = curcumin; DMC = demethoxycurcumin, and BDMC = bisdemethoxycurcumin). The content of C was found to be in the range of 52 to 63%, DMC 19 to 27%, and BDMC 18 to 28%.

Commercially oleoresin turmeric or curcumin is seldom isolated from a specific cultivar; the raw material used is largely a composite of different cultivars, primarily due to availability and

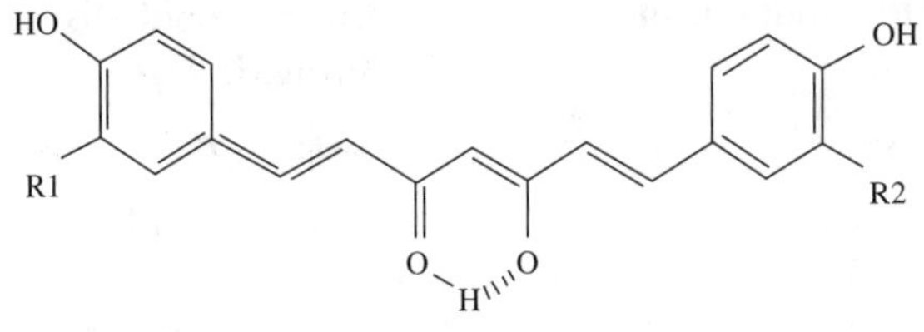

R1	R2	Compound
OMe	OMe	Curcumin
OMe	H	Demethoxycurcumin
H	H	*bis*-Demethoxycurcumin

FIGURE 8.7 Curcuminoids.

TABLE 8.28
Curcuminoid Composition of Indian Turmeric Varieties

Cultivar	C (%)	DMC (%)	BDMC (%)
Alleppy finger turmeric (AFT)	60.7	20.2	19.1
AFT (bulb split)	62.4	19.2	18.4
Barsi Kocha turmeric	52.4	19.7	27.9
Desi Kadappa finger turmeric (polished)	58.7	22.6	18.7
Erode Ghatta turmeric	61.1	18.7	20.1
Erode finger turmeric	59.0	21.8	19.3
Madras finger turmeric	60.1	19.7	20.2
Meghalaya split turmeric	54.5	26.3	19.2
Nizamabad finger turmeric	62.2	20.7	17.1
Panamgali turmeric (Erode)	63.2	18.8	18.0
Rajpuri finger turmeric	52.8	21.5	25.7
Sadashipet finger	52.9	23.7	23.4
Salem Panamgali	59.8	20.6	19.6
Salem finger turmeric (polished)	61.0	19.7	19.4

economic reasons. The distribution of I, II, and III in some turmeric oleoresin samples are given in Table 8.29 (Madsen et al., 2003). The ratio of the curcuminoids in the oleoresin is found to be in line with that of the raw material.

The distribution of curcuminoids in random samples of turmeric dye is displayed in Table 8.30 (Madsen et al., 2003). It is interesting to note that the ratio of the three compounds in the dye differs from that of the oleoresin and the raw material. The content of C is found to be in the range 71 to 88%, DMC 11 to 20%, and BDMC 1 to 10%.

Curcuminoids are recovered from turmeric extract by crystallization. Analysis of the mother extract after harvesting crystalline dye reveals a higher level of DMC and BDMC as compared to the oleoresin (Table 8.31) (Madsen et al., 2003). This shows that the extent of crystallization of the three analogues from the extract follows the order C > DMC > BDMC.

TABLE 8.29
Distribution of Curcuminoids in Turmeric Oleoresin

Sample	C (%)	DMC (%)	BDMC (%)
1	61.9	21.0	17.1
2	64.3	20.1	15.6
3	54.0	23.5	22.4
4	61.5	19.1	19.4
5	57.1	25.0	16.6
6	61.1	22.8	15.2
7	55.1	24.7	20.2
8	62.4	21.6	16.0

TABLE 8.30
Curcuminoid Distribution in Turmeric Dye

Sample	C (%)	DMC (%)	BDMC (%)
1	81.6	15.9	2.5
2	87.4	11.2	1.4
3	77.1	17.7	5.2
4	71.1	19.3	9.6
5	73.8	19.1	7.1
6	81.3	16.2	2.5
7	76.3	19.7	4.1
8	77.1	18.3	4.6
9	80.7	16.5	2.8
10	85.7	12.6	1.7
11	80.4	15.7	4.0
12	80.4	15.8	3.8

TABLE 8.31
Curcuminoid Distribution in the
Mother Extract after Harvesting the
Curcuminoids

Sample	C (%)	DMC (%)	BDMC (%)
1	48.0	24.6	27.4
2	41.0	25.4	33.6
3	28.9	26.6	44.5
4	35.2	26.2	38.6
5	33.8	24.3	41.9

For curcumin dye, it is the cumulative tinctorial strength of components that is of prime importance and not the individual components. So standards usually specify a minimum level of total curcuminoids or gross color value. However, some health food applications specify individual curcuminoid distribution. Curcumin 70 to 80%, DMC 15 to 25%, and BDMC 2.5 to 6.5% is a typical specification. Unless otherwise stated, the term curcumin in commerce refers to the agglomerate of the three constituents. A new compound cyclocurcumin has been identified in turmeric (Supardjan, 2001). The synergistic action of cyclocurcumin and the three curcuminoids is responsible for the nematocidal activity of turmeric. Cyclocurcumin is isolated as a yellow gum by repeated purification from the mother liquor after curcuminoid recovery. It has a molecular formula same as curcumin. Other related components identified include 1,5-*bis*(4-hydroxy-3-methoxyphenyl)-penta-(1E,4E)-1,4-diene-3-one; 1-(4-hydroxy-3-methoxyphenyl)-5-(4-hydroxyphenyl)-penta-(1E,4E)-1,4,-dien-3-one;1-(4-hydroxy-3-methoxyphenyl)-7-(3-4-dihydroxyphenyl)-1,6-heptadiene-3,5-dione; and 1,7-*bis*-(4-hydroxyphenyl)-1,4,6-heptatrien-3-one (Nakayama et al., 1993; Masuda et al., 1993).

8.15 ISOLATION OF TURMERIC DYE (CURCUMIN)

Initial efforts focused on the trapping of curcuminoids by precipitation as lead salts; the coloring matter is liberated from the latter by reaction either with H_2S or H_2SO_4 (Verghese, 1984). Goltman (1957) introduced column chromatography for purifying the crude pigments displaced from the lead salt with H_2S. Later investigators succeeded in upgrading the pigments primarily by recrystallization from ethanol.

Another approach to capture curcuminoids has been through alkali salts. The spice is extracted with an organic solvent, the extract dissolved in alkali and curcuminoids precipitated as salts with acid (Temmler-Werke, 1938). Here the pigments carry oily impurities. An improved method is based on the observation that oil in turmeric can be selectively extracted by an organic solvent like hexane. The extraction process was modified to include two steps — removal of the oil and extraction of pigments (Keil and Dobke, 1940). Firstly, the rhizome is de-oiled by hexane, which has a poor selectivity for the coloring matter. The residue is then freed from the solvent and exhaustively extracted with another solvent like methanol to recover the color matter. After removing the solvent, the concentrated extract is dissolved in alkali, filtered, and acidified with acid to precipitate the pigments.

Curcuminoids can also be extracted from powdered rhizomes using alkaline water (pH about 9) (Ran and Zhou,1988) and recovered by precipitation at pH 3 to 4. In another process, the powder is subjected to distillation with superheated steam to remove the volatile oil, the de-oiled spice is treated with an alkali, followed by precipitation of the curcuminoids with acid (Burger, 1958). The starch present in the spice, however, makes it a clumsy operation.

A key observation in the use of alkali treatment is that the curcuminoids are unstable in alkaline conditions, and the degradation rate rapidly increases with increase in pH (Price and Buescher, 1992).

Several improvements in the extraction method have been suggested by later investigators (Sair and Klee, 1967; Sastry, 1970; Khalique and Amin, 1967; Torres et al., 1998; Anderson et al., 2000, Gaikar and Dandekar, 2001; Madsen et al., 2003). The process for curcumin extraction and purification has now been well standardized by the industry.

8.16 MANUFACTURE OF TURMERIC DYE

Dry turmeric is comminuted and extracted with an organic solvent as discussed earlier. To make turmeric oleoresin, the solvent in the extract is completely evaporated by distillation under vacuum. To recover curcumin, the solvent is only partially removed; the concentrate cooled and allowed to stand for the curcumin to crystallize. Crystallization is enhanced by subcooling with chilled water. The crystals are then separated by centrifuging or vacuum filtration. Crude curcumin so obtained is given a quick wash with the solvent to enhance the purity. A wash with hexane can remove the residual fixed and essential oils, but does not dissolve the curcuminoids. The residual solvent in the product could be removed by injecting live steam and the crystals are then dried in hot air. The yield of curcumin from good quality spice is 2 to 4% depending on the cultivar. The flow diagram for curcumin production is given in Scheme 8.1.

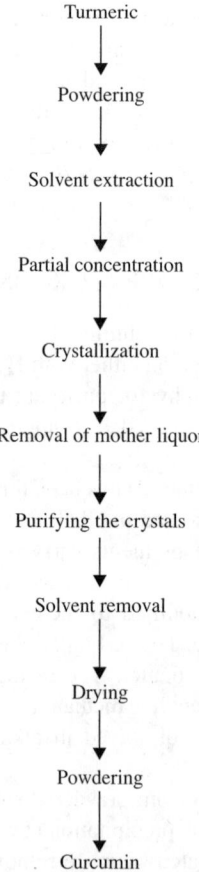

Turmeric

Powdering

Solvent extraction

Partial concentration

Crystallization

Removal of mother liquor

Purifying the crystals

Solvent removal

Drying

Powdering

Curcumin

SCHEME 8.1 Manufacture of curcumin.

8.17 QUALITY EVALUATION OF CURCUMINOIDS

Various methods have evolved over time for the detection and quantification of curcuminoids. In a simple test to detect the presence of curcumin, the sample is treated with ethyl alcohol, then mixed with a few drops of alcoholic H_3BO_3, slightly acidified with tartaric acid, and placed in a small porcelain crucible or on a porcelain plate. The alcohol creeping up in the crucible walls or to the outside ring on the plate carries the characteristic rose-red color to the edge (White, 1930). When turmeric preparations based on turmeric were treated with Ac_2O and a few drops of concentrated H_2SO_4 and exposed to UV light, intense red fluorescence is observed. This analysis is a useful identification tool in pharmacological applications (Schmelzer, 1969). The early quantitative estimation techniques were also based on the color reactions of the dye. A typical method makes use of the reaction of curcumin with H_3BO_3 and H_2SO_4-HOAc mixture. The pink color developed is measured at 500 nm in a spectrophotometer (Satyanarayana et al., 1969).

TLC or paper chromatography provided for the separation of the different curcuminoids, which could then be individually quantified. Rapid and simple TLC methods have been suggested for curcuminoids. The assay usually combines the separation of curcuminoids on silica gel TLC plates, followed by scanning densitometry or spectrometry of the separated components (Punyarajun, 1981; Chen et al., 1983; Yang et al., 1984; Wu, 1995; Cooray et al., 1998; Gupta et al., 1999). If necessary, a more intensely colored complex can be developed by reaction with boric acid to lower the detection limit.

An alternative detection method for the color compounds in turmeric is based on its fluorescent property. Microamounts of curcumin can be determined in a spectrofluorometer (Jasim and Ali, 1992). The process involves dissolving samples in dry Me_2CO, irradiating the resulting clear solution at $\lambda = 424$ nm and measuring the stable intense green yellow fluorescence at $\lambda = 504$ nm.

Curcuminoids exhibit strong absorption between 420 and 430 nm in organic solvents. This property is the basis for spectrophotometric estimations. The primary quality determinant of turmeric extract is the color value and this is best expressed in terms of color units (CU). The color value is essentially the absorbance of a 0.01% ethanolic solution of the extract at the peak maximum, which is approximately at $\lambda = 422$ nm, multiplied by 1000 (EOA, 1958), which is in effect ten times the E1%/1 cm value. Lower dilutions and proportionately higher multiplication factors are employed for analyzing higher color strength products. Color value is a cumulative of the contribution from all the pigment components. Normally the color value of a direct acetone extract of turmeric is in the range of 4000 to 6000 CU, whereas that of commercially pure curcumin is 14,000 to 15,000 CU. Spectrophotometric analysis can also quantify the curcumin content using pure curcumin as standard. The ASTA (ASTA, 1999) has formulated a standard spectrophotometric method for quantification of curcumin in turmeric and its oleoresin. The spectrophotometric method described by Food Chemicals Codex (FCC, 2004) for curcumin estimation makes use of a factor of 165 as absorptivity for curcumin in acetone.

HPLC has emerged as an efficient tool for the quantification of individual curcuminoids. A number of combinations of stationary and mobile phases have been suggested for effective separation of the pigment components (Asakawa et al., 1981; Tonnesen and Karlsen, 1983; Smith and Witowska, 1984; Zhao and Yang, 1986; Rouseff, 1988; Khurana and Ho, 1988; Taylor and McDowell, 1992; Hiserodt et al., 1996; Gupta et al., 1999; Wang et al., 1999; Chauhan et al., 1999; Jayaprakasha et al., 2001; Jayaprakasha et al., 2002). A UV-VIS detector is generally used. Typical HPLC chromatogram of the curcuminoid pool is given in Figure 8.9. High performance thin layer chromatography (HPTLC) (Sankaranarayanan et al., 1993; Khanapure et al., 1993; Lavoine et al., 1998), capillary electrophoresis (Lechtenberg et al., 2004), high-speed countercurrent chromatography (CCC), pH-zone-refining CCC (Patel et al., 2000), on-line high performance liquid chromatography, with UV diode array detector, and electrospray mass spectrometry (He et al., 1998) have also been developed for the quantification of curcuminoids.

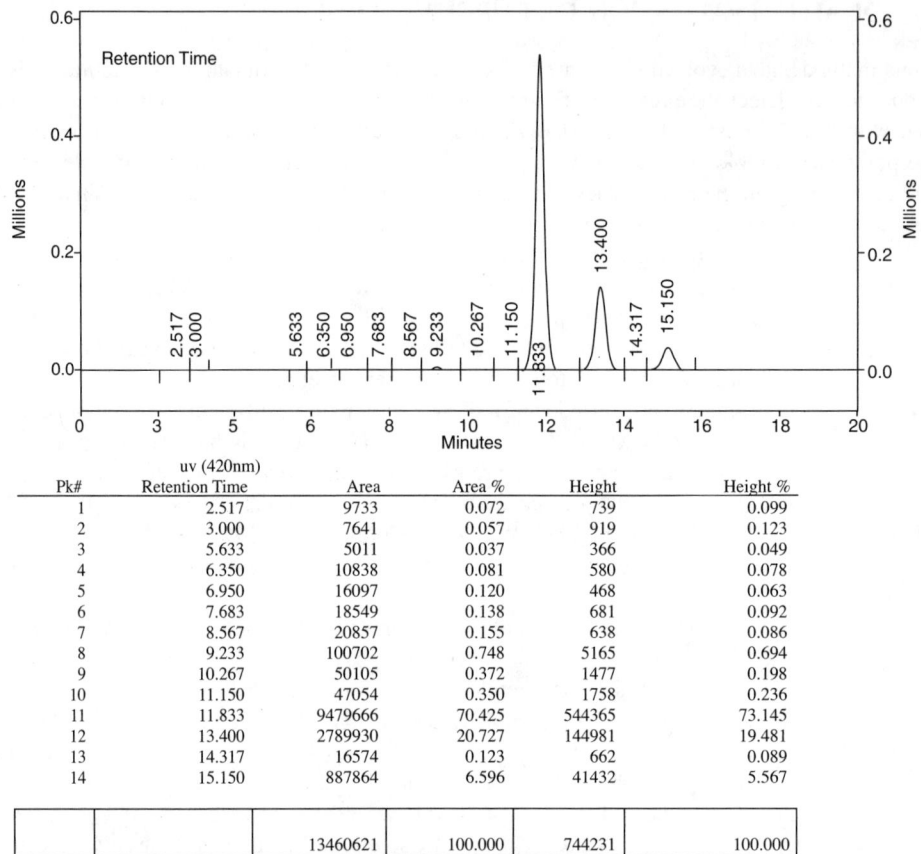

	uv (420nm)				
Pk#	Retention Time	Area	Area %	Height	Height %
1	2.517	9733	0.072	739	0.099
2	3.000	7641	0.057	919	0.123
3	5.633	5011	0.037	366	0.049
4	6.350	10838	0.081	580	0.078
5	6.950	16097	0.120	468	0.063
6	7.683	18549	0.138	681	0.092
7	8.567	20857	0.155	638	0.086
8	9.233	100702	0.748	5165	0.694
9	10.267	50105	0.372	1477	0.198
10	11.150	47054	0.350	1758	0.236
11	11.833	9479666	70.425	544365	73.145
12	13.400	2789930	20.727	144981	19.481
13	14.317	16574	0.123	662	0.089
14	15.150	887864	6.596	41432	5.567
		13460621	100.000	744231	100.000

FIGURE 8.8 Typical HPLC chromatogram of curcuminoids.

Efforts have been made to determine the color value of individual curcuminoids (Tonnesen, 1992). For this, the compounds were separated and the absorbances determined at wavelengths 420, 425, and 430 nm, which correspond to maximum absorption in ethanol for the three curcuminoids. The color values are listed in Table 8.32. The average color value of the curcuminoid complex, therefore, depends on the relative distribution of the three components in the pigment pool. Srinivasan (1953) in his pioneering work found that the content of curcuminoids in turmeric ranged from 2.5 to 6% of which curcumin accounted for about 49% of the total pigments, demethoxycurcumin about 29% and *bis*-demethoxycurcumin about 22%. Madsen et al. (2003) reports the distribution of curcuminoids in Indian varieties (Table 8.26) to be C 52 to 63%, DMC 19 to 27% and

TABLE 8.32
Color Value of Pure Curcuminoids

	Wavelength (nm)		
Curcuminoid	420	425	430
Curcumin (I)	15,280	15,780	15,860
Demethoxycurcumin (II)	15,420	15,460	15,130
bis-Demethoxycurcumin (III)	16,820	16,400	15,650

BDMC 18 to 28%. Verghese (1994) summarized the level of the three analogues by TLC in a number of samples and found that the make up of I, II, and III enjoy random distribution in commercially available products. He found that the samples analyzed had 31 to 57% I, 20 to 40% II, and 7 to 34% III, the averages being in line with the other observations. Taking the distribution reported by Srinivasan, the average color value of the curcuminoid complex in pure form in the wavelength range 420 to 430 nm would be approximately 15,600 to 15,800 CU. Balakrishnan et al. (1983) could obtain a product with a color value of 15,960 CU by repeated purification and suggested a color value of 16,000 CU for pure curcuminoid complex.

A judicious selection of cultivars for extraction enable one to play around with the relative distribution of curcuminoids.

8.18 MODIFIED AND STABILIZED CURCUMINOIDS

Turmeric curcuminoids are important natural pigments. However, they are inherently lipid and water insoluble and are sensitive to light, pH, solvent system, and oxygen (Price and Buescher, 1996). It is common for curcumin to be converted into a convenient application form. Water or oil solubility is achieved using permitted emulsifiers. Polysorbate 80 is the favored emulsifier to confer water solubility. A formulation with 8 to 10% curcumin is clearly water soluble. Mono, diglycerides, and propylene glycol are other common emulsifiers. Curcumin is also available as fine suspensions in vegetable oils. Curcumin as suspended pigment is more stable than the solubilized color.

Numerous patents have been granted for curcumin formulations with improved stability and physical properties. A water-soluble curcumin complex for food coloring is prepared by dissolving curcumin and gelatin in an aqueous acetic acid solution. The complex comprises up to about 15% curcumin by weight (Schranz, 1983). Curcumin can be phosphorylated by phosphorus oxychloride ($POCl_3$) at 0°C in alkaline medium; the phosphorylated product has improved light stability and is water soluble (Yang and Buescher, 1993). This modified curcumin is reported to be about seven times more stable in light than normal curcumin.

Curcumin undergoes degradation and color loss on storage in alkaline conditions and this property limits its use in many products. Curcumin gives a lemon yellow color in acidic aqueous media; in alkaline conditions, the color turns distinctly red and become unstable. A stabilization method reported produces dry curcumin colorant compatible with alkaline and acid dry food mixes. The stability is achieved by encapsulation and spray drying a formulation, where the curcumin in the core is maintained at low, stable pH between 3.5 and 4.5. The formulation comprises curcumin, an organic acid, a buffer, a dispersant for the curcumin, and a film-forming encapsulant. The amount of acid should be effective to maintain the curcumin at the low pH. A preferred use of the colorant is in dry mixes for instant puddings, which are alkaline due to the salts employed to cause setting (Leshik, 1981). Cyclodextrin complexes are said to improve the water solubility as well as the hydrolytic and photochemical stability of curcumin. Complex formation is reported to increase water solubility at pH 5 by over 100 times. The hydrolytic stability of curcumin under alkaline conditions was also strongly improved by complex formation (Tonnesen et al., 2002).

A water soluble coloring complex has been developed by the reaction of curcumin with certain metal ions. The metal-curcumin coloring complex is water soluble and capable of producing varying hues for use in foodstuffs. The metal ion component is selected from the group consisting of stannous ion, zinc ion, and mixtures thereof. A ratio of four to five moles of curcumin to each mole of metal ion is found to have long lasting stability and the ability to provide different hues according to the metal concentration (Maing and Miller, 1981).

In a study on the degradation of curcuminoids under various conditions, it was observed that curcuminoid pigments were sensitive to light; however, the combined effect of air and light was the most deleterious. There was no influence of water activity on the stability of curcuminoid pigments, in curcumin and in turmeric oleoresin-microcrystalline cellulose model systems examined

(Souza et al., 1997). Fatty foods can be colored with microcrystalline dispersions of curcumin. However, because of the crystalline structure, these dispersions are not efficient as tinctorial agents. In addition, curcumin can slowly migrate to the wrapper and stain it. When curcumin is used on a snack, it is sometimes dissolved in hot vegetable oil to increase its solubility, and this in turn is sprayed onto the snack. Some of the curcumin will precipitate on cooling of the snack, due to its lower solubility in the oil. A curcumin complex that surmounts these limitations is prepared from curcumin, a water-soluble branched chain or cyclic polysaccharide and a water-soluble or water-dispersible protein (Todd, 1991). Complexing also increases light stability. Curcuminoids are also stabilized by adding edible acids and/or their water-soluble acidic salts. A combination of curcuminoids, maltitol lactose and citric acid, when mixed and granulated, yielded a product that retained 99% of the color after 40 days (Takagaki and Yamada, 1998).

Water-soluble curcumin coloring agent is produced from ground turmeric by washing it with a soap solution in which the curcumin dissolves. A subsequent treatment of the solution with acid precipitates the curcumin and produces a paste or putty, which is dispersible in fatty-based substances (Stransky, 1979). In another invention, a water soluble curcumin coloring is prepared from turmeric by grinding, washing with aqueous NaCl or KOH to remove dirt, bacteria, sugars, starches, and camphor and extracted with a soap solution containing KOH, castor oil and water at 60°C and pH ~7 to solubilize the pigment. This solution can be mixed with water for use directly in water-based foods. For preparing an oil soluble or oil dispersible curcumin that is still water soluble, the soap solution is mixed with an acid (HCl, HOAc, or a combination) and filtered (Stransky, 1979). Another formulation proposed involves dissolving the ether extract of turmeric in vegetable oil and clarifying with ice cooling to give a scarlet solution that can be used as food dye (Fukazawa, 1970).

8.18.1 MICRO-ENCAPSULATION

Oleoresin turmeric can be microencapsulated to convert to powder form. Microencapsulated oleoresins are micro-fine particles of the product coated with an envelope of an edible medium such as starch, maltodextrin, or natural gums so that the functional components are locked within the tiny capsule. Encapsulated oleoresin is usually prepared by spray drying technique. When incorporated in food, the outer coating dissolves off, thereby releasing the product. Encapsulated oleoresins can be designed to contain a predetermined level of the core material.

Microencapsulation serves the following purposes:

1. control the release of the core material
2. lock the functional components to ensure against loss on storage
3. offer convenience in handling by converting liquids and semisolids into free flowing powder
4. provide uniform dispersibility in the food matrix

Figure 8.9 displays the flow diagram for microencapsulation/spray drying of spice oleoresins (Balakrishnan, 2005).

8.19 USES OF TURMERIC AND ITS EXTRACTS

Turmeric and its extractives have diverse applications. Turmeric is a unique and versatile natural plant produce combining the properties of a spice, food colorant, cosmetic, and drug. In ethnic cuisine, turmeric is a common flavor ingredient. It is one of the most popular natural food colors. Turmeric pigments produce their typical brilliant yellow hue at pH of 2.5 to 7.0. Turmeric has proven antioxidative, anti-inflammatory, anticarcinogenic, antimutagenic, antimicrobial, antiviral, and antiparasitic properties and is a known skin care and health food ingredient capable of pre-

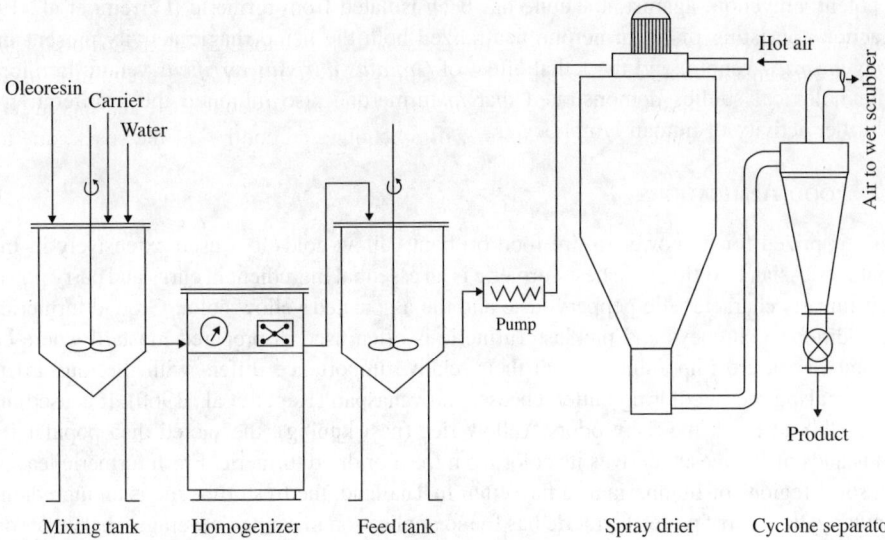

FIGURE 8.9 Schematic layout of microencapsulation/spray drying plant.

venting or retarding a number of illnesses. The unique properties of the spice are attributable to the essential oil and curcuminoid components present therein.

8.19.1 TURMERIC OIL

Unlike the oleoresin, turmeric oil is not highly valued in spice industry. However, there have been extensive investigations on the medicinal properties of turmeric, some of which are attributable to compounds present in the volatile fraction. Oil of turmeric is an antacid, and in small doses, acts as a carminative, stomachic, appetizer, and tonic (CSIR, 1950). The oil given by vapor inhalation is found to have significant effect in removing sputum, relieving coughs, and preventing asthma; it is effective for the treatment of respiratory diseases (Li et al., 1998).

The oil of turmeric has cosmetic applications. It adds a spicy aroma to perfumes and soaps. Moreover, some components of turmeric oil could be converted into compounds of perfumery value through appropriate chemical reactions (Banerjee et al., 1981).

Essential oil of turmeric is reported to possess fungicidal activity comparable with standard antifungal drugs (Singh et al., 1984; Apisariyakul et al., 1995; Saju et al., 1998; Behura et al., 2000; Jayaprakasha et al., 2001; Ramachandraiah et al., 2002). The curcuminoids, however, are devoid of this property (Sawada et al., 1971). The activity of leaf oil on human pathogens is found to be superior to rhizome oil. *ar*-Turmerone and turmerone, the components of turmeric oil, exhibit insect repellency (Su et al., 1982; Lee et al., 2001; Tripathi et al., 2002). Both leaf and rhizome oil possess excellent antimicrobial/antibacterial activity (Banerjee and Nigam, 1978; Iyengar et al., 1995; Negi et al., 1999; Singh et al., 2003; Garg and Jain, 2003), well in line with that of common antibiotic drugs and is especially effective against food borne pathogenic bacteria (Uechi, 2000; Rath et al., 2001).

Turmeric oil is a powerful anti-inflammatory agent. In a study, rats fed orally with a fraction of essential oil of turmeric, boiling range 80 to 110°C, developed anti-inflammatory activity against Freund's adjuvant-induced arthritis. The volatile oil was effective against both early and late inflammation (Chandra and Gupta, 1972). Pharmaceutical composition comprising essential oils extracted from tissues of turmeric or *C. xanthorrhiza* or a combination of both oils and curcumin has been suggested for treating inflammatory condition (Oei, 1992).

A potent antivenom against snakebite has been isolated from turmeric (Ferreira et al., 1992). The fraction consisting of *ar*-turmerone neutralized both the hemorrhagic activity present in the *Bothrops jararaca* venom, and the lethal effect of *Crotalus durissus terrificus* venom in mice.

Immunological studies demonstrated that *ar*-turmerone also inhibited the proliferation and natural killer activity of human lymphocytes.

8.19.2 Food Applications

Turmeric is prized for its power to tint food brilliant yellow-gold. It is used extensively in Indian and Southeast Asian traditional dishes. Turmeric is an essential ingredient in curry and curry powders, where it imparts characteristic peppery taste and the associated yellow color. Ground turmeric is a classic addition to chutneys and pickles. Turmeric is often used in prepared mustard where color, flavor, and aroma are important, though their relative importance differs with the mustard type. Turmeric is also used in coloring butter, cheese, and vanaspati (Kapur et al., 1960). It is used in fish curry, possibly to mask the fishy odors. Yellow rice (nasi kuning), the sacred dish popular in the eastern islands of Indonesia, derives its color from fresh or dried turmeric. Fresh turmeric leaves are used in some regions of Indonesia as a flavoring. In Thailand, the fresh rhizome is an ingredient for the popular yellow curry paste. Turmeric has found application in canned beverages, baked products, dairy products, yogurts, yellow cakes, biscuits, popcorn color, sweets, cake icings, cereals, sauces, gelatins, direct compression tablets, etc. Western cuisine does not use turmeric directly in food, but it forms part of curry powders and sauces to impart the yellow color. Turmeric's flavor has been described as peppery and somewhat bitter, so it is important to be judicious when adding this spice to foods. Curcumin has high color intensity; dosage levels are usually low.

Natural colorants are becoming increasingly important in food industry in view of the mounting safety issues on synthetic dyestuffs. In response to consumer demand, many food manufacturers prefer to keep all their ingredients natural, or derived from natural sources. Traditionally, ground turmeric has been used to enhance both flavor and color, which is now being largely replaced by turmeric oleoresin and curcumin. Cucumin gives a bright yellow color at dosages as low as 5 to 20 ppm. The color shade is comparable to that of tartrazine.

Oleoresin turmeric and curcumin are alcohol soluble and water/lipid insoluble. Oleoresin turmeric can be converted to user friendly versions based on the application. When used in dry mixes, they are precoated on salt or a small portion of the flour or other dry ingredients used in the final product. The precoated blend is then incorporated into the other dry ingredients in a suitable blender. The oleoresin is also microencapsulated and spray dried to convert to powder form. Spray dried version is stable in retort and extrusion processing. Oleoresin turmeric can be rendered water-soluble or dispersible using permitted emulsifiers. Water-soluble and dispersible turmeric colors find application in expanded extruded cereals, boiled sweets, cream fillings, marshmallows, and frosting. Dairy foods such as yogurt, ice cream, and dairy-based beverages can be colored with water-soluble turmeric. Curcumin or defatted oleoresin, essentially devoid of the characteristic turmeric flavor, is preferred for flavor-sensitive products such as cheese and other dairy products, ice cream, lemonade, confections, and baked products. Water-soluble turmeric is used in pickles to provide a bright hue. Chicken soup and broths are often colored with water soluble turmeric. It is also used as coloring matter in pharmacy. Oleoresin turmeric and its microcrystalline suspensions in oil are used to color batters and breadings.

Approximate dosages of turmeric and turmeric oleoresin for typical applications are given in Table 8.33 (Fenaroli, 1975).

Food ingredients permitted for use in Europe are identified with E numbers (Hanssen and Marsden, 1988). Curcumin, listed in the category of colors, has the E number 100. U.S. FDA (1995) considers ground turmeric and turmeric oleoresin as a color additives, which may safely be used for coloring foods generally, in amounts consistent with good manufacturing practice. Turmeric color is exempted from certification.

TABLE 8.33

Approximate Dosages of Turmeric and Turmeric Oleoresin for Typical Applications

Turmeric	
Gelatins and puddings	0.05 ppm
Condiments	760 ppm
Soups	30–50 ppm
Meats	200 ppm
Pickles	690 ppm

Turmeric Oleoresin	
Condiments	640 ppm
Meats	20–100 ppm
Pickles	200 ppm

Combinations of turmeric with other natural colorants yield an array of shades. Oleoresin turmeric is often blended with oleoresin paprika and annatto extract. The addition of oleoresin paprika produces a richer yellow hue. Turmeric and annatto combines as coloring materials for cereal-based products, cheeses, dry mixes, salad dressings, winter butter, and margarine (Freund, 1985). Turmeric is sometimes used in the place of expensive saffron to provide color, though not a substitute.

Curcuminoids are pH sensitive. The kinetics of curcumin degradation in aqueous solution was studied by Tonnesen and Karlsen (1985b). Below pH 7, the solution is yellow. At pH greater than 7.5, curcumin molecules are extremely unstable and gives an orange-red solution. At pH 7.75, the half life of curcumin is 50 min, whereas at pH 10.8, it is about 1 min. Degradation products have been identified to be ferulic acid and feruloylmethane. The photochemical stability of curcumin was studied by the same scientists and published in 1986 (Tonnensen et al., 1986). The curcumin degradation follows first order kinetics with the best stability in methanol (half life being 92.7 h).

Bitterness limits the use of turmeric in some areas. The bitterness of turmeric or turmeric derivatives can be removed by the addition of a glycine (Goldscher, 1979). The turmeric is added to a given food to obtain the color intensity required and sufficient glycine is added to reduce the bitterness to the desired level. In common orange breakfast drink, it took 0.05 g glycene to debitter 0.095 g turmeric. The glycine selected is from the group consisting of purified hydrolyzed vegetable protein, α-glycine, β-glycine, γ-glycine, glycine hydrochloride, a cationic salt of glycine and an anionic salt of glycine. In another process (Inafuku, 1996), dried pulverized turmeric is mixed with cereal husk (or hull) and sugars, inoculated with lactic acid bacteria and heat dried. This method greatly reduces the bitterness in turmeric. Another method suggested for debittering includes adding a vegetable oil to turmeric, mixing with yeast, and retaining at ordinary temperature or slightly heating the mixture for fermentation. A fermented mixture containing turmeric 100, seasame oil 10, and yeast 5 exhibited no bitter taste (Takanashi, 1998).

8.19.3 AS FOOD ANTIOXIDANT

Curcumin possess excellent antioxidant properties. It is reported to be more potent in preventing lipid peroxidation than alpha-tocopherol, pine bark extract, grape seed extract or the commonly used synthetic antioxidant BHT (Sreejayan and Rao, 1993, 1994; Majeed, 1995). A complex of the three curcuminoids was found to be more effective as an antioxidant than each of the components — curcumin, demethoxycurcumin, or bisdemethoxycurcumin — used alone. Turmeric is sometimes added to oils as a preservative. In a study (Ramaswamy and Banerjee, 1948), turmeric dye exhibited excellent antioxidant properties on coconut oil, groundnut oil, cottonseed oil, and sesame oil. The antioxidant property of the dye is attributed to its phenolic structure. This view is supported by the fact that dye of kamala with phenolic group exhibited similar activity, whereas annatto pigment, which is a carotenoid with no phenolic group, behaved as pro-oxidant. Turmeric is reported to inhibit oxidative rancidity in salted cooked fish and is said to be more potent than garlic and onion (Ramanathan and Das, 1993).

In a study (Majeed, 1995), the free-radical scavenging ability of various curcuminoids were evaluated using the DPPH (1.1 diphenyl-2-picrylhydrazyl) radical scavenging method. The results indicated that curcuminoids neutralize free radicals in a dose-dependent manner. Tetrahydrocurcumin (THC) was the most effective, followed by curcumin and bisdemethoxycurcumin. It is suggested that the antioxidant mechanism of curcuminoids may include one or more of the following interactions: (1) scavenging or neutralizing of free radicals, (2) interacting with oxidative cascade and preventing its outcome, (3) oxygen quenching and making it less available for oxidative reactions, (4) inhibition of oxidative enzymes like cytochrome P-450, and (5) chelating or disarming oxidative properties of metal ions like iron (Fe).

Masuda et al. (2001) investigated the antioxidant mechanism of curcumin against peroxide radicals in the presence of ethyl linoleate as one of the polyunsaturated lipids. They found that during the antioxidation process, curcumin reacted with four types of linoleate peroxyl radicals to yield six reaction products. On the basis of the formation pathway for their chemical structures, an antioxidant mechanism of curcumin in polyunsaturated lipids was proposed, which involved an oxidative coupling reaction at the 3′-position of the curcumin with the lipid and a subsequent intramolecular Diels-Alder reaction.

Toda et al. (1985) probed the antioxidative components in the MeOH extract of turmeric rhizomes using a method based on the air-oxidation of linoleic acid. All three curcuminoids were found to be active. Curcumin was the most active component and its 50% inhibitory concentrations for the air oxidation of linoleic acid were 1.83×10^{-2}% (thiobarbituric acid value) and 1.15×10^{-2}% (peroxide value). These values are superior to dl-α-tocopherol.

8.19.4 MEDICINAL PRODUCTS AND USES

Turmeric has been extensively used in medicine throughout Asia. Traditional healers of India and China used it as a remedy for varied conditions from eye infections to intestinal worms to leprosy. It is reported to be alterative, antiperiodic, cholagogue, depurative, haemostatic, stomachic, and tonic; aperient, astringent, carminative, cordial, detergent, diuretic, emmenagogue, maturant, and stimulant. The active components in the spice exhibit a broad spectrum of biological activities viz. antibacterial, antifungal, antiparasitic, antimutagen, anti-inflammatory, hypolipidemic, hepatoprotective; lipoxygenase, cyclooxygenase, and protease-inhibitory; besides being effective active oxygen species scavenger (Khanna, 1999). In folk medicine, turmeric is used in the treatment of abdominal pain, diarrhea, chest pains, colic, dysmenorrhea, epistaxis, fever, flatulence, hematuria, hematemesis, and urinary problems; jaundice, hepatitis, and other affections of the liver; it is applied externally for inflammation, indolent ulcers, itch, ringworm, and sores, boils, bruises, elephantiasis, leucoderma, scabies, smallpox, snakebite, leech bite, and swellings. In Ayurveda, turmeric is commonly administered internally as a stomachic, tonic, and blood purifier, and topically in the prevention and treatment of skin diseases. Mixed with warm milk, it is said to be beneficial in common cold. The juice of fresh rhizome is used as an antiparasitic for many skin infections. A paste made from the powdered rhizome along with lime forms a remedy for inflamed joints. Turmeric is reported to augment the healing process of chronic and acute wounds, ulcers, and sores. Wound healing preparations containing an effective amount of turmeric are reported (Das and Cohly, 1995). A decoction of the rhizome is said to relieve the pain of purulent opthalmia. Aqueous extract of turmeric is useful in the treatment of gall bladder complaints (CSIR, 1950). Hair tonics that retard hair loss, inhibits 5-α-reductase activity and having antidandruff effects have been developed with curcuminoids as the active components (Hamada and Minamino, 1997).

Herbal mixtures containing turmeric as one of the main ingredients are reported to provide therapeutic weight loss as well as lipid reduction and change body composition. The mixture taken as a dietary supplement is effective in the treatment of obesity, including both weight loss and reduction of weight gain (Wei and Xu, 2003). Turmeric extract is used as an antibacterial and anti-inflammatory natural dye for cosmetics like shampoos, lotions, and sprays (Belle, 1980). Curcumin and turmeric extract tested on different strains of bacteria showed that they possess strong antibacterial and antifungal activity (Basu, 1971; Giang, 2002; Wuthi-udomlert et al., 2000; Chauhan et al., 2003). Turmeric is said to discourage the growth of facial hair. Balms containing turmeric powder or extract are effective in treating acne, rashes, eczema, and pimples. Curcumin containing plants and plant extracts are also effective in controlling infections in animals (Kawaguchi and Kida, 1987).

The significance of turmeric in medicine has increased considerably with the discovery of the antioxidant properties of naturally occurring phenolic compounds present therein. Curcumin exhibits strong antioxidant activity (Sharma, 1976; Govindarajan, 1980; Toda et al., 1985; Unnikrishnan and Rao, 1995; Iqbal et al., 2003; Balasubramanyam et al., 2003). A number of patents have been granted for formulations containing curcumin for human health (Oei, 1992; Ammon et al., 1995; Aggarwal, 1999; Majeed et al., 1999; Santhanam et al., 2001; Graus et al., 2002; Saito, 2002; Newmark, 2002; Phan, 2003; Lee et al., 2004; Arbiser, 2004). Analysis of the chemical structure of curcumin in relation to its biological activity have established that the p-hydroxy groups in the molecules are essential for the antioxidant activity of curcuminoids (Majeed, 1995). Molecular structures that contribute for biological activity of curcuminoids are: (1) p-hydroxyl groups — antioxidant activity, (2) keto groups — anti-inflammatory, anticancer, antimutagen, and (3) double bonds — anti-inflammatory, anticancer, antimutagen. In a comparative study on the antioxidant activity of curcumin and demethoxycurcumin by radiation-induced lipid peroxidation in rat liver microsomes (Priyadarsini et al., 2003), it was observed that at equal concentration, the efficiency to inhibit lipid peroxidation changed from 82% with curcumin to 24% with demethoxycurcumin, showing that the phenolic OH was essential for both antioxidant activity and free-radical kinetics.

Curcumin is a potent scavenger of superoxide; the anti-inflammatory activity and superoxide scavenging property can be correlated (Elizabeth and Rao, 1990). Turmeric is effective in protecting animal livers from a variety of hepatotoxic insults (Kiso et al., 1983b; Donatus and Sardjoko, 1990; Deshpande et al., 1998; Park et al., 2000). Administration of curcumin greatly reduced liver injury. The hepatoprotective effect is attributed to the antioxidant properties of the spice. Turmeric and curcumin are reported to reverse biliary hyperplasia, fatty changes, and necrosis induced by aflatoxin production (Soni et al., 1992). Herbal mixtures containing turmeric as one of the main ingredients are reported to provide therapeutic weight loss as well as lipid reduction and change body composition. The mixture taken as a dietary supplement is effective in the treatment of obesity, including both weight loss and reduction of weight gain (Wei and Xu, 2003).

Turmeric oil and curcumin exhibit potent anti-inflammatory effects (Arora et al., 1971; Chandra and Gupta, 1972; Mukhopadhyay et al., 1982; Ammon et al., 1993; Kohli et al., 2005). Oral administration of curcumin in instances of acute inflammation was found to be as effective as cortisone or phenylbutazone, and one-half as effective in cases of chronic inflammation (Mukhopadhyay et al., 1977). Deodhar et al. (1980) studied the anti-inflammatory action of curcumin in patients with rheumatoid arthritis. The study demonstrated a significant improvement with curcumin in the duration of morning stiffness, walking time, and joint swelling, which was almost comparable to phenylbutazone. Satoskar et al. (1986) evaluated the anti-inflammatory property of curcumin in patients with postoperative inflammation. The effect of the drug on individual parameters revealed that curcumin had good anti-inflammatory responses in these patients. Curcumin was found better than phenylbutazone in reducing spermatic cord oedema and tenderness. In various animal studies, a dose range of 100 to 200 mg/kg body weight exhibited good anti-inflammatory activity and seemed to have negligible adverse effect on human systems (Kohli et al., 2005). The most interesting feature of curcumin is the lack of gastrointestinal side effects despite being an anti-inflammatory agent. In rats with Freud's adjuvant-induced arthritis, oral administration of turmeric significantly reduced inflammatory swelling compared to controls (Arora et al., 1971). In monkeys, curcumin was shown to inhibit neutrophil aggregation associated with inflammation (Srivastava, 1989). Turmeric can be applied topically in poultices to relieve pain and inflammation (Leung, 1980). Turmeric's anti-inflammatory properties may be attributed to its ability to inhibit pro-inflammatory arachidonic acid, as well as neutrophil function during inflammatory states. Curcumin may also be applied topically to animal skin to counteract inflammation and irritation associated with inflammatory skin conditions and allergies (Mukhopadhyay et al., 1982; Kawaguchi and Kida, 1987). A toothpaste effective in the treatment of inflammation diseases of the mucous membrane of the oral cavity and periodontitis has been developed with turmeric as one of the ingredients. Dentifrices and other oral compounds containing curcumin, its analogues or derivatives are found to be useful in the prevention and treatment of both dental caries and periodontal diseases (Katyuzhanskaya et al., 1985; Cavanaugh, 1998).

Turmeric and curcumin are capable of suppressing the activity of several common mutagens and carcinogens in a variety of cell types (Soudamini and Kuttan, 1989; Mehta and Moon, 1991; Azuine and Bhide, 1992; Boone et al., 1992). Animal studies as well as *in vitro* studies utilizing human cell lines have demonstrated curcumin's ability to inhibit carcinogenesis at three stages — tumor promotion, angiogenesis, and tumor growth (Limtrakul et al., 1997; Thaloor et al., 1998; Kawamori et al., 1999). In studies on colon and prostate cancer, curcumin was found to inhibit cell proliferation and tumor growth (Hanif et al., 1997; Dorai et al., 2001). Kuttan et al. (1987) reported that an ethanolic extract of turmeric or a curcumin ointment provided symptomatic relief in patients with cancers of oral cavity, breast, vulva, and skin. Curcumin is reported to be effective in the management of chronic anterior uveitis (Lal et al., 1999). Lack of side effects with curcumin is a great advantage when compared with drugs like corticosteroids. Clinical efficacy of curcumin in the treatment of patients suffering from idiopathic inflammatory orbital pseudotumors has also been established (Lal et al., 2000). The anticarcinogenic effects of turmeric and curcumin are partly due to the antioxidant and free-radical scavenging effect; they also enhance the body's natural antioxidant system, increasing glutathione levels, thereby aiding in hepatic detoxification of mutagens and carcinogens, and inhibiting nitrosamine formation (Pizorrno and Murray, 1999).

Turmeric extract is used as an antibacterial, anti-inflammatory natural dye for cosmetics like shampoos, lotions, and sprays (Belle, 1980). Balms containing turmeric powder or extract are effective in treating acne, rashes, eczema, pimples, psoriasis, and fungal infections. Turmeric is said to discourage the growth of facial hair. Women in some parts of India apply turmeric paste on the face and body to enhance fairness and complexion. Tests have shown that turmeric oil and extract possess strong antibacterial and antifungal activity (Giang, 2002; Singh et al., 2002b) against different strains of bacteria, parasites, and pathogenic fungi. In chicks, feed dosed with turmeric reduces intestinal parasites and enhances weight gain (Allen et al., 1998). Topically applied turmeric oil inhibited the dermatophytes and pathogenic fungi in animals (Apisariyakul et al., 1995).

Turmeric's protective effects on the cardiovascular system include lowering cholesterol and triglyceride levels, decreasing susceptibility of low-density lipoprotein to lipid peroxidation, and inhibiting platelet aggregation (Srivastava et al., 1986; Deshpande et al., 1997; Ramirez-Tortosa et al., 1999). These effects have been observed even with low doses of turmeric. Turmeric's effect on cholesterol levels may be due to decreased cholesterol uptake in the intestines and increased conversion of cholesterol to bile acids in the liver (Ramprasad and Sirsi, 1957). The inhibition of platelet aggregation by turmeric constituents could be through their potentiation of prostacyclin synthesis and inhibition of thromboxane synthesis (Srivastava et al., 1986).

Constituents of turmeric exert several protective effects on the gastrointestinal tract. Studies in rats have shown that turmeric inhibits ulcer formation caused by stress, alcohol, indomethacin, pyloric ligation, and reserpine. Turmeric extract significantly increased the gastric wall mucus in rats subjected to these gastrointestinal insults (Rafatulla et al., 1990).

Sodium curcuminate is the water-soluble salt of curcumin. This compound is reported to be better absorbed than the plain curcumin and retains the pharmacological properties of curcumin

(Ghatak and Basu, 1972; Mukhopadhyay et al., 1982; Rao et al., 1984). Sodium curcuminate is reported to exert choleretic effects by increasing biliary excretion of bile salts, cholesterol, and bilirubin, as well as increasing bile solubility, therefore possibly preventing and treating cholelithiasis (Ramprasad and Sirsi, 1957). It is found to inhibit intestinal spasm (Ammon and Wahl, 1991).

Recently THC has been developed from curcumin as an effective antioxidative substance (Pacchetti, 2001). THC is produced by hydrogenating curcumin in an organic solvent with a metallic catalyst (Mimura et al., 1993). The metallic catalyst is selected from the group consisting of nickel, manganese, copper, and zinc. The product has a subdued color compared to the original yellow color of curcumin. THC is of interest in the medicinal and cosmetic industry as an effective skin bioprotectant (Badmaev and Majeed, 2000; Pacchetti, 2001). Formulations containing THC have skin lightening properties and provide protection against UV rays (Majeed and Badmaev, 2003). THC in cosmetic compositions protects keratinous tissue against environmental aggressors such as smoke, smog, and UV radiation (Anderson et al., 2003).

Pharmacokinetic studies in animals demonstrate that 40 to 85% of an oral dose of curcumin passes through the gastrointestinal tract unchanged. The absorbed material is metabolized in the intestinal mucosa and liver (Wahlstrom and Blennow, 1978; Ravindranath and Chandrasekhara, 1980). The medicinal properties of turmeric or curcumin cannot be fully utilized because of its poor bioavailability due to rapid metabolism in the liver and in the intestinal wall. In a study, the effect of combining piperine, the active component in pepper and a known inhibitor of hepatic and intestinal glucuronidation, was evaluated on the bioavailability of curcumin (Shoba et al., 1998). The study showed that piperine enhances the serum concentration, extent of absorption, and bioavailability of curcumin in both rats and humans with no adverse effects. Turmeric is reported to possess anticoagulative properties (Kosuge et al., 1985). Turmeric, curcumin, and curcumin analogs find application in the treatment of patients with HIV infection (Wu, 2004; Sui et al., 1993; Mazumder, 1997).

8.19.5 OTHER USES

Turmeric produces a yellow dye for fabrics and is used for dyeing wool, silk, and cotton. Turmeric is sometimes used in combination with other natural dyes like indigo and safflower to impart different shades (CSIR, 1950). The optimum concentration of the dye, dyeing time, concentration of mordants, mode, and time of mordanting and combination of mordants have been worked out for the dyeing of wool and cotton with turmeric extract (Agrawal et al., 1992; Nishida and Kobayashi, 1992). Dyeing can also be done with chloro, chloronitro, chloroamino, and diazotized chloroamino derivatives and the acetate of curcumin (Faruq et al., 1990). Curcumin is also used for dyeing jute. Curcumin can serve as the starting material for synthetic dyes. In a study, curcumin was chlorinated followed by nitration and reduction of the nitro group to amino. The amino compound was diazotized and coupled with various couplers (including curcumin) to give azo dyes (Faruq et al., 1990). Hair dye compositions containing turmeric extract are reported to impart golden color to light chestnut hair (Manier and Lutz, 1985).

Investigations have been carried out to derive flavoring compounds from curcumin. Curcumin is subjected to hydrolysis under the action of heat and pressure in the presence of water to produce vanillin and other natural flavor products. The proportion of vanillin in the final product varies with the pH of the reaction mixture (Dolfini et al., 1990).

Curcumin can offer protection to photolabile drugs in soft gelatin capsules, thereby replacing synthetic dyes or pigments frequently used. Curcumin is incorporated into the capsule shells by making a curcumin–gelatin complex (Schranz, 1983). Results on accelerated photostability tests have shown that a curcumin content of 0.4% (w/w) in the capsule shell results in a threefold or more increase in the half life of the test compounds. A curcumin content of 0.02% resulted in 20% increase of half life (Tonnesen and Karlsen, 1987). Turmeric extract can also be used to color the outer coating of pharmaceutical tablets (Woznicki et al., 1984). Addition of curcumin can also improve the stability of photolabile drugs in serum samples (Tonnesen and Karlsen, 1988). Acrylic dental filling composition, colored with a low dose of turmeric extract, allows filling process to be followed easily; the filling is colorless or nearly colorless when the process is complete (Kemper, 1981).

Curcumin is employed for the purpose of detecting and warning the presence of cyanide in food, drug, or oral compositions. The curcumin is incorporated into the food packaging material. In the presence of cyanide, the curcumin undergoes a color change causing at least a portion of the packaging material to manifest a color change. The color change is sufficient to be detected by an observer, thereby providing a warning of cyanide adulteration of the product (Raymond et al., 1991).

As a chemical indicator, turmeric changes color in alkaline and acid substrates. Turmeric paper, prepared by soaking unglazed white paper in turmeric tincture and then dried, serves as a reagent for test for alkalinity. Turmeric paper is also used as a test for boric acid (Yoshida et al., 1989). Colorimetric method using curcumin is used in the determination of microquantities of boron in solutions and biological samples (Goltman and Gurevich, 1958; Wimmer and Goldbach, 1999). A diluted tincture of turmeric is suitable for use as a fluorescence indicator even in brown and yellow solutions (CSIR, 1950). An optical chemical sensor has been prepared for the selective detection of *o*-nitrophenol in water, based on the fluorescence quenching of curcumin in PVC (polyvinyl chloride) membrane. The sensor can be used for the detection of *o*-nitrophenol in water samples (Wang et al., 1997).

A fluorescent marking ink has been developed by reaction of a saturated filtered solution of turmeric in alcohol, an aqueous solution of a fluorescent substance and glycerol. The ink, when applied to hand, was invisible in ordinary light and luminisced when exposed to UV radiation (Suryanarayana et al., 1978).

Turmeric is an auspicious article in all religious observances in Hindu households and temples. It is intimately associated with many social and religious customs and beliefs. This aspect of turmeric has recently been summarized by Remadevi and Ravindran (2005).

REFERENCES

Aggarwal, B.B. (1999). Curcumin (diferuloylmethane) inhibition of NF.kappa.B activation. United States Patent 5,891,924, April 6, 1999.

Agrawal, A., Goel, A., and Gupta K.C. (1992). Optimization of dyeing process for wool with natural dye obtained from turmeric (*Curcuma longa*). *Text. Dyer Printer*, 25(22), 28-30.

Allen P.C., Danforth, H.D., and Augustine, P.C. (1998). Dietary modulation of avian coccidiosis. *Int. J. Parasitol.*, 28 (11), 31-1140.

Ammon, H. P. T., Safayhi, H., and Okpanyi, S.N. (1995). Use of preparations of curcuma plants. *United States Patent* 5,401,777, March 28, 1995.

Ammon, H.P.T., Wahl, M.A. (1991). Pharmacology of *Curcuma longa*. *Planta Med.*, 57 (1), 1-7.

Ammon, H.P.T., Safayhi, H., Mack, T., and Sabieraj, J. (1993). Mechanism of antiinflammatory actions of curcumin and boswellic acids. *J. Ethnopharmacol.*, 38(2-3), 113-119.

Anandaraj, M., Devasahayam, S., Zachariah, T.J., Eapen, S.J., Sasikumar, B., and Thankamani, C.K. (2001). Turmeric (Extension Pamphlet). Indian Institute of Spices Research, Calicut, Kerala, India.

Anderson, A.M., Mitchell, M.S., and Mohan, R.S. (2000). Isolation of curcumin from turmeric. *J. Chem. Educ.*, 77(3), 359-360.

AOAC (1998). Bactereological Analytical Manual, 8th Edition, Revision A. Food and Drug Administration. Published by AOAC International, USA.

AOAC (2000a). AOAC Official Method 970.52, Vol 1, 17th Edition, AOAC International, USA.

AOAC (2000b). AOAC Official Method 974.27. Vol I, 17th Edition. AOAC International, USA.

AOAC (2000c). AOAC Official Method 985.35. Vol II, 17th Edition. AOAC International, USA.

Apisariyakul, A., Vanittanakom, N., and Buddhasukh, D. (1995). Antifungal activity of turmeric oil extracted from *Curcuma longa* (Zingiberaceae). *J Ethnopharmacol.*, 49, 163-169.

Arbiser, J.L. (2004). Curcumin and curcuminoid inhibition of angiogenesis. *United States Patent* 6,673,843, January 6, 2004.

Arora, R.B., Basu, N., Kapoor, V. and Jain, A.P. (1971). Anti-inflammatory studies on *Curcuma longa* (Turmeric). *Indian J. Med. Res.*, 59 (8), 1289-1295.

Asakawa, N., Tsuno, M., Hattori, T., Ueyma, M., Shinoda, A., Miyake, Y., and Kagei, K. (1981). Determination of curcumin content of turmeric by high performance liquid chromatography. *Yakugaku Zasshi*, 101(4), 374-377 (CA Vol. 95/1981:12847f).

ASTA (1997a). Analysis of Afltaoxins B1, B2, G1 and G2 by HPLC. Method 24.2. Official Analytical Methods of the American Spice Trade Association, 4th Edition. pp. 149-152.

ASTA (1997b). Determination of Ethylene Oxide (EtO), Method 23.2, and Determination of Ethylene Chlorohydrin residues in spices, Method 23.3. Official Analytical Methods of the American Spice Trade Association, N.J., 4th Edition. pp 127-135.

ASTA (1999). ASTA Cleanliness specifications for spices, seeds and herbs. American Spice Trade Association, N.J. 07632, USA.

ASTA (2002). A concise guide to Spices, Herbs, Seeds, and Extractives. American Spice Trade Association. pp. 48-50.

ASTM (1998). Standard Guide for the Irradiation of Dried Spices, Herbs and Vegetable Seasonings to Control Pathogens and Other Microorganisms. F1885-98, The American Society of Testing Materials.

Azuine, M. and Bhide, S. (1992). Chemopreventive effect of turmeric against stomach and skin tumors induced by chemical carcinogens in Swiss mice. *Nutr. Cancer*, 17, 77-83.

Badmaev, V. and Majeed, M. (2000). Tetrahydrocurcuminoids (THC) as a skin bioprotectant. *Agro Food Industry Hi-Tech*, 11(1), 26-27.

Balakrishnan, K.V. (2005). Postharvest and industrial processing of Ginger. In Ravindran, P.N. and Nirmal Babu, K. (eds) *Ginger-the genus Zingiber'* CRC Press, Boca Raton, Florida. pp. 391-468.

Balakrishnan, K.V., Chandran, C.V., George, K.M., Narayana Pillai, O.G., Mathulla, T., and Verghese, J. (1983). Evaluation of curcumin. *Perfum Flavor*, 8(2), 46-49.

Balakrishnan, K.V., Francis, J.D., George, K.M., and Verghese, J. (1992). *Curcuma longa* L: Profile of dry and wet rhizomes and maturity levels for harvesting. *Indian Spices*, 29(4), 6-8.

Balasubramanyam, M., Koteswari, A.A, Kumar, R.S, Monickaraj S.F, Maheswari, J.U, and Mohan V. (2003). Curcumin-induced inhibition of cellular reactive oxygen species generation: Novel therapeutic implications. *J. Biosc.*, 28, 715-721.

Bambirra, M.L.A., Junqueira, R.G., and Glória, M.B.A. (2002). Influence of post harvest processing conditions on yield and quality of ground turmeric (*Curcuma longa* L.). *Braz. Arch. Biol. Technol.*, 45(4), 423-429.

Banerjee, A. and Nigam, S.S. (1978). Anti-microbial efficacy of the essential oil of *Curcuma longa*. *Indian J. Med. Res.* , 68(11), 864-866.

Banerjee, S., Narayanan, C.S., and Mathew, A.G. (1981). Chemical modification of turmeric oil to more value added products. *Indian Perfum.*, 25(3&4), 25-30.

Bansal, R. P., Bahl, J. R., Garg, S. N., Naqvi, A. A., and Kumar, S.(2002). Differential chemical compositions of the essential oils of the shoot organs, rhizomes and rhizoids in the turmeric *Curcuma longa* grown in Indo-Gangetic Plains. *Pharm. Biol.,* 40(5), 384-389.

Basu, A.P. (1971). Antibacterial activity of *Curcuma longa*. *Indian J. Pharm.*, 33(6), 131.

Baumann, W, Rodrigues, S.V., and Viana, L.M. (2000). Pigments and their solubility in and extractability by supercritical CO_2 – I. The case of curcumin. *Braz. J. Chem. Eng.*, 17(3) p.323-328.

Began, G., Goto, M., Kodama, A., and Hirose, T. (2000). Supercritical Carbon Dioxide Extraction of Turmeric (*Curcuma longa*), *J. Agric. Food Chem.*, 48(6), 2189-2192.

Behura, A., Sahoo, S. and Srivastava, V.K. (2002). Major constituents in leaf essential oils of *Curcuma longa* L. and *Curcuma aromatica* Salisb. *Curr. Sci.*, 83(11), 1312-1313.

Behura, S., Srivastava, V.K. (2004). Essential oils of leaves of *Curcuma* species. *J.Essential Oil Res.*, 16(2), 109-110.

Behura, C., Ray, P., Rath, C.C., Mishra,R.K., Ramachandraiah, O.S., and Charyulu, J.K. (2000). Antifungal activity of essential oils of *Curcuma longa* against five rich pathogens *in vitro*. *J. Essent. Oil-Bearing Plants*, 3(2), 79-84.

Belle, R. (1980). Natural dye for capillary use and cosmetic preparations containing it. *Eur. Pat.* Appl. 20,274 (Cl.A61K7/13), 10 Dec. 1980. Fr. Appl. 79/13,970, 31 May 1979. (CA 94/1981:90041g).

BIS (2002). Spices and Condiments – Turmeric, Whole and Ground – Specification. Indian Standard IS3576:1994. Bureau of Indian Standards, New Delhi.

Boegl, W. and Heide, L. (1985). Chemiluminescence measurements as an identification method for gamma-irradiated foodstuffs. *Radiat. Phys. Chem.*, 25(1-3), 173-85.

Boone C.W., Steele V.E., and Kelloff G.J. (1992). Screening of chemopreventive (anticarcinogenic) compounds in rodents. *Mut. Res.,* 267, 251-255.

Braga, M.E., Leal, P.F., Carvalho, J.E., and Meireles, M.A. (2003). Comparison of yield, composition, and antioxidant activity of turmeric (*Curcuma longa* L.) extracts obtained using various techniques. *J Agric. Food Chem.,* 51(22), 6604-6611.

Brown, D. J. (1970). Determination of ethylene oxide and ethylene chlorohydrin in plastic and rubber surgical equipment sterilized with ethylene oxide. *J. Assoc. Offic. Anal. Chem.,* 53, 263-267.

Buescher, R., and Yang, L. (2000). Turmeric. In *Natural Food Colorants, Science and Technology'*, Marcel Dekker, New York. pp. 205-226.

Burger, A. (1958). Curcuma root and its essential oil. *Perfum.Ess. Oil Rec.,* 49(12), 801-802.

Cavanaugh, P.F., Jr. (1998). Dentifrice compositions containing a curcuminoid. Brit. UK Patent Appl. GB 2,317,339, 25 Mar 1998. (CA 129/1998: 280784r).

Carter, F.D., Copp, F.C., Rao, B.S., Simonsen, J.L., and Subramaniam, K.S. (1939). The constituents of some Indian essential oils. Part XXVI. The structures of l-a- and b-Curcumenes. *J. Chem. Soc. (London),* p. 1504-1509 (1939).

CFR (1995). *Code of Federal Regulations. 21. Food and Drugs.* Part 173C. Office of the Federal Register, National Archives and Records Administration, USA. pp. 117-118.

CFR (2005). Title 40. Part 180. Tolerances and exemptions from tolerances for pesticide chemicals in food. Electronic Code of Federal Regulations (e-CFR). US EPA.

Chandra, D., and Gupta S.S. (1972). Antiinflammatory and antiarthritic activity of volatile oil of *Curcuma longa. Indian J. Med. Res.,* 60(1), 138-42.

Chane-Ming, J., Vera, R., Chalchat, J. C., and Cabassu, P. (2002). Chemical composition of essential oils from rhizomes, leaves and flowers of *Curcuma longa* L. from Reunion Island. *J. Essential Oil Res.,* 14 (4), 249-251.|

Chassagnez, A.L.M., Correa, N.C.F., and Meireles, M.A.A. (1997). Turmeric (*Curcuma longa*) oleoresin extraction with supercritical CO2. *Cienc. Tecnol.Aliment.,* 17(4), 399-404. (CA 129/1998: 215888p).

Charley, N.G. (1938). Preparation of turmeric for the market. *Agric. and Livestock in India,* 8(6), 695-697.

Chatterjee, S., Padwal-Desai, S.R., and Thomas, P. (1999). Effect of gamma-irradiation on the antioxidant activity of turmeric (*Curcuma longa* L.) extracts. *Food Res. Int.,* 32, 487-490.

Chatterjee, S., Variyar, P.S., Gholap, A.S., Padwal-Desai, S.R., and Bongiwar, D.R. (2000). Effect of *gamma*-irradiation on the volatile oil constituents of turmeric (*Curcuma longa*). *Food Res. Int.,* 33, 103-106.

Chauhan, S.K; Singh, B.P.; Agrawala, S. (1999). Estimation of curcuminoids in *Curcuma longa* by HPLC and spectrophotometric methods. *Indian J. Pharm. Sci.,* 61(1), 58-60.

Chauhan, U.K., Soni, P., Shrivastava, R., Mathur, K.C., and Khadikar, P.V. (2003). Antimicrobial activities of the rhizome of *Curcuma longa* Linn. *Oxid. Commun.,* 26(2), 266-270.

Chen, J., Chen, Y., and Yu, J. (1983). Studies on Chinese *Curcuma* plants. IV. Assay of curcuminoids in the root and tuber of *Curcuma* spp. *Zhongcaoyao,* 14(2), 59-63. (CA. 98/1983, 204500w).

Chosdu, R., Hilmy, N., Bagiawati, S. (1985). The effect of gamma radiation on medicinal plants and spices.III. *Curcuma xanthorriza, Curcuma aeruginosa, Curcuma domestica* and *Kaemferia galanga. Maj. BATAN,* 18(2), 37-53. (CA.105/1986, 38271p).

Ciamician, G. and Silber, P. (1897). Zur Kenntniss des Curcumins. *Ber. Dtsch. Chem. Ges.,* (30), 192-195.

Commission of the European Communities (2002). Commission Regulation (EC) No.472/2002. *Official Journal of the European Communities.,* pp. L75/20.

Cooray, N.F., Jansz, E.R., Ranatunga, J., and Wimalasena, S. (1998). Effect of maturity on some chemical constituents of turmeric (*Curcuma longa* L.). *J. Natl. Sci. Counc. Sri Lanka,* 16(1), 39-51.

CSIR (1950). Curcuma - in Wealth of India, Raw Materials, Vol II, Publications and Information Directorate, CSIR, New Delhi, pp. 401-406.

Dahal, K.R. and Idris, S. (1999). *Curcuma longa* L. In *Plant Resources of South-East Asia,* No 13. *Spices.* Backhuys Publishers, Leiden, The Netherlands. pp. 111-116.

Das, S.K. and Cohly; H.P. (1995). Use of *turmeric* in wound healing. *United States Patent* 5,401,504, March 28, 1995.

Daube, F.W. (1870). Ueber Curcumin, den Farbstoff der Curcumawurzel. *Ber. Dtsch. Chem. Ges. Berlin,* 3, 609-613.

Deodhar, S.D., Sethi, R., and Srimal, R.C. (1980). Preliminary study on antirheumatic activity of Curcumin (Diferuloyl methane). *Indian J. Med. Res.,* 71,632-643.

Deshpande, U.R., Joseph, L.J., Manjure, S.S., Samuel, A.M., Pillai, D., and Bhide, S.V. (1997). Effects of turmeric extract on lipid profile in human subjects. *Med. Sci. Res.*, 25(10), 695-698.

Deshpande, U.R, Gadre, S.G., Raste, A.S. (1998). Protective effect of turmeric (*Curcuma longa* L.) extract on carbon tetrachloride-induced liver damage in rats. *Indian J. Exp. Biol.*, 36, 573-577.

Dieterle, H., and Ph.Kaiser. (1932). Constituents of the rhizomes of *Curcuma domestica* (Temoe Lawak). *Arch.Pharm.*, 270, 413-418.

Directorate of Marketing Inspection (1964). Turmeric Grading and Marking Rules, 1964. Ministry of Agriculture, Government of India.

Dolfini, J.E., Glinka, J., and Bosch, A.C. (1990). Hydrolysis of curcumin. *United States Patent* 4,927,805, May 22, 1990.

Donatus, I.A. and Sardjoko, V.N.P. (1990). Cytotoxic and cytoprotective activities of curcumin. Effects on paracetamol-induced cytotoxicity, lipid peroxidation and glutathione depletion in rat hepatocytes. *Biochem. Pharmacol.*, 39, 1869-1875.

Dorai, T., Cao, Y.C., and Dorai B (2001). Therapeutic potential of curcumin in human prostate cancer. III. Curcumin inhibits proliferation, induces apoptosis, and inhibits angiogenesis of LNCaP prostate cancer cells *in vivo. Prostate*, 47, 293-303.

Dung, N.X., Tuyet, N.T.B., and Leclercq, P. A. (1995). Constituents of the leaf oil of *Curcuma domestica* L. from Vietnam. *J.Essent. Oil Res.*, 7(6), 701-703.

EC (2002). SCF/CS/ADD/EMU/186 Final. Opinion of the Scientific Committee on Food on impurities of ethylene oxide in food additives. Health & Consumer Protection Directorate General, European Commission, Brussels, Belgium.

Elizabeth, K., and Rao, M.N.A. (1990). Oxygen radical scavenging activity of curcumin. *Int. J. Pharm.*, 58, 237-240.

EOA (1958). EOA Book of Standards and Specifications. EOA No.271. Oleoresin Turmeric. Essential Oil Association of USA, Inc., N.Y.

FAO (1972). Pesticide Residues in Food. Report of the 1971 Joint Meeting of the FAO Working Party of Experts on Pesticide Residues and the WHO Expert Committee on Pesticide Residues, WHO Techn. Rep. Ser., No. 502, FAO Agricultural Studies, No. 88.

FCC (2004). Monographs/Spice Oleoresins. Oleoresin Turmeric. pp. 446-448 and Appendix VIII/General Tests and Assays. Curcumin Content. P.944. Food Chemicals Codex, Fifth Edition, The National Academies Press, Washington, DC.

Faruq, M. O., Haque, M. Z., Sayeed, M. A., and Sarder, M. A.I. (1990). Chemical investigations on Bangladeshi turmeric. Part II. Substantivity of curcumin and its derivative dyes on jute and wool fibers. *Bangladesh J. Sci. Ind. Res.*, 25(1-4), 142-152.

FDA (1995). Food Defect Action Levels. Department of Health and Human Services, Public Health Service, Food and Drug Administration, Washington, DC 20204.

FDA (2003). Irradiation in the production, processing and handling of food. US Food and Drug Administration (FDA). Federal Register, 21 CFR 179. 26.

Fenaroli, G. (1975) Turmeric. In *Fenaroli's Handbook of Flavor Ingredients.* Second Edition. Vol.1, CRC Press Inc, Ohio, USA. pp. 481-482.

Ferreira, L.A.F., Henriques, O.B., Andreoni, A.A.S., Vital, G.R.F., Campos, M. M.C., Habermehl, G.G., De Moraes, V.L.G. (1992). Antivenom and biological effects of *ar*-turmerone isolated from *Curcuma longa. Toxicon* 30(10), 1211–1218.

Freund, P.R. (1985). Natural colors in cereal-based products. *Cereal Foods World*, 1985, 30(4), 271-273.

Fukazawa, T. (1970). Food dye from turmeric. *Japanese Patent.* Jpn 70 24,672, 17 Aug 1970. (CA. 74/1971:10540t).

Gaikar, V.G., and Dandekar, D.V.(2001). Process for extraction of curcuminoids from *Curcuma* species. *US Patent* 6,224,877. May 1, 2001.

Garg, S.C., and Jain, R.K. (2003). Antimicrobial activity of the essential oil of *Curcuma longa* L. *Indian Perfum.*, 47(2), 199-202.

Garg, S.N., Mengi, N., Patra, N.K., Charles, R., and Kumar, S. (2002). Chemical examination of the leaf essential oil of *Curcuma longa* L. from the North Indian plains. *Flavour Fragrance J.*, 17(2), 103-104.

Ge, F., Shi, Q., Tan, X., Li, Z., and Jing, X. (1997). Technological study on the supercritical-CO2 fluid extraction of *Curcuma longa* oils. *Zhong Yao Cai.* 20(7), 345-350. (Abstract - Entrez PubMed PMID: 12572431).

Ghatak, N. and Basu, N. (1972). Sodium curcuminate as an effective anti-inflammatory agent. *Ind. J. Exp. Biol.*, 10, 235.

Ghosh, P.C. (1919). Curcumin. *J. Chem. Soc.*, 115, 292-299.

Giang, T.S. (2002). Study on chemical components and separation of curcumin from rhizome *Curcuma longa*. *Tap Chi Duoc Hoc.*, 1, 15-17 (CA Vol. 137/2002:182302)

Golding, B.T., Pombo, E., and Samuel, C.J. (1982). Turmerones: isolation from turmeric and their structure determination. *J.Chem. Soc., Chem. Commun.*, 6, 363-364.

Goldscher, K.J. (1979). Turmeric coloring process and composition for foods and beverages. *United States Patent* 4,163, 803, 07 Aug 1979.

Goltman, A.D. (1957). Separation of Curcumin from *Curcuma* Roots. *Ukr. Khim. Zhur.*, 23, 659–661 (CA 52/1958: 8466c).

Goltman, A.D. and Gurevich, V.G. (1958). Colorimetric method of determining microquantities of boron with turmeric. *Ukr. Khim. Zhur.*, 24, 244-250. (CA 52/1958: 15335b).

Gopalam, A. and Ratnambal, M.J. (1987). Gas chromatographic evaluation of turmeric essential oils *Indian Perfum.*, 31(3), 245-248.

Gopalan, B., Goto, M.., Kodama, A., and Hirose, T. (2000). Supercritical carbon dioxide extraction of turmeric (*Curcuma longa*). *J. Agric. Food Chem.*, 48(6), 2189-2192.

Govindarajan, V.S. (1980). Turmeric - Chemistry, technology, and quality. *Crit. Rev. Food Sci. Nutr.*, 12, 199-301.

Graus, I.M.F. and Smit, H.F. (2002). Composition for the treatment of osteoarthritis. *United States Patent* 6,492,429, December 10, 2002.

Guenther, E. (1972). *The Essential Oils*. Robert K. Krieger Publishing Company, N.Y. Vol I, pp. 18, 88-104.

Gupta, A.P., Gupta, M.M., and Kumar, S.(1999). Simultaneous determination of curcuminoids in *Curcuma* samples using high performance thin layer chromatography. *J. Liq. Chromatogr. Relat. Technol.*, 22(10), 1561-1569.

Hamada, K., and Minamino, H. (1997). Hair tonics. *Japanese Patent* Jpn. Kokai Tokkyo Koho JP 09 268,115. (97,268,115), 14 Oct 1997 (CA127/1997: 322650k).

Hanif, R., Qiao, L., Shift, S. J., Rigas, B. (1997). Curcumin, a natural plant phenolic food additive, inhibits cell proliferation and induces cell cycle changes in colon adenocarcinoma cell lines by a prostaglandin-independent pathway. *J. Lab. Clin. Med.*, 130, 576-584.

Hanssen, M. and Marsden, J. (1988). E for Additives. Thorsons, Harper Collins Publishers, London. pp. 63-64.

He, X., Lin, L., Lian, Li. and Lindenmaier, M. (1998). Liquid chromatography-electrospray mass spectrometric analysis of curcuminoids in turmeric (*Curcuma longa*). *J.Chromatogr., A*, 818(1), 127-132.

Heide, L., and Boegl, W. (1985). Chemiluminescence measurements of 20 spices. A method of detecting treatment with ionizing radiation. *Z.Lebensm.-Unters. Forsch*, 181(4), 283-288.

Heide, L., and Boegl, W. (1987). Identification of irridated spices with thermoluminescence and chemoluminescence measurements. *Int. J. Food Sci. Technol.*, 22(2), 93-104.

Heuser, S. G. and Scudamore, K. A. (1967). Determination of ethylene chlorohydrin, ethylene dibromide and other volatile fumigant residues in flour and whole wheat. *Chem.Ind.*, 37, 1557-1560.

Heuser, S. G. and Scudamore, K. A. (1968). Fumigant residues in wheat and flour : Solvent extraction and gas-chromatographic determination of free methyl bromide and ethylene oxide. *Analyst*, 93, 252-258.

Heuser, S. G. and Scudamore, K. A. (1969). Determination of fumigant residues in cereals and other foodstuffs: a multi-detection scheme for gas-chromatography of solvent extracts. *J. Sci. Food Agr.*, 20, 565-57.

Hiserodt, R., Hartman, T. G., Ho, C. T., Rosen, R. T. (1996). Characterization of powdered turmeric by liquid chromatography-mass spectrometry and gas chromatography-mass spectrometry. *J.Chromatogr. A*, 740 (1), 51-63.

Houser, T.J., Biftu, E., Hsieh, P.F. (1975). Extraction rate equations for paprika and turmeric with certain organic solvents. *J.Agric. Food Chem.*, 23(2), 353-355.

Hu,Y., Du, Q., and Tang, Q. (1997). Study of chemical constituents of volatile oil from *Curcuma longa* by GC-MS. *Zhongguo Yaoxue Zazhi (Beijing)*, 32(Suppl.), 35-36. (CA 130/1999: 100339p).

Hu, Y., Du, Q., and Tan, Q. (1998). Determination of chemical constituents of the volatile oil from *Curcuma longa* by gas chromatography-mass spectrometry. *Sepu*, 1998, 16(6), 528-9. (CA 130/1999: 136608q).

Ibrahim, J., Ahmad, A. S., Ali, N. A. M., Ahmad, R. A., and Ibrahim, H. (1999). Chemical composition of the rhizome oils of four *Curcuma* species from Malaysia. *J. Essential Oil Res.*, 11(6),719-723.

Imai, S., Morikiyo, M., Furihata, K., Hayakawa, Y., Seto, H. (1990). Turmeronol A and Turmeronol B, new inhibitors of soyabean lipoxygenase. *Agric. Biol. Chem.*, 54(9), 2367-71.

Inafuku, M. (1996). Ukon (rhizome of *Curcuma longa*) for preparation of food. *Japanese Patent* Jpn. Kokai Tokkyo Koho JP 08,214,825 (96,214,825), 27 Aug 1996. (CA. Vol 125/1996 , 274339a).

Iqbal, M., Sharma, S.D., Okazaki, Y., Fujisawa, M., and Okada, S. (2003). Dietary supplementation of curcumin enhances antioxidant and phase II metabolizing enzymes in ddY male mice: Possible role in protection against chemical carcinogenesis and toxicity. *Pharmacology & Toxicology*, 92(1), 33-38.

ISO (1983). ISO 5562-1983. Turmeric, whole or ground (powdered) – Specification.

Iyengar, M.A., Rama Rao, P.M., Bairy, I., and Kamat, M.S. (1995). Antimicrobial activity of the essential oil of *Curcuma longa* leaves. *Indian Drugs*, 32(6), 249-250.

Iwanof-Gajewsky, Y. (1870). *Ber. Dtsch. Chem. Ges.*, 3, 624-626.

Jackson, C.L., and Clarke, L. (1911). Curcumin. *Am. Chem. J.*, 45, 48-58.

Jasim, F., and Ali, F. (1992). A novel and rapid method for the spectrofluorometric determination of curcumin in curcumin species and flavors. *Microchem. J.*, 46(2), 209-14.

Jayaprakasha, G.K., Negi, P.S., Anandharamakrishnan, C., Sakariah, K.K.(2001). Chemical composition of turmeric oil - a byproduct from turmeric oleoresin industry and its inhibitory activity against different fungi. *J. Biosciences*, 56(1/2), 40-44.

Jayaprakasha, L.G.K., Mohan Rao, L.J. and Sakariah, K.K. (2002). Improved HPLC method for the determination of curcumin, demethoxycurcumin, and bisdemethoxycurcumin. *J. Agric. Food Chem.*, 50(13), 3668-3672.

Kapur, O.P., Srinivasan, M. and Subrahmanyan, V. (1960). Colouring of Vanaspati with Curcumin from turmeric. *Curr. Sci.*, 29(9), 350-351.

Kapoor, L.D. (1990). *Handbook of Ayurvedic and Medicinal Plants*. CRC Press. Boca Raton, Florida. 149-150.

Katyuzhanskaya, A.N., Usalka, L.G., Gondel, V.P., Skylar, V.E., Tereshina T.P., Pekhov A.V., Dyubankova, N.F., Kolobutina V.A, and Danilova, A.P. (1985). Toothpaste. *U.S.S.R Patent* SU 1,132,945, 07 Jan 1985. (CA.102/1985:137609z).

Kawaguchi, T., and Kida, N. (1987). Natural bactericides for the control of infections in animals. *Japanese Patent* Jpn. Kokai Tokkyo Koho JP 62 59,214, 14 Mar 1987. (CA.107/1987, 52004p).

Kawamori, T., Lubet, R., and Steele, V.E. (1999). Chemopreventative effect of curcumin, a naturally occurring anti-inflammatory agent, during the promotion/progression stages of colon cancer. *Cancer Res.*, 59, 597-601.

Keil, F. and Dobke, W. (1940). Extraction of Curcumene. *German. Patent* 700,765, Nov.28, 1940. (CA 35/1941: 7656).

Kelkar, N.C. and Rao, B. S. (1933). Indian Essential Oils. V. Essential oil from the rhizomes of *Curcuma longa* L. (turmeric). *J.Indian Inst. Sci.*, 17A, 7-24.

Kemper, R.N. (1981). Dental filling compositions. *German Patent* 3,038,383, 23 Apr 1981. US Appl. 83,896, 11 Oct.,1979, 12 pp. (CA Vol 95/1981, 68069w).

Khalique, A. and Amin, M.N. (1967). *Curcuma longa*. I. Constituents of the rhizome. *Sci.Res.* 4(4), 193-197.

Khalique, A., and Das, N.R. (1968). Examination of *Curcuma longa* Linn. — Part II. Constituents of the essential oil. *Sci. Res. (Dacca)*, 5(1), 44-49.

Khanapure, V.M., Sankaranarayanan, J., and Jolly, I. (1993). HPTLC fingerprint analysis of *Momordica charantia, Curcuma longa, Emblica officinalis*, and their mixtures. *J. Planar Chromatogr.*, 6(2), 157-160.

Khanna, N.M. (1999). Turmeric-nature's precious gift. *Current Sci.*, 76(10), 1351-1356.

Khurana, A. and Ho, C.T. (1988). High performance liquid chromatographic analysis of curcuminoids and their photo-oxidative decomposition compounds in *Curcuma longa* L. *J.Liq.Chromatogr.*, 11(11), 2295-304.

Kirtikar, K.R. (1912). The use of saffron and turmeric in Hindu marriage ceremonies. *J. Anthropol. Soc. Bombay*, 9(7), 439-454.

Kiso, Y., Suzuki, Y., Oshima, Y., and Hikino, H. (1983a). Stereostructure of Curlone, a sesquiterpenoid of *Curcuma longa* rhizomes. *Phytochemistry*, 22(2), 596-597.

Kiso, Y., Suzuki, Y., and Watanabe, N. (1983b). Antihepatotoxic principles of *Curcuma longa* rhizomes. *Planta Medica*, 49, 185-187.

Kohli, K., Ali, J., Ansari, M.J., and Raheman, Z. (2005). Curcumin: A natural antiinflammatory agent. *Indian J. Pharmacol.*, 37(3), 141-147.

Kojima, H., Yanai, T., Toyota, A., Hanani, E., Saiki,Y. (1998). Essential oils constituents from *Curcuma aromatica*, *C.longa* and *C.xanthorrhiza* rhizomes. In. *International Congress Series: Towards Natural Medicine Research in the 21st Century*, 1157, 531-539. International Symposium on Natural Medicines, Kyoto, Japan, October 28 - 30, 1997. (CA130/1999: 220403b).

Kosuge, T., Ishida, H., Yamazaki, H. (1985). Studies on active substances in the herbs used for Oketsu ("stagnent blood") in Chinese medicine.III. On the anticoagulative principles in curcumae rhizoma. *Chem. Pharm. Bull.*, 33(4), 1499-502.

Krishnamurthy, M.N., Padma Bai, R., Natarajan C.P., and Kuppuswamy S. (1975) Colour content of turmeric varieties and studies of its processing . *J. Food Sci. Technol., (India)*, 12, 12-14.

Krishnamurthy, N., Mathew, A.G., Nambudiri, E.S., Shivasahnkar, S., Lewis, Y.S., and Natarajan C.P. (1976). Oil and Oleoresin of turmeric, *Trop. Sci.*, 18, 37-45.

Kuttan, R., Sudheeran, P.C, and Joseph, C.D. (1987). Turmeric and curcumin as topical agents in cancer therapy. *Tumori.*, 73, 29-31.

Lacroix, M., Marcotte, M., and Ramaswamy, H. (2003). Irradiation of fruits, vegetables, nuts and spices. In: *Handbook of Postharvest Technology. Cereals, Fruits, Vegetables, Tea, and Spices*. Marcel Dekker, New York. pp. 623-652.

Lal, B., Kapoor, A.K., Asthana, O.P., Agrawal, P.K., Prasad, R., and Kumar. P. (1999). Efficacy of curcumin in the management of chronic anterior uveitis. *Phytother. Res.*, 13, 318-322.

Lal, B., Kapoor, A.K., Agrawal, P.K., Asthana, O.P., and Srimal, R.C. (2000). Role of curcumin in idiopathic inflammatory orbital pseudotumours. *Phytother. Res.*, 14, 443-447.

Lampe, V. (1918). Synthese von Curcumin. *Ber. Dtsch. Chem. Ges. Berlin*, 51, 1347-55. (CA Vol. 13/1919: 715).

Lampe, V. and Milobedzka, J. (1913). Studien uber Curcumin. *Ber. Dtsch. Chem. Ges. Berlin*, 46, 2235-40. (CA Vol. 7/1913: 3333).

Lavoine, S., Arnaudo, J.F., Coutiere, D. (1998). Application of high-performance thin layer chromatography (HPTLC). *Rev. Ital. EPPOS*, 580-590. (CA 128/1998: 32079h).

Lechtenberg, M., Quandt, B.., and Nahrstedt, A. (2004). Quantitative determination of curcuminoids in Curcuma rhizomes and rapid differentiation of *Curcuma domestica* Val. and *Curcuma xanthorrhiza* Roxb. by capillary electrophoresis. *Phytochem. Anal.* 15(3), 152-158.

Lee, H.S., Shin W.K., Song, C., Cho, K.Y., and Ahn, Y.J. (2001). Insecticidal activities of ar-turmerone identified in *Curcuma longa* rhizome against *Nilaparvata lugens* (Homoptera: Delphacidae) and *Plutella xylostella* (Lepidoptera: Yponomeutidae) *J. Asia Pacific Entomology*, 4(2), 181-185.

Lee, K. H., Ishida, J., Ohtsu, H., Wang, H.K., Itokawa, H., Chang, C., Shih, C.C.Y. (2004). Curcumin analogues and uses thereof. *United States Patent* 6,790,979 September 14, 2004.

Leela, N.K., Tava, A., Shafi, P.M., John, S. P., Chempakam, B. (2002). Chemical composition of essential oils of turmeric (*Curcuma longa* L.). *Acta Pharm.*, 52, 137-141.

Leshik, R.R. (1981). Stabilized curcumin colorant. *United States Patent* 4,307,117, December 22, 1981.

Leung, A. (1980). *Encyclopedia of Common Natural Ingredients Used in Food, Drugs, and Cosmetics*. John Wiley, New York, NY, pp. 313-314.

Leung, A.Y. and Foster, S. (1996). *Encyclopedia of Common Natural Ingredients Used in Food, Drugs and Cosmetics*, 2nd ed. John Wiley & Sons, Inc. New York. 499–501.

Li, C., Li, L., Luo J., and Huang, N. (1998). Study on the effects of turmeric volatile oil on respiratory tract. *Zhongguo Zhongyao Zazhi*, 23(10), 624-625. (Abstract - Entrez PubMed PMID: 11599365).

Limtrakul, P., Lipigorngoson, S., and Namwong, O. (1997). Inhibitory effect of dietary curcumin on skin carcinogenesis in mice. *Cancer Lett.*,116, 197-203.

Lubis, I. (1968). Phenolic compounds of turmeric. *Ann. Bogor.*, 4(Pt.4), 219-25. (CA 71/1969: 109760q).

Madsen B., Hidalgo, G.V. and Hernandez, V.L. (2003). Purification process for improving total yield of curcuminoid colouring agent. *United States Patent* 6,576,273, June 10, 2003.

Maing, I.Y., and Miller, I. (1981). United States Patent 4,263,333, April 21, 1981.

Majeed, M. (Ed). (1995). *Curcuminoids; antioxidant phytonutrients*. NutriScience Publishers, Inc., N.J.

Majeed, M., and Badmaev, V. (2003). Cross-regulin composition of tumeric-derived tetrahydrocurcuminoids for skin lightening and protection against UVB rays. United States Patent 6,653,327, November 25, 2003.

Majeed, M., Badmaev, V., and Rajendran, R. (1999). Bioprotectant composition, method of use and extraction process of curciminoids. United States Patent 5,861,415, January 19, 1999.

Manier, F., and Lutz, D. (1985). Hair dye compositions containing plant extracts. *Belg.* BE 900,219, 25 Jan, 1985. Fr.Appl. 83/12,458, 28 Jul 1983. (CA Vol. Vol.102/1985, 137585p).

Marongiu, B., Porcedda, S., Caredda, A., De Gioannis, B., and Piras, A. (2002). Supercritical CO_2 extraction of curcumin and essential oil from the rhizomes of turmeric (*Curcuma longa* L.). *J. Essent. Oil-Bear. Plants* , 5(3), 144-153.

Masuda, T., Jitoe, A., Isobe, J., Nakatani, N., Yonemori, S. (1993). Antioxidative and anti-inflammatory curcumin-related phenolics from rhizomes of *Curcuma domestica*. *Phytochemistry*, 32(6), 1557-60.

Masuda, T., Maekawa, T., Hidaka, K., Bando, H., Takeda, Y., and Yamaguchi, H. (2001). Chemical studies on antioxidant mechanism of curcumin: Analysis of oxidative coupling products from curcumin and linoleate. *J. Agric. Food Chem.*, 49(5), 2539-2547.

Maul, A.A., Freitas, B.R., Ribeiro, P.F.S., Polakiewicz, B., and Magalhães, J.F. (2002). Supercritical Fluid Extraction of Turmeric Rhizomes (*Curcuma domestica* Valeton, Zingiberaceae). Poster Presentation 7-16. 24th Symposium on Biotechnology For Fuels and Chemicals, April 28 - May 1, Gatlinburg, TN, USA.

Mazumder, A., Neamati, N., Sunder, S., Schulz, J., Pertz, H., Eich, E., and Pommier, Y. (1997). Curcumin analogs with altered potencies against HIV-1 integrase as probes for biochemical mechanisms of drug action. *J. Med. Chem.*, 40(19), 3057-3063.

McCarron, M., Mills, A.J., Whittaker, D, Sunny, T.P., and Verghese, J. (1995). Comparison of the mono-terpenes derived from green leaves and fresh rhizomes of *Curcuma longa* L. from India. *Flav. Frag. J.*, 10(6), 355-7.

Mehta, R.G., and Moon, R.C. (1991). Characterization of effective chemopreventive agents in mammary gland in vitro using and initiation-promotion protocol. *Anticancer Res.,* 11, 593-596.

Milobedzka, J., Kostanecki, S.V. and Lampe, V. (1910). Zur Kenntniss des Curcumins. *Ber. Dtsch. Chem. Ges. Berlin,* 43, 2163-2170. (CA Vol.4/1910, 2831).

Mimura, A., Takahara, Y., and Osawa, T. (1993). Method for making tetrahydrocurcumin and a substance containing the antioxidative substance tetrahydrocurcumin. *United States Patent* 5,266,344, November 30, 1993.

Moon, C..K, Park, N.S,and Koh, S.K (1976). Studies on the lipid components of *Curcuma longa*. (I). The composition of fatty acids and sterols. *Soul Taehakkyo Yakhak Nonmunjip.,* 1, 105-111. (CA 87/1977: 114582t).

Mukhopadhyay, A., Basu, and N., Ghatak, N. (1982). Anti-inflammatory and irritant activities of curcumin analogues in rats. *Agents Actions,* 12, 508-515.

Mukhopadhy, A., Basu, N., Ghatak, N., and Gujral, P.K. (1977). Local anti-inflammatory effect of curcumin, sodium curcuminate and phenylbutazone. *Indian J. Pharmacol.*, 9(1), 30-31.

Muthu, R.J.R, Jayaram. M., and Majumdar, S.K. (1984). Ethyl formate as a safe general fumigant. *Dev. Agric. Eng.,* 5, 369-93.

Nakayama, R., Tamura, Y., Yamanaka, H., Kikuzaki, H., and Nakatani, N. (1993). Two curcuminoid pigments from *Curcuma domestica*. *Phytochemistry*, 33(2), 501-502.

Natarajan, C.P., and Lewis, Y.S. (1980). Technology of Ginger and Turmeric - In Status Papers and Abstracts, National Seminar on Ginger and Turmeric, Central Plantation Crops Research Institute, Calicut, India, 8-9 April, 1980, pp. 83-89.

Negi, P. S., Jayaprakasha, G. K., Rao, L. J. M., and Sakariah, K. K. (1999). Antibacterial activity of turmeric oil: a byproduct from curcumin manufacture. *J. Agric. Food Chem.*, 47(10), 4297-4300.

Newmark, T. and Schulick, P. (2002). Anti-Inflammatory herbal composition and method of use. *United States Patent* 6,387,416, May 14, 2002.

Nigam, M.C., and Ahmed, A. (1991). *Curcuma longa*: terpenoid composition of its essential oil *Indian Perfum.*, 35(4), 255-257.

Nishida, K, and Kobayashi, K. (1992). Dyeing properties of natural dyes under after-treatment using metallic mordants. *Am. Dyest. Rep.*, 81(5), 61-63.

Oei, B.L. (1992). Combinations of compounds isolated from *Curcuma* spp as anti-inflammatory agents. United States Patent 5,120,538 June 9, 1992.

Oguntimein, B., Weyerstahl, O.P., and Weyerstahl, M.H. (1990). Essential oil of *Curcuma longa* L. leaves. *Flav. Frag. J.*, 5(2), 89-90.

Oshiro, M., Kuroyanagi, M., and Ueno, A. (1990). Structure of sesquiterpenes from *Curcuma longa*. *Phytochemistry*, 29(7), 2201-2205.

Pacchetti, B. (2001). Tetrahydrocurcuminoids. Antioxidants from the roots of *Curcuma longa Cosmetic Technol.*, 4(1), 35-39.

Panduwawala, J.P., Illeperuma, C.D.K., Samarajeewa U. (1998). Iron contamination during commercial grinding of spices. *J. Natl. Sci. Counc. Sri Lanka,* 16(1), 105-114.

Park, E.J., Jeon, C.H., and Ko, G. (2000). Protective effect of curcumin in rat liver injury induced by carbon tetrachloride. *J Pharm Pharmacol.*, 52, 437-440.

Patel, K., Krishna, G., Sokoloski, E., and Ito, Y (2000). Preparative separation of curcuminoids from crude curcumin and turmeric powder by pH-zone-refining countercurrent chromatography. *J. Liq. Chromatogr. & Rel. Technol.*, 23(14), 2209-2218.

Pavolini, T.P., Gambarin, F. and Grinzato, A.M. (1950). Curcumin and Curcuminoids. *Ann.chim. (Rome),* 40, 280-291. (CA Vol.45/1951, 9004i).

Perotti, A. G. (1975). Curcumin--a little known but useful vegetable colour. *Ind. Aliment. Prod. Veg.,* 14, 66.

Perry, R.H and Chilton, C.H (1973) Leaching. In *Chemical Engineer's Handbook.* Fifth Edition. McGraw Hill Inc., N.Y. pp.17.3-17.8.

PFA (2005). The Prevention of Food Adulteration Act, 1954, India. Commercial Law Publishers (India) Pvt Ltd. Delhi. p. 146.

Phan, D. (2003). Compositions and methods of treatment for skin conditions using extracts of turmeric. *United States Patent* 6,521,271, February 18, 2003.

Pizorrno, J.E., Murray, M.T. (1999). *Textbook of Natural Medicine*, 2nd Ed. London: Churchill Livingstone, 689-693.

Price, L. C., and Buescher, R. W. (1992). "Turmeric pigments - Stability characteristics and reaction mechanism of photoxidation and alkaline degradation" in IFT 1992 Annual Meeting Food Expo. New Orleans.

Price, L.C., and Buescher, R.W. (1996). Decomposition of turmeric curcuminoids as affected by light, solvent and oxygen. *J.Food Biochem.*, 20(2), 125-133.

Priyadarsini, K.I., Maity, D.K., Naik, G.H., Kumar, M.S., Unnikrishnan, M.K., Satav, J.G. (2003). Role of phenolic OH and methylene hydrogen on the free radical reactions and antioxidant activity of curcumin. *Free Radic. Biol. Med.,* 35,475-84.

Pruthi, J.S. (1992). Postharvest technology of spices - Pre-treatments, curing, cleaning, grading and packing. *J. Spices and Aromatic Crops,* 1(1), 1-29.

Pruthi, J.S. (1976) *Spices and Condiments.* National Book Trust, New Delhi. p. 269.

Pruthi, J.S. (1980). Quality Control, Packaging and Storage of Turmeric. In Turmeric - Status papers and Abstracts, National Seminar on Ginger and Turmeric, Central Plantation Crops Research Institute, Calicut, India, 8-9 April, 1980, pp. 128-139.

Punyarajun, S. (1981). Determination of the curcuminoid content in Curcuma. *Varasarn Paesachasarthara,* 8(2), 29-31. (CA Vol 95/1981, 179626r).

Purseglove, J.W., Brown, E.G., Green, C.L., and Robbins S.R.J. (1981). "Turmeric" in Spices. Vol 2. Chapter 9. Longman Group Limited, Essex, UK. pp. 532-580.

Rafatulla, S., Tariq, M., and Alyahya, M.A. (1990). Evaluation of turmeric (*Curcuma longa*) for gastric and duodenal antiulcer activity in rats. *J Ethnopharmacol,* 29, 25-34.

Ragelis, E. P., Fisher, B. S., Klimeck, B. A. and Johnson, C. (1968). Isolation and determination of chlorohydrins in foods fumigated with ethylene oxide or with propylene oxide. *J. Assoc. Offic. Anal. Chem.,* 51, 709.

Raina, V.K., Srivastava, S.K., Jain, N., Ahmad, A., Syamasundar,K.V., and Aggarwal, K.K. (2002). Essential oil composition of *Curcuma longa* L. cv. Roma from the plains of northern India. *Flav. Frag. J.,* 17(2), 99-102.

Rajaraman, K., Narayanan, C.S., Sumathy Kutty, M..A., Sankarikutty, B., and Mathew, A.G. (1981). Ethyl acetate as a solvent for extraction of spice oleoresins. *J.Food Sci.Technol.*, 18(3), 101-103.

Ramachandraiah, O.S., Azeemoddin, G., and Charyulu, J.K. (1998). Turmeric (*Curcuma longa* L.) leaf oil, a new essential oil for perfumery industry. *Indian Perfum.*, 42(3), 124-127.

Ramachandraiah, O. S., Azeemoddin, G., Charyulu, J.K., Rath, C. C., Sashi, K., Dash, S. K., Rabindra, K., Mishra, R. K., Behura, C., and Ray, P. (2002). Isolation, characteristics, chemical composition and microbial activity of turmeric (*Curcuma longa* Syn. *C. domestica* valeton) leaf oil *Indian Perfumer,* 46(3), 211-216.

Ramanathan, L., and Das, N..P. (1993). Natural products inhibit oxidative rancidity in salted cooked ground fish. *J. Food Sci.,* 58(2), 318-320, 360.

Ramaswamy, T.S., and Banerjee, B.N. (1948). Vegetable dyes as antioxidants for vegetable oils. *Ann. Biochem. and Exptl. Med. (India),* 8, 55-68.

Ramirez-Tortosa, M.C., Mesa, M.D., and Aguilera, M.C. (1999). Oral administration of a turmeric extract inhibits LDL oxidation and has hypocholesterolemic effects in rabbits with experimental atherosclerosis. *Atherosclerosis,*147, 371-378.

Ramprasad, C., and Sirsi, M. (1957). Curcuma longa and bile secretion. Quantitative changes in the bile constituents induced by sodium curcuminate. *J Sci. Ind. Res.,* 16C, 108-110.

Ran, Q., and Zhou, X. (1988). New methods for isolation of curcumin. *Shipin Kexue (Beijing),* 101, 12-15. (CA Vol.109/1988, 148152r).

Rao, R.M., Reddy, R.C.K., and Subbarayadu, M. (1975). Promising turmeric types of Andhra Pradesh, *Indian Spices,* 12, 2-5.

Rao, S.D., Basu, N., Seth, S.D., and Siddiqui, H.H. (1984). Some aspects of pharmacological profile of sodium curcuminate. *Ind. J.Physiol. Pharmacol.,* 28, 211-215.

Rath, C. C., Dash, S.K., Mishra, R.K., and Charyulu, J.K. (2001). Anti *E.coli* activity of turmeric (*Curcuma longa* L.) essential oil. *Indian Drugs,* 38(3), 106-111.

Ravindranath, V., Chandrasekhara, N. (1980). Absorption and tissue distribution of curcumin in rats. *Toxicol.,*16, 259-265.

Raymond, S., Greenberg, M. J., and Rajani, F. (1991). Curcumin in the detection and warning of cyanide adulterated food products. *United States Patent* 5,047,100 September 10, 1991.

Riaz, M., Iqbal, M.J., and Chaudhary, F.M. (2000). Chemical composition of the volatile oil from rhizomes of *Curcuma longa* Linn. of Pakistan. *Bangladesh J. Sci. Ind. Res.,* 35(1-4), 163-166.

Richmond, R., and Pombo-Villar, E. (1997). Gas chromatography-mass spectrometry coupled with pseudo-Sadtler retention indices, for the identification of components in the essential oil of *Curcuma longa* L. *J. Chromatogr. A* , 760(2), 303-308.

Riddle, J.A., and Ford, J.E. (2000). IFOAM/IOIA International Organic Inspection Manual.

Rouseff, R.L. (1988). High performance liquid chromatographic separation and spectral characterization of pigments in turmeric and annatto. *J. Food Sci.,* 53(6), 1823-1826.

Rupe, H., Luksch, E., and Steinbach, A. (1909). Curcuma Oil. Ber., 42, 2515-20 (CA Vol 3/1909: 2565).

Rustovskii, B.N., and Leonov, P.P. (1924). Some constituents of oil of turmeric. *Chimie et industrie,* 16, 95. (CA Vol.20/1926: 3774).

Sair, L., and Klee, L (1967). Debittering of turmeric. United States Patent 3,340,250 (Cl.260-236.5), Sept 5, 1967.

Saito, K. (2002). Anti-oxidant reducing substance and method of producing the same. *United States Patent* 6,372,265 April 16, 2002.

Saju, K.A., Venugopal, M.N., and Mathew, M.J. (1998). Antifungal and insect-repellent activities of essential oil of turmeric (*Curcuma longa* L.). *Curr. Sci.,* 75(7), 660-662.

Sampathu, S. R., Krishnamurthy, N., Sowbagya, H. B., and Shankaranarayana, M. L. (1988) Studies on quality of turmeric (*Curcuma longa* L.) in relation to curing methods. *J. Food Sci.Technol. (India),* 25(3), 152-155.

Sanagi, M.M., Ahmed, U.K., and Smith, R. M. (1993). Application of supercritical fluid extraction and chromatography to the analysis of turmeric. *J.Chromatogr.Sci.,* 31(1), 20-25.

Sankaracharya, N.B., and Natarajan, C.P. (1975). Technology of spices. *Arecanut and Spices Bull.,* 7(2), 27-43.

Sankaranarayanan, J., Khanapure, V.M., and Jolly, C.I. (1993). HPTLC fingerprint analysis of *Momordica charantia, Curcuma longa, Emblica officinalis* and their mixtures. *J.Planar Chromatogr.,* 6(2), 157-160.

Santhanam, U., Weinkauf, R.L., and Palanker, L.R. (2001). Turmeric as an anti-irritant in compositions containing hydroxy acids or retinoids. *United States Patent* 6,277,881 August 21, 2001.

Sastry, B.S. (1970). Curcumin content of turmeric. *Res. Ind.,* 15(4), 258-60.

Satoskar, R.R., Shah, S.J., and Shenoy, S.G. (1986). Evaluation of anti-inflammatory property of curcumin in patients with post operative inflammation. *Int. J. Clin. Pharmacol. Ther. Toxicol.,* 24(12), 651-654.

Satyanarayana, M.N., Chandrasekhar, A.N, and Rao, D.S. (1969). Estimation of curcumin. *Res. Ind.,* 14(2), 82-83.

Sawada, T., Yamahara, J., Shimazu, S., and Ohta, T. (1971). Evaluation of crude drugs by bioassay. III. Comparison with local varieties of the contents and the fungistatic action of the essential oil from the roots of *Curcuma longa. Shoyakugaku Zasshi,* 25(1), 11-16. (CA 76/1972:21838p).

Schranz, J.L (1983). Water-soluble curcumin complex. United States Patent 4,368,208 January 11, 1983.

Schimmel & Co. (1911). Etherial Oils. Trade Circular, April 1911. thru Chem. Zentr., 1, 1837-1839. (CA. Vol.5/1911:3496).

Schmelzer, U. (1969). Fluorescence analysis and its utilization on examples of various turmeric preparations. *Praktikantenbriefe,* 14(9), 67-69. (CA 73/1970: 102112n).

Scott, P.M., Kennedy, B.P.C. (1975).Analysis of spices and herbs for aflatoxins. *Can. Inst. Food Sci. Technol. J.,* 8(2), 124-125.

Sharifzadeh, M., Sohrabpour, M. (1993). Identification of irradiated spices by the use of thermoluminescence method (TL). *Radia. Phys. Chem.,* 42(1-3), 401-405.

Sharma, O.P. (1976). Antioxidant activity of curcumin and related compounds. *Biochem. Pharmacol.,* 25, 1811-1812.

Sharma, R.K., Misra, B.P., Sarma, T.C., Bordolai, A.K., Pathak, M.G., Leclercq, P.A. (1997). Essential oils of *Curcuma longa* L. from Bhutan. *J.Essent.Oil Res.,* 9(5), 589-592.

Shoba, G., Joy, D., Joseph, T., Majeed, M., Rajendran, R., and Srinivas, P.S.S.R. (1998). Influence of piperine on the pharmacokinetics of curcumin in animals and human volunteers. *Planta Med.,* 64(4), 353-356.

Singh, S., Dube, N.K., Tripathi, S.C., and Singh, S.K. (1984). Fungitoxicity of some essential oils against *Aspergillus flavus. Indian Perfum.,* 28(3-4), 164-166.

Singh, G., OmPrakash, and Maurya, S. (2002 a). Chemical and biocidal investigations on essential oils of some Indian *Curcuma* species. *Progress in Crystal Growth and Characterization of Materials,* 45(1-2), 75-81.

Singh, R., Chandra, R., Bose, M., and Luthra, P.M. (2002 b). Antibacterial activity of *Curcuma longa* rhizome extract on pathogenic bacteria. *Curr. Sci.,* 83(6), 737-740.

Singh, G., Kapoor, I.P.S., Pandey, S.K., and O.P. Singh, O.P. (2003). *Curcuma longa* Linn. - Chemical, antifungal and antibacterial investigation on rhizome oil. *Indian Perfum.,* 47(2), 173-178.

Sit, A.K., and Tiwari, R.S. (1997). Micropropagation of turmeric (*Curcuma longa* L.). *Rec. Horticult.,* 4, 145-148.

Smith, R.M., Witowska, B.A. (1984). Comparison of detectors for the determination of curcumin in turmeric by high-performance liquid chromatography. *Analyst (London),* 109(3), 259-61.

Snyder,J.P., Davis, M.C., Adams, B., Shoji, M., Liotta, D.C., Ferstl, E.M., and Sunay, U.B. (2003). Curcumin analogs with anti-tumor and anti-angiogenic properties. *United States Patent* 6,664,272 December 16, 2003.

Soni, K.B., Rajan, A., and Kuttan, R. (1992). Reversal of aflatoxin induced liver damage by turmeric and curcumin. *Cancer Lett.,* 66,115-121.

Soudamini, N.K., and Kuttan, R. (1989). Inhibition of chemical carcinogenesis by curcumin. *J. Ethnophar-macol.,* 27, 227-233.

Souza, C.R.A., Osme, S.F., Gloria, M.B. (1997). Stability of curcuminoid pigments in model systems. *J.Food Process. Preserv.,* 21(5), 353-363.

Spices Board (1995) Quality Improvement of turmeric. Spices Board, India, October 1995, p.10.

Spices Board (2001). Agmark Grade Specifications for spices. Spices Board, India. January 2001.

Spices Board (2002). Quality requirements of spices for export. Spices Board, India, 2002.

Spices Board (2004). Spices Statistics. Vth Edition. Spices Board, Cochin, India.

Spitz, H. D., and Weinberger, J. (1971). Determination of ethylene oxide, ethylene chlorohydrin and ethylene glycol by gas-chromatography. *J. Pharmac. Sci.,* 60, 271-273.

Sreejayan, N., and Rao, M.N.A. (1993) Curcumin inhibits iron dependent lipid peroxidation. *International J. Pharmaceutics,* 100, 93-97.

Sreejayan, N. and Rao, M.N.A. (1994). Curcuminoids as potent inhibitors of lipid peroxidation. *J. Pharm. Pharmacol.,* 46,1013.

Srinivasan, K.R. (1953). Chromatographic study of the curcuminoids in *Curcuma longa. J.Pharm. Pharmacol.,* 5, 448-57.

Srivastava R. (1989). Inhibition of neutrophil response by curcumin. *Agents Actions,* 28(3-4), 298-303.

Srivastava, R., Puri, V., Srimal, R.C., and Dhawan B.N.(1986). Effect of Curcumin on platelet aggregation and vascular prostacyclin synthesis. *Arzneim.-Forsch.,* 36(4), 715-717.

Stransky, C.E. (1979). Water and Oil soluble curcumin coloring agents. *United States Patent 4,138,212* February 6, 1979.

Su, H.C.F., Horvat, R., and Jilani, G. (1982). Isolation, purification and characterization of insect repellents from *Curcuma longa* L. *J.Agric. Food Chem.,* 30(2), 290-292.

Su, S.L., Wu, Q.N., Ouyang, Z., Wu, D.K., and Chen, J. (2004). Study on the SFE condition for curcumin in *Curcuma longa. Zhongguo Zhong Yao Za Zhi. Sep*,29(9), 857-60. (Abstract — Entrez PubMed PMID: 15575202).

Sui, Z., Salto, R., Li, J., Craik, C., and Ortiz de Montellano, P.R. (1993). Inhibition of the HIV-1 and HIV-2 proteases by curcumin and curcumin boron complexes. *Bioorg. Med. Chem.* 1(6), 415-422.

Suryanarayana, C.V., Siddiqi, M.I.A., Rajaram, N., Gopinath, K.K., Lakshminarayanan, R. (1978). Fluorescent marking ink. *Indian Patent* 144,964, 05 Aug 1978.

Supardjan, A. M. (2001). Chemical content of turmeric curcumin and its derivatives. *Maj. Farm. Indonesia*, 12(3), 115-119.

Tainter, D.R., and Grenis, A.T. (2001). *Spices and Seasonings*. A Food Technology Handbook. 2nd Ed. Wiley-VCH. New York.

Takagaki, A., and Yamada, Y. (1998). Stabilization of curcuminoids and stabilized curcuminoid compositions. Japanese Patent Jpn.Kokai Tokkyo Koho JP 10 191,927 (98 191,927), 28 Jul. 1998. (CA 129/1998:108291s).

Takanashi, T. (1998). Method for removal and suppression of bitter taste of *Curcuma longa. Japanese Patent* Jpn.Kokai Tokkyo Koho JP 10 84,908 (98 84,908), 7 Apr. 1998. (CA. Vol.28/1998, 282131d).

Tang, K., and Chen, G. (2004). Analysis of chemical component of volatile oil from turmeric by gas chromatography-mass spectrometry. *Zhipu Xuebao*, 25(3), 163-165. (CA. Vol 141/2004:212323).

Taylor, S.J., McDowell, I.J. (1992). Determination of the curcuminoid pigments in turmeric (*Curcuma domestica* Val) by reversed-phase high performance liquid chromatography. *Chromatographia*, 34(1-2), 73-77.

Temmler-Werke (1938). Process for extracting the total pigment material of *Curcuma* drug. *German Patent* 658958, 20 April, 1938. Temmler-Worke, Vereinigte Chemische Fabriken, Berlin-Johannisthal.

Thaloor, D., Singh, A.K., and Sidhu, G.S. (1998). Inhibition of angiogenic differentiation of human umbilical vein endothelial cells by curcumin. *Cell Growth Differ.*, 9, 305-312.

Toda, S., Miyase, T., Arichi, H., Tanizawa, H., Takino, Y. (1985). Natural antioxidants.III. Antioxidative components isolated from rhizomes of *Curcuma longa. Chem. Pharm. Bull.*, 33(4), 1725-1728.

Todd, Jr., P.H.(1991). Curcumin complexed on water-dispersible substrates. *United States Patent* 4,999,205, March 12, 1991.

Tonnesen, H.H., and Karlsen, J. (1983). High-performance liquid chromatography of curcumin and related compounds. *J. Chromatogr.*, 259(2), 367-371.

Tonnesen, H. H. and Karlsen, J. (1985a). Studies on curcumin and curcuminoids. V. Alkaline degradation of curcumin. *Z. Lebensm.Unters Forsch.*, 180(2), 132-134.

Tonnesen, H. H., and Karlsen, J. (1985b). Studies on curcumin and curcuminoids. VI. Kinectics of curcumin degradation in aqueous solution. *Z. Lebensm.Unters Forsch.,* 180(5), 402-404.

Tonnesen, H. H., Karlsen, J., and van Henegouwen, G.B. (1986). Studies on Curcumin and Curcuminoids. VIII. Photochemical Stability of Curcumin. *Z Lebensm Unters Forsch.*, 183, 116-122.

Tonnesen, H. H., Karlsen, J. (1987). Studies on Curcumin and Curcuminoids. X. The use of curcumin as a formulation aid to protect light-sensitive drugs in soft gelatin capsules. *Int. J. Pharmaceutics*, 38, 247-249.

Tonnesen, H. H., and Karlsen, J. (1988). Studies on Curcumin and Curcuminoids. XI. Stabilization of photolabile drugs in serum samples by addition of curcumin. *International Journal of Pharmaceutics*, 41, 75-81.

Tonnesen (1992). Studies on curcumin and curcuminoids. XVIII. Evaluation of *Curcuma* products by the use of standardized reference colour values. *Z Lebensm Unters Forsch.*, 194, 129-130.

Tonnesen, H.H., Masson, M., and Loftsson, T. (2002). Studies of curcumin and curcuminoids. XXVII. Cyclodextrin complexation: solubility, chemical and photochemical stability. *Int. J. Pharmaceutics*, 244(1-2), 127-135.

Torres, R.C., Bonifacio, T.S., Herrera, C.L., and Lanto, E.A. (1998). Isolation and spectroscopic studies of curcumin from Philippine *Curcuma longa* L. *Philipp. J.Sci.*, 127(4), 221-228.

Tripathi, A.K., Prajapati, V., Verma, N., Bahl, J.R., Bansal, R.P., Khanuja, S.P.S., and Kumar, S. (2002). Bioactivities of the leaf essential oil of *Curcuma longa* (Var. Ch-66) on three species of stored-product beetles (Coleoptera). *J. Econ. Entomol.*, 95(1), 183-189.

Uechi, S., Y. Ishimine, Y., and F. Hongo, F. (2000). Antibacterial activity of essential oil derived from *Curcuma* sp. (Zingiberaceae) against foodborne pathogenic bacteria and its heat-stability.*Sci. Bull. Fac. Agric. Univ. Ryukyus*, 47, 129-136.

Unnikrishnan, M.K., and Rao, M.N. (1995). Inhibition of nitrite induced oxidation of hemoglobin by cur-cuminoids. *Pharmazie*, 50, 490-492.

USDA (2004). USDA National Nutrient Database for Standard Reference, Release 16-1.

Velappan, E., Thomas, K.G, and Elizabeth, K.G. (1993). New Technologies for on-farm processiong of spices. In: *Post Harvest Technology of Spices*, Proceedings of the National Seminar held at RRL, Trivandrum, 13-14 May, 1993. Spices Board, Cochin, India.

Venkatesha, J., Vanamala, K. R., Khan, M. M. (1997). Effect of method of storage on the viability of seed rhizome in turmeric. *Current Research – University of Agricultural Sciences (Bangalore)*, India. 26 (6/7), 114-115.

Verghese, J. (1993). Isolation of curcumin from *Curcuma longa* L. Rhizome. *Flav. Frag. J.* 8, 315-319.

Verghese, J., and Joy, M.T. (1989). Isolation of the colouring matter from dried turmeric (*Curcuma longa* L.) with Ethyl acetate. *Flav. Frag. J.*, 4, 31-32.

Verghese, J. (1984). On the retrieval of pigments from turmeric. *Indian Spices*, 21(3), 12-13, 16-17.

Verghese, J. (1994). Curcumin – The pigment pool of *C.longa* rhizome. *Indian Spices*, 31(4), 12-16.

Vogel J., and Pelletier, R. (1815). Curcumin. *J.Pharm.*, 2, 50.

Wahlstrom, B., and Blennow, G. (1978). A study on the fate of curcumin in the rat. *Acta Pharmacol. Toxicol.*, 43, 86-92.

Wang, Y., Wang, K.M., Shen, G.L., and Yu, R.Q. (1997). A selective optical chemical sensor for o-nitrophenol based on fluorescence quenching of curcumin. *Talanta*, 44(7), 1319-1327 (CA Vol.127/85651b).

Wang, Y., Hu, W., and Wang, M. (1999). HPLC determination of three curcuminoid constituents in rhizome curcumae. *Yaoxue Xuebao*, 34(6), 467-470. (CA Vol.131/1999: 187799e).

Wei, K., Xu, X. (2003). Herbal composition and method for controlling body weight and composition . United States Patent 6,541,046. April 1, 2003.

Weinberger, J. (1971). GLC determination of ethylene chlorohydrin following co-sweep extraction. *J. Pharmac. Sci.*, 60, 545-547.

Weiss, E.A. (2002) *Spice Crops*. CAB International Publishing, Oxon, UK.

White, W.B. (1930). A modification of the turmeric and annatto tests. Ann. Rept. N.Y. Dept. Agr. and Markets for 1929. Legislative Document No.37, 97-8 (1930). (CA Vol.24/1930, 3726).

Wimmer, M.A., and Goldbach, H.E. (1999). A miniaturized curcumin method for the determination of boron in solutions and biological samples. *J. Plant Nutr. Soil Sci.*, 162(1), 15-18.

Woznicki, E.J., Rosania, L.J., and Marshall, K. (1984). Colored medicinal tablet, natural color pigment and method for using the pigment in coloring food, drug and cosmetic products. *United States Patent*, 4,475,919 October 9, 1984.

Woodrow, J.E., McChesney, M.M., and Seiber, J.N. (1995). Determination of Ethylene oxide in spices using Headspace Gas Chromatography. *J.Agric.Food Chem.*,43(8), 2126-2129.

Wu, G. (1995). Determination of Curcumin in curcuma by TLC-scanning densitometry. *Huaxi Yaoxue Zazhi*, 10(3), 172-174. (CA Vol.123/1995: 266254n).

Wu, T.S. (2004). Herbal pharmaceutical composition for treatment of HIV/AIDS patients. *United States Patent* 6,696,094, February 24, 2004.

Wuthi-udomlert, M., Grisanapan, W., Luanratana, O., and Caichompoo, W. (2000). Antifungal activity of *Curcuma longa* grown in Thailand. *Southeast Asian J. Trop. Med. Public Health*, 31, Suppl.1, 178-182.

Yang, L., and Buescher, R. (1993). Dye and method for making same. *US Patent* 5,210,316, 11 may, 1993.

Yang, M., Dong, X., and Tang, Y. (1984). Studies on the chemical constituents of common turmeric (*Curcuma longa*). *Zhongcaoyao*, 15(5), 197-8. (CA Vol. 101/1984, 167112d).

Yasuda, K., Tsuda, T., Shimizu, H., and and Sugaya, A. (1988). Multiplication of *Curcuma* species by tissue culture. *Planta Med.* 54(1), 75-79.

Yoshida, M., Watabiki, T., Tokiyasu, T., and Ishida, N. (1989). Determination of boric acid in biological materials by curcuma paper. *Nippon Hoigaku Zasshi*, 43(6), 497-501(CA. Vol. 115:225582j).

Zhao, D., and Yang, M. (1986). Separation and determination of curcuminoids in *Curcuma longa* L. and its preparations by HPLC. *Yaoxue Xuebao*, 21(5), 382-5. (CA Vol. 105/1986:102681b).

9 Bioactivity of Turmeric

Satyajit D. Sarker and Lutfun Nahar

CONTENTS

9.1   INTRODUCTION

Turmeric, commonly known as *Haldi* or *Holud*, is the dried rhizome of *Curcuma longa* L. of the ginger family. It is one of the extensively used spices and coloring agents in South and Southeast Asian cuisine, especially in the Indian subcontinent (Grieve, 2005). Turmeric is yellow in color, has a characteristic fragrance and a bitterish, slightly acrid taste, and is a major constituent of curry powder. Turmeric is routinely added to mustard blends and relishes. It is also used in place of saffron to provide color and flavor. The people in South and Southeast Asia, especially in Bangladesh, India, and Indonesia, use turmeric to dye their bodies as part of their wedding rituals. It is still used in rituals of the Hindu religion and as a dye for holy robes.

Turmeric is one of the most important spices in India, which produces nearly the whole world's crop and uses >80% of it. In Southeast Asia, fresh turmeric is much preferred to the dried. In Thailand, the fresh rhizome is grated and added to curry dishes; it is also part of the "yellow curry paste" for Thai curries. Yellow rice prepared with turmeric is still very popular in the Eastern Islands of Indonesia. In Bali, rice cooked with turmeric, coconut milk, Indonesian bay leaves, lemongrass, and pandanus leaves is considered a "cultic dish" in Hindu culture, and sacrificed to the gods.

Previous phytochemical investigations on turmeric revealed that it contains an essential oil (max. 5%) composed of a variety of sesquiterpenes. Most important for the aroma are turmerone (max. 30%), *ar*-turmerone (25%), and zingiberene (25%). Conjugated diarylheptanoids (1,7-diaryl-hepta-1,6-diene-3,5-diones, e.g., curcumin, 3 to 4%) are responsible for the orange color and probably also for the pungent taste.

Over the past few decades, there has been an increased interest in turmeric and its medicinal properties. This is evidenced from the large body of scientific studies published on this topic. The aim of this chapter is certainly not to discuss the myths, cultural practices, or phytochemistry of turmeric, but to present an overview on the biological activities of turmeric, together with a brief discussion on the traditional medicinal uses.

9.2   TRADITIONAL MEDICINAL USES

Turmeric has been used as a yellow dye, medicine, and flavoring agent since 600 B.C. In fact, turmeric has been a traditional remedy in Asian folk medicines for the last 2000 years. In 1280, Marco Polo described turmeric as "a vegetable with the properties of saffron, yet it is not really saffron." Throughout Asia, it has been used internally to treat fevers, stomach problems, allergies, diarrhea, chronic cough, heartburn, wind, bloating, colic, bronchial asthma, flatulence, and jaundice and other liver ailments. Externally, it has been used for reducing inflammation and swelling due to sprains, cuts, and bruises. Turmeric has long been used in both Ayurveda and Chinese medicines as an anti-inflammatory agent, useful for such conditions as arthritis. In India and Bangladesh, turmeric powder or paste is mixed with lime (CaO), warmed, and applied to treat inflammation and swelling of limbs caused by various external injuries. It has also been used externally, to heal sores, and as a cosmetic. Turmeric is a mild aromatic stimulant, carminative, and digestive.

Turmeric water is an Asian cosmetic applied to offer a golden glow to the complexion. The fresh plant juice or a paste is used to treat leprosy and snakebites. The smoke produced by sprinkling powdered turmeric over glowing charcoal is supposed to relieve pain on account of scorpion bites. About 20 drops of the raw turmeric juice, mixed with a pinch of salt, taken just before breakfast, is believed to be an effective remedy for expelling worms. A teaspoon of raw turmeric juice, mixed with honey, can be effective in the treatment of anemia.

Turmeric is beneficial in the treatment of measles. Sun-dried and ground turmeric roots are mixed with a few drops of honey and the juice of a few bitter gourd leaves. The resulting concoction greatly benefits those suffering from measles (Tornetta, 2005). Turmeric, in combination with caraway seeds, is useful for colds in infants. A teaspoon of turmeric powder and a quarter teaspoon of caraway seeds are added to boiling water, and cooled. About 30 ml of this decoction, sweetened with honey, may be taken thrice a day to treat common cold.

Turmeric is effective in the treatment of skin diseases caused by ringworm and scabies. In such cases, the juice of raw turmeric is applied to the affected parts, and turmeric juice, mixed with honey, is taken orally. Turmeric is an excellent natural antibiotic, and one of the best detoxifying herbs by virtue of its beneficial effect on the liver. It is a powerful antioxidant with health-promoting effects on the cardiovascular, skeletal, and digestive systems. Through its beneficial effect on the ligaments, it is highly valued by those who practice Hatha Yoga (McIntyre, 2005). In China it is used to treat the early stages of cervical cancer. Turmeric is used in traditional medicine as a household remedy for various diseases, including biliary disorders, anorexia, cough, diabetic wounds, hepatic disorders, rheumatism, and sinusitis (Chattopadhyay et al., 2004). The people of Ngada use several plants for wound healing, and turmeric is one of them (Sachs et al., 2002).

9.3 COMMERCIAL MEDICINAL PREPARATIONS CONTAINING TURMERIC

Turmeric is commercially available in the forms of capsules containing powder, extracts, and tincture. Because bromelain enhances the absorption and anti-inflammatory effects of curcumin (1), the active ingredient of turmeric, bromelain is often formulated with turmeric products. An ointment based on turmeric is used as an antiseptic in Malaysia. A number of Indian antiseptic creams contain an extract of turmeric. "RA-11 (ARTREX, MENDAR)," a standardized multiplant herbal drug used in the treatment of arthritis, is composed of *Withania somnifera*, *Boswellia serrata*, *Zingiber officinale*, and *C. longa*. "Hyponidd" is a herbomineral formulation composed of the extracts of ten medicinal plants: *Momordica charantia*, *Melia azadirachta*, *Pterocarpus marsupium*, *Tinospora cordifolia*, *Gymnema sylvestre*, *Enicostemma littorale*, *Emblica officinalis*, *Eugenia jambolana*, *Cassia auriculata*, and *C. longa*. "Smoke Shield," a formulation composed of the extracts of *C. longa* and green tea, has long been in the market to reduce smoke-related mutagenicity and toxicity in the population. Ophthacare®, a commercially available eye preparation in India, contains *Carum copticum*, *Terminalia belirica*, *E. officinalis*, *C. longa*, *Ocimum sanctum*, *Cinnamomum camphora*, *Rosa damascena*, and meldespumapum (Biswas et al., 2001). Some of the commercially available medicinal preparations of turmeric or curcumin (1) are presented in Table 9.1.

9.4 BIOACTIVITY

The bioactivity of turmeric is discussed under the categories of crude powder or extract, and isolated active compounds, e.g., curcumin (1). The toxicological aspects of turmeric and its active components are also reviewed.

9.4.1 BIOACTIVITY OF THE CRUDE EXTRACT OR POWDER

Turmeric possesses a great variety of pharmacological activities including anti-inflammatory, anti-HIV, antibacterial, antioxidant properties, and nematocidal activities (Araujo and Leon, 2001; Nagabhusan and Bhide, 1992; Kuttan et al., 1987). It has immune-enhancing properties (Nagabhusan and Bhide, 1992; Kuttan et al., 1987). Turmeric helps to regulate intestinal flora and is well worth taking during and after a course of antibiotics and by those suffering from candida or thrush (McIntyre, 2005). It produces a soothing and bolstering effect on the mucosa of the gut and boosts

TABLE 9.1
Some of the Well-Known Commercially Available Medicinal Preparations of Turmeric or Curcumin

Products	Supplier/Source	Address
Best Curcumin w/Bioperine Capsules, 500 mg	Doctor's Best	http://www.iherb.com/curcumin3.html
Full Spectrum Turmeric Extract Tablets, 450 mg	Planetary Formulas	http://www.iherb.com/curcumin3.html
Curcumin 95 Capsules, 500 mg	Jarrow Formulas	http://www.iherb.com/curcumin3.html
Turmeric Capsules, 250 mg	Paradise Herbs	http://www.iherb.com/curcumin3.html
Ultra Joint Response Tablets, 450 mg	Source Naturals	http://www.iherb.com/glucosamine2.html
Turmeric (Standardized 95% Curcuminoids) Tablets, 450 mg	Nature's Way	http://www.iherb.com/turmericnw.html
Turmeric (Curcumin) Extract Tablets, 400 mg	Source Naturals	http://www.iherb.com/tumeric.html
Turmeric Force Capsules, 500 mg	New Chapter	http://www.iherb.com/turmericforce.html
Curcumin Capsules, 665 mg	Now Foods	http://www.iherb.com/curcumin2.html
Turmeric and Bromelain Capsules, 450 mg	Natural Factors	http://www.iherb.com/tumericbrom.html
Turmeric Catechu Extract, 60 ml	Gaia Herbs	http://www.iherb.com/turmericcat.html
Turmeric Extract, 30 ml	Nature's Apothecary	http://www.iherb.com/turmeric4.html
Turmeric Extract Capsules, 500 mg	Turmeric-Curcumin.com	http://www.turmeric-curcumin.com/

stomach defences against excess acid, drugs, and other irritating substances. It can reduce the risk of gastritis and ulcers.

Turmeric has been reported to be able to protect against the development of cancer and has a long history of use in the treatment of various cancers. It enhances the production of cancer-fighting cells protecting against environmental toxins. An alcohol extract of turmeric applied externally in skin cancer has been shown to reduce itching, relieve pain, and promote healing. Turmeric has been found to be highly effective at inhibiting recurring melanoma in people at high risk. Research has also demonstrated its protective effects against colon and breast cancers (McIntyre, 2005). Its anticancer effect is mainly mediated through induction of apoptosis.

Numerous studies have been carried out with the powder and crude extracts of turmeric for their various biological activities to date. These include its anti-inflammatory, antioxidant, anticarcinogenic, antimutagenic, anticoagulant, antifertility, antidiabetic, antibacterial, antifungal, antiprotozoal, antiviral, antifibrotic, antivenom, antiulcer, hypotensive, and hypocholesteremic activities (Chattopadhyay et al., 2004; Braga et al., 2003). Its anti-inflammatory, anticancer, and antioxidant roles may be clinically exploited to control rheumatism, carcinogenesis, and oxidative stress–related pathogenesis. However, most of these studies have been performed in laboratory conditions, and mostly on animals. In a recent study, it was suggested that most of the beneficial effects of turmeric could be related to its prominent free radical scavenging property (Tilak et al., 2004). The biological activities of the crude extracts or powder of turmeric reported in scientific journals to date are summarized below.

9.4.1.1 Digestive Disorders

Turmeric is beneficial to stomach upset, abdominal cramps, and flatulence. In an animal study, extracts of turmeric reduced secretion of acid from the stomach and protected against injuries such as inflammation along the stomach or intestinal walls, and ulcers caused from certain medications, stress, or alcohol. Bundy et al. (2004) assessed the effects of turmeric extract on irritable bowel syndrome (IBS) in healthy adults, and found that the IBS prevalence decreased remarkably. A poststudy analysis also revealed a significant reduction in abdominal pain/discomfort score. Rafatulla et al. (1990) reported the gastric and duodenal antiulcer activity of turmeric in rats.

9.4.1.2 Osteoarthritis

By virtue of its anti-inflammatory property, turmeric is useful in relieving the symptoms of osteoarthritis. An Ayurvedic formula containing turmeric as well as *Withania somnifera* (winter cherry), *Boswellia serrata* (Boswellia), and zinc could alleviate pain and disability (Kulkarni et al., 1991). Treatment with this herbomineral formulation produced a significant drop in severity of pain ($p < 0.001$) and disability score ($p < 0.05$).

9.4.1.3 Atherosclerosis and Cardiac Diseases

Atherosclerosis is characterized by oxidative damage that affects lipoproteins, the walls of blood vessels, and subcellular membranes (Quiles et al., 1998). The antioxidant property of an ethanolic extract of the rhizomes of *C. longa* on the lipid peroxidation of liver mitochondria and microsome membranes in atherosclerotic rabbits was evaluated. It was concluded that the active compounds present in this extract might protect the prevention of lipoperoxidation of subcellular membranes in a dosage-dependent manner, and thus help with the prevention of atherosclerosis. Mesa et al. (2003) reported that the oral administration of a nutritional dose of turmeric extracts reduced the susceptibility to oxidation of erythrocyte and liver microsome membranes *in vitro*, and could possibly contribute to the prevention of effects caused by a diet high in fat and cholesterol in blood and liver during the development of atherosclerosis. Quile et al. (2002) studied the effect of turmeric extract on the development of experimental atherosclerosis (fatty streak) in rabbits and its interaction with other plasmatic antioxidants. It was found that compared with the turmeric extract–treated group, the control group showed higher plasma lipid peroxide at all experimental times (10, 20, and 30 days) and lower -tocopherol and coenzyme Q levels at 20 and 30 days. Histological results for the fatty streak lesions revealed damage in the thoracic and abdominal aorta that was lower in the turmeric extract–treated group than in the control group at 30 days. It was suggested that supplementation with turmeric could reduce oxidative stress and attenuate the development of fatty streaks in rabbits fed a high cholesterol diet.

Rilantono et al. (2000) demonstrated the antioxidative, anti-inflammatory, and antithrombotic potentials of aqueous extract of turmeric in an experimental photochemical thrombogenesis model using rat femoral artery. A hydroalcoholic extract of turmeric was found to lower the abnormally high values of human plasma fibrinogen, and thus could reduce the risk of heart attacks (Bosca et al., 2000). Olajide (1999) established the antithrombotic effect of the extracts of *Azadiractha indica*, *Bridelia ferruginea*, *Commiphora molmol*, *Garcinia indica*, and *C. longa*. A number of other studies suggested that turmeric could prevent the blockage of arteries that can eventually cause a heart attack or stroke. In animal studies, turmeric extract was found to lower cholesterol levels and inhibit the oxidation of LDL (low density lipoprotein) cholesterol. When too much LDL cholesterol circulates in the blood, it can slowly build up in the inner walls of the arteries that feed the heart and brain. Together with other substances, it can form atherosclerotic plaque, a thick, hard deposit that can clog those arteries. This condition is known as atherosclerosis.

Turmeric could prevent platelet build-up along the walls of an injured blood vessel. Platelets collecting at the site of a damaged blood vessel cause blood clots to form and blockage of the artery as well. The efficacy of turmeric in lowering blood plasma lipid was evaluated in rabbits (Wientarsih et al., 2002). It was found that turmeric did not influence feed, protein, and fat consumption and protein excretion, but increased fat excretion. Cholesterol concentration was decreased by 46.6, 56.4, and 63.2% and HDL (high density lipoprotein) concentration was decreased by 9.9, 14.5, and 21.9% at 2, 3, and 4 g/kg turmeric, respectively. Turmeric decreased LDL and triglyceride concentration by 20.4, 28.5, and 29.5% at 2, 3, and 4 g/kg, respectively. HMG-CoA (3-hydroxy-3-methylglutaryl coenzyme A) reductase inhibitor was increased, and the glucose level was reduced. Lipid peroxidation was prevented at 3 and 4 g/kg turmeric. The enhanced fat excretion could have been mediated through an acceleration of lipid metabolism from extrahepatic tissues to the liver, which would increase the excretion of cholesterol via the bile and into the feces. It was indicated that turmeric could be used as a phytotherapeutic agent under atherosclerosis and cardiovascular disease conditions. Free radical-induced blood lipid peroxidation and especially peroxidized LDL contributes to the pathogenesis of atherosclerosis and related cardiovascular diseases (Ramirez-Bosca et al., 2000). The contribution of apolipoprotein-B (apo-B) to atherogenesis is considered to be the main inductor of one of its earlier steps, i.e., macrophage proliferation. A daily oral administration of a hydroalcoholic extract of *C. longa* decreased LDL and apo-B, and increased HDL and apo-A of healthy subjects. An aqueous ethanolic extract of the rhizomes of turmeric was studied for its effect on LDL oxidation and plasma lipids in atherosclerotic rabbits (Ramirez-Tortosa et al., 1999). It was demonstrated that this extract could be useful in the management of cardiovascular disease in which atherosclerosis is important. It was observed that low dosage but not high dosage decreased the susceptibility of LDL to lipid peroxidation. Both doses had lower levels of total plasma cholesterol than the control group. Moreover, the lower dosage had lower levels of cholesterol, phospholipids, and triglycerides in LDL than the 3.2 mg dosage.

Molecular mechanisms for cardioprotective effects of turmeric on myocardial apoptosis, cardiac function, and antioxidant milieu in an ischemia reperfusion model of myocardial infarction (MI) have recently been described (Gupta, 2004). The use of the anticancer drug doxorubicin is limited by its dose-dependant cardiotoxicity caused by elevated tissue levels of cellular superoxide anion/oxidative stress. Wattanapitayakul et al. (2005) evaluated the cardioprotective effects of an ethanolic extract and an aqueous extract of turmeric against doxorubicin toxicity using crystal violet cytotoxicity assay. Monaty et al. (2004) also have evaluated the cardioprotective potential of turmeric in the ischemia–reperfusion (I/R) model of MI. MI produced after I/R was reduced in the turmeric-treated group. Turmeric treatment restored the myocardial antioxidant status, altered hemodynamic parameters as compared to control I/R, and inhibited the I/R-induced lipid peroxidation. The cardioprotective effect of turmeric possibly resulted from the suppression of oxidative stress and correlated with the improved ventricular function. Histopathological examination further supported these cardioprotective effects of turmeric.

Despande et al. (1997) investigated the unique cholesterol-lowering property of turmeric, and its potential in the prevention of coronary heart disease. Turmeric intake reduced the total cholesterol (9.6 to 12.5%), triglyceride (16.2 to 34.3%), and LDL-C cholesterol (3.5 to 17%) in healthy subjects within 15 days. This hypolipemic effect of turmeric extract was accompanied by decreased plasma lipid peroxidation.

9.4.1.4 Cancer and Tumor

There has been a substantial amount of research on turmeric's anticancer potential against various forms of cancers including colorectal, prostate, oral, blood, and breast cancers (Leal et al., 2003; Pal et al., 2001; Sharma et al., 2001; Soni et al., 1997; Krishnaswamy, 1996; Manoharan et al., 1996; Azuine et al., 1992; Kuttan et al., 1985).

The growth inhibitory effects of turmeric on prostate cancer cell lines were recently studied by Rao et al. (2004). Turmeric was effective on the highly metastatic PC-3M prostate cancer cell line. It was also found that turmeric was most effective against different prostate cancer cell lines of varying metastatic potential and showed only a moderate effect on BG-9 normal skin fibroblasts. Paek et al. (1996) reported that turmeric extract was effective in inducing apoptosis in human myeloid leukemia cells (HL-60). Sharma et al. (2001) studied the pharmacodynamic and pharmacokinetic behavior of oral administration of turmeric extract in patients with colorectal cancer.

The modifying effects of turmeric extracts on 4-nitroquinoline-1-oxide (4NQO)-induced oral carcinoma were investigated in male Wistar strain rats (Manoharan et al., 1996). While all the animals painted with 4NQO on their cheek mucosa for 20 weeks developed oral neoplasms identified histologically as squamous cell carcinoma, and the oral tumor tissue showed a decrease in lipid peroxidation with concomitant changes in antioxidant activity, in the turmeric-treated group of mice, no tumors were observed, and the lipid peroxidation and antioxidant levels showed a pattern similar to that of untreated controls. Later, similar effects of the plant products neem and turmeric during the preinitiation and postinitiation phases of oral carcinogenesis induced by 4NQO were investigated in male Wistar rats (Nagini and Manoharan, 1997). Both neem and turmeric decreased the formation of lipid peroxides and enhanced the levels of enzymic and nonenzymic antioxidants in the preinitiation as well as postinitiation phases. In these studies, the antioxidant activity of turmeric was indicated as the main contributors to its anticarcinogenic effect.

Turmeric was found to inhibit the mutagenesis induced by aflatoxin B-1 (0.5 µg/plate) in *Salmonella* tester strains TA 98 and TA 100 (Soni et al., 1997). Turmeric inhibited the mutation frequency by more than 80% at concentrations of 2 µg/plate. Dietary administration of turmeric (0.05%) to rats reduced the number of γ-glutamyl *trans*-peptidase-positive foci induced by aflatoxin B-1, which is considered as the precursor of hepatocellular neoplasm. These results indicated the usefulness of turmeric in ameliorating aflatoxin-induced mutagenicity and carcinogenicity. "Smoke Shield," which contains an extract of turmeric and green tea, could inhibit mutagenic response *in vitro* and *in vivo* produced by several kinds of mutagens present in our atmosphere (Kuttan et al., 2004). "Smoke Shield" produced inhibition of mutagenicity to *S. typhimurium* induced by sodium azide and 4-nitro-*O*-phenylenediamine (NPD). It was more effective against mutagens needing metabolic activation such as 2-acetamidofluorene (2-AAF) and benzo[a]pyrene. It also inhibited the mutagenicity induced by tobacco extract to *S. typhimuirium* TA 102. The urinary mutagenicity of rats treated with the benzo[a]pyrene and tobacco extract, and also urinary mutagenicity in smokers, were inhibited by "Smoke Shield."

In a recent study, turmeric at concentrations ranging from 0 to 200 µM was applied to human cancer cell cultures (HELA, K-562, and IM-9) with or without X-irradiation (Baatout et al., 2004), and the cell proliferation was monitored by trypan blue exclusion. For the estimation of apoptosis, changes in cell morphology and flow cytometry analysis (DNA content and presence of the sub-G1 peak) were also performed. The microscopic examination of the turmeric-treated cells (with concentrations above 100 µM) demonstrated a characteristic morphology of apoptosis. Furthermore, cells treated with turmeric exhibited a sub-G1 peak from which the magnitude was proportional to the concentration of turmeric. X-irradiation alone induced polyploidization and apoptosis of the three cell lines, proportional to the doses of irradiation with a marked difference in radiation sensitivity between the cell lines (IM-9 < K-562 < HELA). However, when radiation and turmeric were applied together, it was observed that in HELA, K-562, and IM-9, turmeric showed a radiation sensitizing effect only at the dose of 200 µM.

9.4.1.5 Antioxidant/Radical Scavenging Property

Turmeric possesses antioxidant property, and this property has been implicated to its various pharmacological activities (Leal et al., 2003; Miquel et al., 2002; Adegoke et al., 1998; Semwal et al., 1997; Bosca et al., 1995). Scartezzini and Speroni (2000) have reviewed the antioxidant property of *C. longa*.

In a recent study (Ramos et al., 2003), turmeric displayed IC_{50} of <30 µg/ml in the DPPH assay and IC_{50} of <32 µg/ml in lipid peroxidation inhibition testing. It showed a 20% inhibition of the *in vitro*–induced (OH)-O- attack to deoxyglucose. The antioxidant property of "Smoke Shield" was evaluated *in vitro*, in experimental animals as well as in human models by Sreekanth et al. (2003). "Smoke Shield" was found to scavenge superoxide radicals generated by photoreduction of riboflavin (IC_{50} = 91 µg/ml) and hydroxyl radicals generated by Fenton reaction (IC_{50} = 95 µg/ml), and to reduce lipid peroxidation. Administration of "Smoke Shield" to mice increased antioxidant enzymes such as catalase and superoxide dismutase in blood as well as in liver and kidney. Glutathione-*S*-transferase (GST) activity was elevated in liver and kidney. The levels of glutathione and glutathione reductase were elevated, respectively, in blood and kidney. Administration of "Smoke Shield" decreased the lipid peroxidation in serum, liver, and kidney, as well as reduced the levels of conjugated dienes and hydroperoxides. Administration of "Smoke Shield" to smokers increased the superoxide dismutase and glutathione in blood and decreased glutathione peroxidase. It inhibited Phase I enzymes as represented by aniline-hydroxylase and aminopyrene-demethylase *in vitro*. The methanolic extract of *C. longa* showed perxoynitrile scavenging activity having IC_{50} = 1.7 µg/ml (Kim et al., 2003). The peroxynitrite scavenging activity potential of the individual fractions obtained from partitioning was in the order of EtOAc > dichloromethane > water fraction. Turmeric was effective in inhibiting formation of hexanal, (*E*)-2-penenal, (*E*)-2-hexenal, (*E*)-2-heptenal, and (*E*)-2-octenal when slurries of fermented cucumber tissue were exposed to oxygen (Zhou et al., 2000). Turmeric almost completely prevented aldehyde formation at 40 ppm.

Asai et al. (1999) demonstrated the *in vivo* antioxidant potential of turmeric and its ability to prevent the deposition of triacylglycerols in the liver of mice. Phospholipid hydroperoxides in the plasma, red blood cells, and liver were measured after dietary supplementation for 1 week (1% w/w of diet) with a turmeric extract, hexane extract of rosemary, and supercritical CO_2-extracted capsicum pigment (supplemented with cr-tocopherol to prevent fading). A lower phospholipid hydroperoxide level was found in red blood cells of the spice extract–fed mice (65 to 74% of the nonsupplemented control mice). The liver lipid peroxidizability induced with Fe^{2+}/ascorbic acid was effectively suppressed by dietary supplementation with the turmeric and capsicum extracts. The liver triacylglycerol concentration of the turmeric-treated mice was markedly reduced to 50% of the level in the control mice.

Bosca et al. (1995) showed that a 45 days intake (by healthy individuals ranging in age from 27 to 67 years) of turmeric hydroalcoholic extract (at a daily dose equivalent to 20 mg of **1**) resulted in a significant decrease in the levels of serum lipid peroxides, which might play an important pathogenic role in normal senescence and age-related diseases, e.g., atherosclerosis. Reddy and Lokesh (1994) reported that dietary turmeric could lower lipid peroxidation by enhancing the activities of antioxidant enzymes. The effects of pretreatment with tomato, garlic, and turmeric, alone and in combination, against 7,12-dimethylbenz[a]anthracene–induced genetic damage and oxidative stress in male Swiss mice were evaluated by Mohan et al. (2004). The pretreatment reduced the frequencies of 7,12-dimethylbenz[a]anthracene–induced bone marrow micronuclei as well as the extent of lipid peroxidation, possibly mediated by the antioxidant enhancing effects of these dietary agents.

9.4.1.6 Liver Disease and Hepatoprotective Activity

Several animal studies provided evidence that turmeric could protect the liver from a number of damaging substances such as carbon tetrachloride (CCl_4) and paracetamol by clearing such toxins from the body (Luper, 1999). Lin et al. (1995) demonstrated that the extract of *C. xanthorrhiza* could reduce the acute elevation of serum transaminases levels induced by the two kinds of hepatotoxins, and alleviated the degree of liver damage at 24 h after the intraperitoneal (i.p.) administration of two hepatotoxins. Later, Lin et al. (1996) established the hepatoprotective effects of a dose of turmeric on acute hepatotoxicity induced in rats by a single dose of β-D-galactosamine (288 mg/kg, i.p.), and its mechanism of action.

9.4.1.7 Antimicrobial Activity

As one of the traditional medicinal uses of turmeric is attributed to its antimicrobial property, the antimicrobial activities of an ethanolic extract were evaluated against several strains of bacteria and fungi (Chauhan et al., 2003; Saju et al., 1998; Bhavanishankar and Srinivasa, 1979; Banerjee and Nigam, 1978; Lutomski et al., 1974). The rhizome extract was effective against fungi *Fusarium oxysporium, Aspergillus niger, A. nidulans*, and *Alternaria solani* and bacteria *Staphylococcus albus, E. coli*, and *Pseudomonas pyocyanea*. Leal et al. (2003) reported the antimycobacterial activity of turmeric. Kim et al. (2003a) demonstrated the fungicidal activity of turmeric against *Botrytis cineria*,

Erysiphe graminis, *Phytophthora infestans*, *Puccinia recondita*, *Pyricularia oryzae*, and *Rhizocto-nia solani*. The methanolic extract of *C. longa* inhibited the growth of *Helicobacter pylori* (Mahady et al., 2002). Singh et al. (2002b) evaluated the antibacterial potential of *C. longa* rhizome extracts against pathogenic strains of Gram-positive (*Staphylococcus aureus*, *Staph. epidermidis*) and Gram-negative (*E. coli, Pseudomonas aeruginosa, Salmonella typhimurium*) bacteria. The fungitoxic effects of different plant extracts on *Fusarium udum*, which causes wilt disease of *Cajanus cajan in vitro* and *in vivo*, were examined by Singh and Rai (2000). A leaf extract of *Citrus medica*, a root extract of *Asparagus adscendens*, rhizome extracts of *C. longa* and *Z. officinale*, and a bulb extract of *Allium sativum* inhibited up to 100% growth at higher concentrations. Allievi and Gualandris (1984) also reported the antimicrobial activity of the crude extracts of *C. longa*.

9.4.1.8 Wound Healing

In animal studies, turmeric applied to wounds was found to accelerate the healing process. Nitric oxide is one of the important factors in wound healing and its production is regulated by inducible nitric oxide synthase (iNOS). Turmeric was shown to have wound-healing properties (Mani et al., 2002). Since nitric oxide synthetase (NOS) levels are high in wound-related inflammation, an ideal wound-healing and anti-inflammatory agent should be able to bring NOS levels down (Cohly et al., 1999). Turmeric at 0.1 to 10 µg/ml was found to decrease NOS levels in acute and chronic wounds.

9.4.1.9 Eye Disorders

Biswas et al. (2001) studied the potential of Ophthacare® eye drops, which contains turmeric, in the management of various ophthalmic disorders. It was observed that this eye drop could be useful in managing a variety of infective, inflammatory, and degenerative ophthalmic disorders.

9.4.1.10 Alzheimer's Disease

As free radicals are involved in neurodegeneration in Alzheimer's disease, free radical scavengers should be able to reduce the neurodegeneration process in this disease. An aqueous extract of turmeric, having free radical scavenging activity, showed significant neuroprotective property (Koo et al., 2004). It rescued PC12 cells from pyrogallol-induced cell death. Hypoxia/reoxygenation injury of PC12 cells was blocked by this extract. The study also examined the effect of this extract on H_2O_2-induced toxicity in rat pheochromocytoma line PC12 by measuring cell lesion, level of lipid peroxidation, and antioxidant enzyme activities. Following a 30 min exposure of the cells to H_2O_2 (150 µM), a decrease in cell survival, activities of glutathione peroxidase, and catalase as well as increased production of malondialdehyde were observed. Pretreatment of the cells with this extract (0.5 to 10 µg/ml) prior to H_2O_2 exposure increased the cell survival, antioxidant enzyme activities, and decreased the level of malondialdehyde.

9.4.1.11 Antifertility

Ashok and Meenakshi (2004) demonstrated the contraceptive effect of the crude turmeric extract in male albino rats. A reduction in sperm motility and density was observed in both the treated groups. It was suggested that this extract might have affected the androgen synthesis either by inhibiting the Leydig cell function or the hypothalamus pituitary axis and as a result, spermatogenesis was arrested.

9.4.1.12 Anti-inflammatory

Arora et al. (1971) established the anti-inflammatory property of turmeric. Rafi et al. (2003) reviewed the anti-inflammatory property of a number of nutraceuticals including products containing turmeric. Atkinson and Hunter (2003) reported the effect of turmeric extract in the treatment of steroid-dependent inflammatory bowel disease. Ozaki (1990) demonstrated the anti-inflammatory activity of *C. xanthorhiza*. In another review article by Calixto et al. (2003), the anti-inflammatory activity and the possible mechanisms of actions of the constituents of turmeric have been discussed.

Propionibacterium acnes, an anaerobic pathogen, plays an important role in the pathogenesis of acne by inducing certain inflammatory mediators including reactive oxygen species and proinflammatory cytokines. Jain and Basal (2003) utilized reactive oxygen species, interleukin-8 (IL-8), and tumor necrosis factor-α (TNF-α) as the major criteria for the evaluation of anti-inflammatory activity of herbs *Rubia cordifolia*, *C. longa*, *Hemidesmus indicus*, and *Azadirachta indica*. To prove the anti-inflammatory effects of herbs, polymorphonuclear leukocytes and monocytes were treated with culture supernatant of *P. acnes* in the presence or absence of herbs. It was observed that all these herbs caused a suppression of reactive oxygen species from polymorphonuclear leukocytes.

The inhibitors of prostaglandin biosynthesis and nitric oxide production are potential antiinflammatory and cancer chemopreventive agents (Hong et al., 2002). In a study conducted with 170 methanol extracts of natural products including Korean herbal medicines for the inhibition of prostaglandin E-2 production (for COX-2 inhibitors) and nitric oxide formation (for iNOS inhibitors) in lipopolysaccharide-induced mouse macrophages RAW264.7 cells, along with a number of other plant extracts, turmeric was found to inhibit iNOS activity (>70% inhibition at the test concentration of 10 µg/ml).

9.4.1.13   Hypoglycemic Effect and Diabetes

The hypoglycemic and antioxidant potential of "Hyponide" a herbomineral formulation composed of the extracts of ten medicinal plants including turmeric, was investigated on diabetic rats (Babu and Prince, 2004). Oral administration of "Hyponide" (100 to 200 mg/kg) for 45 days lowered the levels of blood glucose, and increased levels of hepatic glycogen and total hemoglobin. "Hyponide" administration also reduced the levels of glycosylated hemoglobin, plasma thiobarbituric acid–reactive substances, hydroperoxides, ceruloplasmin, and α-tocopherol in diabetic rats. Plasma-reduced glutathione and vitamin C were elevated by oral administration of "Hyponide." The hypoglycemic effect of turmeric on rats having aloxan-induced diabetes was also reported by Arun and Nalini (2002).

Yasni et al. (1991) found that *C. xanthorrhiza* could improve the diabetic symptoms such as growth retardation, hyperphagia, polydipsia, elevation of glucose and triglyceride in the serum, and reduction of the ratio of arachidonate to linoleate in the liver phospholipids. It specifically modified the amount and composition of fecal bile acids.

Diabetes is one of the major risk factors for cataractogenesis. Aldose reductase has been reported to play an important role in sugar-induced cataract. The aldose reductase inhibitory activity of *O. sanctum*, *W. somnifera*, *C. longa*, and *A. indica* was studied together with their effect on sugar-induced cataractogenic changes in rat lenses *in vitro* (Halder et al., 2003). All four plants inhibited lens aldose reductase activity. The IC_{50} values of *O. sanctum*, *C. longa*, *A. indica*, and *W. somnifera* were 20, 55, 57, and 89 µg/ml, respectively. *O. sanctum* also showed an inhibition (38.05%) in polyol accumulation followed by *C. longa* and *A. indica* (28.4 and 25.04%, respectively).

9.4.1.14   Others

The effectiveness of turmeric-derived medicinal preparations against *Oketsu*, the so-called blood stasis syndrome, an important pathological conception in Japanese traditional medicine that often accompanies cerebro-vascular disorders, was evaluated by examining their vasomotional effects as one index (Sasaki et al., 2003). The differences in their efficacy were also investigated. Nitric oxide (NO) is the relaxation factor of vascular smooth muscle and an inhibitor of platelet aggregation in blood vessels. Thus any substances that show NO-dependent relaxation are thought to be effective against *Oketsu*. The methanol extracts of the *Curcuma* drugs, derived from various species of *Curcuma*, including *C. longa, C. kwangsiensis, C. phaeocaulis, C. wenyujin,* and *C. zedoaria* exhibited intense effects on relaxation in rings precontracted by prostaglandin F-2-α despite pretreatment with and without *N-G*-nitro-L-arginine methyl ester as an inhibitor of NO synthesis.

Oral administration of an aqueous extract of turmeric to mice elicited dose-dependent relation of immobility reduction in the tail suspension test and the forced swimming test in mice (Yu et al., 2002). The effects of the extracts at the dose of 560 mg/kg were more potent than that of reference antidepressant fluoxetine. It also promoted the adhesion of peripheral neutrophils to human umbilical vein endothelial cells (Madan et al., 2001) and activate nuclear transcription factor κ-B (NF-κ-B), a major transcription factor involved in the transcription of genes encoding ICAM-1, VCAM-1, and E-selectin. These results could be implicated in the usage of aqueous preparation of *C. longa* for upregulation of cell adhesion molecule expression and/or NF-κ-B.

Niederau and Gopfert (1999) reported the beneficial effect of turmeric on pain due to biliary dyskinesia. In the Ayurveda and *Sidha* system of medicine, *A. indica* (neem) and *C. longa* have been used for healing chronic ulcers and scabies. The neem and turmeric were used as a paste for the treatment of scabies in 814 people, and 97% of them were cured within 3 to 15 days of treatment (Charles and Charles, 1992). Rao and Kotagi (1984) demonstrated antiandrogenic activity of turmeric extracts in male rats.

9.4.2 BIOACTIVITY OF ISOLATED COMPOUNDS

Curcumin (**1**), demethoxycurcumin (**2**), *bis*-demethoxycurcumin (**3**), and *ar*-turmerone (**4**) are four major active components of turmeric (Figure 9.1). It also contains high amounts of carotene, equivalent to 50 IU of vitamin A per 100 g (Chopra and Simon, 2000). Curcumin (**1**), the main yellow bioactive component of turmeric, has been reported to possess a variety of medicinal properties including anticancer, antiarthritic, anti-inflammatory, antiedemic, antitumor, antimutagenic, anticoagulant, hepatoprotective, antihypercholesterolemic, nephrotonic, antihypertensive, chemoprotective, carminative, depurative, anti-HIV, antimicrobial, and antiparasitic properties (Turmeri-Crucumin.Com, 2005; Chattopadhyay et al., 2004; Araujo and Leon, 2001). *In vitro*, curcumin exhibits antiparasitic, antispasmodic, anti-inflammatory, and gastrointestinal effects, and also inhibits carcinogenesis and cancer growth. Regarding its *in vivo* effects, there are reports on the antiparasitic and anti-inflammatory activity of **1** in animal models.

Curcumin (**1**) is a powerful anti-inflammatory agent, excellent for treating inflammatory problems such as arthritis, and liver and gall bladder problems. It has been found to block the production of certain prostaglandins and to have effects similar to cortisone and nonsteroidal anti-inflammatory drugs but without the side effects (Srimal and Dhawan, 1973; Ghatak and Basu, 1972). It has long been known to possess antioxidant property (Chirangini et al., 2004). Another unsymmetrically substituted curcuminoid, 5′-methoxycurcumin (**5**), from *C. xanthorrhiza* also possesses potent antioxidant activity (Venkateswarlu et al., 2004). Curcumin (**1**) has also been found to be useful in preventing memory loss and in the prevention of Alzheimer's disease. It possesses anti-inflammatory activity, and is a potent inhibitor of reactive oxygenase rating enzymes such as lipoxygenase/cyclooxygenase, xanthine dehydrogenase/oxiclase, and inducible NO synthase, and an effective inducer of heme oxygenase-1. It is a potent inhibitor of protein kinase C, epidermal growth factor-receptor tyrosine kinase, and I-κ-B kinase (Lin, 2004).

Curcuminoids, a group of phenolic compounds isolated from the rhizomes of turmeric, exhibit a variety of beneficial effects on health and on events that help in preventing certain diseases. A vast majority of these studies were carried out with curcumin (**1**). The most detailed studies using **1** include anti-inflammatory, antioxidant, anticarcinogenic, antiviral, and anti-infectious activities. In addition, the wound-healing and detoxifying properties of **1** have also received considerable attention. The biological activities of various compounds from turmeric, reported in the scientific literature to date, are summarized below.

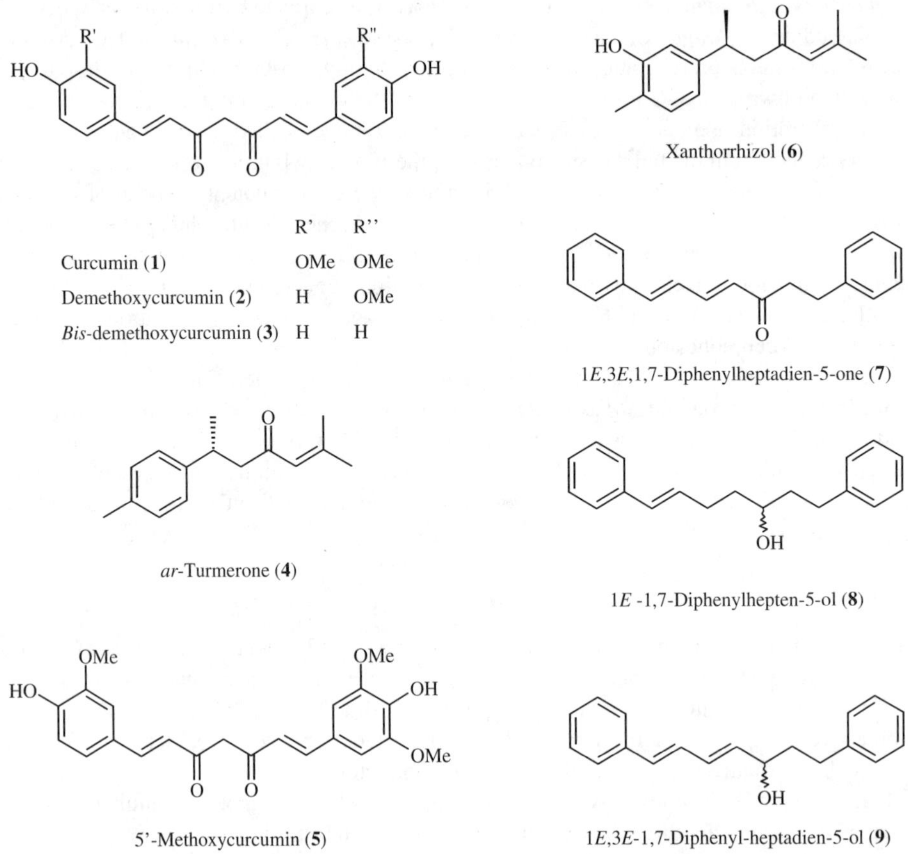

	R'	R''
Curcumin (**1**)	OMe	OMe
Demethoxycurcumin (**2**)	H	OMe
Bis-demethoxycurcumin (**3**)	H	H

Xanthorrhizol (**6**)

ar-Turmerone (**4**)

1*E*,3*E*,1,7-Diphenylheptadien-5-one (**7**)

1*E*-1,7-Diphenylhepten-5-ol (**8**)

5'-Methoxycurcumin (**5**)

1*E*,3*E*-1,7-Diphenyl-heptadien-5-ol (**9**)

FIGURE 9.1 Structures of major bioactive components of turmeric.

9.4.2.1 Antimicrobial Activity

Saju et al. (1998) reported the antifungal activity of the essential oils extracted from turmeric. Later, Singh et al. (2002a) confirmed the antifungal activity of essential oils against the mycelium growth of *Colletotrichum falcatum* and *Fusarium moniliforme* (MIC = 1000 ppm), and *Curvularia palle-scens*, *A. niger*, and *F. oxysporium* (MIC = 2000 ppm). Bioassay-directed fractionation of a hexane extract from the turmeric leaves yielded labda-8(17),12-diene-15,16-dial with antifungal activity against *Candida albicans* at 1 µg/ml, and inhibited the growth of *C. kruseii* and *C. parapsilosis* at 25 µg/ml (Roth et al., 1998). Kim et al. (2003a) demonstrated the antifungal activity of four commercially available compounds derived from *C. longa*.

The antibacterial potential of xanthorrhizol (**6**) against *Streptococcus mutans* was demonstrated by Hwang et al. (2000). Curcumin (**1**) was found to inhibit the growth of *H. pylori* (Mahady et al., 2002). Infection of epithelial cells by the microbial pathogen *H. pylori* leads to activation of the transcription factor NF-κ-B, the induction of proinflammatory cytokine/chernokine genes, and the mitogenic response (cell scattering). In a recent study, it was observed that *H. pylori*–induced NF-κ-B activation and the subsequent release of IL-8 were inhibited by **1** (Foryst-Ludwig et al., 2004). The results demonstrated that **1** could inhibit I-κ-B-α degradation, the activity of I-κ-B kinases α and β (IKK α and β), and NF-κ-B DNA-binding, and block *H. pylori*-induced mitogenic response. This finding suggested that **1** could be a potential therapeutic agent effective against pathogenic processes initiated by *H. pylori* infections. Essential oils of turmeric were active against pathogenic strains of Gram-positive (*Staph aureus*, *Staph. epidermidis*) and Gram-negative (*E. coli*, *P. aerug-*

inosa, S. typhimurium) bacteria (Singh et al., 2002b). The clinical isolate of *Staph aureus* showed more sensitivity toward essential oils than the positive controls. In another study, the mother liquor after isolation of **1** from oleoresin of turmeric was found to contain approximately 40% antibacterial oil (Negi et al., 1999). Various column fractions obtained from the hexane extract were tested for antibacterial activity against *Bacillus cereus*, *B. coagulans*, *B. subtilis*, *Staph. aureus*, *E. coli*, and *P. aeruginosa*. The GC and GC-MS analyses of the active fractions revealed the presence of *ar*-turmerone (**4**), turmerone, and curlone as the major compounds.

Apisariyakul et al. (1995) studied the effect of turmeric oil and curcumin (**1**) against 15 isolates of dermatophytes, 4 isolates of pathogenic molds, and 6 isolates of yeasts. The inhibitory activity of turmeric oil was tested in trichophyton-induced dermatophytosis in guinea pigs. The results showed that all 15 isolates of dermatophytes could be inhibited by turmeric oil at dilutions of 1:40 to 1:320. However, none of the isolates of dermatophytes was inhibited by **1**. The other four isolates of pathogenic fungi were inhibited by turmeric oil at dilutions of 1:40 to 1:80 but none was inhibited by **1**. All six isolates of yeasts tested proved to be insensitive to both turmeric oil and **1**.

9.4.2.2 Nephroprotective Activity

Cisplatin, a widely used anticancer drug, produces undesirable side effects such as nephrotoxicity. The effect of xanthorrhizol (**6**) (Figure 9.1), isolated from the rhizomes of *C. xanthorrhiza*, on cisplatin-induced nephrotoxicity in mice was studied by Kim et al. (2005, 2004). A single dose of cisplatin (45 mg/kg, i.p.) elevated the levels of blood urea nitrogen, serum creatinine, and the kidney-to-body weight ratio, but the pretreatment of **6** (200 mg/kg/day) for 4 d attenuated the cisplatin-induced nephrotoxicity. The preventive effect of **6** was more efficacious than that of **1** with the same amount (200 mg/kg).

Over expression of transforming growth factor (TGF-β) contributes greatly to fibrotic kidney disease. The activator protein-1 (AP-1) inhibition by curcumin (**1**) was shown to reduce collagen accumulation in experimental pulmonary fibrosis (Gaedeke et al., 2004). When applied 30 min before TGF-β, curcumin (**1**) dose-dependently reduced TGF-β–induced increases in plasminogen activator inhibitor-1 (PAI-1), TGF-β-1, fibronectin (FN), and collagen I (Col I) mRNA, and in PAI-1 and fibronectin protein. Prolonged curcumin treatment (6 h) reduced TGF-β receptor type II levels and SMAD2/3 phosphorylation in response to added TGF-β. Depletion of cellular c-jun levels with an RNAi method mimicked the effects of **1** on expression of TGF-β-1, FN, and Col I, but not PAI-1. It was concluded that **1** could block TGF-β's profibrotic actions on renal fibroblasts through downregulation of T-β-RII, and partial inhibition of c-jun activity, and **1** might be an effective candidate for antifibrotic drug development for the treatment of chronic kidney diseases.

Free radicals have been implicated to various pathogenic processes including initiation/promotion stages of carcinogenesis, and antioxidants have been considered to be a protective agent for this reason (Iqbal et al., 2003a,b). An iron chelate, ferric nitrilotriacetate, is a potent nephrotoxic agent and induces acute and subacute renal proximal tubular necrosis by catalyzing the decomposition of hydrogen peroxide–derived production of hydroxyl radicals, which are known to cause lipid peroxidation and DNA damage. The DNA damage is associated with a high incidence of renal adenocarinoma in rodents. Curcumin (**1**) provided protection against lipid peroxidation and DNA damage induced by ferric nitrilotriacetate and hydrogen peroxide *in vitro*.

Babu and Srinivasan (1998) demonstrated that dietary curcumin (**1**) could bring about beneficial modulation of the progression of renal lesions in diabetes. It was inferred that this beneficial ameliorating influence of dietary curcumin on diabetic nephropathy could possibly be mediated through its ability to lower blood cholesterol levels.

9.4.2.3 Hepatoprotective Activity

Several studies demonstrated the hepatoprotective effect of curcumin (**1**) and its analogs (Luper, 1999; Romiti et al., 1998; Kiso et al., 1983). Curcuminoids **1** to **3**, isolated from an EtOAc extract of turmeric rhizomes, showed hepatoprotective effects on tacrine-induced cytotoxicity in human liver-derived Hep G2 cells (Song et al., 2001). The EC_{50} values of **1** to **3** were 86.9, 70.7, and 50.2 μM, respectively. Silybin (EC_{50} = 69.0 μM) and silychristin (EC_{50} = 82.7 μM) were used as positive controls. An administration of 200 or 600 mg/kg of **1** suppressed diethylnitrosamine (DEN)-induced liver inflammation and hyperplasia in rats, as evidenced by histopathological examination (Chuang et al., 2000).

Romiti et al. (1998) investigated **1** for its ability to interact *in vitro* with hepatic P-glycoprotein (Pgp), in a model system represented by primary cultures of rat hepatocytes, in which spontaneous overexpression of multidrug resistance genes occurs. The results suggested that **1** modulated *in vitro* both expression and function of hepatic Pgp, and supported the hypothesis that **1** could reveal itself also as a compound endowed with chemosensitizing properties on multidrug resistance phenotype.

The potential hepatoprotective effect of the turmeric antioxidant protein (TAP) isolated from an aqueous extract of turmeric was evaluated in the CCl_4 treated rats (Subramanian and Selvam, 1999). The increased basal as well as promoter-induced lipid peroxide formation in the tissues of CCl_4-treated rats was inhibited by 40 to 70% by TAP administration. Decreased antioxidant enzyme activities of superoxide dismutase (SOD), catalase (CAT), glutathione peroxidase (GPx), GST, and antioxidant concentrations of reduced glutathione (GSH), total (TSH), protein (PSH), and nonprotein (NPSH) thiols and ascorbic acid in the liver of CCl_4-treated rats were nearly normalized on TAP pretreatment. The increased activities with glucose-6-phosphate dehydrogenase (G6PD), lactate dehydrogenase (LDH), alanine transaminase (ALT), and aspartate transminase (AST) in the liver of CCl_4 treated rats were conferred protection by 50 to 80% on TAP treatment. Glucose-6-phosphatase and the membrane bound ATPase activities were decreased in CCl_4-treated animals and were completely restored on TAP treatment.

9.4.2.4 Cancer and Tumor

Curcumin (**1**) exhibits chemopreventive and growth inhibitory activity against several tumor cell lines (Deeb et al., 2003). The chemopreventive activity of **1**, and its applications in the intervention of carcinogenesis, have been known for a while (Chun et al., 2003). Extensive research over the last five or six decades has indicated that **1** could both prevent and treat various forms of cancers and tumors (Syng-ai et al., 2004; Aggarwal et al., 2003; Anto et al., 2002, 1996; Shukla et al., 2002; Pal et al., 2001; Bhaumik et al., 2000; Singhal et al., 1999; Surh, 1999; Itokawa et al., 1999; Simon et al., 1998; Hanif et al., 1997; Rao et al., 1995; Azuine et al., 1992). Nagabhushan and Bhide (1992) demonstrated that **1** and its analogs could inhibit cancer at initiation, promotion, and progression stages of development. Pharmacological studies have revealed that **1** is an antimutagen as well as an antipromotor for cancer (Han, 1994). Curcumin (**1**) inhibited carcinogenesis of murine skin, stomach, intestine, and liver (Cheng et al., 2001; Chuang et al., 2000; Soni et al., 1997). It suppressed growth of head and neck squamous cell carcinoma (LoTemplo et al., 2004). Huang et al. (1994) reported the inhibitory effects of feeding commercial grade curcumin (77% **1**, 17% **2**, and 3% **3**) in AIN 76A diet on carcinogen-induced tumorigenesis in the forestomach, duodenum, and colon of mice. Curcumin prevented colon cancer in rodents (Sharma et al., 2001; Lin et al., 2000; Kawamori et al., 1999; Rao et al., 1995). The chemopreventive effect of **1** on *N*-nitrosomethylbenzylamine-induced esophageal cancer in rats was reported by Ushida et al. (2000). Dorai et al. (2000, 2001) described the effectiveness of **1** in the treatment of prostate cancer. Curcuminoids **1** to **3** showed activity against leukemia, colon, CNS, melanoma, renal, and breast cancer cell lines (Ramsewak et al., 2000; Mehta et al., 1997; Verma et al., 1997; Ruby et al., 1995). They were found to be potent inhibitors of mutagenesis and croton oil-induced tumor promotion (Anto et al., 1996). These curcuminoids (**1** to **3**) were compared for their cytotoxic, tumor-reducing, and antioxidant activities (Ruby et al., 1995). Curcuminoid **3** was more active than the other two as a cytotoxic agent and in the inhibition of Ehrlich ascites tumor in mice (ILS 74.1%). Xanthorrhizol (**6**) was reported as a novel anticariogenic agent against *Streptococcus mutans* (Hwang et al., 2002).

9.4.2.4.1 Anti-cancer and Anti-tumor Activity

Curcumin (**1**) and genistein inhibited the growth of estrogen-positive human breast MCF-7 cells induced individually or by a mixture of the pesticides endosulfane, DDT, and chlordane or 17-β estradiol (Verma et al., 1997). These compounds could act synergistically to inhibit the induction of MCF-7 cells by the highly estrogenic activity of endosulfane/chlordane/DDT mixtures. Diet containing these compounds could potentially reduce the proliferation of estrogen-positive cells by mixtures of pesticides or estradiol. Later, Simon et al. (1998) examined the effect of **1** to **3** and cyclocurcumin (Cyclocur) on the proliferation of MCF-7 human breast tumor cells. The natural curcuminoids **1** to **3** were potent inhibitors, whereas Cyclocur was less inhibitory. Curcumin (**1**) also exerted a cytostatic effect, which could be implicated to its antiproliferative property. The diketone moiety in **1** was indicated to be essential for the inhibitory activity.

Curcumin (**1**) inhibited the mutagenesis induced by aflatoxin B-1 (0.5 μg/plate) in *Salmonella* tester strains TA 98 and TA 100 (Soni et al., 1997). It inhibited the mutation frequency by more than 80% at concentrations of 2 μg/plate. Dietary administration of **1** (0.005% each) to rats significantly reduced the number of γ-glutamyl transpeptidase–positive foci induced by aflatoxin B-1, which is considered as the precursor of hepatocellular neoplasm. These results indicated the usefulness of **1** in ameliorating aflatoxin-induced mutagenicity and carcinogenicity.

Curcumin (**1**) inhibited tumorigenesis during both initiation and promotion (postinitiation) periods in several experimental animal models (Huang et al., 1997). Topical application of **1** inhibited benzo[a]pyrene (B[a]P)-mediated formation of DNA-B[a]P adducts in the epidermis. It reduced 12-*O*-tetradecanoylphorbol-13-acetate (TPA)-induced increases in skin inflammation, epidermal DNA synthesis, ornithine decarboxylase (ODC) mRNA level, ODC activity, hyperplasia, formation of c-Fos, and c-Jun proteins, hydrogen peroxide, and the oxidized DNA base 5-hydroxymethyl-2′-deoxyuridine (HmdU). Topical application of **1** inhibited TPA-induced increases in the percentage of epidermal cells in synthetic (S) phase of the cell cycle. Curcumin inhibited arachidonic acid-induced edema of mouse ears *in vivo* and epidermal cyclooxygenase and lipoxygenase activities *in vitro*. Other analogs of **1** were also found to be about equipotent as inhibitors of TPA-induced tumor promotion in mouse skin. Topical application of **1** inhibited tumor initiation by B[a]P and tumor promotion by TPA in mouse skin. Dietary curcumin inhibited B[a]P-induced forestomach carcinogenesis, *N*-ethyl-*N*′-nitro-*N*-nitrosoguanidine (ENNG)-induced duodenal carcinogenesis, and azoxymethane (AOM)-induced colon carcinogenesis. Limtrakul et al. (1997) demonstrated the inhibitory effect of dietary curcumin on skin carcinogenesis in mice. Later, the chemopreventive action of dietary curcumin on 7,12-dimethylbenz(a) anthracene initiated and 12,0-tetradecanoylphorbol-13-acetate (TPA)-promoted skin tumor formation in Swiss albino mice was reported (Limtrakul et al., 2001). It was shown that **1** could inhibit a variety of biological activities of TPA. Topical application of **1** was reported to inhibit TPA-induced c-fos, c-jun, and c-myc gene expression in mouse skin. Limtrakul et al. (2001) evaluated the effects of orally administered curcumin (**1**). It was observed that animals in which tumors had been initiated with 7,12-dimethylbenz(a) anthracene and promoted with TPA experienced significantly fewer tumors and less tumor volume if they ingested either 0.2 or 1% curcumin diets. Also, the dietary consumption of **1** resulted in a decreased expression of ras and fos proto-oncogenes in the tumorous skin.

Shukla et al. (2002) reported the antimutagenic potential of **1** using an *in vivo* chromosomal aberration assay in Wistar rats. In cyclophosphamide (mutagen)-treated animals, a significant induction of chromosomal aberration was recorded with decrease in mitotic index. However, in curcumin-supplemented animals, no significant induction in chromosomal damage or change in mitotic index was observed. In different curcumin-supplemented groups, a dose-dependent significant decrease in cyclophosphamide-induced clastogenicity was recorded. The incidence of aberrant cells was found to be reduced by both the doses of **1** when compared to cyclophosphamide-treated group. The anticytotoxic potential of **1** towards cyclophosphamide was also evident as the status

of mitotic index was found to show increment. The study revealed the antigenotoxic potential of
1 against cyclophosphamide induced chromosomal mutations.

Curcumin (**1**), at low concentration, induced differentiation in embryonal carcinoma cell line
PCC4 (Batth et al., 2001). In response to **1**, PCC4 cells ceased to proliferate and showed cell cycle
arrest after 4 h of treatment, followed by their differentiation characterized by increase of
nuclear/cytoplasmic ratio. Bhaumik et al. (2000) demonstrated the differential activation status in
host macrophages and NK cells induced by **1** during the spontaneous regression of subcutaneously
transplanted AK-5 tumors. Closer scrutiny of the cytokine profile and NO production by immune
cells showed an initial downregulation of Th1 cytokine response and WO production by macrophages, and their upregulation in NK cells, which picked up upon prolonged treatment with **1**.

Treatment of highly metastatic murine melanoma cells B16F10 with **1** (15 µM) for 15 days
inhibited matrixmetalloproteinase-2 (MMP-2) activity (Banerji et al., 2004). Expression of membrane type-1 matrixmetalloproteinase (MT1-MMP) and focal adhesion kinase (FAK), an important
component of the intracellular signaling pathway, were also reduced to almost background levels.
MMP-2, MT1–MMP, and FAK did not return to control levels even after 28 days of drug withdrawal.

Ji et al. (2004) evaluated the cytotoxic effect of *ar*-turmerone (**4**) on the K562, L1210, U937,
and RBL-2H3 cell lines using the MTT assay. *ar*-Turmerone (**4**) exhibited potent cytotoxicity (IC_{50}
= 20 to 50 µg/ml) on these cancer cell lines. Pillai et al. (2004) reported the concentration-dependant
inhibitory effect of **1** against human lung cancer cell lines-A549 and H1299.

The role of natural food products in prevention of prostate cancer has been highlighted in recent
epidemiological studies (Deeb et al., 2003). The androgen-sensitive human prostate cancer cell line
LNCaP is only slightly susceptible to TNF-related apoptosis-inducing ligand (TRAIL), a member
of the TNF family of cell death-inducing ligands. Deeb et al. (2003) investigated whether **1** and
TRAIL cooperatively interact to promote death of LNCaP cells. At low concentrations (10 µM of
1 and 20 ng/ml TRAIL), neither of the two agents alone produced significant cytotoxicity in LNCaP
cells, as measured by the 3-(4,5-dimethylthiazol-2-yl)-5-(3-carboxymethoxy-phenyl)-2-(4-sulfo-
nyl)-2H-tetrazolium dye reduction assay. On the other hand, cell death was markedly enhanced
(twofold to threefold) if tumor cells were treated with **1** and TRAIL together.

Shukla and Arora (2003) studied the effect of **1** on the development of altered hepatic foci
(AHF), by using a medium term liver bioassay. A significant protection on diethylnitrosamine
(DEN)-initiated and 2-acetylaminofluorene (AAF)-promoted AHF by **1** was observed. The administration of **1** restored the normal levels of the enzymes GST and γ-glutamyl transferase in rat liver
following DEN–AAF exposure. Similarly, a significant protection was provided by **1** in the enzyme-
deficient foci for the adenosine triphosphatase-, alkaline phosphatase-, and glucose-6-phosphatase-
treated groups in comparison to the DEN–AAF-treated group. These results showed that **1** could
effectively suppress the DEN-induced development of AHF in rat liver.

Cheng et al. (2001) conducted the Phase I trial to evaluate pharmacokinetic behavior of **1** in
relation to its chemopreventive properties in patients with one of the following five high-risk
conditions: (i) recently resected urinary bladder cancer; (ii) arsenic Bowen's disease of the skin;
(iii) uterine cervical intraepithelial neoplasm (CIN); (iv) oral leucoplakia; and (v) intestinal
metaplasia of the stomach. Lal et al. (2000) described the clinical efficacy of **1** in the treatment
of patients suffering from idiopathic inflammatory orbital pseudotumors. It was suggested that **1**
could be used as a safe and effective drug in the treatment of idiopathic inflammatory orbital
pseudotumors.

Hanif et al. (1997) studied the effect of **1** in proliferation and apoptosis in the HT-29 and HCT-
15 human colon cancer cell lines. Curcumin (**1**) dose-dependently reduced the proliferation rate of
both cell lines, causing a 96% decrease by 48 h. The antiproliferative effect was preceded by
accumulation of the cells in the G2/M phase of cell cycle. The effects of dietary curcumin and
ascorbyl palmitate on azoxymethanol-induced hyperproliferation of colonic epithelial cells and the
incidence of focal areas of dysplasia (FADs) were evaluated in female CF-1 mice fed an AIN 76A
diet (Huang et al., 1992b). Subcutaneous injections of AOM (10 mg/kg body wt. once weekly for

6 weeks) caused hyperplasia and the formation of FADs in the colon. Administration of 2% curcumin in the diet inhibited azoxymethanol-induced formation of FADs while administration of 2% ascorbyl palmitate in the diet did not demonstrate inhibition.

9.4.2.4.2　Chemosensitization

Multidrug resistance is often associated with decreased intracellular drug accumulation in patients' tumor cells resulting from enhanced drug efflux. It is related to the overexpression of a membrane protein, P-glycoprotein (Pgp-170), thereby reducing drug cytotoxicity. Therefore, the modulation of multidrug resistance-1 gene expression is an attractive target for new chemosensitizing agents. Limtrakul et al. (2004) found that *bis*-demethoxycurcumin (**3**) was the most active of the curcuminoids present in turmeric for modulation of multidrug resistance-1 gene. Treatment of drug-resistant KB-V1 cells with **1** increased their sensitivity to vinblastine, which was consistent with a decreased multidrug resistance-1 gene product, a P-glycoprotein, on the cell plasma membrane. The radiosensitizing effects of **1** in p53 mutant prostate cancer cell line PC-3 has recently been examined (Chendil et al., 2004). Compared to cells that were irradiated alone (SF2 = 0.635; D-0 = 231 cGy), **1** at 2 and 4 µM concentrations in combination with radiation showed enhancement to radiation-induced clonogenic inhibition (SF2 = 0.224: D-0 = 97 cGy; and SF2 = 0.080: D-0 = 38 cGy) and apoptosis. Curcumin in combination with radiation treatment showed inhibition of TNF-α-mediated NF-κ-B activity resulting in Bcl-2 protein downregulation. Bax protein levels remained constant in these cells after radiation or **1** plus radiation treatments. However, the downregulation of Bcl-2 and no changes in Bax protein levels in **1** plus radiation-treated PC-3 cells, together, altered the Bcl-2:Bax ratio, and this caused the enhanced radiosensitization effect. In addition, significant activation of cytochrome-c and caspase-9 and -3 were observed in **1** plus radiation treatments. It was concluded that **1** was a potent radiosensitizer, and it acted by overcoming the effects of radiation-induced prosurvival gene expression in prostate cancer.

9.4.2.4.3　Mode of Action

A number of studies have been carried out to establish the possible mode of anticancer activity of *Curcumin* and its analogs. Surh (2001, 2002) reviewed the possible molecular mechanisms of chemopreventive actions of **1**. The chemopreventive effect of **1** has been linked to its antioxidative and anti-inflammatory activities. As **1** provides protection against lipid peroxidation and DNA damage induced by Fe-NTA and hydrogen peroxide *in vitro*, it was suggested that **1** might be a suitable candidate for the chemoprevention of Fe-NTA–associated cancer (Iqbal et al., 2003a,b). Previous studies showed that **1** could cause an increase in GST activity in rodent liver, which might be implicated to its anticancer and anti-inflammatory activities (Piper et al., 1998). It was also suggested that the induction of enzymes involved in the detoxification of the electrophilic products of lipid peroxidation might contribute to the anti-inflammatory and anticancer activities of **1**. It was shown that the anti-inflammatory and anticarcinogenic action of **1** could partly be due to the inhibition of G-protein–mediated phospholipase-D (Yamamoto et al., 1997). In another similar study the lipoxygenase-inhibitory activity of **1** was implicated to its anticancer property (Skrzypczak-Jankun et al., 2000). The anticancer potential of **1** is also believed to be related to its ability to suppress proliferation of a wide variety of tumor cells, downregulate transcription factors NF-κ-B, AP-1, and Egr-1; downregulate the expression of COX-2, LOX, NOS, MMP-9, uPA, TNF, chemokines, cell surface adhesion molecules, and cyclin D1; downregulate growth factor receptors (such as EGFR and HER-2); and inhibit the activity of c-Jun N-terminal kinase, protein tyrosine kinases, and protein serine/threonine kinases (Surh, 2001, 2002). It was found that **1** could suppress tumor initiation, promotion, and metastasis. The antimetastatic activity of xanthorrhizol (**6**) has recently been studied using an *in vivo* mouse lung metastasis model and a tumor mass formation assay (Choi et al., 2005). Xanthorrhizol (**6**) inhibited the formation of tumor nodules in the lung tissue and the intra-abdominal tumor mass formation. The antimetastatic activity of **6** could be highly linked to the metastasis-related multiplex signal pathway including ERK, COX-2, and MMP-

9. Curcumin (**1**) inhibited the 12-*O*-tetradecanoylphorbol-13-acetate (TPA)-induced NF-κ-B activation by preventing the degradation of the inhibitory protein I-κ-B-α and the subsequent translocation of the p65 subunit in cultured human promyelocytic leukemia (HL-60) cells (Han et al., 2002). It also repressed the TPA-induced activation of NF-κ-B through direct interruption of the binding of NF-κ-B to its consensus DNA sequences. Similarly, the TPA-induced DNA binding of the activator protein-1 (AP-1) was inhibited by **1** pretreatment.

The most significant mode of anticancer activity of **1** and its analogs is their ability to induce apoptosis (programmed cell death) in cancer cells (Bhaumik et al., 1999; Khar et al., 1999). Curcumin (**1**) decreased the Ehrlich's ascites carcinoma (EAC) cell number by the induction of apoptosis in the tumor cells as evident from flow cytometric analysis of cell cycle phase distribution of nuclear DNA and oligonucleosomal fragmentation (Pal et al., 2001). Probing further into the molecular signals leading to apoptosis of EAC cells, it was found that **1** caused tumor cell death by the upregulation of the proto-oncoprotein Bax, release of cytochrome-c from the mitochondria, and activation of caspase-3. The status of Bcl-2 remained unchanged in EAC, which would signify that **1** was bypassing the Bcl-2 checkpoint and overriding its protective effect on apoptosis. Syngai et al. (2004) studied the cytotoxic effects of **1** on three human tumor cell lines and rat primary hepatocytes. Curcumin (**1**) induced apoptosis in MCF-7, MDAMB, and HepG2 cells in a dose-dependent and time-dependent manner. It was also noted that **1** had no effect on normal rat hepatocytes, which showed no superoxide generation and therefore no cell death. Similar antitumor activity of **1** was also reported by Odot et al. (2004). It was found that the cytotoxic effect observed in the two culture types could be related to the induction of apoptosis. In the *in vivo* studies, the effectiveness of a prophylactic immune preparation of soluble proteins from B16-R cells, or a treatment with **1** as soon as tumoral appearance, alone or in combination, on the murine melanoma B16-R was assessed. The combination treatment inhibited the growth of B16-R melanoma, whereas each treatment by itself showed little effect. Moreover, animals receiving the combination therapy exhibited an enhancement of their humoral antisoluble B16-R protein immune response and a significant increase in their median survival time (>82.8% vs. 48.6% and 45.7%, respectively, for the immunized group and the curcumin-treated group). The chemopreventive action of **1** was predominantly due to its ability to induce apoptosis and to arrest cell cycle (Shim et al., 2001). It was shown that **1** could induce apoptosis in A-431 cells. The curcumin-induced cell death of A-431 exhibited various apoptotic features, including DNA fragmentation and nuclear condensation. Furthermore, the curcumin-induced apoptosis of A-431 cells involved activation of caspase-3-like cysteine protease. The decreased NO production was also shown in A-431 cells treated with **1**, which probably resulted from the inhibition of the iNOS expression by **1**, as in other cell lines. The cell death of A-431 by **1** was due to the induction of apoptosis, which involved caspase-3 activation. The effect of **1** on the activation of the apoptotic pathway in human acute myelogenous leukemia HL-60 cells and in established stable cell lines expressing Bcl-2 and Bcl-xl was also demonstrated by Anto et al. (2002). Curcumin (**1**) inhibited the growth of HL-60 cells (neo) in a dose- and time-dependent manner, whereas Bcl-2 and Bcl-xl-transfected cells were relatively resistant. It activated caspase-8 and caspase-3 in HL-60 neo cells but not in Bcl-2 and Bcl-xl-transfected cells. Similarly, time-dependent poly(ADP)ribose polymerase (PARP) cleavage by **1** was observed in neo cells but not in Bcl-2 and Bcl-xl-transfected cells. Curcumin treatment induced BID cleavage and mitochondrial cytochrome-c release in neo cells but not in Bcl-2 and Bcl-xl-transfected cells. In neo HL-60 cells, **1** also downregulated the expression of cyclooxygenase-2. As DN-FLICE blocked curcumin-induced apoptosis, caspase-8 might have a critical role. It was concluded that **1** could induce apoptosis through mitochondrial pathway involving caspase-8, BID cleavage, cytochrome-c release, and caspase-3 activation.

Bhaumik et al. (2000) reported the induction of apoptosis in AK-5, rat histiocytic cells by **1** leading to the inhibition of tumor growth *in vivo*. Lin (2004) studied the possible mechanism of chemopreventive action of **1**. Curcumin (**1**) inhibited the activation of NF-κ-B and the expressions of oncogenes including c-jun, c-fos, c-myc, NIK, MAPKs, ERK, ELK, PI3K, Akt, CDKs, and

NOS. It was proposed that **1** might be able to suppress tumor promotion through blocking signal transduction pathways in the target cells. The oxidant tumor promoter TPA activated PKC by reacting with zinc thiolates present within the regulatory domain, while the oxidized form of cancer chemopreventive agents such as **1** could inactivate PKC by oxidizing the vicinal thiols present within the catalytic domain. Curcumin-induced apoptosis was mediated through the impairment of ubiquitin-proteasome pathway. Curcumin was first biotransformed to dihydrocurcumin and tetrahydrocurcumin and these compounds subsequently were converted to monoglucuronide conjugates. These results suggested that curcumin-glucuronide, dihydrocurcumin-glucuronide, tetrahydrocurcumin-glucuronide, and tetrahydrocurcumin were the major metabolites of curcumin in mice, rats, and humans. The induction of apoptosis through generation of reactive oxygen species, downregulation of Bcl-X-L and IAP, and the release of cytochrome-c and inhibition of Akt were reported as the possible molecular mechanisms of curcumin-induced cytotoxicity (Woo et al., 2003).

Paek et al. (1996) reported the apoptosis-inducing effect of *ar*-turmerone (**4**) and β-atlantone in human myeloid leukemia cells (HL-60). The exposure of human myeloid leukemia HL-60 cells to clinically achievable concentrations of **4** or β-atlantone produced internucleosomal DNA fragmentation of approximately 200 base-pair multiples, and the morphological changes characteristic of cells undergoing apoptosis. Later, the effects of *ar*-turmerone (**4**) on DNA of human leukemia cell lines, Molt 413, HL-60, and stomach cancer KATO III cells were studied by Aratanechemuge et al. (2002). *ar*-Turmerone (**4**) generated selective induction of apoptosis in human leukemia Molt 4B and HL-60 cells, but not in human stomach cancer KATO III cells. Morphological changes showing apoptotic bodies were observed in the human HL-60 and Molt 4B cells treated with **4**. The fragmentation of DNA by this compound to oligonucleosomal-sized fragments, which is characteristic of apoptosis, was concentration- and time-dependent in Molt 4B and HL-60 cells, but not in KATO III cells. It was suggested that the suppression of growth of these leukemia cell lines by **4** resulted from the induction of apoptosis. Ji et al. (2004) reported that *ar*-turmerone (**4**) could induce the apoptotic activity in the K562, L1210, U937, and RBL-2H3 cells.

Curcumin (**1**) was shown to be responsible for the inhibition of AK-5 tumor (a rat histiocytoma) growth by inducing apoptosis in AK-5 tumor cells via caspase activation (Bhaumik et al., 1999; Khar et al., 1999). It was suggested that redox signaling and caspase activation could be the possible mechanisms responsible for the induction of curcumin-mediated apoptosis in AK-5 tumor cells. Pillai et al. (2004) studied the cellular and molecular changes induced by **1** leading to the induction of apoptosis in human lung cancer cell lines A549 and H1299. It induced apoptosis in both the lung cancer cell lines. A decrease in expression of p53, Bcl-2, and Bcl-X-L was observed after 12 h exposure of 40 μM of **1**. Bak and Caspase genes remained unchanged up to 60 μM of **1** but showed decrease in expression levels at 80 to 160 μM. Previously, the inhibition of caspase-3 by **1** was reported by Piwocka et al. (2001). It was found that in Jurkat cells, **1** prevented glutathione decrease, thus protecting cells against caspase-3 activation and oligonucleosomal DNA fragmentation. On the other hand, it induced nonclassical apoptosis via a still-unrecognized mechanism, which leads to chromatin degradation and high-molecular-weight DNA fragmentation. Deeb et al. (2003) investigated whether **1** and TRAIL cooperatively interact to promote death of LNCaP cells. At low concentrations (10 μM curcumin and 20 ng/ml TRAIL), neither of the two agents alone produced significant cytotoxicity in LNCaP cells, as measured by the 3-(4,5-dimethylthiazol-2-yl)-5-(3-carboxymethoxy-phenyl)-2-(4-sulfonyl)-2H-tetrazolium dye reduction assay. On the other hand, cell death was markedly enhanced (twofold to threefold) if tumor cells were treated with **1** and TRAIL together. The combined **1** and TRAIL treatment increased the number of hypodiploid cells and induced DNA fragmentation in LNCaP cells. The combined treatment induced cleavage of procaspase-3, procaspase-8, and procaspase-9, truncation of Bid, and release of cytochrome-c from the mitochondria, indicating that both the extrinsic (receptor-mediated) and intrinsic (chemical-induced) pathways of apoptosis were triggered in prostate cancer cells treated with a combination of **1** and TRAIL.

Duvoix et al. (2003) studied the induction of apoptosis by **1**. The expression of GST P1-1 (GSTP1-1) is correlated to carcinogenesis and resistance of cancer cells against chemotherapeutic agents. The inhibitory activity of **1** on the expression of GSTP1-1 mRNA as well as protein, and its correlation with the apoptotic effect of **1** on K562 leukemia cells was evaluated. Curcumin inhibited the TNF-α and phorbol ester-induced binding of AP-1 and NF-κ-B transcription factors to sites located on the GSTP1-1 gene promoter. TNF-α-induced GSTP1-1 promoter activity was also inhibited by **1**. Curcumin induced procaspases 8 and 9 as well as poly ADP ribose polymerase cleavage and thus leading to apoptosis in K562 cells. In a similar study (Chan et al., 2003), it was established that **1** could prevent UV irradiation-induced apoptotic changes, including c-jun N-terminal kinase (JNK) activation, loss of mitochondrial membrane potential (MMP), mitochondrial release of cytochrome C, caspase-3 activation, and cleavage/activation of PAK2 in A431 cells. It was concluded that **1** attenuated UV irradiation-induced ROS formation, and ROS triggered JNK activation, which in turn caused MMP change, cytochrome C release, caspase activation, and subsequent apoptotic biochemical changes.

Collett et al. (2001) investigated the effects of **1** on apoptosis and tumorigenesis in male Apc(min) mice treated with the human dietary carcinogen, 2-amino-1-methyl-6-phenylimidazo[4,5-b]pyridine (PhIP). Intestinal epithelial apoptotic index in response to PhIP treatment was approximately twice as great in the wild-type C57BL/6 APC(+/+) strain than in mice. The study demonstrated that the Apc(min) genotype was associated with resistance to PhIP-induced apoptosis in intestinal epithelium; **1** could attenuate Apc(min) resistance to PhIP-induced apoptosis and inhibit PhIP-induced tumorigenesis in proximal Apc(min) mouse small intestine. Curcumin (**1**) could induce apoptosis of several, but not all, cancer cells. Many cancer cells protect themselves against apoptosis by activating NF-κ-B/Rel, a transcription factor that helps in cell survival (Anto et al., 2000). It was suggested that NF-κ-B plays an anti-apoptotic role in curcumin-induced apoptosis.

In addition to the inhibitory effect on proliferation, **1** was shown to block dexamethasone-induced apoptosis of rat thymocytes (Jaruga et al., 1998). Curcumin (**1**) prevented the glutathione loss occurring in dexamethasone-treated thymocytes, enhancing intracellular glutathione content at 8 h to 192% of that of nontreated cells. A 60% increase in acid-soluble sulfhydryl groups was also observed. In the presence of L-buthionine S,R-sulfoximine (BSO, an inhibitor of glutathione synthesis), intracellular glutathione content of thymocytes treated with dexamethasone and **1** fell to 31%, and that of the acid soluble sulfhydryl groups to 23% of control after 8 h. Curcumin treatment elevated the concentrations of glutathione and nonprotein sulfhydryl groups, thus preventing their decrease in apoptotic thymocytes. Coadministration of L-buthionine S,R-sulfoximine and **1** did not affect the antiapoptotic effect of CUR, suggesting a glutathione-independent mechanism of cell protection.

Hanif et al. (1997) studied the effect of **1** in proliferation and apoptosis in the HT-29 and HCT-15 human colon cancer cell lines. The antiproliferative effect was preceded by accumulation of the cells in the G2/M phase of cell cycle. It was concluded that curcumin could inhibit colon cancer cell proliferation *in vitro* mainly by accumulating cells in the G2/M phase and that this effect was independent of its ability to inhibit prostaglandin synthesis.

Banerji et al. (2004) suggested that the antimetastatic property of **1** might be mediated through downregulation of focal adhesion kinase and reduction of matrixmetalloproteinase-2 activity. *H. pylori* is a group 1 carcinogen and is associated with the development of gastric and colon cancer (Mahady et al., 2002). Curcumin (**1**) inhibited the growth of all strains of *H. pylori in vitro* with a minimum inhibitory concentration range of 50–625 µg/ml. It was concluded that the inhibition of the growth of *H. pylori* cagA+ strains *in vitro* by **1** could be one of the mechanisms by which **1** exerts its chemopreventative effects.

9.4.2.5 Diabetes

The curcuminoids and sesquiterpenoids present in the extracts of *C. longa* were found to lower blood glucose levels in type 2 diabetic KK-A(y) mice (6 weeks old, *n* = 5/group) (Nishiyama et al., 2005). The ethanol extract contained curcuminoids and sesquiterpenoids, the hexane extract yielded sesquiterpenoids, and ethanol extract obtained from hexane extract residue provided curcuminoids. Both curcuminoids and sesquiterpenoids in turmeric exhibited significant hypoglycemic effects via PPAR-γ activation as one of the mechanisms. Curcumin (**1**), when administered to rats having alloxan-induced diabetes, reduced the blood sugar, Hb, and glycosylated hemoglobin levels (Arun and Nalini, 2002). It also reduced the oxidative stress encountered by the diabetic rats.

The increased levels of circulating reactive oxygen species as indirectly inferred by the findings of increased lipid peroxidation and decreased antioxidant status has been found in diabetes. Direct measurements of intracellular generation of reactive oxygen species using fluorescent dyes also demonstrated an association of oxidative stress with diabetes (Balasubramanyam et al., 2003). The antioxidant effect of **1** as a function of changes in cellular reactive oxygen species generation was tested, and it was observed that **1** abolished both phorbol-12 myristate-13 acetate and thapsigargin-induced reactive oxygen species generation in cells from control and diabetic subjects. The pattern of these reactive oxygen species inhibitory effects as a function of dose dependency suggested that **1** could mechanistically interfere with protein kinase C and calcium regulation. Simultaneous measurements of reactive oxygen species and Ca^{2+} influx suggested that a rise in cytosolic Ca^{2+} might be a trigger for increased reactive oxygen species generation. The antioxidant and antiangeogenic actions of **1**, as a mechanism of inhibition of Ca^{2+} entry and protein kinase C activity, have the potential for being further exploited for the development of novel antidiabetic drugs.

9.4.2.6 Anti-inflammatory

Curcuminoids (**1** to **3**) have been shown to possess potent anti-inflammatory properties (Lukita-Atmadja et al., 2002; Ozaki, 1990). These compounds are able to inhibit the lipopolysaccharide (LPS)-induced production of TNF-α, IL-1β, and the activation of NF-κ-B in human monocytic derived cells. Protease-activated receptors (PARs) play a pivotal role in inflammation, and human leukemic mast cells (HMC-1) coexpress PAR2 and PAR4. Baek et al. (2003) investigated the effect of **1** on PAR2- and PAR4-mediated HMC-1 activation. Curcumin (10 and 100 μM) inhibited TNF-α secretion from trypsin on activating peptide-stimulated HMC-1. At the same concentration, it also inhibited TNF-α and tryptase mRNA expression in trypsin-stimulated HMC-1. The trypsin-induced extracellular signal–regulated kinase (ERK) phosphorylation was also inhibited by **1**. A number of different molecules involved in inflammation were inhibited by **1** including phospholipase, lipooxygenase, cyclooxygenase 2, leukotrienes, thromboxane, prostaglandins, nitric oxide, collagenase, elastase, hyaluronidase, monocyte chemoattractant protein-1 (MCP-1), interferon-inducible protein, TNF, and IL-12 (Chainani-Wu, 2003). The anti-inflammatory activity of **1** might be due to the inhibition of a number of different molecules that play a role in inflammation. The topical anti-inflammatory activity of three nonphenolic linear 1,7-diarylheptanoids, isolated from turmeric and four new semi-synthetic derivatives of the naturally occurring compounds, were assessed in the murine model of ethyl phenylpropiolate-induced ear edema (Claeson et al., 1996). The naturally occurring compound 1*E*,3*E*,1,7-diphenylheptadien-5-one (**7**) exerted the most potent anti-inflammatory activity, with an ID_{50} value of similar magnitude to that of the reference drug oxyphenbutazone (67 vs. 46 μg/ear, respectively). None of the semi-synthetic diarylheptanoids was more active than **7**. The degree of unsaturation in positions 1 and 3, and the nature of the oxygenated functional group in position 5 of the C-7-chain could play significant roles in determining the *in vivo* activity. The nonphenolic linear 1,7-diarylheptanoids could lead to the discovery of a novel class of topical anti-inflammatory agents. Three nonphenolic diarylheptanoids, alnustone (**7**), *trans*-1,7-diphenyl-1-hepten-5-ol (**8**), and *trans,trans*-1,7-diphenyl-1,3-heptadien-5-ol (**9**), isolated from *C. xanthorrhiza*, showed anti-inflammatory activity in the assay of carrageenin-induced hind paw edema in rats (Claeson et al., 1993). The anti-inflammatory activity of a number of curcuminoids isolated from *C. domestica* was determined on mouse ears by using a tumor promoter, TPA (12-*O*-tetradecanoylphorbol-13-acetate), as an inducer (Masuda et al., 1993). Chandra and Gupta (1972) reported the anti-inflammatory and antiarthritic activity of volatile oil of *C. longa*. The cytokine macrophage migration inhibitory factor has recently emerged as a crucial factor in the pathogenesis of rheumatoid arthritis (Molnar and Garai, 2005). Curcumin (1) and caffeic acid were found to be the most potent inhibitors, exhibiting IC_{50} values in the submicromolar range in the ketonase assay.

To determine the efficacy of curcuminoids in inhibiting the hepatic microvascular inflammatory responses elicited by LPS, Balb/c mice were gavaged intragastrically with curcuminoids [40 mg/kg body weight (bw) or 80 mg/kg bw] 1 h before intravenous injection of LPS (*E. coli*, 0111:B4, 100 mug/kg bw) (Lukita-Atmadja et al., 2002; Ozaki, 1990). The liver was examined 2 h after LPS injection using *in vivo* microscopic methods. LPS-treated mice showed increased phagocytic activity of centrilobular Kupffer cells. The numbers of leukocytes adhering to the sinusoidal wall and swollen endothelial cells increased in both the periportal and centrilobular regions, concomitant with a reduction in the numbers of sinusoids containing flow. Pretreatment with curcuminoids at the doses of 40 or 80 mg/kg bw to endotoxemic mice reduced the phagocytic activity of Kupffer cells, the numbers of adhering leukocytes, and swollen endothelial cells. As a result, the number of sinusoids containing flow was increased in animals treated with 40 mg/kg curcuminoids and restored to control levels with 80 mg/kg curcuminoids. Neutrophil sequestration was reduced when measured in sections stained with naphthol AS-D chloroacetate esterase technique. These results demonstrated that curcuminoids were effective in suppressing the hepatic microvascular inflammatory response to LPS and might be a natural alternative anti-inflammatory substance.

A recent study reported that *in vivo* treatment of mice with 1 reduced the duration and clinical severity of active immunization and adoptive transfer allergic encephalomyelitis (Natarajan and Bright, 2002). Curcumin (1) inhibited encephalomyelitis in association with a decrease in IL-12, a proinflammatory cytokine, production from macrophage/microglial cells and differentiation of neural Ag-specific Th1 cells. *In vitro* treatment of activated T cells with 1 inhibited IL-12-induced tyrosine phosphorylation of Janus kinase 2, tyrosine kinase 2, and STAT3 and STAT4 transcription factors. The inhibition of Janus kinase-STAT pathway by 1 resulted in a decrease in IL-12-induced T cell proliferation and Th1 differentiation. Curcumin could inhibit encephalomyelitis by blocking IL-12 signaling in T cells and be useful in the treatment of MS and other Th1 cell-mediated inflammatory diseases.

While studying the possible mechanism of action for the anti-inflammatory activity of natural products, Srivastava et al. (1995) reported that the effects of 1 on eicosanoid biosynthesis could be a possible explanation for its anti-inflammatory property. Extracts from several spices, including turmeric, and isolated compounds, e.g., curcumin, could inhibit platelet aggregation and modulate eicosanoid biosynthesis. Curcumin (1) inhibited platelet aggregation induced by arachidonate, adrenaline and collagen. This compound inhibited thromboxane B-2 production from exogenous arachidonate in washed platelets with a concomitant increase in the formation of 12-lipoxygenase products. Curcumin also inhibited the incorporation of [C-14]AA into platelet phospholipids and inhibited the deacylation of AA-labeled phospholipids (liberation of free AA) on stimulation with calcium ionophore A23187. Curcumin was studied in a set of *in vitro* experiments in order to elucidate the mechanism of its anti-inflammatory property (Ammon et al., 1993). It inhibited the 5-lipoxygenase activity in rat peritoneal neutrophils as well as the 12-lipoxygenase and the cyclooxygenase activities in human platelets. In a cell free peroxidation system 1 exerted strong antioxidative activity. It was suggested that its effects on the dioxygenases could be owing to its reducing capacity. Curcumin (1) inhibited several types of phospholipases, most effectively G protein-mediated phospholipase D among those tested. It also inhibited 12-*O*-tetradecanoylphorbol-13-acetate-induced G protein-mediated phospholipase D activation in intact J774.1 cells in a dose-dependent manner (Yamamoto et al., 1997). It was suggested that the anti-inflammatory and anticarcinogenic action of 1 could partly be due to the inhibition of G protein-mediated phospholipase D.

9.4.2.7 Antiasthmatic Effect

Curcumin (1) increased mucin release by directly acting on airway mucin-secreting cells (Lee et al., 2003). Antiasthmatic property of 1 was evaluated in a guinea pig model of airway hyperresponsiveness (Ram et al., 2003). Guinea pigs were sensitized with ovalbumin to develop certain characteristic features of asthma: allergen induced airway constriction and airway hyperreactivity to histamine. Guinea pigs were then treated with **1** during sensitization (to examine its preventive effect) or after developing impaired airways features (to examine its therapeutic effect). Status of airway constriction and airway hyperreactivity were determined by measuring specific airway conductance (SGaw) using a noninvasive technique, constant-volume body plethysmography. Treatment with **1** (20 mg/kg) inhibited ovalbumin-induced airway constriction and airway hyperreactivity. The results established the effectiveness of **1** in improving the impaired airways features in the ovalbumin-sensitized guinea pigs.

9.4.2.8 Hypercholesterolemia and Cardiovascular Disorders

Curcuminoids (**1** to **3**) displayed antioxidative, anticarcinogenic, and hypocholesterolemic activities (Asai and Miyazawa, 2001). Curcumin (**1**) is believed to be able to stimulate bile flow and to reduce hypercholesterolemia. The ability of **1** to reduce cyclosporine-induced cholestasis and hypercholesterolemia was investigated in a subchronic bile fistula model (Deters et al., 2003). Male Wistar rats were daily treated with **1** (100 mg/kg p.o.), cyclosporine (10 mg/kg i.p.), and a combination of **1** with cyclosporine. After 2 weeks a bile fistula was installed into the rats to measure bile flow and biliary excretion of bile salts, cholesterol, bilirubin, cyclosporine, and its main metabolites. Blood was taken to determine the concentration of these parameters in serum or blood. Cyclosporine reduced bile flow (−14%) and biliary excretion of bile salts (−10%) and cholesterol (−61%), while increasing serum concentrations of cholesterol and triglycerides by 32 and 82%, respectively. Sole administration of **1** slightly decreased bile flow (−7%) and biliary bile salt excretion (−12%) but showed no effect on biliary excretion of cholesterol and serum lipid concentration. Coadministration of **1** with cyclosporine enhanced the cyclosporine-induced cholestasis, but the cyclosporine-induced hyperlipidemia was not affected. Babu and Srinivasan (1997) demonstrated the blood cholesterol lowering effect of **1** in streptozotocin-induced diabetic rats. In order to understand the mechanism of hypocholesterolemic action of dietary curcumin, activities of hepatic cholesterol-7α-hydroxylase and 3-hydroxy-3-methylglutaryl-CoA (HMG CoA) reductase were measured. Hepatic cholesterol-7α-hydroxylase activity was markedly higher in curcumin-fed diabetic animals suggesting a higher rate of cholesterol catabolism.

Kim et al. (2002) studied the antiangiogenic activity of demethoxycurcumin (**2**) and its effect on genetic reprogramming in cultured human umbilical vein endothelial cells using cDNA microarray analysis. Curcuminoid **2** inhibited the expression of MMP-9, yet showed no direct effect on its activity. These data showed that gene expressional change of MMP-9 was a major mediator for angiogenesis inhibition by **2**. Asai and Miyazawa (2001) studied the effect of curcuminoids on lipid metabolism in rats. It was observed that the liver triacylglycerol and cholesterol concentrations were lower in curcuminoid-treated rats than in control rats. Lipid-lowering potency of dietary curcuminoids *in vivo* could probably be due to alterations in fatty acid metabolism.

A daily intake of turmeric equivalent to 20 mg of **1** for 60 d decreased the high levels of peroxidation of both the HDL and the LDL, *in vivo*, in 30 healthy volunteers ranging in age from 40 to 90 yr (Bosca et al., 1997). Because of the high antioxidant activity of **1** and lack of any considerable toxicity, **1** might be a useful complement to standard hypolipidemic drugs in the prevention and treatment of atherosclerosis. The effects of **1** on the proliferation of blood mononuclear cells and vascular smooth muscle cells were studied (Huang et al., 1992). In human peripheral blood mononuclear cells, **1** dose-dependently inhibited the responses to phytohemagglutinin and mixed lymphocyte reaction at the dose ranges of 10^{-6} to 3×10^{-5} and 3×10^{-6} to 3×10^{-5} M, respectively. Curcumin dose dependently inhibited the proliferation of rabbit vascular smooth muscle cells stimulated by fetal calf serum. It had a greater inhibitory effect on platelet-derived growth factor-stimulated proliferation than on serum-stimulated proliferation. It was suggested that **1** might be useful as a new template for the development of better remedies for the prevention of the pathological changes of atherosclerosis and restenosis. The fraction containing α-curcumene, prepared from the hexane-soluble fraction by silica gel column chromatography, was found to suppress the synthesis of fatty acids from [C-14]acetate in primary cultured rat hepatocytes (Yasni et al., 1994). α-Curcumene was found responsible for exerting triglyceride-lowering activity in *C. xanthorrhiza*.

9.4.2.9 Antioxidant/Radical Scavenging Property

A number of curcuminoids were found to be the radical scavenging principles present in *C. longa* (Kim et al., 2003; Iqbal et al., 2003a,b; Miquel et al., 2002; Song et al., 2001; Nakatani, 2000; Potterat, 1997; Osawa et al., 1995; Ruby et al., 1995; Masuda et al., 1992; Toda et al., 1985). The IC_{50} values of these curcuminoid compounds were in the range of 4 to 30 µg/ml. The structure–activity relationships of diarylheptanoids on peroxynitrite were also reported. Curcuminoids (**1** to **3**) have been shown to be free radical scavengers that suppress the production of superoxide by macrophages (Lukita-Atmadja et al., 2002; Masuda et al., 1992). Srinivas et al. (1992) reported the isolation of a water-soluble antioxidant peptide, turmerin, from turmeric.

The antioxidant properties of **1** were investigated using EPR spectroscopic techniques. Curcumin (**1**) inhibited the *O*-1(2)-dependent 2,2,6,6-tetramethylpiperidine *N*-oxyl (TEMPO) formation in a dose-dependent manner. Curcumin at 2.75 µM caused 50% inhibition of TEMP-*O*-1(2) adduct formation. It was concluded that **1** could only effectively quench singlet oxygen at low concentration in aqueous systems (Das and Das, 2002). Ramsewak et al. (2000) demonstrated the antioxidant potential of **1** and its analogs. The inhibition of liposome peroxidation by curcuminoids **1** to **3** at 100 µg/ml were 58, 40, and 22%, respectively. Curcuminoids protected normal human keratinocytes from hypoxanthine/xanthine oxidase injury (Bonte et al., 1997). Since curcuminoids inhibited nitroblue tetrazolium reduction, a decrease in superoxide radical formation leading to lower levels of cytotoxic hydrogen peroxide was proposed as an explanation for this protective effect. Curcuminoids (**1** to **3**) were compared for their antioxidant activities (Ruby et al., 1995). The IC_{50} values of curcuminoids (**1** to **3**) associated with the inhibition of lipid peroxidation were 20, 14, and 11 µg/ml, respectively. The IC_{50} values for superoxide inhibition were 6.25, 4.25, and 1.9 µg/ml and those for hydroxyl radical were 2.3, 1.8, and 1.8 mg/ml, respectively. The ability of **1** to **3** to suppress the superoxide production by macrophages activated with phorbol-12-myristate-13-acetate (PMA) indicated that all three curcuminoids inhibited superoxide production and **3** produced the maximum effect.

Reddy and Lokesh (1992) reported that **1** could inhibit lipid peroxidation in a dose-dependent manner. However, the inhibition of lipid peroxidation by **1** was reversed by the addition of high concentrations of Fe^{2+}. Sugiyama et al. (1996) demonstrated that tetrahydrocurcumin showed a greater inhibitory effect than **1** on the lipid peroxidation of erythrocyte membrane ghosts induced by *tert*-butylhydroperoxide. It was proposed that tetrahydrocurcumin could scavenge radicals such as *tert*-butoxyl radical and peroxyl radical. To clarify the antioxidative mechanism of **1**, in particular the role of the β-diketone moiety, dimethylated tetrahydrocurcumin was incubated with peroxyl radicals generated by thermolysis of 2,2′-azobis(2,4-dimethylvaleronitrile). Four oxidation products were detected, three of which were identified as 3,4-dimethoxybenzoic acid, 3′,4′-dimethoxyacetophenone, and 3-(3,4-dimethoxyphenyl)-propionic acid. The fourth oxidation product was an unstable intermediate, and its detailed structure has not been determined. These results suggested that the β-diketone moiety of tetrahydrocurcumin must exhibit antioxidative activity by cleavage of the Cr bond at the active methylene carbon between two carbonyls in the β-diketone moiety.

9.4.2.10 Wound Healing

Curcumin (**1**) is known to enhance cutaneous wound healing in normal and diabetic rats (Mani et al., 2002; Sidhu et al., 1999; Phan et al., 2001). Sidhu et al. (1999) evaluated the efficacy of **1** by oral and topical applications on impaired wound healing in diabetic rats and genetically diabetic mice using a full thickness cutaneous punch wound model. Wounds of animals treated with **1** showed earlier re-epithelialization, improved neovascularization, increased migration of various cells including dermal myofibroblasts, fibroblasts, and macrophages into the wound bed, and a higher collagen content. Immunohistochemical localization showed an increase in transforming growth factor-β-1 in curcumin-treated wounds compared to controls. It was concluded that **1** could enhance wound repair in diabetic-impaired healing, and could be developed as a pharmacological agent in such clinical settings. The effect of curcumin treatment by topical application in dexamethasone-impaired cutaneous healing in a full thickness punch wound model in rats was also investigated. The healing in terms of histology, morphometry, and collagenization on the fourth and seventh days postwounding was evaluated, and the regulation of TGF-β-1 and its receptors type I (tIrc) and type II (tIIrc) and iNOS were analyzed. Curcumin (**1**) accelerated healing of wounds with or without dexamethasone treatment as revealed by a reduction in the wound width and gap length compared to controls. Curcumin treatment resulted in the enhanced expression of TGF-β-1 and TGF-β tIIrc in both normal and impaired healing wounds as revealed by immunohistochemistry. Macrophages in the wound bed showed an enhanced expression of TGF-β-1 mRNA in curcumin treated wounds as evidenced by *in situ* hybridization.

9.4.2.11 Neuroprotective Activity and Alzheimer's Disease

Rajakrishnan et al. (1999) investigated the neuroprotective activity of **1** using ethanol as a model of brain injury. Oral administration of **1** to rats caused a reversal in lipid peroxidation and brain lipids, and produced enhancement of glutathione, a nonenzymic antioxidant in ethanol-intoxicated rats, revealing that the antioxidative and hypolipidemic action of **1** could be responsible for its protective role against ethanol-induced brain injury. β-Amyloid-induced oxidative stress is a well-established pathway of neuronal cell death in Alzheimer's disease (Kim et al., 2001). Curcuminoids (**1** to **3**) protected PC12 rat pheochromocytoma and normal human umbilical vein endothelial (HUVEC) cells from β-A(1 to 42) insult, as measured by 3-[4,5-dimethylthiazol-2-yl]-2,5-dipheny bromide reduction assay. The ED_{50} values of **1** to **3** towards PC12 and HUVEC cells were 7.1 ± 0.3, 4.7 ± 0.1, and 3.5 ± 0.2 µg/ml and 6.8 ± 0.4, 4.2 ± 0.3, and 3.0 ± 0.3 µg/ml, respectively. These compounds were better antioxidants than α-tocopherol as determined by the DPPH assay. α-Tocopherol did not protect the cells from β-A(1 to 42) insult even at >50 µg/ml concentration. The results suggested that **1** to **3** might be protecting the cells from β-A(1 to 42) insult through antioxidant pathway. In a similar study (Park and Kim, 2002), curcuminoids **1** to **3**, calebin-A, and 1,7-*bis*-(4-hydroxyphenyl)-1-heptene-3,5-dione protected PC12 cells from β-A insult (ED_{50} = 0.5 to 10 µg/ml) better than Congo red (ED_{50} = 37 to 39 µg/ml).

9.4.2.12 Immunomodulatory and Anti-HIV Activity

Antony et al. (1999) reported the possible immunomodulatory effect of curcumin (**1**) in Balb/c mice. The administration of **1** increased the total WBC count (15,290) significantly on the 12th day. The group of animals treated with vehicle alone showed results similar to that of normal animals (10,130 on 12th day). It increased the circulating antibody titer (512) against SRRC and the plaque-forming cells (PFC) in the spleen, and the maximum number of PFC was observed on the sixth day (1130 PFC/10^6 spleen cells) after immunization with SRRC. Bone marrow cellularity (16.9 × 10^6 cells/femur) and α-esterase positive cells (1622/4000 cells) were also enhanced by **1** administration. An increase in macrophage phagocytic activity was also observed.

IL-12 plays a central role in the immune system by driving the immune response towards T helper 1 (Th1) type responses that are characterized by high IFN-γ and low IL-4 production (Kang et al., 1999). Pretreatment with **1** could inhibit IL-12 production by macrophages stimulated with either lipopolysaccharide (LPS) or head-killed listeria monocytogenes (HKL). Curcumin-pretreated macrophages reduced their ability to induce IFN-γ and increased the ability to induce IL-4 in Ag-primed CD4(+) T cells. Addition of recombinant IL-12 to cultures of curcumin-pretreated macrophages and CD4+ T cells restored IFN-γ production in CD4(+) T cells. The *in vivo* administration of **1** resulted in the inhibition of IL-12 production by macrophages stimulated *in vitro* with either LPS or HKL, leading to the inhibition of Th1 cytokine profile (decreased IFN-γ and increased IL-4 production) in CD4(+) T cells. It was suggested that **1** might inhibit Th1 cytokine profile in CD4(+) T cells by suppressing IL-12 production in macrophages, and thus could be used in the Th1-mediated immune diseases.

The transcription of HIV-1 provirus is regulated by both cellular and viral factors (Barthelemy et al., 1998). It is believed that Tat proteins secreted by HIV1-infected cells may play an additional role in the pathogenesis of AIDS because of its ability to be taken up by noninfected cells. Curcumin (**1**) used at 10 to 100 nM inhibited Tat transactivation of HIV1–LTR lacZ by 70 to 80% in HeLa cells. Several other synthetic analogs of **1** also showed similar properties. Curcumin (**1**), dicaffeoylquinic, and dicaffeoyltartaric acids, L-chicoric acid, and a number of fungal metabolites (equisetin, phomasetin, oteromycin, and integric acid) were proposed as HIV-1 integrase inhibitors (De Clercq, 2000). The anti-HIV effect of **1** was determined on purified HIV-1 integrase (Majumder et al., 1995), and the IC_{50} was 40 μM. Inhibition of an integrase deletion mutant containing only amino acids 50 to 212 suggested that **1** could interact with the integrase catalytic core. Energy minimization studies suggested that the anti-integrase activity of **1** could be due to an intramolecular stacking of two phenyl rings that bring the hydroxyl groups into close proximity. The HIV-1 integrase inhibition activity of **1** may be responsible for its antiviral activity.

9.4.2.13 Eye Diseases

The oral administration of **1** to patients suffering from chronic anterior uveitis at a dose of 375 mg three times a day for 12 weeks showed remarkable improvement of this condition (Lal et al., 1999). The efficacy of **1** and recurrences following treatment were comparable to corticosteroid therapy, which is presently the only available standard treatment for this disease. The lack of side effects with **1** is its greatest advantage compared with corticosteroids. A study of 32 people with uveitis (inflammation of the uvea) suggested that **1** might prove to be as effective as corticosteroids, the type of medication generally prescribed for this eye disorder.

9.4.2.14 Mosquitocidal Activity

ar-Turmerone (**4**) was found to possess mosquitocidal activity with an LD_{100} of 50 μg/ml on *Aedes aegyptii* larvae (Roth et al., 1998). Another compound, labda-8(17),12-diene-15,16 dial, isolated from a hexane extract of turmeric, also displayed 100% mosquitocidal activity on *A. aegyptii* larvae at 10 μg/ml. Volatile oils extracted by steam distillation from turmeric were evaluated in mosquito cages and in a large room for their repellency effects against three mosquito vectors, *A. aegyptii*, *Anopheles dirus*, and *Culex quinquefasciatus* (Tawatsin et al., 2001). The oils from turmeric, with the addition of 5% vanillin, repelled the three species under cage conditions for up to 8 h. The results of large room evaluations confirmed the responses for each repellent treatment obtained under cage conditions. This study demonstrated the potential of volatile oils extracted from turmeric as topical repellents against both day- and night-biting mosquitoes. Saju et al. (1998) also reported the insect repellent property of the essential oils extracted from *C. longa*.

9.4.2.15 Nematocidal Activity

Various laboratory studies suggested that curcuminoids could reduce the destructive activity of parasites or roundworms in the intestine. Nematocidal activity of turmeric and its constituents was reported by Kiuchi et al. (1993). Curcumin (**1**) inhibited chloroquine-resistant *Plasmodium falciparum* growth in culture in a dose-dependent manner with an IC_{50} of ~5 μM (Reddy et al., 2005).

Additionally, oral administration of **1** to mice infected with malaria parasite (*Plasmodium berghei*) reduced blood parasitemia by 80 to 90% and enhanced their survival.

9.4.2.16 Others

Curcumin (**1**) and its derivatives inhibited farnesyl protein transferase with an IC_{50} of 29 to 50 μM (Kang et al., 2004). Three curcuminoids (**1** to **3**) inhibited TNF-α–induced expression of ICAM-1, VCAM-1, and *E*-selectin on human umbilical vein endothelial cells (Gupta and Ghosh, 1999). Among them, **1** was the most active one. It was suggested that as **1** could block the cytokine-induced transcript levels for the leukocyte adhesion molecules, it might be interfering at an early stage of signaling event induced by TNF-α.

Curcumin (**1**), by virtue of its potent antioxidant property, inhibited hydrogen peroxide (H_2O_2)-induced cell damage in NG108-15 cells (Mahakunakorn et al., 2003). When added simultaneously with 500 μM H_2O_2, curcumin (25 to 100 μM) effectively protected cells from oxidative damage. However, when the cells were pretreated with **1** (25 to 100 μM) for 1.5 h before H_2O_2 exposure, **1** was unable to inhibit H_2O_2-induced cell damage.

The impact of **1** on the expression of galectin-3 in glioblastoma cells under basal conditions and under stress invoked by the cell exposure to alkylating agent *N*-methyl-*N'*-nitro-*N*-nitrosoguani-dine (MNNG) and ultraviolet C (UV-C) light was investigated by Dumic et al. (2002). Galectin-3 level was measured by western-blot technique using M3/38 monoclonal antibody. Curcumin decreased the basal level of galectin-3, while the pretreatment of cells with **1** reduced the inducible effect of UV-C radiation and abolished the inducible effect of alkylating agent.

Rasyid et al. (2002) demonstrated that **1** could produce a positive cholekinetic effect. It was reported that **1** (20 mg) was capable of contracting the gall. Later on, the dosage of **1** capable of producing a 50% contraction of the gall bladder was found to be 40 mg. Curcumin stimulated bile flow in rats, whereas its analog bisdemethoxycurcumin (**3**) inhibited bile flow (Deter et al., 1999). It was observed that the choleretic effect of **3** lasted longer than that of **1**. The administration of **1** and **3** transiently increased bile flow to 100% and to 125% of the starting value, respectively. However, only **3** attenuated cyclosporin-induced reduction of bile acid excretion.

The radioprotective action of **1** against the acute and chronic effects and the mortality induced by exposure to radiation was evaluated using female rats (Inano and Onoda, 2002), and it was concluded that **1** could be used as an effective radioprotective agent to inhibit acute and chronic effects, but not mortality, after irradiation.

Rajakrishnan et al. (2000) studied the potential role of **1** and the sulfur-containing amino acid *N*-acetylcysteine against ethanol-induced changes in the levels of prostanoids. Biochemical assessment of liver damage was carried out by measuring the activities of serum enzymes, i.e., aspartate transaminase and alkaline phosphatase, which were increased in rats fed with ethanol, whereas the elevated levels of these enzymes were decreased after **1** and *N*-acetylcysteine treatment. Administration of **1** and *N*-acetylcysteine also decreased the increased levels of prostaglandins E-1, E-2, F-2 α, and D-2 in liver, kidney, and brain tissues of ethanol-fed rats.

Curcumin (**1**) displayed genotoxic effect on human lymphocytes (Kowali et al., 1999). Despite antioxidative properties of **1**, it might, in certain conditions, participate in generating reactive oxygen species, which are able to damage DNA. Bioassay-directed fractionation of EtOAc extract from the rhizomes of *C. longa* yielded three curcuminoids (**1** to **3**), which displayed topoisomerase I and II enzyme inhibition activity (Roth et al., 1998). Xanthorrhizol (**6**) caused mortality of neonate larvae of *S. littoralis* when incorporated into artificial diet, suggesting inactivation of the compounds in the larval gut (Pandji et al., 1993). Yamazaki et al. (1988a,b) reported the isolation of a number of compounds from *C. xanthorhiza* that could prolong the phenobarbital-induced sleeping time, and reduced body temperature. Curcuminoids **1** to **3** were also found to possess insect growth inhibitory activity against *Schistocerca gregaria* and *Dysdercus koenigii* nymphs (Chowdhury et al., 2000).

A potent antivenom against snakebites was isolated from *C. longa* (Ferreira et al., 1992). The fraction consisting of *ar*-turmerone (**4**) neutralized both the hemorrhagic activity present in *Bothrops jararaca* venom, and the lethal effect of *Crotalus durissus terrificus* venom in mice. Immunological studies demonstrated that this fraction also inhibited the proliferation and the natural killer activity of human lymphocytes.

9.5 TOXICOLOGY

Many herbs and medicinal plants contain various biologically active compounds that may trigger side effects and interact with other herbs, supplements, or medications. However, turmeric and the active ingredients, curcuminoids, are considered safe when taken at the recommended doses. Excessive use of pure curcumin (**1**) may produce stomach upset and, in extreme cases, ulcers (Ammon and Wahl, 1991). Turmeric is contraindicated for those who have been diagnosed with gallstones or obstruction of the bile passages without consultation with a qualified practitioner. While studies in pregnant rats, mice, guinea pigs, and monkeys suggested that the use of turmeric or curcumin is safe for those animals in pregnancy, there have been no studies involving pregnant women subjects reported to date.

On the basis of the pharmacological and toxicological studies on possible interactions of turmeric or **1** with other herbs or medicines carried out to date, it can be advised that turmeric or **1** in medicinal forms should not be used in the following circumstances without prior consultation with a qualified medical practitioner:

1. People who are on blood thinning medications, e.g., warfarin, aspirin, etc.
2. People who are on nonsteroidal anti-inflammatory drugs, e.g., indomethacin, ibuprofen, etc.
3. People who are on the hypotensive drug, reserpine.

Safety evaluation studies have indicated that both turmeric and **1** are well tolerated at a very high dose without any toxic effects (Chattopadhyay et al., 2004). Human clinical trials with **1** indicated that there was no dose-limiting toxicity when administered at doses up to 10 g/d (Aggarwal et al., 2003). A Phase 1 human trial with 25 subjects using up to 8000 mg of **1** per day for 3 months found no toxicity (Chainani-Wu, 2003). Five other human trials using 1125 to 2500 mg of **1** per day have also found it to be safe. Cheng et al. (2001) demonstrated that **1** was nontoxic to humans at up to 8000 mg/day when taken by mouth for 3 months. During a pharmacokinetic study on turmeric extracts in cancer patients, it was demonstrated that this extract could be administered safely to patients at doses of up to 2.2 g daily, equivalent to 180 mg of **1** (Sharma et al., 2001). A hydroalcoholic extract of *C. longa* did not show any toxic effects while decreasing the levels of blood lipid peroxides, oxidized lipoproteins, and fibrinogen (Ramirez-Bosca et al., 2000). During the evaluation of the clinical efficacy of **1** in the treatment of patients suffering from idiopathic inflammatory orbital pseudotumors, no side effect was noted in any patients and it demonstrated the safety of **1** (Lal et al., 2000).

Qureshi et al. (1992) conducted the acute (24 h) and chronic (90 d) oral toxicity studies on the ethanolic extracts of the rhizomes of *Alpinia galanga* and *C. longa* in mice. Acute dosages were 0.5, 1.0, and 3 g/kg body weight while the chronic dosage was 100 mg/kg/d as the extract. All external morphological, hematological, and spermatogenic changes, in addition to body weight and vital organ weights, were recorded. During this investigation no significant mortality as compared to the controls was observed. The *C. longa*–treated animals gained no significant weight after chronic treatment. *C. longa*–treatment induced significant changes in heart and lung weights upon chronic treatment. Hematological studies revealed a significant fall in the WBC and RBC levels of the *C. longa*–treated animals as compared to the controls. The gain in weights of sexual organs and increased sperm motility and sperm counts was observed in both groups of extract-treated male mice; however, these changes were highly significant in the *A. galanga*-treated group. Both extracts failed to show any spermatotoxic effects.

9.6 CONCLUSION

Turmeric, curcumin (**1**), and its analogs have the potential for being developed as modern medicine for the treatment of various diseases (Chattopadhyay et al., 2004). From the studies carried out to date, the efficacy of turmeric, curcumin, and its analogs in the treatment of a number of diseases, particularly, various forms of cancers, cardiovascular diseases, and microbial infections has been established. The absence of any significant toxicity associated with turmeric or its components has made it superior to many other contemporary medications.

REFERENCES

Adegoke, G.O., Kumar, M.V., Krishna, A.G.G., Varadaraj, M.C., Sambaiah, K., and Lokesh, B.R. (1998) Antioxidants and lipid oxidation in foods - A critical appraisal. *J. Food Sci. Technol. Mysore,* 35, 283-298.

Aggarwal, B.B., Kumar, A. and Bharti, A.C. (2003) Anticancer potential of curcumin: Preclinical and clinical studies. *Anticancer Research,* 23, 363-398.

Allievi, L. and Gualandris, R. (1984) Study of the antimicrobial action of a turmeric (*Curcuma longa*) extract. *Industrie Alimentari,* 23, 867-870.

Ammon, H.P.T., Safayhi, H., Mack, T. and Sabieraj, J. (1993) Mechanism of antiinflammatory actions of curcumine and boswellic acids. *J. Ethnopharmacol.,* 38, 113-119.

Ammon, H.P.T. and Wahl, M.A. (1991) Pharmacology of *Curcuma longa. Planta Medica,* 57, 1-7.

Anto, R.J., Mukhopadhyay, A., Denning, K. and Aggarwal, B.B. (2002) Curcumin (diferuloylmethane) induces apoptosis through activation of caspase-8, BID cleavage and cytochrome c release: its suppression by ectopic expression of Bcl-2 and Bcl-xl. *Carcinogenesis,* 23, 143-150.

Anto, R.J., Maliekal, T.T. and Karunagaran, D. (2000) L-929 cells harboring ectopically expressed Re1A resist curcumin-induced apoptosis. *J. Biol. Chem.,* 275, 15601-15604.

Anto, R.J., George, J., Babu, K.V., Rajasekharan, K.N. and Kuttan, R. (1996) Antimutagenic and anticarcinogenic activity of natural and synthetic curcuminoids. *Mutation Research-Genetic Toxicology,* 370, 127-131.

Antony, S., Kuttan, R. and Kuttan, G. (1999) Immunomodulatory activity of curcumin. *Immunological Investigations,* 28, 291-303.

Apisariyakul, A., Vanittanakom, N. and Buddhasukh, D. (1995) Antifungal activity of turmeric oil extracted from *Curcuma longa* (Zingiberaceae). *J. Ethnopharmacol.,* 49, 163-169.

Araujo, C.A.C. and Leon, L.L. (2001) Biological activities of *Curcuma longa* L. *Memorias Do Instituto Oswaldo Cruz,* 96, 723-728.

Aratanechemuge, Y., Komiya, T., Moteki, H., Katsuzaki, H., Imai, K. and Hibasami, H. (2002) Selective induction of apoptosis by ar-turmerone isolated from turmeric (*Curcuma longa* L) in two human leukemia cell lines, but not in human stomach cancer cell line. *Int. J. Mol. Med.,* 9, 481-484.

Arora, R.B., Kapoor, V., Basu, N. and Jain, A.P. (1971) Anti-inflammatory studies on *Curcuma longa* (turmeric). *Indian J. Med. Res.,* 59, 1289-1295.

Arun, N. and Nalini, N. (2002) Efficacy of turmeric on blood sugar and polyol pathway in diabetic albino rats. *Plant Foods for Human Nutrition,* 57, 41-52.

Asai, A. and Miyazawa, T. (2001) Dietary curcuminoids prevent high-fat diet-induced lipid accumulation in rat liver and epididymal adipose tissue. *J. Nutrition,* 131, 2932-2935.

Asai, A., Nakagawa, K. and Miyazawa, T. (1999) Antioxidative effects of turmeric, rosemary and capsicum extracts on membrane phospholipid peroxidation and liver lipid metabolism in mice. *Bioscience Biotechnology and Biochemistry,* 63, 2118-2122.

Ashok, P. and Meenakshi, B. (2004) Contraceptive effect of *Curcuma longa* (L.) in male albino rat. *Asian J. Andrology,* 6, 71-74.

Atkinson, R.J. and Hunter, J.O. (2003) A double blind, placebo controlled randomised trial of *Curcuma* extract in the treatment of steroid dependent inflammatory bowel disease. *Gastroenterology,* 124, A205-A205.

Azuine, M.A., Kayal, J.J. and Bhide, S.V. (1992) Protective role of aqueous turmeric extract against mutage-
    nicity of direct-acting carcinogens as well as benzo[a]pyrene-induced genotoxicity and carcinogenic-
    ity. *J. Cancer Res. Clin. Oncol.,* 118, 447-452.

Baatout, S., Derradji, H., Jacquet, P., Ooms, D., Michaux, A. and Mergeay, M. (2004) Effect of *Curcuma* on
    radiation-induced apoptosis in human cancer cells. *Int. J. Oncol.,* 24, 321-329.

Babu, P.S. and Prince, P.S.M. (2004) Antihyperglycaemic and antioxidant effect of hyponidd, an ayurvedic
    herbomineral formulation in streptozotocin-induced diabetic rats. *J. Pharm. Pharmacol.,* 56,
    1435-1442.

Babu, P.S. and Srinivasan, K. (1998) Amelioration of renal lesions associated with diabetes by dietary curcumin
    in streptozotocin diabetic rats. *Mol. Cell. Biochem.,* 181, 87-96.

Babu, P.S. and Srinivasan, K. (1997) Hypolipidemic action of curcumin, the active principle of turmeric
    (*Curcuma longa*) in streptozotocin induced diabetic rats. *Mol. Cell. Biochem.,* 166, 160-175.

Baek, O.S., Kang, O.H., Choi, Y.A., Choi, S.C., Kim, T.H., Nah, Y.H., Kwon, D.Y., Kim, Y.K., Kim, Y.H.,
    Bae, K.H., Lim, J.P. and Lee, Y.M. (2003) Curcurmin inhibits protease-activated receptor-2
    and-4-mediated mast cell activation. *Clinica Chimica Acta,* 338, 135-141.

Balasubramanyam, M., Koteswari, A.A., Kumar, R.S., Monickaraj, S.F., Maheswari, J.U. and Mohan, V. (2003)
    Curcumin-induced inhibition of cellular reactive oxygen species generation: Novel therapeutic impli-
    cations. *J. Biosciences,* 28, 715-721.

Banerjee, A. and Nigam, S.S. (1978) Antimicrobial efficacy of the essential oil of *Curcuma longa. Indian J.
    Med. Res.,* 68, 864-866.

Banerji, A., Chakrabarti, J., Mitra, A. and Chatterjee, A. (2004) Effect of curcurnin on gelatinase A (MMP-2)
    activity in B16F10 melanoma cells. *Cancer Letters,* 211, 235-242.

Barthelemy, S., Vergnes, L., Moynier, M., Guyot, D., Labidalle, S. and Bahraoui, E. (1998) Curcumin and
    curcumin derivatives inhibit Tat-mediated transactivation of type 1 human immunodeficiency virus
    long terminal repeat. *Research in Virology,* 149, 43-52.

Batth, B.K., Tripathi, R. and Srinivas, U.K. (2001) Curcumin-induced differentiation of mouse embryonal
    carcinoma PCC4 cells. *Differentiation,* 68, 133-140.

Bhaumik, S., Jyothi, M.D. and Khar, A. (2000) Differential modulation of nitric oxide production by curcumin
    in host macrophages and NK cells. *FEBS Letters,* 483, 78-82.

Bhaumik, S., Anjum, R., Rangaraj, N., Pardhasaradhi, B.V.V. and Khar, A. (1999) Curcumin mediated apoptosis
    in AK-5 tumor cells involves the production of reactive oxygen intermediates. *FEBS Letters,* 456, 311-314.

Bhavani Shankar, T.N. and Sreenivasa Murthy, V. (1979) Effect of turmeric (*Curcuma longa*) fractions on the
    growth of some intestinal & pathogenic bacteria *in vitro. Indian J. Exp. Biol.,* 17, 1362-1366.

Biswas, N.R., Gupta, S.K., Das, G.K., Kumar, N., Mongre, P.K., Haldar, D. and Beri, S. (2001) Evaluation
    of ophthacare (R) eye drops - A herbal formulation in the management of various ophthalmic disorders.
    Phytotherapy Res., 15, 618-620.

Bonte, F., NoelHudson, M.S., Wepierre, J. and Meybeck, A. (1997) Protective effect of curcuminoids on
    epidermal skin cells under free oxygen radical stress. *Planta Medica,* 63, 265-266.

Bosca, A.R., Soler, A., Carrion-Gutierrez, M.A., Mira, D.P., Zapata, J.P., Diaz-Alperi, J., Bernd, A., Almagro,
    E.Q. and Miquel, J. (2000) An hydroalcoholic extract of *Curcuma longa* lowers the abnormally high
    values of human-plasma fibrinogen. *Mechanisms of Aging and Development,* 114, 207-210.

Bosca, A.R., Gutierrez, M.A.C., Soler, A., Puerta, C., Diez, A., Quintanilla, E., Bernd, A. and Miquel, J.
    (1997) Effects of the antioxidant turmeric on lipoprotein peroxides: Implications for the prevention
    of atherosclerosis. *Age,* 20, 165-168.

Bosca, A.R., Soler, A., Gutierrez, M.A.C., Alvarez, J.L. and Almagro, E.Q. (1995) Antioxidant Curcuma
    extracts decrease the blood lipid peroxide levels of human subjects. *Age,* 18, 167-169.

Braga, M.E.M., Leal, P.F., Carvalho, J.E. and Meireles, M.A.A. (2003) Comparison of yield, composition,
    and antioxidant activity of turmeric (*Curcuma longa* L.) extracts obtained using various techniques.
    J. Agric. Food Chem., 51, 6604-6611.

Bundy, R., Walker, A.F., Middleton, R.W. and Booth, J. (2004) Turmeric extract may improve irritable bowel
    syndrome symptomology in otherwise healthy adults: A pilot study. *J. Alternative and Complementary
    Medicine,* 10, 1015-1018.

Calixto, J.B., Otuki, M.F. and Santos, A.R.S. (2003) Anti-inflammatory compounds of plant origin. Part I.
    Action on arachidonic acid pathway, nitric oxide and nuclear factor kappa B (NF-kappa B). *Planta
    Medica,* 69, 973-983.

Chan, W.H., Wu, C.C. and Yu, J.S. (2003) Curcumin inhibits UV irradiation-induced oxidative stress and apoptotic biochemical changes in human epidermoid carcinoma A431 cells. *J. Cellular Biochem.*, 90, 327-338.

Chandra, D. and Gupta, S.S. (1972) Anti-inflammatory and anti-arthritic activity of volatile oil of *Curcuma longa* (Haldi). *Indian J. Med. Res.*, 60, 138-142.

Charles, V. and Charles, S.X. (1992) The use and efficacy of *Aazadirachta indica* ADR (neem) and *Curcuma longa* (turmeric) in scabies - a pilot-study. *Tropical and Geographical Medicine*, 44, 178-181.

Chattopadhyay, I., Biswas, K., Bandyopadhyay, U. and Banerjee, R.K. (2004) Turmeric and curcumin: Biological actions and medicinal applications. *Current Science*, 87, 44-53.

Chauhan, U.K., Soni, P., Shrivastava, R., Mathur, K.C. and Khadikar, P.V. (2003) Antimicrobial activities of the rhizome of *Curcuma longa* Linn. *Oxidation Commun*, 26, 266-270.

Chendil, D., Ranga, R.S., Meigooni, D., Sathishkumar, S. and Ahmed, M.M. (2004) Curcumin confers radiosensitizing effect in prostate cancer cell line PC-3. *Oncogene*, 23, 1599-1607.

Cheng, A.L., Hsu, C.H., Lin, J.K., Hsu, M.M., Ho, Y.F., Shen, T.S., Ko, J.Y., Lin, J.T., Lin, B.R., Wu, M.S., Yu, H.S., Jee, S.H., Chen, G.S., Chen, T.M., Chen, C.A., Lai, M.K., Pu, Y.S., Pan, M.H., Wang, Y.J., Tsai, C.C. and Hsieh, C.Y. (2001) Phase I clinical trial of curcumin, a chemopreventive agent, in patients with high-risk or pre-malignant lesions. *Anticancer Res.*, 21, 2895-2900.

Chirangini, P., Sharma, G.J. and Sinha, S.K. (2004) Sulfur free radical reactivity with curcumin as reference for evaluating antioxidant properties of medicinal Zingiberales. *J. Environ. Path. Toxicol. Oncol.*, 23, 227-236.

Choi, M.A., Kim, S.H., Chung, W.Y., Hwang, J.K. and Park, K.K. (2005) Xanthorrhizol, a natural sesquiterpenoid from *Curcuma xanthorrhiza*, has an anti-metastatic potential in experimental mouse lung metastasis model. *Biochemical and Biophysical Research Communications*, 326, 210-217.

Chopra, D. and Simon, D. (2000) The Chopra Centre Handbook, India. p. 112.

Chowdhury, H., Walia, S. and Saxena, V.S. (2000) Isolation, characterization and insect growth inhibitory activity of major turmeric constituents and their derivatives against *Schistocerca gregaria* (Forsk) and *Dysdercus koenigii* (Walk). *Pest Management Science*, 56, 1086-1092.

Chuang, S.E., Cheng, A.L., Lin, J.K. and Kuo, M.L. (2000) Inhibition by curcumin of diethylnitrosamine-induced hepatic hyperplasia, inflammation, cellular gene products and cell-cycle-related proteins in rats. *Food Chem. Toxicol.*, 38, 991-995.

Chun, K.S., Keum, Y.S., Han, S.S., Song, Y.S., Kim, S.H. and Surh, Y.J. (2003) Curcumin inhibits phorbol ester-induced expression of cyclooxygenase-2 in mouse skin through suppression of extracellular signal-regulated kinase activity and NF-kappa B activation. *Carcinogenesis*, 24, 1515-1524.

Claeson, P., Pongprayoon, U., Sematong, T., Tuchinda, P., Reutrakul, V., Soontornsaratune, P. and Taylor, W.C. (1996) Non-phenolic linear diarylheptanoids from *Curcuma xanthorrhiza*: A novel type of topical anti-inflammatory agents: Structure-activity relationship. *Planta Medica*, 62, 236-240.

Claeson, P., Panthong, A., Tuchinda, P., Reutrakul, V., Kanjanapothi, D., Taylor, W.C. and Santisuk, T. (1993) 3 Nonphenolic diarylheptanoids with antiinflammatory activity from *Curcuma xanthorrhiza*. *Planta Medica*, 59, 451-454.

Cohly, H.H.P., Rao, M.R., Kanji, V.K., Manisundram, D., Taylor, A., Wilson, M.T., Angel, M.F. and Das, S.K. (1999) Effect of turmeric (chemical plant extract) on in-vitro nitric oxide synthetase (NOS) levels in tissues harvested from acute and chronic wounds. *Wounds - A compendium of Clinical Research and Practice*, 11, 70-76.

Collett, G.P., Robson, C.N., Mathers, J.C. and Campbell, F.C. (2001) Curcumin modifies Apc(min) apoptosis resistance and inhibits 2-amino 1-methyl-6-phenylimidazo[4,5-b]pyridine (PhIP) induced tumour formation in Apc(min) mice. *Carcinogenesis*, 22, 821-825.

Das, K.C. and Das, C.K. (2002) Curcumin (diferuloylmethane), a singlet oxygen (O-1(2)) quencher. *Biochem. Biophys. Res. Commun.*, 295, 62-66.

De Clercq, E. (2000) Current lead natural products for the chemotherapy of human immunodefiency virus (HIV) infection. *Med. Res. Rev.*, 20, 323-349.

Deeb, D., Xu, Y.X., Jiang, H., Gao, X.H., Janakiraman, N., Chapman, R.A. and Gautam, S.C. (2003) Curcumin (diferuloyl-methane) enhances tumor necrosis factor-related apoptosis-inducing ligand-induced apoptosis in LNCaP prostate cancer cells. *Mol. Cancer Therapeutics*, 2, 95-103.

Deshpande, U.R., Joseph, L.J., Manjure, S.S., Samuel, A.M., Pillai, D. and Bhide, S.V. (1997) Effects of turmeric extract on lipid profile in human subjects. *Med. Sci. Res.*, 25, 695-698.

Deters, M., Klabunde, T., Meyer, H., Resch, K. and Kaever, V. (2003) Effects of curcumin on cyclosporine-induced cholestasis and hypercholesterolemia and on cyclosporine metabolism in the rat. *Planta
Medica,* 69, 337-343.

Deters, M., Siegers, C., Muhl, P. and Hansel, W. (1999) Choleretic effects of curcuminoids on an acute
cyclosporin-induced cholestasis in the rat. *Plant Medica,* 65, 610-613.

Dorai, T., Cao, Y.C., Dorai, B., Buttyan, R. and Katz, A.E. (2001) Therapeutic potential of curcumin in human
prostate cancer. III. Curcumin inhibits proliferation, induces apoptosis, and inhibits angiogenesis of
LNCaP prostate cancer cells *in vivo. Prostate,* 47, 293-303.

Dorai, T., Gehani, N. and Katz, A. (2000) Therapeutic potential of curcumin in human prostate cancer. II.
Curcumin inhibits tyrosine kinase activity of epidermal growth factor receptor and depletes the protein.
Mol. Urol., 4, 1-6.

Dumic, J., Dabelic, S. and Flogel, M. (2002) Curcumin - A potent inhibitor of galectin-3 expression. *Food
Technol. Biotechnol.,* 40, 281-287.

Duvoix, A., Morceau, F., Delhalle, S., Schmitz, M., Schnekenburger, M.L., Galteau, M.M., Dicato, M. and
Diederich, M. (2003) Induction of apoptosis by curcumin: mediation by glutathione S-transferase
P1-1 inhibition. *Biochem. Pharmacol.,* 66, 1475-1483.

Ferreira, L.A.F., Henriques, O.B., Andreoni, A.A.S., Vital, G.R.F., Campos, M.M.C., Habermehl, G.G. and
Demoraes, V.L.G. (1992) Antivenom and biological effects of ar-turmerone isolated from *Curcuma
longa* (Zingiberaceae). *Toxicon.,* 30, 1211-1218.

Foryst-Ludwig, A., Neumann, M., Schneider-Brachert, W. and Naumann, M. (2004) Curcumin blocks
NF-kappa B and the motogenic response in *Helicobacter pylori*-infected epithelial cells. *Biochem.
Biophys. Res. Comm.,* 316, 1065-1072.

Gaedeke, J., Noble, N.A. and Border, W.A. (2004) Curcumin blocks multiple sites of the TGF-beta signaling
cascade in renal cells. *Kidney International,* 66, 112-120.

Ghatak, N. and Basu, N. (1972) Sodium curcuminate as an effective anti-inflammatory agent. *Indian J. Exp.
Bio.,* 10, 235-6.

Grieve, M. (2005) A Modern Herbal. Available online at: http://www.botanical.com/botanical/mgmh/t/
turmer30.html.

Gupta, S.K. (2004) Molecular mechanisms for cardioprotective effects of *Curcuma longa* on myocardial
apoptosis, cardiac function and antioxidant milieu in an ischemia reperfusion model of myocardial
infarction. *J. Mol. Cell Cardiology,* 37, 277-277.

Gupta, B. and Ghosh, B. (1999) *Curcuma longa* inhibits TNF-alpha induced expression of adhesion molecules
on human umbilical vein endothelial cells. *Int. J. Immunopharmacol.,* 21, 745-757.

Halder, N., Joshi, S. and Gupta, S.K. (2003) Lens aldose reductase inhibiting potential of some indigenous
medicinal plants. *J. Ethnopharmacol.,* 86, 113-116.

Han, R. (1994) Highlight on the studies of anticancer drugs derived from plants in China. *Stem Cells,* 12, 53-63.

Han, S.S., Keum, Y.S., Seo, H.J. and Surh, Y.J. (2002) Curcumin suppresses activation of NF-kappa B and
AP-1 induced by phorbol ester in cultured human promyelocytic leukemia cells. *J. Biochem. Mol.
Biol.,* 35, 337-342.

Hanif, R., Qiao, L., Shiff, S.J. and Rigas, B. (1997) Curcumin, a natural plant phenolic food additive, inhibits
cell proliferation and induces cell cycle changes in colon adenocarcinoma cell lines by a prostaglandin-independent pathway. *J. Lab Clin. Med.,* 130, 576-584.

Hong, C.H., Hur, S.K., Oh, O.J., Kim, S.S., Nam, K.A. and Lee, S.K. (2002) Evaluation of natural products
on inhibition of inducible cyclooxygenase (COX-2) and nitric oxide synthase (iNOS) in cultured
mouse macrophage cells. *J. Ethnopharmacol.,* 83, 153-159.

Huang, M.T., Newmark, H.L. and Frenkel, K. (1997) Inhibitory effects of Curcumin on tumorigenesis in mice.
J. Cellular Biochem., Suppl., 27, 26-34.

Huang, H.C., Jan, T.R. and Yeh, S.F. (1992a) Inhibitory effect of curcumin, an antiinflammatory agent, on
vascular smooth-muscle cell proliferation. *Eur. J. Pharmacol.,* 221, 381-384.

Huang, M.T., Deschner, E.E., Newmark, H.L., Wang, Z.Y., Ferraro, T.A. and Conney, A.H. (1992b) Effect of
dietary curcumin and ascorbyl palmitate on azoxymethanol-induced colonic epithelial cell proliferation and focal areas of dysplasia. *Cancer Letters,* 64, 117-121.

Huang, M.T., Lou, Y.R., Ma, W., Newmark, H.L., Reuhl, K.R. and Conney, A.H. (1994) Inhibitory effects of
dietary curcumin on forestomach, duodenal, and colon carcinogenesis in mice. *Cancer Res.,* 54,
5841-5847.

Hwang, J.K., Shim, J.S., Park, H.K., Kim, S.N. and Ahn, H.J. (2002) Xanthorrhizol from *Curcuma xanthorrhiza* as a novel anticariogenic agent against *Streptococcus* mutans. *J. Dental Res.,* 81, 0754.

Hwang, J.K., Shim, J.S., Baek, N.I. and Pyun, Y.R. (2000) Xanthorrhizol: A potential antibacterial agent from *Curcuma xanthorrhiza* against *Streptococcus* mutans. *Planta Medica,* 66, 196-197.

Inano, H. and Onoda, M. (2002) Radioprotective action of curcumin extracted from *Curcuma longa* L.: Inhibitory effect on formation of urinary 8-hydroxy-2 '-deoxyguanosine, tumorigenesis, but not mortality, induced by gamma-ray irradiation. *Int. J. Rad. Oncol. Biol. Phys.,* 53, 735-743.

Iqbal, M., Okazaki, Y. and Okada, S. (2003a) *In vitro* curcumin modulates ferric nitrilotriacetate (Fe-NTA) and hydrogen peroxide (H_2O_2)-induced peroxidation of microsomal membrane lipids and DNA damage. *Teratogenesis, Carcinogenesis and Mutagenesis,* Supp 1, 151-160.

Iqbal, M., Sharma, S.D., Okazaki, Y., Fujisawa, M. and Okada, S. (2003b) Dietary supplementation of curcumin enhances antioxidant and phase II metabolizing enzymes in ddY male mice: Possible role in protection against chemical carcinogenesis and toxicity. *Pharmacol. Toxicol.,* 92 33-38.

Itokawa, H., Takeya, K., Hitotsuyanagi, Y. and Morita, H. (1999) Antitumor compounds isolated from higher plants. *Yakugaku Zasshi-Journal of the Pharmaceutical Society of Japan,* 119, 529-583.

Jain, A. and Basal, E. (2003) Inhibition of Propionibacterium acnes-induced mediators of inflammation by Indian herbs. *Phytomedicine,* 10, 34-38.

Jaruga, E., Bielak-Zmijewska, A., Sikora, E., Skierski, J., Radziszewska, E., Piwocka, K. and Bartosz, G. (1998) Glutathione-independent mechanism of apoptosis inhibition by curcumin in rat thymocytes. *Biochem. Pharmacol.,* 56, 961-965.

Ji, M.J., Choi, J., Lee, J. and Lee, Y. (2004) Induction of apoptosis by ar-turmerone on various cell lines. *Int. J. Mol. Med.,* 14, 253-256.

Kang, H.M., Son, K.H., Yang, D.C., Han, D.C., Kim, J.H., Baek, N.I. and Kwon, B.M. (2004) Inhibitory activity of diarylheptanoids on farnesyl protein transferase. *Nat. Prod. Res.,* 18, 295-299.

Kang, B.Y., Song, Y.J., Kim, K.M., Choe, Y.K., Hwang, S.Y. and Kim, T.S. (1999) Curcumin inhibits Th1 cytokine profile in CD4(+) T cells by suppressing interleukin-12 production in macrophages. *British J. Pharmacol.,* 128, 380-384.

Kawamori, T., Lubet, R., Steele, V.E., Kelloff, G.J., Kaskey, R.B., Rao, C.V. and Reddy, B.S. (1999) Chemopreventive effect of curcumin, a naturally occurring anti-inflammatory agent, during the promotion/progression stages of colon cancer. *Cancer Res.,* 59, 597-601.

Khar, A., Ali, A.M., Pardhasaradhi, B.V.V., Begum, Z. and Anjum, R. (1999) Antitumor activity of curcumin is mediated through the induction of apoptosis in AK-5 tumor cells. *FEBS letters,* 445, 165-168.

Kim, M.K., Choi, G.J. and Lee, H.S. (2003a) Fungicidal property of *Curcuma longa* L. rhizome-derived curcumin against phytopathogenic fungi in a greenhouse. *J. Agric. Food Chem.,* 51, 1578-1581.

Kim, D.S.H.L., Park, S.Y. and Kim, J.Y. (2001) Curcuminoids from *Curcuma longa* L. (Zingiberaceae) that protect PC12 rat pheochromocytoma and normal human umbilical vein endothelial cells from PA(1-42) insult. *Neuroscience Letters,* 303, 57-61.

Kim, J.E., Kim, A.R., Chung, H.Y., Han, S.Y., Kim, B.S. and Choi, J.S. (2003b) *In vitro* peroxynitrite scavenging activity of diarylheptanoids from Curcuma longa. *Phytotherapy Res.,* 17, 481-484.

Kim, J.H., Shim, J.S., Lee, S.K., Kim, K.W., Rha, S.Y., Chung, H.C. and Kwon, H.J. (2002) Microarray-based analysis of anti-angiogenic activity of demethoxycurcumin on human umbilical vein endothelial cells: Crucial involvement of the down-regulation of matrix metalloproteinase. *Japanese J. Cancer Research,* 93, 1378-1385.

Kim, S.H., Hong, K.O., Hwang, J.K. and Park, K.K. (2005) Xanthorrhizol has a potential to attenuate the high dose cisplatin-induced nephrotoxicity in mice. *Food and Chem. Toxicol.,* 43, 117-122.

Kim, S.H., Hong, K.O., Chung, W.Y., Hwang, J.K. and Park, K.K. (2004) Abrogation of cisplatin-induced hepatotoxicity in mice by xanthorrhizol is related to its effect on the regulation of gene transcription. *Toxicol. and Applied Pharmacol.,* 196, 346-355.

Kiso, Y., Suzuki, Y., Watanabe, N., Oshima, Y. and Hikino, H. (1983) Validity of the oriental medicines .53. antihepatotoxic principles of *Curcuma longa* rhizomes. *Planta Medica,* 49, 185-187.

Kiuchi, F., Goto, Y., Sugimoto, N., Akao, N., Kondo, K. and Tsuda, Y. (1993) Nematocidal activity of turmeric: synergistic action of curcuminoids. *Chem. Pharm. Bull.,* 41, 1640-1643.

Koo, B.S., Lee, W.C., Chung, K.H., Ko, J.H. and Kim, C.H. (2004) A water extract of *Curcuma longa* L. (Zingiberaceae) rescues PC12 cell death caused by pyrogallol or hypoxia/reoxygenation and attenuates hydrogen peroxide induced injury in PC12 cells. *Life Sciences,* 75, 2363-2375.

Kowalik, J., Trzeciak, A., Wojewodzka, M. and Blasiak, J. (1999) *In vitro* genotoxic effect of curcumin assessed by alkaline single cell gel/comet assay. *Neoplasma,* 46, 64-65.

Krishnaswamy, K. (1996) Indian functional foods: Role in prevention of cancer. *Nutrition Rev.,* 54, S127-S131.

Kulkarni, R.R., Patki, P.S., Jog, V.P., Gandage, S.G. and Patwardhan, B. (1991) Treatment of osteoarthritis with a herbomineral formulation — a double-blind, placebo-controlled, cross-over study. *J. Ethnopharmacol.,* 33, 91-95.

Kuttan, R., Kuttan, G., Joseph, S., Ajith, T.A., Mohan, M. and Srimal, R.C. (2004) Antimutagenicity of herbal detoxification formula smoke shield against environmental mutagens. *J. Experimental and Clinical Cancer Res.,* 23, 61-68.

Kuttan, R., Sudheeran, P.C. and Joseph, C.D. (1987) Turmeric and curcumin as topical agents in cancer therapy. *Tumori genesis,* 73, 29-31.

Kuttan, R., Bhanumathy, P., Nirmala, K. and George, M.C. (1985) Potential anticancer activity of turmeric (*Curcuma longa*). *Cancer Letters,* 29, 197-202.

Lal, B., Kapoor, A.K., Agrawal, P.K., Asthana, O.P. and Srimal, R.C. (2000) Role of curcumin in idiopathic inflammatory orbital pseudo tumours. *Phytotherapy Research,* 14, 443-447.

Lal, B., Kapoor, A.E., Asthana, O.P., Agrawal, P.K., Prasad, R., Kumar, P. and Srimal, R.C. (1999) Efficacy of curcumin in the management of chronic anterior uveitis. *Phytotherapy Res.,* 13, 318-322.

Leal, P.F., Braga, M.E.M., Sato, D.N., Carvalho, J.E., Marques, M.O.M. and Meireles, M.A.A. (2003) Functional properties of spice extracts obtained via supercritical fluid extraction. *J. Agric. Food Chem.,* 69, 523-526.

Lee, C.J., Lee, J.H., Seok, J.H., Hur, G.M., Park, Y.C., Seol, I.C. and Kim, Y.H. (2003) Effects of baicalein, berberine, curcumin and hesperidin on mucin release from airway goblet cells. *Planta Medica,* 69, 523-526.

Limtrakul, P., Anuchapreeda, S. and Buddhasukh, D. (2004) Modulation of human multidrug-resistance MDR-1 gene by natural curcuminoids. *BMC Cancer,* 4, 13.

Limtrakul, P.N., Anuchapreeda, S., Lipigorngoson, S. and Dunn, F.W. (2001) Inhibition of carcinogen induced c-Ha-ras and c-fos proto-oncogenes expression by dietary curcumin. *BMC Cancer,* 1, 1.

Limtrakul, P., Lipigorngoson, S., Namwong, O., Apisariyakul, A. and Dunn, F.W. (1997) Inhibitory effect of dietary curcumin on skin carcinogenesis in mice. *Cancer Letters,* 116, 197-203.

Lin, J.K. (2004) Suppression of protein kinase C and nuclear oncogene expression as possible action mechanisms of cancer Chemoprevention by curcumin. *Arch. Pharm. Res.,* 27, 683-692.

Lin, J.K., Pan, M.H. and Lin-Shiau, S.Y. (2000) Recent studies on the biofunctions and biotransformations of curcumin. *Biofactors,* 13, 153-158.

Lin, S.C., Teng, C.W., Lin, C.C., Lin, Y.H. and Supriyatna, S. (1996) Protective and therapeutic effect of the Indonesian medicinal herb *Curcuma xanthorrhiza* on beta-D-galactosamine-induced liver damage. *Phytother. Res.,* 10, 131-135.

Lin, S.C., Lin, C.C., Lin, Y.H., Supriyatna, S. and Teng, C.W. (1995) Protective and therapeutic effects of curcuma-xanthorrhiza on hepatotoxin-induced liver-damage. *Am. J. Chinese Med.,* 23, 243-254.

LoTempio, M.M., Steele, H.L., Ramamurthy, B., Chakrabarti, R., Calcaterra, T.C., Srivatsan, E.S., Wang, M.B. and Van Waes, C. (2004) Curcumin suppresses growth of head and neck squamous cell carcinoma. *Otolaryngology - Head and Neck Surgery,* 131, 179.

Lukita-Atmadja, W., Ito, Y., Baker, G.L. and McCuskey, R.S. (2002) Effect of curcuminoids as anti-inflammatory agents on the hepatic microvascular response to endotoxin. *Shock,* 17, 399-403.

Luper, S. (1999) A review of plants used in the treatment of liver disease: part two. *Altern. Med. Rev.,* 4, 178-188; 692.

Lutomski, J., Kedzia, B. and Debska, W. (1974) Effect of an alcohol extract and of active ingredients from *Curcuma longa* on bacteria and fungi. *Planta Medica,* 26, 9-19.

Madan, B., Gade, W.N. and Ghosh, B. (2001) *Curcuma longa* activates NF-kappa B and promotes adhesion of neutrophils to human umbilical vein endothelial cells. *J. Ethnopharmacol.,* 75, 25-32.

Mahady, G.B., Pendland, S.L., Yun, G. and Lu, Z. (2002) Turmeric (*Curcuma longa*) and curcumin inhibit the growth of *Helicobacter pylori,* a group 1 carcinogen. *Anticancer Res.,* 22, 4179-4181.

Mahakunakorn, P., Tohda, M., Murakami, Y., Matsumoto, K., Watanabe, H. and Vajaragupta, O. (2003) Cytoprotective and cytotoxic effects of curcumin: Dual action on H_2O_2-induced oxidative cell damage in NG108-15 cells. *Biol. Pharm. Bull.,* 26, 725-728.

Mani, H., Sidhu, G.S., Kumari, R., Gaddipati, J.P., Seth, P. and Maheshwari, R.K. (2002) Curcumin differentially regulates TGF-beta 1, its receptors and nitric oxide synthase during impaired wound healing. *Biofactors,* 16, 29-43.

Manoharan, S., Ramachandran, C.R., Ramachandran, V. and Nagini, S. (1996) Inhibition of 4-nitroquinoline-1-oxide-induced oral carcinogenesis by plant products. *J. Clinical Biochem. Nutrition,* 21, 141-149.

Masuda, T., Jitoe, A., Isobe, J., Nakatani, N. and Yonemori, S. (1993) Antioxidative and antiinflammatory curcumin-related phenolics from rhizomes of *Curcuma domestica. Phytochemistry,* 32, 1557-1560.

Masuda, T., Isobe, J., Jitoe, A. and Nakatani, N. (1992) Antioxidative curcuminoids from rhizomes of *Curcuma xanthorrhiza. Phytochemistry,* 31, 3645-3647.

Mazumder, A., Raghavan, K., Weinstein, J., Kohn, K.W. and Pommier, Y. (1995) Inhibition of human-immunodeficiency-virus type-1 integrase by curcumin. *Biochem. Pharmacol.,* 49, 1165-1170.

McIntyre, A. (2005) Turmeric — an amazing healer. Available on-line at: http://www.positivehealth.com/permit/Articles/Regular/mcintyre73.htm.

Mehta, K., Pantazis, P., McQueen, T. and Aggarwal, B.B. (1997) Antiproliferative effect of curcumin (diferuloylmethane) against human breast tumor cell lines. *Anti-cnacer Drugs,* 8, 470-481.

Mesa, M.D., Aguilera, C.M., Ramirez-Tortosa, C.L., Ramirez-Tortosa, M.C., Quiles, J.L., Baro, L., de Victoria, E.M. and Gil, A. (2003) Oral administration of a turmeric extract inhibits erythrocyte and liver microsome membrane oxidation in rabbits fed with an atherogenic diet. *Nutrition,* 19, 800-804.

Miquel, J., Bernd, A., Sempere, J.M., Diaz-Alperi, J. and Ramirez, A. (2002) The curcuma antioxidants: pharmacological effects and prospects for future clinical use. A review. *Archives of Gerontology and Geriatrics,* 34, 37-46.

Mohan, K.V.P.C., Abraham, S.K. and Nagini, S. (2004) Protective effects of a mixture of dietary agents against 7,12-dimethylbenz[a]anthracene-induced genotoxicity and oxidative stress in mice. *J. Medicinal Food,* 7, 55-60.

Mohanty, I., Arya, D.S., Dinda, A., Joshi, S., Talwar, K.K. and Gupta, S.K. (2004) Protective effects of *Curcuma longa* on ischemia-reperfusion induced myocardial injuries and their mechanisms. *Life Sciences,* 75, 1701-1711.

Molnar, V. and Garai, J. (2005) Plant-derived anti-inflammatory compounds affect MIF tautomerase activity. *Int. Immunopharmacol.,* 5, 849-856.

Nagabhushan, N. and Bhide, S.V. (1992) Curcumin as an inhibitor of cancer. *J. Am. Coll. Nutr.,* 11, 192-198.

Nagini, S. and Manoharan, S. (1997) Biomonitoring the chemopreventive potential of the plant products neem and turmeric in 4-nitroquinoline 1-oxide-induced oral carcinogenesis. *J. Clin. Biochem. Nturition,* 23, 33-40.

Natarajan, C. and Bright, J.J. (2002) Curcumin inhibits experimental allergic encephalomyelitis by blocking IL-12 signalling through Janus kinase-STAT pathway in T lymphocytes. *J. Immunol.,* 168, 6506-6513.

Nakatani, N. (2000) Phenolic antioxidants from herbs and spices. *Biofactors,* 13, 141-146.

Negi, P.S., Jayaprakasha, G.K., Rao, L.J.M. and Sakariah, K.K. (1999) Antibacterial activity of turmeric oil: A by product from curcumin manufacture. *J. Agric. Food Chem.,* 47, 4297-4300.

Niederau, C. and Gopfert, E. (1999) The effect of extracts from Schollkraut and *Curcuma* on upper abdominal pain due to dysfunction of the biliary system: Results of a placebo-controlled, double-blind study. *Medizinische Klinik,* 94, 425-430.

Nishiyama, T., Mae, T., Kishida, H., Tsukagawa, M., Mimaki, Y., Kuroda, M., Sashida, Y., Takahashi, K., Kawada, T., Nakagawa, K. and Kitahara, M. (2005) Curcuminoids and sesquiterpenoids in turmeric (*Curcuma longa* L.) suppress an increase in blood glucose level in type 2 diabetic KK-A(y) mice. *J. Agric. Food Chem.,* 53, 959-963.

Odot, J., Albert, P., Carlier, A., Tarpin, M., Devy, J. and Madoulet, C. (2004) *In vitro* and *in vivo* anti-tumoral effect of curcumin against melanoma cells. *Int. J. Cancer,* 111, 381-387.

Olajide, O.A. (1999) Investigation of the effects of selected medicinal plants on experimental thrombosis. *Phytotherapy Res.,* 13, 231-232.

Osawa, T., Sugiyama, Y., Inayoshi, M. and Kawakishi, S. (1995) Antioxidative activity of tetrahydrocurcuminoids. *Biosci. Biotechnolo. Biochem.,* 59, 1609-1612.

Ozaki, Y. (1990) Antiinflammatory effect of *Curcuma xanthorrhiza* Roxb and its active principles. *Chem. Pharm. Bull.,* 38, 1045-1048.

Paek, S.H., Kim, G.J., Jong, H.S. and Yum, S.K. (1996) Ar-turmerone and beta-atlantone induce internucleo-somal DNA fragmentation associated with programmed cell death in human myeloid leukemia HL-60 cells. *Arch. Pharm. Res.,* 19, 91-94.

Pal, S., Choudhuri, T., Chattopadhyay, S., Bhattacharya, A., Datta, G.K., Das, T. and Sa, G. (2001) Mechanisms of curcumin-induced apoptosis of Ehrlich's ascites carcinoma cells. *Biomed. Biophys. Res. Commun.,* 288, 658-665.

Pandji, C., Grimm, C., Wray, V., Witte, L. and Proksch, P. (1993) Insecticidal constituents from 4 species of the Zingiberaceae. *Phytochemistry,* 34, 415-419.

Park, S.Y. and Kim, D.S.H.L. (2002) Discovery of natural products from *Curcuma longa* that protect cells from beta-amyloid insult: A drug discovery effort against Alzheimer's disease. *J. Nat. Prod.,* 65, 1227-1231.

Phan, T.T., See, P., Lee, S.T. and Chan, S.Y. (2001) Protective effects of curcumin against oxidative damage on skin cells *in vitro*: its implication for wound healing. *J. Trauma.,* 51, 927-931.

Pillai, G.R., Srivastava, A.S., Hassanein, T.I., Chauhan, D.P. and Carrier, E. (2004) Induction of apoptosis in human lung cancer cells by curcumin. *Cancer Letters,* 208, 163-170.

Piper, J.T., Singhal, S.S., Salameh, M.S., Torman, R.T., Awasthi, Y.C. and Awasthi, S. (1998) Mechanisms of anticarcinogenic properties of curcumin: the effect of curcumin on glutathione linked detoxification enzymes in rat liver. *Int. J. Biochem. Cell Biol.,* 30, 445-456.

Piwocka, K., Jaruga, E., Skierski, J., Gradzka, I. and Sikora, E. (2001) Effect of glutathione depletion on caspase-3 independent apoptosis pathway induced by curcumin in Jurkat cells. *Free Radical Biol. Med.,* 31, 670-678.

Potterat, O. (1997) Antioxidants and free radical scavengers of natural origin. *Curr. Org. Chem.,* 1, 415-440.

Quiles, J.L., Mesa, M.D., Ramirez-Tortosa, C.L., Aguilera, C.M., Battino, M., Gil, A. and Ramirez-Tortosa, M.C. (2002) *Curcuma longa* extract supplementation reduces oxidative stress and attenuates aortic fatty streak development in rabbits. *Arteriosclerosis Thrombosis and Vascular Biology,* 22, 1225-1231.

Quiles, J.L., Aguilera, C., Mesa, M.D., Ramirez-Tortosa, M.C., Baro, L. and Gil, A. (1998) An ethanolic-aque-ous extract of *Curcuma longa* decreases the susceptibility of liver microsomes and mitochondria to lipid peroxidation in atherosclerotic rabbits. *Biofactors,* 8, 51-57.

Qureshi, S., Shah, A.H. and Ageel, A.M. (1992) Toxicity studies on *Alpinia galanga* and *Curcuma longa.* *Planta Medica,* 58, 124-127.

Rafatullah, S., Tariq, M., Alyahya, M.A., Mossa, J.S. and Ageel, A.M. (1990) Evaluation of turmeric (*Curcuma longa*) for gastric and duodenal antiulcer activity in rats *J. Ethnopharmacol.,* 29, 25-34.

Rafi, M.M., Yadav, P.N. and Maeng, I.K. (2003) Targeting inflammation using nutraceuticals. *Oriental Foods and Herbs: ACS Symposium Series,* 859, 48-63.

Rajakrishnan, V., Jayadeep, A., Arun, O.S., Sudhakaran, P.R. and Menon, V.P. (2000) Changes in the prostag-landin levels in alcohol toxicity: Effect of curcumin and *N*-acetylcysteine. *J. Nutritional Biochem.,* 11, 509-514.

Rajakrishnan, V., Viswanathan, P., Rajasekharan, K.N. and Menon, V.P. (1999) Neuroprotective role of cur-cumin from *Curcuma longa* on ethanol-induced brain damage. *Phytotherapy Res.,* 13, 571-574.

Ram, A., Das, M. and Ghosh, B. (2003) Curcumin attenuates allergen-induced airway hyperresponsiveness in sensitized guinea pigs. *Biol. Pharm. Bull.,* 26, 1021-1024.

Ramirez-Bosca, A., Soler, A., Carrion, M.A., Diaz-Alperi, J., Bernd, A., Quintanilla, C., Almagro, E.Q. and Miquel, J. (2000) An hydroalcoholic extract of *Curcuma longa* lowers the apo B/apo A ratio - Implications for atherogenesis prevention. *Mechanisms of Ageing and Development,* 119, 41-47.

Ramirez-Tortosa, M.C., Mesa, M.D., Aguilera, M.C., Quiles, J.L., Baro, L., Ramirez-Tortosa, C.L., Mar-tinez-Victoria, E. and Gil, A. (1999) Oral administration of a turmeric extract inhibits LDL oxidation and has hypocholesterolemic effects in rabbits with experimental atherosclerosis. A*therosclerosis,* 147, 371-378.

Ramos, A., Visozo, A., Piloto, J., Garcia, A., Rodriguez, C.A. and Rivero, R. (2003) Screening of antimutage-nicity via antioxidant activity in Cuban medicinal plants. *J. Ethnopharmacol.,* 87, 241-246.

Ramsewak, R.S., DeWitt, D.L. and Nair, M.G. (2000) Cytotoxicity, antioxidant and anti-inflammatory activities of Curcumins I-III from *Curcuma longa. Phytomedicine,* 7, 303-308.

Rao, K.V.K., Schwartz, S.A., Nair, H.K., Aalinkeel, R., Mahajan, S., Chawda, R. and Nair, M.P.N. (2004) Plant derived products as a source of cellular growth inhibitory phytochemicals on PC-3M, DU-145 and LNCaP prostate cancer cell lines. *Current Science,* 87, 1585-1588.

Rao, C.V., Rivenson, A., Simi, B. and Reddy, B.S. (1995) Chemoprevention of colon carcinogenesis by dietary curcumin naturally occurring plant phenolic compound. *Cancer Res.,* 55, 259-266.

Rao, A.J. and Kotagi, S.G. (1984) Antiandrogenic action of the plant *Curcuma longa* root extract in male rats. *IRCS Medical Science Biochemistry,* 12, 500-501.

Rasyid, A., Rahman, A.R.A., Jaalam, K. and Lelo, A. (2002) Effect of different curcumin dosages on human gall bladder. *Asia Pacific J. Clinical Nutrition,* 11, 314-318.

Reddy, A.C.P. and Lokesh, B.R. (1994) Effect of dietary turmeric (*Curcuma longa*) on iron-induced lipid peroxidation in the rat liver. *Food Chem. Toxicol.,* 32, 279-283.

Reddy, A.C.P. and Lokesh, B.R. (1992) Studies on spice principles as antioxidants in the inhibition of lipid peroxidation of rat liver microsomes. *Mol. Cell Biochem.,* 111, 117-124.

Reddy, R.C., Vatsala, P.G., Keshamouni, V.G., Padmanaban, G. and Rangarajan, P.N. (2005) Curcumin for malaria therapy. *Biochemical and Biophysical Research Communications,* 326, 472–474.

Rilantono, L.I., Yuwono, H.S. and Nugrahadi, T. (2000) Dietary antioxidative potential in arteries. *Clinical Hemorheology and Microcirculation,* 23, 113-117.

Romiti, N., Tongiani, R., Cervelli, F. and Chieli, E. (1998) Effects of curcumin on P-glycoprotein in primary cultures of rat hepatocytes. *Life Sciences,* 62, 2349-2358.

Roth, G.N., Chandra, A. and Nair, M.G. (1998) Novel bioactivities of *Curcuma longa* constituents. *J. Nat. Prod.,* 61, 542-545.

Ruby, A.J., Kuttan, G., Babu, K.D., Rajasekharan, K.N. and Kuttan, R. (1995) Antitumor and antioxidant activity of natural curcuminoids. *Cancer Letters,* 94, 79-83.

Sachs, M., von Eichel, J. and Asskali, F. (2002) Wound treatment with coconut oil in Indonesian natives. *Chirurg,* 73, 387-392.

Saju, K.A., Venugopal, M.N. and Mathew, M.J. (1998) Antifungal and insect-repellent activities of essential oil of turmeric (*Curcuma longa* L.). *Current Science,* 75, 660-662.

Sasaki, Y., Goto, H., Tohda, C., Hatanaka, F., Shibahara, N., Shimada, Y., Terasawa, K. and Komatsu, K. (2003) Effects of Curcuma drugs on vasomotion in isolated rat aorta. *Biol. Pharm. Bull.,* 26, 1135-1143.

Scartezzini, P. and Speroni, E. (2000) Review on some plants of Indian traditional medicine with antioxidant activity. *J. Ethnopharmacol.,* 71, 23-43.

Semwal, A.D., Sharma, G.K. and Arya, S.S. (1997) Antioxygenic activity of turmeric (*Curcuma longa*) in sunflower oil and ghee. *J. Food Sci. Technol. Mysore,* 34, 67-69.

Sharma, R.A., McLelland, H.R., Hill, K.A., Ireson, C.R., Euden, S.A., Manson, M.M., Pirmohamed, M., Marnett, L.J., Gescher, A.J. and Steward, W.P. (2001) Pharmacodynamic and pharmacokinetic study of oral Curcuma extract in patients with colorectal cancer. *Clinical Cancer Res.,* 7, 1894-1900.

Shim, J.S., Lee, H.J., Park, S.S., Cha, B.G. and Chang, H.R. (2001) Curcumin-induced apoptosis of A-431 cells involves caspase-3 activation. *J. Biochem. Mol. Biol.,* 34, 189-193.

Shukla, Y., Arora, A. and Taneja, P. (2002) Antimutagenic potential of curcumin on chromosomal aberrations in Wistar rats. *Mutagen Research-Genetic Toxicology and Environmental Mutagenesis,* 515, 197-202.

Shukla, Y. and Arora, A. (2003) Suppression of altered hepatic foci development by curcumin in Wistar rats. *Nutrition and Cancer — An International Journal,* 45, 53-59.

Sidhu, G.S., Manni, H., Gaddipati, J.P., Singh, A.K., Seth, P., Banaudha, K.K., Patnaik, G.K. and Maheshwari, R.K. (1999) Curcumin enhances wound healing in streptozotocin induced diabetic rats and genetically diabetic mice. *Wound Repair and Regeneration,* 7, 362-374.

Simon, A., Allais, D.P., Duroux, J.L., Basly, J.P., Durand-Fontanier, S. and Delage, C. (1998) Inhibitory effect of curcuminoids on MCF-7 cell proliferation and structure-activity relationships. *Cancer Letters,* 129, 111-116.

Singh, G., Singh, O.P. and Maurya, S. (2002a) Chemical and biocidal investigations on essential oils of some Indian *Curcuma* species. *Progress in Crystal Growth and Characterization of Materials,* 45, 75-81.

Singh, R., Chandra, R., Bose, M. and Luthra, P.M. (2002b) Antibacterial activity of *Curcuma longa* rhizome extract on pathogenic bacteria. *Current Science,* 83, 737-740.

Singh, R. and Rai, B. (2000) Antifungal potential of some higher plants against *Fusarium nudum* causing wilt disease of *Cajanus cajan. Microbios,* 102, 165-173.

Singhal, S.S., Awasthi, S., Pandya, U., Piper, J.T., Saini, M.K., Cheng, J.Z. and Awasthi, Y.C. (1999) The effect of curcumin on glutathione-linked enzymes in K562 human leukemia cells. *Toxicol. Letters,* 109, 87-95.

Skrzypczak-Jankun, E., McCabe, N.P., Selman, S.H. and Jankun, J. (2000) Curcumin inhibits lipoxygenase by binding to its central cavity: Theoretical and X-ray evidence. *Int. J. Mol. Med.*, 6, 521-526.

Song, E.K., Cho, H., Kim, J.S., Kim, N.Y., An, N.H., Kim, J.A., Lee, S.H. and Kim, Y.C. (2001) Diarylheptanoids with free radical scavenging and hepatoprotective activity *in vitro* from *Curcuma longa*. *Planta Medica*, 67, 876-877.

Sreekanth, K.S., Sabu, M.C., Varghese, L., Manesh, C., Kuttan, G. and Kuttan, R. (2003) Antioxidant activity of Smoke Shield *in-vitro* and *in-vivo*. *J. Pharm. Pharmacol.*, 55, 847-853.

Srimal, R. and Dhawan, B. (1973) Pharmacology of diferuloyl methane (curcumin), a non-steroidal anti-inflammatory agent. *J. Pharm. Pharmacol.*, 25, 447-52.

Srinivas, L., Shalini, V.K. and Shylaja, M. (1992) Turmerin — a water-soluble antioxidant peptide from turmeric [*Curcuma longa*]. *Arch. Biochem. Biophys.*, 292, 617-623.

Srivastava, K.C., Bordia, A. and Verma, S.K. (1995) Curcumin, a major component of food spice turmeric (*Curcuma longa*) inhibits aggregation and alters eicosanoid metabolism in human blood-platelets. *Prostaglandins Leukotrienes and Essential Fatty Acids*, 52, 223-227.

Subramanian, L. and Selvam, R. (1999) Prevention of CCl_4 - Induced hepatotoxicity by aqueous extract of turmeric. *Nutrition Res.*, 19, 429-441.

Sugiyama, Y., Kawakishi, S. and Osawa, T. (1996) Involvement of the beta-diketone moiety in the antioxidative mechanism of tetrahydrocurcumin. *Biochem. Pharmacol.*, 52, 519-525.

Surh, Y.J. (2002) Anti-tumor promoting potential of selected spice ingredients with antioxidative and anti-inflammatory activities: a short review. *Food Chem. Toxicol.*, 40, 1091-1097.

Surh, Y.J., Chun, K.S., Cha, H.H., Han, S.S., Keum, Y.S., Park, K.K. and Lee, S.S. (2001) Molecular mechanisms underlying chemopreventive activities of anti-inflammatory phytochemicals: down-regulation of COX-2 and iNOS through suppression of NF-kappa B activation. *Mutation Research-Fundamental and Molecular Mechanisms of Mutagenesis*, 480, 243-268.

Surh, Y.J. (1999) Molecular mechanisms of chemopreventive effects of selected dietary and medicinal phenolic substances. *Mutation Research-Fundamental and Molecular Mechanisms of Mutagenesis*, 428, 305-327.

Syng-ai, C., Kumari, A.L. and Khar, A. (2004) Effect of curcumin on normal and tumor cells: Role of glutathione and bcl-2. *Molecular Cancer Therapeutics*, 3, 1101-1108.

Tawatsin, A., Wratten, S.D., Scott, R.R., Thavara, U. and Techadamrongsin, Y. (2001) Repellency of volatile oils from plants against three mosquito vectors. *J. Vector. Ecol.*, 26, 76-82.

Tilak, J.C., Banerjee, M., Mohan, H. and Devasagayam, T.P.A. (2004) Antioxidant availability of turmeric in relation to its medicinal and culinary uses. *Phytotherapy Researh*, 18, 798-904.

Toda, S., Miyase, T., Arichi, H., Tanizawa, H. and Takino, Y. (1985) Natural antioxidants .13. antioxidative components isolated from rhizome of *Curcuma longa* L. *Chem. Pharm. Bull.*, 33, 1725-1728.

Tornetta, G. (2005) The goodness of turmeric, BellaOnline, USA. Available on-line at: http://www.bellaonline.com/articles/art28294.asp.

Turmeric-Curcumin.Com (2005), USA. Available on-line at: http://www.turmeric-curcumin.com/.

Ushida, J., Sugie, S., Kawabata, K., Pham, Q.V., Tanaka, T., Fujii, K., Takeuchi, H., Ito, Y. and Mori, H. (2000) Chemopreventive effect of curcumin on N-nitrosomethylbenzylamine-induced esophageal carcinogenesis in rats. *Japanese J. Cancer Res.*, 91, 893-898.

Venkateswarlu, S., Ramachandra, M.S. and Subbaraju, G.V. (2004) Synthesis and antioxidant activity of 5 '-methoxycurcumin: A yellow pigment from *Curcuma xanthorrhiza*. *Asian J. Chem.*, 16, 827-830.

Verma, S.P., Salamone, E. and Goldin, B. (1997) Curcumin and genistein, plant natural products, show synergistic inhibitory effects on the growth of human breast cancer MCF-7 cells induced by estrogenic pesticides. *Biochem. Biophys. Res. Commun.*, 233, 692-696.

Wattanapitayakul, S.K., Chularojmontri, L., Herunsalee, A., Charuchongkolwongse, S., Niumsakul, S. and Bauer, J.A. (2005) Screening of antioxidants from medicinal plants for cardioprotective effect against doxorubicin toxicity. *Basic and Clinical Pharmacol and Toxicol.*, 96, 80-87.

Wientarsih, L., Chakeredza, S. and ter Meulen, U. (2002) Influence of curcuma (*Curcuma xanthorrhiza* Roxb) on lipid metabolism in rabbits. *J. Science of Food and Agriculture*, 82, 1875-1880.

Woo, J.H., Kim, Y.H., Choi, Y.J., Kim, D.G., Lee, K.S., Bae, J.H., Min, D.S., Chang, J.S., Jeong, Y.J., Lee, Y.H., Park, J.W. and Kwon, T.K. (2003) Molecular mechanisms of curcumin-induced cytotoxicity: induction of apoptosis through generation of reactive oxygen species, down-regulation of Bcl-X-L and IAP, the release of cytochrome c and inhibition of Akt. *Carcinogenesis*, 24, 1199-1208.

Yamamoto, H., Hanada, K., Kawasaki, K. and Nishijima, M. (1997) Inhibitory effect of curcumin on mammalian phospholipase D activity. *FEBS Letters,* 417, 196-198.

Yamazaki, M., Mmaebayashi, Y., Iwase, N. and Kaneko, T. (1988a) Studies on pharmacologically active principles from Indonesian crude drugs .2. hypothermic principle from *Curcuma xanthorrhiza* Roxb. *Chem. Pharm. Bull.,* 36, 2075-2078.

Yamazaki, M., Mmaebayashi, Y., Iwase, N. and Kaneko, T. (1988b) Studies on pharmacologically active principles from Indonesian crude drugs .1. principle prolonging pentobarbital-induced sleeping time from *Curcuma xanthorrhiza* Roxb. *Chem. Pharm. Bull.,* 36, 2070-2074.

Yasni, S., Imaizumi, K., Sin, K., Sugano, M. and Nonaka, G.S. (1994) Identification of an active principle in essential oils and hexane-soluble fractions of *Curcuma xanthorrhiza* Roxb showing triglyceride-lowering action in rats. *Food Chem. Toxicol.,* 32, 273-278.

Yasni, S., Imaizumi, K. and Sugano, M. (1991) Effects of an Indonesian medicinal plant, *Curcuma xanthorrhiza* Roxb, on the levels of serum glucose and triglyceride, fatty-acid desaturation, and bile-acid excretion in streptozotocin induced diabetic rats. *Agric. Biol. Chem.,* 55, 3005-3010.

Yu, Z.F., Kong, L.D. and Chen, Y. (2002) Antidepressant activity of aqueous extracts of *Curcuma longa* in mice. *J. Ethnopharmacol.,* 83, 161-165.

Zhou, A., McFeeters, R.F. and Fleming, H.P. (2000) Inhibition of formation of oxidative volatile components in fermented cucumbers by ascorbic acid and turmeric. *J. Agric. Food Chem.* 48, 4910-4912.

10 Curcumin — Biological and Medicinal Properties

Bharat B. Aggarwal, Indra D. Bhatt, Haruyo Ichikawa,
Kwang Seok Ahn, Gautam Sethi, Santosh K. Sandur,
Chitra Sundaram, Navindra Seeram, and Shishir Shishodia

CONTENTS

10.1   INTRODUCTION

The turmeric (*Curcuma longa*) plant, a perennial herb belonging to the ginger family, is cultivated extensively in south and southeast tropical Asia. The rhizome of this plant is also referred to as the "root" and is the most useful part of the plant for culinary and medicinal purposes. The most active component of turmeric is curcumin, which makes up 2 to 5% of the spice. The characteristic yellow color of turmeric is due to the curcuminoids.

Curcumin is an orange–yellow crystalline powder practically insoluble in water. The structure of curcumin ($C_{21}H_{20}O_6$) was first described in 1913 by Lampe and Milobedeska and shown to be diferuloylmethane (Aggarwal et al., 2003).

Turmeric is used as a dietary spice, coloring agent in foods and textiles, and a treatment for a wide variety of ailments (Figure 10.1). It is widely used in traditional Indian medicine to cure biliary disorders, anorexia, cough, diabetic wounds, hepatic disorders, rheumatism, and sinusitis. Turmeric paste in slaked lime is a popular home remedy for the treatment of inflammation and wounds. For centuries, curcumin has been consumed as a dietary spice at doses up to 100 mg/day. Extensive investigation over the last five decades has indicated that curcumin reduces blood cholesterol (Rao et al., 1970; Patil and Srinivasan, 1971; Keshavarz, 1976; Soudamini et al., 1992; Soni and Kuttan, 1992; Hussain and Chandrasekhara, 1992; Asai and Miyazawa, 2001) prevents LDL oxidation (Ramirez-Tortosa et al., 1999; Naidu and Thippeswamy, 2002; Patro et al., 2002), inhibits platelet aggregation (Srivastava et al., 1986,1995), suppresses thrombosis (Srivastava, 1985) and myocardial infarction (MI) (Dikshit et al., 1995; Nirmala and Puvanakrishnan, 1996a,b; Venkatesan, 1998), suppresses symptoms associated with type II diabetes (Srinivasan, 1972; Babu and Srinivasan, 1995,1997; Arun and Nalini, 2002), rheumatoid arthritis (Deodhar et al., 1980), multiple

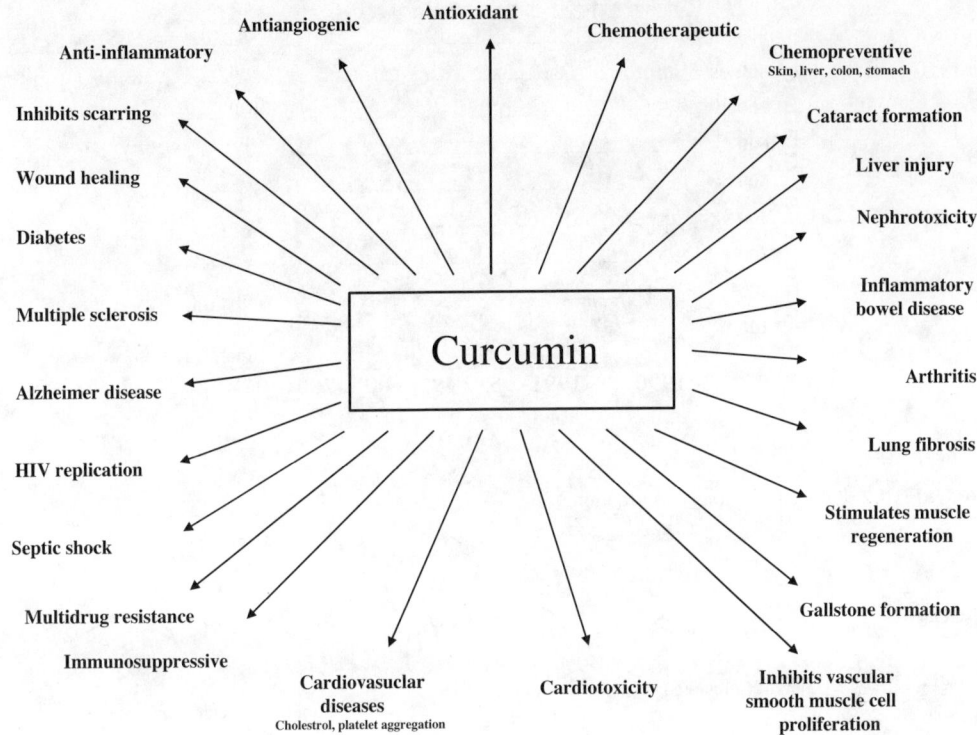

FIGURE 10.1 Medicinal properties of curcumin.

sclerosis (MS) (Natarajan and Bright, 2002), and Alzheimer's disease (Lim et al., 2001; Frautschy et al., 2001), inhibits human immunodeficiency virus (HIV) replication (Sui et al., 1993; Li et al., 1993; Jordan and Drew, 1996; Mazumder et al., 1997; Barthelemy, 1998), enhances wound healing (Sidhu et al., 1998; Phan et al., 2001; Shahed et al., 2001), protects from liver injury (Morikawa et al., 2002), increases bile secretion (Ramprasad and Sirsi, 1956), protects from cataract formation (Awasthi et al., 1996), and protects from pulmonary toxicity and fibrosis (Venkatesan and Chandrakasan, 1995; Venkatesan et al., 1997; Venkatesan, 2000; Punithavathi et al., 2000), is an anti-leishmaniasis (Saleheen et al., 2002; Gomes Dde et al., 2002; Koide et al., 2002) and an antiatherosclerotic (Huang et al., 1992; Chen and Huang, 1998). Additionally, there is extensive literature that suggests that curcumin has potential in the prevention and treatment of a variety of other diseases (Figure 10.2).

10.2 CHEMICAL COMPOSITION OF TURMERIC

Curcumin was first isolated in 1815, obtained in crystalline form in 1870 (Vogel and Pelletier, 1818; Daube, 1870), and identified as 1,6-heptadiene-3,5-dione-1,7-bis(4-hydroxy-3-methoxyphenyl)-(1E,6E) or diferuloylmethane (Figure 10.3). The feruloylmethane skeleton of curcumin was subsequently confirmed in 1910 by the initial work and synthesis by Lampe (Lampe, 1910; Lampe and Milobedzka, 1913). Curcumin is a yellow-orange powder that is insoluble in water and ether but soluble in ethanol, dimethylsulfoxide, and acetone. Curcumin has a melting point of 183°C, molecular formula of $C_{21}H_{20}O_6$, and molecular weight of 368.37 g/mol.

Curcumin (also known as curcumin I) occurs naturally in the rhizome of *Curcuma longa*, which is grown commercially and sold as turmeric, a yellow-orange dye. Turmeric contains curcumin along with other chemical constituents known as the "curcuminoids" (Srinivasan, 1952). The major

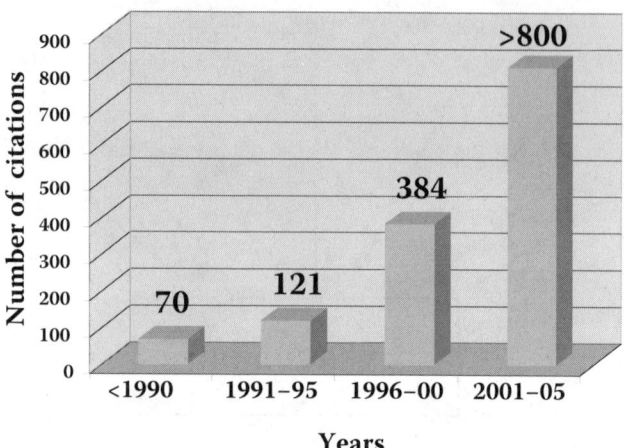

FIGURE 10.2 Pubmed citations on curcumin.

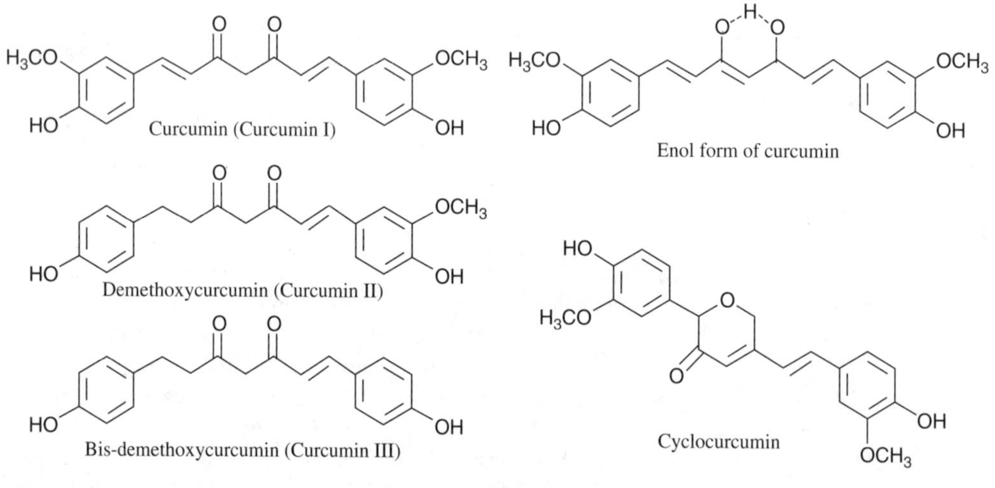

FIGURE 10.3 Structure of curcumin.

curcuminoids present in turmeric are demethoxycurcumin (curcumin II), bisdemethoxycurcumin (curcumin III), and the recently identified cyclocurcumin (Kiuchi et al., 1993). Commercial curcumin contains curcumin I (~77%), curcumin II (~17%), and curcumin III (~3%) as its major components. The curcuminoid complex is also referred to as Indian saffron, yellow ginger, yellow root, *kacha haldi*, ukon, or natural yellow 3.

Spectrophotometrically, curcumin has a maximum absorption (λ_{max}) in methanol at 430 nm, with a Beer's law range from 0.5 to 5 µg/mL (Prasad, 1997). It absorbs maximally at 415 to 420 nm in acetone and a 1% solution of curcumin has 1650 absorbance units. Curcumin has a brilliant yellow hue at pH 2.5 to 7 and takes on a red hue at pH > 7. The spectral and photochemical properties of curcumin have been studied in different solvents by Chignell and coworkers (Chignell et al., 1994). In toluene, the absorption spectrum of curcumin contains some structure, which disappears in more polar solvents such as ethanol and acetonitrile. The fluorescence of curcumin occurs as a broad band in acetonitrile (λ_{max} = 524 nm), ethanol (λ_{max} = 549 nm), or micellar solution (λ_{max} = 557 nm), but has some structure in toluene (λ_{max} = 460, 488 nm) (Chignell et al., 1994). These workers also showed that the fluorescence quantum yield of curcumin is low in sodium dodecyl sulfate solution (phi = 0.011) but higher in acetonitrile (phi = 0.104) (Chignell et al., 1994). In addition, curcumin was observed to produce singlet oxygen upon irradiation ($\lambda_{max} > 400$ nm) in toluene or acetonitrile (phi = 0.11 for 50 μM curcumin). Curcumin quenched singlet oxygen in acetonitrile (kq = 7×10^6/M-s). Singlet oxygen production was about ten times lower in alcohols and was hardly detectable when curcumin was solubilized in an aqueous micellar solution of Triton X-100. However, in sodium dodecyl sulfate solution, no singlet oxygen phosphorescence could be observed for those micelles containing curcumin. Curcumin is also reported to be able to photo-generate superoxide in toluene and ethanol (Chignell et al., 1994).

The interactions between curcumin and biological radical stressors have been studied. For example, Iwunze and coworkers (Iwunze and McEwan, 2004) recently used both fluorescence and absorptiometric techniques to study the interaction between curcumin and peroxynitrite. Using both techniques, these workers observed that signals increased asymptotically until the concentration of peroxynitrite equaled that of the curcumin (held at a constant concentration of 1×10^{-5} M). However, there was a shift in fluorescence wavelength after the initial oxidation of the hydroxyl group, which was attributed to the nitration of the phenoxyl group of curcumin. A second-order reaction rate for the nitration of curcumin by peroxynitrite was concluded with an association constant of 1.2×10^6/M-s and 3.6×10^6/M-s for the fluorescence and absorptiometric techniques, respectively (Iwunze and McEwan, 2004). In another report, Mishra et al. studied the reactions of superoxide–crown ether complex with curcumin (Mishra et al., 2004). Optical absorption spectra showed that on reaction with superoxide, curcumin forms a blue-colored intermediate ($\lambda max = 560$ nm), which subsequently decayed with the development of the absorption band corresponding to the parent curcumin. A 100% regeneration was observed at low superoxide concentrations (1:1–1:3, curcumin:superoxide) and a 60% regeneration at high superoxide concentration (>1:5, curcumin:superoxide). These researchers also determined the rate constant for the reaction of superoxide with curcumin. Based on their studies, the authors concluded that at low superoxide concentrations, curcumin effectively causes superoxide dismutation without itself undergoing any chemical change, but at higher concentrations of superoxide, curcumin inhibits superoxide activity by reacting with it. Toniolo and coworkers also investigated the action of curcumin on superoxide ions (Toniolo et al., 2002). They found that 1 mol of curcumin reacted with 6 mol of anion radical, which provides the perhydroxyl radical and further disproportionates to the anionic form of hydrogen peroxide and oxygen.

The free radical-scavenging mechanism of curcumin has been examined by Ohara and coworkers (Ohara et al., 2005). The second-order rate constants for the radical-scavenging reactions of curcumin and half-curcumin were measured by a stopped-flow spectrophotometer in several organic solvents (methanol, ethanol, acetonitrile, chloroform, and benzene) and in aqueous Triton X-100 (5.0%) micelle solutions at various pH values. The difference in the rate constants and solvent dependence between curcumin and half-curcumin suggested that the enol structure with the intramolecular hydrogen bond of curcumin (Figure 10.3) strongly enhances its radical-scavenging activity. Also, notable pH dependences were observed for the rate constants of both curcumin and half-curcumin in micelle solutions, suggesting that the acid–base dissociation equilibrium of phenol-protons in curcumin and half-curcumin affected their radical-scavenging activities (Ohara et al., 2005).

Priyadarsini and coworkers conducted studies to evaluate the relative importance of the phenolic hydrogens and the –CH$_2$ hydrogens on the antioxidant activity and free radical reactions of curcumin and dimethoxycurcumin (Priyadarsini et al., 2003). They showed that at equal concentrations, the efficiency to inhibit lipid peroxidation (LPO) changed from 82% with curcumin to 24% with dimethoxycurcumin. The kinetics of reaction of 2,2′-diphenyl-1-picrylhydrazyl (DPPH), a stable hydrogen-abstracting free radical, was tested with these two compounds using spectrophotometry. The authors concluded that although the energetics to remove hydrogens from both the phenolic hydrogens and the –CH$_2$ group of the diketo structure were very close, the phenolic hydrogen is essential for both antioxidant activity and free radical kinetics. This was further confirmed by density functional theory (DFT) calculations where it was shown that the phenolic hydrogen is more labile for abstraction, compared to the –CH$_2$ hydrogens in curcumin. Therefore, based on both experimental and theoretical results, this report showed that the phenolic hydrogens play a major role in the antioxidant activity of curcumin.

The photophysical properties of curcumin have been investigated by Khopde et al. (2000), wherein a variety of spectroscopic techniques were used to investigate the photophysical properties of curcumin in different organic solvents and in Triton X-100 aqueous micellar media. The steady-state absorption and fluorescence characteristics of curcumin were found to be sensitive to the solvent characteristics. Curcumin was also found to be a weakly fluorescent molecule and its fluorescence decay properties in most of the solvents could be fitted well to a double-exponential decay function. The shorter component (lifetime in the range 50–350 ps) could be assigned to its enol form (Figure 10.3), whereas the longer component (lifetime in the range 500–1180 ps) was assigned to the diketo form of curcumin (Figure 10.3). These authors also conducted nuclear magnetic resonance (NMR) experiments in CDCl3 and dimethylsulfoxide-D6 and showed that the enol form of curcumin is present in the solution by more than about 95% in these solvents.

The stability of curcumin in aqueous media has been investigated by Bernabe-Pineda et al. (2004). They showed that the stability of curcumin was improved at high pH values (>11.7), fitting a model describable by a pseudo–zero-order rate equation with a rate constant k′ for the disappearance of the curcumin species of 1.39×10^{-9}/Mmin^{-1}. Three acidity constants (pKA) were measured for curcumin, as follows, pKA1 = 8.38 ± 0.04, pKA2 = 9.88 ± 0.02 and pKA3 = 10.51 ±0.01. Formation of quinoid structures played an important role in the tautomeric forms of curcumin in aqueous media, which made the experimental values differ from the theoretically calculated ones, depending on the conditions adopted in this study (Bernabe-Pineda et al., 2004). In a separate report, Souza and coworkers (Souza et al., 1997) also studied the influence of water activity on the stability of curcuminoid pigments in curcumin- and turmeric oleoresin–microcrystalline–cellulose model systems during storage at 21 ± 1°C. Samples were analyzed spectrophotometrically for curcuminoid pigments at specific time intervals and the degradation of the curcuminoids were observed to follow first-order reactions. Although the curcuminoid pigments were sensitive to light, the combined effects of air and light were the most deleterious (Souza et al., 1997). The authors did not observe any influence of water activity on the stability of curcuminoid pigments in the curcumin– and turmeric oleoresin–microcrystalline cellulose model systems. Tonnesen and coworkers (Tonnesen et al., 1986) also investigated the photodecomposition of curcumin on exposure to ultraviolet (UV)/visible radiation and identified the major degradation products. They also examined the photobiological activity of curcumin using bacterial indicator systems (Tonnesen et al., 1987). On irradiation with visible light, curcumin, at low concentrations, was phototoxic for *Salmonella typhimurium* and *Escherichia coli*. The authors concluded that the observed phototoxicity makes curcumin a potential photosensitizing drug that might find application in the phototherapy of psoriasis, cancer, and bacterial and viral diseases (Tonnesen et al., 1987). The same group also prepared cyclodextrin complexes of curcumin to improve the water solubility and the hydrolytic and photochemical stability of the compound (Tonnesen et al., 2002). Complex formation resulted in an increase in water solubility at pH 5 by a factor of at least 10^4. The hydrolytic stability of curcumin under alkaline conditions was strongly improved by complex formation, while the photodecomposition rate was increased compared to a curcumin solution in organic solvents. The cavity size and the charge and bulkiness of the cyclodextrin side chains influenced the stability constant for complexation and the degradation rate of the curcumin molecule (Tonnesen et al., 2002). Wang et al. examined the degradation kinetics of curcumin under various pH conditions and the stability of curcumin in physiological matrices (Wang et al., 1997). When curcumin was incubated in 0.1 M phosphate buffer and serum-free medium at pH 7.2 at 37°C, about 90% decomposed within 30 min. The authors also tested curcumin stability from pH 3 to 10 and showed that the decomposition of curcumin was pH dependent and occurred faster at neutral to basic conditions. Curcumin was more stable in cell culture medium containing 10% fetal calf serum (FCS) and in human blood; less than 20% of curcumin decomposed within 1 hour, and after incubation for 8 hours, about 50% of curcumin still remained. *Trans*-6-(4′-hydroxy-3′-methoxyphenyl)-2,4-dioxo-5-hexenal was predicted as the major degradation product and vanillin, ferulic acid, and feruloyl methane were identified as minor degradation products of curcumin (Wang et al., 1997).

The electrochemical behavior of curcumin has been recently investigated by Wu and coworkers (Wu et al., 2005). In 0.1 mol/L phosphate buffer solution at pH 3, the voltametry behaviors of curcumin at a glassy carbon electrode were studied. The adsorptive potential of curcumin was reported as +0.8V, while its peak potential was +0.386V (Wu et al., 2005).

10.3 ANTIOXIDANT PROPERTIES OF CURCUMIN

Sharma (1976), Ruby et al. (1995), and Sugiyama et al. (1996) studied the antioxidative properties of curcumin and its three derivatives (demethoxy curcumin, bisdemethoxy curcumin, and diacetyl curcumin). The authors demonstrated that these substances provide a protection of hemoglobin from oxidation at a concentration as low as 0.08 mM, except the diacetyl curcumin, which has little effect in the inhibition of nitrite-induced oxidation of hemoglobin. The effect of curcumin on LPO has also been studied in various models by several authors. Curcumin is a good antioxidant and inhibits LPO in rat liver microsomes, erythrocyte membranes, and brain homogenates. The LPO has a main role in the inflammation, in heart diseases, and in cancer.

The antioxidant activity of curcumin could be mediated through antioxidant enzymes such as superoxide dismutase, catalase, and glutathione peroxidase. Curcumin has been shown to serve as a Michael acceptor, reacting with glutathione and thioredoxin (Adams et al., 2005). Reaction of curcumin with these agents reduces intracellular GSH in the cells. The suppression of LPO by curcumin could lead to the suppression of inflammation. In fact, curcumin has been found to be at least ten times more active as an antioxidant than even vitamin E (Khopde et al., 1999). In curcumin, the phenolic and the methoxy group on the phenyl ring and the 1,3-diketone system seem to be important structural features that can contribute to these effects. Another fact proposed in the literature is that the antioxidant activity increases when the phenolic group with a methoxy group is at the ortho position (Motterlini et al., 2000)

10.4 MOLECULAR TARGETS OF CURCUMIN

Various studies have shown that curcumin modulates numerous targets (Figure 10.4, Table 10.1, and Table 10.2). These include the growth factors, growth factor receptors, transcription factors, cytokines, enzymes, and genes regulating apoptosis.

10.4.1 CURCUMIN DOWNREGULATES THE ACTIVITY OF EGFR AND EXPRESSION OF HER2/NEU

HER2/neu and epithelial growth factor receptor (EGFR) activity represent one possible mechanism by which curcumin suppresses the growth of breast cancer cells. Almost 30% of the breast cancer cases have been shown to overexpress the HER2/neu protooncogene (Slamon et al., 1987), and both HER2 and EGF receptors stimulate proliferation of breast cancer cells. Overexpression of these two proteins correlates with progression of human breast cancer and poor patient prognosis (Slamon et al., 1987). Curcumin has been shown to downregulate the activity of EGFR and HER2/neu (Korutla and Kumar, 1994; Korutla et al., 1995) and to deplete the cells of HER2/neu protein (Hong et al., 1999). Additionally, we have recently found that curcumin can downregulate bcl-2 expression, which may contribute to its antiproliferative activity (Mukhopadhyay et al., 2001).

Like geldanamycin, curcumin has been shown to provoke the intracellular degradation of HER2 (Tikhomirov and Carpenter, 2003). HER2 mutations, however, limit the capacity of geldanamycin to disrupt the tyrosine kinase activity of HER2. Thus, these HER2 mutants are resistant to geldanamycin-induced degradation, but they maintain their sensitivity to curcumin through ErbB-2 degradation.

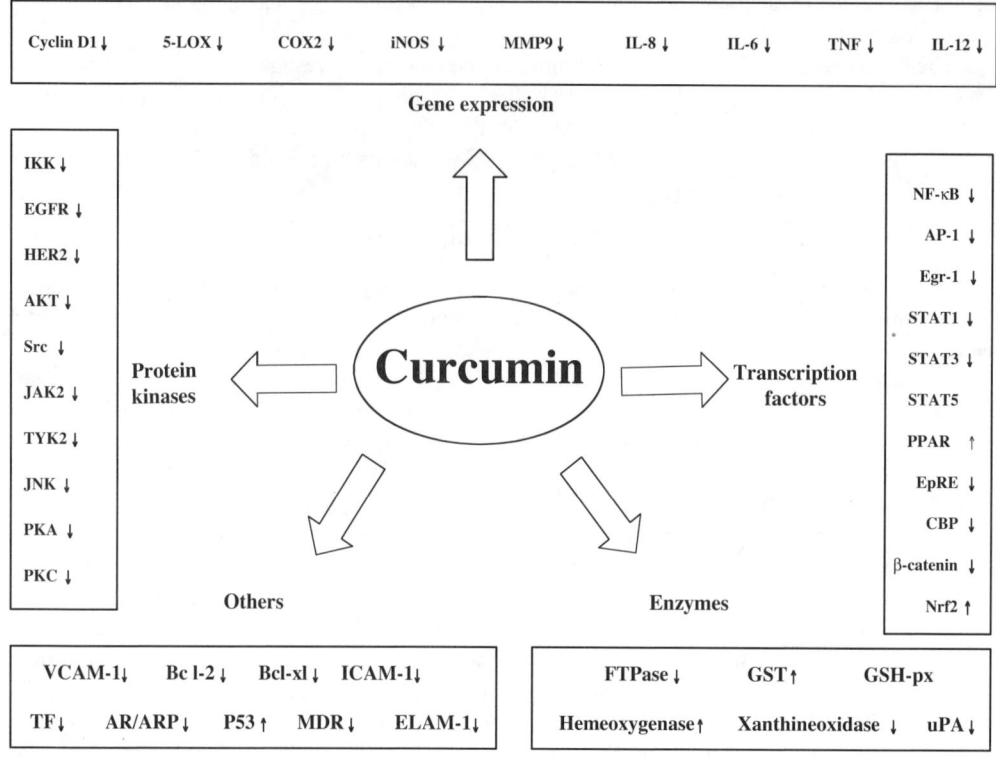

FIGURE 10.4 Molecular targets of curcumin.

10.4.2 CURCUMIN DOWNREGULATES THE ACTIVATION OF NF-κB

Curcumin may also operate through the suppression of nuclear factor-κB (NF-κB) activation. NF-κB is a nuclear transcription factor required for the expression of genes involved in cell proliferation, cell invasion, metastasis, angiogenesis, and resistance to chemotherapy (Baldwin, 2001). This factor is activated in response to inflammatory stimuli, carcinogens, tumor promoters, and hypoxia, which is frequently encountered in tumor tissues (Pahl, 1999). Several groups, including ours, have shown that activated NF-κB suppresses apoptosis in a wide variety of tumor cells (Wang et al., 1996; Lee et al., 1995; Giri and Aggarwal, 1998), and it has been implicated in chemoresistance (Wang et al., 1996). Furthermore, the constitutively active form of NF-κB has been reported in human breast cancer cell lines in culture (Nakshatri et al., 1997), carcinogen-induced mouse mammary tumors (Kim et al., 2000), and biopsies from patients with breast cancer (Sovak et al., 1997). Our laboratory has shown that various tumor promoters, including phorbol ester, tumor necrosis factor (TNF), and H_2O_2 activate NF-κB and that curcumin downregulates the activation (Singh and Aggarwal, 1995). Subsequently, others showed that curcumin-induced downregulation of NF-κB is mediated through suppression of IκBα kinase activation (Jobin et al., 1999; Plummer et al., 1999). Recently, we have shown that curcumin downregulated cigarette smoke-induced NF-κB activation through inhibition of IκBα kinase in human lung epithelial cells (Shishodia et al., 2003). We also found that curcumin suppresses the constitutively active NF-κB activation in mantle cell lymphoma (Shishodia et al., 2005). This led to the downregulation of cyclin D1, COX-2, and matrix metalloproteinase (MMP)-9 by curcumin. Philip and Kundu (2003) have recently reported that curcumin downregulates osteopontin (OPN)-induced NF-κB–mediated promatrix metalloproteinase-2 activation through IκBα/IKK signaling (Philip and Kundu, 2003). Zheng et al. demonstrated that curcumin arrested cell growth at the G(2)/M phase and induced apoptosis in human

TABLE 10.1
Effect of Curcumin on Different Cell Signaling Pathways

Refs.

Inhibition of NF-κB Signaling Pathway

Suppresses the activation of transcription factor NF-κB	Singh and Aggarwal, 1995
Inhibits IL-1α and TNF-induced NF-κB	Xu et al., 1997
Inhibits TPA-induced activation of NF-κB	Surh et al., 2000; Han et al., 2002
Inhibits anticancer drug-induced activation of NF-κB	Chuang et al., 2002
Inhibits TNF production and release	Chan, 1995; Jang et al., 2001
Inhibits inflammatory cytokine production by peripheral blood monocytes and alveolar macrophages	Abe et al., 1999
Regulation of proinflammatory cytokine expression	Literat et al., 2001
Blocks NF-κB activation and proinflammatory gene expression by inhibiting IκB kinase activity	Jobin et al., 1999
Downregulates chemokine expression and release	Xu et al., 1997; Cipriani et al., 2001; Hidaka et al., 2002
Inhibit the angiogenic response stimulated by FGF-2, including expression of MMP-9	Mohan et al., 2000; Lin et al., 1998
Inhibits IL-1–stimulated NF-κB and downregulates MMP gene expression	Liacini et al., 2002; Onodera et al., 2002
Inhibits TNF-mediated cell surface expression of adhesion molecules and of NF-κB activation	Kumar et al., 1998; Gupta & Ghosh, 1999
Reduces endothelial tissue factor gene expression	Bierhaus et al., 1997
Inhibits COX-2 transcription and expression	Plummer et al., 1999; Zhang et al., 1999; Goel et al., 2001; Surh et al., 2001
Inhibits NOS expression and nitrite production	Pan et al., 2000; Surh et al., 2001; Chan et al., 1998; Brouet and Ohshima, 1995; Chan et al., 1995; Onoda and Inano, 2000
Induces p21 (WAF1/CIP1) and C/EBPβ expression	Hour et al., 2002

Inhibition of AP-1 Signaling Pathway

Suppresses PMA-induced c-Jun/AP-1 activation	Han et al., 2002
Inhibits TNF-induced expression of monocyte chemoattractant JE via fos and jun genes	Hanazawa et al., 1993
Inhibits TPA-induced expression of c-fos, c-jun, and c-myc proto-oncogenes mRNAs	Kakar and Roy, 1994
Inhibits TPA- and UV-B light-induced expression of c-Jun and c-Fos	Lu et al., 1994
Reduces endothelial tissue factor gene expression	Bierhaus et al., 1997
Inhibits IL1α and TNF-induced AP-1	Xu et al., 1997
Inhibits TPA-induced activation of AP-1	Surh et al., 2000
Inhibits thrombin-induced, AP-1–mediated, plasminogen activator inhibitor 1 expression	Chen et al., 2000
Inhibits release of MIP-1α, MIP-1β, and RANTES, and AP-1	Cipriani et al., 2001
Inhibits IL-1–stimulated AP-1 and downregulates MMP gene expression	Liacini et al., 2002
Suppresses transcription factor Egr-1	Pendurthi and Rao, 2000
Downregulates transactivation and gene expression of androgen receptors	Nakamura et al., 2002

Inhibition of MAPK Pathway

Inhibits JNK signaling pathway	Chen and Huang, 1998

Continued

TABLE 10.1 *(Continued)*
Effect of Curcumin on Different Cell Signaling Pathways

                                                                              Refs.

Inhibits IL-1-stimulated MAP kinases and downregulates   Liacini et al., 2002
  MMP gene expression

Inhibition of Growth Factor Pathway

Inhibits EGF receptor kinase activity   Korutla and Kumar, 1994
Inhibits ligand-induced activation of EGF receptor tyrosine   Korutla et al., 1995
  phosphorylation
Inhibits PTK activity of p185neu and also depletes p185neu   Hong et al., 1999
Inhibits PTK activity of EGF receptor and depletes the   Dorai et al., 2000
  protein
Inhibition of serine protein kinase pathway
Inhibits protein kinase C activity induced by PMA   Liu et al., 1993
Inhibits phosphorylase kinase   Reddy and Aggarwal, 1994
Inhibits cyclic AMP-dependent protein kinase   Hasmeda and Polya, 1996

Others

Inhibits LOX and COX activities   Huang et al., 1991; Ramsewak et al., 2000; Skrzypczak-
                                                     Jankun et al., 2000
Induces GST activity   Susan and Rao, 1992; Oetari et al., 1996; Awasthi et al., 2000
Inhibits HIV-1 and HIV-2 proteases   Sui et al., 1993
Inhibits PMA-induced xanthine dehydrogenase/oxidase Lin et al., 1994
Modulates brain Na$^+$/K$^+$ ATPase activity   Kaul and Krishnakanth, 1994
Modulates cytochrome P450 activity   Oetari et al., 1996; Thapliyal et al., 2001; Ciolino et al., 1998
Inhibits the Ca^{2+}-ATPase of sarcoplasmic reticulum Logan-Smith et al., 2001; Logan-Smith et al., 2002
Increases the rate of accumulation of Ca^{2+} Logan-Smith et al., 2001
Inhibits SERCA Ca^{2+} pumps   Bilmen et al., 2001; Sumbilla et al., 2002
Inhibits mammalian phospholipase D activity   Yamamoto et al., 1997
Inhibits of IL-12 production in LPS-activated macrophages Kang et al., 1999; Kang et al., 1999
Blocks TGF-β1–induced uPA expression   Santibanez et al., 2000
Induces cell migration in nontumorigenic murine colon Fenton et al., 2002
  epithelial cells through MT-MMP expression
Inhibits heme oxygenase-1   Motterlini et al., 2000; Scapagnini et al., 2002
Modulates aryl hydrocarbon receptor   Ciolino et al., 1998
Modulates *P*-glycoprotein in primary cultures of rat Romiti et al., 1998
  hepatocytes
Intercalates in DNA and poison Topo II isomerase   Snyder and Arnone, 2002
Stimulates the stress-induced expression of stress proteins Kato et al., 1998

Abbreviations: AP-1, activating protein-1; NF-κB, nuclear-factor kappa B, LPS, lipopolysaccharide; IL, interleukin; MMP, matrix metalloproteinase; EGF, epidermal growth factor; PTK, protein tyrosine kinase; COX, cyclooxygenase; LOX, lipoxygenase; GST, glutathione S-transferase; NOS, nitric oxide synthase; TNF, tumor necrosis factor; ATPase, adenosine triphosphatase; FGF, fibroblast growth factor; JNK, c-Jun N-terminal kinase.

TABLE 10.2
Proteins/Enzymes That Physically Interact with Curcumin

Protein/Enzyme	IC$_{50}$	Refs.
Xanthine oxidase	—	Lin and Shih, 1994
Lipooxygenase	—	Skrzypczak-Jankun et al., 2000
Cyclooxygenase-2	—	Ramsewak et al., 2000
IκBα kinase	—	Bharti et al., 2003
P-Glycoprotein	—	Anuchapreeda et al., 2002; Romiti et al., 1998
Glutathione S-transferase	1.79–2.29 μM	Oetari et al., 1996
Protein kinase A	—	Reddy and Aggarwal, 1994
Protein kinase C	—	Reddy and Aggarwal, 1994
Protamine kinase	—	Reddy and Aggarwal, 1994
Phosphorylase kinase	—	Reddy and Aggarwal, 1994
Autophosphorylation-activated protein kinase	—	Reddy and Aggarwal, 1994
pp60c-src tyrosine kinase	—	Reddy and Aggarwal, 1994
Ca^{2+}dependent protein kinase	41 μM	Hasmeda and Polya, 1996
Ca^{2+}-ATPase of sarcoplasmic reticulum	—	Logan-Smith et al., 2001
Aryl hydrocarbon receptor	—	Ciolino et al, 1998
Rat liver cytochrome P450s	—	Oetari et al., 1996
Topo II isomerase	—	Synder and Arnone, 2002
Inositol 1,4,5-triphosphate receptor	10 μM	Dyer et al., 2002
Glutathione	—	Awasthi et al., 2000

melanoma cells by inhibiting NF-κB activation and thus depletion of endogenous nitric oxide (Zheng et al., 2004). Kim et al. recently reported that curcumin inhibited lipopolysaccharide (LPS)-induced mitogen-activated protein kinase (MAPK) activation and the translocation of NF-κB p65 in dendritic cells (Kim et al., 2005).

10.4.3 CURCUMIN DOWNREGULATES THE ACTIVATION OF STAT3 PATHWAY

Numerous reports suggest that interleukin-6 (IL-6) promotes survival and proliferation of various tumors, including multiple myeloma (MM) cells, through the phosphorylation of a cell-signaling protein, signal transducers, and activators of transcription (STAT)-3. Thus, agents that suppress STAT3 phosphorylation have the potential for the treatment of MM (Bharti et al., 2003). Bharti et al. demonstrated that curcumin inhibited IL-6–induced STAT3 phosphorylation and consequent STAT3 nuclear translocation. Curcumin had no effect on STAT5 phosphorylation but inhibited the interferon (IFN)-α–induced STAT1 phosphorylation. The constitutive phosphorylation of STAT3 found in certain MM cells was also abrogated by treatment with curcumin. Curcumin-induced inhibition of STAT3 phosphorylation was reversible. Compared with AG490, a well-characterized Janus kinase (JAK)-2 inhibitor, curcumin was a more rapid (30 min vs. 8 h) and more potent (10 μM vs. 100 μM) inhibitor of STAT3 phosphorylation. Similarly, the dose of curcumin that completely suppressed proliferation of MM cells, AG490 had no effect. In contrast, STAT3-inhibitory peptide that can inhibit the STAT3 phosphorylation mediated by Src blocked the constitutive phosphorylation of STAT3 and also suppressed the growth of myeloma cells. TNF-α and lymphotoxin (LT) also induced the proliferation of MM cells, but through a mechanism independent of STAT3 phosphorylation. In addition, dexamethasone-resistant MM cells were found to be sensitive to curcumin. Overall, these results demonstrated that curcumin was a potent inhibitor of STAT3 phosphorylation and this plays a role in curcumin's suppression of proliferation of MM.

Kim et al. (2003) investigated the inhibitory action of curcumin on JAK-STAT signaling in the brain. Curcumin markedly inhibited the phosphorylation of STAT1 and 3 as well as JAK1 and 2 in rat primary microglia activated with gangliosides, LPS, or IFN-γ (Kim et al., 2003). Li et al. (2001) showed that curcumin suppressed oncostatin-M-stimulated STAT1 phosphorylation, Deoxy ribonucleic acid (DNA)-binding activity of STAT1, and c-Jun N-terminal kinase activation without affecting JAK1, JAK2, JAK3, ERK1/2, and p38 phosphorylation. Curcumin also inhibited Oncostatin M-induced MMP-1, MMP-3, MMP-13, and TIMP-3 gene expression.

Natarajan et al (2002) showed that treatment of activated T-cells with curcumin inhibited IL-12–induced tyrosine phosphorylation of JAK2, tyrosine kinase 2, and STAT3 and STAT4 transcription factors. The inhibition of the JAK–STAT pathway by curcumin resulted in a decrease in IL-12–induced T-cell proliferation and Th1 differentiation.

10.4.4 CURCUMIN ACTIVATE PEROXISOME PROLIFERATOR–ACTIVATED RECEPTOR-γ (PPAR)

Activation of PPAR-γ inhibits the proliferation of nonadipocytes. The level of PPAR-γ is dramatically diminished along with activation of hepatic stellate cells (HSC). Xu et al. (2003) demonstrated that curcumin dramatically induced the gene expression of PPAR-γ and activated PPAR-γ in activated HSC. Blocking its *trans*-activating activity by a PPAR-γ antagonist markedly decreased the effects of curcumin on the inhibition of cell proliferation. Zheng and Chen (2004) reported that curcumin stimulated PPAR γ activity in activated HSC *in vitro*, which was required for curcumin to reduce cell proliferation, induce apoptosis, and suppress extracellular matrix gene expression. Chen and Xu (2005) recently reported that curcumin activation of PPAR γ inhibited Moser cell (human colon cancer cell line) growth and mediated the suppression of the gene expression of cyclin D1 and EGFR.

10.4.5 CURCUMIN DOWNREGULATES THE ACTIVATION OF AP-1 AND JNK

Activated protein-1 (AP-1) is another transcription factor that has been closely linked with proliferation and transformation of tumor cells (Karin et al., 1997). The activation of AP-1 requires the phosphorylation of c-jun through activation of stress-activated kinase JNK (Xia et al., 2000). The activation of JNK is also involved in cellular transformation (Huang et al., 1999). Curcumin has been shown to inhibit the activation of AP-1 induced by tumor promoters (Huang et al., 1991) and JNK activation induced by carcinogens (Chen and Tan, 1998). Bierhaus et al. (1997) demonstrated that curcumin caused inhibition of AP-1 due to its direct interaction with AP-1DNA binding motif (Bierhaus et al., 1997). Prusty and Das (2005) recently reported that curcumin downregulated AP-1–binding activity in tumorigenic HeLa cells.

Dickinson et al. (2003) have demonstrated that the beneficial effects elicited by curcumin appear to be due to changes in the pool of transcription factors that compose EpRE and AP-1 complexes, affecting gene expression of glutamate-cysteine ligase and other phase II enzymes (Dickinson et al., 2003). Squires et al have demonstrated that curcumin suppresses the proliferation of tumor cells through inhibition of Akt/PKB (protein kinase B) activation (Squires et al., 2003).

10.4.6 CURCUMIN SUPPRESSES THE INDUCTION OF ADHESION MOLECULES

The expression of various cell surface adhesion molecules such as intercellular cell adhesion molecule-1, vascular cell adhesion molecule-1, and endothelial leukocyte adhesion molecule-1 on endothelial cells is absolutely critical for tumor metastasis (Ohene-Abuakwa and Pignatelli, 2000). The expression of these molecules is in part regulated by NF-κB (Iademarco et al., 1995). We have shown that the treatment of endothelial cells with curcumin blocks the cell surface expression of adhesion molecules, and this accompanies the suppression of tumor cell adhesion to endothelial cells (Kumar et al., 1998). We have demonstrated that downregulation of these adhesion molecules is mediated through the downregulation of NF-κB activation (Kumar et al., 1998). Gupta and Ghosh (1999) reported that curcumin inhibits TNF-induced expression of adhesion molecules on human umbilical vein endothelial cells (HUVECs). Jaiswal et al. (2002) showed that curcumin treatment causes p53- and p21-independent G(2)/M phase arrest and apoptosis in colon cancer cell lines. Their results suggest that curcumin treatment impairs both Wnt signaling and cell–cell adhesion pathways, resulting in G(2)/M phase arrest and apoptosis in HCT-116 cells.

10.4.7 CURCUMIN DOWNREGULATES COX-2 EXPRESSION

Overexpression of COX-2 has been shown to be associated with a wide variety of cancers, including that of colon (Fournier and Gordon, 2000), lung (Hida et al., 1998), and breast (Harris et al., 2000) cancers. The role of COX-2 in the suppression of apoptosis and tumor cell proliferation has been demonstrated (Williams et al., 1999). Furthermore, celebrex, a specific inhibitor of COX-2, has been shown to suppress mammary carcinogenesis in animals (Reddy et al., 2000). Several groups have shown that curcumin downregulates the expression of COX-2 protein in different tumor cells (Plummer et al., 1999; Chen et al., 1999), most likely through the downregulation of NF-κB activation (Plummer et al., 1999), which is needed for COX-2 expression. Chun et al. (2003) reported that curcumin inhibited phorbol ester–induced expression of COX-2 in mouse skin through suppression of extracellular signal-regulated kinase activity and NF-κB activation. COX-2 has been implicated in the development of many human cancers.

Plummer et al. explored the inhibition of COX-2 activity as a systemic biomarker of drug efficacy, a biomarker of potential use in the clinical trials of many chemopreventive drugs known to inhibit this enzyme. They measured COX-2 protein induction and PGE_2 production in human blood after incubation with LPS. When 1 μM curcumin was added *in vitro* to blood from healthy volunteers, LPS-induced COX-2 protein levels and concomitant PGE_2 production were reduced by 24% and 41%, respectively (Plummer et al., 2001).

10.4.8 CURCUMIN INHIBITS ANGIOGENESIS

For most solid tumors, including breast cancer, angiogenesis (blood vessel formation) is essential for tumor growth and metastasis (Folkman, 2001). The precise mechanism that leads to angiogenesis is not fully understood, but growth factors that cause proliferation of endothelial cells have been shown to play a critical role in this process. Curcumin has been shown to suppress the proliferation of human vascular endothelial cells *in vitro* (Singh et al., 1996) and abrogate the fibroblast growth factor-2–induced angiogenic response *in vivo* (Mohan et al., 2000), thus suggesting that curcumin is also an antiangiogenic factor. CD13/aminopeptidase N (APN) is a membrane-bound, zinc-dependent metalloproteinase that plays a key role in tumor invasion and angiogenesis. Shim et al. (2003) observed that curcumin binds to APN and irreversibly inhibits its activity. Indeed curcumin has been shown to suppress angiogenesis *in vivo* (Arbiser et al., 1998). Dorai et al. (2001) also reported that curcumin inhibits angiogenesis of LNCaP prostate cancer cells *in vivo*.

To elucidate the possible mechanisms of antiangiogenic activity by curcumin, Park et al. (2002) performed cDNA microarray analysis and found that curcumin modulated cell-cycle–related gene expression. Specifically, curcumin induced G0/G1 and/or G2/M phase cell cycle arrest, upregulated CDKIs, p21WAF1/CIP1, p27KIP1, and p53, and slightly downregulated cyclin B1 and cdc2 in ECV304 cells. The upregulation of CDKIs by curcumin played a critical role in the regulation of cell cycle distribution in these cells, which may underlie the antiangiogenic activity of curcumin.

10.4.9 CURCUMIN SUPPRESSES THE EXPRESSION OF MMP-9 AND iNOS

The MMPs make up a family of proteases that play a critical role in tumor metastasis (Kumar et al., 1999). One of them, MMP-9 has been shown to be regulated by NF-κB activation (Lin et al., 1998), and curcumin has been shown to suppress its expression (Lin et al., 1998). Swarnakar et al. (2005) recently reported that curcumin attenuates the activity of MMP-9 during prevention and healing of indomethacin-induced gastric ulcer. Chan et al. (2003) reported that curcumin reduced the production of iNOS mRNA in a concentration-dependent manner in *ex vivo* cultured BALB /c mouse peritoneal macrophages.

Curcumin has also been demonstrated to downregulate iNOS expression, also regulated by NF-κB and involved in tumor metastasis (Pan et al., 2000). These observations suggest that curcumin must have antimetastatic activity. Indeed, there is a report suggesting that curcumin inhibits tumor metastasis (Menon et al., 1999).

10.4.10   CURCUMIN DOWNREGULATES CYCLIN D1 EXPRESSION

Cyclin D1, a component subunit of cyclin-dependent kinase (Cdk)-4 and Cdk6, is the rate-limiting factor in the progression of cells through the first gap (G1) phase of the cell cycle (Baldin et al., 1993). Cyclin D1 has been shown to be overexpressed in many cancers including breast, esophagus, head and neck, and prostate (Bartkova et al., 1994; Adelaide et al., 1995; Caputi et al., 1999; Nishida et al., 1994; Gumbiner et al., 1999; Drobnjak et al., 2000). It is possible that the antiproliferative effects of curcumin are due to inhibition of cyclin D1 expression. We found that curcumin can indeed downregulate cyclin D1 expression (Mukhopadhyay et al., 2001; Bharti et al., 2003; Mukhopadhyay et al., 2002) and this downregulation occurred at the transcriptional and posttranscriptional level. Choudhuri et al. (2005) recently reported that curcumin reversibly inhibits normal mammary epithelial cell cycle progression by downregulating cyclin D1 expression and blocking its association with Cdk4/Cdk6, as well as by inhibiting phosphorylation and inactivation of retinoblastoma protein.

10.4.11   CURCUMIN INHIBITS ANDROGEN RECEPTORS AND AR-RELATED COFACTORS

Nakamura et al. (2002) have evaluated the effects of curcumin in cell growth, activation of signal transduction, and transforming activities of both androgen-dependent and -independent cell lines. The prostate cancer cell lines LNCaP and PC-3 were treated with curcumin, and its effects on signal transduction and expression of androgen receptor (AR) and AR-related cofactors were analyzed. Their results showed that curcumin downregulates transactivation and expression of AR, AP-1, NF-κB, and cAMP response element-binding protein (CREB)-binding protein (CBP). It also inhibited the transforming activities of both cell lines as evidenced by reduced colony-forming ability in soft agar. These studies suggest that curcumin has a potential therapeutic effect on prostate cancer cells through downregulation of AR and AR-related cofactors, AP-1, NF-κB, and CBP (Nakamura et al., 2002).

10.4.12   CURCUMIN INHIBITS FPTASE

Ras proteins must be isoprenylated at a conserved cysteine residue near the carboxyl terminus (Cys-186 in mammalian Ras p21 proteins) in order to extend their biological activity. Previous studies indicate an intermediate in the mevalonate pathway, most likely farnesyl pyrophosphate, is the donor of this isoprenyl group, and that using inhibitors of the mevalonate pathway could block the transforming properties of the ras oncogene. Chen et al. (1997) examined the effects of curcumin on farnesyl protein transferase (FPTase). They found that partially purified FPTase capable of catalyzing the farnesylation of unprocessed Ras p21 proteins *in vitro* was inhibited by curcumin and its derivatives. This is another potential mechanism by which curcumin could suppress cellular growth.

10.4.13   SUPPRESSION OF EGR-1 BY CURCUMIN

The transcription factor, early growth response-1 gene product (Egr-1), is a member of the family of immediate early response genes and regulates a number of pathophysiologically relevant genes in vasculature, which are involved in growth, differentiation, immune response, wound healing, and blood clotting. Pendurthi et al. (2000) investigated the effect of curcumin on Egr-1 expression in endothelial cells and fibroblasts. Gel mobility shift assays showed that pretreatment of endothelial cells and fibroblasts with curcumin suppressed tumor promoting agent (TPA) and serum-induced Egr-1 binding to the consensus Egr-1 binding site and also to the Egr-1 binding site present in the promoter of the tissue factor gene. Western blot analysis revealed that curcumin inhibited TPA-induced *de novo* synthesis of Egr-1 protein in endothelial cells. Suppression of Egr-1 protein expression in curcumin-treated cells stemmed from the suppression of Egr-1 mRNA. Northern blot analysis showed that curcumin inhibited serum and TPA-induced expression of tissue factor and urokinase-type plasminogen activator receptor mRNA in fibroblasts. These results showed that curcumin suppresses the induction of Egr-1 and thereby modulates the expression of Egr-1–regulated genes in endothelial cells and fibroblasts. The downregulation of tissue factor by curcumin has also been demonstrated by another group (Bierhaus et al., 1997).

10.4.14 SUPPRESSION OF MAPKs BY CURCUMIN

Most inflammatory stimuli are known to activate three independent MAPK pathways, leading to activation of p44/42 MAPK (also called ERK1/ERK2), JNK, and p38 MAPK pathway. Chen et al. (1999) found that curcumin inhibits JNK activation induced by various agonists including PMA plus ionomycin, anisomycin, UV-C, gamma radiation, TNF, and sodium orthovanadate. Although both JNK and ERK activation by PMA plus ionomycin were suppressed by curcumin, the JNK pathway was more sensitive. The IC50 (50% inhibition concentration) of curcumin was between 5 and 10 µM for JNK activation and was 20 µM for ERK activation. In transfection assays, curcumin moderately suppressed mitogen-activated protein kinase kinase (MEKK)-1–induced JNK activation; however, it effectively blocked JNK activation caused by cotransfection of TAK1, GCK, or human progenitor kinase 1 (HPK1). Curcumin did not directly inhibit JNK, SEK 1 (SAPK/Erk kinase), MEKK1, or HPK1 activity. Although curcumin suppressed TAK1 and GCK activities at high concentrations, this inhibition cannot fully account for the JNK inhibition by curcumin *in vivo*. Thus, these results suggested that curcumin affected the JNK pathway by interfering with the signaling molecule(s) at the same level or proximally upstream of the mitogen-activated protein kinase kinase kinase (MAPKKK) level. The inhibition of the MEKK1–JNK pathway reveals a possible mechanism of suppression of AP-1 and NF-κB signaling by curcumin and may explain the potent anti-inflammatory and anticarcinogenic effects of this chemical.

10.4.15 SUPPRESSION OF PROTEIN KINASES BY CURCUMIN

Curcumin could also mediate its effects through inhibition of various other serine/threonine protein kinases. Our group showed that treatment of highly purified protein kinase A (PKA), protein kinase C (PKC), protamine kinase (cPK), phosphorylase kinase (PhK), autophosphorylation-activated protein kinase (AK), and pp60c-src tyrosine kinase with curcumin inhibited all kinases. PhK was completely inhibited at low concentration of curcumin (Reddy and Aggarwal, 1994). At around 0.1 mM curcumin, PhK, pp60c-src, PKC, PKA, AK, and cPK were inhibited by 98, 40, 15, 10, 1, and 0.5%, respectively. Lineweaver–Burke plot analysis indicated that curcumin is a noncompetitive inhibitor of PhK, with a Ki of 0.075 mM.

Other investigators have shown suppression of PMA-induced activation of cellular PKC by curcumin (Liu et al., 1993). Treatment of cells with 15 or 20 µM curcumin inhibited TPA-induced PKC activity in the particulate fraction by 26 or 60%, respectively, and did not affect the level of PKC. Curcumin also inhibited PKC activity in both cytosolic and particulate fractions *in vitro* by competing with phosphatidylserine. However, the inhibitory effect of curcumin was reduced after preincubation with the thiol compounds. These findings suggested that the suppression of PKC activity may contribute to the molecular mechanism of inhibition of TPA-induced tumor promotion by curcumin.

Besides *in vitro* suppression, curcumin could also inhibit PKC in the cells (Hasmeda and Polya, 1996). Hasmeda and Polya (1996) showed that curcumin inhibits Ca^{2+}- and phospholipid-dependent PKC and of the catalytic subunit of cyclic AMP-dependent protein kinase (cAK; IC50 values 15 and 4.8 µM, respectively). Curcumin inhibits plant Ca^{2+}-dependent protein kinase (CDPK) (IC50 41 µM), but does not inhibit myosin light chain kinase or a high-affinity 3′,5′-cyclic AMP–binding phosphatase. Curcumin inhibits cAK, PKC, and CDPK in a fashion that is competitive with respect to both ATP (adenosine triphosphatase) and the synthetic peptide substrate employed. The IC50 values for inhibition of cAK by curcumin are very similar when measured with kemptide (LRRASLG) (in the presence or absence of ovalbumin) or with casein or histone III-S as substrates. However, the presence of bovine serum albumin (0.8 mg ml⁻¹) largely overcomes inhibition of cAK by curcumin.

10.5 ANTICANCER PROPERTIES OF CURCUMIN

Several studies indicate that curcumin is a potent anticancer agent (Figure 10.5). The tumorigenesis of the skin, mammary gland, oral cavity, forestomach, oesophagus, stomach, intestine, colon, lung, and liver have been shown to be suppressed by curcumin (Huang et al., 1988; Kuttan et al., 1985, 1987; Rao et al., 1984; Lee et al., 2005; Chuang et al., 2000; Deshpande et al., 1998; Ushida et al., 2000; Limtrakul et al., 1997; Piper et al., 1998) (Table 10.3).

To explain the anticarcinogenic effects of curcumin on different tumors, a wide variety of mechanisms have been implicated, including inhibition of ROI, suppression of inflammation, down-regulation of ODC, inhibition of cell proliferation, inhibition of cytochrome P450 isoenzymes, induction of GSH, suppression of certain oncogenes (e.g., cHa-ras, c-jun, and c-fos), inhibition of transcription factors NF-κB and AP-1, suppression of COX2, inhibition of cell-cycle–related proteins (PCNA, cyclin E, p34 cdc2), inhibition of chromosomal damage, inhibition of oxidation of DNA bases, inhibition of malondialdehyde (MDA) DNA adduct formation, inhibition of tumor implantation, inhibition of protein tyrosine kinase and protein kinase C activity, inhibition of biotransformation of carcinogens, and induction of gluthathione S-transferase (GST) activity (Huang et al., 1988; Kuttan et al., 1985, 1987; Rao et al., 1984; Deshpande et al., 1998; Sharma et al., 2001; Chuang et al., 2002; Tanaka et al., 2004; Inano et al., 1999).

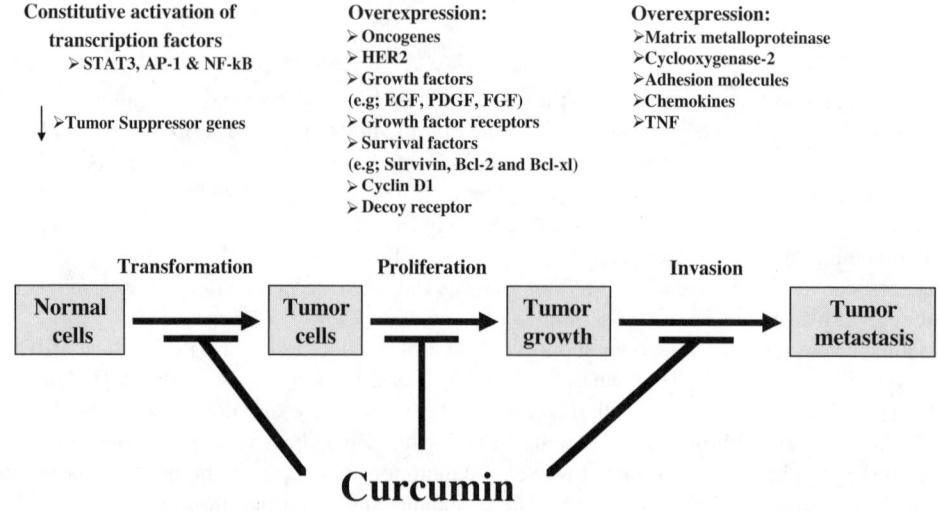

FIGURE 10.5 Antitumor properties of curcumin.

TABLE 10.3
Effects of Curcumin on Survival and Proliferation of Different Cell Types

Cell Type and Effect	Mechanism	Refs.
Inhibits proliferation of breast tumor cells	Inhibition of ODC; arrested cells at G2/S	Mehta et al., 1997; Ramachandran and You, 1999; Simon et al., 1998; Verma et al., 1997
Induction of apoptosis in H-ras transformed MCF10A cells	P53-dependent bax induction	Kim et al., 2001; Choudhari et al., 2002
Induces apoptosis of AK-5 cells	Production of ROI Activation of caspase-3	Bhaumik et al., 1999; Khar et al., 2001
Inhibits proliferation of colon cancer (HT29; HCT-15) cells	Accumulation of cells in G2/M phase	Hanif et al., 1997
Induces apoptosis in colon (LoVo) cancer cells	Accumulates in S, G2/M phase of cell cycle	Chen et al., 1999; Mori et al., 2001; Moragoda et al., 2001
	Induction of HSP70 and p53	Chen et al., 1996; Samaha et al., 1997
Transformed cells	—	Gautam et al., 1998
Fibroblast (NIH3T3), colon cancer (HT29)	Induces cell shrinkage	Ramsewak et al., 2000; Jiang et al., 1996; Nogaki et al., 1998; Pal et al., 2001
Kidney cancer (293), hepatocellular Carcinoma (HepG2)	Chromatin condensation DNA fragmentation	Park et al., 2005
Induces apoptosis of leukemia cells	—	Kuo et al., 1996
Induces apoptosis of T-cell leukemia (Jurkat) modulation by GSH	Independent of mitochondria and caspases	Piwocka et al., 1999; Piwocka et al., 2001
Induces growth arrest and apoptosis of B-cell lymphoma	Downregulation of Egr-1 c-myc, bcl-xl, NF-κB and p53	Han et al., 1999
Induces apoptosis in HL-60 cells	Increases Sub-G1; activates caspase-3	Bielak-Zmijewska et al., 2000
Induces apoptosis of myeloid (HL-60) cells	Cytochrome C release; caspase-3 activation; loss of MMTP; caspase-9 activation	Pan et al., 2001
Induces apoptosis of myeloid (HL-60) cells	Caspase-8 activation; BID cleavage; cytochrome C release	Anto et al., 2002
Induces apoptosis of basal cell carcinoma cells	p53-dependent	Jee et al., 1998
Inhibits proliferation of prostate cancer cells	Downregulation of bcl-2 and NF-κB	Mukhopadhyay et al., 2001
Induces apoptosis of melanoma cells	Fas/FLICE pathway; p53-independent	Bush et al., 2001
Inhibits the proliferation of HUVEC	Accumulation of cells in S-phase	Singh et al., 1996; Park et al., 2002
Inhibits proliferation of oral epithelial cells	—	Khafif et al., 1998; Elattar and Virji, 2000
Induces apoptosis of T-lymphocytes	Loss of mitochondrial membrane potential, plasma membrane asymmetry & permeability; GSH-independent	Sikora et al., 1997; Jaruga et al., 1998
Induces apoptosis of γδT-cells	Increase in annexin V reactivity; nuclear expression of active caspase-3; cleavage of PARP; nuclear disintegration; translocation of AIF to the nucleus; large-scale DNA chromatolysis	Cipriani et al., 2001
Induced apoptosis of breast cancer cells	P53-dependent Bax induction	Chowdhuri et al., 2002

Continued

TABLE 10.3 *(Continued)*
Effects of Curcumin on Survival and Proliferation of Different Cell Types

Cell Type and Effect	Mechanism	Refs.
Induces apoptosis of osteoclasts	—	Ozaki et al., 2000
Induces apoptosis in VSMC	Reduction in the S-phase; increase in G0/G1 phase; increase in TUNNEL-positive cells; DNA fragmentation; decrease in mRNA for c-myc and bcl-2 but not p53; inhibition of PKC and PTK	Chen and Huang, 1998
Inhibits the PDGF-induced proliferation of VSMC	—	Huang et al., 1992a
Inhibits the PHA-induced proliferation of PBMC	—	Huang et al., 1992b
Induces apoptosis in hepatocytes	Increase MMTP, loss of MMP, mitochondrial swelling; inhibition of ATP synthesis, oxidation of membrane thiol	Morin et al., 2001; Gomez-Lechon et al., 2002

Abbreviations: ROI, reactive oxygen intermediates; HUVEC, human umbilical vein vascular endothelial cells; MMTP, mitochondrial membrane permeability transition pore; GSH, glutathione; VSMC, vascular smooth muscle cells; PARP, poly(ADP-ribose) polymerase; AIF, apoptosis-inducing factor; MMP, mitochondrial membrane potential; TUNEL, TdT-mediated dUTP nick end labeling; PKC, protein kinase C; PTK, protein tyrosine kinase; PBMC, peripheral blood mono-nuclear cells; PHA, phytohemagglutinin.

10.5.1 *In Vitro Studies*

Curcumin has been shown to inhibit the proliferation of a wide variety of tumor cells, including B-cell and T-cell leukemia (Kuo et al., 1996; Ranjan et al., 1999; Piwocka et al., 1999; Han et al., 1999), colon carcinoma (Chen et al., 1999), epidermoid carcinoma (Korutla and Kumar, 1994), head and neck squamous cell carcinoma (Aggarwal et al., 2004), MM (Bharti et al., 2003), and mantle cell lymphoma (Shishodia et al., 2005). It has also been shown to suppress the proliferation of various breast carcinoma cell lines in culture (Mehta et al., 1997; Ramachandran and You, 1999; Simon et al., 1998).

Mehta et al. (1997) examined the antiproliferative effects of curcumin against several breast tumor cell lines, including hormone-dependent and -independent and multidrug-resistance (MDR) lines. Cell growth inhibition was monitored by [3H] thymidine incorporation, Trypan blue exclusion, crystal violet dye uptake, and flow cytometry. All the cell lines tested, including the MDR-positive ones, were highly sensitive to curcumin. The growth inhibitory effect of curcumin was time- and dose dependent, and correlated with its inhibition of ornithine decarboxylase activity. Curcumin preferentially arrested cells in the G2/S phase of the cell cycle.

Thioredoxin reductases (TrxR) have been found to be overexpressed by a number of human tumors. Fang et al. (2005) reported that rat TrxR1 activity in Trx-dependent disulfide reduction was inhibited by curcumin. The IC50 value for the enzyme was 3.6 µM after incubation at room temperature for 2 h *in vitro*. The inhibition occurred with enzyme only in the presence of NADPH (nicotinamide adenine dinucleotide phosphate, reduced form) and persisted after removal of curcumin. By using mass spectrometry and blotting analysis, they showed that this irreversible inhibition by curcumin was caused by alkylation of both residues in the catalytically active site (Cys (496)/Sec (497)) of the enzyme. Inhibition of TrxR by curcumin added to cultured HeLa cells was also observed with an IC50 of around 15 µM. Modification of TrxR by curcumin provides a possible mechanistic explanation for its cancer-preventive activity.

Kang et al. reported that the exposure of human hepatoma cells to curcumin led to a significant decrease in histone acetylation. Curcumin treatment resulted in a comparable inhibition of histone acetylation in the absence or presence of trichostatin A and showed no effect on the *in vitro* activity of HDAC (histone deacetylase). Curcumin treatment significantly inhibited the HAT (histone acetyltraseferase) activity both *in vivo* and *in vitro*. Curcumin-induced hypoacetylation led to the loss of cell viability in human hepatoma cells (Kang et al., 2005).

The transcription factor AP-1 plays a central role in the transcriptional regulation of specific types of high-risk human papillomaviruses (HPVs) such as HPV16 and HPV18, which are etiologically associated with the development of cancer of the uterine cervix in women. Prusty et al. (2005) showed that curcumin can selectively downregulate HPV18 transcription as well as the AP-1–binding activity in HeLa cells. Most interestingly, curcumin can reverse the expression dynamics of c-fos and fra-1 in this tumorigenic cell line.

Curcumin synergized with the chemotherapeutic agent Vinorelbine in suppressing the growth of human squamous cell lung carcinoma H520 cells. Both the agents caused apoptosis by increasing the protein expression of Bax and Bcl-Xl while decreasing that of Bcl-2 and Bcl-X (L), releasing apoptogenic cytochrome c and augmenting the activity of caspase-9 and caspase-3. Curcumin treatment induced 23.7% apoptosis in the H520 cells, while Vinorelbine caused 38% apoptosis. Pretreatment with curcumin enhanced the Vinorelbine-induced apoptosis to 61.3%. These findings suggest that curcumin has the potential to act as an adjuvant chemotherapeutic agent and enhance chemotherapeutic efficacy of Vinorelbine in H520 cells *in vitro* (Sen et al., 2005).

Curcumin significantly inhibited the growth of human gastric carcinoma (AGS) cells in a dose- and time-dependent manner (Koo et al., 2004). Curcumin caused a 34% decrease in AGS proliferation at 5 µmol/L, 51% at 10 µmol/L, and 92% at 25 µmol/L after 4 d of treatment. When curcumin (10 µmol/L) was removed after a 24-h exposure, the growth pattern of curcumin-treated AGS cells was similar to that of control cells, suggesting reversibility of curcumin on the growth of AGS cells. Combining curcumin with 5-FU (5-fluorouracil) significantly increased growth inhibition of AGS cells compared with either curcumin or 5-FU alone, suggesting synergistic actions of the two drugs. After 4 d of treatment with 10 µmol/L of curcumin, the G2/M phase fraction of cells was 60.5% compared to 22.0% of the control group, suggesting a G2/M block by curcumin treatment. The curcumin concentrations (5 µmol/L) used in this study were similar to steady-state concentrations (1.77 ± 1.87 µmol/L) in human serum of subjects receiving chronic administration of a commonly recommended dose (8 g/d). Thus, curcumin may be useful for the treatment of gastric carcinoma, especially in conjunction with 5-FU.

Using time-lapse video and immunofluorescence-labeling methods, Holy (2004) demonstrated that curcumin significantly alters microfilament organization and cell motility in PC-3 and LNCaP human prostate cancer cells *in vitro*. Curcumin rapidly arrests cell movements and subsequently alters cell shape in the highly motile PC-3 cell line, but has a less noticeable effect on the relatively immobile LNCaP cell line. Stress fibers are augmented, and the overall quantity of f-actin appears to increase in both types of cells following curcumin treatment. At least some of the effects of curcumin appear to be mediated by protein kinase C (PKC), as treatment with the PKC inhibitor, bisindolylmaleimide, inhibits the ability of curcumin to block CB (cytochalasin B)-induced membrane blebbing. These findings demonstrate that curcumin exerts significant effects on the actin cytoskeleton in prostate cancer cells, including altering microfilament organization and function. This may represent an important mechanism by which curcumin functions as a chemopreventative agent, and as an inhibitor of angiogenesis and metastasis.

Chemoresistance is a major problem in the treatment of patients with MM due to constitutive expression of NF-κB and STAT-3. Bharti et al. (2003, 2004) showed that the suppression of NF-κB and STAT3 activation in MM cells by *ex vivo* treatment with curcumin resulted in a decrease in adhesion to bone marrow stromal cells, cytokine secretion, and the viability of cells.

Helicobacter pylori is a Group 1 carcinogen and is associated with the development of gastric and colon cancer. A methanol extract of the dried powdered turmeric rhizome and curcumin were tested against 19 strains of *H. pylori*, including 5 cagA+ strains. Both the methanol extract and curcumin inhibited the growth of all strains of *H. pylori in vitro*, with a minimum inhibitory concentration range of 6.25 to 50 μg/ml (Mahady et al., 2002). These data demonstrate that curcumin inhibits the growth of *H. pylori* cagA+ strains *in vitro*, and this may be one of the mechanisms by which curcumin exerts its chemopreventative effects.

Chen et al. (2004) used microarray analysis of gene expression profiles to characterize the anti-invasive mechanisms of curcumin in highly invasive lung adenocarcinoma cells (CL1-5). Results showed that curcumin significantly reduces the invasive capacity of CL1-5 cells in a concentration range far below its levels of cytotoxicity (20 μM) and that this anti-invasive effect was concentration dependent (10.17 ± 0.76 x 10^3 cells at 0 μM; 5.67 ± 1.53 x 10^3 cells at 1 μM; 2.67 ± 0.58 x 10^3 cells at 5 μM; $1.15 \pm 1.03 \times 10^3$ cells at 10 μM; $p < 0.05$) in the Transwell cell culture chamber assay. Using microarray analysis, 81 genes were downregulated and 71 genes were upregulated after curcumin treatment. Below sublethal concentrations of curcumin (10 μM), several invasion-related genes were suppressed, including MMP14 (0.65-fold), neuronal cell adhesion molecule (0.54-fold), and integrins alpha6 (0.67-fold) and beta4 (0.63-fold). In addition, several heat-shock proteins (Hsp) (Hsp27 [2.78-fold], Hsp70 [3.75-fold], and Hsp40-like protein [3.21-fold]) were induced by curcumin. Real-time quantitative reverse transcription-polymerase chain reaction (RT-PCR), Western blotting, and immunohistochemistry confirmed these results in both RNA and protein levels. Curcumin (1 to 10 μM) reduced the MMP14 expression in both mRNA and protein levels and also inhibited the activity of MMP2, the downstream gelatinase of MMP14, by gelatin zymographic analysis.

Kim et al. (2002) evaluated the antiangiogenic activity of demethoxycurcumin (DC), a structural analog of curcumin, and investigated the effect of DC on genetic reprogramming in cultured HUVECs using cDNA microarray analysis. Of the 1024 human cancer-focused genes arrayed, 187 genes were upregulated and 72 genes were downregulated at least twofold by DC. Interestingly, nine angiogenesis-related genes were downregulated over fivefold in response to DC, suggesting that the genetic reprogramming was crucially involved in antiangiogenesis by the compound. MMP-9, the product of one of the angiogenesis-related genes downregulated over fivefold by DC, was investigated using gelatin zymography. DC potently inhibited the expression of MMP-9, yet showed no direct effect on its activity.

10.5.2 *In Vivo* Studies

Numerous studies have been performed to evaluate the chemopreventive properties of curcumin (Table 10.4). Kuttan et al (1985) examined the anticancer potential of curcumin *in vivo* in mice using Dalton's lymphoma cells grown as ascites. Initial experiments indicated that curcumin reduced the development of animal tumors. They encapsulated curcumin (5 mg/ml) into neutral and unil-amelar liposomes prepared by sonication of phosphatidylcholine and cholesterol. An aliquot of liposomes (50 mg/kg) was given i.p. to mice the day after giving the Dalton's lymphoma cells and continued for 10 days. After 30 and 60 days, the surviving animals were counted. When curcumin was used in liposomal formulations at concentration of 1 mg/animal, all animals survived 30 days and only two of the animals developed tumors and died before 60 days.

Busquets et al. (2001) showed that systemic administration of curcumin (20 μg/kg body weight) for six consecutive days to rats bearing the highly cachectic Yoshida AH-130 ascites hepatoma resulted in an important inhibition of tumor growth (31% of total cell number). Interestingly, curcumin was also able to reduce by 24% *in vitro* tumor cell content at concentrations as low as 0.5 μM without promoting any apoptotic events. Although systemic administration of curcumin has previously been shown to facilitate muscle regeneration, administration of the compound to tumor-bearing rats did not result in any changes in muscle wasting, when compared with the untreated tumor-bearing animals. Indeed, both the weight and protein content of the gastrocnemius muscle significantly decreased as a result of tumor growth, and curcumin was unable to reverse

TABLE 10.4
Chemopreventive Effects of Curcumin

Effects	Refs.
Inhibits tumor promotion in mouse skin by TPA	Huang et al., 1988
Inhibits TPA-induced increase in mRNA for ODC in mouse epidermis	Lu et al., 1993
Inhibits chemical carcinogen-induced stomach and skin tumors in Swiss mice	Azuine and Bhide, 1992
Inhibits TPA- and UV-B light-induced expression of c-Jun and c-Fos in mouse epidermis	Lu et al., 1994
Inhibits TPA-induced tumor promotion by curcumin analogues	Huang et al., 1995
Inhibits UV-A-induced ODC induction and dermatitis induced by TPA in mouse skin	Ishizaki et al., 1996
Inhibits TPA-induced tumor promotion and oxidized DNA bases in mouse epidermis	Huang et al., 1997
Inhibits skin carcinogenesis in mice	Limtrakul et al., 1997
Inhibits chemical carcinogenesis by curcumin	Soudamini and Kuttan, 1989
Protection from fuel smoke condensate-induced DNA damage in human lymphocytes	Shalini and Srinivas, 1990
Reverses aflatoxin-induced liver damage	Soni and Kuttan, 1992
Inhibits BPDE-induced tumor initiation	Huang et al., 1992a
Inhibits azoxymethanol-induced colonic epithelial cell proliferation and focal areas of dysplasia	Huang et al., 1992b
Inhibits azoxymethane-induced aberrant crypt foci formation in the rat colon	Rao e al., 1993
Inhibits forestomach, duodenal, and colon carcinogenesis in mice	Huang et al.,1994
Prevents colon carcinogenesis	Rao et al., 1995
Prevents 1,2-dimethylhydrazine initiated mouse colon carcinogenesis	Kim et al., 1998
Inhibits the promotion/progression stages of colon cancer	Kawamori et al., 1999
Induces genetic reprogramming in pathways of colonic cell maturation	Mariadason et al., 2000
Inhibits PhIP-induced tumor formation in Apc(min) mice	Collett et al., 2001
Inhibits GST and MDA–DNA adducts in rat liver and colon mucosa	Sharma et al., 2001
Prevents familial adenomatous polyposis	Perkins et al., 2002
Suppresses methyl (acetoxymethyl) nitrosamine-induced hamster oral carcinogenesis	Azuine and Bhide, 1992b
Inhibits 4-nitroquinoline 1-oxide-induced oral carcinogenesis	Tanaka et al., 1994
Inhibits experimental cancer: in forestomach and oral cancer models	Azuine and Bhide, 1994
Inhibits benzo[a]pyrene-induced forestomach cancer in mice	Singh et al., 1998
Inhibits N-nitrosomethylbenzylamine-induced esophageal carcinogenesis in rats.	Ushida et al., 2000
Prevents N-methyl-N'-nitro-N-nitrosoguanidine and NaCl-induced glandular stomach carcinogenesis	Ikezaki et al.,2001
Inhibits azoxymethane-induced colon cancer and DMBA-induced mammary cancer in rats	Pereira et al., 1996
Inhibits DMBA-induced mammary tumorigenesis and DMBA-DNA adduct formation	Singletary et al., 1996
Prevents DMBA-induced rat mammary tumorigenesis	Deshpande et al., 1998
Inhibits formation of DMBA-induced mammary tumors and lymphomas/leukemias in Sencar mice	Huang et al., 1998
Inhibits the promotion stage of tumorigenesis of mammary gland in rats irradiated with γ-rays	Inano et al., 1999
Prevents radiation-induced initiation of mammary tumorigenesis in rats	Inano and Onoda, 2002; Inano et al., 2000
Inhibits DMBA-induced mammary tumorigenesis by dibenzoylmethane, a analogue of curcumin	Lin et al., 2001

Continued

TABLE 10.4 *(Continued)*
Chemopreventive Effects of Curcumin

Effects	Refs.
Inhibits mammary gland proliferation, formation of DMBA–DNA adducts in mammary glands and mammary tumorigenesis in Sencar mice by dietary dibenzoylmethane	Lin et al., 2001
Inhibits B[a]P plus NNK-induced lung tumorigenesis in A/J mice	Hecht et al., 1999
Inhibits the initiation stage in a rat multiorgan carcinogenesis model	Takaba et al., 1997
Retards experimental tumorigenesis and reduction in DNA adducts	Krishnaswamy et al., 1998
Induces glutathione linked detoxification enzymes in rat liver	Piper et al., 1998
Inhibits TPA-induced ODC activity	Lee and Pezzuto, 1999
Structurally related natural diarylheptanoids antagonize tumor promotion	Chun et al., 1999
Inhibits aflatoxin B(1) biotransformation	Lee et al., 2001
Inhibits B[a]P-induced cytochrome P-450 isozymes	Thapliyal et al., 2001
Inhibits diethylnitrosamine-induced murine hepatocarcinogenesis	Chuang et al., 2000
Inhibits diethylnitrosamine-induced hepatic hyperplasia in rats	Chuang et al., 2000
Inhibits Purnark against benzo(a)pyrene-induced chromosomal damage in human lymphocytes	Ghaisas and Bhide, 1994
Inhibits carcinogen induced c-Ha-ras and c-fos proto-oncogenes expression	Limtrakul et al., 2001
Prevents intravesical tumor growth of the MBT-2 tumor cell line following implantation in C3H mice	Sindhwani et al., 2001
Suppresses growth of hamster flank organs by topical application	Liao et al., 2001
Inhibits Epstein–Barr virus BZLF1 transcription in Raji DR-LUC cells	Hergenhahn et al., 2002
Prevents radiation-induced mammary tumors	Inano and Onoda, 2002
Induces GST activity by curcumin in mice	Susan and Rao, 1992

Abbreviations: BPDE, benzo[a]pyrene diolepoxide; TPA, 12-*O*-tetradecanoylphorbol-13-acetate; ODC, ornithine decarboxylase; DMBA, 7,12-dimethylbenz[a]anthracene; GST, glutathione *S*-transferase; MDA, malondialdehyde; PhIP, 2-amino 1-methyl-6-phenylimidazo[4,5-b]pyridine; BaP, benzo[a]pyrene; NNK, 4-(methyl-nitrosamino)-1-(3-pyridyl)-1-butanone.

this tendency. It was concluded that curcumin, in spite of having clear antitumoral effects, has little potential as an anticachectic drug in the tumor model used in the study.

Menon et al. (1995) reported curcumin-induced inhibition of B16F10 melanoma lung metastasis in mice. Oral administration of curcumin at concentrations of 200 nmol/kg body weight reduced the number of lung tumor nodules by 80%. The life span of the animals treated with curcumin was increased by 143.85% (Menon et al., 1995). Moreover, lung collagen hydroxyproline and serum sialic acid levels were significantly lower in treated animals than in the untreated controls. Curcumin treatment (10 µg/ml) significantly inhibited the invasion of B16F-10 melanoma cells across the collagen matrix of a Boyden chamber. Gelatin zymographic analysis of the trypsin-activated B16F-10 melanoma cells sonicate revealed no metalloproteinase activity. Curcumin treatment did not inhibit the motility of B16F-10 melanoma cells across a polycarbonate filter *in vitro*. These findings suggest that curcumin inhibits the invasion of B16F-10 melanoma cells by inhibition of MMPs, thereby inhibiting lung metastasis.

Curcumin decreases the proliferative potential and increases apoptotic potential of both androgen-dependent and androgen-independent prostate cancer cells *in vitro*, largely by modulating the apoptosis suppressor proteins and by interfering with the growth factor receptor signaling pathways as exemplified by the EGF receptor. To extend these observations, Dorai et al. (2001) investigated the anticancer potential of curcumin in a nude mouse prostate cancer model. The androgen-dependent LNCaP prostate cancer cells were grown, mixed with Matrigel, and injected subcutaneously. The experimental group received a synthetic diet containing 2% curcumin for up to 6 weeks. At the end point, mice were killed, and sections taken from the excised tumors were evaluated for pathology, cell proliferation, apoptosis, and vascularity. Results showed that curcumin induced a marked decrease in the extent of cell proliferation as measured by the BrdU incorporation assay and a significant increase in the extent of apoptosis as measured by an *in situ* cell death assay. Moreover, a significant decrease in the microvessel density as measured by CD31 antigen staining was also seen. It was concluded that curcumin was a potentially therapeutic anticancer agent, as it significantly inhibited prostate cancer growth, as exemplified by LNCaP *in vivo*, and it had the potential to prevent the progression of this cancer to its hormone refractory state. Aggarwal et al. (2004) recently reported that curcumin inhibits growth and survival of human head and neck squamous cell carcinoma cells via modulation of NF-κB signaling.

The chemopreventive activity of curcumin was observed when it was administered prior to, during, and after carcinogen treatment, as well as when it was given only during the promotion/progression phase of colon carcinogenesis (Kawamori et al., 1999). Collett et al. (2001) investigated the effects of curcumin on apoptosis and tumorigenesis in male apc (min) mice treated with the human dietary carcinogen PhIP. The intestinal epithelial apoptotic index in response to PhIP treatment was approximately twice as great in wild type C57BL/6 APC (+/+) strain as in Apc (min) mice. Curcumin enhanced PhIP-induced apoptosis and inhibited PhIP-induced tumorigenesis in the proximal small intestine of Apc (min) mice. Mahmoud et al. (2000) investigated the effect of curcumin for the prevention of tumors in C57BL/6J-Min/+ (Min/+) mice that bear germline mutation in the apc gene and spontaneously develop numerous intestinal adenomas by 15 weeks of age. At a dietary level of 0.15%, curcumin decreased tumor formation in Min-/- mice by 63%. Examination of intestinal tissue from the treated animals showed the tumor prevention by curcumin was associated with increased enterocyte apoptosis and proliferation. Curcumin also decreased expression of the oncoprotein β-catenin in the erythrocytes of the Min/+mouse, an observation previously associated with an antitumor effect.

Recently, Perkins et al. (2002) also examined the preventive effect of curcumin on the development of adenomas in the intestinal tract of the C57BL/6J-Min/+ mouse, a model of human familial APC. These investigators explored the link between its chemopreventive potency in the Min/+ mouse and levels of drug and metabolites in target tissue and plasma. Mice received dietary curcumin for 15 weeks, after which adenomas were enumerated. Levels of curcumin and metabolites were determined by high-performance liquid chromatography in plasma, tissues, and feces of mice after either long-term ingestion of dietary curcumin or a single dose of [^{14}C] curcumin (100 mg/kg) intraperitoneally. Whereas curcumin at 0.1% in the diet was without effect, however, at 0.2 and 0.5%, it reduced adenoma multiplicity by 39 and 40%, respectively.

Odot et al. (2004) showed that curcumin was cytotoxic to B16-R melanoma cells resistant to doxorubicin. They demonstrated that the cytotoxic effect observed was due to the induction of programmed cell death. They examined the effectiveness of a prophylactic immune preparation of soluble proteins from B16-R cells, or a treatment with curcumin as soon as tumoral appearance, alone or in combination, on the murine melanoma B16-R. The combination treatment resulted in substantial inhibition of growth of B16-R melanoma, whereas each treatment by itself showed little effect. Moreover, animals receiving the combination therapy exhibited an enhancement of their humoral antisoluble B16-R protein immune response and a significant increase in their median survival time (> 82.8% vs. 48.6% and 45.7%, respectively for the immunized group and the curcumin-treated group).

10.6 CURCUMIN AND CHEMOSENSITIVITY

Chemosensitivity is the susceptibility of tumor cells to the cell-killing effects of anticancer drugs. Most of the chemotherapeutic agents frequently induce drug resistance. HER2, a growth factor receptor overexpressed in breast cancer, has been implicated in Taxol-induced resistance, probably through the activation of NF-κB. Acquired resistance to chemotherapeutic agents is most likely mediated through a number of mechanisms including the gene product MDR protein. Multidrug resistance is a phenomenon that is often associated with decreased intracellular drug accumulation in the tumor cells of a patient, resulting from enhanced drug efflux. It is often related to the overexpression of P-glycoprotein on the surface of tumor cells, thereby reducing drug cytotoxicity. Curcumin has been shown to augment the cytotoxic effects of chemotherapeutic drugs, including doxorubicin (Harbottle et al., 2001), tamoxifen (Verma et al., 1998), cisplatin and camptothecin, daunorubicin, vincristine, and melphalan (Bharti et al., 2003). Taxol has a major disadvantage in its dose-limiting toxicity. Bava et al. (2005) reported that combination of Taxol with curcumin augments anticancer effects more efficiently than Taxol alone. This combination at the cellular level augments activation of caspases and cytochrome c release. Similarly, the combination of curcumin with cisplatin resulted in a synergistic antitumor activity in hepatic cancer HA22T/VGH cell line, which constitutively expresses activated NF-κB. Combination of curcumin with cisplatin led to an additive decrease in the expression of c-myc, Bcl-X_L, c-IAP-2, and XIAP (Notarbartolo et al., 2005)

Curcumin was found to be cytotoxic to B16-R melanoma cells resistant to doxorubicin either cultivated as monolayers or grown in three-dimensional (3D) cultures (spheroids). A prophylactic immune preparation of soluble proteins from B16-R cells, or a treatment with curcumin upon tumor appearance, alone or in combination, on the murine melanoma B16-R resulted in substantial inhibition of growth of B16-R melanoma, whereas each treatment by itself showed little effect. Moreover, animals receiving the combination therapy exhibited an enhancement of their humoral antisoluble B16-R protein immune response and a significant increase in their median survival time (Odot et al., 2004).

NF-κB has been implicated in the development of drug resistance in cancer cells. The basal level of NF-κB activity has been found to be heterogeneous in various cancer cells and roughly correlated with drug resistance. Curcumin has been shown to downregulate doxorubicin-induced NF-κB activation. (Chuang et al., 2002). Multidrug resistance (MDR) is a major cause of chemotherapy failure in cancer patients. One of the resistance mechanisms is the overexpression of drug efflux pumps such as P-glycoprotein and multidrug resistance protein 1 (MRP1, [ABCC1]). Upon treating the cells with etoposide in the presence of 10 μM curcuminoids, the sensitivity of etoposide was increased by several folds in MRP1 expressing HEK 293 cells (Limtrakul et al., 2004). Curcumin also decreased P-glycoprotein function and expression, and the promotion of caspase-3 activation in MDR gastric cancer cells. Gastric cancer cells treated with curcumin decreased the IC50 value of vincristine and promoted vincristine-mediated apoptosis in a dose-dependent manner. Curcumin reversed the MDR of the human gastric carcinoma SGC7901/VCR cell line (Tang et al., 2005). Curcumin decreased P glycoprotein expression in a concentration-dependent manner and was also found to have the same effect on MDR1 mRNA levels (Anuchapreeda et al., 2002; Romiti et al., 1998). The effect of curcumin on apoptosis in multidrug resistant cell lines has been reported. Piwocka et al. demonstrated that curcumin induced cell death in multidrug-resistant CEM(P-gp4) and LoVo(P-gp4) cells in a caspase-3–independent manner (Piwocka et al., 2002). Mehta et al. (1997) also examined the antiproliferative effects of curcumin against multidrug-resistant (MDR) lines, which were found to be highly sensitive to curcumin. The growth-inhibitory effect of curcumin was time- and dose dependent and correlated with its inhibition of ornithine decarboxylase activity. Curcumin preferentially arrested cells in the G2/S phase of the cell cycle.

Subtoxic concentrations of curcumin sensitize human renal cancer cells to TRAIL (TNF-related apoptosis inducing ligand)-mediated apoptosis. Apoptosis induced by the combination of curcumin and TRAIL is not interrupted by Bcl-2 overexpression. Treatment with curcumin significantly induces death receptor 5 expression accompanying the generation of the reactive oxygen species (ROS) (Jung et al., 2005). Curcumin causes cell death in melanoma cell lines with mutant p53. Since melanoma cells with mutant p53 are strongly resistant to conventional chemotherapy, curcumin overcomes the chemoresistance of these cells and provides potential new avenues for treatment (Bush et al., 2001).

10.7 RADIOSENSITIZING EFFECTS OF CURCUMIN

Radiotherapy plays an important role in the management of cancers. While its role in achieving local control following surgery in patients with early stage cancer is well established, there is still unclear evidence to explain the factors governing radioresistance in patients who develop recurrences.

Chendil et al. (2004) investigated the radiosensitizing effects of curcumin in p53 mutant prostate cancer cell line PC-3. Compared to cells that were irradiated alone, curcumin at 2 and 4 µM concentrations in combination with radiation showed significant enhancement to radiation-induced clonogenic inhibition and apoptosis. In PC-3 cells, radiation upregulated TNFα protein, leading to an increase in NF-κB activity resulting, in the induction of Bcl-2 protein. However, curcumin in combination with radiation treated showed inhibition of TNFα-mediated NF-κB activity, resulting in bcl-2 protein downregulation. The results suggested that curcumin is a potent radiosensitizer, and it acts by overcoming the effects of radiation-induced prosurvival gene expression in prostate cancer.

Khafif et al. (2005) investigated whether curcumin can sensitize squamous cell carcinoma (SCC) cells to the ionizing effects of irradiation. Incubation with curcumin only (3.75 µM) for 48 hours did not decrease the number of cells or the ability to form colonies in the absence of radiation. However, in plates that were exposed to 1 to 5 Gy of radiation, cell counts dropped significantly if pretreated with curcumin with a maximal effect at 2.5 Gy. The colonogenic assay revealed a significant decrease in the ability to form colonies following pretreatment with curcumin at all radiation doses. Thus, curcumin may serve as an adjuvant in radiotherapy.

10.7.1 CURCUMIN CAN INDUCE RADIOPROTECTION

There are several studies that suggest that curcumin is radioprotective. Thresiamma et al. (1996) showed that curcumin protects from radiation-induced toxicity. They showed that whole body irradiation of rats (10 Gy as five fractions) produces lung fibrosis within 2 months as seen from increased lung collagen hydroxyproline and histopathology. Oral administration of curcumin (200 µmole/kg body weight) significantly reduced the lung collagen hydroxyproline. In serum and liver LPO increased by irradiation was reduced significantly by curcumin treatment. The liver superoxide dismutase (SOD) and GSH peroxidase activity increased was reduced significantly by curcumin. Curcumin also significantly reduced the whole body irradiation-induced increased frequency of micronucleated polychromatic erythrocytes in mice. In another study, Thresiamma et al. (1998) later investigated the protective effect of curcumin on radiation-induced genotoxicity. They showed that induction of micronuclei and chromosomal aberrations produced by whole body exposure of γ-radiation (1.5–3 Gy) in mice was significantly inhibited by oral administration of curcumin (400 µmol/kg body weight), inhibited micronucleated polychromatic and normochromatic erythrocytes, significantly reduced the number of bone marrow cells with chromosomal aberrations and chromosomal fragments, and inhibited the DNA strand breaks produced in rat lymphocytes upon radiation as seen from the DNA unwinding studies (Thresiamma et al., 1998). In contrast to these studies, Araujo et al. (1999) showed potentiation by curcumin of γ-radiation–induced chromosome aberrations in Chinese hamster ovary cells. They treated the cells with curcumin (2.5, 5, and 10 µg/ml), and then irradiated (2.5 Gy) during different phases of the cell cycle. Curcumin at 10 µg/ml enhanced the chromosomal damage frequency. Curcumin did not show protective effect against the clastogenicity of γ-radiation. Instead, an obvious increase in the frequencies of chromosome aberrations was observed with curcumin at 10 µg/ml plus γ-radiation during S and G2/S phases of the cell cycle. Why there is a difference in the results of Thresiamma et al. (1998) and that of Araujo et al. (1999) is unclear.

Inano and Onoda (2002) investigated the radioprotective action of curcumin on the formation of urinary 8-hydroxy-2′-deoxyguanosine (8-OHdG), tumorigenesis, and mortality induced by gamma-ray irradiation. The evaluation of the protective action of dietary curcumin (1%, w/w) against the long-term effects revealed that curcumin (1%, w/w) significantly decreased the incidence of mammary and pituitary tumors. However, the experiments on survival revealed that curcumin was not effective when administered for 3 days before and/or 3 days after irradiation (9.6 Gy). These findings demonstrate that curcumin can be used as an effective radioprotective agent to inhibit acute and chronic effects, but not mortality, after irradiation.

How curcumin exactly provides radioprotection is not fully understood. There are studies, however, that indicate that curcumin can inhibit the radiation-induced damage of specific proteins (Kapoor and Priyadarsini, 2001). Varadkar et al. (2001) examined the effect of curcumin on radiation-induced PKC activity isolated from the liver cytosol and the particulate fraction of unirradiated mice and mice irradiated at 5 Gy. Following irradiation, the PKC activity was increased in both cytosolic and particulate fractions. Curcumin was found to inhibit the activated cytosolic and particulate PKC at very low concentrations. Since activation of PKC is one of the means of conferring radioresistance on a tumor cell, suppression of PKC activity by curcumin may be one of the means of preventing the development of radioresistance following radiotherapy.

Another potential mechanism of radioprotection involves suppression of radiation-induced gene expression. Oguro and Yoshida (2001) examined the effect of curcumin on UVA-induced ODC and metallothionein gene expression in mouse skin. They showed that UVA induced metallothionein mRNA in mouse skin, and 1,4-Diazabicylo-[2,2,2]-octane (DABCO), a singlet oxygen scavenger, reduced UVA-mediated induction of MT mRNA (by 40%). UVA slightly enhanced TPA-mediated ODC mRNA induction, while it enhanced ODC enzyme activity by 70%. UVA additively intensified TPA-mediated MT mRNA induction. Curcumin dramatically inhibited both TPA- and TPA + UVA-induced expression of ODC and MT genes.

10.8 CARDIOVASCULAR DISEASES

10.8.1 Effect of Curcumin on Atherosclerosis and MI

The effect of curcumin on MI in the cat and the rat has been investigated (Dikshit et al., 1995; Nirmala and Puvanakrishnan, 1996; Nirmala et al., 1999). Dikshit et al. (1995) examined the prevention of ischemia-induced biochemical changes by curcumin in the cat heart. Myocardial ischemia was induced by the ligation of the left descending coronary artery. Curcumin (100 mg/kg, i.p.) was given 30 minutes before ligation. Cats were killed and hearts were removed 4 h after coronary artery ligation. Levels of GSH, malonaldelhyde (MDA), myeloperoxidase (MPO), SOD, catalase, and lactate dehydrogenase (LDH) were estimated in the ischemic and nonischemic zones. Curcumin protected the animals against decrease in the heart rate and blood pressure following ischemia. In the ischemic zone, after 4 hours of ligation, an increase in the level of MDA and activities of MPO and SOD (cytosolic fraction) were observed. Curcumin pretreatment prevented the ischemia-induced elevation in MDA contents and LDH release but did not affect the increase in MPO activity. Thus, curcumin prevented ischemia-induced changes in the cat heart.

10.8.2 Curcumin Inhibits Proliferation of Vascular Smooth Muscle Cells

The proliferation of peripheral blood mononuclear cells (PBMC) and vascular smooth muscle cells (VSMC) is a hallmark of atherosclerosis. Huang et al. (1992a) investigated the effects of curcumin on the proliferation of PBMC and VSMC from the uptake of [³H] thymidine. Curcumin dose dependently inhibited the response to phytohemagglutinin and the mixed lymphocyte reaction in human PBMC at dose ranges of 1 to 30 μM and 3 to 30 μM, respectively. Curcumin (1–100 μM) dose dependently inhibited the proliferation of rabbit VSMC stimulated by fetal calf serum. Curcumin had a greater inhibitory effect on platelet-derived growth factor-stimulated proliferation than on serum-stimulated proliferation. Analogues of curcumin (cinnamic acid, coumaric acid, and ferulic acid) were much less effective than curcumin as inhibitors of serum-induced smooth muscle cell proliferation. This suggested that curcumin may be useful for the prevention of the pathological changes associated with atherosclerosis and restenosis.

Chen and Huang (1998) examined the possible mechanisms underlying curcumin's antiproliferative and apoptotic effects using the rat VSMC cell line A7r5. Curcumin (1–100 μM) inhibited serum-stimulated [^{3}H] thymidine incorporation of both A7r5 cells and rabbit VSMC. Cell viability, as determined by the trypan blue dye exclusion method, was unaffected by curcumin at the concentration range 1 to 10 μM in A7r5 cells. However, the number of viable cells after 100 μM curcumin treatment was less than the basal value. Following curcumin (1–100 μM) treatment, cell cycle analysis revealed a G0/G1 arrest and a reduction in the percentage of cells in S phase. Curcumin at 100 μM also induced cell apoptosis, as demonstrated by hematoxylin-eosin staining, TdT-mediated dUTP nick end labeling, DNA laddering, cell shrinkage, chromatin condensation, and DNA fragmentation. The membranous protein tyrosine kinase activity stimulated by serum in A7r5 cells was significantly reduced by curcumin (10–100 μM). On the other hand, PMA-stimulated cytosolic PKC activity was reduced by 100 μM curcumin. The levels of c-myc mRNA and bcl-2 mRNA were significantly reduced by curcumin but had little effect on the p53 mRNA level. These results demonstrate that curcumin inhibited cell proliferation, arrested cell cycle progression, and induced cell apoptosis in VSMC. These results may explain how curcumin prevents the pathological changes of atherosclerosis and postangioplasty restenosis.

10.8.3 CURCUMIN LOWERS SERUM CHOLESTEROL LEVEL

Numerous studies suggest that curcumin lowers serum cholesterol levels (Rao et al., 1970; Patil and Srinivasan, 1971; Keshavarz, 1976; Soudamini et al., 1992; Hussain and Chandrasekhara, 1992; Soni and Kuttan, 1992; Hussain and Chandrasekhara, 1994). Soudamini et al. (1992) investigated the effect of oral administration of curcumin on serum cholesterol levels and on LPO in the liver, lung, kidney, and brain of mice treated with carbon tetrachloride, paraquat, and cyclophosphamide. Oral administration of curcumin significantly lowered the increased peroxidation of lipids in these tissues, produced by these chemicals. Administration of curcumin also significantly lowered the serum and tissue cholesterol levels in these animals, indicating that the use of curcumin helps in conditions associated with peroxide-induced injury, such as liver damage and arterial diseases. Soni and Kuttan (1992) examined the effect of curcumin administration in reducing the serum levels of cholesterol and lipid peroxides in ten healthy human volunteers receiving 500 mg of curcumin per day for 7 days. A significant decrease in the level of serum lipid peroxides (33%), an increase in high-density lipoproteins (HDL)-cholesterol (29%), and a decrease in total serum cholesterol (12%) were noted. Because curcumin reduced serum lipid peroxides and serum cholesterol, the study of curcumin as a chemopreventive substance against arterial diseases was suggested.

Curcuma xanthorrhiza Roxb., a medicinal plant used in Indonesia (known as *temu lawak* or Javanese turmeric), has been shown to exert diverse physiological effect. However, little attention has been paid to its effect on lipid metabolism. Yasni et al. (1993) investigated the effects of *C. xanthorrhiza* on serum and liver lipids, serum HDL cholesterol, apolipoprotein, and liver lipogenic enzymes. In rats given a cholesterol-free diet, *C. xanthorrhiza* decreased the concentrations of serum triglycerides, phospholipids, and liver cholesterol and increased the concentrations of serum HDL-cholesterol and apolipoproteins. The activity of liver fatty acid synthase, but not glycerophosphate dehydrogenase, was decreased by the medicinal plant. In rats on a high-cholesterol diet, *C. xanthorrhiza* did not suppress the elevation of serum cholesterol, although it did decrease liver cholesterol. Curcuminoids prepared from *C. xanthorrhiza* had no significant effects on the serum and liver lipids. These studies, therefore, indicate that *C. xanthorrhiza* contains an active principle other than the curcuminoids that can modify the metabolism of lipids and lipoproteins.

In later studies Yasni et al., (1994) identified the major component (approximately 65%) of the essential oil as alpha-curcumene. Addition of essential oils (0.02%), prepared by steam distillation, to a purified diet lowered hepatic triglyceride concentration without influencing serum triglyceride levels, whereas addition of the hexane-soluble fraction (0.5%) lowered the concentration of serum and hepatic triglycerides. Rats fed with the essential oil and hexane-soluble fraction had lower hepatic fatty acid synthase activity. The fraction containing β-curcumene, prepared from the hexane-soluble fraction by silica gel column chromatography, suppressed the synthesis of fatty acids from [^{14}C]-acetate in primary cultured rat hepatocytes.

Skrzypczak-Jankun et al. (2003) showed the 3D structural data and explained how curcumin interacts with the fatty acid–metabolizing enzyme, soybean lipoxygenase. Curcumin binds lipoxygenase in a noncompetitive manner. Trapped in that complex, it undergoes photodegradation in response to x-rays, but utilizes enzyme catalytic ability to form the peroxy complex Enz-Fe-O-O-R as 4-hydroperoxy-2-methoxy-phenol, which later is transformed into 2-methoxycyclohexa-2, 5-diene-1,4-dione. However, when Rukkumani et al. (2002) compared the effects of curcumin and photo-irradiated curcumin on alcohol and polyunsaturated fatty acid-induced hyperlipidemia, they found that photo-irradiated curcumin was more effective than curcumin in treating the above pathological conditions.

10.8.4 Curcumin Inhibits LDL Oxidation

The oxidation of low-density lipoproteins (LDL) plays an important role in the development of atherosclerosis. Atherosclerosis is characterized by oxidative damage, which affects lipoproteins, the walls of blood vessels, and subcellular membranes. Several studies suggest that curcumin inhibits oxidation of LDL (Asai et al., 2001; Ramirez-Tortosa et al., 1999; Naidu and Thippeswamy, 2002; Quiles et al., 1998). Naidu and Thippeswamy (2002) examined the effect of curcumin on copper ion-induced LPO of human LDL by measuring the formation of thiobarbituric acid reactive substance (TBARS) and relative electrophoretic mobility of LDL on agarose gel. Curcumin inhibited the formation of TBARS effectively throughout the incubation period of 12 hours and decreased the relative electrophoretic mobility of LDL. Curcumin at 10 μM produced 40 to 85% inhibition of LDL oxidation. The inhibitory effect of curcumin was comparable to that of BHA (butylated hydroxyanisole), but more potent than ascorbic acid. Further, curcumin significantly inhibited both initiation and propagation phases of LDL oxidation.

Ramirez-Tortosa et al. (1999) evaluated the effect of curcumin on LDL oxidation susceptibility and plasma lipids in atherosclerotic rabbits. A total of 18 rabbits were fed for 7 weeks on a diet containing 95.7% standard chow, 3% lard, and 1.3% cholesterol, to induce atherosclerosis. The rabbits were divided into groups, two of which were also orally treated with turmeric extract at doses of 1.66 (group A) and 3.2 (group B) mg/kg body weight. A third group (group C) acted as an untreated control. Plasma and LDL lipid composition, plasma alpha-tocopherol, plasma retinol, LDL TBARS, LDL lipid hydroperoxides were assayed, and aortic atherosclerotic lesions were evaluated. The low but not the high dosage of turmeric extracts decreased the susceptibility of rabbit LDL to LPO. Both doses produced lower levels of total plasma cholesterol than the control group. Moreover, the lower dosage group had lower levels of cholesterol, phospholipids, and triglycerides than the 3.2-mg dosage.

Quiles et al. (1998) evaluated the antioxidant capacity of a turmeric extract on the LPO of liver mitochondria and microsome membranes in atherosclerotic rabbits. Male rabbits fed 3% (w/w) lard and 1.3% (w/w) cholesterol diet was randomly assigned to three groups. Two groups were treated with different dosages of a turmeric extract (A and B) and the third group (control) with a curcumin-free solution. Basal and in vitro 2,2′-azobis (2-amidinopropane) dihydrochloride–induced hydroperoxide and TBARS production in liver mitochondria and microsomes were analyzed. Group A had the lowest concentration of mitochondrial hydroperoxides. In microsomes, the basal hydroperoxide levels were similar in all groups but, after the induction of oxidation, group C registered the highest value; TBARS production followed the same trend in mitochondria. These findings suggest that active compounds in turmeric extract may be protective against lipoperoxidation of subcellular membranes in a dosage-dependent manner.

Asai et al. (2001) examined the effect of curcumin on lipid metabolism in rats fed a control, moderately high-fat diet (15 g soybean oil/100 g diet), and those given supplements of 0.2 g curcuminoids/100 g diet. Liver triacylglycerol and cholesterol concentrations were significantly lower in rats fed curcumin than in control rats. Plasma triacylglycerols in the very low-density lipoproteins fraction were also lower in curcumin-fed rats than in control ($P < 0.05$). Hepatic acyl-CoA oxidase activity of the curcumin group was significantly higher than that of the control. Furthermore, epididymal adipose tissue weight was significantly reduced with curcuminoid intake in a dose-dependent manner. These results indicated that dietary curcuminoids have lipid-lowering potency in vivo, probably due to alterations in fatty acid metabolism.

10.8.5 CURCUMIN INHIBITS PLATELET AGGREGATION

Shah et al. (1999) studied the mechanism of the antiplatelet action of curcumin. They found that curcumin inhibited platelet aggregation mediated by the platelet agonist's epinephrine (200 μM), ADP (4 μM), platelet-activating factor (PAF: 800 nM), collagen (20 μg/ml), and arachidonic acid (AA: 0.75 mM). Curcumin preferentially inhibited PAF- and AA-induced aggregation (IC50: 25-20 μM), whereas much higher concentrations of curcumin were required to inhibit aggregation induced by other platelet agonists. Pretreatment of platelets with curcumin resulted in inhibition of platelet aggregation induced by calcium ionophore A-23187 (IC50: 100 μM), but curcumin up to 250 μM had no inhibitory effect on aggregation induced by the PKC activator phorbol myrsitate acetate (1 μM). Curcumin (100 μM) inhibited the A-23187–induced mobilization of intracellular Ca_2^+ as determined by using fura-2 acetoxymethyl ester. Curcumin also inhibited the formation of thromboxane A2 (TXA2) by platelets (IC50: 70 μM). These results suggest that the curcumin-mediated preferential inhibition of PAF- and AA-induced platelet aggregation involves inhibitory effects on TXA2 synthesis and Ca_2^+ signaling, but without the involvement of PKC.

10.9 CURCUMIN STIMULATES MUSCLE REGENERATION

Skeletal muscle is often the site of tissue injury due to trauma, disease, developmental defects, or surgery. Yet, to date, no effective treatment is available to stimulate the repair of skeletal muscle. Thaloor et al. (1999) investigated the kinetics and extent of muscle regeneration in vivo after trauma, following systemic administration of curcumin to mice. Biochemical and histological analyses indicated faster restoration of normal tissue architecture in mice treated with curcumin after only 4 days of daily intraperitoneal injection, whereas controls required over 2 weeks to restore normal tissue architecture. Curcumin acted directly on cultured muscle precursor cells to stimulate both cell proliferation and differentiation under appropriate conditions. The authors suggested that this effect of curcumin was mediated through suppression of NF-κB; inhibition of NF-κB–mediated transcription was confirmed using reporter gene assays. They concluded that NF-κB exerts a role in regulating myogenesis, and that modulation of NF-κB activity within muscle tissue is beneficial for muscle repair. The striking effects of curcumin on myogenesis suggest therapeutic applications for treating muscle injuries.

10.10 CURCUMIN ENHANCES WOUND HEALING

Tissue repair and wound healing are complex processes that involve inflammation, granulation, and remodeling of the tissue. Perhaps, the earliest report that curcumin has wound-healing activity was reported by Gujral and coworkers (Srimal and Dhawan, 1973). Sidhu et al. (1998) examined the wound-healing capacity of curcumin in rats and guinea pigs. Punch wounds in curcumin-treated animals closed faster in treated than in untreated animals. Biopsies of the wound showed reepithelialization of the epidermis and increased migration of various cells including myofibroblasts, fibroblasts, and macrophages in the wound bed. Multiple areas within the dermis showed extensive neovascularization, and Masson's trichrome staining showed greater collagen deposition in curcumin-treated wounds. Immunohistochemical localization showed an increase of transforming growth factor beta 1 (TGF-β1) in curcumin-treated wounds as compared with untreated wounds. *In situ* hybridization and PCR analysis also showed an increase in the mRNA transcripts of TGF-β1 and fibronectin in curcumin-treated wounds. Because TGF-β1 is known to enhance wound healing, it is possible that curcumin modulates TGF-β1 activity.

To further understand its therapeutic effect on wound healing, the antioxidant effects of curcumin on H_2O_2 and hypoxanthine-xanthine oxidase-induced damage to cultured human keratinocytes and fibroblasts were investigated by Phan et al. (2001). Cell viability was assessed by colorimetric assay and quantification of LDH release. Exposure of human keratinocytes to curcumin at 10 μg/mL significantly protected against the keratinocytes from H_2O_2-induced oxidative damage. Interestingly, exposure of human dermal fibroblasts to curcumin at 2.5 μg/ml showed significant protective effects against H_2O_2. No protective effects of curcumin on either fibroblasts or keratinocytes against hypoxanthine-xanthine oxidase-induced damage were found. These investigators thus concluded that curcumin indeed possessed powerful inhibitory capacity against H_2O_2-induced damage in human keratinocytes and fibroblasts and that this protection may contribute to wound healing.

Mani et al. (2002) investigated the effect of curcumin treatment by topical application in dexamethasone-impaired cutaneous healing in a full-thickness punch wound model in rats. They assessed healing in terms of histology, morphometry, and collagenization on the fourth and seventh days postwounding and analyzed the regulation of TGF-β1, its receptors type I (tIrc) and type II (tIIrc) and iNOS. Curcumin significantly accelerated healing of wounds with or without dexamethasone treatment as revealed by a reduction in the wound width and gap length compared to controls. Curcumin treatment enhanced expression of TGF-β1 and TGF-β tIIrc in both normal and impaired healing wounds. Macrophages in the wound bed showed an enhanced expression of TGF-β1 mRNA in curcumin-treated wounds as evidenced by *in situ* hybridization. iNOS levels were increased, following curcumin treatment in unimpaired wounds, but not so in the dexamethasone-impaired wounds. Their study indicated an enhancement in dexamethasone-impaired wound repair by topical curcumin and its differential regulatory effect on TGF-β1, its receptors, and iNOS in this cutaneous wound-healing model.

10.11 CURCUMIN SUPPRESSES SYMPTOMS ASSOCIATED WITH ARTHRITIS

Deodhar et al. (1980) were the first to report on the antirheumatic activity of curcumin in human subjects. They performed a short-term double blind crossover study in 18 patients with "definite" rheumatoid arthritis to compare the antirheumatic activity of curcumin (1200 mg/day) with phenylbutazone (300 mg/day). Subjective and objective assessment in patients who were taking corticosteroids just prior to the study showed significant ($P < 0.05$) improvements in morning stiffness, walking time, and joint swelling, following two weeks of curcumin therapy.

Liacini et al. (2002) examined the effect of curcumin in articular chondrocytes. IL-1, the main cytokine instigator of cartilage degeneration in arthritis, induces MMP-3 and MMP-13 RNA and protein in chondrocytes through the activation of mitogen-activated protein kinase (MAPK), AP-1, and NF-κB transcription factors. Curcumin achieved 48 to 99% suppression of MMP-3 and 45 to 97% of MMP-13 in human and 8 to 100% (MMP-3) and 32 to 100% (MMP-13) in bovine chondrocytes. Inhibition of IL-1 signal transduction by these agents could be useful for reducing cartilage resorption by MMPs in arthritis.

10.12 CURCUMIN REDUCES THE INCIDENCE OF CHOLESTEROL GALLSTONE FORMATION

Hussain and Chandrasekhara (1992) studied the efficacy of curcumin in reducing the incidence of cholesterol gallstones induced by feeding a lithogenic diet in young male mice. Feeding a lithogenic diet supplemented with 0.5% curcumin for 10 weeks reduced the incidence of gallstone formation to 26%, as compared to 100% incidence in the group fed with the lithogenic diet alone. Biliary cholesterol concentration was also significantly reduced by curcumin feeding. The lithogenic index, which was 1.09 in the cholesterol-fed group, was reduced to 0.43 in the 0.5% curcumin-supplemented group. Further, the cholesterol:phospholipid ratio of bile was also reduced significantly when 0.5% curcumin supplemented diet was fed. A dose–response study with 0.2%, 0.5%, and 1% curcumin-supplemented lithogenic diets showed that 0.5% curcumin was more effective than a diet with 0.2% or 1% curcumin. How curcumin mediates antilithogenic effects in mice was further investigated by this group (Hussain and Chandrasekhara, 1994). For this purpose, the hepatic bile of rats was fractionated by gel filtration chromatography, and the low-molecular weight (LMW) protein fractions were tested for their ability to influence cholesterol crystal growth in model bile. The LMW protein fraction from the lithogenic agent-fed control group's bile shortened the nucleation time and increased the crystal growth rate and final crystal concentration. But with the LMW protein fractions from the bile of rats given curcumin, the nucleation times were prolonged, and the crystal growth rates and final crystal concentrations were decreased. The LMW fractions were further purified into three different sugar-specific proteins by affinity chromatography. A higher proportion of LMW proteins from the control group bile was bound to Con-A, whereas higher proportions of LMW proteins from the groups fed with curcumin were bound to wheat germ agglutinin (WGA) and *Helix pomatia* lectin. The Con-A–bound fraction obtained from the control group showed a pronucleating effect. In contrast, the WGA-bound fraction obtained from curcumin group showed a potent antinucleating activity.

10.13 CURCUMIN MODULATES MS

MS is an inflammatory disease of the central nervous system (CNS), which afflicts more than 1 million people worldwide. The destruction of oligodendrocytes and myelin sheath in the CNS is the pathological hallmark of MS. MS is an inflammatory autoimmune disease of the CNS resulting from myelin antigen-sensitized T-cells in the CNS. Experimental allergic encephalomyelitis (EAE), a CD4+ Th1 cell-mediated inflammatory demyelinating autoimmune disease of the CNS, serves as an animal model for MS. IL-12 plays a crucial proinflammatory role in the induction of neural antigen-specific Th1 differentiation and pathogenesis of CNS demyelination in EAE and MS.

Natarajan and Bright (2002) investigated the effect of curcumin on the pathogenesis of CNS demyelination in EAE. *In vivo* treatment of SJL/J mice with curcumin significantly reduced the duration and clinical severity of active immunization and adoptive transfer EAE (Natarajan and Bright, 2002). Curcumin inhibited EAE in association with a decrease in IL-12 production from macrophage/microglial cells and differentiation of neural antigen-specific Th1 cells. *In vitro* treatment of activated T-cells with curcumin inhibited IL-12–induced tyrosine phosphorylation of Janus kinase 2, tyrosine kinase 2, and STAT3 and STAT4 transcription factors. The inhibition of Janus kinase–STAT pathway by curcumin resulted in a decrease in IL-12–induced T-cell proliferation and Th1 differentiation. These findings show that curcumin inhibits EAE by blocking IL-12 signaling in T-cells and suggest its use in the treatment of MS and other Th1 cell-mediated inflammatory diseases.

Verbeek and coworkers (2005) examined the effects of oral flavonoids as well as of curcumin on autoimmune T-cell reactivity in mice and on the course of experimental autoimmune encephalomyelitis (EAE), a model for MS. Continuous oral administration of flavonoids significantly affected antigen-specific proliferation and IFN-gamma production by lymph node–derived T-cells following immunization with an EAE-inducing peptide (Verbeek et al., 2005). The effects of curcumin on EAE were assessed using either passive transfer of autoimmune T-cells or active disease induction. In passive EAE, curcumin led to delayed recovery of clinical symptoms rather than to any reduction in disease. Oral curcumin had overall mild but beneficial effects.

10.14 CURCUMIN BLOCKS THE REPLICATION OF HIV

Transcription of type 1 HIV-1 provirus is governed by the viral long-terminal repeat (LTR). Drugs can block HIV-1 replication by inhibiting the activity of its LTR. Li et al. examined the effect of curcumin on HIV-1 LTR-directed gene expression and virus replication (Abraham et al., 1993). Curcumin was found to be a potent and selective inhibitor of HIV-1 LTR-directed gene expression, at concentrations that have minor effects on cells. Curcumin inhibited p24 antigen production in cells either acutely or chronically infected with HIV-1 through transcriptional repression of the LTR. Sui et al. (1993) examined the effect on the HIV-1 and HIV-2 proteases by curcumin and curcumin–boron complexes. Curcumin was a modest inhibitor of HIV-1 (IC50 = 100 µM) and HIV-2 (IC_{50} = 250 µM) proteases. Simple modifications of the curcumin structure raised the IC50 value, but complexes of the central dihydroxy groups of curcumin with boron lowered the IC_{50} to a value as low as 6 µM. The boron complexes were also time-dependent inactivators of the HIV proteases. The increased affinity of the boron complexes may reflect binding of the orthogonal domains of the inhibitor in intersecting sites within the substrate-binding cavity of the enzyme, while activation of the α, β-unsaturated carbonyl group of curcumin by chelation to boron probably accounts for time-dependent inhibition of the enzyme.

Mazumder et al. (1997) examined the effect of curcumin analogs with altered potencies against HIV-1 integrase. They reported that curcumin inhibited HIV-1 integrase activity. They also synthesized and tested analogs of curcumin to explore the structure–activity relationships and mechanism of action of this family of compounds in more detail. They found that two curcumin analogs, dicaffeoylmethane and rosmarinic acid, inhibited both activities of integrase, for IC50 values below 10 µM. They demonstrated that lysine 136 may play a role in viral DNA binding and that two curcumin analogs had equivalent potencies against both an integrase mutant and a wild-type integrase, suggesting that the curcumin-binding site and the substrate-binding site may not overlap. Combining one curcumin analog with the recently described integrase inhibitor NSC 158393 resulted in integrase inhibition that was synergistic, again suggesting that drug-binding sites may not overlap. They also determined that these analogs could inhibit binding of the enzyme to the viral DNA, but that this inhibition is independent of divalent metal ion. Furthermore, kinetic studies of these analogs suggest that they bound to the enzyme at a slow rate. These studies can provide mechanistic and structural information to guide the future design of integrase inhibitors.

The transcription of HIV-1 provirus is regulated by both cellular and viral factors. Various pieces of evidence suggest that Tat protein secreted by HIV1-infected cells may have additional activity in the pathogenesis of AIDS because of its ability to also be taken up by noninfected cells. Barthelemy et al. (1998) showed that curcumin used at 10 to 100 nM inhibited Tat transactivation of HIV1-LTR lacZ by 70 to 80% in He La cells. To develop more efficient curcumin derivatives, the researchers synthesized and tested in the same experimental system the inhibitory activity of reduced curcumin (C1), which lacks the spatial structure of curcumin; allyl-curcumin (C2), which possesses a condensed allyl derivative on curcumin that plays the role of metal chelator; and tocopheryl-curcumin (C3), whose structural alterations enhances the antioxidant activity of the molecule. Results obtained with the C1, C2, and C3 curcumin derivatives showed a significant inhibition (70 to 85%) of Tat transactivation. Despite the fact that tocopheryl–curcumin (C3) failed to scavenge O^{2-}, this curcumin derivative exhibited the most activity; 70% inhibition was obtained at 1 nM, whereas only 35% inhibition was obtained with the curcumin.

Acetylation of histones and nonhistone proteins is an important posttranslational modification involved in the regulation of gene expression in eukaryotes and all viral DNA that integrates into the human genome (e.g., the HIV). Dysfunction of histone acetyltransferases (HATs) is often associated with the manifestation of several diseases. In this respect, HATs are the new potential targets for the design of therapeutics. Balasubramanyam and coworkers (2004) report that curcumin is a specific inhibitor of the p300/CBP HAT activity but not of p300/CBP-associated factor, *in vitro* and *in vivo*. Furthermore, curcumin could also inhibit the p300-mediated acetylation of p53 *in vivo*.

It specifically represses the p300/CBP HAT activity-dependent transcriptional activation from chromatin but not a DNA template Balasubramanyam et al., 2004). It is significant that curcumin could inhibit the acetylation of HIV–Tat protein*in vitro* by p300 as well as proliferation of the virus, as revealed by the repression in syncytia formation upon curcumin treatment in SupT1 cells. Thus, nontoxic curcumin, which targets p300/CBP, may serve as a lead compound in combinatorial HIV therapeutics.

10.15 CURCUMIN AFFECTS ALZHEIMER'S DISEASE

Brain inflammation in Alzheimer's disease (AD) patients is characterized by increased cytokines and activated microglia. Epidemiological studies suggest reduced AD risk is associated with long-term use of nonsteroidal anti-inflammatory drugs (NSAIDs). Whereas chronic ibuprofen suppressed inflammation and plaque-related pathology in an Alzheimer transgenic APPSw mouse model (Tg2576), excessive use of NSAIDs targeting cyclooxygenase I can cause gastrointestinal, liver, and renal toxicity. One alternative NSAID is curcumin. Lim et al. (2001) found that curcumin reduces oxidative damage and amyloid pathology in an Alzheimer transgenic mouse model. To evaluate whether it could affect Alzheimer-like pathology in the APPSw mice, they tested the effect of a low (160 ppm) and a high (5000 ppm) dose of dietary curcumin on inflammation, oxidative damage, and plaque pathology. Low and high doses significantly lowered oxidized proteins and IL-1β, a proin-flammatory cytokine usually elevated in the brains of these mice. With low-dose, but not high-dose, curcumin treatment, the astrocytic marker glial fibrillary acidic protein was reduced, and insoluble beta-amyloid (Aβ), soluble Aβ, and plaque burden were significantly decreased, by 43 to 50%. However, levels of amyloid precursor in the membrane fraction were not reduced. Microgliosis was also suppressed in neuronal layers but not adjacent to plaques. In view of its efficacy and apparent low toxicity, this Indian spice component has promise for the prevention of Alzheimer's disease.

Ono's group (2004) reported previously that nordihydroguaiaretic acid (NDGA) inhibits fAbeta formation from Abeta(1–40) and Abeta(1–42) and destabilize preformed fAbeta(1–40) and fAbeta(1–42) dose dependently *in vitro*. Using fluorescence spectroscopic analysis with thioflavin T and electron microscopic studies, they examined the effects of curcumin on the formation, extension, and destabilization of fAbeta(1-40) and fAbeta(1-42) at pH 7.5 at 37°C *in vitro*. They next compared the antiamyloidogenic activities of curcumin with NDGA. Curcumin dose depen-dently inhibited fAbeta formation from Abeta(1-40) and Abeta(1-42) as well as their extension. In addition, it dose dependently destabilized preformed fAbetas. The overall activities of curcumin and NDGA were similar. The effective concentrations EC50 of curcumin and NDGA for the formation, extension, and destabilization of fAbetas were in the order of 0.1 to 1 μM. Although the mechanism by which curcumin inhibits fAbeta formation from Abeta and destabilize preformed fAbeta *in vitro* remains unclear; they could be a key molecule for the development of therapeutics for AD (Ono et al., 2004).

Yang and coworkers (2005) investigated whether its efficacy in AD models could be explained by effects on Abeta aggregation. Under aggregating conditions *in vitro*, curcumin inhibited aggre-gation IC50 = 0.8 μM] as well as disaggregated fibrillar Abeta40 (IC50 = 1 μM)], indicating favorable stoichiometry for inhibition. Curcumin was a better Abeta40 aggregation inhibitor than ibuprofen and naproxen, and prevented Abeta42 oligomer formation and toxicity between 0.1 and 1.0 μM. Under EM, curcumin decreased dose dependently Abeta fibril formation beginning with 0.125 μM. The effects of curcumin did not depend on Abeta sequence but on fibril-related confor-mation. AD and Tg2576 mice brain sections incubated with curcumin revealed preferential labeling of amyloid plaques. *In vivo* studies showed that curcumin injected peripherally into aged Tg mice crossed the blood–brain barrier and bound plaques. When fed to aged Tg2576 mice with advanced amyloid accumulation, curcumin labeled plaques and reduced amyloid levels and plaque burden. Hence, curcumin directly binds small beta-amyloid species to block aggregation and fibril formation *in vitro* and *in vivo*. These data suggest that low-dose curcumin effectively disaggregates Abeta as well as prevents fibril and oligomer formation, supporting the rationale for curcumin use in clinical trials preventing or treating AD.

10.16 CURCUMIN PROTECTS AGAINST CATARACT FORMATION IN LENSES

Age-related cataractogenesis is a significant health problem worldwide. Oxidative stress has been suggested to be a common underlying mechanism of cataractogenesis, and augmentation of the antioxidant defenses of the ocular lens has been shown to prevent or delay cataractogenesis. Awasthi et al. (1996) tested the efficacy of curcumin in preventing cataractogenesis in an *in vitro* rat model. Rats were maintained on an AIN-76 diet for 2 weeks, after which they were given a daily dose of corn oil alone or 75 mg curcumin/kg in corn oil for 14 days. Their lenses were removed and cultured for 72 h in vitro in the presence or absence of 100 µmol 4-hydroxy-2-nonenal (4-HNE)/L, a highly electrophilic product of LPO. The results of these studies showed that 4-HNE led to opacification of cultured lenses as indicated by the measurements of transmitted light intensity using digital image analysis. However, the lenses from curcumin-treated rats were resistant to 4-HNE-induced opacification. Curcumin treatment significantly induced the GST isozyme rGST8-8 in rat lens epithelium. Because rGST8-8 utilizes 4-HNE as a preferred substrate, we suggest that the protective effect of curcumin may be mediated through the induction of this GST isozyme. These studies suggest that curcumin may be an effective protective agent against cataractogenesis induced by LPO.

Suryanarayana and coworkers (2003) investigated the effect of curcumin on galactose-induced cataractogenesis in rats. The data indicated that curcumin at 0.002% (group C) delayed the onset and maturation of cataract. In contrast, even though there was a slight delay in the onset of cataract at the 0.01% level (group D), maturation of cataract was faster when compared to group B. Biochemical analysis showed that curcumin at the 0.002% level appeared to exert antioxidant and antiglycating effects, as it inhibited LPO, AGE (advanced glycated end products)-fluorescence, and protein aggregation. Though the reasons for faster onset and maturation of cataract in group D rats were not clear, the data suggested that, under hyperglycemic conditions, higher levels of curcumin (0.01%) in the diet may increase oxidative stress, AGE formation, and protein aggregation. However, feeding of curcumin to normal rats up to a 0.01% level did not result in any changes in lens morphology or biochemical parameters. These results suggest that curcumin is effective against galactose-induced cataract only at very low amounts (0.002%) in the diet. On the other hand, at and above a 0.01% level, curcumin seems to not be beneficial under hyperglycemic conditions, at least with the model of galactose-cataract.

Suryanarayana and coworkers (2005) reported that turmeric and curcumin are effective against the development of diabetic cataract in rats. Although, both curcumin and turmeric did not prevent streptozotocin-induced hyperglycemia, as assessed by blood glucose and insulin levels, slit lamp microscope observations indicated that these supplements delayed the progression and maturation of cataract. The present studies suggest that curcumin and turmeric treatment appear to have countered the hyperglycemia-induced oxidative stress, because there was a reversal of changes with respect to LPO, reduced glutathione, protein carbonyl content, and activities of antioxidant enzymes in a significant manner. Also, treatment with turmeric or curcumin appears to have minimized osmotic stress, as assessed by polyol pathway enzymes. Most important, aggregation and insolubilization of lens proteins due to hyperglycemia was prevented by turmeric and curcumin. Turmeric was more effective than its corresponding levels of curcumin. Further, these results imply that ingredients in the study's dietary sources, such as turmeric, may be explored for anticataractogenic agents that prevent or delay the development of cataract.

Padmaja's group investigated that Wistar rat pups treated with curcumin before being administered with selenium showed no opacities in the lens. The LPO, xanthine oxidase enzyme levels in the lenses of curcumin, and selenium-cotreated animals were significantly less when compared to selenium-treated animals. The superoxidase dismutase and catalase enzyme activities of curcumin and selenium-cotreated animal lenses showed an enhancement. Curcumin cotreatment seems to prevent oxidative damage and found to delay the development of cataract (Padmaja et al., 2004).

10.17 CURCUMIN PROTECTS FROM DRUG-INDUCED MYOCARDIAL TOXICITY

Cardiotoxicity is one of the major problems associated with administration of many chemotherapeutic agents. Venkatesan (1998) examined the protective effect of curcumin on acute Adriamycin (ADR) myocardial toxicity in rats. ADR toxicity, induced by a single intraperitoneal injection (30 mg/kg), was revealed by elevated serum creatine kinase (CK) and LDH. The levels of the LPO products, conjugated dienes, and malondialdehyde were markedly elevated by ADR. ADR also caused a decrease in myocardial glutathione content and glutathione peroxidase activity and an increase in cardiac catalase activity. Curcumin treatment (200 mg/kg) 7 days before and 2 days following ADR significantly ameliorated the early manifestation of cardiotoxicity (ST segment elevation and an increase in heart rate) and prevented the rise in serum CK and LDH exerted by ADR. ADR-treated rats that received curcumin displayed a significant inhibition of LPO and augmentation of endogenous antioxidants. These results suggest that curcumin inhibits ADR cardiotoxicity and might serve as novel combination of chemotherapeutic agent with ADR to limit free-radical-mediated organ injury.

10.18 CURCUMIN PROTECTS FROM ALCOHOL-INDUCED LIVER INJURY

Because induction of NF-κB–mediated gene expression has been implicated in the pathogenesis of alcoholic liver disease (ALD) and curcumin inhibits the activation of NF-κB, Nanji et al. (2003) determined whether treatment with curcumin would prevent experimental ALD and elucidated the underlying mechanism. Four groups of rats (six rats/group) were treated by intragastric infusion for 4 weeks. One group received fish oil plus ethanol (FE) and a second group received fish oil plus dextrose (FD). The third and fourth groups received FE or FD supplemented with 75 mg/kg/day of curcumin. Liver samples were analyzed for histopathology, LPO, NF–κB binding, TNFα, IL-12, monocyte chemotactic protein-1, macrophage inflammatory protein-2, COX-2, iNOS, and nitrotyrosine. Rats fed FE developed fatty liver, necrosis, and inflammation, which was accompanied by activation of NF-κB and the induction of cytokines, chemokines, COX-2, iNOS, and nitrotyrosine formation. Treatment with curcumin prevented both the pathological and the biochemical changes induced by alcohol. Because endotoxin and the Kupffer cell are implicated in the pathogenesis of ALD, they also investigated whether curcumin suppressed the stimulatory effects of endotoxin in isolated Kupffer cells. Curcumin blocked endotoxin-mediated activation of NF-kappaB and suppressed the expression of cytokines, chemokines, COX-2, and iNOS in Kupffer cells. Thus curcumin prevented experimental ALD, in part, by suppressing induction of NF-κB-dependent genes.

Hepatic fibrogenesis occurs as a wound-healing process after many forms of chronic liver injury. Hepatic fibrosis ultimately leads to cirrhosis if not treated effectively. During liver injury, quiescent HSC, the most relevant cell type, become active and proliferative. Oxidative stress is a major and critical factor for HSC activation. Activation of PPAR-γ inhibits the proliferation of nonadipocytes. The level of PPAR-γ is dramatically diminished along with the activation of HSC during liver injury. Xu et al. (2003) examined the effect of curcumin on HSC proliferation. They hypothesized that curcumin inhibits the proliferation of activated HSC by inducing PPAR-γ gene expression and reviving PPAR-γ activation. Their results indicated that curcumin significantly inhibited the proliferation of activated HSC and induced apoptosis *in vitro*. They also demonstrated, for the first time, that curcumin dramatically induced the expression of the PPAR-γ gene and activated PPAR-γ in activated HSC. Blocking its transactivating activity by a PPAR-γ antagonist markedly abrogated the effects of curcumin on inhibition of cell proliferation. These results provided a novel insight into mechanisms underlying the inhibition of activated HSC growth by curcumin. The character-istics of curcumin, including antioxidant potential, reduction of activated HSC growth, and no adverse health effects, make it a potential candidate for prevention and treatment of hepatic fibrosis. Recently, Van der Logt et al. (2003) demonstrated that curcumin exerted its anticarcinogenic effects in gastrointestinal cancers through the induction of UDP-glucuronosyltransferase enzymes.

10.19 CURCUMIN PROTECTS FROM DRUG-INDUCED LUNG INJURY

Cyclophosphamide causes lung injury in rats through its ability to generate free radicals with subse-quent endothelial and epithelial cell damage. Venkatesan and Chandrakasan (1995) examined the effect of curcumin on cyclophosphamide-induced early lung injury. In order to observe the protective effects of curcumin on cyclophosphamide-induced early lung injury, healthy, pathogen-free male Wistar rats were exposed to 20 mg/100 g body weight of cyclophosphamide, given intraperitoneally as a single injection. Prior to cyclophosphamide intoxication, curcumin was administered orally daily for 7 days. At various times (2, 3, 5, and 7 days after insult), serum and lung samples were analyzed for angiotensin-converting enzyme (ACE), LPO, reduced glutathione, and ascorbic acid. Bronchoal-veolar lavage fluid (BALF) was analyzed for biochemical constituents. The lavage cells were examined for LPO and glutathione content. Excised lungs were analyzed for antioxidant enzyme levels. Bio-chemical analyses revealed increased lavage fluid total protein, albumin, ACE, LDH, N-acetyl-beta-D-glucosaminidase (NAG), alkaline phosphatase, acid phosphatase, lipid peroxide, GSH, and ascorbic acid levels 2, 3, 5, and 7 days after cyclophosphamide intoxication. Increased levels of LPO and decreased levels of GSH and ascorbic acid were seen in serum, lung tissue, and lavage cells of cyclophosphamide-treated groups. Serum ACE activity increased, which coincided with the decrease in lung tissue levels. Activities of antioxidant enzymes were reduced with time in the lungs of cyclophosphamide-treated groups. A significant reduction in the lavage fluid biochemical constituents and in LPO products in the serum, lung and lavage cells occurred concomitantly with an increase in antioxidant defense mechanisms in curcumin-fed cyclophosphamide rats. Therefore, the study indi-cated that curcumin is effective in moderating the cyclophosphamide-induced early lung injury.

In another study, Venkatesan et al. (1997) investigated the effect of curcumin on bleomycin (BLM)-induced lung injury. The data indicated that BLM-mediated lung injury resulted in increases in lung lavage fluid biomarkers such as total protein, ACE, LDH, NAG, LPO products, SOD, and catalase. BLM administration also increased the levels of malondialdehyde in BALF and broncho-alveolar lavage (BAL) cells and greater amounts of alveolar macrophage (AM) SOD activity. By contrast, lower levels of reduced GSH were observed in lung lavage fluid, BAL cells, and AM. Stimulated superoxide anion and H_2O_2 release by AM from BLM-treated rats were higher. Curcumin treatment significantly reduced lavage fluid biomarkers. In addition, it restored the antioxidant status in BLM rats. These data suggested that curcumin treatment reduces the development of BLM-induced inflammatory and oxidant activity. Therefore, curcumin offers the potential for a novel pharmacological approach in the suppression of drug- or chemical-induced lung injury.

Punithavathi et al., (2000) also evaluated the ability of curcumin to suppress BLM-induced pulmonary fibrosis in rats. A single intratracheal instillation of BLM (0.75 U/100 g, sacrificed 3, 5, 7, 14, and 28 days post-BLM) resulted in significant increases in total cell numbers, total protein, and ACE, and in alkaline phosphatase activities in BALF. Animals with fibrosis had a significant increase in lung hydroxyproline content. AM from BLM-treated rats elaborated significant increases in TNF-α, and superoxide and nitric oxide production in culture medium. Interestingly, oral admin-istration of curcumin (300 mg/kg) 10 days before and daily thereafter throughout the experimental time period inhibited BLM-induced increases in total cell counts and biomarkers of inflammatory responses in BALF. In addition, curcumin significantly reduced the total lung hydroxyproline in BLM-treated rats. Furthermore, curcumin remarkably suppressed the BLM-induced AM production of TNFα, SOD, and nitric oxide. These findings suggest curcumin is a potent anti-inflammatory and antifibrotic agent against BLM-induced pulmonary fibrosis in rats. Punithavathi et al. (2003) also examined whether curcumin prevented amiodarone-induced lung fibrosis in rats. They found that curcumin had a protective effect on amiodarone-induced pulmonary fibrosis. Curcumin inhibited the increases in lung myeloperoxidade activity, TGF-β1 expression, lung hydroxyproline content, and expression of type I collagen and c-Jun protein in amiodarone-treated rats.

Paraquat (PQ), a broad-spectrum herbicide, can cause lung injury in humans and animals. An early feature of PQ toxicity is the influx of inflammatory cells, releasing proteolytic enzymes and oxygen free radicals, which can destroy the lung epithelium and cause pulmonary fibrosis. Suppressing early lung injury before the development of irreversible fibrosis is critical to effective therapy. Venkatesan (2000) showed that curcumin confers remarkable protection against PQ-induced lung injury. A single intraperitoneal injection of PQ (50 mg/kg) significantly increased the levels of protein, angiotensin converting enzyme (ACE), alkaline phosphatase, NAG, and TBARS, and neutrophils in the BALF, while it decreased GSH levels. In PQ-treated rats BAL cells, TBARS concentration was increased at the same time as glutathione content was decreased. In addition, PQ caused a decrease in ACE and glutathione levels and an increase in levels of TBARS and myeloperoxidase activity in the lung. Interestingly, curcumin prevented the general toxicity and mortality induced by PQ and blocked the rise in BALF protein, ACE, alkaline phophatase, NAG, TBARS, and neutrophils. Likewise, it prevented the rise in TBARS content in both BAL cell and lung tissue and MPO activity of the lung, reduced lung ACE, and abolished BAL cell and lung glutathione levels. These findings indicate that curcumin has important therapeutic potential in suppressing PQ lung injury.

Nicotine, a pharmacologically active substance in tobacco, has been identified as a major risk factor for lung diseases. Kalpana's group (Kalpana and Menon, 2004) showed that curcumin exerted its protective effect against nicotine-induced lung toxicity by modulating the biochemical marker enzymes, LPO, and augmenting antioxidant defense system. They evaluated the protective effects of curcumin on LPO and antioxidants status in BALF and BAL of nicotine-treated Wistar rats. Lung toxicity was induced by subcutaneous injection of nicotine at a dose of 2.5 mg/kg body weight (5 days a week, for 22 weeks) and curcumin (80 mg/kg body weight) was given simultaneously by intragastric intubation for 22 weeks. Measurements of biochemical marker enzymes — alkaline phosphatase, LDH, LPO, and antioxidants — were used to monitor the antiperoxidative effects of curcumin. The increased biochemical marker enzymes as well as lipid peroxides in BALF and BAL of nicotine treated rats was accompanied by a significant decrease in the levels of glutathione, glutathione peroxidase, superoxide dismutase, and catalase. Administration of curcumin significantly lowered the biochemical marker enzymes, LPO, and enhanced the antioxidant status.

They also showed that curcumin exerts its protective effect against nicotine-induced lung toxicity by modulating the extent of LPO and augmenting antioxidant defense system. They evaluated the protective effects of curcumin on tissue LPO and antioxidants in nicotine-treated Wistar rats. Lung toxicity was induced by subcutaneous injection of nicotine at a dose of 2.5 mg/kg (5 days a week, for 22 weeks). Curcumin (80 mg/kg) was given simultaneously by intragastric intubation for 22 weeks. The enhanced level of tissue lipid peroxides in nicotine-treated rats was accompanied by a significant decrease in the levels of ascorbic acid, vitamin E, reduced glutathione, glutathione peroxidase, superoxide dismutase, and catalase. Administration of curcumin significantly lowered the level of LPO and enhanced the antioxidant status.

10.20 CURCUMIN PROTECTS FROM DRUG-INDUCED NEPHROTOXICITY

Nephrotoxicity is another problem observed in patients given chemotherapeutic agents. Venkatesan et al. (Venkatesan, 1998; Venkatesan et al., 2000) showed that curcumin prevents ADR-induced nephrotoxicity in rats. Treatment with curcumin markedly protected against ADR-induced proteinuria, albuminuria, hypoalbuminemia, and hyperlipidemia. Similarly, curcumin inhibited ADR-induced increase in urinary excretion of NAG (a marker of renal tubular injury), fibronectin and glycosaminoglycan, and plasma cholesterol. It restored renal function in ADR-treated rats, as judged by the increase in glomerular filteration rate (GFR). The data also demonstrated that curcumin protected against ADR-induced renal injury by suppressing oxidative stress and increasing kidney glutathione content and glutathione peroxidase activity. In like manner, curcumin abolished ADR-stimulated kidney microsomal and mitochondrial LPO. These data suggest that administration of curcumin is a promising approach in the treatment of nephrosis caused by ADR.

10.21 CURCUMIN INHIBITS SCARRING

Keloid and hypertrophic scars commonly occur after injuries. Overproliferation of fibroblasts, overproduction of collagen, and contraction characterize these pathologic scars. Current treatment of excessive scars with intralesional corticosteroid injections used individually or in combination with other methods often have unsatisfactory outcomes, frustrating both the patient and the clinician. Phan et al. (2003) investigated the inhibitory effects of curcumin on keloid fibroblasts (KF) and hypertrophic scar-derived fibroblasts (HSF) by proliferation assays, fibroblast-populated collagen lattice contraction, and electron microscopy. Curcumin significantly inhibited KF and HSF proliferation in a dose- and time-dependent manner. Curcumin seemed to have potent effects in inhibiting proliferation and contraction of excessive scar-derived fibroblasts.

10.22 CURCUMIN PROTECTS FROM INFLAMMATORY BOWEL DISEASE

Inflammmatory bowel disease (IBD) is characterized by oxidative and nitrosative stress, leucocyte infiltration, and upregulation of proinflammatory cytokines. Ukil et al. (2003) recently investigated the protective effects of curcumin on 2,4,6- trinitrobenzene sulphonic acid–induced colitis in mice, a model for IBD. Reports showed that curcumin could prevent and improve experimental colitis in murine model with inflammatory bowel disease (IBD) and could be a potential target for the patients with IBD (Jian et al., 2004; 2005).

Intestinal lesions were associated with neutrophil infiltration, increased serine protease activity (may be involved in the degradation of colonic tissue), and high levels of malondialdehyde. Dose–response studies revealed that pretreatment of mice with curcumin at 50 mg/kg daily i.g. for 10 days significantly ameliorated diarrhea and the disruption of colonic architecture. Higher doses (100 and 300 mg/kg) had comparable effects. In curcumin-pretreated mice, there was a significant reduction in the degree of both neutrophil infiltration and LPO in the inflamed colon as well as decreased serine protease activity. Curcumin also reduced the levels of NO and O_2^- associated with the favorable expression of Th1 and Th2 cytokines and inducible NO synthase. Consistent with these observations, NF-κB activation in colonic mucosa was suppressed in the curcumin-treated mice. These findings suggested that curcumin exerts beneficial effects in experimental colitis and may, therefore, be useful in the treatment of IBD.

Salh et al. (2003) also showed that curcumin is able to attenuate colitis in the dinitrobenzene (DNB) sulfonic acid-induced murine model of colitis. When given before the induction of colitis, it reduced macroscopic damage scores and NF-κB activation, reduced myeloperoxidase activity, and attenuated the DNB-induced message for IL-1β. Western blotting analysis revealed a reproducible DNB-induced activation of p38 MAPK in intestinal lysates detected by a phosphospecific antibody. This signal was significantly attenuated by curcumin. Furthermore, the above workers showed that the immunohistochemical signal is dramatically attenuated at the level of the mucosa by curcumin. Thus they concluded that curcumin attenuates experimental colitis through a mechanism that also inhibits the activation of NF-κB and effects a reduction in the activity of p38 MAPK. They proposed that this agent may have therapeutic implications for human IBD.

10.23 CURCUMIN ENHANCES THE IMMUNOSUPPRESSIVE ACTIVITY

Chueh et al. (2003) have demonstrated that curcumin enhances the immunosuppressive activity of cyclosporine in rat cardiac allografts and in mixed lymphocyte reactions. Their study demonstrated for the first time the effectiveness of curumin as a novel adjuvant immunosuppressant with cyclosporine both *in vivo* and *in vitro*. The immunosuppressive effects of curcumin were studied in rat heterotopic cardiac transplant models, using Brown–Norway hearts transplanted to WKY hosts. In the Brown–Norway-to-WKY (Wistar-Kyoto) model, curcumin alone significantly increased the mean survival time, to 20.5 to 24.5 days as compared to 9.1 days in nontreated controls. The combination of curcumin and subtherapeutic doses of cyclosporine further prolonged the mean survival time to 28.5 to 35.6 days, better than that of curcumin or cyclosporine alone. Cytokine analysis revealed significantly reduced expression of IL-2, IFNγ and granzyme B in the day 3 specimens of the curcumin and curcumin plus cyclosporine-treated allografts compared with the nontreated allograft controls.

10.24 CURCUMIN PROTECTS AGAINST VARIOUS FORMS OF STRESS

Curcumin has been identified as a potent inducer of hemoxygenase-1 (HO-1), a redox-sensitive inducible protein that provides protection against various forms of stress. Curcumin stimulated the expression of Nrf2, an increase associated with a significant increase in HO-1 protein expression and HO-1 activity (Balogun et al., 2005). Chan et al. (2005) reported that curcumin prevented MG-induced cell death and apoptotic biochemical changes such as mitochondrial release of cytochrome-c, caspase-3 activation, and cleavage of poly [ADP-ribose] polymerase (PARP). Using the cell-permeable dye 2′,7′-dichlorofluorescein diacetate (DCF-DA) as an indicator of ROS generation, curcumin abolished MG-stimulated intracellular oxidative stress. The study demonstrates that curcumin significantly attenuates MG-induced ROS formation, and suggest that ROS triggers cytochrome-c release, caspase activation, and subsequent apoptotic biochemical changes. Besides, curcumin effectively blocks the detrimental effects of RTV, which is associated with many cardio-vascular complications and causes vascular dysfunction through oxidative stress (Chai et al., 2005).

10.25 CURCUMIN PROTECTS AGAINST ENDOTOXIN SHOCK

Madan and Ghosh (2003) have demonstrated that curcumin exerts protective effects in high-dose endotoxin shock by improving survival and reducing the severity of endotoxin shock symptoms such as lethargy, diarrhea, and watery eyes following a challenge with lipopolysaccharide. They demonstrated that curcumin inhibits the transmigration and infiltration of neutrophils from blood vessels to the underlying liver tissue and, hence, inhibits the damage to the tissue. Curcumin blocks the induced expression of ICAM-1 and VCAM-1 in liver and lungs.

10.26 CURCUMIN PROTECTS AGAINST PANCREATITIS

Gukovsky et al. (2003) reported that curcumin ameliorates pancreatitis in two rat models. In both, cerulein pancreatitis and pancreatitis induced by a combination of ethanol diet and low-dose curcumin, curcumin decreased the severity of the disease. Curcumin markedly inhibited NF-κB and AP-1, IL-6, TNFα, and iNOS in the pancreas. Based on these studies, Gukovsky et al. (2003) suggested that curcumin may be useful for the treatment of pancreatitis. Pathologic activation of both digestive zymogens and the transcription factor NF-κB are early events in acute pancreatitis; these pathologic processes are inhibited in experimental pancreatitis by curcumin and the pH modulator chloroquine (Nagar and Gorelick, 2004).

10.27 CURCUMIN CORRECTS CYSTIC FIBROSIS DEFECTS

Cystic fibrosis is caused by mutations in the gene encoding the cystic fibrosis transmembrane conductance regulator (CFTR). The most common mutation, DeltaF508, results in the production of a misfolded CFTR protein that is retained in the endoplasmic reticulum and targeted for degradation. Curcumin is a nontoxic Ca-adenosine triphosphatase pump inhibitor that can be administered to humans safely. Egan et al. (2004) demonstrated that oral administration of curcumin to homozygous DeltaF508 CFTR mice in doses comparable, on a weight-per-weight basis, to those well tolerated by humans corrected these animals' characteristic nasal potential difference defect. These effects were not observed in mice homozygous for a complete knockout of the CFTR gene. Curcumin also induced the functional appearance of DeltaF508 CFTR protein in the plasma membranes of transfected baby hamster kidney cells. Thus, curcumin treatment may be able to correct defects associated with the homozygous expression of DeltaF508 CFTR.

10.28 CURCUMIN BIOAVAILABILITY, PHARMACODYNAMICS, PHARMACOKINETICS, AND METABOLISM

Numerous studies have been performed on the biotransformation of curcumin (Table 10.5). Lin et al. (2000) showed that curcumin was first biotransformed to dihydrocurcumin and tetrahydrocurcumin and that these compounds subsequently were converted to monoglucuronide conjugates. Thus, curcumin–glucuronide, dihydrocurcumin–glucuronide, tetrahydrocurcumin–glucuronide, and tetrahydrocurcumin are major metabolites of curcumin in mice.

Since the systemic bioavailability of curcumin is low, its pharmacological activity may be mediated, in part, by its metabolites. To investigate this possibility, Ireson et al. (2002) compared curcumin metabolism in human and rat hepatocytes in suspension with that in rats *in vivo*. Analysis by high-performance liquid chromatography with detection at 420 and 280 nm permitted characterization of metabolites with both intact diferoylmethane structure and increased saturation of the heptatrienone chain. Chromatographic inferences were corroborated by mass spectrometry. The major metabolites in suspensions of human or rat hepatocytes were identified as hexahydrocurcumin and hexahydrocurcuminol. In rats, *in vivo*, curcumin administered i.v. (40 mg/kg) disappeared from the plasma within 1 hour of dosing. After p.o. administration (500 mg/kg), parent drug was present in plasma at levels near the detection limit. The major products of curcumin biotransformation identified in rat plasma were curcumin glucuronide and curcumin sulfate, whereas hexahydrocurcumin, hexahydrocurcuminol, and hexahydrocurcumin glucuronide were present in small amounts (Ireson et al., 2002). To test the hypothesis that curcumin metabolites resemble their progenitor in that they can inhibit COX-2 expression, curcumin and four of its metabolites at a concentration of 20 μM were compared in terms of their ability to inhibit phorbol ester–induced prostaglandin E2 (PGE2) production in human colonic epithelial cells. Curcumin reduced PGE2 levels to preinduction levels, whereas tetrahydrocurcumin, hexahydrocurcumin, and curcumin sulfate had only weak PGE2 inhibitory activity, and hexahydrocurcuminol was inactive. The results suggested that (1) the major products of curcumin biotransformation by hepatocytes occurred only at low abundance in rat plasma after curcumin administration and (2) metabolism of curcumin by reduction or conjugation generates species with reduced ability to inhibit COX-2 expression. Because the gastrointestinal tract seems to be exposed more prominently to unmetabolized curcumin than any other tissue, the results support the clinical evaluation of curcumin as a colorectal cancer chemopreventive agent.

Curcumin has very poor bioavailability. In Ayurveda, black pepper (*Piper nigrum*), long pepper, (*Piper longum*) and ginger (*Zingiber officinalis*) are collectively termed Trikatu, and are essential ingredients of numerous prescriptions, used for a wide range of disorders. Numerous studies suggest that Trikatu possesses bioavailability enhancing effect (Johri and Zutshi, 1992). Since curcumin belong to the same family as ginger, it has similar enhancer activity (Chuang et al., 2002).

TABLE 10.5
Pharmacokinetics, Biotransformation, Tissue Distribution, and Metabolic Clearance Rates of Curcumin

Animal	Route	Dose	Remarks	Refs.
Mice	i.p.	0.1 g/kg	2.25 µg/ml in plasma in first 15 min At 1 hour, intestine, spleen, liver, and kidney are 177, 26, 27, 8 µg/kg; 0.4 µg/g brain Biotransformed from DHC to THC, and then converted to monoglucuronide conjugates	Pan et al., 1999
Mice	i.p.	100 mg/kg[a]	39–240 nmol/g tissue small intestine	Perkins et al., 2002
Rats	Oral	1 g/kg	75% excreted in the feces Negligible in urine Poorly absorbed in the gut No toxicity at 5 g/kg	Wahlstrom and Blennow, 1978
	i.v.	—	Transported into bile Major part metabolized	
Rats	Oral, i.v., and i.p.	a	Mostly fecal excretion Excreted in the bile Major biliary metabolite THC/HHC glucuronides	Holder et al., 1978
Rats	Oral	400 mg	60% absorbed None in urine; conjugated glucuronides and sulfates None in heart blood Less than 5 µg/ml in portal blood Negligible in liver /kidney (<20 µg/tissue) for 24 hours At 24 hours, 38% in lower part of the gut	Ravindranath and Chandrashekara, 1980
Rats	—	2 g/kg	Low serum levels Piperine increased bioavailability by 154%	Shoba et al., 1998
Rats	Oral, i.g.	2% of diet	Plasma levels 12 nM Liver (0.1–0.9 nmol/g; colon (0.2–1.8 µmol/g) Increased hepatic GST (16%) Decreased colon MDH–DNA adduct (36%) More in plasma, less in colon	Sharma et al., 2001
Rats	i.v.	40 mg/kg	Disappeared from plasma in 1 hour	Ireson et al., 2001
	p.o.	500 mg/kg	Detectable in plasma Biotransformed to curcumin glucuronide, and sulfate	
Human	Oral	2 g	Serum level not detectable Piperine increased bioavailability by 2000%	Shoba et al., 1998
Human	Oral	375 mg × 3/d	Well tolerated for 12 weeks Significant benefit	Lal et al., 1999
Human	Oral	375 mg × 3/d	Well tolerated for 6–22 months Significant benefit	Lal et al., 1999
Human	Oral	1–12 g/d	Well tolerated up to 8 g/d up to 3 months Serum levels peaked at 1–2 hours and declined at 12 hours Serum levels 0.51 ± 0.11; 0.63 ± 0.06; 1.77 ± 1.87 µM	Cheng et al., 2001
Human	Oral	36–180 mg/d	Most in feces, none in blood or urine 59% decrease in lymphocytic GST after 14 days	Sharma et al., 2001

[a] Combined with [^{3}H] curcumin.

Abbreviations: DHC, dihydrocurcumin; THC, tetrahydrocurcumin; HHC, hexahydrocurcumin; i.v., intravenous; i.p., intraperitoneal; i.g., intragastric; p.o., post-oral; d, day; GST, glutathione S-transferase.

Shoba et al. (1998) found that curcumin has a poor bioavailability due to its rapid metabolism in the liver and intestinal wall. In this study, the effect of combining piperine, a known inhibitor of hepatic and intestinal glucuronidation, was evaluated on the bioavailability of curcumin in rats and healthy human volunteers. When curcumin was given alone, in the dose 2 g/kg to rats, moderate serum concentrations were achieved over a period of 4 hours. Concomitant administration of piperine 20 mg/kg increased the serum concentration of curcumin for a short period of 1 to 2 hours. Time to maximum was significantly increased, while elimination half-life and clearance significantly decreased, and the bioavailability was increased by 154%. On the other hand, in humans, after a dose of 2 g curcumin alone, serum levels were either undetectable or very low. Concomitant administration of piperine 20 mg produced much higher concentrations from 0.25 to 1 hour later, and the increase in bioavailability was 2000%. The study shows that in the dosages used, piperine enhances the serum concentration, extent of absorption, and bioavailability of curcumin in both rats and humans with no adverse effects (Shoba et al., 1998).

In the studies by Kumar et al. (2002), natural biodegradable polymers, namely bovine serum albumin and chitosan, were used to encapsulate curcumin to form a depot drug delivery system. Microspheres were prepared by emulsion–solvent evaporation method coupled with chemical cross-linking of the natural polymers. As much as of 79.49 and 39.66% of curcumin could be encapsulated into the biodegradable carriers with albumin and chitosan, respectively. *In vitro* release studies indicated a biphasic drug release pattern, characterized by a typical burst-effect followed by a slow release, which continued for several days. It was evident from Kumar's study that the curcumin biodegradable microspheres could be successfully employed as prolonged release drug delivery system for better therapeutic management of inflammation as compared to oral or subcutaneous administration of curcumin. Kumar et al. (2000) synthesized bioconjugates of curcumin to improve its systemic delivery. Di-O-glycinoyl curcumin (I) and 2′-deoxy-2′-curcuminyl uridine (2′-cur-U) (IV) were quite potent against multiresistant microorganisms. These bioconjugates served dual purpose of systemic delivery as well as therapeutic agents against viral diseases.

Garcea et al. recently measured curcumin levels in normal and malignant human liver tissue after oral administration of the compound. In total, 12 patients with hepatic metastases from colorectal cancer received 450 to 3600 mg of curcumin daily, for one week prior to surgery (Garcea et al., 2004). The levels of curcumin and its metabolites in portal and peripheral blood, bile, and liver tissue of patients with hepatic metastases from colorectal cancer were measured. Curcumin was poorly available, following oral administration, with low nanomolar levels of the parent compound and its glucuronide and sulphate conjugates found in the peripheral or portal circulation. While curcumin was not found in liver tissue, trace levels of products of its metabolic reduction were detected. In patients who had received curcumin, levels of malondialdehyde–DNA (M(1)G) adduct, which reflect oxidative DNA changes, were not decreased in posttreatment normal and malignant liver tissue when compared to pretreatment samples. The results suggest that doses of curcumin required to furnish hepatic levels sufficient to exert pharmacological activity are probably not feasible in humans.

10.29 CLINICAL EXPERIENCE WITH CURCUMIN

Twelve different studies of the safety and efficacy of curcumin in humans have been reported (Figure 10.6). For example, Deodhar et al. (2004) performed a short-term, double blind, cross-over study in 18 patients (22 to 48 year) to compare the antirheumatic activity of curcumin and phenylbutazone (Deodhar et al., 2004). They administered 1200 mg curcumin/day or 300 mg phenylbutazone/day for 2 weeks. These investigators reported that curcumin was well tolerated, had no side effects, and showed comparable antirheumatic activity.

Lal et al. (1999) administered curcumin orally to patients suffering from chronic anterior uveitis (CAU) at a dose of 375 mg three times a day for 12 weeks. Of 53 patients enrolled, 32 completed the 12-week study (Lal et al., 2000). They were divided into two groups: one group of 18 patients

Table 10.6 Clinical studies with curcumin in human subjects

Study	Patients	Dose	Comments	Ref.
Double blind, cross-over study	18 pts. (22-48 yrs)	1200 mg /day x 2wks	Antirheumatic	Deodhar et al. (2004)
	46 male pts. (15-68)	400 mg; 3x/day x5 days	Inguinal hernia	Satosakar et al. (1986)
	10 volun.	500 mg/day x7 days	Serum cholesterol & LPO	Soni & Kuttan (1992)
	40 pts.	625 mg; 4x/day x 8 wks	well-tolerated	James et al. (1996)
	53 pts.	375 mg; 3x/day x12 wks	Chronic anterior uveitis	Lal et al. (1999)
	8 pts.	375 mg; 3x/day 6-22 months	Idiopathic inflamm. orbital pseudotumors	Lal et al. (2000)
Prospective Phase I	25 pts.	500 mg-12,000mg/day x3 months	H&N cancers	Cheng et al. (2001)
	15 pts.	36-180 mg 4 months	Colorectal Serum GST-down	Sharma et al. (2001)
	12 pts.	450-3600 mg/day x1 wks	Hepatic metastasis of colorectal cancer	Garcea et al. (2004)
	15 pts.	450-3600 mg/day x4 months	Advanced colorectal cancer	Sharma et al. (2004)
	12 pts. (47-72 yrs)	3.6, 1.8 or 0.45 g/day 7 days	Colorectal cancer	Garcea et al. (2005)
	10 (20-40 yrs)	18 mg	Colon cancer chemoprevention	Plummer et al. (2001)

received curcumin alone, whereas the other group of 14 patients, who had a strong PPD reaction, in addition, received antitubercular treatment. The patients in both the groups started improving after two weeks of treatment. All the patients who received curcumin alone improved, whereas the group receiving antitubercular therapy along with curcumin had a response rate of 86%. Follow-up of all the patients for the next 3 yr indicated a recurrence rate of 55% in the first group and of 36% in the second group. Four of 18 (22%) patients in the first group and 3 of 14 patients (21%) in the second group lost their vision in the follow-up period because of various complications, e.g., vitritis, macular edema, central venous block, cataract formation, and glaucomatous optic nerve damage, etc. None of the patients reported any side effects. The efficacy of curcumin and recurrences following treatment are comparable to corticosteroid therapy, which is at present considered the only available standard treatment for this disease. The lack of side effects with curcumin is its greatest advantage compared with corticosteroids. A double blind multicenter clinical trial of this drug for CAU is highly desirable to further validate the results of the study.

Satoskar et al. (1986) evaluated the anti-inflammatory properties of curcumin in patients with postoperative inflammation. They studied 46 male patients (between the ages of 15 and 68 years) having inguinal hernia and/or hydrocoele. After the hernia operation, spermatic cord edema and tenderness were evaluated. Either curcumin (400 mg) or placebo (250 mg lactose) or phenylbutazone (100 mg) was administered three times a day for a period of 5 d from the first postoperative day. Curcumin was found to be quite safe, and phenylbutazone and curcumin produced a better anti-inflammatory response than placebo (Satoskar et al., 1986).

Soni and Kuttan (1992) examined the effect of curcumin on serum levels of cholesterol and lipid peroxides in ten healthy human volunteers. A dose of 500 mg of curcumin per day for 7 days significantly decreased the level of serum lipid peroxides (33%), increased HDL cholesterol (29%), and decreased total serum cholesterol (11.63%). The results suggest curcumin as a chemopreventive substance against arterial diseases.

James (1996) led a New England clinical trial of curcumin's effectiveness as an antiviral agent in 40 participants. Two dropped out; 23 were randomized to high-dose group (four capsules, four times a day) and 15 to lose-dose (three capsules, three times a day) for 8 weeks. Though it had no antiviral effects, curcumin was well tolerated, and most participants liked taking curcumin and felt better.

Lal et al. (2000) described for the first time the clinical efficacy of curcumin in the treatment of patients suffering from idiopathic inflammatory orbital pseudotumors. Curcumin was administered orally at a dose of 375 mg/three times/day for a period of 6 to 22 months in eight patients. They were followed up for a period of 2 years at three-monthly intervals. Five patients completed the study, of which four recovered completely. In the remaining patient, the swelling regressed completely, but some limitation of movement persisted. No side effect was noted in any patient, and there was no recurrence. Thus curcumin could be used as a safe and effective drug in the treatment of idiopathic inflammatory orbital pseudotumors.

Cheng et al. (2001) examined the toxicology, pharmacokinetics, and biologically effective dose of curcumin in humans. This prospective phase I study evaluated curcumin in patients with one of the following five high-risk conditions: (1) recently resected urinary bladder cancer; (2) arsenic Bowen's disease of the skin; (3) uterine cervical intraepithelial neoplasm (CIN); (4) oral leucoplakia; and (5) intestinal metaplasia of the stomach. Curcumin was taken orally for 3 months. Biopsy of the lesion sites was done immediately before and 3 months after starting curcumin treatment. The starting dose was 500 mg/day. If no toxicity of grade II or higher was noted in at least three successive patients, the dose was escalated to 1000, 2000, 4000, 8000, or 12000 mg/day in order. The concentration of curcumin in serum and urine was determined by high-pressure liquid chromatography (HPLC). A total of 25 patients were enrolled in this study. There was no treatment-related toxicity for doses up to 8000 mg/day. Beyond 8000 mg/day, the bulky volume of the drug was unacceptable to the patients. The serum concentration of curcumin usually peaked at 1 to 2 hours after oral intake of curcumin and gradually declined within 12 hours. The average peak serum concentrations after taking 4000 mg, 6000 mg, and 8000 mg of curcumin were 0.51 ± 0.11 μM, 0.63 ± 0.06 μM, and 1.77 ± 1.87 μM, respectively. Urinary excretion of curcumin was undetectable. One of four patients with CIN and one of seven patients with oral leucoplakia developed frank malignancies in spite of curcumin treatment. In contrast, histologic improvement of precancerous lesions was seen in one of two patients with recently resected bladder cancer, two of seven patients with oral leucoplakia, one of six patients with intestinal metaplasia of the stomach, one of four patients with CIN, and two of six patients with Bowen's disease. In conclusion, this study demonstrated that curcumin is not toxic to humans at doses up to 8000 mg/day when taken by mouth for three months. These results also suggested a biologic effect of curcumin in the chemoprevention of cancer.

Sharma et al. (2001) examined the pharmacodynamics and pharmacokinetics of curcumin in humans in a dose-escalation pilot study. A novel standardized turmeric extract in proprietary capsule form was given at doses between 440 and 2200 mg/day, containing 36 to 180 mg of curcumin (Sharma et al., 2001). Fifteen patients with advanced colorectal cancer refractory to standard chemotherapies received turmeric extract daily for up to 4 months. The activity of GST and levels of a DNA adduct (M(1)G) formed by malondialdehyde, a product of LPO and prostaglandin biosynthesis, were measured in patients' blood cells. Oral turmeric extract was well tolerated, and dose-limiting toxicity was not observed. Neither curcumin nor its metabolites were detected in blood or urine, but curcumin was recovered from feces. Curcumin sulfate was identified in the feces of one patient. Ingestion of 440 mg of turmeric extract for 29 days was accompanied by a 59% decrease in lymphocytic GST activity. At higher dose levels, this effect was not observed. Leukocytic M(1)G levels were constant within each patient and unaffected by treatment. Radiologically stable disease was demonstrated in five patients for 2 to 4 months of treatment. The results suggested that: (1) tumeric extract can be administered safely to patients at doses of up to 2.2 g daily, equivalent to 180 mg of curcumin; (2) curcumin has low oral bioavailability in humans and may undergo intestinal metabolism; and (3) larger clinical trials of curcuma extract are merited.

Sharma et al. (2004) designed a dose-escalation study to explore the pharmacology of curcumin in humans. Fifteen patients with advanced colorectal cancer refractory to standard chemotherapies consumed capsules compatible with curcumin doses between 0.45 and 3.6 g daily for up to 4 months. Levels of curcumin and its metabolites in plasma, urine, and feces were analyzed by HPLC and mass spectrometry. Three biomarkers of the potential activity of curcumin were translated from preclinical models and measured in patient blood leukocytes: GST activity, levels of deoxyguanosine adduct M (1) G, and PGE (2) production induced *ex vivo*. Dose-limiting toxicity was not observed. Curcumin and its glucuronide and sulfate metabolites were detected in plasma in the 10 nmol/l range and in urine. A daily dose of 3.6 g curcumin engendered 62% and 57% decreases in inducible PGE (2) production in blood samples taken 1 hour after dose on days 1 and 29, respectively, of treatment compared with levels observed immediately predose ($P < 0.05$). A daily oral dose of 3.6 g of curcumin is advocated for Phase II evaluation in the prevention or treatment of cancers outside the gastrointestinal tract. PGE (2) production in blood and target tissue may indicate biological activity. Levels of curcumin and its metabolites in the urine can be used to assess general compliance (Sharma et al., 2004).

Garcea et al. (2004) investigated whether oral administration of curcumin results in concentrations of the agent in normal and malignant human liver tissue, which are sufficient to elicit pharmacological activity. In total, 12 patients with hepatic metastases from colorectal cancer received 450 to 3600 mg of curcumin daily for one week prior to surgery. Curcumin was poorly available, following oral administration, with low nanomolar levels of the parent compound and its glucuronide and sulphate conjugates found in the peripheral or portal circulation. The results suggest that doses of curcumin required to furnish hepatic levels sufficient to exert pharmacological activity are probably not feasible in humans (Garcea et al., 2004).

Garcea et al. (2005) also tested the hypothesis that pharmacologically active levels of curcumin that can be achieved in the colorectum of humans as measured by effects on levels of M(1)G and COX-2 protein. Patients with colorectal cancer ingested curcumin capsules (3,600, 1,800, or 450 mg daily) for 7 days. Biopsy samples of normal and malignant colorectal tissue, respectively, were obtained at diagnosis and at 6 to 7 hours after the last dose of curcumin. Blood was taken 1 hour after the last dose of curcumin. Curcumin and its metabolites were detected and quantitated by high-performance liquid chromatography with detection by UV spectrophotometry or mass spectrometry. The concentrations of curcumin in normal and malignant colorectal tissue of patients receiving 3600 mg of curcumin were 12.7 ± 5.7 and 7.7 ± 1.8 nmol/g, respectively. Curcumin sulfate and curcumin glucuronide were identified in the tissue of these patients. Trace levels of curcumin were found in the peripheral circulation. M(1)G levels were 2.5-fold higher in malignant tissue as compared with normal tissue. Administration of curcumin (3600 mg) decreased M(1)G levels from 4.8 ± 2.9 adducts per 107 nucleotides in malignant colorectal tissue to 2.0 ± 1.8 adducts per 107 nucleotides. COX-2 protein levels in malignant colorectal tissue were not affected by curcumin. The results suggest that a daily dose of 3.6 g curcumin achieves pharmacologically efficacious levels in the colorectum with negligible distribution of curcumin outside the gut (Garcea et al., 2005).

A Phase I study of curcumin for the chemoprevention of colon cancer has recently been concluded (http://clinicaltrials.gov/ct/gui/show/NCT00027495?order=5). This study aimed at determining the maximum tolerated dose of curcumin as a chemopreventive agent of colon cancer in healthy subjects.

10.30 NATURAL ANALOGUES OF CURCUMIN

Natural curcumin contains three major curcuminoids, namely curcumin, demethoxycurcumin, and bisdemethoxycurcumin (Figure 10.6). Several analogues of curcumin have been identified from other plant sources. These include 6- and 8-gingerol, 6-paradol, cassumunin, galangals, diarylheptanoids, yakuchinones, isoeugenol, and dibenzoylmethane. Like curcumin, gingerol, paradol, cassumunin, shogaol, and diarylheptanoids are also derived from the roots of the plant (Table 10.7)

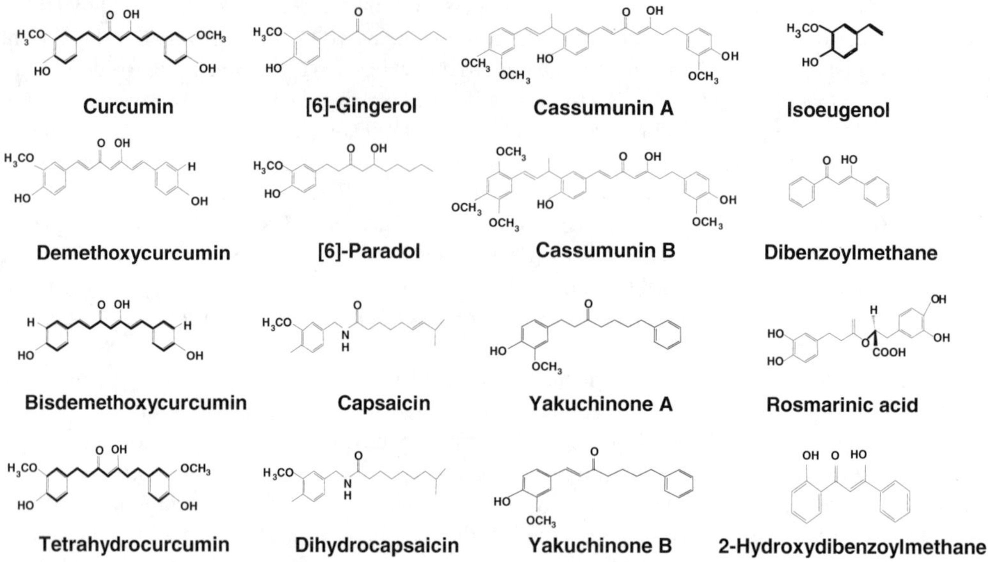

FIGURE 10.6 Natural analogues of curcumin.

TABLE 10.7
Sources and Site of Action of Natural Analogues of Curcumin

Analogues	Source	Target	Refs.
6-Gingerol	Ginger (*Z. officinale* Roscoe)	TNF, NF-κB, AP-1, COX2, ODC, iNOS, p38MAPK, antifungal	Kim et al., 2005
8-Gingerol	Ginger (*Z. officinale* Roscoe)		Kim et al., 2005
6-Paradol	Ginger (*Z. officinale* Roscoe)	Caspase activation	Keum et al., 2002
Shogaol	Ginger (*Z. officinale* Roscoe)	Helicobacter pylori	Mahady et al., 2003
Cassumunin A and B	Ginger (*Z. cassumunar*)	Antioxidant	Mosuda et al., 1998
Diarylheptanoids	Ginger (*Zingiber* sps.)	PGE2 and LT	Hong et al., 2004
Dibenzoylmethane	Licorice (*Glycyrrhiza echinata*)	COX1-1, LOX, HIF, VEGF	Hong et al., 2004
Galanals A and B	Zingiber (*Zingiber mioga* Roscoe)	Caspase 3, Bcl-2	Miyoshi et al., 2003
Garcinol	Kokum (*Garcinia indica*)	NF-κB, COX-2, iNOS, HAT	Pan et al., 2001
Isoeugenol	Cloves (*Eugenia caryophyllus*)	NF-κB, antioxidant β-amyloid	Chainy et al., 2000
Yakuchinone A and B	Galanga (*Alpinia officinarum*)	PG synthetase, COX1-1, iNOS, NF-κB, insecticidal, adhesion molecules, TNF, AP-1, 5-HETE	Chun et al., 2002

Note: For structure of these analogues, see Figure 10.6.

(Kim et al., 2005; Keum et al., 2002; Mahady et al., 2003; Masuda et al., 1998; Hong et al., 2004; Miyoshi et al., 2003; Pan et al., 2001; Chainy et al., 2000; Chun et al., 2002). Although most of these analogues exhibit activities very similar to curcumin, whether they are more potent or less potent than curcumin has not been established. Yakuchinones have been shown to be more potent inhibitors of 5-HETE production than curcumin (Flynn et al., 1986). Synthetic cassumunins also show stronger protective activity than curcumin against oxidative cell death induced by hydrogen peroxide (Masuda et al., 1998). Garcinol is more potent than curcumin in inhibiting tumor cells (Pan et al., 2001). The anticancer potential of galangals, however, is comparable to that of curcumin (Miyoshi et al., 2003). Curcumin has been shown to be more cytotoxic than isoeugenol, bis-eugenol, and eugenol (Fujisawa et al., 2004).

10.31 SYNTHETIC ANALOGS OF CURCUMIN

Commercial curcumin isolated from the rhizome of *Curcuma longa* Linn. contains three major curcuminoids (approximately 77% curcumin, 17% demethoxycurcumin, and 3% bisdemethoxy-curcumin) (Figure 10.7). Commercial curcumin, pure curcumin, and demethoxycurcumin are about equipotent as inhibitors of TPA-induced tumor promotion in mouse skin, whereas bis-demethoxycurcumin is somewhat less active (Huang et al., 1997). Besides curcumin, several analogues of curcumin have been synthesized and tested (Ishida et al., 2002; Dinkova-Kostova and Talalay, 1999). Tetrahydrocurcumin, an antioxidative substance, which is derived from cur-cumin by hydrogenation, has been shown to have a protective effect on oxidative stress in cholesterol-fed rabbits (Naito et al., 2004). Kumar et al. (2003) have developed an analogue of curcumin, 4-hydroxy-3-methoxybenzoic acid methyl ester (HMBME), which targets the Akt/NF-

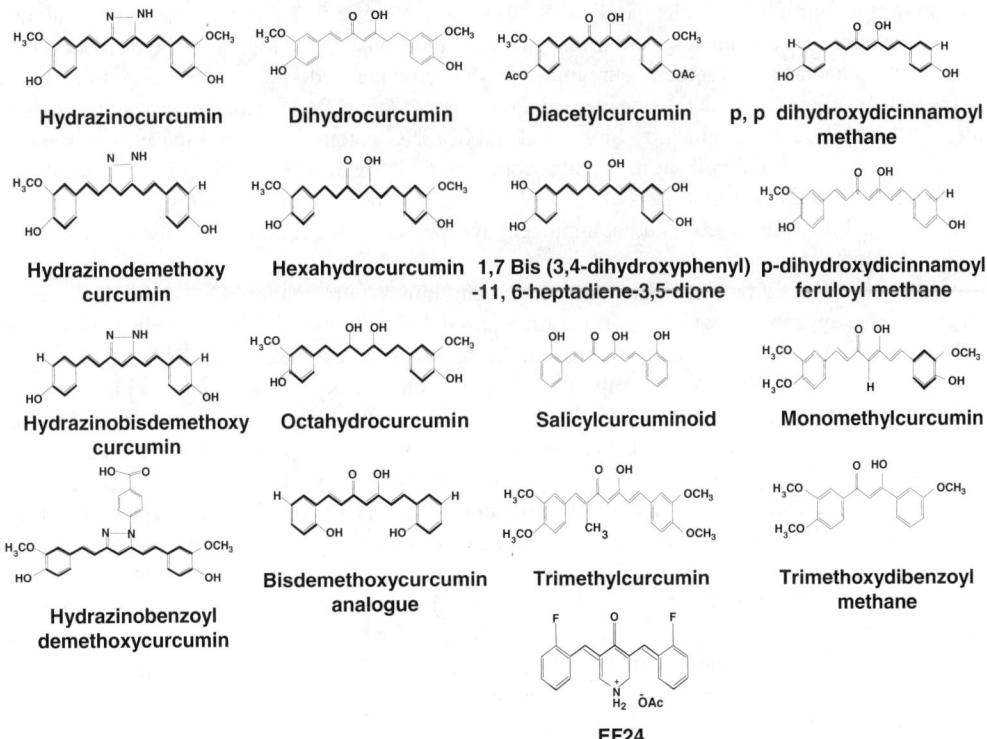

FIGURE 10.7 Synthetic analogues of curcumin.

κB signaling pathway (Kumar et al., 2003). They demonstrated the ability of this novel compound to inhibit the proliferation of human and mouse PCA cells. Overexpression of constitutively active Akt reversed the HMBME-induced growth inhibition and apoptosis, illustrating the direct role of Akt signaling in HMBME-mediated growth inhibition and apoptosis. HMBME-mediated inhibition of Akt kinase activity may have a potential in suppressing/decreasing the activity of major survival/antiapoptotic pathways.

Using an *in vitro* SVR assay, Robinson et al. (2003) have demonstrated potent antiangiogenic properties in aromatic enone and dieneone analogues of curcumin. Based on a simple pharmacophore model, the aromatic enone and aromatic dienone analogues of curcumin were prepared using standard drug design concepts.

Devasena et al. (2002) examined the protective effect of a curcumin analogue [bis-1,7-(2-hydroxyphenyl)-hepta-1,6-diene-3,5-dione] on hepatic LPO and antioxidant status during 1,2-dimethylhydrazine-induced colon carcinogenesis in male Wistar rats. They observed that the curcumin analog exerted chemopreventive effects against cancer development at extrahepatic sites by modulating hepatic biotransformation enzymes and antioxidant status. The effect was comparable with that of curcumin. They proposed that the hydroxyl group in the aromatic ring is responsible for the protective effect rather than the methoxy group. Mishra et al. (2002) synthesized a novel curcumin conjugate viz. 1,7-bis (4-O-glycinoyl-3-methoxyphenyl)-1,6- heptadiene-3, 5, dione (I), which was attached to the deoxy-11 mer, 5′-GTT AGG GTT AG-3′, a complementary sequence of telomerase RNA template. This novel anticancer prodrug has the potential to target the telomerase sequence.

The antitumor properties of metal chelates of synthetic curcuminoids (John et al., 2002) have also been investigated. John et al. (2002) examined four synthetic curcuminoids, 1,7-bis(4-hydroxy-3-methoxyphenyl)-1, 6-heptadiene-3, 5-dione (curcumin1), 1,7-bis(piperonyl)-1,6-heptadiene-3, 5-dione (piperonyl curcumin), 1, 7-bis(2-hydroxy naphthyl)-1, 6-heptadiene-2, 5-dione (2-hydroxy naphthyl curcumin), 1,1-bis(phenyl)-1, 3, 8, 10-undecatetraene-5, 7-dione (cinnamyl curcumin), and their copper(II) complexes were investigated for their possible cytotoxic and antitumor activities. Copper chelates of synthetic curcuminoids showed enhanced antitumor activity. In addition to these, a novel curcumin derivative, named hydrazinocurcumin (HC), was synthesized and examined for its biological activities by Shim et al. (2002). HC potently inhibited the proliferation of bovine aortic endothelial cells at nanomolar concentrations (IC50 = 520 nM without cytotoxicity. Snyder et al. (2002) reported the synthesis of several different structural analogues of curcumin and examined their antitumor and antiangiogenic properties. They found analogues that are more potent than native curcumin.

Limtrakul et al. (2004) investigated natural curcuminoids, pure curcumin, demethoxycurcumin, and bisdemethoxycurcumin, isolated from turmeric, for their potential ability to modulate the human MDR-1 gene expression in multidrug-resistant human cervical carcinoma cell line, KB-V1. Multidrug resistance (MDR) is a very important phenomenon that is often associated with decreased intracellular drug accumulation in patient's tumor cells resulting from enhanced drug efflux (Limtrakul et al., 2004). It is related to the overexpression of a membrane protein, P-glycoprotein (Pgp-170), thereby reducing drug cytotoxicity. They found that all the three curcuminoids inhibited MDR-1 gene expression, and bisdemethoxycurcumin produced maximum effect. The commercial grade curcuminoid (approximately 77% curcumin, 17% demethoxycurcumin, and 3% bisdemthoxycurcumin) decreased MDR-1 gene expression in a dose-dependent manner and had about the same potent inhibitory effect on MDR-1 gene expression as natural curcuminoid mixtures. Their results indicate that bisdemethoxycurcumin is the most active of the curcuminoids present in turmeric for modulation of MDR-1 gene. Treatment of drug resistant KB-V1 cells with curcumin increased their sensitivity to vinblastine, which was consistent with a decreased MDR-1 gene product, a P-glycoprotein, on the cell plasma membrane.

In a recent study, Selvam et al. (2005), isolated curcuminoids from turmeric and their pyrazole and isoxazole analogues were synthesized. They compared for antioxidant and COX-1/COX-2 inhibitory and anti-inflammatory activities. The designed analogues exhibited significant

COX-2/COX-1 selectivity and also better antioxidant activity. Molecular docking studies revealed that these compounds were able to dock the active site of COX-2/COX-1. This approach would help in designing novel potent inhibitors. Furness et al. (2005) showed the efficacy of different curcumin analogs to exhibit antiangiogenic properties.

Gafner et al. (2004) investigated curcumin and 22 of its derivatives for their chemopreventive potential. Based on COX-2 inhibition, curcumin (IC50 = 15.9 μM), 1,7-bis (3-fluoro-4-hydroxyphenyl)-1,6-heptadiene-3,5-dione (IC50 = 23.7 μM) and 2,6-bis (3-fluoro-4-hydroxybenzylidene) cyclohexanone (IC50 = 5.5 μM) were found to be most potent. Tricyclic derivatives 2,6-bis (4-hydroxy-3-methoxybenzylidene)cyclohexanone, 2,6-bis (4-hydroxy-3,5-dimethoxybenzylidene) cyclohexanone and 2,5-bis (4-hydroxy-3,5-dimethoxybenzylidene)cyclopentanone inhibited LPS-induced COX-2 and iNOS gene expression in murine macrophages with potency equal to curcumin. RT-PCR experiments demonstrated suppression of COX-2 and iNOS gene expression occurred at the transcriptional level (Gafner et al., 2004).

Kalpana and Menon (2004) showed that curcumin and its analogues inhibited nicotine-mediated imbalances in oxidant–antioxidant status in male Wistar rats. In this study, nicotine was injected subcutaneously at a dose of 2.5 mg/kg of body weight (5 days a week, for 22 weeks). The enhanced circulatory lipid peroxides in nicotine-treated rats was accompanied by a significant decrease in the levels of ascorbic acid, vitamin E, reduced glutathione, glutathione peroxidase, superoxide dismutase, and catalase. Administration of curcumin and curcumin analogue significantly lowered the LPO and enhanced the antioxidant status with modulation in the levels of zinc, copper, and ferritin. However, the effect was more significant in curcumin analogue-treated rats than in curcumin-treated rats.

The inhibitory mechanism of curcumin and its derivative (CHC007) against beta-catenin/T-cell factor (Tcf) signaling in various cancer cell lines was investigated by Park et al. (2005). β-Catenin presents two different facets. One aspect is that it contributes to the cell–cell adhesion in cooperation with E-cadherin. The other aspect is that it possesses transcriptional activity in cooperation with T-cell factor (Tcf)/lymphoid enhancer factor (Lef) transcription factor in a nucleus. β-catenin gene is mutated in many cancer cells including colorectal cancer, melanoma, hepatocellular carcinoma, and gastric carcinoma; the transcriptional activity of β-catenin is upregulated in these cancer cells. Therefore, if β-catenin's transcriptional activity can be markedly downregulated, tumor growth will be suppressed. However, there exist few β-catenin inhibitors and one of them is aspirin. Curcumin and its derivatives showed excellent inhibition of β-catenin/Tcf signaling in different cancer cell lines and the reduced β-catenin/Tcf transcriptional activity is due to the decreased nuclear β-catenin and Tcf-4.

Synthetic curcumin analogs inhibited complex formations between Fos-Jun heterodimer and activator protein-1 (AP-1) DNA as reported by Hahm et al. (2002). These curcumin analogs have been observed to repress the AP-1 transcription in AP-1-transfected cells, and they also inhibited the increased expression of Jun/AP-1 protein by 12-O-tetradecanoylphorbol-13-acetate (TPA) in the same cells. Curcumin analogs downregulated the expression of MMP-9 (gelatinase-B), correlating with inhibition of cellular invasion and migration in conditions such as tumor invasion and metastasis.

10.32 STRUCTURE–ACTIVITY RELATIONSHIP OF CURCUMIN

To elucidate which portion of the molecule is critical for the activity, a large number of structural analogs of curcumin have been synthesized. Some analogues are more active than native curcumin, while others are less active (Table 10.7), (Bartherlemy et al., 1998; Gomes Dde et al., 2002; Tonnesen et al., 2002; Punithavathi et al., 2003; Dinkova-Kostova and Talalay, 1999; Kumar et al., 2003; Mishra et al., 2002; John et al., 2002; Shim et al., 2002; Selvam et al., 2005; Khopde et al., 1999; Rao et al., 1982; Hahm et al., 2002; Douglas, 1993; Rukkumani et al., 2004; Vajragupta et al., 2004). It was found that the phenolic analogues were more active than the nonphenolic analogues

TABLE 10.8
Relative Potency of Curcumin and Its Synthetic Analogues

Effects	Refs.
Analogues More Potent than Curcumin	
THC: lipid peroxidation under aqueous condition by pulse radiolysis technique	Khopde et al., 1999
THC: preventing nitrite-induced oxidation of hemoglobin	Venkatesan et al., 2003
NaC: carrageenin-induced rat hind paw edema	Rao et al., 1982
HMBME: inhibition of prostate cancer	Kumar et al., 2003
BJC005, CHC011, and CHC007: formation of Fos-/Jun- DNA complex	Hahm et al., 2002
Tocopheryl curcumin: inhibiting Tat transactivation of HIV-LTR	Barthelemy et al., 1998
4, 4′-DAC: histamine blocking activity	Douglas, 1993
Copper chelates of 2-hydroxynapthyl curcumin: antitumor activity	John et al., 2002
Hydrazinocurcumin: BAECs proliferation	Shim et al., 2002
o-hydroxy substituted analog: inhibiting alcohol and PUFA induced oxidative stress	Rukkumani et al., 2004
Di-O-glycinoyl curcumin and 2′-deoxy-2′-curcuminyl uridine: antiviral activity	Mishra et al., 2002
Pyrazole and isoxazole analogues: Cox-2 inhibitory activity	Selvam et al., 2005
1,7-bis-(2-hydroxy-4-methoxyphenyl)-1,6-heptadiene-3,5-dione): AL activity	Gomes et al., 2002
Salicylcurcuminoid: antioxidant	
Analogues Less Potent than Curcumin	
THC: lipid peroxidation under aerated condition by pulse radiolysis technique	Khopde et al., 2000
THC: TPA-induced mouse ear edema and skin carcinogenesis	Dinkova-koztova and Talalay, 1999
Analogues as Potent as Curcumin	
5-hydroxy-1,7-diphenyl-1,4,6-heptatriene-3-one: scavenge hydroxyl radicals	Tonnesen and Greenhill, 1992
Manganese complexes of curcumin and diacetylcurcumin: scavenge hydroxyl radicals	Vajragupta et al., 2004

Abbreviations: THC, tetrahydrocurcumin; NaC, sodium curcuminate; HMBME, 4-hydroxy-3-methoxybenzoic acid methyl ester; DAC, diacetylcurcumin; BAEC, bovine aortic endothelial cells; PUFA, poly-unsaturated fatty acids; Cox-2: cyclooxygenase-2; AL, anti-leishmanial.

(Venkatesan, 2000). The highest antioxidant activity was obtained when the phenolic group was sterically hindered by the introduction of two methyl groups at the ortho position. The phenolic group is essential for the free-radical scavenging activity, and the presence of the methoxy group further increases the activity (Sreejayan and Rao, 1996). Curcumin shows both antioxidant and pro-oxidant effects. Ahsan et al. have shown that both these effects are determined by the same structural moieties of the curcuminoids (Ahsan et al., 1999).

Dinkova–Kostova showed that the presence of hydroxyl groups at the ortho position on the aromatic rings and that beta-diketone functionality was required for high potency in inducing Phase 2 detoxification enzymes (Dinkova–Kostova and Talalay, 1999). Curcumin is a noncompetitive inhibitor of rat liver microsomal delta 5 desaturase and delta 6 desaturase. Kawashima et al. (1996) have shown that only half the structure is essential for the desaturase inhibition. A 3-hydroxy group of the aromatic ring is essential for the inhibition and a free carboxyl group at the end opposite to the aromatic ring interferes with the inhibitory effect.

Simon et al. found that the presence of the diketone moiety in the curcumin molecule seems to be essential for its ability to inhibit the proliferation of MCF-7 human breast tumor cells (Simon et al., 1998). The aromatic enone and dienone analogs of curcumin have been demonstrated to have potent antiangiogenic property in an *in vitro* SVR assay (Robinson et al., 2003).

TABLE 10.9

Cancer Incidence in India and the U.S.

Cancer	U.S. Cases	U.S. Deaths	India Cases	India Deaths
Breast	660	160	79	41
Prostate	690	130	20	9
Colon/rectum	530	220	30	18
Lung	660	580	38	37
Head and neck SCC	140	44	153	103
Liver	44	41	13	12
Pancreas	108	103	8	8
Stomach	81	50	33	30
Melanoma	145	27	18	1
Testis	21	1	3	1
Bladder	202	43	15	11
Kidney	115	44	6	4
Brain, nervous system	65	47	19	14
Thyroid	55	5	12	3
Endometrial cancers	163	41	132	72
Ovary	76	50	20	12
Multiple myeloma	50	40	6	5
Leukemia	100	70	19	17
Non-Hodgkin lymphoma	180	90	17	15
Hodgkin's disease	20	5	7	4

Showing cases per 1 million persons calculated on the basis of current consensus; endometrial cancers include Cervix uteri and Corpus uteri.

GLOBOCAN 2000: Cancer Incidence, Mortality and Prevalence Worldwide, Version 1.0. IARC Cancer Base No. 5. Lyon, IARC Press, 2001.

10.33 CONCLUSION

All evidences accumulated so far clearly indicate that curcumin protects against cancer, cardiovascular diseases, and diabetes, the major ailments in the U.S. This drug has also shown preventive as well as therapeutic effects against Alzheimer's disease, MS, cataract formation, AIDS, and drug-induced nonspecific toxicity in the heart, lung, and kidney. Several of the studies establishing curcumin's potential were carried out in animals. Further testing of curcumin in humans is underway to confirm these observations. A clinical development plan for using curcumin to treat cancer was recently described by the NCI. Studies also show that in countries such as India, which consume curcumin, the profile of cancer incidence is very different than those that do not, such as in the U.S. (Table 10.9). How curcumin produces its therapeutic effects is not fully understood, but they are probably mediated in part through the antioxidant and anti-inflammatory action of curcumin. It is quite likely that curcumin mediates its effects through other mechanisms as well. Over a dozen different cellular proteins and enzymes have been identified to which curcumin binds. High-throughput ligand-interacting technology and microarray technology have begun to reveal more molecular targets and genes affected by curcumin.

Some of the sources of curcumin are given in Table 10.10.

TABLE 10.10
Commercial Sources of Curcumin

Human use:

Sabinsa (http:www.sabinsa.com/products/circumin_book.htm; Piscataway, NJ)

Synthite Industrial Chemicals (www.synthite.com/health.html)

Kalsec (http://www.kalsec.com/products/turmeric_over.cfm; Kalamazoo, MI)

Life Extension (http://www.lef.org/newshop/items/item00552.html?source=WebProtProd)

Turmeric Curcumin (http://www.turmeric-curcumin.com/)

Iherb (http://www.iherb.com/curcuminl.html)

Club Natural (http://www.clubnatural.com/curex9550180.html, Irvine, CA)

American Nutrition (www.AmericanNutrition.com)

Amerifit (www.amerifit.geomerx.com/items/categories.cfm?categoryid=2, Bloomfield, CT)

XKMS (www.xkms.org/Webvitamins-32/Curcumin-Power-60C.htm)

Immune Support (https://www.Immunesupport.com/shop/prodlisting.cfm?NOTE=NOC)

Nature's (www.naturesnutrition.com/SKU/55114.htm)

Big Fitness (www.bfwse.com/jr-021.html)

Power house Gym (http://store.yahoo.com/musclespot/curcumin95.html), MMS

MMS Pro (http://www.mmspro.com/)

Herbal Fields (http://www.herbalfields.com/curcumin.html)

Amazon.com

Research use:

Sigma Aldrich (http://www.sigmaaldrich.com/cgibin/hsrun/Distributed/HahtShop/HAHTpage/HS_CatalogSearch)

Calbiochem (http://www.calbiochem.com/Products/ProductDetail_CBCB.asp?catNO=239802)

LKT laboratories (www.lktlabs.com)

REFERENCES

Abe, Y., Hashimoto, S. and Horie, T. (1999) Curcumin inhibition of inflammatory cytokine production by human peripheral blood monocytes and alveolar macrophages. *Pharmacol Res* 39(1), 41-47.

Abraham, S.K., Sarma, L. and Kesavan, P.C. (1993) Protective effects of chlorogenic acid, curcumin and beta-carotene against gamma-radiation-induced in vivo chromosomal damage. *Mutat Res* 303(3), 109-112.

Adams, B.K., Cai, J., Armstrong, J., Herold, M., Lu, Y.J., Sun, A., Snyder, J.P., Liotta, D.C., Jones, D.P. and Shoji, M. (2005) EF24, a novel synthetic curcumin analog, induces apoptosis in cancer cells *via* a redox-dependent mechanism. *Anticancer Drugs* 16(3), 263-275.

Adelaide, J., Monges, G., Derderian, C., Seitz, J.F. and Birnbaum, D. (1995) Oesophageal cancer and amplification of the human cyclin D gene CCND1/PRAD1. *Br J Cancer* 71(1), 64-68.

Aggarwal, B.B., Kumar, A. and Bharti, A.C. (2003) Anticancer potential of curcumin: preclinical and clinical studies. *Anticancer Res* 23(1A), 363-398.

Aggarwal, S., Takada, Y., Singh, S., Myers, J.N. and Aggarwal, B.B. (2004) Inhibition of growth and survival of human head and neck squamous cell carcinoma cells by curcumin via modulation of nuclear factor-kappaB signaling. *Int J Cancer* 111(5), 679-692.

Ahsan, H., Parveen, N., Khan, N.U. and Hadi, S.M. (1999) Pro-oxidant, anti-oxidant and cleavage activities on DNA of curcumin and its derivatives demethoxycurcumin and bisdemethoxycurcumin. *Chem Biol Interact* 121(2), 161-175.

Anto, R.J., Mukhopadhyay, A., Denning, K. and Aggarwal, B.B. (2002) Curcumin (diferuloylmethane) induces apoptosis through activation of caspase-8, BID cleavage and cytochrome c release: its suppression by ectopic expression of Bcl-2 and Bcl-xl. *Carcinogenesis* 23(1), 143-150.

Anuchapreeda, S., Leechanachai, P., Smith, M.M., Ambudkar, S.V. and Limtrakul, P.N., (2002) Modulation of P-glycoprotein expression and function by curcumin in multidrug-resistant human KB cells. *Biochem Pharmacol* 64(4), 573-582.

Araujo, C.A., Alegrio, L.V., Gomes, D.C., Lima, M.E., Gomes-Cardoso, L. and Leon, L.L. (1999) Studies on the effectiveness of diarylheptanoids derivatives against Leishmania amazonensis. *Mem Inst Oswaldo Cruz* 94(6), 791-794.

Arbiser, J.L., Klauber, N., Rohan, R., van Leeuwen, R., Huang, M.T., Fisher, C., Flynn, E. and Byers, H.R. (1998) Curcumin is an *in vivo* inhibitor of angiogenesis. *Mol Med* 4(6): 376-383.

Arun, N. and Nalini, N. (2002) Efficacy of turmeric on blood sugar and polyol pathway in diabetic albino rats. *Plant Foods Hum Nutr* 57(1), 41-52.

Asai, A. and Miyazawa, T. (2001) Dietary curcuminoids prevent high-fat diet-induced lipid accumulation in rat liver and epididymal adipose tissue. *J Nutr* 131(11), 2932-2935.

Awasthi, S., Pandya, U., Singhal, S.S., Lin, J.T., Thiviyanathan, V., Seifert, W.E.Jr., Awasthi, Y.C. and Ansari, G.A. (2000) Curcumin-glutathione interactions and the role of human glutathione S-transferase P1-1. *Chem Biol Interact* 128(1), 19-38.

Awasthi, S., Srivatava, S.K., Piper, J.T., Singhal, S.S., Chaubey, M. and Awasthi, Y.C. (1996) Curcumin protects against 4-hydroxy-2-trans-nonenal-induced cataract formation in rat lenses. *Am J Clin Nutr* 64(5): 761-766.

Azuine, M.A. and Bhide, S.V. (1992) Chemopreventive effect of turmeric against stomach and skin tumors induced by chemical carcinogens in Swiss mice. *Nutr Cancer* 17(1), 77-83.

Azuine, M.A. and Bhide, S.V. (1992) Protective single/combined treatment with betel leaf and turmeric against methyl (acetoxymethyl) nitrosamine-induced hamster oral carcinogenesis. *Int J Cancer* 51(3), 412-415.

Azuine, M.A. and Bhide, S.V. (1994) Adjuvant chemoprevention of experimental cancer: catechin and dietary turmeric in forestomach and oral cancer models. *J Ethnopharmacol* 44(3), 211-217.

Babu, P.S. and Srinivasan, K. (1995) Influence of dietary curcumin and cholesterol on the progression of experimentally induced diabetes in albino rat. *Mol Cell Biochem* 152(1), 13-21.

Babu, P.S. and Srinivasan, K. (1997) Hypolipidemic action of curcumin, the active principle of turmeric (*Curcuma longa*) in streptozotocin induced diabetic rats. *Mol Cell Biochem* 166(1-2), 169-175.

Balasubramanyam, K., Varier, R.A., Altaf, M., Swaminathan, V., Siddappa, N.B., Ranga, U. and Kundu, T.K. (2004) Curcumin, a novel p300/CREB-binding protein-specific inhibitor of acetyltransferase, represses the acetylation of histone/nonhistone proteins and histone acetyltransferase-dependent chromatin transcription. *J Biol Chem* 279(49), 51163-51171.

Baldin, V., Lukas, J., Marcote, M.J., Pagano, M. and Draetta, G. (1993) Cyclin D1 is a nuclear protein required for cell cycle progression in G1. *Genes Dev* 7(5), 812-821.

Baldwin, A.S. (2001) Control of oncogenesis and cancer therapy resistance by the transcription factor NF-kappaB. *J Clin Invest* 107(3), 241-246.

Balogun, E., Hoque, M., Gong, P., Killeen, E., Green, C.J., Forest,i R., Alam, J. and Motterlini, R.(2005) Curcumin activates the haem oxygenase-1 gene via regulation of Nrf2 and the antioxidant-responsive element. *Biochem J* 371(Pt 3), 887-895.

Barthelemy, S., Vergnes, L., Moynier, M., Guyot, D., Labidalle, S. and Bahraoui, E. (1998) Curcumin and curcumin derivatives inhibit Tat-mediated transactivation of type 1 human immunodeficiency virus long terminal repeat. *Res Virol* 149(1), 43-52.

Bartkova, J., Lukas, J., Muller, H., Lutzhoft, D., Strauss, M. and Bartek, J. (1994) Cyclin D1 protein expression and function in human breast cancer. *Int J Cancer* 57(3), 353-361.

Bava, S.V., Puliappadamba, V.T., Deepti, A., Nair, A., Karunagaran, D. and Anto, R.J.. (2005) Sensitization of taxol-induced apoptosis by curcumin involves down-regulation of nuclear factor-kappaB and the serine/threonine kinase Akt and is independent of tubulin polymerization. *J. Biol. Chem.*, 280(8), 6301-6308.

Bernabe-Pineda, M., Ramirez-Silva, M.T., Romero-Romo, M., Gonzalez-Vergara, E. and Rojas-Hernandez, A. (2004) Determination of acidity constants of curcumin in aqueous solution and apparent rate constant of its decomposition. *Spectrochim Acta A Mol Biomol Spectrosc* 60(5): 1091-1097.

Bharti, A.C., Donato, N. and Aggarwal, B.B. (2003) Curcumin (diferuloylmethane) inhibits constitutive and IL-6-inducible STAT3 phosphorylation in human multiple myeloma cells. *J Immunol* 171(7), 3863-3871.

Bharti, A.C., Donato, N., Singh, S. and Aggarwal, B.B. (2002) Curcumin (diferuloylmethane) downregulates the constitutive activation of nuclear factor-kappaB and I kappa B alpha kinase in human multiple myeloma cells leading to suppression of proliferation and induction of apoptosis. *Blood.* 101 (3): 1053-1062.

Bharti, A.C., Donato, N., Singh, S. and Aggarwal, B.B. (2003) Curcumin (diferuloylmethane) down-regulates the constitutive activation of nuclear factor-kappa B and IkappaBalpha kinase in human multiple myeloma cells, leading to suppression of proliferation and induction of apoptosis. *Blood* 101(3), 1053-1062.

Bharti, A.C., Shishodia, S., Reuben, J.M., Weber, D., Alexanian, R., Raj-Vadhan, S., Estrov, Z., Talpaz, M. and Aggarwal, B.B. (2004) Nuclear factor-kappaB and STAT3 are constitutively active in CD138+ cells derived from multiple myeloma patients, and suppression of these transcription factors leads to apoptosis. *Blood* 103(8), 3175-3184.

Bhaumik, S., Anjum, R., Rangaraj, N., Pardhasaradhi, B.V. and Khar, A. (1999) Curcumin mediated apoptosis in AK-5 tumor cells involves the production of reactive oxygen intermediates. *FEBS Lett* 456(2), 311-314.

Bielak-Zmijewska, A., Koronkiewicz, M., Skierski, J., Piwocka, K., Radziszewska, E. and Sikora, E. (2000) Effect of curcumin on the apoptosis of rodent and human nonproliferating and proliferating lymphoid cells. *Nutr Cancer* 38(1), 131-138.

Bierhaus, A., Zhang, Y., Quehenberger, P., Luther, T., Haase, M., Muller, M., Mackman, N., Ziegler, R. and Nawroth, P.P. (1997) The dietary pigment curcumin reduces endothelial tissue factor gene expression by inhibiting binding of AP-1 to the DNA and activation of NF-kappa B. *Thromb Haemost* 77(4), 772-782.

Bilmen, J.G., Khan, S.Z., Javed, M.H. and Michelangeli, F. (2001) Inhibition of the SERCA Ca2+ pumps by curcumin. Curcumin putatively stabilizes the interaction between the nucleotide-binding and phosphorylation domains in the absence of ATP. *Eur J Biochem* 268(23), 6318-27.

Brouet, I. and Ohshima, H. (1995) Curcumin, an anti-tumour promoter and anti-inflammatory agent, inhibits induction of nitric oxide synthase in activated macrophages. *Biochem Biophys Res Commun* 206(2), 533-540.

Bush, J.A., Cheung, K.J.Jr. and Li, G. (2001) Curcumin induces apoptosis in human melanoma cells through a Fas receptor/caspase-8 pathway independent of p53. *Exp Cell Res* 271(2), 305-314

Busquets, S., Carbo, N., Almendro, V., Quiles, M.T., Lopez-Soriano, F.J. and Argiles, J.M. (2001) Curcumin, a natural product present in turmeric, decreases tumor growth but does not behave as an anticachectic compound in a rat model. *Cancer Lett* 167(1), 33-38.

Caputi, M., Groeger, A.M., Esposito, V., Dean, C., De Luca, A., Pacilio, C., Muller, M.R., Giordano, G.G., Baldi, F., Wolner, E. and Giordano, A. (1999) Prognostic role of cyclin D1 in lung cancer. Relationship to proliferating cell nuclear antigen. *Am J Respir Cell Mol Biol* 20(4), 746-750.

Chai, H., Yan, S., Lin, P., Lumsden, A.B., Yao, Q. and Chen, C.(2005) Curcumin blocks HIV protease inhibitor ritonavir-induced vascular dysfunction in porcine coronary arteries. *J Am Coll Surg* 200(6), 820-830.

Chainy, G.B., Manna, S.K., Chaturvedi, M.M. and Aggarwal, B.B. (2000) Anethole blocks both early and late cellular responses transduced by tumor necrosis factor: effect on NF-kappaB, AP-1, JNK, MAPKK and apoptosis. *Oncogene* 19(25), 2943-2950.

Chan, M.M. (1995) Inhibition of tumor necrosis factor by curcumin, a phytochemical. *Biochem Pharmacol* 49(11), 1551-6.

Chan, M.M., Fong, D., Soprano, K.J., Holmes, W.F. and Heverling, H. (2003) Inhibition of growth and sensitization to cisplatin-mediated killing of ovarian cancer cells by polyphenolic chemopreventive agents. *J Cell Physiol* 194(1), 63-70.

Chan, M.M., Ho, C.T. and Huang, H.I. (1995) Effects of three dietary phytochemicals from tea, rosemary and turmeric on inflammation-induced nitrite production. *Cancer Lett* 96(1), 23-29.

Chan, M.M., Huang, H.I., Fenton, M.R. and Fong, D. (1998) *In vivo* inhibition of nitric oxide synthase gene expression by curcumin, a cancer preventive natural product with anti-inflammatory properties. *Biochem Pharmacol* 55(12), 1955-1962.

Chan, W.H., Wu H.J. and Hsuuw, Y.D.(2005) Curcumin Inhibits ROS Formation and Apoptosis in Methylglyoxal-Treated Human Hepatoma G2 Cells. *Ann N Y Acad Sci* 1042, 372-378.

Chen, A. and Xu, J. (2005) Activation of PPAR{gamma} by curcumin inhibits Moser cell growth and mediates suppression of gene expression of cyclin D1 and EGFR. *Am J Physiol Gastrointest Liver Physiol* 288(3), G447-456.

Chen, H., Zhang, Z.S., Zhang, Y.L. and Zhou, D.Y. (1999) Curcumin inhibits cell proliferation by interfering with the cell cycle and inducing apoptosis in colon carcinoma cells. *Anticancer Res* 19(5A), 3675-3680.

Chen, H.W. and Huang, H.C. (1998) Effect of curcumin on cell cycle progression and apoptosis in vascular smooth muscle cells. *Br J Pharmacol* 124(6): 1029-1040.

Chen, H.W., Yu ,S.L., Chen, J.J., Li, H.N., Lin, Y.C., Yao, P.L., Chou, H.Y., Chien, C.T., Chen, W.J., Lee, Y.T. and Yang, P.C. (2004) Anti-invasive gene expression profile of curcumin in lung adenocarcinoma based on a high throughput microarray analysis. *Mol Pharmacol* 65(1), 99-110.

Chen, X., Hasuma, T., Yano, Y., Yoshimata, T., Morishima, Y., Wang, Y. and Otani, S. (1997) Inhibition of farnesyl protein transferase by monoterpene, curcumin derivatives and gallotannin. *Anticancer Res* 17(4A), 2555-2564.

Chen, X., He, Q., Liu, W., Xu, Q., Ye, Y., Fu, B. and Yu, L. (2000) [AP-1 mediated signal transduction in thrombin-induced regulation of PAL-1 expression in human mesangial cells]. *Chin Med J (Engl)* 113(6), 514-519.

Chen, Y.C., Kuo, T.C., Lin-Shiau, S.Y. and Lin, J.K. (1996) Induction of HSP70 gene expression by modulation of Ca(+2) ion and cellular p53 protein by curcumin in colorectal carcinoma cells. *Mol Carcinog* 17(4), 224-234.

Chen, Y.R. and Tan, T.H. (1998) Inhibition of the c-Jun N-terminal kinase (JNK) signaling pathway by curcumin. *Oncogene* 17(2), 173-178.

Chen, Y.R., Zhou, G. and Tan, T.H. (1999) c-Jun N-terminal kinase mediates apoptotic signaling induced by N-(4-hydroxyphenyl)retinamide. *Mol Pharmacol* 56(6), 1271-1279.

Chendil, D., Ranga, R.S., Meigooni, D., Sathishkumar, S. and Ahmed, M.M. (2004) Curcumin confers radiosensitizing effect in prostate cancer cell line PC-3. *Oncogene* 23(8), 1599-1607.

Cheng, A.L., Hsu, C.H., Lin, J.K., Hsu, M.M., Ho, Y.F., Shen, T.S., Ko, J.Y., Lin, J.T., Lin, B.R., Ming-Shiang, W., Yu, H.S., Jee, S.H., Chen, G.S., Chen, T.M., Chen, C.A., Lai, M.K., Pu, Y.S., Pan, M.H., Wang, Y.J., Tsai, C.C. and Hsieh, C.Y. (2001) Phase I clinical trial of curcumin, a chemopreventive agent, in patients with high-risk or pre-malignant lesions. *Anticancer Res* 21(4B), 2895-2900.

Chignell, C.F., Bilski, P., Reszka, K.J., Motten, A.G., Sik, R.H. and Dahl, T.A. (1994) Spectral and photo-chemical properties of curcumin. *Photochem Photobiol* 59(3): 295-302.

Choudhuri, T., Pal, S., Agrwarwal, M.L., Das, T. and Sa, G. (2002) Curcumin induces apoptosis in human breast cancer cells through p53-dependent Bax induction. *FEBS Lett* 512(1-3), 334-340.

Choudhuri, T., Pal, S., Das, T., and Sa, G. (2005) Curcumin selectively induces apoptosis in deregulated cyelin-D$_1$ expressed cells at G$_2$ phase of cell cycle in a P$_{53}$- dependent manner. J Biol Chem, 280(20): 20059-20062.

Chuang, S.E., Cheng, A.L., Lin, J.K. and Kuo, M.L. (2000) Inhibition by curcumin of diethylnitro-samine-induced hepatic hyperplasia, inflammation, cellular gene products and cell-cycle-related pro-teins in rats. *Food Chem Toxicol* 38(11), 991-995.

Chuang, S.E., Kuo, M.L., Hsu ,C.H., Chen, C.R., Lin, J.K., Lai, G.M., Hsieh, C.Y. and Cheng, A.L. (2000) Curcumin-containing diet inhibits diethylnitrosamine-induced murine hepatocarcinogenesis. *Carcino-genesis* 21(2), 331-335.

Chuang, S.E., Yeh, P.Y., Lu, Y.S., Lai, G.M., Liao, C.M., Gao, M. and Cheng, A.L. (2002) Basal levels and patterns of anticancer drug-induced activation of nuclear factor-kappaB (NF-kappaB), and its atten-uation by tamoxifen, dexamethasone, and curcumin in carcinoma cells. *Biochem Pharmacol* 63(9), 1709-1716.

Chueh, S.C., Lai, M.K., Liu, I.S., Teng, F.C. and Chen, J.(2003) Curcumin enhances the immunosuppressive activity of cyclosporine in rat cardiac allografts and in mixed lymphocyte reactions. *Transplant Proc* 35(4), 1603-1605.

Chun, K.S., Kang, J.Y., Kim, O.H., Kang, H. and Surh. Y.J. (2002) Effects of yakuchinone A and yakuchinone B on the phorbol ester-induced expression of COX-2 and iNOS and activation of NF-kappaB in mouse skin. *J Environ Pathol Toxicol Oncol* 21(2), 131-139.

Chun, K.S., Keum, Y.S., Han, S.S., Song, Y.S., Kim, S.H. and Surh, Y.J. (2003) Curcumin inhibits phorbol ester-induced expression of cyclooxygenase-2 in mouse skin through suppression of extracellular signal-regulated kinase activity and NF-kappaB activation. *Carcinogenesis* 24(9), 1515-1524.

Chun, K.S., Sohn, Y., Kim, H.S., Kim, O.H., Park, K.K., Lee, J.M., Moon, A., Lee, S.S. and Surh, Y.J (1999) Anti-tumor promoting potential of naturally occurring diarylheptanoids structurally related to cur-cumin. *Mutat Res* 428 (1-2), 49-57.

Ciolino, H.P., Daschner, P.J., Wang, T.T. and Yeh, G.C. (1998) Effect of curcumin on the aryl hydrocarbon receptor and cytochrome P450 1A1 in MCF-7 human breast carcinoma cells. *Biochem Pharmacol* 56(2), 197-206.

Cipriani, B., Borsellino, G., Knowles, H., Tramonti, D., Cavaliere, F., Bernardi, G., Battistini, L. and Brosnan, C.F. (2001) Curcumin inhibits activation of Vgamma9Vdelta2 T cells by phosphoantigens and induces apoptosis involving apoptosis-inducing factor and large scale DNA fragmentation. *J Immunol* 167(6), 3454-3462.

Collett, G.P., Robson, C.N., Mathers, J.C. and Campbell, F.C. (2001) Curcumin modifies Apc(min) apoptosis resistance and inhibits 2-amino 1-methyl-6-phenylimidazo[4,5-b]pyridine (PhIP) induced tumour formation in Apc(min) mice. *Carcinogenesis* 22(5), 821-825.

Daube, F.V. (1870) Uber Curcumin, den Farbstoff der Curcumawurzel. *Ber.* 3: 609.

Deodhar, S.D., Sethi, R. and Srimal, R.C. (1980) Preliminary study on antirheumatic activity of curcumin (diferuloyl methane). *Indian J Med Res* 71, 632-634.

Deshpande, S.S., Ingle, A.D. and Maru, G.B. (1998) Chemopreventive efficacy of curcumin-free aqueous turmeric extract in 7,12-dimethylbenz[a]anthracene-induced rat mammary tumorigenesis. *Cancer Lett* 123(1), 35-40.

Devasena, T. Rajasekaran, K.N. and Menon, V.P. (2002) Bis-1,7-(2-hydroxyphenyl)-hepta-1,6-diene-3,5-dione (a curcumin analog) ameliorates DMH-induced hepatic oxidative stress during colon carcinogenesis. *Pharmacol Res* 46(1), 39-45.

Dickinson, D.A., Iles, K.E., Zhang, H., Blank, V. and Formanm H.J. (2003) Curcumin alters EpRE and AP-1 binding complexes and elevates glutamate-cysteine ligase gene expression. *Faseb J* 17(3), 473-475.

Dikshit, M., Rastogi, L., Shukla, R. and Srimal, R.C. (1995) Prevention of ischaemia-induced biochemical changes by curcumin & quinidine in the cat heart. *Indian J Med Res* 101, 31-35.

Dinkova-Kostova, A.T. and Talalay, P. (1999) Relation of structure of curcumin analogs to their potencies as inducers of Phase 2 detoxification enzymes. *Carcinogenesis* 20(5), 911-914.

Dorai, T., Cao, Y.C., Dorai, B., Buttyan, R. and Katz, A.E. (2001) Therapeutic potential of curcumin in human prostate cancer. III. Curcumin inhibits proliferation, induces apoptosis, and inhibits angiogenesis of LNCaP prostate cancer cells in vivo. *Prostate* 47(4), 293-303.

Dorai, T., Gehani, N. and Katz, A. (2000) Therapeutic potential of curcumin in human prostate cancer. II. Curcumin inhibits tyrosine kinase activity of epidermal growth factor receptor and depletes the protein. *Mol Urol* 4(1), 1-6.

Douglas, D.E. (1993) 4,4'-Diacetyl curcumin--in-vitro histamine-blocking activity. *J Pharm Pharmacol* 45(8), 766.

Drobnjak, M., Osman, I., Scher, H.I., Fazzari, M. and Cordon-Cardo, C. (2000) Overexpression of cyclin D1 is associated with metastatic prostate cancer to bone. *Clin Cancer Res* 6(5), 1891-1895.

Dyer, J.L., Khan, S.Z., Bilmen, J.G., Hawtin, S.R., Wheatley, M., Javed, M.U. and Michelangeli, F. (2002) Curcumin: a new cell-permeant inhibitor of the inositol 1,4,5-trisphosphate receptor. *Cell Calcium* 31(1), 45-52.

Egan, M.E., Pearson, M., Weiner, S.A., Rajendran, V., Rubin, D., Glockner-Pagel, J., Canny, S., Du, K., Lukacs, G.L. and Caplan, M.J.(2004) Curcumin, a major constituent of turmeric, corrects cystic fibrosis defects. *Science* 304(5670), 600-602.

Elattar, T.M. and Virji, A.S. (2000) The inhibitory effect of curcumin, genistein, quercetin and cisplatin on the growth of oral cancer cells in vitro. *Anticancer Res* 20(3A), 1733-1738.

Fang, J., Lu, J. and Holmgren, A. (2005) Thioredoxin reductase is irreversibly modified by curcumin: a novel molecular mechanism for its anticancer activity. *J Biol Chem* 280(26), 25284-290.

Fenton, J.I., Wolff, M.S., Orth, M.W. and Hord, N.G. (2002) Membrane-type matrix metalloproteinases mediate curcumin-induced cell migration in non-tumorigenic colon epithelial cells differing in Apc genotype. *Carcinogenesis* 23(6), 1065-1070.

Flynn, D.L., Rafferty, M.F. and Boctor, A.M. (1986) Inhibition of 5-hydroxy-eicosatetraenoic acid (5-HETE) formation in intact human neutrophils by naturally-occurring diarylheptanoids: inhibitory activities of curcuminoids and yakuchinones. *Prostaglandins Leukot Med* 22(3), 357-360.

Folkman, J. (2001) Can mosaic tumor vessels facilitate molecular diagnosis of cancer? *Proc Natl Acad Sci U S A* 98(2), 398-400.

Fournier, D.B. and Gordon, G.B. (2000) COX-2 and colon cancer: Potential targets for chemoprevention. *J Cell Biochem* 77(S34), 97-102.

Frautschy, S.A., Hu, W., Kim, P., Miller, S.A., Chu, T., Harris-White, M.E. and Cole, G.M. (2001) Phenolic anti-inflammatory antioxidant reversal of Abeta-induced cognitive deficits and neuropathology. *Neurobiol Aging* 22(6), 993-1005.

Fujisawa, S., Atsumi, T., Ishihara, M. and Kadoma, Y. (2004) Cytotoxicity, ROS-generation activity and radical-scavenging activity of curcumin and related compounds. *Anticancer Res* 24(2B), 563-569.

Furness, M.S., Robinson, T.P., Ehlers, T., Hubbard, R.Bt, Arbiser, J.L., Goldsmith, D.J. and Bowen, J.P. (2005) Antiangiogenic agents: studies on fumagillin and curcumin analogs. *Curr Pharm Des* 11(3), 357-373.

Gafner, S., Lee, S.K., Cuendet, M., Barthelemy, S., Vergnes, L., Labidalle, S., Mehta, R.G., Boone, C.W. and Pezzuto, J.M. (2004) Biologic evaluation of curcumin and structural derivatives in cancer chemoprevention model systems. *Phytochemistry* 65(21), 2849-2859.

Garcea, G., Berry, D.P., Jones, D.J., Singh, R., Dennison, A.R., Farmer, P.B., Sharma, R.A., Steward, W.P. and Gescher, A.J.(2005) Consumption of the putative chemopreventive agent curcumin by cancer patients: assessment of curcumin levels in the colorectum and their pharmacodynamic consequences. *Cancer Epidemiol Biomarkers Prev* 14(1), 120-125.

Garcea, G., Jones, D.J., Singh, R., Dennison, A.R., Farmer, P.B., Sharma, R.A., Steward, W.P., Gescher, A.J. and Berry, D.P. (2004) Detection of curcumin and its metabolites in hepatic tissue and portal blood of patients following oral administration. *Br J Cancer* 90(5), 1011-1015.

Gautam, S.C., Xu, Y.X., Pindolia, K.R., Janakiraman, N. and Chapman, R.A. (1998) Nonselective inhibition of proliferation of transformed and nontransformed cells by the anticancer agent curcumin (diferuloylmethane). *Biochem Pharmacol* 55(8), 1333-1337.

Ghaisas, S.D. and Bhide, S.V. (1994) *In vitro* studies on chemoprotective effect of Purnark against benzo(a)pyrene-induced chromosomal damage in human lymphocytes. *Cell Biol Int* 18(1), 21-27.

Giri, D.K. and Aggarwal, B.B. (1998) Constitutive activation of NF-kappaB causes resistance to apoptosis in human cutaneous T cell lymphoma HuT-78 cells. Autocrine role of tumor necrosis factor and reactive oxygen intermediates. *J Biol Chem* 273(22), 14008-14014.

Goel, A., Boland, C.R. and Chauhan, D.P. (2001) Specific inhibition of cyclooxygenase-2 (COX-2) expression by dietary curcumin in HT-29 human colon cancer cells. *Cancer Lett* 172(2), 111-118.

Gomes Dde, C., Alegrio, L.V., de Lima, M.E., Leon, L.L. and Araujo, C.A. (2002) Synthetic derivatives of curcumin and their activity against Leishmania amazonensis. *Arzneimittelforschung* 52(2), 120-124.

Gomez-Lechon, M.J., O'Connor, E., Castel,l J.V. and Jover, R. (2002) Sensitive markers used to identify compounds that trigger apoptosis in cultured hepatocytes. *Toxicol Sci* 65(2), 299-308.

Gukovsky, I., Reyes C.N., Vaquero, E.C., Gukovskaya, A.S. and Pandol, S.J.(2003) Curcumin ameliorates ethanol and nonethanol experimental pancreatitis. *Am J Physiol Gastrointest Liver Physiol* 284(1), G85-95.

Gumbiner, L.M., Gumerlock, P.H., Mack, P.C., Chi, S.G., deVere White, R.W., Mohler, J.L., Pretlow, T.G. and Tricoli, J.V. (1999) Overexpression of cyclin D1 is rare in human prostate carcinoma. *Prostate* 38(1), 40-45.

Gupta, B. and Ghosh, B. (1999) Curcuma longa inhibits TNF-alpha induced expression of adhesion molecules on human umbilical vein endothelial cells. *Int J Immunopharmacol* 21(11), 745-57.

Hahm, E.R., Cheon, G., Lee, J., Kim, B., Park, C. and Yang, C.H. (2002) New and known symmetrical curcumin derivatives inhibit the formation of Fos-Jun-DNA complex. *Cancer Lett* 184(1), 89-96.

Han, S.S., Chung, S.T., Robertson, D.A., Ranjan, D. and Bondada, S. (1999) Curcumin causes the growth arrest and apoptosis of B cell lymphoma by downregulation of egr-1, c-myc, bcl-XL, NF-kappa B, and p53. *Clin Immunol* 93(2), 152-161.

Han, S.S., Keum, Y.S., Seo, H.J. and Surh, Y.J. (2002) Curcumin Suppresses Activation of NF-kappaB and AP-1 Induced by Phorbol Ester in Cultured Human Promyelocytic Leukemia Cells. *J Biochem Mol Biol* 35(3), 337-342.

Hanazawa, S., Takeshita, A., Amano, S., Semba, T., Nirazuka, T., Katoh, H. and Kitano, S. (1993) Tumor necrosis factor-alpha induces expression of monocyte chemoattractant JE via fos and jun genes in clonal osteoblastic MC3T3-E1 cells. *J Biol Chem* 268(13), 9526-9532.

Hanif, R., Qiao, L., Shiff, S.J. and Rigas, B. (1997) Curcumin, a natural plant phenolic food additive, inhibits cell proliferation and induces cell cycle changes in colon adenocarcinoma cell lines by a prostaglandin-independent pathway. *J Lab Clin Med* 130(6), 576-584.

Harbottle, A., Daly, A.K., Atherton, K. and Campbell, F.C. (2001) Role of glutathione S-transferase P1, P-glycoprotein and multidrug resistance-associated protein 1 in acquired doxorubicin resistance. *Int J Cancer* 92(6), 777-783.

Harris, R.E., Alshafie ,G.A., Abou-Issa, H. and Seibert, K. (2000) Chemoprevention of breast cancer in rats by celecoxib, a cyclooxygenase 2 inhibitor. *Cancer Res* 60(8), 2101-2103.

Hasmeda, M. and Polya, G.M. (1996) Inhibition of cyclic AMP-dependent protein kinase by curcumin. *Phytochemistry* 42(3), 599-605.

Hecht, S.S., Kenney, P.M., Wang, M., Trushin, N., Agarwal, S., Rao, A.V. and Upadhyaya, P. (1999) Evaluation of butylated hydroxyanisole, myo-inositol, curcumin, esculetin, resveratrol and lycopene as inhibitors of benzo[a]pyrene plus 4-(methylnitrosamino)-1-(3-pyridyl)-1-butanone-induced lung tumorigenesis in A/J mice. *Cancer Lett* 137(2), 123-130

Hergenhahn, M., Soto, U., Weninger, A., Polack, A., Hsu, C.H., Cheng, A.L. and Rosl, F. (2002) The chemopreventive compound curcumin is an efficient inhibitor of Epstein-Barr virus BZLF1 transcription in Raji DR-LUC cells. *Mol Carcinog* 33(3), 137-145.

Hida, T., Yatabe, Y., Achiwa, H., Muramatsu, H., Kozaki, K., Nakamura, S., Ogawa, M., Mitsudomi, T., Sugiura, T. and Takahashi, T. (1998) Increased expression of cyclooxygenase 2 occurs frequently in human lung cancers, specifically in adenocarcinomas. *Cancer Res* 58(17), 3761-3764..

Hidaka, H., Ishiko, T., Furuhashi, T., Kamohara, H., Suzuki, S., Miyazaki, M., Ikeda, O., Mita, S., Setoguchi, T. and Ogawa, M. (2002) Curcumin inhibits interleukin 8 production and enhances interleukin 8 receptor expression on the cell surface:impact on human pancreatic carcinoma cell growth by autocrine regulation. *Cancer* 95(6), 1206-1214.

Holder, G.M., Plummer, J.L. and Ryan, A.J. (1978) The metabolism and excretion of curcumin (1,7-bis-(4-hydroxy-3-methoxyphenyl)-1,6-heptadiene-3,5-dione) in the rat. *Xenobiotica* 8(12), 76176-8.

Holy, J. (2004) Curcumin inhibits cell motility and alters microfilament organization and function in prostate cancer cells. *Cell Motil Cytoskeleton* 58(4), 253-268.

Hong, J., Bose, M., Ju, J., Ryu, J.H., Chen, X., Sang, S., Lee, M.J. and Yang, C.S. (2004) Modulation of arachidonic acid metabolism by curcumin and related beta-diketone derivatives: effects on cytosolic phospholipase A(2), cyclooxygenases and 5-lipoxygenase. *Carcinogenesis* 25(9), 1671-1679.

Hong, R.L., Spohn, W.H. and Hung, M.C. (1999) Curcumin inhibits tyrosine kinase activity of p185neu and also depletes p185neu. *Clin Cancer Res* 5(7): 1884-891.

Hour, T.C., Chen, J., Huang, C.Y., Guan, J.Y., Lu, S.H. and Pu, Y.S. (2002) Curcumin enhances cytotoxicity of chemotherapeutic agents in prostate cancer cells by inducing p21(WAF1/CIP1) and C/EBPbeta expressions and suppressing NF-kappaB activation. *Prostate* 51(3): 211-218.

Huang, C., Li, J., Ma, W.Y. and Dong, Z. (1999) JNK activation is required for JB6 cell transformation induced by tumor necrosis factor-alpha but not by 12-O-tetradecanoylphorbol-13-acetate. *J Biol Chem* 274(42), 29672-29676.

Huang, H.C., Jan, T.R. and Yeh, S.F. (1992) Inhibitory effect of curcumin, an anti-inflammatory agent, on vascular smooth muscle cell proliferation. *Eur J Pharmacol* 221(2-3): 381-384.

Huang, M.T., Deschner, E.E., Newmark, H.L., Wang, Z.Y., Ferraro, T.A. and Conney, A.H. (1992b) Effect of dietary curcumin and ascorbyl palmitate on azoxymethanol-induced colonic epithelial cell proliferation and focal areas of dysplasia. *Cancer Lett* 64(2), 117-121.

Huang, M.T., Lou, Y.R., Ma, W., Newmark, H.L., Reuhl, K.R. and Conney, A.H. (1994) Inhibitory effects of dietary curcumin on forestomach, duodenal, and colon carcinogenesis in mice. *Cancer Res* 54(22), 5841-5847.

Huang, M.T., Lou, Y.R., Xie, J.G., Ma, W., Lu, Y.P., Yen, P., Zhu, B.T., Newmark, H. and Ho, C.T. (1998) Effect of dietary curcumin and dibenzoylmethane on formation of 7,12-dimethyl-benz[a]anthracene-induced mammary tumors and lymphomas/leukemias in Sencar mice. *Carcinogenesis* 19(9), 1697-1700.

Huang, M.T., Lysz, T., Ferraro, T., Abidi, T.F., Laskin, J.D. and Conney, A.H. (1991) Inhibitory effects of curcumin on *in vitro* lipoxygenase and cyclooxygenase activities in mouse epidermis. *Cancer Res* 51(3), 813-819.

Huang, M.T., Ma, W., Lu, Y.P., Chang, R.L., Fisher, C., Manchand, P.S., Newmark, H,L. and Conney, A.H. (1995) Effects of curcumin, demethoxycurcumin, bisdemethoxycurcumin and tetrahydrocurcumin on 12-O-tetradecanoylphorbol-13-acetate-induced tumor promotion. *Carcinogenesis* 16(10), 2493-2497.

Huang, M.T., Ma, W., Yen, P., Xie, J.G., Han, J., Frenkel, K., Grunberger, D. and Conney, A.H. (1997) Inhibitory effects of topical application of low doses of curcumin on 12-O-tetradecanoylphor-bol-13-acetate-induced tumor promotion and oxidized DNA bases in mouse epidermis. *Carcinogenesis* 18(1), 83-88.

Huang, M.T., Newmark, H.L. and Frenkel, K. (1997) Inhibitory effects of curcumin on tumorigenesis in mice. *J Cell Biochem Suppl* 27, 26-34.

Huang, M.T., Smart, R.C., Wong, C.Q. and Conney, A.H. (1988) Inhibitory effect of curcumin, chlorogenic acid, caffeic acid, and ferulic acid on tumor promotion in mouse skin by 12-O-tetradecanoylphorbol-13-acetate. *Cancer Res* 48(21), 5941-5946.

Huang, M.T., Wang, Z.Y., Georgiadis, C.A., Laskin, J.D. and Conney, A.H. (1992) Inhibitory effects of curcumin on tumor initiation by benzo[a]pyrene and 7,12-dimethylbenz[a]anthracene. *Carcinogenesis* 13(11), 2183-2186.

Huang, T.S., Lee ,S.C. and Lin, J.K. (1991) Suppression of c-Jun/AP-1 activation by an inhibitor of tumor promotion in mouse fibroblast cells. *Proc Natl Acad Sci U S A* 88(12), 5292-6.

Hussain, M.S. and Chandrasekhara, N. (1992) Effect on curcumin on cholesterol gall-stone induction in mice. *Indian J Med Res* 96, 288-291.

Hussain, M.S. and Chandrasekhara, N. (1994) Biliary proteins from hepatic bile of rats fed curcumin or capsaicin inhibit cholesterol crystal nucleation in supersaturated model bile. *Indian J Biochem Biophys* 31(5), 407-412.

Iademarco, M.F., Barks, J.L. and Dean, D.C. (1995) Regulation of vascular cell adhesion molecule-1 expression by IL-4 and TNF-alpha in cultured endothelial cells. *J Clin Invest* 95(1), 264-271.

Ikezaki, S., Nishikawa, A., Furukawa, F., Kudo, K., Nakamura, H., Tamura, K. and Mori, H. (2001) Chemopreventive effects of curcumin on glandular stomach carcinogenesis induced by N-methyl-N'-nitro-N-nitrosoguanidine and sodium chloride in rats. *Anticancer Res* 21(5), 3407-3411.

Inano, H. and Onoda, M. (2002) Prevention of radiation-induced mammary tumors. *Int J Radiat Oncol Biol Phys* 52(1), 212-223.

Inano, H. and Onoda, M. (2002) Radioprotective action of curcumin extracted from Curcuma longa LINN: inhibitory effect on formation of urinary 8-hydroxy-2'-deoxyguanosine, tumorigenesis, but not mortality, induced by gamma-ray irradiation. *Int J Radiat Oncol Biol Phys* 53(3), 735-743.

Inano, H., Onoda, M., Inafuku, N., Kubota, M., Kamada, Y., Osawa, T., Kobayashi, H. and Wakabayashi, K. (1999) Chemoprevention by curcumin during the promotion stage of tumorigenesis of mammary gland in rats irradiated with gamma-rays. *Carcinogenesis* 20(6), 1011-1018.

Inano, H., Onoda, M., Inafuku, N., Kubota, M., Kamada, Y., Osawa, T., Kobayashi, H. and Wakabayashi, K. (2000) Potent preventive action of curcumin on radiation-induced initiation of mammary tumorigenesis in rats. *Carcinogenesis* 21(10), 1835-1841.

Ireson, C., Orr, S., Jones, D.J., Verschoyle, R., Lim, C.K., Luo, J.L., Howells, L., Plummer, S., Jukes, R., Williams, M., Steward, W.P. and Gescher, A. (2001) Characterization of metabolites of the chemopreventive agent curcumin in human and rat hepatocytes and in the rat in vivo, and evaluation of their ability to inhibit phorbol ester-induced prostaglandin E2 production. *Cancer Res* 61(3), 1058-1064.

Ireson, C.R., Jones, D.J., Orr, S., Coughtrie, M.W., Boocock, D.J., Williams, M.L., Farmer, P.B., Steward, W.P. and Gescher, A.J. (2002) Metabolism of the cancer chemopreventive agent curcumin in human and rat intestine. *Cancer Epidemiol Biomarkers Prev* 11(1), 105-111.

Ishida, J., Ohtsu, H., Tachibana, Y., Nakanishi, Y., Bastow, K.F., Nagai, M., Wang, H.K., Itokawa, H. and Lee, K.H. (2002) Antitumor agents. Part 214: synthesis and evaluation of curcumin analogues as cytotoxic agents. *Bioorg Med Chem* 10(11), 3481-3487.

Ishizaki, C., Oguro, T., Yoshida, T., Wen, C.Q., Sueki, H. and Iijima, M. (1996) Enhancing effect of ultraviolet A on ornithine decarboxylase induction and dermatitis evoked by 12-o-tetradecanoylphorbol-13-acetate and its inhibition by curcumin in mouse skin. *Dermatology* 193(4), 311-317.

Iwunze, M.O. and McEwan, D. (2004) Peroxynitrite interaction with curcumin solubilized in ethanolic solution. *Cell Mol Biol (Noisy-le-grand)* 50(6): 749-752.

Jaiswal, A.S., Marlow, B.P., Gupta, N. and Narayan, S. (2002) Beta-catenin-mediated transactivation and cell-cell adhesion pathways are important in curcumin (diferuylmethane)-induced growth arrest and apoptosis in colon cancer cells. *Oncogene* 21(55), 8414-8427.

James, J.S. (1996) Curcumin: clinical trial finds no antiviral effect. *AIDS Treat News* (no 242), 1-2.

Jang, M.K., Sohn, D.H. and Ryu, J.H. (2001) A curcuminoid and sesquiterpenes as inhibitors of macrophage TNF-alpha release from Curcuma zedoaria. *Planta Med* 67(6), 550-552.

Jaruga, E., Bielak-Zmijewska, A., Sikora, E., Skierski, J., Radziszewska, E., Piwocka, K. and Bartosz, G. (1998) Glutathione-independent mechanism of apoptosis inhibition by curcumin in rat thymocytes. *Biochem Pharmacol* 56(8), 961-965.

Jaruga, E., Salvioli, S., Dobrucki, J., Chrul, S., Bandorowicz-Pikula, J., Sikora, E., Franceschi, C., Cossarizza, A. and Bartosz, G. (1998) Apoptosis-like, reversible changes in plasma membrane asymmetry and permeability, and transient modifications in mitochondrial membrane potential induced by curcumin in rat thymocytes. *FEBS Lett* 433(3), 287-293.

Jaruga, E., Sokal, A., Chrul, S. and Bartosz, G. (1998) Apoptosis-independent alterations in membrane dynamics induced by curcumin. *Exp Cell Res* 245(2), 303-312.

Jee, S.H., Shen, S.C., Tseng, C.R., Chiu, H.C. and Kuo, M.L. (1998) Curcumin induces a p53-dependent apoptosis in human basal cell carcinoma cells. *J Invest Dermatol* 111(4), 656-661.

Jian, Y.T., Mai, G.F., Wang, J.D., Zhang, Y.L., Luo, R.C. and Fang, Y.X. (2005) Preventive and therapeutic effects of NF-kappaB inhibitor curcumin in rats colitis induced by trinitrobenzene sulfonic acid. *World J Gastroenterol* 11(12), 1747-1752.

Jian, Y.T., Wang, J.D., Mai, G.F., Zhang, Y.L. and Lai, Z.S.(2004) [Modulation of intestinal mucosal inflammatory factors by curcumin in rats with colitis.]. *Di Yi Jun Yi Da Xue Xue Bao* 24(12), 1353-1358.

Jiang, M.C., Yang-Yen, H.F., Yen, J.J. and Lin, J.K. (1996) Curcumin induces apoptosis in immortalized NIH 3T3 and malignant cancer cell lines. *Nutr Cancer* 26(1), 111-1120.

Jobin, C., Bradham, C.A., Russo, M.P., Juma, B., Narula, A.S., Brenner, D.A. and Sartor. R.B. (1999) Curcumin blocks cytokine-mediated NF-kappa B activation and proinflammatory gene expression by inhibiting inhibitory factor I-kappa B kinase activity. *J Immunol* 163(6), 3474-3483.

John, V.D., Kuttan, G. and Krishnankutty, K. (2002) Anti-tumour studies of metal chelates of synthetic curcuminoids. *J Exp Clin Cancer Res* 21(2), 219-224.

Johri, R.K. and Zutshi U. (1992) An Ayurvedic formulation 'Trikatu' and its constituents. *J Ethnopharmacol* 37(2), 85-91.

Jordan, W.C. and Drew, C.R. (1996) Curcumin — a natural herb with anti-HIV activity. *J Natl Med Assoc* 88(6), 333.

Jung, E.M., Lim, J.H., Lee, T.J., Park, J.W., Choi, K.S. and Kwon, T.K. (2005) Curcumin sensitizes tumor necrosis factor-related apoptosis-inducing ligand (TRAIL)-induced apoptosis through reactive oxygen species-mediated up-regulation of death receptor 5 (DR5). *Carcinogenesis.*

Kakar, S.S. and Roy. D (1994) Curcumin inhibits TPA induced expression of c-fos, c-jun and c-myc proto-oncogenes messenger RNAs in mouse skin. *Cancer Lett* 87(1), 858-9.

Kalpana, C. and Menon, V.P. (2004) Curcumin ameliorates oxidative stress during nicotine-induced lung toxicity in Wistar rats. *Ital J Biochem* 53(2), 82-86.

Kalpana, C. and Menon, V.P. (2004) Inhibition of nicotine-induced toxicity by curcumin and curcumin analog: a comparative study. *J Med Food* 7(4), 467-471.

Kang, B.Y., Chung, S.W., Chung, W., Im, S., Hwang, S.Y, and Kim, T.S. (1999) Inhibition of interleukin-12 production in lipopolysaccharide-activated macrophages by curcumin. *Eur J Pharmacol* 384(2-3), 191-5.

Kang, B.Y., Song, Y.J., Kim, K.M., Choe, Y.K., Hwang, S.Y. and Kim, T.S. (1999) Curcumin inhibits Th1 cytokine profile in CD4+ T cells by suppressing interleukin-12 production in macrophages. *Br J Pharmacol* 128(2), 380-384.

Kang, J., Chen, J., Shi, Y., Jia, J. and Zhang, Y. (2005) Curcumin-induced histone hypoacetylation: the role of reactive oxygen species. *Biochem Pharmacol* 69(8), 1205-1213.

Kapoor, S. and Priyadarsini, K.I. (2001) Protection of radiation-induced protein damage by curcumin. *Biophys Chem* 92(1-2), 119-126.

Karin, M., Liu, Z. and Zandi, E. (1997) AP-1 function and regulation. *Curr Opin Cell Biol* 9(2), 240-6.

Kato, K., Ito, H., Kamei, K. and Iwamoto, I. (1998) Stimulation of the stress-induced expression of stress proteins by curcumin in cultured cells and in rat tissues in vivo. *Cell Stress Chaperones* 3(3), 152-160.

Kaul, S. and Krishnakanth, T.P. (1994) Effect of retinol deficiency and curcumin or turmeric feeding on brain Na(+)-K+ adenosine triphosphatase activity. *Mol Cell Biochem* 137(2), 101-107.

Kawamori, T., Lubet, R., Steele, V.E., Kelloff, G.J., Kaskey, R.B., Rao, C.V. and Reddy, B.S. (1999) Chemo-preventive effect of curcumin, a naturally occurring anti-inflammatory agent, during the promotion/progression stages of colon cancer. *Cancer Res* 59(3), 597-601.

Kawashima, H., Akimoto, K., Jareonkitmongkol, S., Shirasaka, N. and Shimizu, S. (1996) Inhibition of rat liver microsomal desaturases by curcumin and related compounds. *Biosci Biotechnol Biochem* 60(1), 108-110.

Keshavarz, K. (1976) The influence of turmeric and curcumin on cholesterol concentration of eggs and tissues. *Poult Sci* 55(3), 1077-1083.

Keum, Y.S., Kim, J., Lee, K.H., Park, KK., Surh, Y.J., Lee, J.M., Lee, S.S., Yoon, J.H., Joo, S.Y., Cha, I.H. and Yook, J.I. (2002) Induction of apoptosis and caspase-3 activation by chemopreventive [6]-paradol and structurally related compounds in KB cells. *Cancer Lett* 177(1), 41-47.

Khafif, A., Hurst, R., Kyker, K., Fliss, D.M., Gil, Z. and Medina, J.E. (2005) Curcumin: a new radio-sensitizer of squamous cell carcinoma cells. *Otolaryngol Head Neck Surg* 132(2), 317-321.

Khafif, A., Schantz, S.P., Chou, T.C., Edelstein, D. and Sacks, P.G. (1998) Quantitation of chemopreventive synergism between (-)-epigallocatechin-3-gallate and curcumin in normal, premalignant and malignant human oral epithelial cells. *Carcinogenesis* 19(3), 419-424.

Khar, A., Ali, A.M., Pardhasaradhi, B.V., Varalakshmi, C.H., Anjum, R. and Kumari , A.L. (2001) Induction of stress response renders human tumor cell lines resistant to curcumin-mediated apoptosis: role of reactive oxygen intermediates. *Cell Stress Chaperones* 6(4), 368-376.

Khopde, S.M., Priyadarsini, K.I., Guha, S.N., Satav, J.G., Venkatesan, P. and Rao, M.N. (2000) Inhibition of radiation-induced lipid peroxidation by tetrahydrocurcumin: possible mechanisms by pulse radiolysis. *Biosci Biotechnol Biochem* 64(3), 503-509.

Khopde, S.M., Priyadarsini, K.I., Palit, D.K. and Mukherjee, T. (2000) Effect of solvent on the excited-state photophysical properties of curcumin. *Photochem Photobiol* 72(5): 625-631.

Khopde, S.M., Priyadarsini, K.I., Venkatesan, N. and Rao, M.N.A. (1999) Free radical scavenging ability and antioxidant efficiency of curcumin and its substituted analogue. *Biophysical Chemistry* 80(2), 83-89.

Kim, D.W., Sovak, M.A., Zanieski, G., Nonet, G., Romieu-Mourez, R., Lau, A.W., Hafer, L.J., Yaswen, P., Stampfer, M., Rogers, A.E., Russo, J. and Sonenshein, G.E. (2000) Activation of NF-kappaB/Rel occurs early during neoplastic transformation of mammary cells. *Carcinogenesis* 21(5), 871-879.

Kim, H.Y., Park, E.J., Joe, E.H. and Jou, I. (2003) Curcumin suppresses Janus kinase-STAT inflammatory signaling through activation of Src homology 2 domain-containing tyrosine phosphatase 2 in brain microglia. *J Immunol* 171(11), 6072-6079.

Kim, J.H., Shim, J.S., Lee, S.K., Kim, K.W., Rha, S.Y., Chung, H.C. and Kwon, H.J. (2002) Microarray-based analysis of anti-angiogenic activity of demethoxycurcumin on human umbilical vein endothelial cells: crucial involvement of the down-regulation of matrix metalloproteinase. *Jpn J Cancer Res* 93(12), 1378-1385.

Kim, J.M., Araki, S., Kim, D.J., Park, C.B., Takasuka, N., Baba-Toriyama, H., Ota, T., Nir, Z., Khachik, F., Shimidzu, N., Tanaka, Y., Osawa, T., Uraji, T., Murakoshi, M., Nishino, H. and Tsuda, H. (1998) Chemopreventive effects of carotenoids and curcumins on mouse colon carcinogenesis after 1,2-dim-ethylhydrazine initiation. *Carcinogenesis* 19(1), 81-85.

Kim, K.H., Park, H.Y., Nam, J.H., Park, J.E., Kim, J.Y., Park, M.I., Chung, K.O., Park, K.Y. and Koo, J.Y. (2005) The inhibitory effect of curcumin on the growth of human colon cancer cells (HT-29, WiDr) *in vitro*. *Korean J Gastroenterol* 45(4), 277-284.

Kim, M.S., Kang, H.J. and Moon, A. Inhibition of invasion and induction of apoptosis by curcumin in H-ras-transformed MCF10A human breast epithelial cells. *Arch Pharm Res* 24(4), 349-354.

Kim, S.O., Kundu, J.K., Shin, Y.K., Park, J.H., Cho, M.H., Kim, T.Y. and Surh, Y.J. (2005) [6]-Gingerol inhibits COX-2 expression by blocking the activation of p38 MAP kinase and NF-kappaB in phorbol ester-stimulated mouse skin. *Oncogene*.

Kiuchi, F., Goto, Y., Sugimoto, N., Akao, N., Kondo, K. and Tsuda, Y. (1993) Nematocidal activity of turmeric: synergistic action of curcuminoids. *Chem Pharm Bull (Tokyo)* 41(9): 1640-1643.

Koide, T., Nose, M., Ogihara, Y., Yabu, Y. and Ohta, N. (2002) Leishmanicidal effect of curcumin *in vitro*. *Biol Pharm Bull* 25(1): 131-133.

Koo, J.Y., Kim, H.J., Jung, K.O. and Park, K.Y. (2004) Curcumin inhibits the growth of AGS human gastric carcinoma cells *in vitro* and shows synergism with 5-fluorouracil. *J Med Food* 7(2), 117-121.

Korutla, L. and Kumar, R. (1994) Inhibitory effect of curcumin on epidermal growth factor receptor kinase activity in A431 cells. *Biochim Biophys Acta* 1224(3), 597-600.

Korutla, L., Cheung, J.Y., Mendelsohn, J. and Kumar, R. (1995) Inhibition of ligand-induced activation of epidermal growth factor receptor tyrosine phosphorylation by curcumin. *Carcinogenesis* 16(8): 1741-1745.

Krishnaswamy, K., Goud, V.K., Sesikeran, B., Mukundan, M.A. and Krishna, T.P. (1998) Retardation of experimental tumorigenesis and reduction in DNA adducts by turmeric and curcumin. *Nutr Cancer* 30(2), 163-166.

Kumar, A., Dhawan, S., Hardegen, N.J. and Aggarwal, B.B. (1998) Curcumin (Diferuloylmethane) inhibition of tumor necrosis factor (TNF)-mediated adhesion of monocytes to endothelial cells by suppression of cell surface expression of adhesion molecules and of nuclear factor-kappaB activation. *Biochem Pharmacol* 55(6), 775-783.

Kumar, A., Dhawan, S., Mukhopadhyay, A. and Aggarwal, B.B. (1999) Human immunodeficiency virus-1-tat induces matrix metalloproteinase-9 in monocytes through protein tyrosine phosphatase-mediated activation of nuclear transcription factor NF-kappaB. *FEBS Lett* 462(1-2), 140-4.

Kumar, A.P., Garcia, G.E., Ghosh, R., Rajnarayanan, R.V., Alworth, W.L. and Slaga, T.J. (2003) 4-Hydroxy-3-methoxybenzoic acid methyl ester: a curcumin derivative targets Akt/NF kappa B cell survival signaling pathway: potential for prostate cancer management. *Neoplasia* 5(3), 255-66.

Kumar, S., Dubey K.K., Tripathi, S., Fujii, M. and Misra, K.(2000) Design and synthesis of curcumin-bio-conjugates to improve systemic delivery. *Nucleic Acids Symp Ser* (44), 75-76.

Kumar, V., Lewis, S.A., Mutalik, S., Shenoy D.B., Venkatesh and Udupa, N. (2002) Biodegradable micro-spheres of curcumin for treatment of inflammation. *Indian J Physiol Pharmacol* 46(2), 209-217.

Kuo, M.L., Huang, T.S. and Lin, J.K. (1996) Curcumin, an antioxidant and anti-tumor promoter, induces apoptosis in human leukemia cells. *Biochem Biophys Acta* 1317(2), 95-100.

Kuttan, R., Bhanumathy, P., Nirmala, K. and George, M.C. (1985) Potential anticancer activity of turmeric (*Curcuma longa*). *Cancer Lett* 29(2), 197-202.

Kuttan, R., Sudheeran, P.C. and Joseph, C.D. (1987) Turmeric and curcumin as topical agents in cancer therapy. *Tumori* 73(1), 29-31.

Lal, B., Kapoor, A.K., Agrawal, P.K., Asthana, O.P. and Srimal, R.C.(2000) Role of curcumin in idiopathic inflammatory orbital pseudotumours. *Phytother Res* 14(6), 443-447.

Lal, B., Kapoor, A.K., Asthana, O.P., Agrawal, P.K., Prasad, R., Kumar, P. and Srimal, R.C.(1999) Efficacy of curcumin in the management of chronic anterior uveitis. *Phytother Res* 13(4), 318-322.

Lampe, V. and Milobedzka, J. (1913) Ver. Dtsch.Chem. *Ges.* 46: 2235.

Lee, H., Arsura, M., Wu, M., Duyao, M., Buckler, A.J. and Sonenshein, G.E. (1995) Role of Rel-related factors in control of c-myc gene transcription in receptor-mediated apoptosis of the murine B cell WEHI 231 line. *J Exp Med* 181(3), 1169-1177.

Lee, S.E., Campbell, B.C., Molyneux, R.J., Hasegawa, S. and Lee, H.S. (2001) Inhibitory effects of naturally occurring compounds on aflatoxin B(1) biotransformation. *J Agric Food Chem* 49(11), 5171-5177.

Lee, S.K. and Pezzuto, J.M. (1999) Evaluation of the potential of cancer chemopreventive activity mediated by inhibition of 12-O-tetradecanoyl phorbol 13-acetate-induced ornithine decarboxylase activity. *Arch Pharm Res* 22(6), 559-564.

Lee, S.L., Huang, W.J., Lin, W.W., Lee, S.S. and Chen, C.H. (2005) Preparation and anti-inflammatory activities of diarylheptanoid and diarylheptylamine analogs. *Bioorg Med Chem*.

Li, C.J., Zhang, L.J., Dezube, B.J., Crumpacker, C.S. and Pardee, A.B. (1993) Three inhibitors of type 1 human immunodeficiency virus long terminal repeat-directed gene expression and virus replication. *Proc Natl Acad Sci U S A* 90(5), 1839-1842.

Li, W.Q., Dehnade, F. and Zafarullah, M. (2001) Oncostatin M-induced matrix metalloproteinase and tissue inhibitor of metalloproteinase-3 genes expression in chondrocytes requires Janus kinase/STAT signaling pathway. *J Immunol* 166(5), 3491-3498.

Liacini, A., Sylvester, J., Li, W.Q. and Zafarullah, M. (2002) Inhibition of interleukin-1-stimulated MAP kinases, activating protein-1 (AP-1) and nuclear factor kappa B (NF-kappaB) transcription factors down-regulates matrix metalloproteinase gene expression in articular chondrocytes. *Matrix Biol* 21(3), 251-262.

Liao, S., Lin, J., Dang, M.T., Zhang, H., Kao, Y.H., Fukuchi, J. and Hiipakka, R.A. (2001) Growth suppression of hamster flank organs by topical application of catechins, alizarin, curcumin, and myristoleic acid. *Arch Dermatol Res* 293(4), 200-205.

Lim, G.P., Chu, T., Yang, F., Beech, W., Frautschy, S.A. and Cole, G.M. (2001) The curry spice curcumin reduces oxidative damage and amyloid pathology in an Alzheimer transgenic mouse. *J Neurosci* 21(21), 8370-7.

Limtrakul, P., Anuchapreeda, S. and Buddhasukh, D. (2004) Modulation of human multidrug-resistance MDR-1 gene by natural curcuminoids. *BMC Cancer* 4, 13.

Limtrakul, P., Anuchapreeda, S., Lipigorngoson, S. and Dunn, F.W. (2001) Inhibition of carcinogen induced c-Ha-ras and c-fos proto-oncogenes expression by dietary curcumin. *BMC Cancer* 1, 1.

Limtrakul, P., Lipigorngoson, S., Namwong, O., Apisariyakul, A. and Dunn, F.W. (1997) Inhibitory effect of dietary curcumin on skin carcinogenesis in mice. *Cancer Lett* 116(2), 197-203.

Lin, C.C., Ho, C.T. and Huang, M.T. (2001) Mechanistic studies on the inhibitory action of dietary dibenzoylmethane, a beta-diketone analogue of curcumin, on 7,12-dimethylbenz[a]anthracene-induced mammary tumorigenesis. *Proc Natl Sci Counc Repub China B* 25(3), 158-165.

Lin, C.C., Lu, Y.P., Lou, Y.R., Ho, C.T., Newmark, H.H., MacDonald, C., Singletary, K.W. and Huang, M.T. (2001) Inhibition by dietary dibenzoylmethane of mammary gland proliferation, formation of DMBA-DNA adducts in mammary glands, and mammary tumorigenesis in Sencar mice. *Cancer Lett* 168(2), 125-132.

Lin, J.K. and Shih, C.A. (1994) Inhibitory effect of curcumin on xanthine dehydrogenase/oxidase induced by phorbol-12-myristate-13-acetate in NIH3T3 cells. *Carcinogenesis* 15(8), 1717-21.

Lin, J.K., Pan, M.H. and Lin-Shiau, S.Y.(2000) Recent studies on the biofunctions and biotransformations of curcumin. *Biofactors* 13(1-4), 153-158.

Lin, L.I., Ke, Y.F., Ko, Y.C. and Lin, J.K. (1998) Curcumin inhibits SK-Hep-1 hepatocellular carcinoma cell invasion *in vitro* and suppresses matrix metalloproteinase-9 secretion. *Oncology* 55(4), 349-353.

Literat, A., Su, F., Norwicki, M., Durand, M., Ramanathan, R., Jones, C.A., Minoo, P. and Kwong, K.Y. (2001) Regulation of pro-inflammatory cytokine expression by curcumin in hyaline membrane disease (HMD). *Life Sci* 70(3), 253-267.

Liu, J.Y., Lin ,S.J. and Lin, J.K. (1993) Inhibitory effects of curcumin on protein kinase C activity induced by 12-O-tetradecanoyl-phorbol-13-acetate in NIH 3T3 cells. *Carcinogenesis* 14(5), 857-61.

Logan-Smith, M.J., East, J.M. and Lee, A.G. (2002) Evidence for a global inhibitor-induced conformation change on the Ca(2+)-ATPase of sarcoplasmic reticulum from paired inhibitor studies. *Biochemistry* 41(8), 2869-2875.

Logan-Smith, M.J., Lockyer, P.J., East, J.M. and Lee, A.G. (2001) Curcumin, a molecule that inhibits the Ca2+-ATPase of sarcoplasmic reticulum but increases the rate of accumulation of Ca2+. *J Biol Chem* 276(50), 46905-46911.

Lu, Y.P., Chang, R.L., Huang, M.T. and Conney, A.H. (1993) Inhibitory effect of curcumin on 12-O-tetradecanoylphorbol-13-acetate-induced increase in ornithine decarboxylase mRNA in mouse epidermis. *Carcinogenesis* 14(2), 293-297.

Lu, Y.P., Chang, R.L., Lou, Y.R., Huang, M.T., Newmark, H.L., Reuhl, K.R. and Conney, A.H. (1994) Effect of curcumin on 12-O-tetradecanoylphorbol-13-acetate- and ultraviolet B light-induced expression of c-Jun and c-Fos in JB6 cells and in mouse epidermis. *Carcinogenesis* 15(10), 2363-2370.

Madan, B. and Ghosh, B.(2003) Diferuloylmethane inhibits neutrophil infiltration and improves survival of mice in high-dose endotoxin shock. *Shock* 19(1), 91-96..

Mahady, G.B., Pendland, S.L., Yun, G. and Lu, Z.Z. (2002) Turmeric (*Curcuma longa*) and curcumin inhibit the growth of Helicobacter pylori, a group 1 carcinogen. *Anticancer Res* 22(6C), 4179-81.

Mahady, G.B., Pendland, S.L., Yun, G.S., Lu, Z.Z. and Stoia, A. (2003) Ginger (Zingiber officinale Roscoe) and the gingerols inhibit the growth of Cag A+ strains of Helicobacter pylori. *Anticancer Res* 23(5A), 3699-3702.

Mahmoud, N.N., Carothers, A.M., Grunberger, D., Bilinski, R.T., Churchill, M.R., Martucci, C., Newmark, H.L. and Bertagnolli, M.M. (2000) Plant phenolics decrease intestinal tumors in an animal model of familial adenomatous polyposis. *Carcinogenesis* 21(5), 921-927.

Mani, H., Sidhu, G.S., Kumari ,R., Gaddipati ,J.P., Seth, P. and Maheshwari, R.K. (2002) Curcumin differentially regulates TGF-beta1, its receptors and nitric oxide synthase during impaired wound healing. *Biofactors* 16(1-2), 29-43.

Mariadason, J.M., Corner, G.A. and Augenlicht, L.H. (2000) Genetic reprogramming in pathways of colonic cell maturation induced by short chain fatty acids: comparison with trichostatin A, sulindac, and curcumin and implications for chemoprevention of colon cancer. *Cancer Res* 60(16), 4561-4572.

Masuda, T., Matsumura, H., Oyama, Y., Takeda, Y., Jitoe, A., Kida, A. and Hidaka, K. (1998) Synthesis of (+/-)-cassumunins A and B, new curcuminoid antioxidants having protective activity of the living cell against oxidative damage. *J Nat Prod* 61(5), 609-613.

Mazumder, A., Neamati, N., Sunder, S., Schulz, J., Pertz, H., Eich, E. and Pommier, Y. (1997) Curcumin analogs with altered potencies against HIV-1 integrase as probes for biochemical mechanisms of drug action. *J Med Chem* 40(19), 3057-3063.

Mehta, K., Pantazis, P., McQueen, T. and Aggarwal, B.B. (1997) Antiproliferative effect of curcumin (difer-uloylmethane) against human breast tumor cell lines. *Anticancer Drugs* 8(5), 470-81.

Menon, L.G., Kuttan, R. and Kuttan, G. (1995) Inhibition of lung metastasis in mice induced by B16F10 melanoma cells by polyphenolic compounds. *Cancer Lett* 95(1-2), 221-225.

Menon, L.G., Kuttan, R. and Kuttan, G. (1999) Anti-metastatic activity of curcumin and catechin. *Cancer Lett* 141(1-2), 159-165.

Mishra, B., Priyadarsini, K.I., Bhide, M.K., Kadam, R.M. and Mohan, H. (2004) Reactions of superoxide radicals with curcumin: probable mechanisms by optical spectroscopy and EPR. *Free Radic Res* 38(4): 355-362.

Mishra, S., Tripathi, S. and Misra, K. (2002) Synthesis of a novel anticancer prodrug designed to target telomerase sequence. *Nucleic Acids Res Suppl* (2), 277-278.

Miyoshi, N., Nakamura, Y., Ueda, Y., Abe, M., Ozawa, Y., Uchida, K. and Osawa, T. (2003) Dietary ginger constituents, galanals A and B, are potent apoptosis inducers in Human T lymphoma Jurkat cells. *Cancer Lett* 199(2), 113-119.

Mohan. R., Sivak, J., Ashton, P., Russo, L.A., Pham, B.Q., Kasahara, N., Raizman, M.B. and Fini, M.E. (2000) Curcuminoids inhibit the angiogenic response stimulated by fibroblast growth factor-2, including expression of matrix metalloproteinase gelatinase B. *J Biol Chem* 275(14), 10405-10412.

Moragoda, L., Jaszewski, R. and Majumdar, A.P. (2001) Curcumin induced modulation of cell cycle and apoptosis in gastric and colon cancer cells. *Anticancer Res* 21(2A), 873-8.

Mori, H., Niwa, K., Zheng, Q., Yamada, Y., Sakata, K. and Yoshimi, N. (2001) Cell proliferation in cancer prevention; effects of preventive agents on estrogen-related endometrial carcinogenesis model and on an in vitro model in human colorectal cells. *Mutat Res* 480-481, 201-7.

Morikawa, T., Matsuda, H., Ninomiya, K. and Yoshikawa, M. (2002) Medicinal foodstuffs. XXIX. Potent protective effects of sesquiterpenes and curcumin from Zedoariae Rhizoma on liver injury induced by D-galactosamine/lipopolysaccharide or tumor necrosis factor-alpha. *Biol Pharm Bull* 25(5), 627-631.

Morin, D., Barthelemy, S., Zini, R., Labidalle, S. and Tillement, J.P. (2001) Curcumin induces the mitochon-drial permeability transition pore mediated by membrane protein thiol oxidation. *FEBS Lett* 495(1-2), 131-136.

Motterlini, R., Foresti, R., Bassi, R. and Green, C.J. (2000) Curcumin, an antioxidant and anti-inflammatory agent, induces heme oxygenase-1 and protects endothelial cells against oxidative stress. *Free Radic Biol Med* 28(8), 1303-1312.

Mukhopadhyay, A. Banerjee, S. Stafford, L.J., Xia, C.X., Liu, M. and Aggarwal, B.B. (2002) Cur-cumin-induced suppression of cell proliferation correlates with donregulation of cyclin D1 expression and CDK4-mediated retinoblastoma protein phosphorylation. *Oncogene* 21(57), 8852-8862.

Mukhopadhyay, A., Bueso-Ramos, C., Chatterjee, D., Pantazis, P. and Aggarwal, B.B. (2001) Curcumin downregulates cell survival mechanisms in human prostate cancer cell lines. *Oncogene* 20(52): 7597-7609.

Nagar, A.B. and Gorelick, F.S. (2004) Acute pancreatitis. *Curr Opin Gastroenterol* 20(5), 439-443.

Naidu, K.A. and Thippeswamy, N.B. (2002) Inhibition of human low density lipoprotein oxidation by active principles from spices. *Mol Cell Biochem* 229(1-2), 19-23,.

Naito, M., Wu, X., Nomura, H., Kodama, M., Kato, Y. and Osawa, T. (2002) The protective effects of tetrahydrocurcumin on oxidative stress in cholesterol-fed rabbits. *J Atheroscler Thromb* 9(5), 243-250.

Nakamura, K., Yasunaga, Y., Segawa, T., Ko, D., Moul, J.W., Srivastava, S. and Rhim, J.S. (2002) Curcumin down-regulates AR gene expression and activation in prostate cancer cell lines. *Int J Oncol* 21(4), 825-830.

Nakshatri, H., Bhat-Nakshatri, P., Martin, D.A., Goulet, R.J.Jr., and Sledge, G.W. Jr., (1997) Constitutive activation of NF-kappaB during progression of breast cancer to hormone-independent growth. *Mol Cell Biol* 17(7), 3629-3639.

Nanji, A.A., Jokelainen, K., Tipoe, G.L., Rahemtulla, A., Thomas, P. and Dannenberg, A.J. (2003) Curcumin prevents alcohol-induced liver disease in rats by inhibiting the expression of NF-kappa B-dependent genes. *Am J Physiol Gastrointest Liver Physiol* 284(2), G321-327.

Natarajan, C. and Bright, J.J. (2002) Curcumin inhibits experimental allergic encephalomyelitis by blocking IL-12 signaling through Janus kinase-STAT pathway in T lymphocytes. *J Immunol* 168(12), 6506-6513.

Nirmala, C. and Puvanakrishnan, R. (1996) Effect of curcumin on certain lysosomal hydrolases in isoproter-enol-induced myocardial infarction in rats. *Biochem Pharmacol* 51(1), 47-51.

Nirmala, C. and Puvanakrishnan, R. (1996) Protective role of curcumin against isoproterenol induced myo-cardial infarction in rats. *Mol Cell Biochem* 159(2), 85-93.

Nirmala, C., Anand, S. and Puvanakrishnan, R. (1999) Curcumin treatment modulates collagen metabolism in isoproterenol induced myocardial necrosis in rats. *Mol Cell Biochem* 197(1-2): 31-37.

Nishida, N., Fukuda, Y., Komeda, T., Kita, R., Sando, T., Furukawa, M., Amenomori, M., Shibagaki, I., Nakao, K., Ikenaga, M. and et al., (1994) Amplification and overexpression of the cyclin D1 gene in aggressive human hepatocellular carcinoma. *Cancer Res* 54(12), 3107-3110.

Nogaki, A., Satoh, K., Iwasaka, K., Takano, H., Takahama, M., Ida, Y. and Sakagami, H. (1998) Radical intensity and cytotoxic activity of curcumin and gallic acid. *Anticancer Res* 18(5A), 3487-3491.

Notarbartolo, M., Poma, P., Perri, D., Dusonchet, L., Cervello, M. and D'Alessandro, N. (2005) Antitumor effects of curcumin, alone or in combination with cisplatin or doxorubicin, on human hepatic cancer cells. Analysis of their possible relationship to changes in NF-kB activation levels and in IAP gene expression. *Cancer Lett* 224(1), 53-65.

Odot, J., Albert, P., Carlier, A., Tarpin, M., Devy, J. and Madoulet, C. (2004) *In vitro* and *in vivo* anti-tumoral effect of curcumin against melanoma cells. *Int J Cancer* 111(3), 381-387.

Oetari. S., Sudibyo, M., Commandeur, J.N., Samhoedi, R. and Vermeulen, N.P. (1996) Effects of curcumin on cytochrome P450 and glutathione S-transferase activities in rat liver. *Biochem Pharmacol* 51(1), 39-45.

Oguro, T. and Yoshida, T. (2001) Effect of ultraviolet A on ornithine decarboxylase and metallothionein gene expression in mouse skin. *Photodermatol Photoimmunol Photomed* 17(2), 71-78.

Ohara, K., Mizukami, W., Tokunaga, A., Nagaoka, S., Uno, H. and Mukai, K (2005) Solvent and pH. *Bulletin of the Chemical Society of Japan* 78(4): 615-621.

Ohene-Abuakwa, Y. and Pignatelli, M. (2000) Adhesion molecules in cancer biology. *Adv Exp Med Biol* 465, 115-126.

Ono, K., Hasegawa, K., Naiki, H. and Yamada, M. (2004) Curcumin has potent anti-amyloidogenic effects for Alzheimer's beta-amyloid fibrils *in vitro*. *J Neurosci Res* 75(6), 742-750.

Onoda, M. and Inano, H. Effect of curcumin on the production of nitric oxide by cultured rat mammary gland. *Nitric Oxide* 4(5), 505-515.

Onodera S, Nishihira J, Iwabuchi K, Koyama Y, Yoshida K, Tanaka S and Minami A. (2002) Macrophage migration inhibitory factor up-regulates matrix metalloproteinase-9 and -13 in rat osteoblasts. Rele-vance to intracellular signaling pathways. *J Biol Chem* 277(10), 7865-3874.

Ozaki, K., Kawata, Y., Amano, S. and Hanazawa, S. (2000) Stimulatory effect of curcumin on osteoclast apoptosis. *Biochem Pharmacol* 59(12), 1577-1581.

Padmaja, S. and Raju ,T.N. (2004) Antioxidant effect of curcumin in selenium induced cataract of Wistar rats. *Indian J Exp Biol* 42(6), 601-3.

Pahl, H.L. (1999) Activators and target genes of Rel/NF-kappaB transcription factors. *Oncogene* 18(49), 6853-6866.

Pal, S., Choudhuri, T., Chattopadhyay, S., Bhattacharya, A., Datta, G.K., Das, T. and Sa, G. (2001) Mechanisms of curcumin-induced apoptosis of Ehrlich's ascites carcinoma cells. *Biochem Biophys Res Commun* 288(3), 658-665.

Pan, M.H., Chang, W.L., Lin-Shiau, S.Y., Ho, C.T. and Lin, J.K. (2001) Induction of apoptosis by garcinol and curcumin through cytochrome c release and activation of caspases in human leukemia HL-60 cells. *J Agric Food Chem* 49(3), 1464-1474.

Pan, M.H., Huang, T.M. and Lin, J.K. (1999) Biotransformation of curcumin through reduction and glucu-ronidation in mice. *Drug Metab Dispos* 27(4), 486-494.

Pan, M.H., Lin-Shiau, S.Y. and Lin, J.K. (2000) Comparative studies on the suppression of nitric oxide synthase by curcumin and its hydrogenated metabolites through down-regulation of IkappaB kinase and NFkap-paB activation in macrophages. *Biochem Pharmacol* 60(11), 1665-1676.

Park, M.J., Kim, E.H., Park, I.C., Lee, H.C., Woo, S.H., Lee, J.Y., Hong, Y.J., Rhee, C.H., Choi, S.H., Shim, B.S., Lee, S.H. and Hong, S.I. (2002) Curcumin inhibits cell cycle progression of immortalized human umbilical vein endothelial (ECV304) cells by up-regulating cyclin-dependent kinase inhibitor, p21WAF1/CIP1, p27KIP1 and p53. *Int J Oncol* 21(2), 379-383.

Park, S.D., Jung, J.H., Lee, H.W., Kwon, Y.M., Chung, K.H., Kim, M.G. and Kim, C.H. (2005) Zedoariae rhizoma and curcumin inhibits platelet-derived growth factor-induced proliferation of human hepatic myofibroblasts. *Int Immunopharmacol* 5(3), 555-569.

Patil, T.N. and Srinivasan, M. (1971) Hypocholesteremic effect of curcumin in induced hypercholesteremic rats. *Indian J Exp Biol* 9(2), 167-169.

Patro, B.S., Rele, S., Chintalwar, G.J., Chattopadhyay, S., Adhikari, S. and Mukherjee ,T. (2002) Protective activities of some phenolic 1,3-diketones against lipid peroxidation: possible involvement of the 1,3-diketone moiety. *Chembiochem* 3(4), 364-370.

Pendurthi, U.R. and Rao, L.V. (2000) Suppression of transcription factor Egr-1 by curcumin. *Thromb Res* 97(4), 179-189.

Pereira, M.A., Grubbs, C.J., Barnes, L.H., Li, H., Olson, G.R., Eto, I., Juliana, M., Whitaker, L.M., Kelloff, G.J., Steele, V.E. and Lubet, R.A. (1996) Effects of the phytochemicals, curcumin and quercetin, upon azoxymethane-induced colon cancer and 7,12-dimethylbenz[a]anthracene-induced mammary cancer in rats. *Carcinogenesis* 17(6), 1305-1311.

Perkins, S., Verschoyle, R.D., Hill, K., Parveen, I., Threadgill, M.D., Sharma, R.A., Williams, M.L., Steward, W.P. and Gescher, A.J. (2002) Chemopreventive efficacy and pharmacokinetics of curcumin in the min/+ mouse, a model of familial adenomatous polyposis. *Cancer Epidemiol Biomarkers Prev* 11(6), 535-540.

Phan, T.T., See, P., Lee, S.T. and Chan, S.Y. (2001) Protective effects of curcumin against oxidative damage on skin cells *in vitro*: its implication for wound healing. *J Trauma* 51(5), 927-31.

Phan, T.T., Sun, L., Bay, B.H., Chan, S.Y. and Lee, S.T. (2003) Dietary compounds inhibit proliferation and contraction of keloid and hypertrophic scar-derived fibroblasts in vitro: therapeutic implication for excessive scarring. *J Trauma* 54(6), 1212-1224.

Philip, S. and Kundu, G.C. (2003) Osteopontin induces nuclear factor kappa B-mediated promatrix metallo-proteinase-2 activation through I kappa B alpha /IKK signaling pathways, and curcumin (diferulolyl-methane) down-regulates these pathways. *J Biol Chem* 278(16), 14487-14497.

Piper, J.T., Singhal, S.S., Salameh, M.S., Torman, R.T., Awasthi, Y.C. and Awasthi, S. (1998) Mechanisms of anticarcinogenic properties of curcumin: the effect of curcumin on glutathione linked detoxification enzymes in rat liver. *Int J Biochem Cell Biol* 30(4), 445-456.

Piwocka, K., Bielak-Mijewska, A. and Sikora, E. (2002) Curcumin induces caspase-3-independent apoptosis in human multidrug-resistant cells. *Ann N Y Acad Sci* 973, 250-254.

Piwocka, K., Jaruga, E., Skierski, J., Gradzka, I. and Sikora, E. (2001) Effect of glutathione depletion on caspase-3 independent apoptosis pathway induced by curcumin in Jurkat cells. *Free Radic Biol Med* 31(5), 670-678.

Piwocka, K., Zablocki, K., Wieckowski, M.R., Skierski, J., Feiga, I., Szopa, J., Drela, N., Wojtczak, L. and Sikora, E. (1999) A novel apoptosis-like pathway, independent of mitochondria and caspases, induced by curcumin in human lymphoblastoid T (Jurkat) cells. *Exp Cell Res* 249(2), 299-307.

Plummer, S.M., Hill, K.A., Festing, M.F., Steward, W.P., Gescher, A.J. and Sharma, R.A. (2001) Clinical development of leukocyte cyclooxygenase 2 activity as a systemic biomarker for cancer chemopreventive agents. *Cancer Epidemiol Biomarkers Prev* 10(12), 1295-1299.

Plummer, S.M., Holloway, K.A., Manson, M.M., Munks, R.J., Kaptein, A., Farrow, S. and Howells, L. (1999) Inhibition of cyclo-oxygenase 2 expression in colon cells by the chemopreventive agent curcumin involves inhibition of NF-kappaB activation via the NIK/IKK signalling complex. *Oncogene* 18(44), 6013-6020.

Prasad, N.S. (1997) Spectrophotometric estimation of curcumin. *Indian Drugs* 34(4): 227-228.

Priyadarsini, K.I., Maity, D.K., Naik, G.H., Kumar, M.S., Unnikrishnan, M.K., Satav, J.G. and Mohan, H. (2003) Role of phenolic O-H and methylene hydrogen on the free radical reactions and antioxidant activity of curcumin. *Free Radic Biol Med* 35(5): 475-484.

Prusty, B.K. and Das, B.C. (2005) Constitutive activation of transcription factor AP-1 in cervical cancer and suppression of human papillomavirus (HPV) transcription and AP-1 activity in HeLa cells by curcumin. *Int J Cancer* 113(6), 951-960.

Punithavathi, D., Venkatesan, N. and Babu, M. (2000) Curcumin inhibition of bleomycin-induced pulmonary fibrosis in rats. *Br J Pharmacol* 131(2): 169-172.

Punithavathi, D., Venkatesan, N. and Babu, M. (2003) Protective effects of curcumin against amiodarone-induced pulmonary fibrosis in rats. *Br J Pharmacol* 139(7), 1342-1350.

Quiles, J.L., Aguilera, C., Mesa, M.D., Ramirez-Tortosa, M.C., Baro, L. and Gil, A. (1998) An ethanolic-aqueous extract of *Curcuma longa* decreases the susceptibility of liver microsomes and mitochondria to lipid peroxidation in atherosclerotic rabbits. *Biofactors* 8(1-2), 51-57.

Ramachandran, C. and You, W. (1999) Differential sensitivity of human mammary epithelial and breast carcinoma cell lines to curcumin. *Breast Cancer Res Treat* 54(3), 269-278.

Ramirez-Tortosa, M.C., Mesa, M.D., Aguilera, M.C., Quiles, J.L., Baro, L., Ramirez-Tortosa, C.L., Martinez-Victoria, E. and Gil, A. (1999) Oral administration of a turmeric extract inhibits LDL oxidation and has hypocholesterolemic effects in rabbits with experimental atherosclerosis. *Atherosclerosis* 147(2), 371-378.

Ramprasad, C. and Sirsi, M. (1956) Studies on Indian medicinal plants: Curcuma longa Linn.-effect of curcumin & the essential oils of C. longa on bile secretion. *J Sci Industr Res* 15(C): 262-265.

Ramsewak, R.S., DeWitt, D.L. and Nair, M.G. (2000) Cytotoxicity, antioxidant and anti-inflammatory activities of curcumins I-III from Curcuma longa. *Phytomedicine* 7(4), 303-308.

Ranjan, D., Johnston, T.D., Reddy, K.S., Wu, G., Bondada, S. and Chen, C. (1999) Enhanced apoptosis mediates inhibition of EBV-transformed lymphoblastoid cell line proliferation by curcumin. *J Surg Res* 87(1), 1-5.

Rao, C.V., Rivenson, A., Simi, B. and Reddy, B.S. (1995) Chemoprevention of colon carcinogenesis by dietary curcumin, a naturally occurring plant phenolic compound. *Cancer Res* 55(2), 259-66.

Rao, C.V., Rivenson, A., Simi, B. and Reddy, B.S. (1995) Chemoprevention of colon cancer by dietary curcumin. *Ann N Y Acad Sci* 768, 201-204.

Rao, C.V., Simi, B. and Reddy, B.S. (1993) Inhibition by dietary curcumin of azoxymethane-induced ornithine decarboxylase, tyrosine protein kinase, arachidonic acid metabolism and aberrant crypt foci formation in the rat colon. *Carcinogenesis* 14(11), 2219-2225.

Rao, D.S., Sekhara, N.C., Satyanarayana, M.N. and Srinivasan, M. (1970) Effect of curcumin on serum and liver cholesterol levels in the rat. *J Nutr* 100 (11), 1307-1315.

Rao, T.S., Basu, N. and Siddiqui, H.H. (1982) Anti-inflammatory activity of curcumin analogues. *Indian J Med Res* 75, 574-578.

Rao, T.S., Basu, N., Seth, S.D. and Siddiqui, H.H. (1984) Some aspects of pharmacological profile of sodium curcuminate. *Indian J Physiol Pharmacol* 28(3), 211-215.

Ravindranath, V. and Chandrasekhara, N. (1980) Absorption and tissue distribution of curcumin in rats. *Toxicology* 16(3), 259-265.

Reddy, B.S., Hirose, Y., Lubet, R., Steele, V., Kelloff, G., Paulson, S., Seibert, K. and Rao, C.V. (2000) Chemoprevention of colon cancer by specific cyclooxygenase-2 inhibitor, celecoxib, administered during different stages of carcinogenesis. *Cancer Res* 60(2), 293-297.

Reddy, S. and Aggarwal, B.B. (1994) Curcumin is a non-competitive and selective inhibitor of phosphorylase kinase. *FEBS Lett* 341(1), 19-22.

Robinson, T.P., Ehlers, T., Hubbard, I.R., Bai, X., Arbiser, J.L., Goldsmith, D.J. and Bowen, J.P. (2003) Design, synthesis, and biological evaluation of angiogenesis inhibitors: aromatic enone and dienone analogues of curcumin. *Bioorg Med Chem Lett* 13(1), 115-117.

Romiti, N., Tongiani, R., Cervelli, F. and Chieli, E. (1998) Effects of curcumin on P-glycoprotein in primary cultures of rat hepatocytes. *Life Sci* 62(25), 2349-2358.

Ruby, A.J., Kuttan, G., Babu, K.D., Rajasekharan, K.N. and Kuttan, R. (1995) Anti-tumour and antioxidant activity of natural curcuminoids. *Cancer Lett* 94(1), 79-83.

Rukkumani, R., Aruna, K., Varma, P.S., Rajasekaran, K.N. and Menon, V.P. (2004) Comparative effects of curcumin and an analog of curcumin on alcohol and PUFA induced oxidative stress. *J Pharm Pharm Sci* 7(2), 274-283.

Rukkumani, R., Sri Balasubashini, M., Vishwanathan, P. and Menon, V.P. (2002) Comparative effects of curcumin and photo-irradiated curcumin on alcohol- and polyunsaturated fatty acid-induced hyperlipidemia. *Pharmacol Res* 46(3), 257-264.

Saleheen, D., Ali, S.A., Ashfaq, K., Siddiqui, A.A., Agha, A. and Yasinzai, M.M. (2002) Latent activity of curcumin against leishmaniasis *in vitro*. *Biol Pharm Bull* 25(3): 386-389.

Salh, B., Assi, K., Templeman, V., Parhar, K., Owen, D., Gomez-Munoz, A. and Jacobson, K. (2003) Curcumin attenuates DNB-induced murine colitis. *Am J Physiol Gastrointest Liver Physiol* 285(1), G235-243.

Samaha, H.S., Kelloff, G.J., Steele, V., Rao, C.V. and Reddy, B.S. (1997) Modulation of apoptosis by sulindac, curcumin, phenylethyl-3-methylcaffeate, and 6-phenylhexyl isothiocyanate: apoptotic index as a biomarker in colon cancer chemoprevention and promotion. *Cancer Res* 57(7), 1301-1305.

Santibanez, J.F., Quintanilla, M. and Martinez, J. (2000) Genistein and curcumin block TGF-beta 1-induced u-PA expression and migratory and invasive phenotype in mouse epidermal keratinocytes. *Nutr Cancer* 37(1), 49-54.

Satoskar, R.R., Shah, S.J. and Shenoy, S.G. (1986) Evaluation of anti-inflammatory property of curcumin (diferuloyl methane) in patients with postoperative inflammation. *Int J Clin Pharmacol Ther Toxicol* 24(12), 651-654.

Scapagnini, G., Foresti, R., Calabrese, V., Giuffrida Stella, A.M., Green, C.J. and Motterlini, R. (2002) Caffeic acid phenethyl ester and curcumin: a novel class of heme oxygenase-1 inducers. *Mol Pharmacol* 61(3), 554-561.

Selvam, C., Jachak, S.M., Thilagavathi, R. and Chakraborti, A.K. (2005) Design, synthesis, biological evaluation and molecular docking of curcumin analogues as antioxidant, cyclooxygenase inhibitory and anti-inflammatory agents. *Bioorg Med Chem Lett* 15(7), 1793-1797.

Sen, S., Sharma, H. and Singh, N. (2005) Curcumin enhances Vinorelbine mediated apoptosis in NSCLC cells by the mitochondrial pathway. *Biochem Biophys Res Commun* 331(4), 1245-1252.

Shah, B.H., Nawaz, Z., Pertani, S.A., Roomi, A., Mahmood, H., Saeed, S.A. and Gilani, A.H. (1999) Inhibitory effect of curcumin, a food spice from turmeric, on platelet-activating factor- and arachidonic acid-mediated platelet aggregation through inhibition of thromboxane formation and Ca2+ signaling. *Biochem Pharmacol* 58(7), 1167-1172.

Shahed, A.R., Jones, E. and Shoskes, D. (2001) Quercetin and curcumin up-regulate antioxidant gene expression in rat kidney after ureteral obstruction or ischemia/reperfusion injury. *Transplant Proc* 33(6), 2988.

Shalini, V.K. and Srinivas, L. (1990) Fuel smoke condensate induced DNA damage in human lymphocytes and protection by turmeric (Curcuma longa). *Mol Cell Biochem* 95(1), 21-30.

Sharma, O.P. (1976) Antioxidant activity of curcumin and related compounds. *Biochem Pharmacol* 25(15): 1811-1812.

Sharma, R.A., Euden, S.A., Platton, S.L., Cooke,D.N., Shafayat, A., Hewitt, H.R., Marczylo, T.H., Morgan, B., Hemingway, D., Plummer, S.M., Pirmohamed, M., Gescher, A.J. and Steward, W.P. (2004) Phase I clinical trial of oral curcumin: biomarkers of systemic activity and compliance. *Clin.Cancer Res* 10(20), 6847-6854.

Sharma, R.A., Ireson, C.R., Verschoyle, R.D., Hill, K.A., Williams, M.L., Leuratti, C., Manson, M.M., Marnett, L.J., Steward, W.P. and Gescher, A. (2001) Effects of dietary curcumin on glutathione S-transferase and malondialdehyde-DNA adducts in rat liver and colon mucosa: relationship with drug levels. *Clin Cancer Res* 7(5), 1452-1458.

Sharma, R.A., McLelland, H.R., Hill, K.A., Ireson, C.R., Euden, S.A., Manson, M.M., Pirmohamed, M., Marnett, L.J., Gescher, A.J. and Steward, W.P. (2001) Pharmacodynamic and pharmacokinetic study of oral Curcuma extract in patients with colorectal cancer. *Clin Cancer Res* 7(7), 1894-900.

Shim, J.S., Kim, D.H., Jung, H.J., Kim, J.H., Lim, D., Lee, S.K., Kim, K.W., Ahn, J.W., Yoo, J.S., Rho, J.R., Shin, J. and Kwon, H.J (2002) Hydrazinocurcumin, a novel synthetic curcumin derivative, is a potent inhibitor of endothelial cell proliferation. *Bioorg Med Chem* 10(9), 2987-2992.

Shim, J.S., Kim, J.H., Cho, H.Y., Yum, Y.N., Kim, S.H., Park, H.J., Shim, B.S., Choi, S.H. and Kwon, H.J. (2003) Irreversible inhibition of CD13/aminopeptidase N by the antiangiogenic agent curcumin. *Chem Biol* 10(8), 695-704.

Shishodia, S., Amin, H.M., Lai, R. and Aggarwal, B.B. (2005) Curcumin (diferuloylmethane) inhibits constitutive NF-kappaB activation, induces G1/S arrest, suppresses proliferation, and induces apoptosis in mantle cell lymphoma. *Biochem Pharmacol* 70(5), 700-713.

Shishodia, S., Potdar, P., Gairola, C.G. and Aggarwal, B.B. (2003) Curcumin (diferuloylmethane) down-regulates cigarette smoke-induced NF-kappaB activation through inhibition of IkappaBalpha kinase in human lung epithelial cells: correlation with suppression of COX-2, MMP-9 and cyclin D1. *Carcinogenesis* 24(7), 1269-1279.

Shoba, G., Joy, D., Joseph, T., Majeed, M., Rajendran R. and Srinivas, P.S.(1998) Influence of piperine on the pharmacokinetics of curcumin in animals and human volunteers. *Planta Med* 64(4), 353-356.

Sidhu, G.S., Singh, A.K., Thaloor, D., Banaudha, K.K., Patnaik, G.K., Srimal, R.C. and Maheshwari, R.K (1998) Enhancement of wound healing by curcumin in animals. *Wound Repair Regen* 6(2), 167-177.

Sikora, E., Bielak-Zmijewska, A., Piwocka, K., Skierski, J. and Radziszewska, E. (1997) Inhibition of proliferation and apoptosis of human and rat T lymphocytes by curcumin, a curry pigment. *Biochem Pharmacol* 54(8), 899-907.

Simon, A., Allais, D.P., Duroux, J.L., Basly, J.P., Durand-Fontanier, S. and Delage, C. (1998) Inhibitory effect of curcuminoids on MCF-7 cell proliferation and structure-activity relationships. *Cancer Lett* 129(1), 111-116.

Sindhwani, P., Hampton, J.A., Baig, M.M., Keck, R. and Selman, S.H. (2001) Curcumin prevents intravesical tumor implantation of the MBT-2 tumor cell line in C3H mice. *J Urol* 166(4), 1498-1501.

Singh, A.K., Sidhu, G.S., Deepa, T. and Maheshwari, R.K. (1996) Curcumin inhibits the proliferation and cell cycle progression of human umbilical vein endothelial cell. *Cancer Lett* 107(1), 109-115.

Singh, S. and Aggarwal, B.B. (1995) Activation of transcription factor NF-kappa B is suppressed by curcumin (diferuloylmethane) [corrected]. *J Biol Chem* 270(42), 24995-5000.

Singh, S.V., Hu, X., Srivastava, S.K., Singh, M., Xia, H., Orchard, J.L. and Zaren, H.A. (1998) Mechanism of inhibition of benzo[a]pyrene-induced forestomach cancer in mice by dietary curcumin. *Carcinogenesis* 19(8), 1357-1360.

Singletary, K., MacDonald, C., Wallig, M. and Fisher, C. (1996) Inhibition of 7,12-dimethylbenz[a]anthracene (DMBA)-induced mammary tumorigenesis and DMBA-DNA adduct formation by curcumin. *Cancer Lett* 103(2), 137-141.

Skrzypczak-Jankun, E., McCabe, N.P., Selman, S.H. and Jankun, J. (2000) Curcumin inhibits lipoxygenase by binding to its central cavity: theoretical and X-ray evidence. *Int J Mol Med* 6(5), 521-526.

Skrzypczak-Jankun, E., Zhou, K., McCabe, N.P., Selman, S.H. and Jankun, J. (2003) Structure of curcumin in complex with lipoxygenase and its significance in cancer. *Int J Mol Med* 12(1), 17-24.

Slamon, D.J., Clark, G.M., Wong, S.G., Levin, W.J., Ullrich, A. and McGuire, W.L. (1987) Human breast cancer: correlation of relapse and survival with amplification of the HER-2/neu oncogene. *Science* 235(4785), 177-182.

Snyder, J.P., Davis, M.C. and Adams Bea. (2002) Curcumin analogs with anti-tumor and anti-angiogenic properties. *United States Patent Application Publication* US 2002/0019382.

Snyder, R.D. and Arnone, M.R. (2002) Putative identification of functional interactions between DNA intercalating agents and topoisomerase II using the V79 in vitro micronucleus assay. *Mutat Res* 503(1-2), 21-35.

Soni, K.B. and Kuttan, R. (1992) Effect of oral curcumin administration on serum peroxides and cholesterol levels in human volunteers. *Indian J Physiol Pharmacol* 36(4), 273-275.

Soni, K.B., Rajan, A. and Kuttan, R. (1992) Reversal of aflatoxin induced liver damage by turmeric and curcumin. *Cancer Lett* 66(2), 115-121.

Soudamini, K.K. and Kuttan, R. (1989) Inhibition of chemical carcinogenesis by curcumin. *J Ethnopharmacol* 27(1-2), 227-233.

Soudamini, K.K., Unnikrishnan, M.C., Soni, K.B. and Kuttan, R. (1992) Inhibition of lipid peroxidation and cholesterol levels in mice by curcumin. *Indian J Physiol Pharmacol* 36(4), 239-243.

Souza, C.R.A., Osme, S.F. and Gloria, M.B.A. (1997) Stability of curcuminoid pigments in model systems. *Journal of Food Processing and Preservation* 21(5): 353-363.

Sovak, M.A., Bellas, R.E., Kim, D.W., Zanieski, G.J., Rogers, A.E., Traish, A.M. and Sonenshein, G.E (1997) Aberrant nuclear factor-kappaB/Rel expression and the pathogenesis of breast cancer. *J Clin Invest* 100(12), 2952-2960.

Squires, M.S., Hudson, E.A., Howells, L., Sale, S., Houghton, C.E., Jones, J.L., Fox, L.H., Dickens, M., Prigent, S.A. and Manson, M.M. (2003) Relevance of mitogen activated protein kinase (MAPK) and phosphotidylinositol-3-kinase/protein kinase B (PI3K/PKB) pathways to induction of apoptosis by curcumin in breast cells. *Biochem Pharmacol* 65(3), 361-376.

Sreejayan, N. and Rao, M.N. (1996) Free radical scavenging activity of curcuminoids. *Arzneimittelforschung* 46(2), 169-171.

Srimal, R.C. and Dhawan, B.N. (1973) Pharmacology of diferuloyl methane (curcumin), a non-steroidal anti-inflammatory agent. *J Pharm Pharmacol* 25(6), 447-452.

Srinivasan, K.R. (1952) The coloring matter in Turmeric. *Current Science*: 311.

Srinivasan, M. (1972) Effect of curcumin on blood sugar as seen in a diabetic subject. *Indian J Med Sci* 26(4), 269-270.

Srivastava, K.C., Bordia, A. and Verma, S.K. (1995) Curcumin, a major component of food spice turmeric (*Curcuma longa*) inhibits aggregation and alters eicosanoid metabolism in human blood platelets. *Prostaglandins Leukot Essent Fatty Acids* 52(4), 223-227.

Srivastava, R., Dikshit, M., Srimal, R.C. and Dhawan, B.N. (1985) Anti-thrombotic effect of curcumin. *Thromb Res* 40(3), 413-7.

Srivastava, R., Puri, V., Srimal, R.C. and Dhawan, B.N. (1986) Effect of curcumin on platelet aggregation and vascular prostacyclin synthesis. *Arzneimittelforschung* 36(4), 715-717.

Sugiyama, Y., Kawakishi, S. and Osawa, T. (1996) Involvement of the beta-diketone moiety in the antioxidative mechanism of tetrahydrocurcumin. *Biochem Pharmacol* 52(4), 519-525.

Sui, Z., Salto, R., Li, J., Craik, C. and Ortiz de Montellano, P.R. (1993) Inhibition of the HIV-1 and HIV-2 proteases by curcumin and curcumin boron complexes. *Bioorg Med Chem* 1(6), 415-422.

Sumbilla, C., Lewis, D., Hammerschmidt, T. and Inesi, G. (2002) The slippage of the Ca2+ pump and its control by anions and curcumin in skeletal and cardiac sarcoplasmic reticulum. *J Biol Chem* 277(16), 13900-13906.

Suresh Babu, P. and Srinivasan, K. (1998) Amelioration of renal lesions associated with diabetes by dietary curcumin in streptozotocin diabetic rats. *Mol Cell Biochem* 181(1-2), 87-96.

Surh, Y.J., Chun, K.S., Cha, H.H., Han, S.S., Keum, Y.S., Park, K.K. and Lee, S.S. (2001) Molecular mechanisms underlying chemopreventive activities of anti-inflammatory phytochemicals: down-regulation of COX-2 and iNOS through suppression of NF-kappa B activation. *Mutat Res* 480-481, 243-268.

Surh, Y.J., Han, S.S., Keum, Y.S., Seo, H.J. and Lee, S.S. (2000) Inhibitory effects of curcumin and capsaicin on phorbol ester-induced activation of eukaryotic transcription factors, NF-kappaB and AP-1. *Biofactors* 12(1-4), 107-112.

Suryanarayana, P., Krishnaswamy, K. and Reddy, G.B. (2003) Effect of curcumin on galactose-induced cataractogenesis in rats. *Mol Vis* 9, 223-230.

Suryanarayana, P., Saraswat, M., Mrudula, T., Krishna, T.P., Krishnaswamy, K. and Reddym, G.B. (2005) Curcumin and turmeric delay streptozotocin-induced diabetic cataract in rats. *Invest Ophthalmol Vis Sci* 46(6), 2092-2099.

Susan, M. and Rao, M.N. (1992) Induction of glutathione S-transferase activity by curcumin in mice. *Arzneimittelforschung* 42(7), 962-964.

Swarnakar, S., Ganguly, K., Kundu, P., Banerjee, A., Maity, P. and Sharma, A.V. (2005) Curcumin regulates expression and activity of matrix metalloproteinases 9 and 2 during prevention and healing of indomethacin-induced gastric ulcer. *J Biol Chem* 280(10), 9409-9415.

Takaba, K., Hirose, M., Yoshida, Y., Kimura, J., Ito, N. and Shirai, T. (1997) Effects of n-tritriacontane-16,18-dione, curcumin, chlorphyllin, dihydroguaiaretic acid, tannic acid and phytic acid on the initiation stage in a rat multi-organ carcinogenesis model. *Cancer Lett* 113(1-2), 39-46.

Tanaka, T., Makita, H., Ohnishi, M., Hirose, Y., Wang, A., Mori, H., Satoh, K., Hara, A. and Ogawa, H. (1994) Chemoprevention of 4-nitroquinoline 1-oxide-induced oral carcinogenesis by dietary curcumin and hesperidin: comparison with the protective effect of beta-carotene. *Cancer Res* 54(17), 4653-4659.

Tanaka, Y., Kobayashi, H., Suzuki, M., Kanayama, N. and Terao, T. (2004) Transforming growth factor-beta1-dependent urokinase up-regulation and promotion of invasion are involved in Src-MAPK-dependent signaling in human ovarian cancer cells. *J Biol Chem* 279(10), 8567-8576.

Tang ,X.Q., Bi, H., Feng, J.Q. and Cao, J.G. (2005) Effect of curcumin on multidrug resistance in resistant human gastric carcinoma cell line SGC7901/VCR. *Acta Pharmacol Sin* 26(8), 1009-1016.

Thaloor, D., Miller, K.J., Gephart, J., Mitchell, P.O. and Pavlath, G.K. (1999) Systemic administration of the NF-kappaB inhibitor curcumin stimulates muscle regeneration after traumatic injury. *Am J Physiol* 277(2 Pt 1), C320-329.

Thapliyal, R. and Maru, G.B. (2001) Inhibition of cytochrome P450 isozymes by curcumins *in vitro* and *in vivo*. *Food Chem Toxicol* 39(6), 541-547.

Thapliyal, R., Deshpande, S.S. and Maru, G.B. (2001) Effects of turmeric on the activities of benzo(a)pyrene-induced cytochrome P-450 isozymes. *J Environ Pathol Toxicol Oncol* 20(1), 59-63.

Thresiamma, K.C., George, J. and Kuttan, R. (1996) Protective effect of curcumin, ellagic acid and bixin on radiation induced toxicity. *Indian J Exp Biol* 34(9), 845-847.

Thresiamma, K.C., George, J. and Kuttan, R. (1998) Protective effect of curcumin, ellagic acid and bixin on radiation induced genotoxicity. *J Exp Clin Cancer Res* 17(4), 431-434.

Tikhomirov, O. and Carpenter, G. (2003) Identification of ErbB-2 kinase domain motifs required for geldanamycin-induced degradation. *Cancer Res* 63(1), 39-43.

Toniolo, R., Di Narda, F., Susmel, S., Martelli, M., Martelli, L. and Bontempelli, G. (2002) Quenching of superoxide ions by curcumin. A mechanistic study in acetonitrile. *Ann Chim* 92(3): 281.

Tonnesen, H.H. and Greenhill, J.V. (1992) Studies on curcumin and curcuminoids XXII: curcumin as a reducing agent and as a radical scavenger. *Int.J. Pharm* 87, 79-87.

Tonnesen, H.H., de Vries, H., Karlsen, J. and Beijersbergen van Henegouwen, G. (1987) Studies on curcumin and curcuminoids. IX: Investigation of the photobiological activity of curcumin using bacterial indicator systems. *J Pharm Sci* 76(5): 371-373.

Tonnesen, H.H., Karlsen, J. and van Henegouwen, G.B. (1986) Studies on curcumin and curcuminoids. VIII. Photochemical stability of curcumin. *Z Lebensm Unters Forsch* 183(2): 116-122.

Tonnesen, H.H., Masson, M. and Loftsson, T. (2002) Studies of curcumin and curcuminoids. XXVII. Cyclodextrin complexation: solubility, chemical and photochemical stability. *Int J Pharm* 244(1-2): 127-135.

Ukil, A., Maity, S., Karmakar, S., Datta, N., Vedasiromoni, J.R. and Das, P.K. (2003) Curcumin, the major component of food flavour turmeric, reduces mucosal injury in trinitrobenzene sulphonic acid-induced colitis. *Br J Pharmacol* 139(2), 209-218.

Ushida, J., Sugie, S., Kawabata, K., Pham, Q.V., Tanaka, T., Fujii, K., Takeuchi, H., Ito, Y. and Mori, H. (2000) Chemopreventive effect of curcumin on N-nitrosomethylbenzylamine-induced esophageal carcinogenesis in rats. *Jpn J Cancer Res* 91(9), 893-898.

Vajragupta, O., Boonchoong, P. and Berliner, L.J. (2004) Manganese complexes of curcumin analogues: evaluation of hydroxyl radical scavenging ability, superoxide dismutase activity and stability towards hydrolysis. *Free Radic Res* 38(3), 303-314.

Van Der Logt, E.M., Roelofs, H.M., Nagengast, F.M. and Peters, W.H. (2003) Induction of rat hepatic and intestinal UDP-glucuronosyltransferases by naturally occurring dietary anticarcinogens. *Carcinogenesis*, 2003.

Varadkar, P., Dubey, P., Krishna, M. and Verma, N. (2001) Modulation of radiation-induced protein kinase C activity by phenolics. *J Radiol Prot* 21(4), 361-370.

Venkatesan, N. (1998) Curcumin attenuation of acute adriamycin myocardial toxicity in rats. *Br J Pharmacol* 124(3), 425-427.

Venkatesan, N. (200) Pulmonary protective effects of curcumin against paraquat toxicity. *Life Sci* 66(2): PL21-28.

Venkatesan, N. and Chandrakasan, G. (1995) Modulation of cyclophosphamide-induced early lung injury by curcumin, an anti-inflammatory antioxidant. *Mol Cell Biochem* 142(1): 79-87.

Venkatesan, N., Punithavathi, D. and Arumugam, V. (2000) Curcumin prevents adriamycin nephrotoxicity in rats. *Br J Pharmacol* 129(2), 231-234.

Venkatesan, N., Punithavathi, V. and Chandrakasan, G. (1997) Curcumin protects bleomycin-induced lung injury in rats. *Life Sci* 61(6): PL51-58.

Venkatesan, P., Unnikrishnan, M.K., Kumar, S.M. and Rao, M.N.A. (2003) Effect of curcumin analogues on oxidation of haemoglobin and lysis of erythrocytes. *Current Science* 84(1), 74-78.

Verbeek, R., van Tol, E.A. and van Noort, J.M. (2005) Oral flavonoids delay recovery from experimental autoimmune encephalomyelitis in SJL mice. *Biochem Pharmacol* 70(2), 220-228.

Verma, S.P., Goldin, B.R. and Lin, P.S. (1998) The inhibition of the estrogenic effects of pesticides and environmental chemicals by curcumin and isoflavonoids. *Environ Health Perspect* 106(12), 807-812.

Verma, S.P., Salamone, E. and Goldin, B. (1997) Curcumin and genistein, plant natural products, show synergistic inhibitory effects on the growth of human breast cancer MCF-7 cells induced by estrogenic pesticides. *Biochem Biophys Res Commun* 233(3), 692-6.

Vogel, and Pelletier, (1818) *J. Pharm.* 2: 50.

Wahlstrom, B. and Blennow, G (1978) A study on the fate of curcumin in the rat. *Acta Pharmacol Toxicol (Copenh)* 43(2), 86-92.

Wang, C.Y., Mayo, M.W. and Baldwin, A.S., Jr. (1996) TNF- and cancer therapy-induced apoptosis: potentiation by inhibition of NF-kappaB. *Science* 274(5288), 784-787.

Wang, Y.J., Pan, M.H., Cheng, A.L., Lin, L.I., Ho, Y.S., Hsieh, C.Y. and Lin, J.K. (1997) Stability of curcumin in buffer solutions and characterization of its degradation products. *J Pharm Biomed Anal* 15(12): 1867-1876.

Williams, C.S., Mann, M. and DuBois, R.N. (1999) The role of cyclooxygenases in inflammation, cancer, and development. *Oncogene* 18(55), 7908-16.

Wu, P., Chen, W., Zhang,Y. and Lin, X. (2005) Electrochemical behavior and determination of curcumin. *Dianhuaxue* 11(3): 346-349.

Xia, Y., Makris, C., Su, B., Li, E., Yang, J., Nemerow, G.R. and Karin, M. (2000) MEK kinase 1 is critically required for c-Jun N-terminal kinase activation by proinflammatory stimuli and growth factor-induced cell migration. *Proc Natl Acad Sci U S A* 97(10), 5243-5248.

Xu, J., Fu, Y. and Chen, A. (2003) Activation of peroxisome proliferator-activated receptor-gamma contributes to the inhibitory effects of curcumin on rat hepatic stellate cell growth. *Am J Physiol Gastrointest Liver Physiol* 285(1), G20-30.

Xu, Y.X., Pindolia, K.R., Janakiraman, N., Chapman, R.A. and Gautam, S.C. (1997) Curcumin inhibits IL1 alpha and TNF-alpha induction of AP-1 and NF-kB DNA-binding activity in bone marrow stromal cells. *Hematopathol Mol Hematol* 11(1), 49-62.

Xu, Y.X., Pindolia, K.R., Janakiraman, N., Noth, C.J., Chapman, R.A. and Gautam, S.C. (1997) Curcumin, a compound with anti-inflammatory and anti-oxidant properties, down-regulates chemokine expression in bone marrow stromal cells. *Exp Hematol* 25(5), 413-422.

Yamamoto, H., Hanada, K., Kawasaki, K. and Nishijima, M. (1997) Inhibitory effect on curcumin on mammalian phospholipase D activity. *FEBS Lett* 417(2), 196-198.

Yang, F., Lim, G.P., Begum, A.N., Ubeda ,O.J., Simmons, M.R., Ambegaokar, S.S., Chen, P.P., Kayed, R., Glabe, C.G., Frautschy, S.A. and Cole, G.M. (2005) Curcumin inhibits formation of amyloid beta oligomers and fibrils, binds plaques, and reduces amyloid *in vivo*. *J Biol Chem* 280(7), 5892-5901.

Yasni, S., Imaizumi, K., Nakamura, M., Aimoto, J. and Sugano, M. (1993) Effects of Curcuma xanthorrhiza Roxb. and curcuminoids on the level of serum and liver lipids, serum apolipoprotein A-I and lipogenic enzymes in rats. *Food Chem Toxicol* 31(3), 213-218.

Yasni, S., Imaizumi, K., Sin, K., Sugano, M., Nonaka, G. and Sidik. (1994) Identification of an active principle in essential oils and hexane-soluble fractions of Curcuma xanthorrhiza Roxb. showing triglyceride-lowering action in rats. *Food Chem Toxicol* 32(3): 273-278.

Zhang, F., Altorki, N.K., Mestre, J.R. (1999) Subbaramaiah K and Dannenberg AJ, Curcumin inhibits cyclooxygenase-2 transcription in bile acid- and phorbol ester-treated human gastrointestinal epithelial cells. *Carcinogenesis* 20(3), 445-451.

Zheng, M., Ekmekcioglu, S., Walch, E.T., Tang, C.H. and Grimm, E.A. (2004) Inhibition of nuclear factor-kappaB and nitric oxide by curcumin induces G2/M cell cycle arrest and apoptosis in human melanoma cells. *Melanoma Res* 14(3), 165-171.

Zheng, O.S. and Chen, A. (2004) Activation of PPAR gamma is required for curcumin to induce apoptosis and to inhibit the expression of extra-cellular matrix genes in hepatic stellate cells *in vitro*. *Biochem J*, 384(1): 149-157.

11 Turmeric — Production, Marketing, and Economics

M.S. Madan

CONTENTS

11.1 INTRODUCTION

Turmeric is an important commercial crop grown for its aromatic rhizomes used for culinary and cosmetic purposes since antiquity. Turmeric is the basic ingredient in almost all curry powders and a major source of natural coloring for foodstuffs, pharmaceutical, and cosmetic applications. The color ingredient in turmeric, known as curcumin, is gaining wider use world over to replace the artificial yellow color. Turmeric is reported to be the native of South or Southeast Asia; its center of domestication is probably the Indian subcontinent (UNCTAD/GATT, 1982). Currently India is the largest producer, exporter, and consumer of this commodity in the world. Other producers in Asia include Bangladesh, Pakistan, Sri Lanka, Taiwan, China, Burma (Myanmar), and Indonesia. Turmeric is also produced in the Caribbean and Latin America: Jamaica, Haiti, Costa Rica, Peru, and Brazil. Although there are no authentic figures available to show the actual world production; as of now the world trade in turmeric is around 37,000 t valued at US$40,160 million. This quantity excludes the quantity consumed by the producing countries. India exports hardly 7 to 10% of the total production of 527,980 t (2002 to 2003). The major importing countries are the Japan, Iran, United Arab Emirates (UAE), Bangladesh, Singapore, Netherlands, and Sri Lanka accounting for nearly 80% of turmeric traded the world over.

Official statistics on area, production, and productivity are not available for all the producing countries, and the available data are conflicting. However, the trade-related figures are comparatively comprehensive. Multitude of processed spice mixtures with turmeric as a major component entering into the world market is not accounted separately. Despite certain limitations in the availability of data, this chapter makes use of available published data from the Spices Board, Directorate of Spices and Areca Nut Development, Govt. of India, International Trade Centre, Geneva, USDA (United States Developmental Agency), FAO (Food and Agricultural Organization) Circular series etc. The aim of this effort is to get some broad indications on the possible changes that have taken place in the crop economy during the last 2 decades and ascertain the prospects based on observed trends.

11.2 PRODUCTION

11.2.1 World Scenario

Although there is no source of data to show country-wise actual production of the crop, all available literature accept the fact that India produces more than 80% of the total world production. Table 11.1 provides the available data on the share of major producers in total world production in two different periods. As it can be seen from the table, India's share in total production compared to other traditional producing countries reflects its dominance in total world turmeric industry.

During the 1980s there were very few countries engaged in production of turmeric. Turmeric cultivation and processing is labor intensive; countries have hence neglected this crop and consequently they are not very active now in the world market. However, many other Asian, Latin American, and Caribbean countries have entered into turmeric production. Chinese turmeric is considered as good as the Indian turmeric and is accepted in the world market when compared to the Caribbean varieties. Production in Vietnam and Pakistan is also picking up, while traditionally producing countries like Peru have increased the production level. Production from these countries may pose a threat to the Indian monopoly in the world turmeric industry, though the combined export of all other countries producing turmeric presently amounts to less than 20% of the total world imports.

In this chapter, an effort is being made to analyze the salient features of major producing and consumption countries individually for which data are available.

TABLE 11.1
Production in Major Turmeric-Producing Countries
(1982–1983 and 2002–2003)

	1982–1983			2002–2003		
Country	Area × 1000 ha	Production × 1000 t	Yield kg/ha	Area × 1000 ha	Production × 1000 t	Yield kg/ha
India	86.40	167.50	1939	149.41	527.96	3534
Pakistan	4.23	31.30	7392	4.21	39.8	9400
Bangladesh	12.00	21.00	1750	NA	NA	NA

NA: Data not available.

11.2.1.1 Pakistan

Pakistan is traditionally a turmeric-producing country. Punjab, Sindh, and NWFP (North-West Frontier of Pakistan) are the major provinces contributing in turmeric production. Table 11.2 provides the state-wise area, production, and productivity of turmeric for the period from 1986–1987 to 2003–2004. As it can be seen from the table, Pakistan is producing around 38,400 t of turmeric from more than 4000 ha of area under the crop. An all time high production level of 45,000 t was achieved during 1994 to 1995. As it can be seen from the table, the increased production level over the period was achieved mainly because of the improved productivity level rather than area expansion. The highest annual compound growth rate for area (2.36%), production (2.98%), and productivity (1.25%) was achieved during that period.

TABLE 11.2
Turmeric Production in Pakistan (1986–1987 to 2003–2004)

	Punjab		Sindh		NWFP		All Pakistan		
Year	Area (ha)	Production (t)	Area (ha)	Production (t)	Area (ha)	Production (t)	Area (ha)	Production (t)	Yield (kg/Ha)
1986–1987	2,400	20,600	(a)	(b)	1,400	8,600	3,800	29,200	7,700
1987–1988	1,900	17,000	(a)	(b)	1,200	7,200	2,100	24,200	7,800
1988–1989	2,000	17,000	(a)	(b)	1,200	7,400	3,200	24,400	7,600
1989–1990	2,100	20,300	100	(b)	1,100	6,600	3,300	26,900	8,300
1990–1991	2,000	20,600	100	100	1,100	6,500	3,200	27,200	8,400
1991–1992	2,300	22,500	100	100	1,200	10,700	3,600	33,300	9,200
1992–1993	2,900	27,900	200	100	1,200	10,900	4,300	38,900	9,000
1993–1994	3,100	28,400	200	100	1,700	16,400	5,000	44,900	9,000
1994–1995	3,100	29,400	100	(b)	1,600	15,600	4,800	45,000	9,300
1995–1996	3,100	29,900	100	(b)	900	8,600	4,100	38,500	9,300
1996–1997	3,200	30,600	100	(b)	1,000	3,100	4,100	33,700	9,400
1997–1998	3,300	31,500	100	(b)	800	7,800	4,200	39,300	9,300
1998–1999	3,400	32,900	100	(b)	900	8,800	4,400	41,700	9,500
1999–2000	3,500	34,000	100	(b)	1,000	10,000	4,600	44,000	9,500
2000–2001	3,500	34,400	100	100	800	7,800	4,400	42,300	9,600
2001–2002	3,600	33,900	100	100	600	5,900	4,300	39,900	9,300
2002–2003	3,700	35,300	100	100	400	4,400	4,200	39,800	9,400
2003–2004	3,600	34,800	100	100	400	3,500	4,100	38,400	9,400
ACGR							2.36	2.98	1.25

11.2.1.2 India

In India, more than 200 districts spread throughout 20 states are cultivating turmeric in an area of 162,950 ha with a production of 552,300 t. Major turmeric-producing states are Andhra Pradesh, Orissa, Tamil Nadu, Kerala, West Bengal, Gujarat, and Karnataka. Figure 11.1 presents the spatial distribution of turmeric production and productivity in the country. As it can be seen from the Figure 11.1, turmeric production in the country is concentrated in south and central India. It is cultivated as a rainfed or irrigated crop, and in the latter case, mostly confined to Andhra Pradesh and Tamil Nadu.

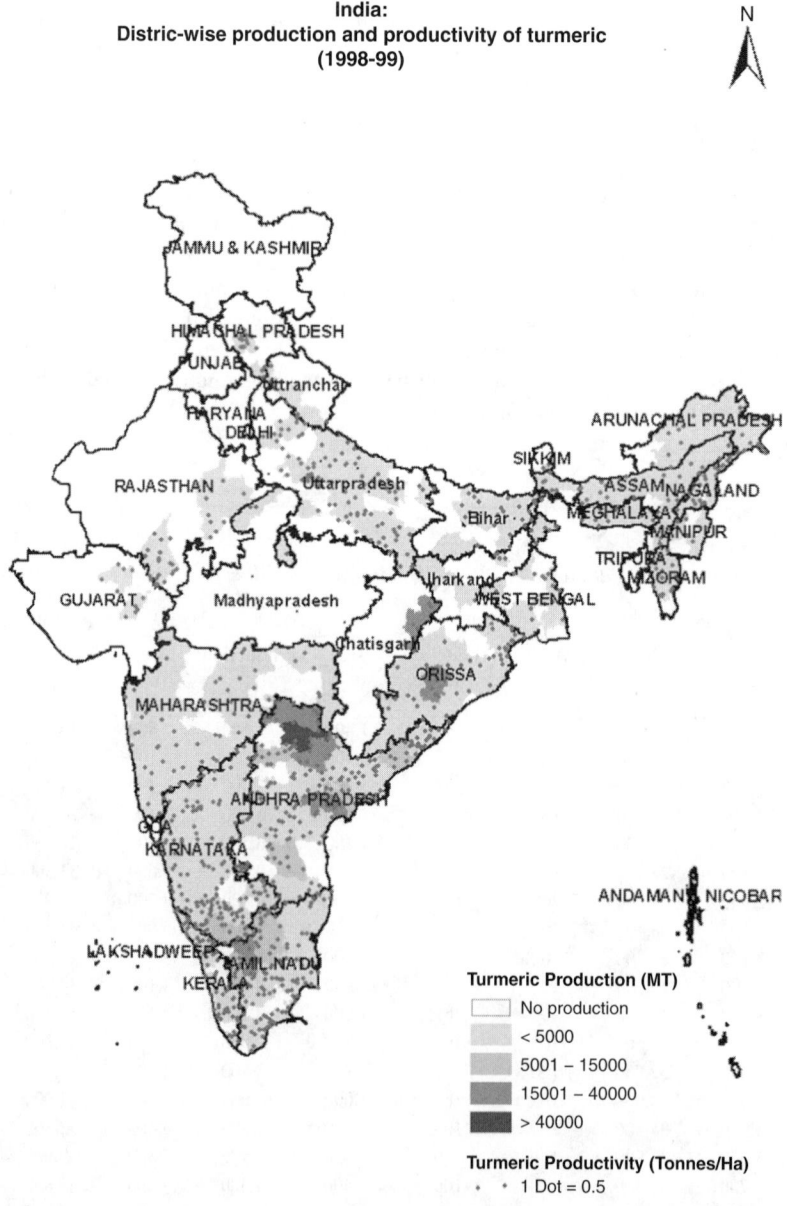

India:
Distric-wise production and productivity of turmeric
(1998-99)

N

Turmeric Production (MT)

☐ No production
< 5000
5001 – 15000
15001 – 40000
> 40000

Turmeric Productivity (Tonnes/Ha)
· · 1 Dot = 0.5

FIGURE 11.1 Spatial distribution of turmeric production and productivity in India (1998–1999).

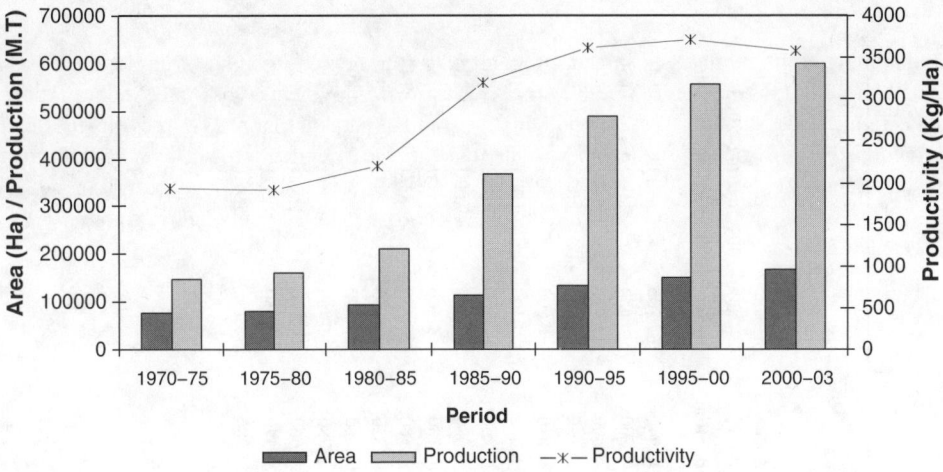

FIGURE 11.2 Area, production, and average yield per hectare of turmeric in India.

Andhra Pradesh accounts for the major share of both area (38.64%) and production (58.98%) of turmeric in India and has maintained the position during the last 3 decades followed by Tamil Nadu. These two States along with Orissa, the other traditional turmeric-growing State, account for nearly 70% of the area under the crop and nearly 90% of the total production in the country.

11.2.1.2.1 Area and Production

Increase in area and production, along with improvement in productivity of turmeric, increased steadily (Figure 11.2), though there was a year-to-year fluctuation. Similarly, the productivity also varied widely during the period. Highest annual production of 719,600 t was achieved during the 2000 to 2001 crop yr from an area of 187,430 ha with a record average productivity level of 3839 kg/ha.

11.2.1.2.2 Production in States

State-wise area, production, and yield per ha of turmeric are given in Table 11.3 with five yearly average. Although the crop cultivation has spread to many new areas, the traditional turmeric-producing States like Andhra Pradesh, Tamil Nadu, and Orissa still dominate the industry. Production scenario in the major producing States are analyzed here.

11.2.1.2.2.1 Andhra Pradesh

Turmeric in Andhra Pradesh is cultivated mainly in Nizamabad, Guntur, Cuddapah, and Karimnagar districts. In Guntur district, which had the largest area under the crop during 1981 to 1982, the cultivation was restricted to alluvial soils of Krishna river delta. Duggirala is the commercial cultivar of this region. It has bold rhizomes with thick fingers. The yield ranges from 12 to 15 t/ha (Rao, 1985). The seed rate is 10 to 12 quintals/ha. However, the reported average yield as per the official statistics is 4912 kg/ha, which has come down from more than 6000 kg/ha in the past. Mulching is not practiced in this area and for providing shade a few castor plants are grown in the field.

Nizamabad–Karimnagar belt, which accounts for 42.25% of the area under the crop in the State, produces 45.93% of the total production (Table 11.4). The main cultivar grown in this region is Armoor. The rhizomes are thinner than those of Duggirala and have a light orange brownish color. The quality of the produce is considered superior compared to Duggirala. It is generally grown under irrigation as a mixed crop with cowpea and maize.

In Cuddapa district, Tekurpet and Mydukur types of turmeric are cultivated. Both have bright orange color. The fingers are intermediate between Duggirala and Armoor in thickness and the

TABLE 11.3
State-Wise Area, Production, and Yield of Turmeric in India

States	1980–1981 to 1984–1985		1985–1986 to 1989–1990		1990–1991 to 1994–1995		1995–1996 to 1999–2000		2000–2001 to 2002–2003		
	Area[a]	Production[b]	Area[a]	Production[b]	Area[a]	Production[b]	Area[a]	Production[b]	Area[a]	Production[b]	Yield[c]
Andhra Pradesh	23,767	74,333	33,775	128,225	51,240	241,880	55,340	292,940	67,700	375,100	5,541
Arunachal Pradesh	133	333	275	550	320	960	380	1,200	550	2,000	3,636
Assam	8,367	5,133	7,550	5,500	8,580	5,540	8,060	5,600	11,700	8,100	692
Bihar	4,233	6,533	3,975	5,975	3,040	3,740	3,060	3,120	3,000	3,000	1,000
Karnataka	2,300	9,333	3,325	18,775	4,280	50,136	5,120	25,040	8,000	38,300	4,788
Kerala	3,100	5,600	3,200	6,225	3,080	5,780	3,780	8,660	4,100	9,000	2,195
Madhya Pradesh	400	367	1,725	2,075	680	1,820	740	760	350	350	1,000
Maharashtra	8,200	12,867	6,925	9,000	6,840	8,660	7,180	9,080	7,000	9,000	1,286
Meghalaya	1,167	1,700	1,250	1,750	1,340	1,800	1,420	7,100	1,500	8,600	5,733
Orissa	25,867	32,867	24,350	29,275	24,580	51,000	25,460	55,580	26,350	56,600	2,148
Rajasthan	133	400	100	325	160	500	200	820	150	500	3,333
Tamil Nadu	8,867	56,267	17,950	113,925	15,260	85,840	20,580	111,660	28,300	138450	4,892
Tripura	1,033	1,567	1,200	1,800	1,500	3,580	1,440	2,780	1,450	6,400	4,414
Utter Pradesh	633	733	375	575	920	1,520	820	1,660	1,100	1,900	1,727
West Bengal	2,850	7,350	5,000	10,375	10,680	20,020	12,520	21,320	13,350	22,200	1,663
India	94,167	214,933	112,825	345,325	133,460	490,540	150,230	557,086	175,190	635,950	3,630

[a] Area (in ha).
[b] Production (in Mt).
[c] Yield (kg/ha).

TABLE 11.4
District-Wise Area and Production of Turmeric in Andhra Pradesh

District	1981–1982			1993–1994 to 1997–1998			1998–1999 and 2003–2004			% Share in Total	
	Area[a]	Production[b]	Yield[c]	Area[a]	Production[b]	Yield[c]	Area[a]	Production[b]	Yield[c]	Area[a]	Production[b]
Adilabad	1,900	4,700	2,474	5,564	26,072	4,686	3,732	15,386	4,123	9.69	8.27
Chittoor	NA	NA	NA	73	320	4,378	37	160	4,378	0.10	0.09
Cuddapah	1,900	5,600	2,947	2,405	12,042	5,006	2,153	8,821	4,098	5.59	4.74
East Godavari	800	2,000	2,500	527	3,112	5,906	664	2,556	3,853	1.72	1.37
Guntur	3,300	10,500	3,182	4,939	29,967	6,068	4,119	20,234	4,912	10.70	10.88
Karimnagar	5,000	10,200	2,040	12,659	84,262	6,656	8,830	47,231	5,349	22.94	25.40
Khammam	NA	NA	NA	392	2,352	6,006	196	1,176	6,006	0.51	0.63
Krishna	900	2,400	2,667	1,760	9,333	5,303	1,330	5,866	4,411	3.45	3.15
Kurnool	NA	NA	NA	788	4,520	5,735	394	2,260	5,735	1.02	1.22
Medak	NA	NA	NA	1,240	7,239	5,839	620	3,620	5,839	1.61	1.95
Nalgoda	NA	NA	NA	80	516	6,424	40	258	6,424	0.10	0.14
Nellore	NA	NA	NA	172	1,029	5,991	86	515	5,991	0.22	0.28
Nizamabad	5,300	13,000	2,453	9,570	63,341	6,619	7,435	38,170	5,134	19.31	20.53
Prakasam	NA	NA	NA	302	1,796	5,948	151	898	5,948	0.39	0.48
Rangareddi	1,400	4,200	3,000	3,357	19,399	5,779	2,379	11,800	4,961	6.18	6.35
Srikakulam	NA	NA	NA	161	909	5,663	80	455	5,663	0.21	0.24
Vijianagaram	NA	NA	NA	23	132	5,823	11	66	5,823	0.03	0.04
Vishakhapatanam	600	1,200	2,000	1,403	7,529	5,367	1,002	4,365	4,358	2.60	2.35
Warangal	NA	NA	NA	7,367	31,091	4,221	3,683	15,546	4,221	9.57	8.36
West Godaveri	800	2,300	2,875	688	4,154	6,034	744	3,227	4,336	1.93	1.74
AP State	23,500	60,100	2,557	53,500	311,780	5,828	38,500	185,940	4,830	100	100

[a] Area (in ha).
[b] Production (in t).
[c] Yield (kg/ha).

TABLE 11.5
District-Wise Turmeric Production in Tamil Nadu
(2001–2002)

District	Area (ha)	Production (Mt)	Yield (t/ha)
Villupuram	1,361	4,263	3.13
Vellore	625	3,127	5
Thiruvannamalai	234	1,171	5
Salem	3,443	10,982	3.19
Namakkal	2,249	20,185	8.98
Dharmapuri	2,174	6,355	2.92
Coimbatore	2,467	11,774	4.77
Erode	10,045	55,197	5.49
Thiruchirapalli	275	1,376	5
Karur	344	1,721	5
Perambalur	223	1,116	5
Others	198	990	5
State total	23,638	118,257	5

Source: Directorate of Horticulture, Government of Tamil Nadu, Chennai, India.

gross yield is comparable to Duggirala. In the Rayalaseema region, it is grown mostly as a pure crop, and in the East Godawari a short duration cultivar Kasturi is grown.

11.2.1.2.2.2   Tamil Nadu

Tamil Nadu is the second largest producer of turmeric in the country. Although the turmeric grown in Erode region is preferred for grinding, cross-contamination of different varieties and the improper post harvest practices followed has led to less acceptability of Erode turmeric in recent years. Currently, most of the turmeric produce is used for domestic consumption and the unit value realization by the farmers is, therefore, comparatively lower. The quality of Salem turmeric is comparatively better and has acceptance in the international market for grinding and blending purposes. Turmeric grown in these two districts is called Madras turmeric in the international market. Table 11.5 provides district-wise area, production, and productivity for turmeric in the State.

During 1980 to 1981, Tamil Nadu was producing 18,960 t from 9470 ha area under cultivation. Yield was mere 2.0 t/ha. Coimbatore, Erode, Salem, and Tiruchi were the districts producing turmeric during the period. By 1992 to 1993, the crop has spread to almost all the districts in the State. Change in production during the period was 470%, while the area expansion achieved was 87.19% and the productivity increase was more than 200%. The trend continued in subsequent years as well, though the level of increase was not as dramatic as witnessed in the previous decade. Newly emerged districts such as Namakkal have achieved the highest productivity of 8.98 t/ha. The State as a whole has achieved a positive change (56.38%) in turmeric production between the period from 1980–1981 to 1989–1990 and 1990–1991 to 2001–2002. This achievement was mainly due to the change in productivity (483.72%) and area expansion (89.45%) owing to introduction of many high-yielding varieties during this period.

Salem, Erode, Dharmapuri, and Coimbatore are the major marketing centers for turmeric in Tamil Nadu. An Agriculture Export Zone (AEZ) for turmeric in the districts of Salem and Erode is being set up by Government of Tamil Nadu based on the recommendation of Spices Board, which had identified these two districts as a potential area for setting up an AEZ for turmeric.

11.2.1.2.2.3   Karnataka

Turmeric is grown in isolated pockets in a variety of soils in the districts of Dakshin Kannada, Belgaum, Gulbarga, Mysore, Chamrajnagar, and Bijapur. It is being cultivated both under rainfed

TABLE 11.6
District-Wise Area and Production of Turmeric in Karnataka State (2001–2002)

District	Area (ha)	Production (Mt)	Yield (kg/ha)	Value in Lakhs*
Bangalore (urban)	158	632	4,000	51
Bangalore (rural)	102	508	4,980.4	12
Mysore	2,217	18,058	8,145.2	1,505
Chamarajanagar	2,537	11,222	4,423.3	1,014
Kodagu	150	1,200	8,000	24
Udupi	153	918	6,000	349
Hassan	122	806	6,606.6	32
Belgaum	2,594	14,618	5,635.3	1,841
Bijapur	174	1,392	8,000	22
Bagalkot	384	7,560	19,687.5	619
Haveri	105	695	6,619	104
Gulbarga	749	2,971	3,966.6	440
Bidar	222	1,201	5,409.9	147
Others[a]	522	3,908	100,522.8	719
State Total	10,189	65,689	64,47.1	6,879

[a] Districts with <100 t production.

* One Lakh = 100,000 = 00 (0.1 million).

Source: Directorate of Horticulture, Govt. of Karnataka, Bangalore, India.

TABLE 11.7
Trend in Production of Turmeric in Karnataka

Year	Area (ha)	Production (Mt)	Yield (t/ha)
1992–1993	3,400	48,800	14.35
1993–1994	4,600	74,700	16.24
1994–1995	5,600	20,880	3.73
1995–1996	4,700	20,600	4.38
1996–1997	4,800	26,200	5.46
1997–1998	4,300	24,100	5.60
1998–1999	5,000	25,400	5.08
1999–2000	4,300	24,100	5.61
2000–2001	5,000	25,400	5.08
2001–2002	6,800	28,900	4.25
2002–2003	9,300	41,000	4.41
2003–2004	6,700	35,600	5.31

and irrigated conditions both as a pure and inter/mixed crop in coconut and arecanut gardens (Vasanthakumar, 1985). Table 11.6 presents the distirict-wise area, production, and productivity in the State during 2001 to 2002.

As it can be seen from the Table 11.7, area under the crop in Karnataka State during 1993 to 1994 was 4600 ha producing 74,700 t. The productivity level achieved was 16.24 t/ha, which has come down to 5.46 t/ha and has been stagnant over the years. The area under the crop has been fluctuating over the years due to variations in price of turmeric and other competing crops like banana, sugarcane, and paddy.

TABLE 11.8

Labor Requirement in Cultivation of Turmeric (Per Acre)

Operation	Labor Requirement (Numbers)	Type of Labor
Digging	38	Men
Peg marking and bed preparation	72	Men
FYM application	1 + 4	Men/women
Pit making	20	Women
Fertilizer application	5	Women
Seeding	4	Women
Covering	4	Women
Mulching	15	Women
Stick removal and weeding	30	Women
Fertilizer application and raking	10	Women
Mulching	22	Women
Earthing up	19	Men
Plant protection	3 + 1	Men/women
Steps 9 to 12 repeated	19 + 62	Men/women
Harvesting	27	Men
Total	352	Men/women

Source: IISR, 2002.

11.2.1.2.3 Production Economics

Time series data indicate that the coefficient of variation of farm price was 72.4% higher than that of the production (55%) over a period indicating the violent fluctuation in price of turmeric in the country. This fluctuating prospect had greater impact on production economics of the farming community. The problem can be better understood from the fact that, farmers buy seed rhizomes for prices as high as Rs. 25/kg at times, but their harvested crop could fetch them only less than one fifth of this price. In order to avoid the price-related risk, the farmer cultivates turmeric as mixed or inter crop under various cropping systems, though pure crop is not uncommon. In the major turmeric-growing state of Kerala, nearly one-fourth of the area under the crop is in the uplands as pure crop, whereas the major area (45%) is in the garden land category and the rest is under mixed cropping system.

The estimated per kg production cost in Kerala for turmeric during 2002 to 2003 was around Rs. 37/kg and the benefit:cost ratio was less than 1.5. Production cost was comparatively more than that in other States owing to higher labor cost and other added costs toward chemical fertilizers accounting nearly 72% of the total cost. Turmeric is a high labor and input-demanding crop. According to Vinning (1990) turmeric cultivation requires 309 man d per ha and is a processing-intensive crop. As regards labor requirement, the actual enumeration done to estimate operation-wise labor requirement indicates that turmeric requires nearly 337 men and women d/ha for the entire period of cultivation excluding marketing (Table 11.8).

Table 11.9 gives the estimated input requirement for turmeric cultivation in the States of Karnataka and Kerala following the recommended package of practice, whereas Table 11.10 provides the input–output budget for turmeric for the crop year 2002 to 2003.

The estimated standard cost–return budget for turmeric in India (Table 11.10) reflects the fact that more than 65% of the total cost incurred is toward labor and seed material purchase. It can be further observed that the turmeric farmer gets a marginal benefit, which can be wiped off easily due to unexpected loss in production and slight fall in price.

A study conducted to estimate the economics of turmeric cultivation in Chamrajnagar district of Karnataka by Lokesh and Chandrakanth (2004) brought out the fact that the benefit:cost ratio is more favorable for improved varieties (1.29), when compared to local variety (1.06). This higher

TABLE 11.9
Input Utilization in Turmeric Production (Per Hectare)

Particulars	Quantity	Value (Rs)	Share in Total (%)
Seed rhizome[a] (kg) at Rs. 20/kg[a]	2,000	40,000	32.97
Manure (t). at Rs. 800/t	30	24,000	19.78
Neem cake/ground nut cakes (t)	1	6,000	4.95
Chemical fertilizer (kg)			
N	60	626	0.52
P	50	978	0.81
K	120	888	0.73
Green leaves for mulching (t)	20	6,000	4.75
Mancozeb for seed treatment (kg)	2.4	912	0.75
Labor	364	29,120	24.00
Tractor units	18	1,440	1.19
Irrigation (h)	224	3,360	2.77
Plant protection (Rs.)		8,000	6.59
Total (Rs.)		12,1324	100

[a] The present price of seed rhizome is Rs. 6/kg.

Source: IISR, 2002.

TABLE 11.10
Cost of Cultivation of Turmeric (Rs/ha)

Cost Items	Cost (Rs/ha)	Share in Total (%)
Labor	29,120	24.00
Tractor units	1,440	1.19
Seed rhizome	40,000	32.97
Manure	24,000	19.78
Neem cake	6,000	4.95
Chemical fertilizer		
N	626	0.52
P	978	0.81
K	888	0.73
Mulching	6,000	4.95
Plant protection (Rs.)	8,000	6.59
Seed treatment	912	0.75
Irrigation	3,360	2.77
Total cost	121,324	100
Production (yield, kgs)	30,000	
Value of production:		
Seed rhizome 25% of the production at Rs 20/kg	150,000	
Bulk sale 75% of the production at Rs 4/kg	90,000	
Gross returns (Rs)	240,000	
Net returns (Rs)	118,676	
BC ratio	1.98	

Source: IISR, 2002.

profitability is mainly due to higher productivity (3100 kg/ha) achieved in improved varieties when compared to the yield level of 2500 kg/ha in local varieties. In Belgaum district of the State, Kerutagi et al., (2000) has estimated the benefit:cost ratio of 2.76 for both Salem and Cuddpah varieties of turmeric.

Economics worked out by Kandiannan (2002) in an onfarm experiment at Bhavanisagar (BSR) in Tamil Nadu, indicate that the gross, net returns/ha, and benefit:cost ratio was Rs. 132,500, Rs. 74,000, and 2.26, respectively, for a popular variety of BSR2. According to the study, closer spacing and higher input of fertilizer in the form of N resulted in higher gross and net returns and benefit:cost ratio up to 2.69.

Sikka et al. (1985) have estimated Rs. 2.26/kg and 2.70/kg as production cost in Nizamabad and Erode, respectively, during 1985. The prevailed market price during the same period was Rs. 2.47/kg and 2.81/kg in Nizamabad and Erode markets, respectively. Thus, in the past the farmer was getting only a nominal profit and if the price falls further, he might be in loss. During the same year a survey conducted by Murthy and Naidu (1989) to appraise the profitability of turmeric farming in Guntur district of Andhra Pradesh, indicated that the break-even output was 43.39, 40.56, and 41.71 quintals for marginal, medium, and large farms, respectively.

11.2.1.2.4 Trends in Area, Production and Productivity

The time series data on area, production, and productivity of turmeric along with growth index worked out for the period from 1970–1971 to 2002–2003 are presented in Table 11.11. Looking to the near constant production between the years 1984–1985 to 1987–1988, year 1985 to 1986 was taken as the base year to estimate the growth index.

11.2.1.2.4.1 Production

Indian production has been showing a steadily increasing trend from 150,600 t in 1970 to 1971 to 5,274,600 t during 2002 to 2003. An increase of nearly 250% in production is due to the combined improvement in both area (49.3%) and productivity (97.25%). Andhra Pradesh and Tamil Nadu together accounted for more than 80% of total production in the country. In a region-wise grouping, the southern region comprising Tamil Nadu, Kerala, Karnataka, and Andhra Pradesh accounted for 88.2% production with 61.8% area during 2000–2001 to 2002–2003. Distribution of turmeric-producing area as a percent of total cropped area and production (district-wise) worked out clearly shows the concentration of turmeric cultivation in states of Andhra Pradesh, Orissa, and Tamil Nadu, indicating their dominance in turmeric production in the country. Data on turmeric production along with trend line for the period 1970–1971 to 2002–2003 is presented in Figure 11.3. A perusal of the figure indicates significant increase in production over the years.

State-wise average area, production, and productivity for the periods 1980–1981 to 1984–1985, 1985–1986 to 1989–1990, 1990–1991, 1994–1995, 1995–1996 to 1999–2000, and 2000–2001 to 2002–2003 are given in Table 11.12. As it can be seen from the table, against the national average yield of around 2.28t/ha achieved during 1980–1981 to 1984–1985, States like Andhra Pradesh, Karnataka, and Tamil Nadu have been consistently recording a higher levels of yield. Tamil Nadu achieved highest productivity level of 6.35t/ha during the period and has maintained the productivity along with increase in area during the subsequent periods. Andhra Pradesh has maintained its supremacy in turmeric production in all the periods except during 2000–2001 to 2002–2003.

11.2.1.2.4.2 Productivity

Productivity of turmeric in the country has increased over the years from 1.87 t/ha during 1970 to 1971 to 3.53 t/ha during 2002 to 2003. Productivity registered during 2002 to 2003 is nearly two times the productivity of 1970 to 1971 level. The yield level has shown a steady improvement until the middle of 1980s except for occasional fluctuations toward the lower side during 1976 to 1977 and 1977 to 1978. It seems that the yield increase (79.53%) during this period contributed much to the increase in production (143.76%). Thus, the productivity level improved slowly and steadily from 1980 to 1981 onward and reached an average of 3.534t/ha during 2002 to 2003 with an

TABLE 11.11
Area, Production, and Average Yield Per Hectare of Turmeric
in India

Year	Area (ha)	Growth Index	Production (t)	Growth Index	Yield (kg/ha)	Growth Index
1970–1971	80,500	73.65	150,600	41.02	1,871	55.70
1971–1972	75,900	69.44	178,400	48.60	2,350	69.98
1972–1973	69,000	63.13	121,100	32.99	1,755	52.26
1973–1974	74,300	67.98	133,900	36.48	1,802	53.66
1974–1975	78,100	71.45	145,700	39.69	1,866	55.54
1975–1976	71,800	65.69	135,200	36.83	1,883	56.06
1976–1977	66,800	61.12	109,700	29.88	1,642	48.90
1977–1978	75,700	69.26	126,300	34.40	1,668	49.68
1978–1979	89,700	82.07	190,400	51.87	2,123	63.20
1979–1980	105,000	96.07	235,400	64.12	2,242	66.75
1980–1981	101,500	92.86	216,900	59.08	2,137	63.63
1981–1982	90,700	82.98	191,300	52.11	2,109	62.80
1982–1983	85,900	78.59	173,100	47.15	2,015	60.00
1983–1984	94,300	86.28	212,500	57.89	2,253	67.09
1984–1985	102,300	93.60	259,200	70.61	2,534	75.44
1985–1986	109,300	100.00	367,100	100.00	3,359	100.00
1986–1987	109,900	100.55	319,900	87.14	2,911	86.67
1987–1988	108,700	99.45	303,900	82.78	2,796	83.24
1988–1989	123,400	112.90	390,400	106.35	3,164	94.20
1989–1990	124,000	113.45	459,500	125.17	3,706	110.33
1990–1991	119,000	108.87	342,400	93.27	2,877	85.67
1991–1992	120,300	110.06	373,200	101.66	3,102	92.37
1992–1993	130,200	119.12	407,700	111.06	3,131	93.23
1993–1994	148,400	135.77	707,400	192.70	4,767	141.93
1994–1995	149,400	136.69	622,000	169.44	4,163	123.96
1995–1996	139,300	127.45	462,900	126.10	3,323	98.94
1996–1997	135,200	123.70	528,900	144.08	3,912	116.47
1997–1998	139,700	127.81	549,200	149.61	3,931	117.05
1998–1999	160,700	147.03	598,260	162.97	3,723	110.84
1999–2000	176,250	161.25	646,170	176.02	3,666	109.16
2000–2001	187,430	171.48	719,600	196.02	3,839	114.31
2001–2002	162,950	149.09	552,300	150.45	3,389	100.92
2002–2003	149,410	136.70	527,960	143.82	3,534	105.21

Source: Anonymous, 1985; Directorate of Arecanut and Spices Development (DASD),
Government of India, 1970, 1983, 1985, 2004.

occasional fluctuation in both the sides. The estimated growth index for the year 2002 to 2003 in production is only 105.21 over the base year (1980 to 1981), while the highest of 141.93% was achieved during 1993 to 1994.

11.2.1.2.4.3   Growth Estimates

In order to get the long-term trends in area, production, and productivity in major producing states in the country, semilogarithmic growth equations were used. The resultant estimates presented in Table 11.12 indicate that the overall trend in area under turmeric registered an average annual growth rate of 2.91% for the period from 1980–1981 to 2002–2003. Growth in production was at

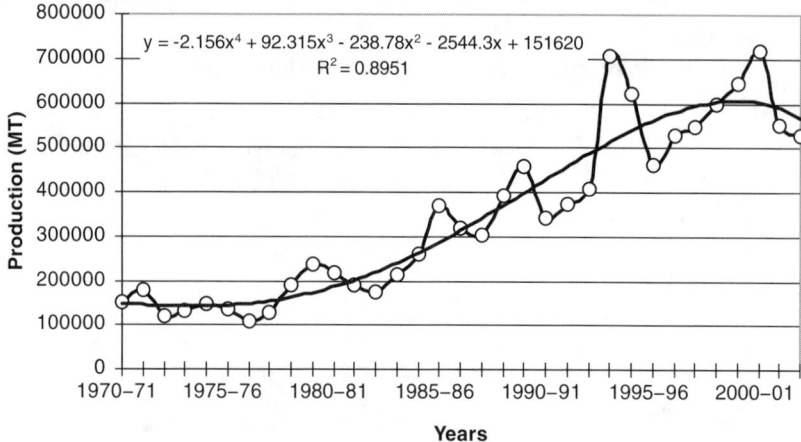

FIGURE 11.3 Trend in turmeric production in India (1970–1971 to 2002–2003).

the rate of 5.36% during the same period indicating a slight improvement in productivity, which was around 2.38% for the period.

As regards the period-wise performance, all the turmeric-producing States except Maharashtra have recorded better growth rate in all the three parameters during period I followed by period II. The State-wise results were reflected in the estimated national level growth rates for these periods. However, the period III recorded a negative growth rate in terms of area and production, while there was a positive-growth rate of 2.38% in productivity at National level.

11.2.1.2.4.4   Impact of Change in Area and Productivity on Production

To ascertain the impact of area expansion and productivity on production, period-wise data for major producing States were analyzed using a simple technique that was followed by Aida et al., (1988). Results presented in Table 11.13 show that, there is a positive sign in all the three parameters in all the three periods indicating the steady improvement in production due to both area expansion and productivity increase for the country as a whole. The detailed component analysis revealed that the change in productivity had more positive role in all the three periods, while in the last period improved productivity played major role in production increase.

Orissa State, which has the largest area under the crop, next to Andhra Pradesh, had performed better in periods II and III. Change in production during period II (38.10%) and period III (25.28%) was contributed by improvement in both area and productivity improvement. However, contribution by productivity increase of 11.21% and 54.14% during the periods II and III, respectively, was more pronouncing than the contribution by area expansion.

The same type of analysis that was carried out for other major producing States is given in Table 11.13. In Andhra Pradesh, the situation is almost similar to the national scenario. In the case of Tamil Nadu the other major producing State, increase in production during period III was largely due to area expansion. The role of productivity is negative. Changes in production, area, and yield were negative during period II in Kerala, and during period I and II in the case of Maharashtra.

11.2.1.2.4.5   Production Constraints

Status paper prepared by Spices Board (1990) on the crop highlights the fact that mostly small and marginal farmers cultivate turmeric. Nonavailability of good planting materials in sufficient quantities, lack of scientific know-how among farmers, incidence of diseases, failure of monsoon rains in the rainfed areas, nonavailability of mulch materials, unscientific postharvest operations and price/supply fluctuations are the major constraints identified on the production front. Mixing of

TABLE 11.12
Annual Average Compound Growth Rate (%) in Turmeric Area, Production, and Yield

State	Period	Area	Production	Yield
India	I	4.05	9.82	5.55
	II	3.88	5.00	1.08
	III	−4.24	−2.23	2.11
	Overall	2.91	5.36	2.38
Tamil Nadu	I	6.63	12.11	5.13
	II	11.32	12.85	1.38
	III	−27.60	−36.22	−11.91
	Overall	3.67	3.60	−0.07
Orissa	I	0.78	3.33	2.53
	II	1.24	1.80	0.55
	III	−3.36	−3.92	−0.58
	Overall	0.26	4.83	4.56
Karnataka	I	7.39	13.40	5.59
	II	5.09	−1.60	−6.37
	III	−18.46	−14.29	5.12
	Overall	5.57	6.30	0.69
Kerala	I	0.16	1.33	1.16
	II	4.03	6.37	2.25
	III	−12.81	−12.38	0.49
	Overall	1.02	2.18	1.15
Maharashtra	I	−2.72	−9.25	−6.71
	II	0.59	0.59	−0.01
	III	−1.90	−2.69	−0.80
	Overall	−0.65	−1.63	−0.99
Assam	I	0.64	1.18	0.54
	II	3.76	5.37	1.56
	III	1.87	1.79	−0.08
	Overall	1.77	2.54	0.75
West Bengal	I	6.16	9.83	3.45
	II	3.42	1.54	−5.47
	III	−8.10	−2.48	6.12
	Overall	5.52	6.70	1.54
Andhra Pradesh	I	9.33	14.02	4.29
	II	2.67	8.53	5.7
	III	−12.33	−13.13	−0.92
	Overall	5.42	8.60	3.01

Note: 1980–1981 to 1989–1990 — Period I; 1990–1991 to 1999–1900 — Period II; 2000–2001 to 2002–2003 — Period III; 1980–1981 to 2002–2003 — Overall.

TABLE 11.13
Change in Turmeric Production, Area, and the Relative Contribution of Changes in Area and Yield for Selected Periods

States	1980–1981/1984–1985 to 1985–1986/1989–1990	1985–1986/1989–1990 to 1990–1991/1994–1995	1990–1991/1994–1995 to 1995–1996/1999–1900
	All India: Change in:		
Production	71.41	43.20	10.06
Area	22.72	17.66	11.59
Productivity	38.66	20.58	0.45
	Change in Production Due to Change in:		
Area	39.49	45.30	114.44
Productivity	65.12	86.91	2430.71
	Andhra Pradesh: Change in:		
Production	83.73	113.38	11.90
Area	60.41	42.03	13.62
Productivity	17.25	44.86	0.57
	Change in Production Due to Change in:		
Area	77.69	46.30	113.58
Productivity	296.93	94.69	2262.69
	Orissa: Change in:		
Production	9.61	38.10	25.28
Area	−5.19	3.26	8.29
Productivity	16.60	33.15	15.84
	Change in Production Due to Change in:		
Area	−58.05	9.94	35.35
Productivity	−34.68	11.21	54.19
	Tamil Nadu: Change in:		
Production	57.33	10.36	22.49
Area	37.03	6.47	29.33
Productivity	17.05	0.66	−1.22
	Change in Production Due to Change in:		
Area	69.52	63.54	126.79
Productivity	200.07	949.01	−2090.54
	Kerala: Change in:		
Production	19.71	−7.23	47.62
Area	10.34	−2.34	20.00
Productivity	8.55	−4.03	21.63
	Change in Production Due to Change in:		
Area	54.71	31.61	46.81
Productivity	120.03	57.61	93.11
	Maharashtra: Change in:		
Production	−24.12	−10.00	2.85
Area	−10.61	−7.12	4.38
Productivity	−14.82	−3.45	−1.41

Continued

TABLE 11.13 *(Continued)*
Change in Turmeric Production, Area, and the Relative Contribution of Changes in Area and Yield for Selected Periods

States	1980–1981/1984–1985 to 1985–1986/1989–1990	1985–1986/1989–1990 to 1990–1991/1994–1995	1990–1991/1994–1995 to 1995–1996/1999–1900
	Change in Production Due to Change in:		
Area	40.61	70.09	152.58
Productivity	69.90	21.36	−302.78
	Assam: Change in:		
Production	5.88	5.09	24.67
Area	−12.20	18.31	17.19
Productivity	36.36	−21.74	6.80
	Change in Production Due to Change in:		
Area	−227.68	338.42	71.95
Productivity	−41.92	−68.57	241.26

Note: Based on time-series data from Spices Board, India, Cochin, Kerala, India.

varieties and mixed output is another problem faced by the market. Major production constraints in turmeric cultivation as given by many authors including the Spices Board of India are as follows:

- Low productivity (3630 kg/ha) against an achieved average productivity of more than 40,000 kg/ha elsewhere within the country and more than 9500 kg/ha in Pakistan
- Prevalence of low-yielding traditional varieties in the absence of inadequate supply of quality planting materials of improved varieties
- Under rainfed cultivation, failure of rains and increased labor costs are some of the factors responsible for higher cost of cultivation
- Nonadoption of improved varieties and application of low level of fertilizer and micronutrients are responsible reasons for low productivity
- Nonadoption of recommended postharvest processing by farmers and poor marketing facilities especially in northeastern states of the country result in poor returns to farming community
- Lack of remunerative prices in subsequent years leads to less enthusiasm to take up turmeric cultivation or neglect of the crop
- Keeping the above facts in mind, there is an urgent need to develop cropping systems with turmeric as a component. Although it is being cultivated as an intercrop in coconut and arecanut plantations, we are yet to develop ideal systems with attention on cost–benefit factor, soil disturbance, shade and root effect, and other factors.

11.3 MARKETING

11.3.1 PRODUCTS OF COMMERCE

Three primary products of turmeric traded in the world market are dried rhizome, turmeric powder, oils, and oleoresin. Apart from this, turmeric powder becomes an essential and major component of the various types of curry powder mixes traded. Fresh turmeric is of less importance; dried rhizomes are the major form in which it is internationally traded. Dried turmeric is used directly as a spice and for the preparation of its extractives — turmeric oleoresin and turmeric oil.

Indian turmeric (especially the Alleppey Finger Turmeric) with its high quality is well known in international trade. Chinese turmeric is also considered as good as Indian turmeric, unlike the West Indian turmeric. Chinese turmeric is expected to pose a threat to Indian turmeric in the future (Shukla, 1985).

11.3.1.1 Dried Rhizome

Turmeric is mostly traded as whole rhizome, which is then used in production of other value-added products like turmeric powder, oils, and oleoresin. The dried rhizomes are traded in three categories: fingers, bulbs, and splits. Fingers are the secondary branches from the mother rhizome. Bulbs and splits are the bulbs cut into halves and quarters before curing. The fingers command a higher price than the bulbs and splits.

11.3.1.2 Turmeric Powder

Ground turmeric is mostly used in the retail market and by the food processors. Rhizomes are ground to approximately 60- to 80-mesh particle size. Turmeric powder is a major ingredient in curry powders and pastes. In the food industry, it is mostly used as coloring and flavoring agent, in chicken bouillon, soups, sauces, gravies, and dry seasonings, and in a variety of other products. Recently the powder has also been used as a colorant in cereals.

11.3.1.3 Oils and Oleoresin of Turmeric

Turmeric contains a volatile oil, which gives the characteristic flavor of the spice, but which is not attractively priced for the merit of its commercial distillation. The principal application of turmeric oil is in some confectionary products and aerated waters.

Turmeric oleoresin obtained by solvent extraction of the ground spice, contains coloring matter, volatile oil, fatty oil, and bitter principles. Manufacturers offer various turmeric oleoresins with curcumin content varying from 3.5 to 4.2% or more.

11.3.2 FACTORS OF DEMAND/EXPORT

Major factor that contributes to export demand/potential of a commodity is its quality. In turmeric, the quality parameters are its yellow color and characteristic flavor. Yellow color is due to the orange–yellow pigment curcumin and its analogues. Curcumin content of turmeric varies from 2.5 to 8%, depending on the variety. Price also varies depending on the curcumin content of the dried turmeric rhizome as it can be seen from Table 11.14; Alleppey turmeric obtained the best price in the New York market when compared to the other traded varieties.

The Multi-Commodity Exchange of India Ltd. has defined the quality requirement of turmeric as follows: quality of turmeric is assessed on the basis of several factors, which include the pigment (curcumin) content, the organoleptic character, the general appearance, size, and physical form of the rhizome (Anonymous, 2003). The relative importance of these various quality attributes is dependent upon the intended end use of the product. In the U.K. and the U.S., most consumers prefer turmeric in the form of polished fingers for spice applications (Anon, 1996).

The chief factors of good quality in finger turmeric are a high content of pigment, giving a deep yellow color, and low bitter-principle content. The rhizomes should also be rough, hard, and brittle, with numerous encircling, ridge-like annulations. Length and thickness of the rhizomes and internal color are also important characteristics in the differentiation of cultivars. When manually fractured, the break should be clean — not splintery, or fibrous — and the broken surfaces waxy or horny and resinous in appearance. The endodermis, separating the cortex from the central cylinder (stele), should be clearly visible. When used as spice or condiment, the aroma and flavor imparted

TABLE 11.14
Annual Average Price of Turmeric at New York Market (1995–2005)

Year	Indian Alleppey Type (Curcumin%)				Other Types	
	Curcumin (5.0)	Curcumin (5.5)	Curcumin (5.5–6.0)	Curcumin (6.0)	Madras	Peruvian
1995	0.62	0.45	0.65	NA	0.55	0.44
1996	0.64	0.97	0.67	NA	0.55	0.44
1997	NA	0.78	NA	0.80	0.55	0.44
1998	NA	0.97	NA	0.99	0.56	0.44
1999	0.64	0.89	NA	0.99	0.62	NA
2000	0.71	0.74	NA	0.95	0.60	NA
2001	0.61	0.63	NA	0.98	0.56	NA
2002	0.65	0.67	NA	NA	0.55	NA
2003	0.75	0.79	NA	NA	0.56	NA
2004	0.75	0.87	NA	NA	0.58	NA
2005	0.87	0.90	NA	NA	NA	NA

Source: FAS, USDA, Circular Series FTROP 1–99, March 1999, U.S.

by the volatile oil are important. The aroma should have a musky, pepper-like character and the flavor should be slightly aromatic and somewhat bitter (Pruthi, 1989).

When turmeric is intended for use specifically as a coloring agent, either in the powdered form or as an oleoresin extract, the general appearance and physical form of the whole rhizome is less important. In this case, high curcumin content is essential and low volatile oil content is desirable. Bulbs, splits, and old rhizomes are often suitable for this purpose.

The principal quality determinants of cured turmeric, i.e., pigment, volatile-oil, and bitter principle contents, are mainly governed by the intrinsic characteristics of the cultivars grown and can be improved by appropriate selection of planting material. The quality is also influenced by the state of maturity of the rhizome at harvest and by the care taken in its handling, curing, and grading (Jaiswal, 1980).

11.3.3 INDIAN DRIED TURMERIC

Indian turmeric entering the international market is the high curcumin Alleppey turmeric and the other types known as Madras turmeric, Rajpuri turmeric, Sangli, Duggirala, Nizamabad, etc. named after the major production areas in the country. The bulk of Indian exports are of the Madras turmeric type.

11.3.3.1 Alleppey Turmeric

Alleppey Turmeric is produced in Kerala, particularly in Thoduphuzha and Muvattupuzha areas (central Kerala), and marketed in Alleppey and exported through the Cochin port. This turmeric is deep yellow to orange-yellow in color and has higher curcumin content — up to 6.5%. Almost the entire production of Alleppey turmeric is exported to the U.S. market in the unpolished form and used largely as food colorant.

11.3.3.2 Madras Turmeric

Madras Turmeric is produced in predominantly turmeric-growing districts of Salem, Erode, and Coimbatore districts in Tamil Nadu and exported through Madras and Tuticorin ports. The rhizomes are mustard yellow in color and have curcumin content of about 3.5%. Madras turmeric is the most common type used by curry powder/masala manufacturers in U.K. and Europe (Kannan, 1995).

Apart from these two types of dried turmeric from India, the third type that is internationally traded is the West Indian turmeric. This type is exported from the Caribbean, and the Central and South American countries. The rhizomes are dull yellowish-brown in color, mostly small, and of poor appearance.

11.3.4 MARKET STRUCTURE

Regarding the market structure, there are a number of firms and individuals actively participating in the trade especially in the case of dried turmeric. Large number of dealers, brokers, and various other intermediaries between dealer and user or even between dealer and dealer exist both in exporting and in importing countries. The village merchants play a vital role in fixing the price of turmeric, which depends upon quality factors, such as hardiness, intensity of the core, polished or unpolished, etc.

The prevailing marketing channel for turmeric in India is like the one given in Figure 11.4, with slight variation among regions. To begin with, farmers, after retaining the needed quantity for seeding purpose and for domestic consumption, sell off the rest to commission agents/village traders who collect the produce at farm gate or transport it to nearby market place. The produce collected in the assembly markets in the *taluk*/block level is transported to the regional/district-level main marketing centers. Farmers having a large production base often take their produce to the local and/or regional market directly. From the regional markets, it is moved to terminal/distribution markets like Cochin, Chennai, Bombay, Bangalore, Kolkata, and New Delhi. Table 11.15 provides the list of major markets (assembling/distribution) for turmeric in each producing States, and Table 11.16 provides the market-wise quantity handled during 2000. From the table it is evident that the Erode market is the biggest, handling 37% of the total arrival in all the markets. Bombay, Chennai, Cochin, and Tuticorin are the other major ports through which turmeric is exported.

In Karnataka, farmers take an advance payment for production of the crop from the commission agent/wholesalers from the regulated market of Erode in Tamil Nadu. In turn, the farmers are forced to sell their produce to the commission agents at the farm gate itself (Lokesh and Chandrakant, 2004). In Tamil Nadu, the market in Erode is a regulated one with better facilities; here the farmers sell both polished and unpolished turmeric rhizomes (Srinivasan et al., 2004). The turmeric market

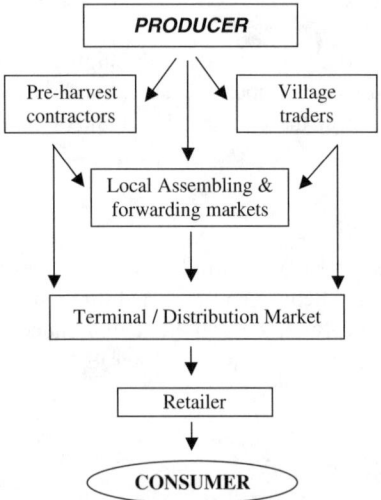

FIGURE 11.4 Commodity-distribution system for turmeric in India.

TABLE 11.15
Important Assembling and Distributing/Terminal Markets for Turmeric in India

State	Assembling Markets	Terminal/Distribution Market
Andhra Pradesh	Duggirala, Cuddapah, Nizamabad, Rajahmundry, Vijayawada, Kodur, Tenali	Hyderabad
Maharashtra	Bombay, Sangli, Karad, Poona, Tekkari, Kolhapur, Tasgaon, Nagpur	Bombay, Sangli
Tamil Nadu	Mettupalayam, Erode, Salem, Coimbatore, Karur, Tiruchirapally, Madras, Madurai	Chennai, Erode, Madurai
Orissa	Berhampur, Parlekimedi, Tikkabali	
Kerala	Cochin, Calicut, Alleppey, Telicherry, Muvattupuzha, Kalpetta, Baliapattam, Thodupuzha, Wayanad, Koduvally, Kodencherry, Badagara, Talibaramba	Cochin, Calicut, Alleppey, Tellicherry
Bihar	Patna	Patna
Other distribution/terminal markets		Bangalore, Amritsar, Calcutta, Kanpur

TABLE 11.16
Estimated Market-Wise Arrival During 2000

Market	Quantity Handled	Percent Share
Erode	1,800,000 Bags	37.89
Nizamabad	1,200,000 Bags	25.26
Cuddapah	250,000 Bags	5.26
Duggirala	400,000 Bags	8.42
Maharashtra	400,000 Bags	8.42
Warangal	400,000 Bags	8.42
Others	300,000 Bags	6.32
Total	4,750,000 Bags	100

at Duggirala is also brought under the regulated marketing act, and intermediaries are avoided. Sangli market in Maharashtra is the only futures market in India exclusively for turmeric.

Naidu and Hanumanthaiah (1988) analyzed the price spread in turmeric marketing in Duggirala market in Andhra Pradesh. According to them, in this regulated market the grower gets 56.36% of the consumer's price with an index of market efficiency of 1.29.

In Karnataka, the study conducted by Lokesh and Chandrakanth (2004) identified the following three marketing channels:

1. Producer → Erode APMC (Agricultural Produce Marketing Committee) market
2. Producer → Commission Agent → Erode APMC market
3. Producer → Local Trader → Erode APMC market

11.3.5 MARKET INTEGRATION FOR TURMERIC IN INDIA

In order to measure the relative influence of past market price on present local market price, Timmer's Index of Market Connection (IMC) was estimated. The method was based on the Ravallion Model. Arshad (1990) used this technique to research the integration of the palm oil market in Malaysia. In general, the closer the index to zero, the greater the degree of market

TABLE 11.17
Market Integration for Turmeric in India

Reference Market	Local Market	b1	b2	b3	IMC
Delhi	Mumbai	−0.016	0.651	1.058	−0.02
Delhi	Kolkata	0.567	0.662	0.365	1.55
Delhi	Cochin	0.024	0.957	1.198	0.02
Mumbai	Cuddapah	0.495	0.719	0.262	1.88
Mumbai	Duggirala	0.449	0.885	0.295	1.53
Mumbai	Nizamabad	0.402	0.732	0.318	1.27
Mumbai	Chennai (Erode)	0.117	0.914	0.767	0.15
Mumbai	Chennai (Salem)	−0.146	1.843	1.245	−0.12
Cochin	Chennai (Erode)	0.322	0.666	0.505	0.64
Cochin	Chennai (Salem)	0.119	0.782	0.828	0.14
Chennai	Cuddapah	0.399	0.530	0.382	1.05
Chennai	Duggirala	−0.068	0.532	0.729	−0.09
Chennai	Nizamabad	−1.038	0.486	1.346	−0.77

Note: Integration of Market Connection (IMC) worked out following Ravallion (1986).

integration. However, Timmer, allowing for assumption of stability in central market prices and the absence of local characteristics, considered that value of IMC less than 1 reflects a high degree of short-run integration. Market integration study done in turmeric marketing between different markets is given in Table 11.17. As it can be seen from the table, with IMC less than 1, Cochin and Chennai with values nearer to 1 or <1 are highly integrated.

Price trends of turmeric in three markets of Andhra Pradesh (Duggirala, Nizamabad, and Cuddapah) were analyzed by Sudhakar (1996), and the result indicated cyclic fluctuations in market price with a peak price in approximately every 8 yr. He observed similar price trend for all the three markets (i.e., they are highly integrated). Causes of price fluctuations were also discussed.

Amritsar, Delhi, Kanpur, Hyderabad, Sangli, and Bombay are the important markets for turmeric from Nizamabad, whereas Calcutta, Patna, Ranchi, Madras, Ahamadabad, and Varanasi are important markets for the Erode turmeric.

Largest quantity (about 80%) of the produce is marketed through the second channel, where the producer gets 81.38% of the wholesale price, while the commission agent gets the remaining 18.62% for his service. The price received by the farmer amounts to 52% of the price of the processed end product (turmeric powder) purchased by the consumer. According to them, the major value addition of converting turmeric rhizomes to turmeric powder fetches an addition of 48% to the base value of turmeric.

11.3.6 ECONOMICS OF DRY TURMERIC PRODUCTION

Fresh turmeric requires processing before being sold in the market by the farmer. The processing of turmeric basically involves washing, sorting, bleaching, drying, polishing, and coloring (Joseph and Sethumadhavan, 1985; Velappan et al., 1993; George, 1995). The method and equipment used for processing varies from region to region depending on the variety harvested leading to variation in the end product. According to the survey conducted by Srinivasan et al. (2004), hardly 30% of the farmers of Tamil Nadu follow the recommended postharvest technology including processing. Senthilnathan et al., (2001) reported that, while most farmers do curing and drying, only 58% farmers polish the dried rhizome before sale. The remaining 42% farmers sold their produce without polishing to the intermediaries in the marketing system. The cost of polishing comes to Rs. 2/kg on a hire basis. In Karnataka, dried turmeric rhizomes are polished at the farm itself by hiring a power-operated rotary drum. The hiring charge was Rs. 20/quintal (100 kg). According to Lokesh and Chandrakanth (2004), there are around 40 such machines available for hire in the Chamrajnagar district alone. Facility to make turmeric powder is also available in the district to make use of the available raw material.

11.3.6.1 Curing

Fresh turmeric is processed for obtaining dry turmeric. The fingers are separated from mother rhizomes. Mother rhizomes are usually kept as seed material. Curing involves boiling of fresh rhizomes in water and drying in the sun.

In the traditional method of curing, the cleaned rhizomes are boiled in water just enough to immerse them. Boiling is stopped when froth comes out and white fumes appear giving out a typical odor. The boiling lasts for 45 to 60 min, until the rhizomes turn soft. The stage at which boiling is stopped largely influences the color and aroma of the final product. Overcooking spoils the color of the final product, while undercooking renders the dried product brittle.

In the improved method of curing, the cleaned fingers (approximately 50 kg) are taken in a perforated trough of $0.9 \rightarrow 0.5 \rightarrow 0.4$ m size made of galvanized iron or mild steel sheet with extended parallel handle. The perforated trough containing the fingers is then immersed in a pan; 100 l of water is poured into the trough to immerse the turmeric fingers. The whole mass is boiled until the fingers become soft. The cooked fingers are taken out of the pan by lifting the trough and draining the water into the pan. The water used for boiling turmeric rhizomes can be used for curing fresh samples. The processing of turmeric is to be done 2 or 3 d after harvesting. If there is delay in processing, the rhizomes should be stored under shade or covered with sawdust or coir dust.

11.3.6.2 Drying

The cooked fingers are dried in sun by spreading them in 5 to 7 cm thick layers on bamboo mats or drying floor. A thinner layer is not desirable, as the color of the dried product may be adversely affected. During nighttime the rhizomes should be heaped or covered with a material, which provides aeration. It may take 10 to 15 d for the rhizomes to become completely dry. Artificial drying, using cross-flow hot air at a maximum temperature of 60°C also gives a satisfactory product. In the case of sliced turmeric, artificial drying has clear advantages in giving a brighter colored product than sun drying, which tends to bleach the surface. The yield of dry product varies from 20 to 30%, depending upon the variety and the location where the crop is grown (Devakaran and Nair, 1982; Krishnamurthy, 1993; George, 1995).

11.3.6.3 Polishing

Dried turmeric has poor appearance and rough dull outer surface with scales and root bits. Smoothening and polishing the outer surface by manual or mechanical rubbing improve the appearance (George and Velappan, 1995; Shankaracharya and Natarajan, 1973; Narayanan et al., 1978; Lakshmanachar, 1985).

Manual polishing consists of rubbing the dried turmeric fingers on a hard surface. The improved method is using a hand-operated barrel or drum mounted on a central axis, the sides of which are made of expanded metal mesh. When the drum filled with turmeric is rotated, polishing is effected by abrasion of the surface against the mesh as well as by mutual rubbing against each other as they roll inside the drum. Turmeric is also polished in power-operated drums. The units are suitably mounted on a truck so that it can be moved from place to place. Capacity of the unit varies from 100 to 800 kg/batch and the cost ranges up to Rs. 10,000/unit (Natarajan and Lewis, 1985). Economics of installing a turmeric-polishing unit in Erode district of Tamil Nadu worked out by Senthilnathan et al. (2001) is given in Table 11.18.

TABLE 11.18
Economics of Installing Turmeric Polishing Unit

Particulars	Cost in Rupees
Capital Investment	
Cost of machinery with the capacity of 3000 t/yr	160,500.00
Cost of site/land value	150,000.00
Cost of building	300,000.00
Total	610,500.00
Fixed Costs	
Depreciation of machinery and building	31,050.00
Interest on investment amount on machinery, building, and land	103,785.00
Insurance and premium	51,000.00
Repair and maintenance	75,000.00
Tax	6,000.00
Total fixed costs	266,835.00
Variable Cost	
Labor charge	24,600.00
Electricity charge	30,000.00
Incidental charge	3,300.00
Miscellaneous	30,000.00
Total	87,900.00
Total costs (266,835 + 87,900)	354,735.00
Income through Turmeric Polishing at Rs 200/t	
3000 t i.e., 200 × 3000	600,000.00
Net profit/yr (600,000–354,735)	245,265.00
Break-even volume in t	1,563.18
Benefit/cost ratio	1.26
Net present value in rupees	645,084.00
Internal rate of return (%)	68.25

Source: Senthilnathan et al. (2001).

This facility can be created in the premises of regulated markets for turmeric, so that the farmers can get the benefit and attract more farmers to the market. In Karnataka, power-operated rotary drums with the capacity to polish 100 kg/h costs Rs. 50,000 only. The hire charge is Rs.20/quintal (100 kg).

11.3.6.4 Coloring

The color of the processed turmeric influences the price of the produce. For an attractive product, turmeric powder (mixed in water) may be sprinkled during the last phase of polishing.

11.3.6.5 Grading

In India, Agmark grading is carried out both for internal trade as well as for export. Turmeric is one among the 41 commodities compulsorily graded before export. Agmark grading of turmeric consists of lot inspection, test sampling, packing, labeling, sealing, and check sampling.

11.4 WORLD SCENARIO

World trade in turmeric amounted to more than US$ 40 million in 2004. Quantity traded has gone up from a mere 3771 t in 1994 to an average of 41,673 t during 2004; an increase of more than 1000%.

Major exporters of turmeric and turmeric-based products to the international market are India, China, Peru, Pakistan, Vietnam, Myanmar, and Brazil, with Singapore and Netherlands reexporting. In order to analyze the issues related to export and import trade, the study has distinguished two groups of countries. They are as follows:

1. Producer-exporters (countries engaged in cultivation of turmeric and usually exporting the surplus over domestic consumption occasionally, however, they may be importing turmeric as well from some other countries.)
2. Reexporters

11.4.1 EXPORT

In 1994, India had the share of 92.5% in the total world export and has been able to maintain its monopoly, though the percent share has come down to less than 80% in the recent past. As it can be seen from Table 11.19, during 1999 to 2001, the share of producer-exporters in the total export was nearly 86% and the rest was shared by countries reexporting turmeric. Indian contribution to total world trade in turmeric during 1999 to 2001 was 87.1% of the total quantity and 78.8% of the total value. China, with 3.8% and 5.5%, respectively, in quantity and value terms, stood second among the turmeric exporting countries. China's exports in 2001 declined steeply when compared to its performance in 1999. Among the reexporting countries, The Netherlands, with 3.3% share in value and 1.7% share in quantity leads the list. Germany, U.K., Belgium, and U.S. are the other reexporting countries for turmeric. However, in recent years Singapore has taken the lead among the reexporting countries.

Further analysis of Table 11.19 indicates that the unit price (US$ 613/t) earned by Indian turmeric export is much lower than the world average of US$ 705/t and is lowest among all indicating the price competitiveness of Indian turmeric in international market. The export price of reexporting countries ranging from US$ 1400/t (The Netherlands) to US$ 2422/t (U.K.) is much more than the average unit price (US$ 859/t) earned by other producer exporters.

11.4.1.1 Export Performance by India

The world scenario viewed from the Indian perspective provides a complex situation for turmeric economy. India, being the largest producer of turmeric in the world, has the potential to play the major role in the world trade of turmeric. Table 11.20 shows export of turmeric from India along with estimated growth index for the period from 1970–1971 to 2004–2005. Looking to the stable export performance in terms of quantity during the period from 1975–1976 to 1978–1979, 1976 to 1977 was taken as base year for estimation of growth rate. During 1970 to 1971, India exported only 11,109 t earning a foreign exchange worth Rs. 38.3 million. Record export of 44,627 t was achieved during 2001 with a growth index of 401.72% over the base production. However, the record earning (Rs. 1565 million) through turmeric export achieved in 2004 to 2005 with growth index of more than 4000% may be due to the increased unit price over the period leading to 1061% growth index. Figure 11.5 shows the increasing trend of quantity exported and value realized. As it can be observed from the figure, up to 1990 there were violent fluctuations in turmeric export from the country. Though there were fluctuations after 1990 also, the cycle was long. The quantity exported as percentage of the total production has gone up to 11.3% in 1980 before going down to less than 3% in 1988. However, in recent years, the percent export is being maintained around 7% of the total production.

TABLE 11.19
Total World Exports of Turmeric

Country	1999		2000		2001		Average 1999–2001		Price (US$/kg)	% share in	
	Quantity[a]	Value[b]	Quantity[a]	Value[b]	Quantity[a]	Value[b]	Quantity[a]	Value[b]		Quantity[a]	Value[b]
Producer-Exporters											
India	35,791	27,249	38,149	21,023	35,000	21,156	36,313	23,143	0.64	87.1	78.8
China	2,100	2,199	1,575	1660	1,112	1,002	1,596	1,620	1.02	3.8	5.5
Peru	585	557	527	447	465	457	526	487	0.93	1.3	1.7
Reexporters											
The Netherlands	616	1,023	649	902	835	1,015	700	980	1.40	1.7	3.3
Germany	260	531	269	505	248	446	259	494	1.91	0.6	1.7
U.S.	98	260	138	308	164	360	133	309	2.32	0.3	1.1
U.K.	168	391	154	436	162	343	161	390	2.42	0.4	1.3
Belgium	32	57	46	85	67	118	49	87	1.78	0.1	0.3
Others	2,084	2,012	1,995	2,026	1,730	1,510	1,936	1,849	0.96	4.6	6.3
World	41,734	34,278	43,501	27,391	39,783	26,406	41,673	29,359	0.70	100	100

[a] Quantity (in Mt).
[b] Value (US$ 1000).

TABLE 11.20
Growth Index for Turmeric Export from India
(1970–1971 to 2004–2005)

Year	Quanity (Mt)	Growth Index	Value (Rs Lakhs)	Growth Index	Unit Price (Rs/kg)	Growth Index
1970–1971	11,109	100.00	383	100.00	3.43	100.00
1971–1972	14,173	127.58	290	75.73	2.05	59.77
1972–1973	6,731	60.59	182	47.48	2.70	78.72
1973–1974	7,921	71.30	365	95.21	4.61	134.40
1974–1975	9,227	83.06	414	108.07	4.49	130.90
1975–1976	11,755	105.82	421	109.84	3.58	104.37
1976–1977	11,109	100.00	383	100.00	3.43	100.00
1977–1978	11,253	101.30	830	216.43	7.38	215.16
1978–1979	11,978	107.82	1,241	323.69	10.36	302.04
1979–1980	26,610	239.54	1,981	516.50	7.44	216.91
1980–1981	14,517	130.68	788	205.55	5.43	158.31
1981–1982	11,986	107.89	517	134.93	4.32	125.95
1982–1983	7,595	68.37	424	110.45	5.58	162.68
1983–1984	10,892	98.05	1,106	288.29	10.15	295.92
1984–1985	12,802	115.24	1,716	447.41	13.4	390.67
1985–1986	8,562	77.07	1,209	315.39	14.13	411.95
1986–1987	19,529	175.79	1,918	500.25	9.82	286.30
1987–1988	8,747	78.74	923	240.62	10.55	307.58
1988–1989	18,967	170.74	1,939	505.72	10.22	297.96
1989–1990	16,900	152.13	1,614	420.87	9.55	278.43
1990–1991	13,624	122.64	1,548	403.81	11.37	331.49
1991–1992	19,661	176.98	3,776	984.75	19.21	560.06
1992–1993	19,726	177.57	4,885	1274.01	24.77	722.16
1993–1994	25,436	228.97	5,256	1,370.64	20.66	602.33
1994–1995	28,286	254.62	4,518	1,178.18	15.97	465.60
1995–1996	27,050	243.50	4,620	1,204.87	17.08	497.96
1996–1997	23,019	207.21	5,845	1,524.14	25.39	740.23
1997–1998	28,875	259.92	8,307	2,166.14	28.77	838.78
1998–1999	37,297	335.74	12,914	3,367.80	34.63	1,009.62
1999–2000	37,776	340.05	12,352	3,221.06	32.7	953.35
2000–2001	44,627	401.72	11,558	3,013.96	25.9	755.10
2001–2002	37,778	340.07	9,074	2,366.21	24.02	700.29
2002–2003	32,402	291.67	10,338	2,695.91	31.91	930.32
2003–2004	37,044	333.46	13,112	3,419.23	36.96	1,077.55
2004–2005	43,000	387.07	15,650	4,081.15	36.40	1,061.09

Source: Spices Board, Cochin, Ministry of Commerce, Government of India.

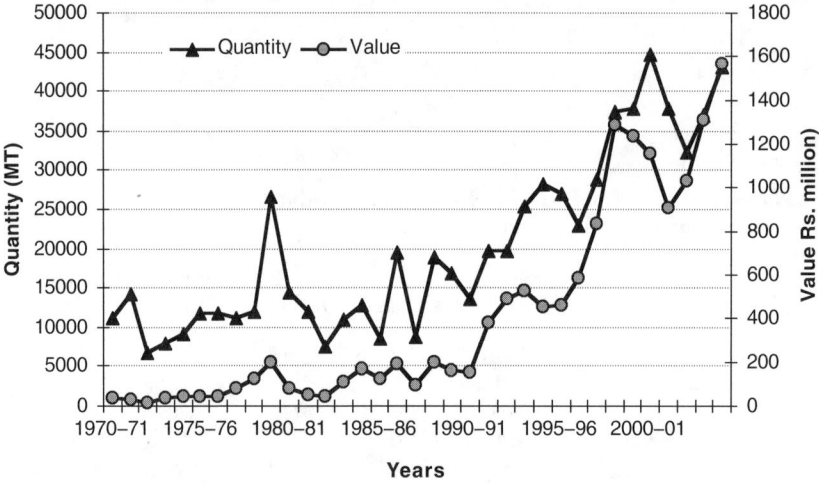

FIGURE 11.5 Trend in export of turmeric from India (1970–1971 to 2004–2005).

TABLE 11.21
Compound Annual Growth Rate in Turmeric Export from India

Period	Qunatity	Value	Price
1980–1981 to 1989–1990	4.10	13.50	9.02
1990–1991 to 1999–1900	9.66	20.15	9.56
2000–2001 to 2004–2005	–0.93	10.23	11.75
1995–1996 to 2000–2001	12.90	23.09	9.03
1980–1981 to 2004–2005	6.78	15.41	8.12
1990–1991 to 1994–1995	18.75	28.05	7.81

Raveendran and Aiyasamy (1982) examined the factors influencing the export of turmeric from India for the period 1960 to 1980 using linear regression. The estimated coefficient of variation in the quantity exported was about 50, while that of export prices was 70.7, indicating a larger variation in export prices of turmeric. The regression analysis showed that 63% of the variation in the export could be explained by the production lagged by 1 yr, export price relative to domestic price of turmeric and time, and proxy for export promotion measures. Only the time variable is found to be significant, suggesting the nonresponsiveness of exports to the price variable.

Analysis of export performance of Indian turmeric economy between 1980 to 1981 and 2004 to 2005 by estimating growth rates (Table 11.21) reflects the following features:

- The physical volume of exports has increased by around 6.78% annually, whereas annual growth in value terms works out to be around 15.41%. The annual growth in unit price realization over this period works out to be around 8.12%.
- At a decadal disaggregated level, performance of exports of turmeric from the country does not look encouraging. In terms of average annual growth in unit value realization, there is a slow upward drifting over the decades to achieve the highest growth rate of 11.75% during 2000–2001 to 2004–2005 (i.e., after complete removal of Quantitative Restrictions on imports as per WTO agreement). This growth in unit value realization was against the background of the fact that there was a negative growth (–0.93%) in the

TABLE 11.22
Instability Index in Turmeric Exports from India (1970–1971 to 2004–2005)

Period	Instability Index in		
	Quantity	Value	Price
1970–1971 to 1979–1980	45.85	50.06	48.89
1980–1981 to 1989–1990	69.35	73.41	31.19
1990–1991 to 1999–1900	17.74	34.39	27.46
2000–2001 to 2004–2005	16.29	20.6	15.11
1970–1971 to 2004–2005	43.67	52.62	32.33

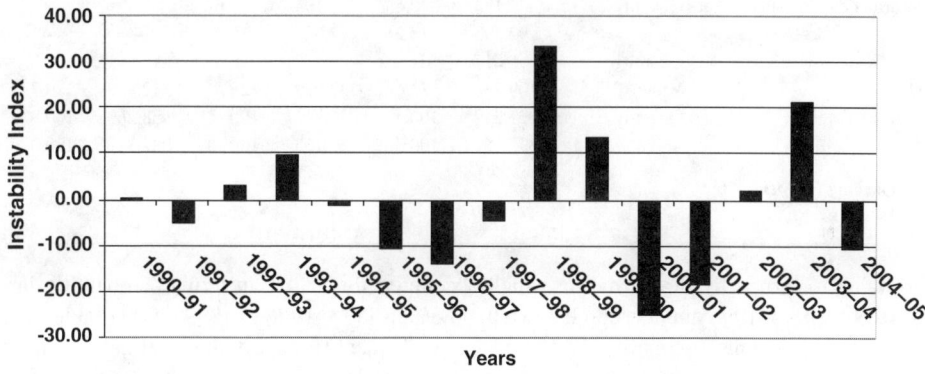

FIGURE 11.6 Instability index in turmeric export from India (1990–1991 to 2004–2005).

physical volume of exports. During the first half of the 1990s, however, there was a spurt in the growth of physical exports, accompanied by better performance in unit value realization, in spite of considerable devaluation of Indian rupee over this period. The trend continued in the second half of the 1900s (post-WTO period) as well.

11.4.1.2 Export Instability

In order to estimate the observed instability in turmeric export in terms of quantity, value, and price instability index estimation was done using the time series data, and the results are presented in Table 11.22. It could be observed from the table that there was an instability in case of quantity, value, and unit price of turmeric exports, and the instability was relatively higher in case of export value (52.62%) compared to quantity (43.67%) and unit price (32.33%) for the entire period from 1970–1971 to 2004–2005. The estimated instability index for the periods 1970–1971 to 1979–1980 and 1980–1981 to 1999–2000 was more than the accepted level of 40%. However, instability in turmeric export has come down well below the 40% level in the subsequent periods. The above instability index was a close approximation of the average year-to-year percentage variation adjusted for trend. Figure 11.6 presents the instability index in graphical presentation for export value earned during 1990–1991 to 2004–2005. As it can be seen from the figure there is a year-to-year violent fluctuation in turmeric export during this period. The negative trend is more frequent than the positive.

11.4.1.3 Composition of Indian Exports

As far as item-wise export of turmeric from India is concerned, there is a marked change in recent years. Traditionally, the country was exporting dried rhizomes only. More than half of the total

TABLE 11.23
Item-Wise Export of Turmeric from India

Item	Quantity[a]/Value[b]	1990–1991 to 1994–1995		1995–1996 to 1999–2000		2000–2001 to 2002–2003	
		Average	Share (%)	Average	Share (%)	Average	Share (%)
Turmeric dry	Quantity	13,033	58.4	18,214	53.3	22,668	59.2
	Value	227,529	44.8	476,326	41.0	519,603	50.3
Turmeric fresh	Quantity	4,121	18.5	5,989	17.5	—	—
	Value	78,028	15.4	164,054	14.1	—	—
Turmeric oil	Quantity	—	—	1	0.0	—	—
	Value	42	0.0	1,033	0.1	—	—
Turmeric oleoresins	Quantity	134	0.6	195	0.6	—	—
	Value	92,393	18.2	177,920	15.3	—	—
Turmeric powder	Quantity	5,017	22.5	9,762	28.6	15,601	40.8
	Value	109,730	21.6	342,089	29.5	512,708	49.7
Turmeric total	Quantity	21,347	100	30,803	100	38,269	100.0
	Value	399,682	100	880,755	100	1032,311	100.0

[a] Quantity (in Mt).
[b] Value (Rs 1,000).

export value is earned by dry turmeric, which accounted for 58.4% in terms of quantity during 1990–1991 to 1994–1995 and has come down to 53.3% in 1995–1996 to 1999–2000 (Table 11.23). Fresh turmeric accounts for hardly 17.5% of the total quantity exported; in terms of value the percent share is 14.1% only. Turmeric oil and oleoresin are the other products exported that have returned high value. Although there is not much demand and remunerative price for turmeric oil, oleoresin from turmeric has great demand from the importing developed country markets. Turmeric powder and curry powder with turmeric powder as a major component are the other value-added products of turmeric exported from the country. Turmeric powder accounts for more than 40% of the total export in terms of quantity. As in the case of reexporting countries, especially the EU countries, India has the potential to strengthen the processing industry to add more value-added products into its export basket. Figure 11.7 and Figure 11.8 depict the trend of product components in total export of turmeric during the period 1991–1992 to 2002–2003 in terms of quantity and value, respectively.

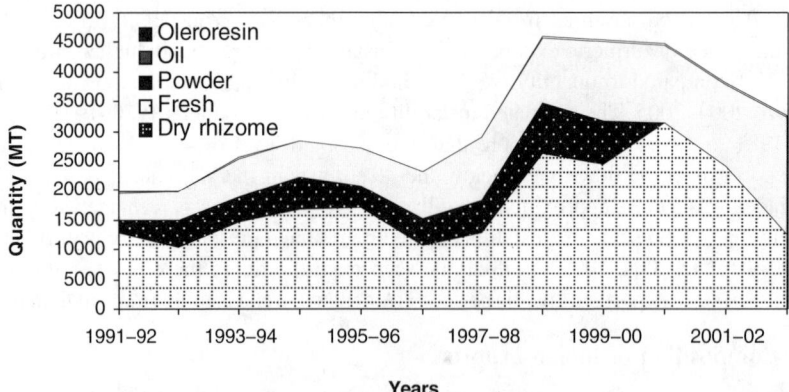

FIGURE 11.7 Item-wise export of turmeric (quantity) from India (1991–1992 to 2002–2003).

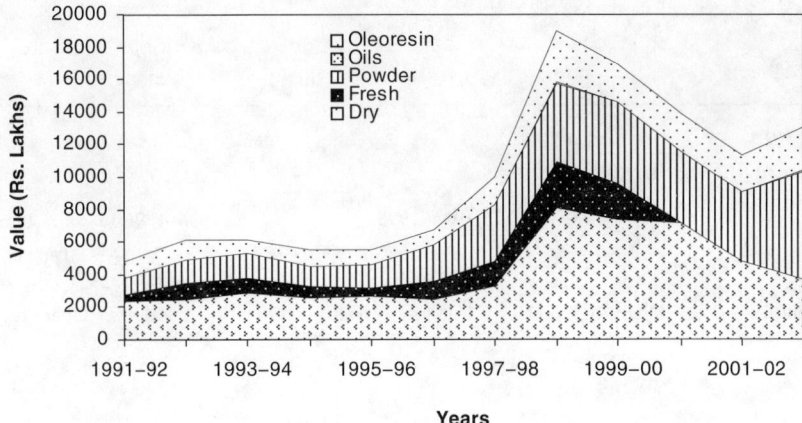

FIGURE 11.8 Item-wise export of turmeric (value) from India (1991–1992 to 2002–2003).

11.4.1.4 Direction of Indian Exports

Zone-wise export of turmeric from India during 1960–1961 to 1977–1978 in percentage of total is given in Table 11.24. As it can be seen from the table, East Asia and Middle East together shared more than 50% of the total quantity exported from India. America and Europe are the other blocks to where the Indian turmeric was exported.

During 1988–1989 to 1994–1995, nearly 40% of the Indian export was to West Asian countries. However, during the period 1995–1996 to 2003, the share of those traditional buyers of Indian turmeric has declined and the country has found new markets in neighboring countries like Sri Lanka and Bangladesh. UAE is the single largest market for turmeric from India both in terms of quantity and value. During 1988–1989 to 1994–1995 period, India has exported more than 29% of its total turmeric export to UAE (Table 11.24). However, during 1999–1996 to 2002–2003, share in Indian exports of turmeric to UAE has declined to 15% and during the same period, Bangladesh and Sri Lanka have increased their import share from India from 6.1% to 19.6%. During the same period, traditional buyers of Indian turmeric, such as U.K., Japan, The Netherlands, Singapore, etc. have reduced their share considerably. Increased share under the category of other countries indicates the fact that, market for Indian turmeric has opened up in many other new countries. As it can be seen from the Table 11.25, the concentration index has come down steadily over the period to indicate the fact that India does not depend on a few countries as its market for turmeric export.

11.4.2 Imports

In International market, the major importers of turmeric are Iran, Japan, U.K., Sri Lanka, Malaysia, U.S., etc. where an increase in imports in terms of volume and value has been recorded over the years, these countries are importing mainly from India, China, Peru, Thailand, etc. (ITC, 1995).

Iran occupied the major share (19.6%) among the importers of turmeric during 1999 to 2001 period, followed by Japan (12.8%), Sri Lanka (8.5%), U.S. (8.36%), U.K. (6.9%), Malaysia (6.48%), and the rest (29.9%) by other turmeric importing countries (Table 11.26). Trend in import of turmeric by major importing countries has not changed much except during 1999 and 2003 in the recent past.

11.4.2.1 Japan

Japan, one of the major importers of turmeric in the world, imports around 3000 t of turmeric worth US$ 2.9 million (2003). More than half of its imports was from India (Table 11.27). China,

TABLE 11.24
Direction of Indian Turmeric Exports

| Countries | 1988–1989 to 1994–1995 | | | | 1995–1996 to 2002–2003 | | | |
| | Quantity | Value | Percent Share in Total | | Quantity | Value | Percent Share in Total | |
			Quantity	Value			Quantity	Value
U.S.	1,822	429	9.2	13.6	2,444	893	6.8	8.8
Canada	205	46	1.0	1.4	212	67	0.6	0.7
France	300	50	1.5	1.6	578	130	1.6	1.3
Germany	565	104	2.8	3.3	827	221	2.3	2.2
U.K.	1,667	294	8.4	9.3	1,999	485	5.5	4.8
The Netherlands	415	76	2.1	2.4	438	86	1.2	0.9
Islamic Representative of Iran	735	113	3.7	3.6	3,081	519	8.5	5.1
Bahrain	146	28	0.7	0.9	198	45	0.5	0.4
Kuwait	221	27	1.1	0.9	172	49	0.5	0.5
Liberia	250	9	1.3	0.3	35	7	0.1	0.1
Libyan Arab Jamahriya	132	12	0.7	0.4	26	7	0.1	0.1
Israel	226	33	1.1	1.1	331	68	0.9	0.7
UAE	5,841	763	29.4	24.2	5,419	876	15.0	8.6
Oman	125	21	0.6	0.7	128	34	0.4	0.3
Saudi Arabia	325	47	1.6	1.5	572	141	1.6	1.4
Singapore	906	146	4.6	4.6	932	202	2.6	2.0
Sri Lanka	395	49	2.0	1.5	2,535	389	7.0	3.8
Bangladesh	809	101	4.1	3.2	4,549	885	12.6	8.7
Malaysia	343	61	1.7	1.9	902	257	2.5	2.5
Japan	1,804	275	9.1	8.7	2,294	560	6.3	5.5
Yemen	454	58	2.3	1.8	196	20	0.5	0.2
Others	2,215	379	11.1	12.0	5,864	1,415	16.2	14.0
Total	19,887	3,155	100.0	100.0	36,155	10,143	100.0	100.0

TABLE 11.25
Country-Wise Export of Turmeric and Concentration Index

| | 1994–1995 | | | 1998–1999 | | | 2003–2004 | | |
	Quantity	Value	Country	Quantity	Value	Country	Quantity	Value	Country
Share (%) by major importers	30.30	21.87	UAE	23.74	23.81	UAE	18.35	15.77	UAE
	10.35	10.56	U.K.	12.23	8.02	Bangladesh	11.44	10.84	U.K.
	10.02	10.37	Japan	11.84	4.85	Pakistan	10.26	10.84	Bangladesh
Concentration index	37.99	34.79		33.8	33.05		28.36	26.87	

TABLE 11.26
Total World Imports of Turmeric

Country	1999–2001 Quantity[a]	Value[b]	Percent share in Quantity[a]	Value[b]	2003 Quantity[a]	Value[b]	Percent share in Quantity[a]	Value[b]
Japan	4,515	3,737	12.8	12.8	4,184	4,752	9.28	11.18
U.S.	2,952	2,950	8.3	10.1	2,449	3,622	5.43	6.54
Iran	6,927	4,399	19.6	15.1	6,596	2,996	14.63	17.63
U.K.	2,453	2,258	6.9	7.8	2,770	3,101	6.15	7.40
Germany	1,461	1,305	4.1	4.5	1,112	1,574	2.47	2.97
The Netherlands	1,235	1,193	3.5	4.1	1,290	1,379	2.86	3.45
Sri Lanka	3,000	1,531	8.5	5.3	3,857	1,587	8.56	10.31
Malaysia	2,254	1,436	6.4	4.9	2,817	2,242	6.25	7.53
Bangladesh	—	—	—	—	2,599	1,267	5.77	6.95
India	—	—	—	—	3,015	1,811	6.69	8.06
Others	10,583	10,227	29.9	35.1	14,251	6,727	31.92	17.98
World	35,379	29,135	100	100	44,940	374,161	100	100

[a] Quantity (in Mt).
[b] Value (US$ 1000).

TABLE 11.27
Turmeric Imports by Japan (2002–2004)

Country	2002 Quantity[a]	Value[b]	Share (%) in Quantity[a]	Value[b]	2003 Quantity[a]	Value[b]	Share (%) in Quantity[a]	Value[b]
India	2958	2128	67.9	54.3	2885	2859	68.9	60.2
China	1085	1277	24.9	32.6	869	1181	20.8	24.9
Indonesia	46	135	1.1	3.4	140	253	3.3	5.3
Viet Nam	56	103	1.3	2.6	92	167	2.2	3.5
Malaysia	24	59	0.5	1.5	41	63	1.0	1.3
Myanmar	110	53	2.5	1.4	75	57	1.8	1.2
Thailand	44	52	1.0	1.3	22	34	0.5	0.7
U.S.	2	45	0.0	1.1	2	41	0.0	0.9
Pakistan	10	19	0.2	0.5	6	13	0.2	0.3
Fiji	10	18	0.2	0.5	0	2	0.0	0.0
Philippines	3	6	0.1	0.2	2	4	0.0	0.1
Bangladesh	48	66	1.1	1.4	0	0	0.0	0.0
Other	7	22	0.2	0.6	3	11	0.1	0.2
Total	4354	3919	100	100	4184	4752	100	100

[a] Quantity (in Mt).
[b] Value (US$ 1000).

TABLE 11.28
Imports by EU Countries (2003)

Origin	Value (US$ 1000)	Share (%)
World	7266	100
India	6528	89.84
Indonesia	246	3.39
Brazil	132	1.82
Peru	94	1.29
Pakistan	76	1.04
China	50	0.69
Others	141	1.94

with nearly 25% share stands second, while Indonesia, Vietnam, and Malaysia are other major exporters of turmeric to Japan.

11.4.2.2 European Union

Imports of turmeric to the EU market are relatively small compared to Iran and Middle East, where turmeric is used to add flavor and color to many rice-based dishes. EU imports are between 3000 and 3500 t per annum. The U.K. is the largest market, with imports having risen from around 1500 to 2500 in the first half of 1990s. In 2003, U.K.'s import was 2770 t. The import is mainly to meet the growing demand from the Asian community living there. Nearly 70% of the consumption is for curry powder and the rest is for coloring extracts.

Germany is the second largest market, with imports rising from around 485 to 880 t in the first half of 1990s. During the period from 1999 to 2001, the average import of turmeric was 1461 t worth US$ 1.305 million. In Germany these are used in mustard and curry sauces. Some special forms of turmeric are imported for use in natural pharmaceutical products (Anon, 1996). France and The Netherlands imported around 1000 to 1200 t each during 1999 to 2001. Source country-wise imports into EU countries given in Table 11.28 reveals the fact that nearly 90% of the total imports in terms of value are from India.

11.4.2.3 U.S. Imports of Turmeric

U.S. is the major market for the Alleppey turmeric from India. It is learned that the entire production of Alleppey turmeric produced in Kerala State is exported to U.S. market. Quantity of turmeric imported (2176 tons) into U.S. has almost doubled between 1996 and 2004 (Table 11.29). But the value of turmeric imports has not changed much during this period. Regarding the source of imports, India, with more than 96% share, was the largest exporter of turmeric to U.S. However, in recent years it has come down to 90%. China, Indonesia, and Vietnam are the emerging sources of import for U.S. (Garner, 1995).

Singapore, the major entreport for spices, is one of the largest importers and reexporters of turmeric in the world market. Singapore's annual import is around 1500 to 2500 t. Myanmar is the major source for Singapore's imports followed by India and Malaysia (Table 11.30).

11.4.2.4 India

India was one of the major importers of turmeric during 2003. It has imported US$ 1.8 billion worth of turmeric. Of the total imports, more than 77% was from Vietnam and Myanmar (Table 11.31). Nigeria and Morocco were the other countries from where India imported turmeric. In fact, there was a decline in turmeric production in the country leading to less export surplus.

TABLE 11.29
Turmeric Imports by Country of Origin U.S. (1997 and 2004) (Qty: tons; Value: US$'000)

| | 1996 | | | | 2004 | | | |
| | | | Share (%) in | | | | Share (%) in | |
Countries	Quantity	Value	Quantity	Value	Quantity	Value	Quantity	Value
China	1	4	0.1	0.2	6	1	0.1	0.0
India	2091	2082	96.1	92.0	3981	2395	90.9	91.0
Indonesia	17	21	0.8	0.9	7	7	0.2	0.2
Thailand	7	16	0.3	0.7	43	23	1.0	0.9
Germany					54	24	1.2	0.9
Fiji					97	56	2.2	2.1
Peru					72	62	1.6	2.4
Others	59	141	2.7	6.2	121	65	0.2	0.1
World	2176	2264	100	100	4381	2631	100	100

TABLE 11.30
Turmeric Import by Singapore (2002–2004) (Qty: tons; Value: US$'000)

| | 2002 | | | | 2003 | | | |
| | | | Share (%) in | | | | Share (%) in | |
Country	Quantity	Value	Quantity	Value	Quantity	Value	Quantity	Value
India	802	527	30.6	46.2	660	588	45.2	61.9
Myanmar	1423	472	54.3	41.4	512	211	35.0	22.2
Malaysia	393	131	15.0	11.5	269	119	18.4	12.6
U.S.	0.191	6	0.0	0.5	0.3	5	0.0	0.5
China	2	2	0.1	0.2	3.0	3	0.2	0.3
Pakistan	0.453	1	0.0	0.1			0.0	0.0
Thailand	0.535	1	0.0	0.1	0.7	9	0.0	1.0
Areas, nes	0.078	0	0.0	0.0	0.1	1	0.0	0.1
Vietnam							0.0	0.0
World	2621	1140	100	100	1460	950	100	100

11.4.3 MARKET OPPORTUNITIES

According to the ITC market development paper (1995), consumption of spices is likely to increase, due to an augmented production of high flavored food by the food industry. In addition, increasing interest in health food and, consequently, "natural" instead of "artificially" flavored and colored food, will also increase consumption of spices.

The turmeric market is likely to grow along with the spread and demand for Asian foods the world over. With the increasing use of curry powder and spice mixtures in the world and the trend in the developing countries toward shifting to natural colors in food, from artificial coloring matter, the prospects of increased trade in turmeric appears to be quite good. The market for turmeric as a natural coloring material is expected to grow throughout the world and hence increase the demand for turmeric oleoresin and spray-dried turmeric extracts. Awareness about the commodity's medicinal properties will create a separate market in the health and cosmetics fields. The growing population and flow of money into the fast growing middle income group in the developing countries, including India, will lead to increased demand for spices in general and turmeric in particular.

TABLE 11.31
Indian Imports of Turmeric During 2003

Origin	Value (US$ 1000)	Share (%)
World	1811	100
Vietnam	718	39.64
Myanmar	677	37.4
Nigeria	201	11.09
Ethiopia	53	2.91
Morocco	48	2.64
Malaysia	33	1.81
Indonesia	32	1.75
Nepal	28	1.56
Thailand	21	1.16
U.K.	1	0.04

11.4.3.1 Competitiveness of Indian Turmeric Industry

In order to understand the position and competitiveness of individual exporters in the world trade of turmeric, market shares and unit value ratios were calculated. In the absence of time series data on prices for individual products from various countries, the unit price was worked out from the value of export and quantity exported. While calculating the unit price, individual items of export were not taken into account. Therefore, there was bound to be slight variation depending upon the share of value-added products in the export basket of individual countries. However, the estimated unit value ratios help in comparing the prices of each exporting country with others and with the average of total imports. The ratio is computed by dividing the price received by a country's export by the world average price. When the unit price ratio is less than 1, it is considered that the country possess competitiveness in the export market for its product. Accordingly, as it can be observed that India, with an average unit price ratio of 0.9, is considered highly competitive in the world market when compared to other competing countries.

Any country's competitive power in exporting a commodity depends crucially on its relative price and quality of that commodity over the competing countries. India is having comparatively a better competitive position in the international market for turmeric from unit price point of view. Percent share in total quantity exported is also being maintained over the years. The country has the potential.

Productivity level is up to 40,000 kg/ha (fresh) against the present level of productivity of 15,000 to 18,000 t (fresh). Within the country, there is variation in productivity among the producing States. There are improved high-yielding varieties with acceptable quality parameters (Sasikumar, 2004; Kallupurakkal and Ravindran, 2002). However, increased cost of production and less productivity may work against India in the future, as countries like China, Vietnam, and Myanmar are increasing their production level. Therefore attempts to increase production and productivity that reduce the cost of production must assume prime importance. The country has enough potential to increase the productivity as there is a wide gap between the average productivity and achievable productivity level. To be successful in the changing environment, it would be imperative to be innovative and proactive. It is pertinent to mention that export demand of turmeric is inelastic. The quantity exported is not affected by increase or decrease in the price. Only the quality of the produce can have impact on world import. India, being the major producer of turmeric in the world, has a large domestic market to serve as cushion against occasional unfavorable price level in the international market. Therefore, India stands to gain a lot from the world market for turmeric by improving the quality and generating enough exportable surplus.

11.4.3.2 Risk and Uncertainty

A review of prices and yields reveals considerable price and yield volatility with relatively little correlation between the two variables. The crop's exceptional vulnerability to diseases increases the yield risk substantially. In addition to abruptly fluctuating prices, turmeric root is relatively susceptible to serious disease problems providing an ever-present possibility for a production problem that leads to sharp reduction in yields. A sustainable turmeric economy is possible only when these risks are minimized.

Although different production scenarios are being followed the world over (for which details are available) and appear to have earned adequate profit, in reality, turmeric industry has to face risk and uncertainty of different forms. Further, each country has to face considerable competition from other turmeric-producing countries, because many new countries have entered into the industry in recent years. Of late, India is losing its market share to other producing countries not because of the price competitiveness but because of changing world order in trade relations. From price point of view, the FOB (Free on Board) prices have been generally good since 1991 to 1992, though there was a fall in 1994 to 1995 and 2001 to 2002 crop year. During 2003 to 2004 onward, export is earning better returns; the price received is more than double over the price prevailed in the 1980s. The price for dry turmeric was well below the break-even point in many crop year forcing the farmers to reduce production in subsequent years. In addition to this abrupt fluctuation in price, the crop is also highly susceptible to serious disease problems leading to reduction in yield and unmarketable production. At times the farmer may have to lose up to 80% crop toward the end of the crop cycle. Thus, the turmeric crop industry is influenced by the risk factors of yield and price, though they are not related as per the analysis of long-term data. However, the analysis of variance indicates that price variability of turmeric is greater than yield variability.

11.4.3.3 Prospects and Policy Measures

1. India, though being the major producer of turmeric, consumes more than 80% of total share in production and exports less than 10% to the world market; however, it has enough surplus to meet the export demand.
2. There is a definite pattern of cyclical fluctuation in production mainly due to the producers' response to price. Price stabilization measures can give better returns to predominantly small and marginal turmeric growers.
3. In the past, Iran, UAE, and other Middle East countries were the major market for Indian turmeric accounting for more than 60% of its total export. In the changed scenario in the recent past, their share has been reducing slowly and neighboring countries like Sri Lanka and Bangladesh are emerging as the new markets.

While considering policy measures to strengthen the turmeric economy of the country, it is imperative to undertake action plans at disaggregated levels — the regional level to begin with, followed by the handling of national issues. Following are the suggested policy measures to overcome the constraints faced by the country:

1. Healthy seed production through the implementation of the "seed village concept" by regular field monitoring and developing seed certification procedures
2. Impose quarantine regulation to restrict seed transportation from one state to other especially where disease is a major problem
3. There is a need to educate the farmers not only about postharvest technology for the crop, but also to avoid mixing of varieties. Mixed consignments have poor demand in the market.

Varieties with high curcumin content and potential productivity up to 40,000 kg/ha (fresh) are already available with research organizations in the country. However, inadequate extension activities have not allowed the technology to reach farmers' fields. Although the developmental plans of the country address all these issues, its implementation in true sense, is still lacking.

1. With sweeping changes occurring in the standard of life, life style, and consumption pattern in the buying countries, and with the focus shifting toward value addition and branded consumer packs, the market development activities, too, need to undergo modifications.
2. Importing countries show definite preference to uncontaminated and clean product. There is a need for collective efforts on the part of farmers, traders, and exporters to upgrade quality through improved preharvesting practices, postharvest handling, processing and packaging, and storage to keep up with the grade specifications, pesticide residues, aflatoxin, and microbial load.
3. Indian farmers need to be educated and trained to effectively stand up to the challenges. The need to adopt measures to be more competitive in terms of both quality and productivity assumes greater significance in view of the opening up of agricultural sector and lowering of Agricultural tariffs in accordance with WTO.
4. Since a high-value product line is emerging through organic farming, efforts should be made to popularize organic farming in turmeric, so that it fetches higher prices in foreign markets.
5. Value addition does not mean production of turmeric derivatives alone. One EU document reveals that in the export market, "buyers are looking for clean, well flavored, artificially dried product with high hygiene levels, in contrast to the bulk of the materials which has been sun dried on the ground" (Commonwealth Secretariat, 1996, p.45).

REFERENCES

Aida, L., Nineveth, E., Emlano, E. and Ocompo, H.B. (1988) Estimating returns to research investment in mango in the Philippines. Los Banos, Laguna; Philippino Council for Agriculture, Forestry and Natural Resources Research and Development, 1988, 114p.

Anonymous (1985) Cocoa, Arecanut and Spices Statistics 1970-1983. Directorate of Cocoa, Arecanut & Spices Development, Government of India, Calicut, Kerala.

Anonymous (1996) Guidelines for Exporters of Spices to the European Market. Commonwealth Secretariat, 106pp, ISBN:85092-528-2 EIDD.

Anonymous (2003) Turmeric Futures. Feasibility study, Multi Commodity Exchange of India Ltd., Mumbai – 400099.

Arshad, F.M. (1990) The Integration of Palm Oil Market in Pesisular Malaysia, *Indian Journal of Agricultural Economics,* 45:21-30.

DASD (Directorate of Areacanut and Spices Development) Cocoa, Arecanut and Spices Statistics (1970, 1983, 1985, 2003) DASD, Calicut, Kerala.

Devakaran, D. and Balaraman Nair, B.M. (1985) India's export trade in ginger and turmeric. *In* Proc. National Seminar on Ginger and Turmeric, (eds) Nair, M.K., Premkumar, T., Ravindran, P.N. and Sarma, Y.R. Central Plantation Crops Research Institute, Kasaragod.

FAO (2000) Spices. In. Definition and Classification of Commodities (Draft). http://www.fao.org/WAICENT/faoinfo/economic/faodef/fdefioe.htm.

FAS/USDA (1999) Tropical Products: World Markets and Trade. United States Department of Agriculture, Foreign Agricultural Service, Circular Series FTROP 1-99, March, 1999.

Garner, J.F. (1995) Market Expectations. World Spice Congress, Feb 15-18, 1995, Cochin, India.

George, C.K. (1995) Turmeric – Quality improvement at farm level. *In* Quality improvement of turmeric, (eds) Sivadasan, C.R. and N.A.Devandra Shenoy. Spices Board, Cochin.

George, C.K. and Velappan, E. (1985) Production and development of ginger and turmeric in India. *In* Proc. National Seminar on Ginger and Turmeric, (eds) Nair, M.K., Premkumar, T., Ravindran, P.N. and Sarma, Y.R. Central Plantation Crops Research Institute, Kasaragod.

IISR (2002) *Annual Report*. Indian Institute of Spices Research, PB No.1701, Marikunnu PO, Calicut, Kerala (India).

International Trade Centre UNCTAD/GATT (1982) Spices: a survey of the world market. Geneva, 1982. 2vols.

International Trade Centre UNCTAD/GATT (1996) The Global Spice Trade and the Uruguay Round agreements Geneva: ITC/CS, 1996, xi, 99p.

International Trade Centre UNCTAD/WTO (1998) Market Research File on Spices: Overview of the European Union, Poland, Hungary, Czech Republic, Russian Federation. M.DPMD/98/0103/Rev.1, 83p.

ITC (1982) Spices: A Survey of the World Market, Vol.I, International Trade Centre, UNCTAD/GATT; Geneva, 1982.

ITC (1995) Market Development: Market brief on turmeric, overview of the world market. International Trade Centre, UNCTAD/WTO; Geneva, July 1995.

Jaiswal, P.C. (1980) Hand book of Agriculture. Indian Council of Agril. Research, New Delhi:1179-1184.

Joseph, P. and Sethumadhavan, P. (1985) Curing of turmeric. *In* Proc. National Seminar on Ginger and Turmeric, (eds) Nair, M.K., Premkumar, T., Ravindran, P.N. and Sarma, Y.R. Central Plantation Crops Research Institute, Kasaragod.

Kallupurakkal, J.A. and Ravindran, P.N. (2002) Turmeric: hints for cultivation. *Spice India, August, 2002.*

Kandiannan, K. (2002) Influence of varieties, times of planting, spacing and nitrogen levels on growth, yield and quality, crop-weather and growth simulation modelling and yield forecast in turmeric. Unpublished Ph.D thesis submitted to TNAU, Coimbatore.

Kannan, S. (1995) Turmeric – Indian exports and international markets. *In* Quality improvement of turmeric, (eds) Sivadasan, C.R. and Devandra Shenoy, N.A. Spices Board, Cochin.

Kerutagi, M.G., Kotikal, Y.K., Hulamani, N.C., and Hiremath, G.K. (2000) Costs and returns of turmeric production in Belgaum district of Karnataka; *Karnataka J. Agricultural Sci.,* 13 (1):209-211.

Krishnamurthy, N., Sampathu, S.R. and Sowbhagya, H.B. (1993) Farm processing of some spices – Pepper, turmeric & ginger. *In* Proc. Post Harvest Technology of Spices, Indian Society for Spices, Calicut.

Lakshmanachar, M.S. (1985) Marketing of ginger and turmeric in India. *In* Proc. National Seminar on Ginger and Turmeric, (eds) Nair, M.K., Premkumar, T., Ravindran, P.N. and Sarma, Y.R. Central Plantation Crops Research Institute, Kasaragod.

Lokesh, G.B. and Chandrakanth, M.G. (2004) Economics of production, marketing and processing of turmeric in Karnataka, *Ind. Jour. Agril. Mktg., 18(2).*

Murthy, C.S. and Naidu, M.R. (1989) Break-even analysis for appraising profitability in turmeric farming in Guntur district of Andhra Pradesh. *Indian Cocoa, Arecanut and Spices J.,* 12(4): 113-115.

Naidu, M.R. and Hanumanthaiah, C.V. (1988) Price spreads of turmeric and chillies regulated marketing in Guntur district, Andhra Pradesh – A comparative study. *Ind. J. Agril. Mktg., 2(1).*

Natarajan, C.P., and Lewis, Y.S. (1985) Technology of ginger and turmeric. *In* Proc. National Seminar on Ginger and Turmeric, (eds) Nair, M.K., Premkumar, T., Ravindran, P.N. and Sarma, Y.R. Central Plantation Crops Research Institute, Kasaragod.

Pruthi, J.S. (1989) Post –harvest technology of spices and Condiments – Pre-treatments, Curing, cleaning, grading and packing. Report of the Second Meeting of the International Spice Group, Singapore, 6-11 March 1989.

Rao, V.C. (1985) Problems and prosepects of cultivation and marketing of turmeric in Andhra Pradesh. *In* Proc. National Seminar on Ginger and Turmeric, (eds) Nair, M.K., Premkumar, T., Ravindran, P.N. and Sarma, Y.R. Central Plantation Crops Research Institute, Kasaragod.

Ravallion, M. (1986) Testing Market Integration. *Asian J. Agricultural Economics,* 68:102-9.

Raveendran, N. (1982) An analysis of domestic trade of turmeric in Tamil Nadu. *Agricultural Situation in India, 37(8): 511-515.*

Raveendran, N. and Aiyasamy, P.K. (1982) An analysis of export growth and export prices of turmeric in India. *Indian J. Agricultural Economics,* 37(3): 323-325.

Sasikumar, B. (2004) Ginger and Turmeric – Improved varieties and preservation of seed rhizomes. *Spice India, February,* 2004.

Senthilnathan,S., Srinivasa, N., Govindarajan, K., and Karunakaran, K.R. (2001) Better returns in turmeric through processing. *Spice India, August 2001, pp.2-6.*

Shukla, G.S. (1985) Quality control on Ginger and Turmeric. *In* Proc. National Seminar on Ginger and Turmeric, (eds) Nair, M.K., Premkumar, T., Ravindran, P.N. and Sarma, Y.R. Central Plantation Crops Research Institute, Kasaragod.

Sikka, R.K., Lakshmanachar, M.S. and George, C.K. (1985) Price fall in turmeric. *In* Proc. National Seminar on Ginger and Turmeric, (eds.) Nair, M.K., Premkumar, T., Ravindran, P.N. and Sarma, Y.R. Central Plantation Crops Research Institute, Kasaragod.

Spices Board (1988) Report on the Domestic Survey of Spices (Part-1). Spices Board, Ministry of Commerce, Govt. of India., Cochin-18.

Spices Board (1990) Production and Export of Turmeric. Status paper on spices, pp.37-39. Spices Board, Cochin-18.

Spices Board (1992) Report of the Forum for Increasing Export of Spices. Spices Board, Ministry of Commerce, Govt. of India., Cochin-18.

Spices Board (2000) Draft Annual Report (1999–2000). Spices Board, Ministry of Commerce, Govt. of India, Cochin, p. 22.Spices Board (2004) Spices Statistics. Spice board, Cochin.

Srinivasan, K.R., Vetriselvan, J., Vasanthakumar, J. and Subbiah, A. (2004) Improved techniques for turmeric growers in Erode. *Spice India, February, 2004, pp. 9-12.*

Sriramrao, T. (1985) Commercial varieties of turmeric in Andhra Pradesh. *In* Proc. National Seminar on Ginger and Turmeric, (eds.) Nair, M.K., Premkumar, T., Ravindran, P.N. and Sarma, Y.R. Central Plantation Crops Research Institute, Kasaragod.

Sudhakar, G. (1996) Price trends of turmeric in Andhra Pradesh markets. *Indian Cocoa, Arecanut and Spices Journal,* 20(2): 52-56.

UNCTAD/GATT (1982) Spices: a survey of the world market, International Trade Centre, Geneva.

United States Department of Agriculture/Foreign Agricultural Service (1998) Tropical Products: World Markets and Trade. Circular Series, FTROP 1-98, March, 1998.

United States Department of Agriculture/Foreign Agricultural Service (1999) Tropical Products: World Markets and Trade. Circular Series, FTROP 1-99, March, 1999.

Vasanthakumar, G.K. (1985) Problems and prospects of ginger and turmeric cultivation in Karnataka. *In* Proc. National Seminar on Ginger and Turmeric, (eds.) Nair, M.K., Premkumar, T., Ravindran, P.N. and Sarma, Y.R. Central Plantation Crops Research Institute, Kasaragod.

Velappan, E., Thomas, K.G. and Elizabeth, K.G. (1993) New technologies for on-farm processing of spices. *In* Proc. Post Harvest Technology of Spices, Indian Society for Spices, Calicut.

Vinning, G. (1990) Marketing Perspectives on a Potential Pacific Spice Industry. ACIAR Technical Reports No.15.60p.

12 Turmeric in Traditional Medicine

R. Remadevi, E. Surendran, and Takeatsu Kimura

CONTENTS

12.1 INTRODUCTION

Turmeric is found only under cultivatation. Its original home is believed to be either the South or Southeast Asian (SEA) region. In India, Ayurveda and Unani systems of medicines have used turmeric from time immemorial. Kirtikar and Basu (1984) mentioned that the Mohammedans use turmeric medicinally in the same manner as the Hindus. There are many Arabic names such as *Uruk-es-Sufr* meaning gold root; *Uruk-es-Sabaghin* i.e., dyers root, and *Zard-Chubah* meaning stick saffron. According to Dymock et al. (1867), turmeric came into use in India as a substitute for saffron and other yellow dyes, which were used by the ancient Aryans before they invaded the country.

Turmeric rhizome is used externally in China and Cambodia for cutaneous afflictions and internally against colic, amenorrhoea, and congestions. In Cambodia, the leaves are considered antipyretic. In Japan, China, and Papua New Guinea, the root is used as a stimulant. In Madagasker, the rhizome is used as tonic, stimulant, laxative, carminative, cordial, emmenagogue, astringent, diuretic, and maturant (Kirtikar and Basu, 1984). Soft tubers are rubbed on insect bites. In Chinese traditional medicine, the rhizome with some other ingredients is used in cases of retrograde biliary gastritis (Anon, 2001)

It is considered as an auspicious substance and has been intimately associated with the Hindu customs and rituals from ancient times. This aspect has recently been reviewed by Remadevi and Ravindran (2005).

12.2 TURMERIC IN INDIAN TRADITIONAL MEDICINE

The first evidences of the use of turmeric, known as *Haridra,* are found in *Atharvaveda* and it was considered a curative drug for skin disease, graying of hair, and for charming away jaundice. In Tibetan medicine also, the term "Haridra" is given for turmeric.

Ayurveda is the offshoot of *Atharvaveda* and is succeeded by the treatises called *Samhitas*, of which *Charaka Samhita* and *Susruta Samhita* are the renowned ones. References on *haridra* are found in them. Charaka included turmeric as one of the drugs in the following *ganas* (groups): *Lekhaneeya* (reducing corpulence), *Kushtaghna* (remedy for skin diseases), *Kandooghna* (remedy for itching), and *Vishaghna gana* (remedy for poisonous affections). Susruta also included *haridra* in the groups of drugs (*Vargas*) indicated for: *Yoniroga* (vaginal and uterine diseases), *Stanya Aamaya* (disorders of breast milk), *Deepaneeya* (digestive), etc. The above-given references are from the main classifications of drugs in both the texts, and there are many indications obtainable from the descriptions given in different contexts of treatment. Along with turmeric, Emblic myrobalan (*Emblica officinalis*) is indicated as the specific drug of choice for diabetes by Vagbhata. *Samhita* period was followed by the *Nighantu* period and during that time, several lexicons were written and published, which contain mainly the descriptions about drugs, i.e., their synonyms, properties, and uses.

There are about 55 synonyms for turmeric in Sanskrit, indicating mostly some of its qualifying properties, out of which 35 have been identified etymologically and philologically (Shah, 1997). These names give us an insight into the type of usages turmeric was put to. For example: *Jwaraanthik* (alleviating fever), *Krimighni,* (killing worms, wormicidal) *Mehaghni* (antibiotic activity), *Vishaghni* (destroying poison), etc. (Shaw, 1997; Remadevi and Ravindran, 2005).

Four types of turmeric are mentioned in *Bhavaprakasa* (a famous Ayurvedic lexicon). They are:

Haridra (Turmeric) — *Curcuma longa*
Vana Haridra (Wild turmeric) — *Curcuma aromatica*
Karpoora Haridra (Camphor turmeric) — *Curcuma amada*
Daru Haridra (Tree turmeric) — *Berberis aristata*

Among these four varieties, the majority of authorities consider turmeric and wild turmeric to be the same plants indicated above. The other two varieties are considered different, having separate identity and uses; indeed, the last one belongs to a primitive dicot family entirely different in properties and chemical composition.

In *Bhaishajyaratnavali* (an Ayurvedic treatise of the 11th century, containing several herbomineral medicines), the purification method suggested for turmeric is steam cooking of the rhizomes. The liquid used should be cow's urine or a decoction of specific five leaves or a decoction of an aromatic plant group described in Ayurveda or the juice of the plant *Biophytum sensitivum*. This type of purification is not practiced anywhere in India now. Turmeric is now processed by boiling rhizomes in clean water for about 30 min and then sun drying. The identity test prescribed is that good turmeric should be reddish orange in appearance when broken or cut into two and should also have a moist feeling (Nadkarni, 1976).

12.3 PROPERTIES OF TURMERIC

According to *Ayurveda,* turmeric has the following properties:

Rasa (taste) — *Thikta* (Bitter) and *Katu* (pungent)
Guna (property) — *Rooksha* (irritant, to make dry, rough)
Veerya (potency) — *Ushna* (hot)
Vipaka (metabolic property) — *Katu* (pungent)

Turmeric is bitter in taste and its action is "pungent-like" after digestion and metabolism. Being hot, light, acrid, and irritant, it is able to reduce corpulence; stimulate all functions, and clear channels. Bhavamisra (an *Ayurvedic* scholar, the author of the ancient lexicon *Bhavaprakasa Nighantu*) denotes turmeric as a curing agent for *Kapha* (phlegmatic disorders) and *Pitta* (digestive, metabolic, and related diseases). It is very good for skin afflictions and acts as an enhancer of complexion. It is effective in all types of skin diseases, diabetes, bleeding, and other blood-related diseases, inflammations, anemia, and abscess. In *Rajanighantu* (another ancient lexicon by Narahari), *Haridra* (turmeric) is stated to be an effective remedy for rheumatoid arthritis and itching, in addition to the above. *Nighanturatnakara* (yet another ancient lexicon of *Ayurveda*) points out more actions such as anthelmintic property, antipoisonous effects, and curative property in catarrhal affections, anorexia (absence of appetite), and enlargement of neck glands. Indications for the use of turmeric as a specific single drug are available in *Charaka samhita, Susruta samhita, Ashtanga sangraha,* and the lexicons of Chakradatta and Vangasena (all of which are ancient treatises of Ayurveda), for diabetes, leprosy, extreme thirst, elephantiasis, and calculus.

The use of turmeric in India generally comes under different headings as shown below:

*   As a spice
*   As an auspicious substance in Indian religious rituals
*   As a dye
*   As a cosmetic
*   As a medicine
    *   In tribal medicine
    *   In Ayurvedic medicine
*   As a home remedy (folk medicine)
    *   In other traditional systems

The use of turmeric as a spice, a dye, or a cosmetic is well known the world over. The socioreligious aspect is also interesting, and it reveals how strongly turmeric is related to the Indian tradition (Remadevi and Ravindran, 2005). It is indeed regarded as the golden boon to mankind. Investigations on its unlimited uses are going on even now, although the saga of turmeric started probably about 4000 yr ago.

12.4 INDICATIONS

Turmeric has got a wide range of activities, properties, and uses as per the ancient traditional medicine texts, some of which are as aromatic, stimulant, tonic, carminative, and anthelmintic. It is effective in treating liver obstruction and dropsy, is externally used for ulcers and inflammation, cures flatulence, dyspepsia, anorexia, intermittent fevers, prurigo, eczema, sprain, bruises, wounds, inflammatory troubles of joints, small pox, chicken pox, catarrhal and purulent ophthalmia, conjunctivitis, cough, ring worm and other parasitic skin diseases, piles, common cold, catarrh, coryza, hysterical fits, relieves pain in scorpion sting, chronic otorrhoea, reduces indolent swellings, and is used in the treatment of urinary diseases, leucoderma, diseases of blood, bad taste in mouth, elephantiasis, diarrhoea, bronchitis, vertigo, and gonorrhoea, (Nadkarni 1976; Kritikar and Basu 1984). It is intellect-promoting (*Sayana*), antidote for snake venom (*Kausika Sutra*), in cardiac complaints and jaundice (*Atharva veda samhita*).

Duke (2003) has made an exhaustive list of the known and reported uses of turmeric in the treatment of illnesses. Turmeric is indicated against a variety of health problems and pathological conditions and used traditionally by a large number of ethnic communities in a variety of conditions. Some of the properties are well documented and validated by pharmacological and clinical trials, while many remain to be validated (Duke, 2003). Jager (1997) compiled 114 biological properties of turmeric from the USDA database.

> Abscess, Adenoma, Adenosis, Allergy, Alzheiner's, Amenorrhea, Anorexia, Arthrosis, Asthma, Atherosclerosis, Athlete's foot, *Bacillus*, Bacteria, Bite, Bleeding, Boil, Bowen's disease, Bronchosis, Bruise, Bursitis, Cancer, Cancer – abdomen, Cancer – bladder, Cancer – breast, Cancer – cervix, Cancer – colon, Cancer – duodenum, Cancer – esophagus, Cancer – joint, Cancer – liver, Cancer – mouth, Cancer – skin, Cancer – stomach, Cancer – uterus, Cardiopathy, Cataract, Catarrh, Chest ache, Childbirth, Cholecystosis, Circulosis, Cold, Colic, Coma, Congestion, Conjuctivosis, Constipation, Coryza, Cramp, Cystosis, Dermatosis, Diabetes, Diarrhea, Dropsy, Duodenosis, Dysgeusia, Dysmenorrhea, Dyspepsia, Dysuria, Eczemia, Edema, Elephantiasis, Enterosis, Epilepsy, Epistaxis, Esophagosis, Fever, Fibrosis, Fungus, Gallstone, Gastrosis, Gonorrhea, Gray hair, Headache, Hematemesis, Hematuria, Hemorroid, Hepatosis, High blood pressure, High cholesterol, High triglycerides, Hyper lipidemia, Hysteria, Immunodepression, Infection, Inflammation, Itch, Jaundice, Laryngosis, Leprosy, Leukemia, Leishmania, Leukoderma, Leukoplakia, Lymphoma, Malaria, Mania, Morning sickness, Mucososis, Mycosis, Nematode, Nephrosis, Nervousness, Ophthalmia, Osteoarthrosis, Ozena, Pain, Parasite, Polyp, Psoriasis, Puerperium, Radiation injury, Restenosis, Rheumatism, Rhinosis, Ring worm, Scabies, Small pox, Sore, Sore throat, Sprain, Stone, Staphylococcus, Stroke, Swelling, Syphilis, Trauma, Ulcer, Uveosis, Vertigo, Vomiting, Wart, Water retention, Whitelow, Worm, Wound, Yeast (Duke 2003).

In Chinese medicine, turmeric rhizomes and tubers (root tubers) are used for different purposes. Turmeric rhizome is said to be a "blood" and *Qi* (vital energy) stimulant, with analgesic properties. It is used to treat chest and abdominal pain and destention, jaundice, frozen shoulder, amenorrhoea due to blood stasis, and postpartem abdominal pain due to stasis. It is also used for injuries (Chang and But, 1987). The "tuber" has properties more or less similar, but is used in hot conditions as it is more cooling and has been used to treat viral hepatitis (Bensky and Gamble, 1986).

Other uses of turmeric in traditional system are:

1. It is an essential substance to purify the gum resin of *Commiphora mukul* (Guggul) before it is made use of in Ayurvedic formulations.
2. Turmeric powder is mixed with the latex of *Snuhi* (*Euphorbia nerifolia*) plant and is then coated over the surgical thread repeatedly. This thread is known as *Ksharasoothra*, which is tied on piles and fistula to cure them effectively.
3. In veterinary medicine, turmeric is used to heal wounds or ulcers of animals.

4. In "leech therapy," turmeric powder is sprinkled over the leech to detach it from the biting site. Again turmeric powder is added to the water, in which the leech is kept, to make it vomit the sucked blood.
5. Turmeric powder is used as an insect and ant repellant and sprinkled around the vessels to be protected.
6. Turmeric is included in the group of yellow substances (*Peetha varga*) in *Rasa sastra* (Alchemy), used in the processing of Mercury.

12.5 PROPERTIES OF TURMERIC COMPONENTS

Turmeric has extensive use as mentioned above. Turmeric, as a therapeutic measure, is discussed below as per the indications available from *Ayurvedic* texts, using modern terminology. Mills and Bone (2000) and Remadevi and Surendran (2006) have recently summarized the properties and uses of turmeric.

12.5.1 ANALGESIC ACTION

Turmeric is indicated in painful traumatic conditions and deep-rooted pains (Sarma, 2005). The powdered rhizome is effective in the treatment of sprain and inflammation (Khare, 2000). Turmeric paste mixed with a little lime and saltpeter and applied hot is a popular application to sprains (Nadkarni, 1976). A decoction of turmeric in purulent conjunctivitis is very effective in relieving pain.

12.5.2 ANTI-INFLAMMATORY ACTION

Inflammatory changes of joints are often associated with rheumatic complaints. Turmeric is attributed with hot potency and anti-inflammatory action. It cures the etiological factors and pathological changes of inflammation. The anti-inflammatory activity of curcumin was first reported in 1971 (Srimal et al., 1971). It was further reported that oral doses of curcumin possess significant anti-inflammatory action in both acute and chronic animal models. Curcumin had been proved to be safe in human trials and had demonstrated anti-inflammatory activity, as reported by Chandra and Gupta (1972). In clinical trials, curcumin was reported to be effective in rheumatoid arthritis. (Deodhar et al., 1980). A clinical trial in eight patients with definite rheumatoid arthritis showed significant improvement in morning stiffness and joint swelling after two week-therapy (Chattopadhyaya et al., 2004). Sathyavathi et al. (1978) found, on experimental study, that turmeric has significant anti-inflammatory activity in rats, compared favorably with hydrocortisone acetate and phenylbutazone. The volatile oil obtained from the rhizome, given orally, was compared with cortisone acetate against adjuvant arthritis in rats. It suppressed the primary swelling on the third day. The protective effect of volatile oil of turmeric in early inflammatory lesions has been attributed to the antihistaminic effect (Chandra and Gupta, 1972). Anti-inflammatory action of turmeric was studied by Tripathi et al. (1973), and the volatile oil was found to inhibit trypsin as well as hyaluronidase enzymes. Srimal et al. (1971) found that curcumin inhibited carrageenin-induced edema in rats and mice, as well as subacute arthritis in rats and cotton pellet-induced granuloma formation in rats. Razga et al. (1995) studied experimentally the effect of curcumin and found that it significantly reduced croton oil-induced ear edema in mouse. The anti-inflammatory activity is supposed to be related to its known inhibitory effect on leukotriene B4 formation.

Iyengar et al. (1994), after experimental studies, reported that the anti-inflammatory effect of the volatile oil of turmeric is similar to that of phenylbutazone. The oil was found to be potent in cotton pellet granuloma study also. Srivastava et al. (1995) reported that turmeric inhibited aggregation of platelet and altered eicosanoid metabolism in human blood platelets. These effects on eicosanoid biosynthesis might explain the anti-inflammatory properties of curcumin. Alcohol and aqueous extract of *Vanaharidra* (*Curcuma aromatica*) showed anti-inflammatory activity on the paw edema of mice in a study conducted by Jangde et al. (1998). Alcoholic extract was slightly more effective than the aqueous one. Thus, *Sodhahara* (anti-inflammatory) and *Vataasra roga samana* (antirheumatoid arthritis) action of *haridra* mentioned by ancient scholars of Indian medicine is now well proven through experimental and clinical trials.

Slaked lime is traditionally mixed with turmeric powder for topical application as an anti-inflammatory agent (Nadkarni, 1976), and the paste is used against scorpion sting for fast relief. The Hindu women of India apply "kumkum" (a mixture of turmeric powder and slaked lime, which produces a brilliant red colour) on their forehead and the same is regarded as a sacred mark of the happily married. The process of mixing slaked lime and turmeric powder may probably increase the water solubility of curcumin through salt formation (Mills and Bone, 2000). The anti-inflammatory activity of sodium curcuminate was investigated in rats in an experimental model of this traditional use. Sodium curcuminate exhibited considerably higher anti-inflammatory activity than either curcumin or hydrocortisone in acute and chronic tests. This was confirmed in a later study in which curcumin and sodium curcuminate were more potent than phenyl butazone in acute and chronic models. However, they were much less effective than ibuprofen in reducing subacute inflammation (Mills and Bone, 2000). But curcumin was found to have a lower ulcerogenic index (0.60) than a nearly equivalent active dose of phenylbutazone (1.70). Lower doses of curcumin given to guinea pigs protected them from gastric ulceration from phenylbutazone. Ulceration caused by high doses of curcumin is associated with a marked reduction in mucin secretion. The probable mode of action of curcumin has been discussed recently by Mills and Bone (2000), Khanna (1999), Joe et al. (2004), and Chattopadhyay et al. (2004).

12.5.3 HEALING PROPERTY, SKIN CARE

According to Ayurveda, turmeric is *Vranahara* (ulcer healing), *Varnya* (improve complexion), *Tvakdoshahara* (cure skin diseases), and *Kandoohara* (cure itching). Till recently, before the onslaught of synthetic and herbal skin care products in the market, womenfolk were dependent more on turmeric, and they used to smear their bodies with a mixture of turmeric–sandal paste for gaining a golden glow to their skin, (Remadevi and Ravindran, 2005). Turmeric helps to remove hairs and impart colour and improve complexion of skin. Several Sanskrit synoyms of turmeric indicate its color-improving property (such as: *varna-datri* — one who gives color, idicates its use as enhancer of body complexion; *hemaragi* and *hemaragini* — both indicate golden color, meaning that it is used by womenfolk to get a golden complexion; *yoshti priya*, meaning favourite of young women, indicating its use for enhancing beauty; *hridayavilasini*, meaning giving delight to heart, charming; etc.). It is considered as an effective wound-healing medicine and is strongly related to the social customs of India. If a wound occurs as a part of a ritual, only turmeric powder is used for healing.The wounds are usually caused by old, rusty, unclean iron sword or hooks while performing certain rituals; even in such cases the wounds get healed without any pus formation or infection.

The fresh juice of turmeric is believed to have antiparasitic property in many skin afflictions. Externally, it is used for indolent ulcers (Chopra et al., 1958). A paste of turmeric and the leaves of *Justicia adhatoda* with cow's urine are rubbed on the skin in case of prurigo. Similar combinations such as turmeric and neem leaves, turmeric and ashes of plantain tree, etc. are used in the same manner. Turmeric powder with cow's urine is given internally also in prurigo and eczema. Turmeric mixed with gingelly oil is applied over the body to prevent skin eruptions. A coating of turmeric powder or a thin paste is applied on small pox and chicken pox patients to facilitate the process of scabbing (Nadkarni, 1976). *Haridrakhanda,* a traditional confectionary preparation described in *Bhaishjya ratnavali*, is very effective in prurigo, boils, urticaria, and chronic skin eruptions.

Oil of turmeric and its ether and chloroform extracts have proved to be antifungal, antiprotozoan, antiviral, and antibacterial (Chattopadhyaya et al., 2004). Experimental studies proved that curcumin enhances cutaneous wound healing in rats and guinea pigs by increasing the formation of granulation tissue and biosynthesis of extracellular matrix proteins (Joe et al., 2004). The bactericidal properties of turmeric have been proved by clinical testing (Khanna, 1999). Systemic treatment with curcumin in local muscle injury led to faster restoration (Joe et al., 2004).

In an experimental study, Rafatullah et al. (1990) proved significant antiulcerogenic activity of the ethanol extract of turmeric in rats. The oral dose of extract 500 mg/kg showed highly significant protective effect against cytodestructive agents. A paste of fresh rhizome of turmeric, on applying over the cut end of the umbilical cord of a newborn, showed quick healing effect (Bhat et al., 1993).

The wound-healing property of turmeric was investigated by Cohly et al. (1999), who observed that turmeric decreased the nitric oxide synthetase (NOS) levels and proved effective in chronic and acute wounds. In a screening for antibiotic property, turmeric showed broad-spectrum antibacterial activity (Omoloso and Vagi, 2001). Turmeric oil obtained as a by-product from curcumin manufacture was subjected to antibacterial study by Negi et al. (1999) and found effective against *Bacillus cereus, Bacillus coagulans, Bacillus subtilis, Staphylococcus aureus, Escherichia coli,* and *Pseudomonas aeruginosa.*

Jaekwan et al. (2000) reported that xanthorrhizol from *Curcuma xanthorrhiza* is a potential antibacterial agent against *Streptococcus nutans* and proved that it is also a potential anticarcinogenic agent.

12.5.4 ANTIDIABETIC PROPERTY

From the *Samhita* period itself (ca. 4000 yrs), turmeric was famous for its antidiabetic property. Vagbhata (an ancient Ayurvedic scholar who wrote *Ashtanga sangraha*) in his treatise gives specific drugs of choices for various diseases. Turmeric along with Emblic myrobalan (*E. officinalis*) is the specific combination for diabetes. The irritant or drying property, antiphlegm action, etc. make turmeric effective against all diseases associated with metabolism and fat deposition. The action is denoted as *Lekhaneeya* by Charaka.

Experimental study reports also prove the efficacy of turmeric in diabetes. Arun and Nalini (2002) studied the efficacy of turmeric on blood sugar and polyol pathway in albino rats and found that both turmeric and curcumin decreased blood sugar level in alloxan-induced diabetes. Curcumin was found to be capable of decreasing the complications in diabetes mellitus (Sajithlal et al., 1998). The report suggests that the antidiabetic action of turmeric may be mainly through the vitalization of pancreatic cells and by stimulation of insulin production. The ethanolic extract of turmeric was found to lower blood glucose level when given as injection to experimental rats. The lowering effect was 37.2% after 3 h and 59.5% after 6 h (Tank et al., 1990). Feeding curcumin to diabetic rats improved their metabolic status (Balm and Srinivasan, 1995, 1997). Diabetic rat maintained on a 0.5% curcumin diet for 8 weeks excerted comparatively lower amounts of albumin, urea, creatinine, inorganic phosphorous, sodium, and potassium. Glucose excretion or the fasting sugar level was unaffected by dietary curcumin, and the body weight was not improved.

12.5.5 ANTHELMINTIC PROPERTY

Turmeric is said to be *Krimihara* (anthelmintic) and *Krimighna* (destroyer of worms) in Ayurvedic lexicons. The juice of turmeric has anthelmintic property on internal use. In the rural areas of Nepal, turmeric powder or paste boiled in water with a little common salt is taken as an anthelmintic (Nadkarni, 1976). Alcoholic extract of rhizomes was found to have antiprotozoal activity against *Entamoeba histolytica* (Dhar et al., 1968). Curcumin has antileishmania activity *in vitro* (Koide et al., 2002)

12.5.6 Turmeric in Respiratory Diseases

Turmeric is well accepted as a *Kaphahara* drug (phlegmatic conditions are termed as "Kapha" and that which cures it is *Kaphahara*). Turmeric is anti-inflammatory and antipurulent in nature. Thus, it is useful in case of respiratory system diseases. Traditionally, it is prescribed for rhinitis, bronchitis, and cough. Turmeric powder mixed with warm milk is a household remedy for common cold and allergic rhinitis. It is reported that volatile oil of turmeric as oral drug in a clinical trial was found very effective in the treatment of bronchial asthma (Jain et al., 1990). Oral administration of turmeric powder provided significant relief from asthma and cough (Jain et al., 1979). Fresh rhizome proved effective against whooping cough and other coughs and in dyspnea (Khare, 2000).

In catarrh and coryza, the inhalation of burning turmeric fumes causes copious mucous discharge and gives instant relief (Nadkarni, 1976). The root, parched and powdered, is given in bronchitis (Kirtikar and Basu, 1984). A report of clinical trials in respiratory diseases such as bronchial asthma, bronchitis, bronchiectasis, and tropical eosinophilia revealed that turmeric could play a vital role as an adjuvant in improving the airway resistance. A trial conducted by Katiyar et al. (1999) in 114 patients led to the conclusion that turmeric has no role in the treatment of tropical pulmonary eosinophilia. Antiasthmatic property of curcumin had been tested in guinea pig model by Ram et al. (2003).

12.5.7 Turmeric in Urinary Diseases

In the Ayurvedic system of medicine, turmeric is an effective remedy in urinary disorders, on account of its pungent, bitter, and hot properties, as well as its specific antiseptic property (Nadkarni, 1976). Some recent experimental studies suggested that the administration of curcumin is a promising approach in the treatment of renal disorders (Venkatesan et al., 2000, and Jones and Shoskes, 2000). In Brunes (Darussalam), turmeric rhizome is used to cure urinary infection, as a traditional method. Kolammal (1979) quoted Vangasena (an ancient Ayurvedic expert, who had written his own treatise) that turmeric is good for calculus. Leskover (1982) reported that curcumin and curcumioids as oral drug prevent the formation of urinary calculi. The nephroprotective effect of curcumin was analyzed by Venkatesan et al. (2000) in rats. They studied the effect of curcumin on adriamycin (ADR)-induced nephrosis in rats and found that the injury was prevented by curcumin treatment. Curcumin protected ADR induced proteinuria, albuminuria, hypoalbuminaemia, hyperlipemia, and urinary excretion. Curcumin restored renal function.

12.5.8 Turmeric in Liver Diseases

From ancient times, turmeric was linked very much with the traditional ways of treatments based on sorcery. For curing jaundice, turmeric paste was applied over the body of the patient, and the sorcerer carried out magical expulsion of the disease. After that, the turmeric was washed off and the people believed that the disease also got washed off together with the turmeric (Remadevi and Ravindran, 2005). Turmeric is considered good for afflictions of the liver (Chopra et al., 1958; Kirtikar and Basu 1984). Turmeric is effective in treating jaundice and is recommended in the diet of patients suffering from jaundice or even infective hepatitis (Anon, 2001). Clinical trial with turmeric and *Phyllanthus fraternus* for treating infective hepatitis has proved very effective, without any side effects (Anon, 2001). In Japan, crude turmeric rhizomes were tested in experimental animals against CCl_4-induced hepatotoxicity. The curcuminoids showed significant antihepatotoxic action (Anon, 2001). Ethanolic extract of turmeric showed significant hepatoprotective effect. Curcumin in combination with *Eclipta alba* and *P. fraternus* was a promising combination against liver injuries, which normalized the level of lipid accumulated in the liver and brought down the level of serum bilirubin in CCl_4-induced hepatotoxicity in experimental rats. The level of serum triglycerides, pre-β-lipoproteins and cholesterol improved and that of glycogen normalized after treatment (Anon, 2001). Turmeric is reported to cause contraction of the gallbladder. Turmeric and curcumin are especially effective in increasing bile flow in infected bile ducts. Curcumin has antilithogenic property and a dose of 0.5 g reduced the incidence of cholesterol gallstones in experimental animals (Anon, 2001). Hussain and Chandrasekhara (1992) reported that curcumin reduced the rate of gallstone formation in response to a lithogenic diet. Rasyid et al. (1999) investigated the effect of curcumin on gallbladder volume of healthy volunteers using ultrasonography and found that curcumin induced contraction of human gallbladder. According to them, sodium curcuminate was found to stimulate bile flow by action as a hydrocholagogue. Pretreatment of rats with curcumin 1 h before the ingestion of paracetamol protected them from the vascular changes and necrosis in liver tissue. Pulla Reddy and Lokesh (1994) studied the effect of dietary turmeric on iron-induced lipid peroxidation in rat liver and reported lowered lipid peroxide level by enhancing the activities of antioxidant enzymes. Chow et al. (1995) investigated the hepatoprotective properties of an aqueous extract of rhizomes of Javanese turmeric (*C. xanthorrhiza*) against acute liver damage induced by acetaminophen or CCl_4 in mice and found substantial alleviation in liver damage. Rajakrishnan et al. (1998, 1999) reported elevated serum cholesterol, phospholipids and free fatty acids in ethanol-fed rats, but on curcumin treatment, they decreased significantly, thus establishing the protective effect of curcumin in ethanol toxicity. Protective effect of turmeric extract on CCl_4-induced liver damage in rats was reported by Deshpande et al. (1998) who observed reduced serum levels of cholesterol and bilirubin, reduced activities of aspartate aminotransferase (AST), alanine aminotransferase (ALT), and alkaline phosphatase. Jeon et al. (2000) also found curcumin very effective in an assay on its protective effect in rat liver.

12.5.9 TURMERIC IN DIGESTIVE SYSTEM

Turmeric is a traditionally used spice and has formed an essential ingredient in Indian recipes from time immemorial. Ayurveda specifies it as an antinoxious substance, which eliminates the toxicity in any dietary substance and protects from its aftereffects. In the digestive system, turmeric acts as a carminative and protective against intestinal gas formation. The hot potency of turmeric (as per Ayurveda) enables it as a digestive and stimulant. Turmeric is an important constituent of the group of drugs indicated for diarrhea, in *Ashtanga hridaya* and *Susruta samhita*, two of the most respected lexicons in *Ayurveda*. Turmeric is antiflatulent, digestive, and stimulant due to its hot potency. It is reported to have antispasmodic activity, inhibiting excessive peristaltic movements of the intestine (Chopra et al., 1958; Anon., 2001)

Bhavanisankar and Srinivasa Murthy (1979) reported the antiflatulent activity of turmeric/curcumin in experimental animals. Curcumin enhanced intestinal lipase, sucrase, and maltase activity (Patel and Srinivasan, 1996). Turmeric powder increased mucin secretion in rabbits and thus acted as a protecting agent against irritants (Lee et al., 2003). In experimental studies, curcumin showed protective effects from ulcerogenic effects of phenylbutazone (Dasgupta et al., 1969; Sinha et al., 1974), but 0.5% curcumin failed to protect, and at higher doses of 50 mg or 100 mg/kg it produced ulcers (Gupta et al., 1980). Curcumin blocked indomethacin; ethanol and stress-induced gastric ulcers in experimental rats (Chattopadhyaya et al., 2004). Van Dau et al. (1998) studied turmeric clinically in 118 patients on the healing of duodenal ulcer. The dosage was fixed as 6 g daily, as suggested in the Vietnamese pharmacopeia and reported as not superior to the placebo in healing duodenal ulcer after 4 to 8 weeks.

Mukherjee et al. (1961) found that turmeric powder appreciably increases the mucin content of gastric juice in rabbits. The report suggested that the beneficial effects of turmeric powder as a therapeutic agent in gastric disorders might possibly be due to its mucus-stimulatory effect. Srihari Rao et al. (1982) observed the antispasmodic activity of sodium curcuminate in isolated guinea pig ileum.

12.5.10 TURMERIC IN OPHTHALMIC CARE

Turmeric is indicated in traditional medicine in catarrahal and purulent ophthalmia, conjunctivitis, etc. A decoction made from the rhizomes of turmeric is said to relieve pain of purulent ophthalmia and, in India, it is a common practice to use a piece of cloth soaked in turmeric solution for wiping away discharges of acute conjunctivitis and ophthalmia (Chopra et al., 1958; Nadkarni, 1976; Kritikar and Basu 1984). Central Food Technological Research Institute, Mysore, isolated a water-soluble peptide (0.1% of dry weight) from turmeric, having antioxidant activity. It inhibited deoxyribonucleic acid (DNA) damage, especially produced by wood smoke, and reported that it can reduce the opacity on eye lens, produced by smoke condensate and thereby prevent loss of vision (Anon., 2001). Awasthi et al. (1996), after experimental study, concluded that curcumin might be an effective protective agent against cataractogenesis induced by lipid peroxidation. Efficacy of curcumin in the management of chronic anterior uveitis (CAU) was investigated clinically by Lal et al. (1999). Curcumin was administered orally to patients suffering from CAU at a dose of 375 mg tds for 12 weeks and found that the efficacy of curcumin in curing CAU, and the recurrences following treatment were comparable to that of corticosteroid therapy. The lack of side effects with curcumin forms the greatest advantage, compared to corticosteroids.

Halder et al. (2003) screened some indigenous plants for their lens aldose reductase (LAR)-inhibiting potential. Turmeric and three other indigenous plants were found effective in inhibiting LAR activity.

12.5.11 ANTITUMOR, ANTICANCEROUS ACTIVITY

Ancient Ayurvedic lexicons indicate turmeric as *arbudahara*, meaning curative of tumors or cancerous afflictions. *Nighanturatnakara*, *Nighantusangraha*, and *Rajavallabhanighantu* (ancient texts) point out turmeric as an effective remedy to *Apachi*, a Sanskrit term to denote enlargement of neck glands. Dietary turmeric could be effectively used as a chemopreventive agent in benzo-(alpha)-pyrene-induced forestomach tumors in Swiss mice. An ethanolic extract of turmeric, as well as an ointment containing curcumin, is reported to produce remarkable symptomatic relief in patients with external cancerous lesions (Anon, 2001). It is now proved that the antioxidants present in turmeric neutralize carcinogenic free radicals (Lehrh et al., 1992). Curcuminoids possess anti-carcinogenic property due to their oxygen radical-scavenging property (Saudamini et al., 1989). Kuttan et al. (1985) evaluated and proved the anticancer activity of turmeric.

The antioxidant and antitumor-promotimg effects of curcumin were shown to be due to the induction of apoptosis in human leukemia cells, and this aspect was studied and positively proved by Kuo et al. (1996). Goel et al. (2001) investigated the specific inhibitory effect of cyclooxygenase (cox)-2 by dietary curcumin in human colon cancer cells. The suppressive effect of curcumin on human breast carcinoma cells was proved by Shao et al. (2002). Choudhari et al. (2002) also found that curcumin induced apoptosis in human breast cancer. It has been shown that the inhibition of arachidonic acid metabolism, modulation of cellular signal transduction pathways, inhibition of hormone, growth factor, and oncogene activity are some of the mechanisms by which curcumin causes tumor suppression (Gesher et al., 1998). Mutagenesis induced by ultraviolet (UV) irradiation is suppressed by the presence of cucumin (Oda, 1995). A Chinese traditional medicine containing turmeric is reported to have inhibitory action on induced Ehrlich ascitis tumor cells in experimental animals (Anon, 2001). Azuine et al. (1994) studied the effect of catechin and dietary turmeric in forestomach and oral cancer models and found that they are effective chemopreventive agents when regularly consumed. Catechin in drinking water and dietary turmeric significantly inhibited the tumor burden and incidence in tumor models. In a comparative study of curcuminoids for their free radical-scavenging activity, Nair and Rao (1996) found turmeric to be the most potent free radical scavenger, followed by dimethoxycurcumin and bis-demethoxy curcumin. Acetyl curcumin was found inactive. Parashar et al. (1995) reported the use of turmeric preparations in the treatment of cancer. In the course of a search for antitumor agents, the extract of turmeric was found to be effective in inducing apoptosis or programmed cell death (PCD) in human myeloid leukaemia cells (HL — 60) by Sang Hyun et al. (1996). The active compounds for PCD were isolated from hexane extract of the rhizome and identified as ar-turmerone and beta-atlantone. Singletary et al. (1996)

evaluated curcumin for its capacity to inhibit mammary tumor, initiating the activity of dimethyl-benz(a)anthracene (DMBA) and the *in vivo* formation of mammary DMBA–DNA adducts in female rats. It is concluded that curcumin can act as an effective chemoprotective agent toward DMBA-induced mammary tumerogenesis and mammary adduct formation.

Protective effects of curcumin and two other plant constituents were studied by Thresiamma et al. (1995) on radiation-induced toxicity, and the results indicated that oral administration of curcumin, ellagic acid, bixin, and alpha tocopherol significantly reduced the lung collagen hydrox-yproline, lipid peroxidation, serum lipid, and superoxide dismutase activity. The effect of topical application of curcumin was investigated by Mou Tuan et al. (1997) and found that all doses of curcumin inhibited the tumor promotion. Curcumin had a strong inhibitory effect on DNA and ribonucleic acid (RNA) synthesis in cultured Hela cells.

In another study, Bonte et al. (1997) observed that curcuminoids protected normal human keratinocytes from hypoxanthine/xanthine oxidase injury. Further, they proposed that since curcumi-noids synergistically inhibited nitrobluetetrazolium reduction, a decrease in superoxide radical for-mation, leading to lower levels of cytotoxic hydrogen peroxide, might explain the protective effect.

The chemopreventive effect of curcumin was assayed by Kawamori et al. (1999) during the promotion/progression stages of colon cancer. The inhibition of adenocarcinomas of the colon was reported as dose dependent. Curcumin treatment during the initiation and postinitiation stages as well as throughout the promotion/progression stage increased apoptosis in colon tumors, compared with groups receiving azoxymethane (AOM) and the control diet. Khar et al. (1999) revealed that the antitumor activity of curcumin is mediated through the induction of apoptosis in AK -5 tumor cells. Gupta et al. (1999) reported that turmeric inhibited tumor necrosis factor (TNF)-α-induced expression of adhesion molecules on human umbilical vein endothelial cells. Curcumin was the most potent among the three compounds tested, as inhibiting TNF-α induced expression of intercellular adhesion molecule-1(ICAM-1), vascular cell adhesion molecule-1 (VCAM-1), and E-selectin by human umbilical vein endothelial cells. Curcumin-I, II, and III from turmeric were assayed by Ramsewak et al. (2000) for their cytotoxicity and antioxidant, and anti-inflammatory activities. These compounds were reported to have potent activity against leukemia and colon, central nervous system (CNS), melanoma, renal, and breast cancer cell lines. Aggarwal et al. (2003) evaluated the anticancer potential of curcumin. Human clinical trials indicated no dose-limiting toxicity when administered at doses up to 10 g/d. The available evidences indicate that turmeric and curcumin can inhibit cancer at the initiation, promotion, and progression stages of TPA (12-O-tetradecanoylphorbol-13-acetate)-induced tumor promotion in mouse skin (Mills and Bone, 2000). All the studies thus suggest that curcumin has enormous potential in the prevention and therapy of cancer.

The mechanism of anticancerous action of curcumin is considered in different angles. Curcumin inhibits the expression of nuclear factor kappa beta (NF-κB) and block the estrogen-mimiking chemicals that promote cancer growth. Curcumin inhibits cox and Lypoxygenase (Lox), two enzymes that promote inflammation and are believed to play a significant role in the development and progression of squamous cell carcinoma and colon cancer. Curcumin protects the body by scavenging free radicals; inducing apoptosis, and stopping cancer cells from multiplying. Curcumin enhances immunity by stimulating CD_4 + T-helper and B type immune cells. It inhibits angiogenesis by blocking AP-1 enzyme and removing iron and copper from the bloodstream.

12.5.12 ANTICHOLESTEROL ACTION

According to *Ayurveda*, due to the hot potency, the drying or irritant property, the bitter and pungent taste, and the pungency as metabolic property together make turmeric a very effective remedy in all phlegmatic afflictions of body. Hence, it is used in all cases of fat-related diseases. Turmeric, as well as curcumin, is reported to reduce the uptake of cholesterol from the gut and increase the high-density lipids (HDL) cholesterol and decrease low-density lipids (LDL) type. It can also inhibit the peroxidation of serum LDL, which can lead to atherosclerotic lesions. Thus, turmeric can prevent coronary problems and heart diseases (Anon, 2001). Khanna (1999) reported that the levels of serum cholesterol and liver cholesterol decreased to one-half, while cholesterol-fed rats were treated with curcumin. Deposition of cholesterol was found to be high in liver sections of rats fed with cholesterol and least in specimens from animals concurrently fed with curcumin. Curcumin increased fecal excretion of bile acids and cholesterol, in normal and hypercholesterolemic rats. This biliary drainage explains the reduction of tissue cholesterol (Patil and Srinivasan, 1971). Kumar et al. (1998) evaluated the antiatherosclerotic effect of "Lipo tab forte," a herbal preparation in cholesterol-fed rabbits. Lipo tab forte is a turmeric-containing herbal formulation, developed on the principles of the Unani system of medicine. The cholesterol levels were significantly decreased by the test drug and also lowered the aorta thiobarbituric acid-reactive substance levels very significantly. Hepatic cholesterol level also reported to be reduced significantly.

Arafa (2005) investigated the possible hypolipidemic effect of curcumin in rats fed on a high-cholesterol diet (HCD). He found an obvious hypocholesterolemic effect that is supposed to be due to an effect on cholesterol absorption, degradation, or elimination, but not due to an antioxidant mechanism. The report suggests that the ingestion of curcumin-containing spices in diet, especially rich in fat, could have a lipid-lowering effect.

12.5.13 ANTIFERTILITY

In Bihar, India, the tribals use turmeric as an effective antifertility agent. Turmeric is reported to possess antifertility activity, as observed in experimental animals. Petroleum ether and aqueous extracts produced 100% anti-implantation effects in rats at a dose of 200 mg/kg body weight fed orally on the first to seventh day of pregnancy (Garg et al., 1978). Garg (1974) further reported incomplete inhibition of implantation by turmeric extracts. Curcumin is reported to inhibit testosterone 5-reductase, which converts testosterone to 5-dihydrotestosterone, and thereby inhibit the growth of flank organs in hamster (Liao et al., 2001). Rithaporn et al. (2003) studied the effect of curcumin as a potential vaginal contraceptive and found that it inhibited human sperm motility and had the potential for the development of novel intravaginal contraceptive. The test results indicated that curcumin had a selective sperm-immobilizing effect in addition to a previously studied anti-human immunodeficiency virus (HIV) property.

Purohit and Bhagat (2004) investigated the contraceptive effect of turmeric in male albino rats and observed a reduction in sperm motility and density in treated group. Turmeric is supposed to have affected the androgen synthesis, either by inhibiting the Ley dig cell function or hypothalamus pituitary axis, thereby inhibiting the spermatogenesis.

The effect of hexane extract of rhizomes of *Curcuma comosa* (from Thailand) on fertility in adult male rats was investigated by Piyachaturawat et al. (1999). The alterations produced by *C.comosa* extract were similar to the effects of estradiol. A 7-d treatment did not significantly affect fertility, but suggested that the suppressing effect on male reproductive organs was mediated through the estrogen-like action of the plant extract.

12.5.14 TURMERIC ON THE CNS

Turmeric has intellect-promoting action, according to Ayurvedic descriptions. Turmeric along with honey and ghee is claimed to be *Medhya* (a Sanskrit term used to indicate intellect promoting activity), according to an ancient Ayurvedic scholar Sayana (Sastry, 2005). Vajragupta et al. (2003) and Thyagarajan et al. (2004) investigated the neuroprotective effect of curcumin and reported its efficacy against vascular dementia by exerting antioxidant activity. Rajakrishnan et al. (1999) studied the neruroprotective role of curcumin on ethanol-induced brain damage. Oral administration of curcumin caused a significant reversal in lipid peroxidation and enhanced glutathione content, revealing that the antioxidative and hypolipemic action of curcumin were responsible for its protective role against ethanol-induced brain injury.

A recent study conducted at the University of California, Los Angeles, has found that turmeric is a potential drug against Alzheimer's disease. These studies were conducted on genetically altered mice, and the results suggested that turmeric not only inhibited accumulation of beta amyloid in the brain of Alzheimer's patients, but also broke up the existing plaques (Anon. 2005). In a drug discovery effort against Alzheimer's disease, Park and Kim (2002) were able to isolate compounds from turmeric, which effectively protect PC 1 2 cells from Beta-A insult.

The antidepressant activity of aqueous extract of turmeric in mice was investigated by Yu et al. (2002). The extracts of turmeric, at 140 mg/kg or above for 14 d, significantly inhibited monoamine oxidase (MAO) activity in mouse. The effect of retinol deficiency and turmeric and curcumin intake on brain microsomal Na$^+$K$^+$–ATPase and its kinetic parameters were studied in rats by Kaul and Krsihnakanth (1994). The results demonstrated that retinol deficiency caused changes in the brain microsomal membrane-bound Na$^+$K$^+$–ATPase activity, and curcumin or turmeric feeding could reverse the induced changes.

Xu et al. (2005) investigated the effect of curcumin on depression-like behavior in mice. They studied a traditional Chinese formula, *Xiaoyaosan*, in which turmeric is the major component, which has been used effectively to treat depression-related diseases in China. Curcumin treatment significantly reduced the duration of immobility in test animals. The neurochemical assays showed that curcumin produced marked increase in serotonin levls but not in adrenalin or dopamin levels.

12.5.15 TURMERIC IN CARDIAC COMPLAINTS

In *Sounakeeya Atharva Veda Samhita* (an ancient treatise on *Ayurveda)*, turmeric powder is proposed for dry massage in *Hridroga* (cardiac complaints). The attributes of turmeric as per Ayurvedic science is self-explanatory for its cardioprotective effects that are amply supported by many experimental studies. Curcumin protected experimental mice against thrombotic change (Kosuge et al., 1985). Liu et al. (1984) reported anticoagulant property of curcumin and some other principles isolated from turmeric. The antioxidant properties of phenolics enable turmeric to protect against cardiovascular diseases precipitated by free radical–mediated damage to biological systems (Joe et al., 2004). Nirmala and Puvanakrishnan (1996) found that curcumin decreases the severity of pathological changes and thus protects from damage caused by myocardial infarction. Sumbilla et al. (2002) investigated the slippage of Ca^{2+} pump in skeletal and cardiac sarcoplasmic reticulum and found that curcumin improved Ca^{2+} transport, thereby raising the possibility of pharmacological interventions to correct the defective Ca^{2+} homeostasis in the cardiac muscle. Dikshit et al. (1995) after an experimental study reported that curcumin prevented ischemia-induced changes in the cat heart. Nirmala et al. (1999) conducted a study to evaluate whether curcumin — a potent antioxidant — has any specific role in the synthesis and degradation of collagen in the rat heart with myocardial necrosis induced by Isoproterenol HCL (ISO). Curcumin pre- and cotreatment with ISO decreased the degree of degradation of the existing collagen matrix and newly synthesized collagen. The observed effects are suggested to be due to the free radical-scavenging capacity and inhibition of lysosomal enzyme release by curcumin.

Ashraf et al. (2005) studied the atheroscleroprotective potential of diet supplementation with garlic and turmeric, by measuring serum lipid profile and changes in cardiovascular parameters i.e., arterial block, pressure, electrocardiogram (ECG), and heart rate. This study demonstrated that garlic and turmeric are potent vasorelaxant and also reduce the atherogenic properties of cholesterol.

12.5.16 IMMUNE PROTECTIVE ACTION

Ayurveda specifies turmeric as an alleviator of *Vata*, *Pitta,* and *Kapha*. According to the Ayurvedic system of medicine, all the physiological functions of the human body are governed by three basic biological parameters — the *thridosha*, three *doshas*, or the three basic qualities: *Vata*, *Pitta*, and *Kapha; Vata* is responsible for all voluntary and involuntary movements in the human body, *Pitta* is responsible for all digestive and metabolic activities, and *Kapha* provides the static energy (strength) for holding tissues together, and also provides lubrications at various joints of friction. When these three qualities (*dosha*) are in a normal state of equilibrium, the human body is healthy and sound, but when they lose equilibrium and become vitiated due to various internal and external factors, they produce varied diseases. Ayurvedic treatment of any disease is aimed at restoring the equilibrium of the three *doshas*, or qualities.

Lele (2001) had given a list of plants clinically proven as immunomodulatory drugs, and among them, turmeric is a prominent one. He also praises the wisdom of ancient Indians in choosing the most appropriate antioxidants for routine use to maintain positive health through drugs, such as turmeric, garlic, ginger, gooseberry, etc., which are a part of the normal Indian vegetarian diet. Turmeric is reported to have an antioxidant activity at various levels; it has the capability to scavenge the primary radical, break chain propagation, and repair reconstruction of lipid membrane, aided by glutathione peroxidase or by increase in level of glutathione. Churchill et al. (2000) studied the effect of curcumin on lymphocytes and found an increase in mucosal CD4 (+) T-cells and B-cells in pretreated animals. This finding suggests that curcumin modulate lymphocyte-mediated immune functions. Kang et al. (1999) reported that curcumin inhibited Th 1-cytokine profile in CD4 < Sup (+) > T-cells by suppressing interleukin-12 production in macrophages. The investigation pointed to a possible therapeutic use of curcumin in Th 1-mediated immune disease.

On an experimental analysis of curcumin on mice, Antony et al. (1999) observed a significant increase in macrophage phagocyte activity, and the results indicated the immunomodulatory activity of curcumin. Taher et al. (2003) found that curcumin inhibited UV light–induced HIV activation. The investigators concluded that the mechanism by which curcumin modulates UV activation of HIV–LTR gene expression mainly involves the inhibition of NF-κB activation.

12.5.17 OTHER ACTIVITIES AND USES

Turmeric is also used in the health care of cattle. Pradhan (1999) evaluated the utility of an herbal topical gel containing turmeric in mastitis control and udder health improvement. It was found to be an effective prophylactic in preventing mastitis by maintaining udder health.

Curcumin inhibited human HIV-1 integrase *in vitro* and it is a modest inhibitor of HIV-1 and HIV-2 proteases. Curcumin also has anti–Epstein–Barr virus activity. Many turmeric components exert various biological effects, which are summarized in Table 12.1.

12.6 THERAPEUTICS

As a medicine, turmeric is used in different fields of health care. In tribal medicine, it is used as single or as a compound drug for several ailments. In Ayurveda, it forms a major or important ingredient in many traditional recipes. It is used alone as a single drug remedy also. As a folk medicine or home remedy, it occupies the foremost place. Other traditional systems of medicine such as Chinese, Arabian, etc. also use turmeric as a potent drug. Some of the therapeutic uses are listed below.

12.6.1 AYURVEDA

The system of Ayurvedic medicine relies mainly on compound medications, though many single usages are also available. It is not possible to list the names of all the Ayurvedic preparations containing turmeric. Some of the important ones are mentioned in Table 12.2.

12.6.2 USAGES IN PRIMARY HEALTH CARE

Turmeric plays a very important role in upkeeping the health of the rural Indian people. A great many home remedies having turmeric as the main component are being used routinely by millions

TABLE 12.1
Biological Properties of the Major Components of Turmeric

Component	Pharmacological Property
Curcumin	Anti-HIV., anti-EBV, antiadenoma – carcinogenic., antiaflatoxin, antiatherosclerotic, antiaggregant, antiangiogenic, antiarachidonate, anticancer, antiedemic, anti-ischemic, anti-inflammatory, antileukemic, antileukotrene, antilymphomic, antimelanomic, antimetastatic, antimutagenic, antinitrososaminic, antioxidant, antiperoxidant, antiprostaglandin, antisarcomic, metal chelator, antithromboxane, antitumor agent, antiviral, apoptotic, cox-2 inhibitor, fibrinolytic, hepatoprotective, immunostimulant, ornithine decarboxylase inhibitor, protease inhibitor, protein kinase inhibitor.
Bis-desmethoxycurcumin:	Antiangiogenic, anti-inflammatory, cytotoxic, anticancer
Demethoxy curcumin	Antiangiogenic, anti-inflammatory, anticancer
Tetrahydro curcumin:	Antioxidant, anti-inflammatory
Alpha curcumene	Antitumour, anti-inflammatory
Ar-turmerone	Anti-inflammatory, antitumour, cox-2 inhibitor, choleretic, hepatotonic
Turmerine	Antimutagenic, antioxidant, DNA protectant
Curcumol	Anticancer, antisarcomic, antitumour (cervix)
Curdione	Antileukopenic, antisarcomic, antitumor, anti-X-radiation
Dehydrocurdione	Analgesic, antiarthritic, antiedemic, anti-inflammatory, antioxidant, antipyretic, calcium channel blocker
Zingiberene	Antirhinoviral, antiulcer, carminative

Source: From various sources and Duke (2003).

of Indians for their primary health care. Some such uses of turmeric in primary health care are listed below.

1. Cold: Milk boiled with turmeric rhizome or powder, sweetened with sugar, is a popular remedy for cold and other allergic affections.
2. Leech therapy: Turmeric paste is applied over bruises, insect bites, leech bites, etc. Turmeric powder is sprinkled on the bite to repel the leech during leech therapy.
3. Conjunctivitis: A decoction of turmeric as external wash is effective in relieving pain of purulent conjunctivitis.
4. Coryza: The fumes of burning turmeric, if inhaled, cause a copious mucous discharge and relieve the congestion in coryza.
5. Scorpion sting: The smoke produced by sprinkling turmeric powder over burnt charcoal will relieve scorpion sting if the affected part is exposed to the smoke for a few minutes.
6. Vertigo: A paste of fresh turmeric rhizome, if applied on the head, is good in cases of vertigo.
7. Joint inflammations: A paste of turmeric powder along with lime is a good remedy for joint inflammations.
8. Diabetes: Half ounce (15 ml) of fresh turmeric juice with equal quantity of fresh goose-berry juice (*E. officinalis*) in empty stomach can control diabetes. Turmeric powder, juice of gooseberry, and honey, if taken together, cure diabetes.
9. Fever: A mixture of a quarter teaspoon of turmeric powder, a small piece of ginger, and 10 black peppercorns boiled in one-and-a-half glasses of water (approx. 300 ml) and concentrated to half a glass. To this, honey or sugar candy is added and given three or four times to cure fever.
10. Stomachache: Fresh turmeric juice of about 2 Oz, with a pinch of salt, cures stomachache. Turmeric powder is ground with honey to make pills weighing 10 to 12 g,

TABLE 12.2
Traditional Ayurvedic Formulations Containing Turmeric

Choorna (Powder)

Nimbadi choornam	Skin diseases, rheumatoid arthritis.
Rajanyaadi choornam	Diarrhea, jaundice, anemia, fever, teething trouble
Raasnaadi choornam	Catarrh, coryza, headache
Saraswatha choornam	Intellect promoter, disorders of speech, mind, and memory
Sudarsana choornam	Hepatopathy, spleenomagali, fever

Kwatha (Decoction)

Mahaamanjishthadi kwatha	Skin diseases, filariasis
Nisaakathakaadi kwatha	Diabetes
Nisoseeraadi kwatha	Diabetes
Pathyadi kvaatham	Disorders of head, ear, and nose
Raasnadi kvatham	Rheumatic complaints
Vaasakaadi kvaatha	Cough, bleeding disorders

Thaila (Oil)

Arimedadi thailam	Eye disease, earache, headache
Arka thailam	Skin affictions
Chandanaadi thaila	Bleeding disorder, epilepsy, insanity
Jaatyaadi thailam	Wounds, skin disorder, fistula
Karanja thaila	Skin afflictions
Kumkumaadi thaila	Acne, discoloration of face
Laakshaadi thaila	Fever, wasting, catarrh in children
Mahaalaakshaadi thaila	Fever, wasting, catarrh in children
Mahaanaaraayana thaila	Rheumatic disorders, wasting disorders, colic,
Nalpaamaradi thailam	Skin afflictions
Parinithakeraksheeraadi thailam	Brachial palsy
Pramehamihira thaila —	Diabetes, fever, rheumatic diseases
Somaraajee thaila	Fistula, rheumatoid arthritis, itching
Thriphalaadi thaila	Headache, catarrh, tubercular glands

Lepa (External Application — Poultice)

Dasaanga lepa	Fever, edema, erecepalus, skin disorders
Haridraadi lepa	Piles
Kritamaalaadi kalka	Skin disorders
Nisaadi lepa	Skin disorders
Rajanee lepa	Piles

Avaleha (Confectionary)

Haridra khanda	Urticaria, allergic complaints
Kalyaana avaleha	Speech disorders

Ghritha (Ghee)

Arjunaadi ghritam	Cardiac diseases,
Jaatyaadi ghritam	Fistula, ulcer, wounds
Kaaseesaadi ghritam	Skin disorders
Mahaachaitasa ghrita	Mental disorders
Mahakalyanaka ghritam	Insanity, loss of memory
Mahapanchagavya ghritam	Epilepsy, edema, fistula
Phala ghritam	Infertility, gynecological diseases
Thriphala ghritam	Tumour, jaundice, cataract

Continued

TABLE 12.2 *(Continued)*
Traditional Ayurvedic Formulations Containing Turmeric

Kshaara (Ash)

Haridraadi kshaara	Peptic ulcer

Gutika (Pills)

Chandraprabhaa gutika	Diabetes, calculus, piles, urinary disorders
Chandraprabhaa varti	Eye disorders
Vilwaadi gulika	Insect bites, indigestion

Anjana (Collyrium)

Muktaadi mahaa anjana	Eye diseases
Nayana shona anjana	Cataract

Arishta and Asava (Medicated Alcohols)

Aswagandhaarishta	Fainting, mental disorders, loss of memory, epilepsy, insanity
Dasamoolaarishtam	Hiccough, cough, dysnpnoe, sprain
Devadaarvarishta	Diabetes, rheumatic disorders
Kooshmaandaasava	Cough
Kumaaryaasava	Hepatopathy, splenomegaly, anemia, heart diseases, dysmenorrhoea
Pippalyasava	Indigestion, anemia, gastric diseases
Saraswathaarishtam	Epilepsy, mental disorders, insanity

and, if administered one pill daily for one month, can cure frequently recurring stomach pain.

11. Flatulence: A glass of milk boiled with turmeric powder and garlic will relieve flatulence.

12. Constipation: A paste of two chebulic myrobalan (*Terminalia chebula*) and a piece of dried turmeric, daily morning and evening, is effective in chronic constipation.

13. Irregular menstruation: The juice of a large piece of fresh turmeric along with a little salt and cumin seed powder daily in the morning is effective in treating irregular menstruation.

14. Ringworm, itching, or eczema: A paste of fresh turmeric rhizome and fresh neem leaves can be applied.

15. Wounds: The fine powder of dried turmeric should be applied and bandaged to arrest bleeding in wounds.

16. Scars of chicken pox: A paste of fresh turmeric rhizome and red sandalwood (*Pterocarpus santalinus*) can be applied for scars of chicken and smallpox.

17. Cough: 1 to 2 g of turmeric powder with honey or ghee.

18. Menorrhagia: Turmeric powder with guggulu (powdered gum resin of *Commiphora mukul*) can be used to treat menorrhagia.

19. Skin afflictions: Turmeric powder along with cow's urine as internal medicine and turmeric powder and butter for external application can be used to treat skin afflictions.

20. Purulent eye infections: Boil one part of turmeric powder in 20 parts of water. The strained liquid can be used frequently as eye drops.

21. Piles with inflammation – Apply externally turmeric powder with aloe.

22. Filariasis: Turmeric mixed with treacle in cow's urine cures elephantiasis.

23. Sprains, wounds, bruises, inflammation: — external application with hot turmeric paste, lime and salt petre mixture.

24. Prurigo: A paste of turmeric and leaves of *Adhatoda vasika* with cow's urine should be rubbed on the skin. A paste of turmeric and neem leaves is good for external application. External application of turmeric powder and ashes of plantain tree is effective in prurigo.

25. Piles: An ointment of turmeric, hemp leaves, onion, and warm mustard oil gives great relief when the piles are painful and protruding.
26. Otorrhoea: Turmeric powder and alum powder in a proportion of 1:20, if blown into the ear, will cure chronic otorrhoea.
27. Diarrhea and dysentery: Buttermilk is boiled with turmeric powder, curry leaves, dried ginger, and some salt. It is used as a curry and for drinking to arrest diarrhea and correct digestion.
28. Skin diseases: A decoction made up of curry leaves and turmeric, taken daily, is effective in skin afflictions.
29. Skin afflictions, after chicken pox and after delivery: Water boiled with turmeric and neem leaves is used for washing wounds and ulcers and for bathing.
30. Caterpillar itching or urticaria: Turmeric powder mixed with sour buttermilk and applied over the affected parts is found to be very effective in relieving itching and redness.
31. Chicken pox: roasted turmeric powder is sprinkled over the open ulcers for rapid healing.
32. Insect bites: Turmeric powder mixed with holy basil leaf juice is applied over to the affected area to relieve pain, itching, and inflammation.
33. Deep ulcers: They are dressed with cooked rice powder and turmeric powder, filling the cavity with it for raising and healing the deep ulcer.
34. Inflammation and pain: Fresh turmeric rhizome and the whole plant of *Cyathula prostrata* (*Apamarga*) are made into a paste and applied over sprain to relieve inflammation and pain.
35. Common cold: Turmeric powder is mixed with bruised black pepper and spread over a small piece of cotton cloth. It is then rolled to make a wick. Soak it in sesame oil and burn one end of it. The smoke that comes out is good for inhalation for relieving nasal congestion in common cold.
36. Worm infections: A paste of fresh turmeric rhizome and garlic is an effective remedy in worm infections of the foot, especially around the toes. Instead of garlic, henna leaf can be used for better results.
37. Allergic reactions: A pinch of turmeric powder mixed with a small quantity of fresh cow's milk early in the morning on an empty stomach is very effective in allergic reactions such as allergic rhinitis, asthma, skin affection, etc.
38. Common cold, sinusitis: Rub turmeric powder overhead during common cold, sinusitis, etc.
39. Mouth ulcers: Turmeric powder with honey, as external application, is an effective remedy for ulcers in the mouth.
40. Lice on head: Apply turmeric and neem leaves paste along with kerosene oil over the scalp and wash after half an hour using traditional herbal shampoo. A paste of turmeric and custard apple leaves is also said to have similar effect.

In tribal medicine

1. Otorrhoea: Fresh turmeric rhizome is covered with a cloth smeared with oil and is burnt. The fumes are directed to the ear through a cone made up of Jack tree leaves. The fumes are said to be effective in healing otorrhoea.
2. Ulcers in mouth: Fresh turmeric juice along with honey is applied over the ulcers in mouth in infants.
3. Ulcers in mouth and intestine: Take seven neem leaves, seven black peppercorns, and equal weight of fresh turmeric rhizome, grind together, and administer internally.

4. Snake bite: Cut a fresh rhizome of turmeric and rub the cut surface over the bite.
5. Asthma: Make a paste of fresh turmeric and mustard seeds and apply it over the body. Administer internally a teaspoon of this paste.
6. Rat bite: Grind Holy basil (*Ocimum tenuifolium*) and fresh turmeric mix with a glass of water and administer for 3 d.
7. Spider bite: Grind holy Basil (*Ocimum tenuifolium*), *Ruta graveolens*, and fresh turmeric rhizome and collect the juice. Administer it one teaspoon two times a day for 2 d and also apply the juice over the affected part.
8. Blood formation: Grind the leaves of *Ichnocarpus frutescence* and fresh turmeric rhizome into a paste. Administer the same in a dose of 10 g every day.
9. For improving complexion and to remove body smell: Apply fresh turmeric paste all over the body before bath.
10. Burns: Boil fresh turmeric in coconut oil and grind it to a paste along with the oil and apply it over the affected part.
11. Concussion and swelling: Grind fresh turmeric rhizome with drumstick tree bark and apply it over the affected area. Apply fresh turmeric paste. Make a paste of Doob grass (Bermuda grass) and fresh turmeric rhizome and apply over the affected part. Apply a paste of fresh turmeric rhizome and bamboo shoots over the bruise.
12. Dental pain: Make a paste of Holy basil leaves and fresh turmeric rhizome and apply it near the site of pain.
13. Chronic diarrhea: Administer 10 g of the paste made up of turmeric and tender leaves of guava tree.
14. Diabetes: Drink 1 oz of gooseberry juice with turmeric powder and honey.
15. Fracture: Make a fine paste with mango tree bark, scrapings of bamboo, and fresh turmeric and apply over the area of fracture and tie it up.
16. Headache: Make a paste of fresh turmeric in a bronze vessel and apply it on the forehead.
17. Stone injury in the leg: Grind stem bark of *Erythrena variegata*, bamboo shoots covering, and fresh turmeric into a fine paste. Apply it over the affected part. Apply a paste of fresh turmeric and Bermuda grass (*Cynodon dactylon*).
18. Ring worm infection: Grind henna leaves and turmeric, heat the paste with coconut oil, and apply over the infected part.
19. Skin diseases: Rub a paste of *Leucas aspera* leaves and turmeric over the body before bath.
20. Sprain: Make a paste of turmeric and sweet flag (*Acorus calamus*) rhizomes and apply it over the affected area.
21. Worm infestation: Grind fresh turmeric and cumin seed and drink a teaspoon full.
22. Fever: Grind boiled turmeric and shoots of castor plants and apply it on the head.
23. Headache: Prepare a paste with fresh turmeric and touch-me-not plant (*Mimosa pudica*) and apply it on the forehead.

12.7 TURMERIC — USE IN CHINESE AND OTHER SOUTHEAST ASIAN TRADITIONAL MEDICINE

In the Southeast Asian (SEA) region, the most ancient and the most significant traditional medicine system is the Chinese medicine, often called traditional Chinese medicine (TCM), which has influenced the traditional medical systems of the neighboring countries. TCM has been influenced very much by the various Chinese systems of philosophy, notably Taoism, Budhism, and NeoConfucianism. TCM is based on the philosophical concept that the human body is a small universe with a set of complete and sophisticated interconnected system. The system is based on the concepts such as the *Yin* and *Yang* and *Qi*. TCM also includes the concept of the meridian system.

The understanding of the *Yin* and *Yang* is basic to the Chinese medicine. This concept holds that all things have two opposite aspects, *Yin* and *Yang*, which are both opposite and at the same time interdependent. The ancient Chinese used water and fire to symbolize *Yin* and *Yang*; anything moving, hot, bright, and hyperactive is *Yang*, and anything quiescent, cold, dim, and hypoactive is *Yin*. Each of the *Yin* and *Yang* properties is a condition for the existence of the other; neither can exist in isolation. Each organ has an element of *Yin* and *Yang* within it. The histological structures and nutrients are *Yin* and the functional activites are *Yang*. In a healthy body, the two are kept in balance; when the balance is vitiated, the body becomes ill. In such a condition, *Qi* gets affected and a disease develops. The concept of *Qi* is also a philosophical one, it can be best considered as the totality of the vital force circulating in the body (Lewith, 2006).

The Chinese system uses a variety of medicinal plant combinations to treat the diseased conditions of the body, which is primarily meant for regaining the balance between the *Yin* and *Yang* and for correcting the blockages or incorrect direction in the flow of *Qi*, disorders located in a specific organ, and the emotional problems that accompany physical illness. In addition to medicinal plants, the TCM also makes use of the acupuncture procedure for correcting the flow of *Qi* in the body. Turmeric and some of the related species are among the medicinal plants commonly used in Chinese medicine for the above-mentioned purposes.

In Chinese medicine, four species of curcuma are used. They are:

C. aromatica (C. wenyujin) – Yujin, Wenyujin (gentle *yujin, pian jianghuang* (Slice turmeric), *guangyjin (Guangzhou yujin)*. This is the main source of *yujin* and is the alternate source for *ezhu* (in place of *C. zedoaria*); very rarely used as *jianghuang* (in place of *C. longa*). Grown widely.

C. kwarigsiensis – guangxi ezhu, maoezhu (fibrous *ezhu*), *guiezhu* (cinnamon *ezhu*). This is the alternate source for *ezhu*, and sometimes used as *yujin*. Grown in the Guangzi province.

C. longa – jianghuang, chuan yujin (Sichuan *yujin*). This is the primary source for the drug *juanghuang*; it is also used as *yujin* in the Sichuan province. Grown in many areas in China as a spice, medicinal plant, and as a cash crop.

C. zedoaria – ezhu, huopeng ezhu, peng ezhu. Main source of *ezhu*, alternative source of *yujin*. Grown in many parts (also occur in the wild). Occasionally some related species are also substituted (such as *C. aeruginosa, C. phaeocaulis, C. pallida* etc.) (Dharmananda, 2005)

The Chinese pharmacopeal texts, such as the Pharmacopoeia of the Peoples Republic of China, mention the salient features of the three drugs.

Yujin: Root tubers harvested in winter, cleaned, steamed or boiled thoroughly, and sundried. The product (*yujin*) is ovoid to long fusiform, some of them slightly compressed, or curved, 2–6 × 0.5–2 cm diameter; greyish-yellow, brown to greyish-brown; longitudinally and disorderly wrinkled; fractured surface horny, greyish to greyish black, and exhibiting a pale-colored endodermis ring in the central part. Odor — slight, taste — bland.

Jianghuang: Rhizomes harvested in winter, cleaned, boiled or steamed, dried in sun. Irregularly ovate, cylindrical, or fusiform; frequently curved, branched 2 to 5 cm long and 1 to 3 cm diameter; deep yellow, rough with wrinkled striations and distinct rings (leaf scars), cut surface — yellow to deep orange-yellow, endodermis ring — distinct, with scattered dots representing vascular bundles. Fragrant, characteristic odor, taste — bitter and acrid.

Ezhu: Harvested in winter, cleaned, steamed or boiled, dried. The dried produce in ovate, conical or long fusiform, 2–8 × 1.5–4 cm in diameter. Externally greyish-yellow to yellow-brown, annular nodes distinct, with round slightly sunken scars. Fractured surface yellowish green or dark brown; waxy, endodermis ring yellowish-white, odor slightly fragrant, taste slightly bitter and pungent.

The Chinese pharmacopeia provides the following properties (Bensky and Gamble, 1986).

Yujin: " To promote flow of *Qi*, to eliminate, blood stasis, to calm the nerves and ease the mind, and increase the flow of bile. It is indicated for amenorrhea, dysmenorrhea, distending or pricking pain in the chest and abdomen; impairment of consciousness in febrile diseases; epilepsy, mania, jaundice with dark urine."

Jiang huang: "To eliminate blood stasis, promote the flow of *Qi*, stimulate menstrual discharge, and relieve pain. It is indicated in pricking pain in the chest and hypochondriac regions; amenorrhea, mass formation in the abdomen, rheumatic pain of the shoulders and arms, traumatic swelling and pain."

Ezhu: "To promote the flow of *Qi*, and eliminate blood stasis with powerful effect, and to relieve pain of removing the stagnation of undigested food. It is indicated for mass in abdomen, amenorrhea due to stagnation of undigested food, carcinoma of cervix at early stage."

The Illustrated *Chinese Materia Medica* mentions that *yujin* is good for the treatment of bleeding (such as blood ejection, spontaneous external bleeding, blood in urine, blood stranguary, and vicarious menstruation); that *Jiang huang* treats kidney pain, and that *Ezhu* dispels wind and clears heat and alleviates menstrual blockage due to blood stasis and pain due to traumatic injury (Dharmananda, 2005).

Yujin is included in the following Chinese medicines:
Guarvan Tang for treating coronary heart disease
Babao Risheng Dan for treating cold in the spleen and stomach causing pain and food retention
Yujin san for heat accumulation in the small intestine discharging to the bladder with blood in the urine
Augong Nichuang Wan for febrile diseases causing mental disorientation
Biajin Wan, for epilepsy and other neurological symptoms due to phlegm accumulation
Yujin Yinzi, for febrile disease with clouding of the spout

Ezhu is included in the following formulations:

Sanleng Heshang Tang — used for treating injuries, especially to the hypochondrum
Da Qiqi Tang — used in the treatment of masses due to *Qi* stagnancy and blood stasis, manifesting as soft and fixed abdominal masses
Neixiao Wan — used for *Qi* stagnancy and blood accumulation
Xiaoshi San — used for alleviating chronic food retention in children
Muxiang Binlang Wan — used for distention, fullness, and aching in the abdomen due to stagnation of *Qi* and accumulation of dampness and heat
Sanleg Wan — used for abdominal pain due to *Qi* and blood stagnation in women
Wengjing Tang — used for alleviating abdominal coldness that causes lengthened menstrual cycle and menstrual blood clots with painful menstruation

Jianghuang is used in preparations such as:

Zhixue Yuqi Zhi Fang — for regulating *Qi* and blood circulation in cases of hypochondriac pain

Zhongman Fenxiao Wan — for treating fullness and pain in the abdomen due to stagnation
   of *Qi* and accumulation of moisture
Juanbi Tang — for relieving pain (*bi* syndrome) in the shoulders, back, and neck
Jianghuang San — for alleviating pain in the heart

The formula *Jiuqi Niantong Wan* includes both *Ezhu* and *Yujin*, and it is used for treating stomach cold and abdominal pain, with *Qi* stagnation and distention (Dharmananda, 2005).

According to the Pharmacology and Applications of Chinese Materia Medica and Chinese Herbal Medicine Materia Medica (Bensky and Gamble, 1993), turmeric has been shown to have the following pharmacological properties in experimental animal systems:

- Reduces blood lipids
- Improves blood circulation to heart
- Lowers blood pressure
- Reduces platelet accumulation and promotes fibrinolysis
- Increases bile formulation and secretion
- Reduces inflammation
- Alleviates pain
- Stimulates uterine contraction

In clinical trials, turmeric was used to reduce blood lipids, treat angina pectoris, alleviate stomachache, remove gallstones, treat jaundice, and relieve postpartum pain.

C. zedoaria was reported to have the following effects in laboratory animal studies:

Antineoplastic effect
Prevention of leucopenia
Inhibition of platelet aggregation
Stimulation of the smooth muscles of the gastrointestinal tract

C. zedoaria has been used to treat coronary heart disease, liver cancer, anemia, and chronic pelvic inflammation. In another publication on Chinese herbal medicine, *C. zedoaria* is indicated for cancer, not only of the cervix and liver, but also of the ovary, lung, and thyroid and for lymphosarcoma and uterine fibroids. It not only inhibits cancer but also helps prevent leucopenia due to cancer therapies. It is used in the anticancer pill *Pingxiao Dan* and in several other anticancer drugs (Chang and But, 1987).

In Chinese Herbal Medicine Materia Medica (Bensky and Gamble, 1986), it was reported that *yujin* could lower cholesterol in rabbits fed on atherosclerotic diet, and that it could alleviate the symptoms of viral hepatitis in humans. All the three species of *Curcuma* have potential value for preventing and treating cardiovascular diseases, as might be expected of blood-vitalizing drugs. Turmeric and *C. zedoaria* in large doses would induce abortion in pregnant rats; traditional texts recommend that *yujin* should be used with caution during pregnancy.

12.8 USE OF TURMERIC IN OTHER SEA COUNTRIES

In Japanese medicine, turmeric is regarded as stomachic, stimulant, carminative, hematinic, styptic in all kinds of hemorrhages and a remedy for certain type of jaundice and other liver troubles. In Japan and Korea, it is indicated in cholecystidis, choleithiasis, and hepatitis. In Thai medicine, turmeric is considered as an antidiarrheal and stomachic and is used in the treatment of skin diseases. The juice of fresh turmeric is used in sore eyes.

Other *Curcuma* species are also rarely used in medicines. *C. aurantiaca* is used in local medicine in Indonesia. *C. caesia* is used as a rubefacient and as an alternative for *C. zedoaria*. *C. comosa* is used in Thailand as a tonic for the abnormalities of the uterus after birth. *C. euchronia* and *C. petiolata* are used in Indonesia as skin medicine and applied externally. *C. viridifolia* and *C. purpurescens* are used as antitussives in Indonesia. *C. xanthorrhiza* is used in Indonesia as a laxative and for hepatitis and cholelithasis.

C. zedoaria is more widely used. In Thai medicine, it is considered antidiarrheal and antipyretic, in Burma, it is regarded as a tonic to the brain and heart and is widely used for removing foul breath. It is used as an emmanagogue in Indonesia. In Japan and Korea, it is used in indigestion, abdominal pain, and blood stasis.

12.8.1　Formulations Containing Turmeric and Other Species

In Japan, the Ministry of Health and Welfare adopts only one traditional preparation (containing turmeric) composed by Seishu Hamaoka (1760–1835). This commercial drug, *Chuoko,* is an ointment meant for acute wounds and skin eruptions, bruises, and sprains. This formulation consists of: sesame oil, beeswax, turmeric rhizome, and phellodendron bark. In China, there are many traditional formulations that contain turmeric, *C. aromatica,* and *C. zedoaria.* Some of these are:

Yujin Jiudiaosan (Inkai, 1983) — contains *C. aromatica* (*Yujin*) together with eight other plant raw drugs. The formulation is used for pain in the eyes.

Chensho Yilijindan (Xinyixueyuan, 1977) — contains *C. aromatica* rhizome and three other plants. It is used for chest pain, abdominal pain, and urinary bladder pain.

Baiju Wan (Xinyixueyuan, 1977) — *C. aromatica* together with alum is used for hypochondria and epilepsy.

Yujin dan (Xinyixueyuan, 1977) — *C. aromatica* rhizome together with alum used for other raw drugs. It is used in infant epilepsy.

Jianghuang San (Perry, 1980) — contains turmeric powder and *C. zerumbet* along with six other herbs. It is used for pains, in general, menstrual disorder, toothache, and pains after childbirth.

Jingqinsiwutang (Nordal, 1963, 1965) — contains turmeric and eight other herbs. It is used for short interval of menstruation.

Yujun Yinzi — Root tubers of *C. aromatica,* together with five other herbs. It is used for fidget and chest pain.

Kimura (1996, 1998, 1999, 2001) has collected and collated the traditional and folk medicines, which also gives the many and varied medicinal uses of turmeric as practiced by people of the orient.

12.9　CONCLUSION

From the Vedic era onward, turmeric has been used as a medicine. There is no other oriental plant that has such a diversity of usage. As a spice, it is used all over the world. It is a medicine par exellence for healing a variety of illnesses. As a cosmetic, its history dates back to time immemorial. As a natural dye, it is being used from ancient times to the present. It is gaining more attraction in recent times due to its protective attributes in cases of cancer and Alzheimer's disease. Above all, the sacredness attached to this plant and the role that it has acquired in Indian social traditions is remarkable. Its medicinal properties are traditionally utilized in veterinary science also. With such a broad spectrum of use, turmeric, the golden rhizome, becomes a "living gold" that brings *mangala* — good luck. This golden spice is a divine boon, and no wonder it is held in high esteem in all oriental cultures.

REFERENCES

Aggarwal, B. B., Kumar, A. and Bharti, A. C. (2003) Anti-cancer potential of Curcumin: preclinical and clinical studies. *Anticancer Research,* 231, 363-398.

Anonymous (2005) News report in Hindu, April 25, 2005.

Anonymus. (2001) *Wealth of India,* Supplement series.Vol.2. National Institute of Science Communication, Council of Scientific &Industrial Research, New Delhi.

Antony, M.B. (2003) Indigenous Medicinal Plants; their extracts and isolates as a value added export product. Agrobios Pub., Jodhpur, India.

Antony, S., Kuttan, R. and Kuttan, G. (1999) Immunomodulatory activity of curcumin. *Immunological Investigations,* 28, 291-303

Apisariyakul, A., Vanittanakom, N., and Buddhasukh, D. (1995) Antifungal activity of turmeric oil extracted from *Curcuma longa* (Zingiberaceae). *J. Ethnopharmacology* , 49, 163-169.

Arafa, H.M. (2005) Curcumin attenuates diet-induced hypercholesterolemia in rats. *Med. Sci. Monit.,* 11(7), 228-234.

Arun, N. and Nalini, N. (2002) Efficacy of turmeric on blood sugar and polyol pathway in diabetic albino rats. *Plant Foods for Human Nutri.,* 57, 41-52.

Atharvaveda samhitha (Ancient Sanskrit text) - Cited from Sastri (2005).

Awasthi, S., Srivatava, S.K., Piper, J.T., Singhal, S.S., Chaubey, M. and Awasthi, Y.C. (1996) Curcumin protects against 4-hydroxy-2-trans-nonenal-induced cataract formation in rat lenses. *American J. Clinical Nutrition,* 64, 761-766.

Azuine, M.A. and Bhide, S.V. (1994) Adjuvant chemoprevention of experimental cancer: catechin and dietary turmeric in fore stomach and oral cancer models. *J. Ethnopharmacology,* 44, 211-217.

Bensky, D. and Gamble, A. (1986) *Chinese Herbal Medicine Materia Medica* 390-391, East land press, Seattle.

Bhavamisra. *Bhavaprakasa,* (San) 7 ᵗʰ edn. 1990, Chaukhambha Viswabharathi, Varanasi (ancient text reprint, Hindi commentary).

Bhavanisankar,T.N. and Srinivasa Murthy,V. (1979) Effect of turmeric (*Curcuma longa*) fractions on the growth of some intestinal and pathogenic bacteria *in vitro. Ind. J. Exp.Biol.,* 17,1363-1366.

Bonte, F., Noel-Hudson, M.S., Wepierre, J. and Meybeck, A. (1997) Protective effect of curcuminoids on epidermal skin cells under free oxygen radical stress. *Planta Medica,* 63, 265-266.

Chakrapanidatta– *Chakradatta,* commentary by Tripadhi, J.P. (1976) Chowkhmbha Sanskrit Series Office, Varanasi (ancient text reprint).

Chandra, D. and Gupta, S.S. (1972) Anti- inflammatory and anti-arthritic activity of volatile oil of *Curcuma longa* (Haldi) : *Ind. J.Med.Res.,* 60.131-142

Chang, H.M. and But, P.P. (1987) *Pharmacology and Applications of Chinese Materia Medica,* 2, 936-939.

Charaka samhitha, Revised by Charaka and Dridhabala, Chaukhambha Visvabharati, Varanasi (1984-reprint) (ancient text).

Chattopadhyaya, I., Biswas, K., Bandopadhyay U. and Banerjee, R.K. (2004) Turmeric and curcumin: bio-logical actions and medicinal applications. *Current Science,* 87, 44-53.

Chopra, R.N., Chopra, I.C., Handa, K.L. and Kapur, L.D. (1958) *Indigenous Drugs of India.* Academic Publishers, Calcutta-700073.

Choudhari, T., Pal, S., Aggarwal, M.L., Das, T. and Sa, G. (2002) Curcumin induces apoptosis in human breast cancer cells through p53-dependent Bax induction. *FEBS Lett.,* 512.334-340.

Chow, L.S., Lin, C.C., Lin, Y.H., Supriyatna, S., and Teng C.W. (1995) Protective and therapeutic effects of *Curcuma xanthorrhiza* on hepatotoxin-induced liver damage. *American J.Chinese Medicine,* 23, 243-254.

Churchill, M., Chadburn, A., Bilinski, R.T. and Bergagnolli, U. (2000) Inhibition of intestinal tumours by curcumin is associated with changes in the intestinal immune cell profile. *J.Surg. Res.,* 89, 169-175.

Cohly, H. H. P., Rao, M. R., Kanji, V. K., Manisundram, D., Taylor, A., Wilson, M. T., Angel, M. F. and Das, S. K. (1999) Effect of turmeric (Chemical Plant Extract) on *in-vitro* nitric oxide synthetase (NOS) levels in tissues harvested from acute and chronic wounds, *Wounds,* 11, 3, 70-76.

Dasgupta, S.R., Sinha, M., Sahana, C.C. and Mukherjee, B.P. (1969) A study of the effect of an extract of *Curcuma longa* Linn. on experimental gastric ulcers in animals. *Ind. J. Pharmacol.,* 1, 49-54

Deodhar, S.D., Sethi, R. and Srimal, R.C. (1980) Preliminary study on anti rheumatic activity of Curcumin (diferuloylmethane). *Ind .J. Med Res.,* 71, 632-634

Deshpande, U. R., Gadre, S. G., Raste, A. S., Pillai, D., Bhide, S. V. and Samuel, A. M. (1998) Protective effect of turmeric (*Curcuma longa* L.) extract on carbon tetrachloride-induced liver damage in rats. *Ind. J. of Experimental Biology*, 36, 573-577.

Dhar, M.L., Dhar, M.M., Dhavan, B.N., Mehrotra, B.N. and Ray, C. (1968) Screening of Indian plants for biological activity. Part 1 *Ind. J. Exptl. Biol.*, 6, 232.

Dikshit, M., Rastogi, L., Sukla, R., Srimal, R.C. (1995) Prevention of ischaemia- induced biochemical changes by curcumin and quinidine in the cat heart. *Ind.J.Med. Res.*, 101, 31-35.

Duke, J.A. (2003) *CRC Handbook of Medicinal Spices*. CRC Press, Boca Raton, USA.

Dymock, W., Warden, C.J.H.and Hooper, D. (1867) *Pharmacographia Indica*, Part VI. Education Society's Press, Byculla. Bombay.

Garg, S.K. (1974) Effect of *Curcuma longa* (rhizomes) on fertility in experimental animals. *Planta Med.*, 26.225-227.

Garg, S.K., Mathur, V.S. and Chaudhury, R.R. (1978) Screening of Indian plants for antiferility activity. *Indian J.Exp.Biol.*, 16.1077-1079.

Gesher, A., Pastorino.U., Plummer.S.M. and Manson, M.M. (1998) Suppression of tumour development by substances derived from the mechanisms and clinical implications. *Br. J. Clin.Pharmacol.*. 45, 1-12.

Goel, A., Boland, C.R. and Chauhan, D.P. (2001) Specific inhibition of cyclooxygenase –2(COX-2) expression by dietary curcumin in HT 29 human colon cancer cells. *Cancer Lett.*, 172, 111-118.

Gupta, B., Kulsreshtha, V.K., Srivastava, R.K. and Prasad, D.N. (1980) Mechanisms of curcumin induced gastric ulcer in rats. *Indian J.Med.Res.*, 71, 806-814.

Gupta, B. and Ghosh, B. (1999) *Curcuma longa* inhibits TNF-<alpha> induced expression of adhesion molecules on human umbilical vein endothelial cells. *International J. Immunopharmacology*, 21, 11, 745-757.

Halder, N., Joshi, S. and Gupta, S. K. (2003) Lens aldose reductase inhibiting potential of some indigenous plants.. *J. Ethnopharmacology*. 86, 1.113-116.

Hussain, M.S. and Chandrasekhara, N. (1992) Effect of curcumin on cholesterol gall stone induction in mouse. *Indian J.Med.Re.*, .96.288-291.

Inkai, K.I.D.H. (1983) (ed.) *Kampo Igaku Daijiten* Vol2, Yakuho Hen, Renmin Weisheng Chuban, Yukonsha, Beijing, Kyoto.

Iyengar, M.A., Rao, M.P., Rao, S.G. and Kamath, M.S. (1994) Antiinflammatory activity of volatile oil of *Curcuma longa* leaves. *Indian Drugs,* 31, 528-531.

JaeKwan, H., JaeSeok, S., NamIn,B., ansd YuRyang, P. (2000) Xanthorrhizol a potential antibacterial agent from *Curcuma xanthorrhiza* against *Streptococcus* mutans. *Planta Medica*, 66, 2, 196-197.

Jager, P. de. (1997) Turmeric, Vidyasagar Pub., California, USA. pp.67.

Jain J.P. *et al.* (1990) *J.Res. ayur Siddha* VII (1-4). 20-30). Cited from Khare.(2000)

Jangde, C. R., Phadnaik, B. S. and Bisen, V. V. (1998) Anti-inflammatory activity of extracts of *Curcuma aromatica* Salisb. *Indian Veterinary J.*, 75, 76-77.

Japanese Pharmacopoea (2001) 14th edn., Hirokawa Pub. Co. Ltd., Tokyo.

Japanese Pharmacopoea-14. (2005) Supplement2, Horokawa Publ. Co.,Tokyo.

Jeon, P.E., and Jeon Chul Hyun,J., GeoNil, K., Kim JaeBaek,K., and DongHwan, S. (2000) Protective effect of curcumin in rat liver injury induced by carbon tetrachloride. *J. Pharmacy and Pharmacology*, 52, 437-440

Joe, B., Vijayakumar, M. and Lokesh B.R. (2004) Biological properties of Curcumin –cellular and molecular mechanisms of action. *Critical Reviews in Food Science and Nutrition,* 47, 97-111.

Jones, E.A. and Shoskes, D.A. (2000) The effect of mycophenolate mofetil and polyphenolic bio- flavonoids on renal ishchemia reperfusion injury and repair. *J.Urol.*, 163, 999-1004.

Kang, B. Y., Song, Y. J., Kim, K. M., Choe, Y. K., Hwang, S. Y., and Kim, T. S.(1999) Curcumin inhibits Th1 cytokine profile in CD4<sup(+)> T cells by suppressing interleukin-12 production in macrophages. *British J.Pharmacology,* 128, 2, 380-384

Kaul, S. and Krishnakanth, T.P. (1994) Effect of retinol deficiency and curcumin or turmeric feeding on brain Na+-K+ adenosine triphosphatase activity. *Molecular and Cellular Biochemistry,* 137, 101-107.

Kausika Sutra – (ancient text) Cited from Sastry (2005).

Kawamori, T., Lubet, R., Steele, V. E., Kelloff, G. J., Kaskey, R. B., Rao, C. V., Reddy, B. S. (1999) Chemopreventive effect of curcumin, a naturally occurring anti-inflammatory agent, during the promotion/progression stages of colon cancer. *Cancer Research (Baltimore),* 59, 3, 597-601, 5C00504

Khanna, N.M. (1999), Turmeric : Nature's Precious gift. *Current Science*, 76, 1351- 1356.

Khar, A., Ali, A.M.., Pardhasaradhi, B.V.V., Begum, Z. and Rana Anjum, R., (1999) Antitumor activity of curcumin is mediated through the induction of apoptosis in AK-5 tumor cells. *FEBS Letters*, 445, 1, 165-168

Khare, C.P. (2000) *Indian Herbal therapies*. Vishv vijay Private Ltd. New Delhi.

Kimura, T. (1996, 1998, 1999, 2001) International collation of traditional and folk medicine, Northeast Asia, Part1, 2, 3, and 4. World Scientific, Singapore.

Kirtikar, K. R. and Basu, B.D. (1984) *Indian Medicinal Plants*, Bishensing Mahendrapal Singh, Dehra Dun, India (Reprint).

Koide, T., Nose, M., Ogihara, Y., Yabu, Y. and Ohta, N. (2002) Leishmanicidal effect of curcumin *in vitro*. *Biological & Pharmaceutical Bulletin*, 25, 131-133.

Kolammal, M. (1979) *Pharmacognosy of Ayurvedic Drugs*. Vol.10. Pharmacognosy unit, Govt Ayurveda College, Trivandrum,

Kosuge, T., Ishida, H. and Yamazaki.H. (1985) Studies on active substances in the herbs used for 'Oketsu' (stangnent blood) in Chinese medicine on the coagulative principles in *Curcuma longa* rhizome. *Chem. Pharm. Bull.(Tokyo)*, 33,1499-1502.

Kumar, S.R., Pillai, K.K., Balani, D.K., Hussain, S.Z. (1998) Anti-atherosclerotic effect of Lipotab Forte in cholesterol-fed rabbits. *J.Ethnopharmacology*, 59, 3 125-130.

Kuo, M.L., Huang, T.S. and Lin, J.K. (1996) Curcumin, an antioxidant and antitumor promoter, induces apoptosis in human leukemia cells. *Biochim.Biophys. Acta*, 1317, 95-100.

Kuttan, R., Bhanumati, P., Nirmala, K. and George, M.C. (1985) Potential anticancer activity of turmeric (*Curcuma longa*). *Cancer Lett.*, 29,197-202.

Lal, B., Kapoor, A. K., Asthana, O. P., Agrawal, P. K., Prasad, R., Kumar, P. and Srimal, R.C. (1999) Efficacy of curcumin in the management of chronic anterior uveitis. *Phytotherapy Research*,13,4,318-322.

Lee, C.J., Lee, J. H. Scok, J.H.,, Hur, G.M., Park, Y.C., Scol, I.C. and Kim, Y.H. (2003), Effects of baicalein, berberine, curcumin and hespiridin on mucin release from airway goblet cells. *Planta Med.*, 69,523-526.

Lehrh et al. (1992) – Cited from Khanna et al.(1999)

Lele. R.D. (2001) *Ayurveda and Modern Medicine*. Bharatiya Vidya Bhavan, Mumbai, 400007

Leskover. P. (1982) German patent 3.046, 580; *chem. Abstr.*, 97,120.487

Lewith, G.T. (2006) The Basic practice of Chinese traditional medicine. Accessed from web on http://www.healthy.net/Library/Books/Medacu/Mod1.htm, 9.2.2006.

Liao, S., Lin, J., Dang, M.T., Zhang, H., Kao, Y.H., Fukuchi, J. and Hiipakka, R.A (2001) Growth suppression of hamster flank organs by topical application of catechins, alizarin, curcumin, and myristoleic acid. *Arch.Dermatol Res.*, 293, 200-205.

Liu.Y.(1984)– US patent, 4, 842-859 Chem Abst.111, 160200.

Mills, S. and Bone, K. (2000) *Principles and Practice of Phytotherapy*. Churchil Livingston, London.

MouTuan, H., Ma, W., Yen, P., JianGuo, X., JingKang, H., Frenkel, K., Grunberger, D., and Conney, A.H. (1997) Inhibitory effects of topical application of low doses of curcumin on 12-O-tetradecanoylphor-bol-13-acetate-induced tumor promotion and oxidized DNA bases in mouse epidermis. *Carcinogenesis*, 18, 83-88.

Mukherjee, B., Zaidi, S.H., and Singh, G.B. (1961) Spices and Gastric function Part I Effect of *Curcuma longa* on the gastric secretion in rabbits, *J..Sci. Industr. Res.*, 20 C, 25.

Nadkarni,, K.M. (1976) *Indian Materia Medica* 1.3rd ed., Popular Prakasan, Bombay.

Nair, S., and Rao, M.N.A. (1996) Free radical scavenging activity of curcuminoids. *Arzneimittel Forschung*, 46, 169-171.

Negi, P. S., Jayaprakasha, G. K., Rao, L. J. M. and Sakariah, K. K. (1999) Antibacterial activity of turmeric oil: a byproduct from curcumin manufacture. *J. Agricultural and Food Chemistry*, 47, 4243-4297

Nirmala, C. and Puvanakrishnan, R. (1996) Protective role of curcumin against isoproterenol - induced myocardial infarction in rats. *Mol.Cell.Biochem*, 159, 85-93.

Nirmala, C., Anand, S., and Rengarajulu, P. (1999) Curcumin treatment modulates collagen metabolism in isoproterenol induced myocardial necrosis in rats., *Molecular and Cellular Biochemistry*,197, 31-37.

Oda (1995) – Cited from Joe et al. (2004).

Omoloso, A. D. and Vagi, J. K. (2001) Broad spectrum antibacterial activity of *Allium cepa*, *Allium roseum*, *Trigonella foenum graecum* and *Curcuma domestica*. *Natural Product Sciences*, 7, 13-16.

Parashar, K.K., Bhauser, K.N. and Soni, H.K. (1995) *Curcuma longa* Linn. (haridra) enters into new horizon for remedy of cancer. *Vaniki Sandesh,* 19(3), 1-4.

Park, S. Y. and Kim, D. S. H. L. (2002) Discovery of natural products from *Curcuma longa* that protect cells from beta-amyloid insult: a drug discovery effort against Alzheimer's Disease .*J. Natural Products,* 265, 1227-1231.

Patel, K. and Srinivasan, K. (1996) Influence of dietary spices or their active principles on digestive enzymes of small intestinal mucosa in rats. *Int. J.Food Sci.Nutr.,* 47, 55-59.

Patil, T.N. and Srinivasan, M. (1971) Hypo cholesteremic effect of Curcumin in induced-hyper cholesteremic rats. *Ind. J. Exp. Biol..* 9, 167-169.

Perry, L.M. (1980*) Medicinal plants of East and Southeast Asia,* MIT Press, Cambridge, MS. USA.

Piyachaturawat, P., Charoenpiboonsin, J., Toskulkao, C. and Suksamrarn, A. (1999) Reduction of plasma cholesterol by *Curcuma comosa* extract in hypercholesterolaemic hamsters. *J. Ethnopharmacology,* 66, 199-204.

Pradhan, N. R. (1999) Utility of herbal topical gel in mastitis control and udder health improvement. *Indian Veterinary J.,* 76, 546-548.

Pulla Reddy, A.C. and Lokesh, B.R. (1994) Effect of dietary turmeric (*Curcuma longa*) on iron-induced lipid peroxidation in the rat liver. *Food and Chemical Toxicology,* 32, 279-283.

Purohit, A. and Meenakshi, B. (2004) Contraceptive effect of *Curcuma longa* (L.) in male albino rat. *Asian J. Andrology.* 6, 71-74.

Rafatullah, S. (1990*) Medicinal, Aromatic and Poisonous Plants.* King Saud University, Riyadh. Saudi Arabia.

Rajakrishnan, V., Menon, V. P. and Rajashekaran, K. N. (1998) Protective role of curcumin in ethanol toxicity. *Phytotherapy Research,* 12, 55-56.

Rajakrishnan, V., Viswanathan, P., Rajasekharan, K. N. and Menon, V. P. (1999) Neuroprotective role of curcumin from *Curcuma longa* on ethanol-induced brain damage. *Phytotherapy Research,* 13, 571-574

Ram, A., Das, M. and Ghosh, B. (2003) Curcumin attenuates allergen induced airway hyperresponsiveness in sensitized guinea pigs. *Biological & Pharmaceutical Bulletin,* 26, 1021-1024.

Ram, P.R. and Mehrota, B.N. (1994) *Compendium of Indian Medicinal Plants.*CSIR, New Delhi.

Rasyid *et al.* (1999) – Cited from B.Joe *et al.* (2004)

Razga, Z. and Gabor, M. (1995) Effects of curcumin and nordihydroguaiaretic acid on mouse ear oedema induced by Croton oil or dithranol. *Pharmazie,* 50, 156-157.

Remadevi R. and Ravindran P.N. (2005) Turmeric : Myths and Traditions *Spice India.* 18 (8), 11-17

Remadevi, R. and Surendran, E. (2006) Turmeric : medicinal properties and uses in traditional medicine. *Spice India,* 19 (3), 2-14.

Rithaporn, T., Monga, M., Rajasekaran, M. (2003) Curcumin: a potential vaginal Contraceptive – *Contraception,* 68, 219-223.

Sajithlal, G.B., Chittra, P, and Chandrakesan, G. (1998) Effect of Curcumin on the advanced glycation and cross-linking of collagen in diabetic rats. *Biochem Pharmacol.,* 56, 1607 – 1614.

Sang Hyun, P., Kim, G.J., Jeong, H.S., Yum, S.K. (1996) Ar-turmerone and beta-atlantone induce internucleosomal DNA fragmentation associated with programmed cell death in human myeloid leukemia HL-60 cells. *Archives of Pharmacal Research,* 19, 91-94.

Sarma P.V. (2005) *Dravyaguna Vijnana* Vol.2, Chaukhambha Bharati Academy Varanasi, India (Reprint)

Sastry, J.L.N. (2005) -*Dravyaguna Vijnana.* Chaukhambha orientalia, Varanasi.

Sathyavathy, G.V., Raina, M.K. and Sarma M. (1978) *Medicinal Plants of India,* Vol.1. Indian Council of Medical Research, New Delhi.

Sayana – Cited by J.L.N.Sastri

Shah, N.C. (1997) Traditional uses of turmeric (*Curcuma longa*) in India. *J. Med. Aromatic Plants,* 19, 948-954.

Shao, Z.M., Shen, Z.Z., Liu.C.H., Sartippour, M.R., Go, V.L., Hever, D. and Nguyen, M., (2002) Curcumin exerts multiple suppressive effects on human breast carcinoma cells. *Int.J.Cancer,* 98, 234-240.

Singletary, K., MacDonald, C., Wallig, M., Fisher, C. (1996) Inhibition of 7,12-dimethylbenz [a] anthracene (DMBA)-induced mammary tumorigenesis and DMBA-DNA adduct formation by curcumin. *Cancer Letters,* 103, 137-141.

Sinha, M., Mukherjee, B.P., Muherjee, B. and Dasgupta, S.R. (1974) Study on the 5-hydroxytryptamine contents in guinea pig stomach with relation to phenylbutazone induced gastric ulcers and the effects of curcumin thereon. *Indian J. Pharmacol.,* 6, 87-96.

Srihari Rao, T., Basu, N., and Siddiqui,. H.H. (1982) Anti inflammatory activity of curcumin analogues. *Ind. J.Med. Res.,* 75, 574 -578.

Srimal, R.C., Khanna, K.M. and Dhawan, B.N. (1971) A preliminary report on anti inflammatory activity of curcumin. *Ind. J. Pharmacol.,* 3, 10.

Srivastava, K.C., Bordia, A. and Verma, S.K. (1995) Curcumin, a major component of food spice turmeric (*Curcuma longa*) inhibits aggregation and alters eicosanoid metabolism in human blood platelets. *Prostaglandins Leukotrienes and Essential Fatty Acids,* 52, 223-227.

Sudhadevi, P.K. (1998) *Adivasi chikitsareetikal, Sasyajam Mruthyunjayam,* All India Radio, Trichur.

Sumbilla, C., Lewis, D., Hammerschmidt, T. and Inesi, G. (2002) The slippage of the Ca2+ pump and its control by anions and curcumin in skelaetal and cardiac sarcoplasmic reticulum. *Biol.Chem.,* 277, 13900-13906.

Taher, M. M., Lammering, G., Hershey, C. and Valerie, K. (2003) Curcumin inhibits ultraviolet light induced human immunodeficiency virus gene expression. *Molecular and Cellular Biochemistry,* 254, 289-297.

Tank R. et al. (1990) *Indian Drugs* V.27 (11) 587 -589 – Cited by.Khare, (2000)

Thresiamma, K.C., George, J and Kuttan, R. (1995). Protective effect of curcumin, ellagic acid and bixin on radiation-induced toxicity. International symposium on radiomodifiers in human health, Manipal, India, 28-31 December 1995 [edited by Uma Devi, P.; Bisht,K.S.] *Indian Journal of Experimental Biology* 34(9), 845-847.

Tripathi, R.M., Guptha, S.S. and Chandra, D (1973) Anti- trypsin and anti- hyaluronidase activity of *Curcuma longa* (Haldi). *Ind.J.Pharmacol.,* 5, 260.

Vajragupta, O., Boonchoong, P., Watanabe, H., Tohda, M., Kummasud, N. and Sumanaont, Y. (2003) Manganese complexes of curcumin and its derivatives evaluation for the radical acavenging ability and neuroprotective activity. *Free Radic. Bio. Med.,* 35, 1632-1644.

Van Dau, N., Ngoc Ham, N., Huy Khac, D., Thi Lam, N., Tong Son, P., Thi Tan, N., Duc Van, D., Dahlgren, S., Grabe, M., Johansson, R., Lindgren, G., Stjernstrom, N. (1998) The effects of a traditional drug, turmeric (*Curcuma longa*), and placebo on the healing of duodenal ulcer. *Phytomedicine,* 51, 29-34.

Vangasena – Cited from Kolammal (1979).

Venkatesan (2000) Cited from B. Joe et al.

Xinyixueyuan, J. (ed.) (1977) *Zhongyao Dacidian,* Shanghai Kexue Jishu Chuban, Shanghai.

Xu, Y., Ku, B.S., Yao, H.Y., Lin, Y.H., Ma, X., Zhang, X.H. and Li, XJ. (2005) The effect of curcumin on depressive like behaviours in Mice. *Eur J. Pharmacol.,* . 518, 40-46.

Yakumukyoku, K. (1995) *Ippanyo Kamposhosho no Tebiki,* 181, Yakugyo Jihou sha, Tokyo.

Yu, Z. F., Kong, L. D., Chen, Y. (2002) Antidepressant activity of aqueous extracts of *Curcuma longa* in mice. *J. Ethnopharmacology,* 83, 61-165.

13 Turmeric as Spice and Flavorant

K.S. Premavalli

CONTENTS

13.1 INTRODUCTION

Spices are the plant products or a mixture thereof free from extraneous matter, cultivated, and processed for their aroma, pungency, flavor and fragrance, natural color, and medicinal qualities or otherwise desirable properties. They consist of rhizomes, bulbs, barks, flower buds, stigmata, fruits, seeds, and leaves of plant origin. Spices are food adjuncts, which have been in use for thousands of years, to impart flavor and aroma or piquancy to foods (Billing and Sherman, 1998; Sherman and Billing, 1999; Sherman and Hash, 2001). They are used to prepare culinary dishes and have little or no nutritive value, but they stimulate the appetite, add zest for food, enhance the taste, and delight the gourmet. As there is a need to reduce the fat, salt, and sugar used in food preparation for health reasons, it becomes critical to pay attention to alternative ways to enhance the natural flavors of foods. Value can also be added to meals by enhancing and improving presentation and by using appropriate garnishes. The primary function of a spice in food is to improve its sensory appeal to the consumer. Food presentation is the arrangement of food on a plate, tray, or steam line in a simple appetizing way. This is generally accomplished by imparting its own characteristic color, flavor, aroma, and mouth feel to the food.

Perception of flavor is comprised of the sensory combination and integration of odors, tastes, oral irritations, thermal sensations, and mouth feels that arise from a particular food (Breslin, 2001). When the flavors are perceived as a food is taken, their sensation immediately invokes feelings about the degree of pleasure of the immediate moment, at the same time strongly influencing intentions about consumption of that type at a later date. In this fashion, flavor plays a prominent role in the delivery of nutrition even though the majority of flavor compounds provide few or no calories in food. A spiced diet is likely to make life not only spicier but also healthier. Spices have also been recognized to possess medicinal properties and their use in traditional systems of medicine has been on record for a long time. With the advancement in the technology of spices and in the knowledge of the chemistry and pharmacology of their active principles, their health benefit effects were investigated more thoroughly in recent decades. Many health benefit attributes of these common food adjuncts have been recognized in the past few decades by pioneering experimental research involving both animal studies and human trials (Srinivasan, 2005).

The rhizome of turmeric has been used as a medicine, spice, and coloring agent for more than 2 millennia. It is principally known for its yellow-orange coloring power, having a musky flavor and aroma, which necessitates classifying it as a spice (Purseglove et al., 1981). Although the use of natural colors in food is an ancient practice, it is currently gaining increasing importance because consumers are wary of food industries using synthetic colors, whereas use of natural colors is seen as an ecologically sustainable and nonhazardous process (Downham and Collins, 2000; Roy et al., 2004). In ancient Indian literature, turmeric is referred to as "Haridra" which is being used for coloring, flavoring, and digestive properties (Patel and Srinivasan, 2004). The rhizome (underground stem, often referred as root) of turmeric has been used in Asian cookery, medicine, cosmetics, and fabric dying for more than 2000 years. Marco Polo wrote about turmeric in his memoirs, fostering its popularity in Europe during medieval times as a colorant and medicine. Turmeric is extensively used in India, the Middle East, and the Far East in food preparations. Besides, turmeric is traditionally used as the women's cosmetics for its characteristic fragrance and color, due to its "cosmeceutical" and "antiaging" benefits. Use of turmeric is also very auspicious in the form of whole plant, rhizome, as well as powder in religious functions (Saavala, 2003). Turmeric is often used as an inexpensive alternative to saffron. The primary product of *C. longa*, true turmeric, is the cured, dried rhizome. Cured and dried turmeric of commerce are available in the form of bulbs and fingers. The external appearance of rhizomes is bright yellow to dull yellow with a polished or rough surface.

13.2 MAJOR CHEMICAL CONSTITUENTS

Turmeric contains gum, starch, minerals, cellulose, volatile oil, and a yellow colorant. The chemical composition of turmeric is given in Table 13.1 and the nutritional composition in Table 13.2. The essential oil is a pale yellow to orange-yellow volatile oil (6%) composed of a number of monoterpenes and sesquiterpenes, including zingiberene, curcumene, α- and β-turmerone among others. The coloring principles (about 5%) are curcuminoids, 50 to 60% of which are a mixture of curcumin, monodesmethoxycurcumin, and bisdesmethoxycurcumin. Representative structures of curcuminoids are presented in Figure 13.1 (Pfeiffer et al., 2003).

TABLE 13.1
Chemical Composition (%) of Turmeric

Source	Moisture	Starch	Protein	Fiber	Ash	Fixed Oil	Volatile Oil	Alcohol Extractives
China	9.0	48.7	10.8	4.4	6.7	8.8	2.0	9.2
Pulena	9.1	50.0	6.1	5.8	8.5	7.6	4.4	7.3
Alleppey (finger)	11.0	30.8	—	4.0	—	—	3.4	24.2
Alleppey (bulbs)	12.0	26.3	—	4.6	—	—	3.4	16.2
Kadur	19.0	32.1	—	3.7	—	—	4.5	16.3
Duggirala	11.0	32.8	—	1.8	—	—	2.9	13.9

Source: Govindarajan, V.S. (1980).

TABLE 13.2
Nutritional Composition of Turmeric per 100 g

Composition	USDA Handbook 8–2[a] (Ground)	ASTA[b]
Water (g)	11.36	6.0
Food energy (kcal)	354	390
Protein (g)	7.83	8.5
Fat (g)	9.88	8.9
Carbohydrates (g)	64.93	69.9
Ash (g)	6.02	6.8
Calcium (g)	0.182	0.2
Phosphorous (mg)	268	260
Sodium (mg)	38	10
Potassium (mg)	2525	2500
Iron (mg)	41.42	47.5
Thiamine (mg)	0.152	0.090
Riboflavin (mg)	0.233	0.190
Niacin (mg)	5.140	4.8
Ascorbic acid (mg)	25.85	50
Vitamin A activity (RE)	Trace	ND[c]

[a] Composition of foods, spices and herbs, USDA Agricultural Handbook 8-2 January, 1977.

[b] The nutritional composition of spices. ASTA Research Committee.

[c] ND = not detected.

Source: Tainter and Grenis (2001).

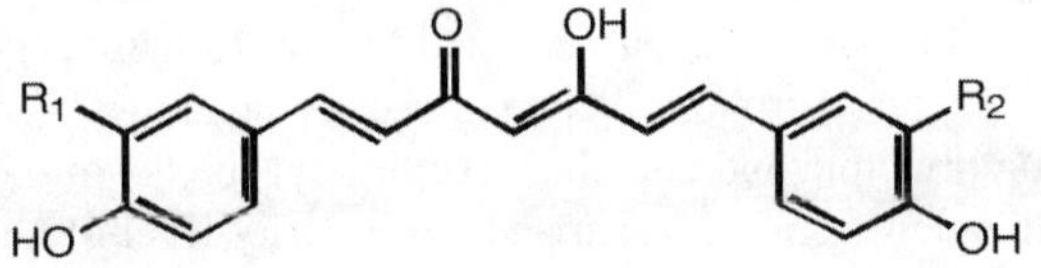

		cCUR	CUR F4
CUR I	$R_1 = OCH_3$; $R_2 = OCH_3$	71.5%	38.7%
CUR II	$R_1 = OCH_3$; $R_2 = H$	19.4%	25.7%
CUR III	$R_1 = H$; $R_2 = H$	9.1%	35.6%

cCUR: Commercial curcumin
CUR F4: Fraction of turmeric extract in aqueous media.

FIGURE 13.1 Representative structures of curcuminoids.

Essential oils and the diarylheptanoid curcumin, which are the major secondary metabolites of turmeric, have been shown to be largely responsible for the pharmacological activities of turmeric powder, extracts, and oleoresins (Garg et al., 1999). Turmeric is widely consumed for a variety of uses, including as a dietary spice, a dietary pigment, and an Indian folk medicine for the treatment of various illnesses. It is used in the textile and pharmaceutical industries as a natural dye (Srimal and Dhawan, 1973; Lokhande and Doruugade, 1999: Aggarwal, et al., 2004). Turmeric in its varied form is extensively used in Ayurveda, Unani, and Siddha medicine as home remedy for various diseases. Recent research worldwide is mainly focused on the medicinal and pharmacological aspects of its constituents specifically with reference to curcumin (Srimal and Dhawan, 1973; WHO, 1999; Vargas, et al., 2000; Dweck, 2002; Billing and Sherman, 1998; Khanna, 1999; Lewis and Lewis, 2003; Aggarwal et al., 2004; Chattopadhyay et al., 2004; Joe, et al., 2004: Scalbert et al., 2005; Finley, 2005).

13.3 TURMERIC EXTRACTS

There are three principal types of turmeric products, namely, essential oil, oleoresin, and curcumin. Essential oil of turmeric is obtained by the steam distillation of turmeric powder and contains all the volatile flavor components of the spice and none of the color. There is only a small commercial demand for this product. Turmeric oleoresin is the commonly produced extract that contains the flavor compounds and color in the same relative proportion as that present in the spice. It is obtained by solvent extraction of the ground turmeric, a process identical to that used in the production of other spice oleoresins. They ensure storage stability in the final product and are free from contamination. Custom-made blends are also offered to suit the specific requirement of the buyer. Spice oleoresins are mainly used in processed meat, fish and vegetables, soups, sauces, chutneys and dressings, cheeses and other dairy products, baked foods, confectionery, and snacks and beverages. India enjoys the distinction of being the single largest supplier of spice oleoresins to the world. Oleoresins are used at very low concentrations because they are highly concentrated. They have greater heat stability than essential oils. Turmeric oleoresin, from its use pattern, chiefly functions as a food color, and secondarily in some of the products to impart a characteristic mild spicy flavor. Cleanliness of rhizomes and high content of color of the curcuminoids determine the selection of preferred varieties such as Alleppey finger turmeric and Madras turmeric, as these types are known for their high curcumin contents. Turmeric oleoresin usually contains 25 to 55% curcumin. Curcumin is the pure coloring principle and contains very little of the flavor components of turmeric. It is produced by crystallization from oleoresin and has a purity level of about 95%, which is the standard commercially available. The new European specifications for curcumin states that the dye content must be not less than 90% when measured spectrophotometrically at 426 nm in ethanol.

It is important to note that the distinction between these three products lies in the ratio of color to flavor. A spice oleoresin contains the total sapid odorous and related characterizing principles normally associated with the spice. Thus, the ratio of flavor components to curcumin in ground spice and oleoresin is the same. A color however, results when an attempt is made to reduce the flavor of the product and increase the relative concentration of color such that the ratio of flavor to color is altered in favor of the color. Thus in the case of turmeric, the ratio of curcumin to volatile oil is of the order 50:50, usually lying within the range of 40:60 to 60:40 (Henry, 1998).

Turmeric is valued for its characteristic color and flavor. The principal component of color is curcumin, and generally, it is present in the range of 0.3 to 7% depending on the variety (Magda, 1994). Among the 100 turmeric cultivars evaluated at Indian Institute of Spices Research, Calicut, Kerala, India, PCT 14 was found best with respect to curcumin (7.9%), essential oil (7%), and oleoresin (15%) (Ratnambal et al., 1992). Analysis of essential oils and curcumin from turmeric accessions from the plains of northern India indicated that the oil content of rhizomes varied between 0.16 and 1.94% on a fresh weight basis. The accessions were classified into two categories: (1) those in whose essential oil, the sum of major terpenes (β-pinene, *p*-cymene, α-curcumene, β-curcumene, ar-turmerone, α-turmerone, and β-turmerone) was in the range 58 to 79%, and (2) those in whose oil the sum was 10 to 22%. The rhizomes of all the accessions were also evaluated for their curcumin content, which was found to vary from 0.61 to 1.45% on a dry weight basis (Garg et al., 1999). Analysis of rhizomes and leaves of turmeric variety Roma grown under the agroclimatic conditions of the North Indian plains at Lucknow, on hydrodistillation, gave 2.2% of oils, which were analyzed by GC and GC–MS. The rhizome oil contained 84 constituents, comprising 100% of the oil, of which the major ones were 1,8-cineole (11.2%), -turmerone (11.1%), -caryophyllene (9.8%), ar-turmerone (7.3%), and -sesquiphellandrene (7.1%). The leaf oil contained 83 components, comprising 97.4% of the total oil, of which the main constituents were terpipolene (26.4%), 1,8-cineole (9.5%), -phellandrene (8%), and terpinen-4-ol (7.4%) (Raina et al., 2002), whereas 20 compounds were identified in fresh turmeric leaves collected at flowering stage of the crop, accounting for 72% of the contents. The oil of turmeric consisted mainly of monoterpenoids, monoterpene hydrocarbons (57%), oxygenated monoterpenes (10%), sesquiterpene hydrocarbons (3.3%), and oxygenated sesquiterpenes (2.1%). The major constituents of the oil were p-cymene (25.4%) and 1,8-cineole (18%), followed by cis-sabinol (7.4%) and β-pinene (6.3%) (Garg et al., 2002).

13.4 TURMERIC PRODUCTS

13.4.1 TURMERIC AS A COLORANT

Turmeric is one among the three natural spice colorants, i.e., paprika, saffron, and turmeric. The principal functional property of turmeric is being a food colorant. The characteristic deep yellow color is due to the group of pigments, curcuminoids; the important ones being curcumin, demethoxy-curcumin and bis-demethoxy-curcumin. All the three compounds exhibit fluorescence under ultraviolet light, and this feature can be used for detecting turmeric in presence of other yellow colors. These three compounds can be separated by TLC along with their geometrical isomers, which are also expected to form in traces. The estimation of curcuminoids is easily achieved by extraction with alcohol and absorption spectra measured at 429, 424, and 419 nm for curcumin, demethoxy-curcumin, and bis-demethoxy-curcumin, respectively. Curcumin with boric acid gives a characteristic color, the formation of rubro-curcumin and rosocyanin. Curcumin is used as a colorant for improving the color of broiler meat and overall performance (Awang et al., 1992; Sultan, 2003). Chatterjee et al. (1999) have reported that gamma irradiation does not affect the composition or stability of natural pigments present in turmeric.

13.4.2 TURMERIC AS SPICE

Food is one of the most intimate and important components of our environment. Flavor is considered an important attribute determining the acceptance of food by the consumer. Whether we accept or reject food depends mainly on its flavor (Carterette and Friedman, 1989; Guichard, 2002), which plays a very important role in the palatability of food and is one of the key parameters determining the overall quality of a food product (Dattatreya, et al., 2002). Turmeric as a spice not only imparts color to food but also enriches the flavor. The pleasant flavor of turmeric is due to the volatile oil fraction. The oil content ranges from 0.3 to 7.2% (is orange-yellow), and contains mainly turmerone, alpha-turmerone, and zingiberene (Magda, 1994). Prolonged distillation of turmeric can convert minor amounts of alcohols into ketones, i.e., the formation of turmerones. Hence, a standard heating process has to be established for the fixing of color and formation of aroma, since curing and drying are followed in the processing of turmeric.

The use of spices varies greatly according to the region and culture of the people. Hirasa and Takemasa (1998) developed a patterning theory on the use of spices. According to this theory, the sustainability of a spice depends on the interaction of the spice with other materials that occurs in the mouth, which results in the "synthesis" of a new taste and flavor perception. A preference for

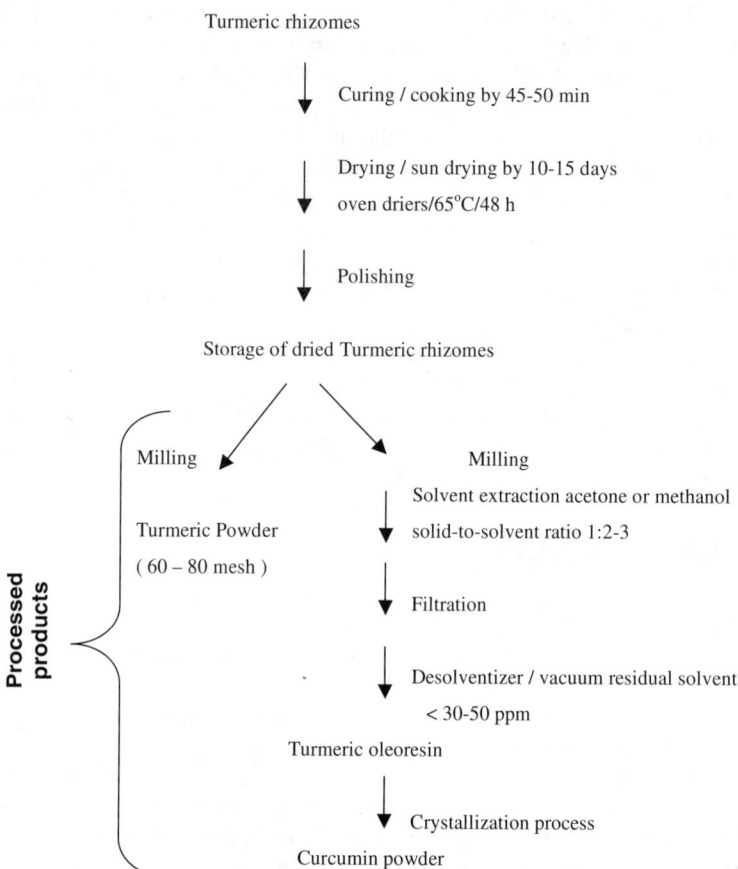

Turmeric rhizomes

Curing / cooking by 45-50 min

Drying / sun drying by 10-15 days
oven driers/65°C/48 h

Polishing

Storage of dried Turmeric rhizomes

Milling   Milling

                                           Solvent extraction acetone or methanol
Turmeric Powder   solid-to-solvent ratio 1:2-3

(60 – 80 mesh)

                                           Filtration

                                           Desolventizer / vacuum residual solvent
                                              < 30-50 ppm

Turmeric oleoresin

                                           Crystallization process

Curcumin powder

FIGURE 13.2 Preparation of turmeric products. *Source:* Govindarajan (1980).

a specific spice is determined by individual judgement. Hirasa and Takemasa (1998) developed a frequency patterning analysis, in which the frequency of each of 40 spices is analyzed so that a pattern of spice use for each nation, each cooking ingredient, and each cooking technique can be evolved. Govindarajan (1980) has depicted the preparation of various turmeric products (Figure 13.2). The study by the above workers led to the following conclusions:

- Turmeric is more suitable for Indian, Southeast Asian, and British cuisines, the suitability decreasing in that order.
- Turmeric is suitable for preparation of (increasing order): Beans and seeds, vegetables, meat, seafood, milk, egg, grains, and fruits.
- Turmeric is more suitable for: (1) fried, (2) steamed, and (3) food dressed with sauce. Less suitable for boiled, baked, deep fried, and pickled foods.

However, the above patterning analysis study probably did not include the South Indian vegetarian and nonvegetarian dishes. In South India, turmeric is added to about every dish, and it is an essential ingredient in all curry masala mixes. Turmeric is common with baked or fried items especially in the case of deep fried banana chips. In this case, banana slices are steeped in turmeric water for about 1/2 to 1 h, which imparts a more attractive yellow color to the chips.

Turmeric is used primarily to impart color to the food in most regional cookings. In Eastern cooking, turmeric, red chillies or paprika, and saffron (rarely) are the coloring spices. This is also the

TABLE 13.3
Deodorizing Rate of Turmeric in Comparison
with Other Common Spices[a]

Spice	Deodorizing Rate (%)
Turmeric	5
Ginger	4
Cardamom	9
Pepper	30
Star anise	39
Allspice	61
Clove	79
Coriander	3
Fennel	0
Cumin	11
Anise	27
Celery	44
Mint	90
Rosemary	97
Thyme	99

[a] Deodorizing rate-percent of methyl mercaptan (500 mg) captured by methanol extract of each spice.

Source: Tokita et al. (1984).

case with Western, Indian, and Southeast Asian cooking. In U.S., Germany, and other West European regions, paprika is the only coloring spice used. In U.K., turmeric is commonly used, whereas in Italian dishes mostly saffron is used. It is also interesting to note that in Chinese and Japanese dishes coloring spices are not used, or when used Perilla is the preferred herb that gives a red color to the dish. Turmeric (and also paprika/chillies) is useful in meat, seafood, milk products, egg, grains and seeds, and vegetable dishes, but it is not used for imparting color to beverages. Turmeric is useful in all forms of cooking, with heat or without heat, but it is not used with fresh food such as fresh salads.

Spices also perform a deodorizing function in food. In many food preparations spices are used for masking or deodorizing unpleasant odors or flavors. Turmeric has very low deodorizing property; its deodorizing rate is only 5% (Table 13.3).

13.5 PRESERVATIVE ACTION AND HEALTH BENEFITS

Curcuminoids from turmeric are reported to possess antioxidative, antibacterial properties. Besides, they also exert anti-inflammatory and anticarcinogenic properties. Turmerene and curlone present in turmeric oil are the major compounds responsible for the antibacterial activity. Negi et al. (1999) studied the antibacterial activity using a by-product of curcumin manufacture. Curcumin, which is the main yellow pigment of turmeric, exhibits antioxidant activity and the chloroform extract had higher antioxidative activity than the other solvent systems (Lim et al., 1996). However, the curcuminoid-derived compounds, which are colorless, also exert antioxidant activity (Osawa et al., 1995). The antioxidative activity of turmeric as such in comparison to other spices is given in Table 13.4. Chatterjee et al. (1999) have reported that γ-irradiation at a dose of 10 kGy does not affect the antioxidative activity of turmeric extract. The antioxidant principle of turmeric as reported by Hirasa and Takemasa (1998) includes curcumin, 4-hydroxy cinnmoyl (feruloyl)methane, and *bis*(4-hydroxy cinnamoyl) methame. The 50% inhibhitory concentration values of these and related compounds are

TABLE 13.4
Antioxidative Activity of Turmeric in Comparison with Other Common Spices Against Lard

Spice	Ground Spice POV (meq/kg)	Petroleum Ether Soluble Fractions POV (meq/kg)	Petroleum Ether Insoluble Fraction POV (meq/kg)
Turmeric	399.3	430.6	293.7
Ginger	40.9	24.5	35.5
Black pepper	364.5	31.3	486.5
Chillies	108.3	369.1	46.2
Cardamom	423.8	711.8	458.6
Cinnamon	324.0	36.4	448.9
Clove	22.6	33.8	12.8
Mace	13.7	29.0	11.3
Nutmeg	205.6	31.1	66.7
Rosemary	3.4	6.2	6.2
Sage	2.9	5.0	5.0

TABLE 13.5
50% Inhibitory Concentration (IC$_{50}$) of Antioxidative Compounds of Turmeric and Related Compounds on Air Oxidation of linoleic Acid

Sample	50% Inhibitory Concentration TBAV (%)	POV (%)
Methanol ex.	1.22×10^{-2}	1.21×10^{-2}
Curcumin	1.83×10^{-2}	1.15×10^{-2}
4-Hydroxycinnamoyl (feruloyl) methane	1.88×10^{-2}	2.79×10^{-2}
bis(4-Hydroxycinnamoyl) methane	2.80×10^{-2}	3.17×10^{-2}
Caffeic acid	5.63×10^{-3}	5.30×10^{-3}
Ferulic acid	8.95×10^{-3}	5.41×10^{-3}
Protocatechuic acid	1.85×10^{-2}	1.54×10^{-2}
Vanillic acid	2.01×10^{-2}	1.83×10^{-2}
BHA	3.37×10^{-3}	3.75×10^{-3}
BHT	1.92×10^{-3}	2.24×10^{-3}
dl-α-Tocopherol	1.95×10^{-1}	2.4×10^{-1}

Source: Hirasa and Takemasa (1998).

given in Table 13.5. IC$_{50}$ values of these compounds were lower than that of α-tocopherol meaning that they are more effective antioxidants. Besides food systems, curcumin has been studied for its antioxidant activity in rats as well as under *in vitro* system. Noguchi et al. (1994) reported that curcumin reacted with stable radicals such as galvinoxyl and *N,N*-diphenyl-1-picryl hydrazyl, suggesting that it can serve as a hydrogen donor and is a strong antioxidant as compared to eugenol. Song et al. (2001) reported free radical scavenging and hepato-protective activity of turmeric rhizomes in *in vitro* system, and the activity was much better than that of ascorbic acid. The mechanism of antioxidative activity of curcumin has been reported by Masuda et al. (2002). He found that curcumin formed dimers as radical termination products especially at 2′-position of curcumin molecule and oxidative coupling reaction at 3′-position (Masuda et al., 2001). Watanabe and Fukai (1997, 2000) have reported that curcumin suppresses the oxidative stress by scavenging various free radicals and

TABLE 13.6
Typical Curry Powder Formulation Containing Turmeric

Ingredients	Typical Range (%)
Coriander	10–50
Turmeric	10–35
Cumin	5–20
Fenugreek	5–20
Ginger	5–20
Celery	0–15
Black pepper	0–10
Red pepper	0–10
Cinnamon	0–15
Nutmeg	0–15
Cloves	0–15
Caraway	0–15
Fennel	0–15
Cardamom	0–15
Salt	0–10

Source: Tainter and Grenis (2001).

its antioxidative activity seems to be derived from its suppressive effects. Asai and Miyazawa (2001) showed that phenolic yellowish pigments of turmeric display antioxidative activity in rats while Okada et al. (2001) in his study on induced oxidative renal damage in male mice showed that curcumin is an effective protectant against oxidative stress. Curcumin, having antioxidative property, may act as anticancer agent, but also inhibits the regulatory enzymes and exhibits anticarcinogenic action.

13.6 APPLICATIONS

Turmeric as a spice has properties of sensory, physiological, functional, and preservative action. It is widely used in the preparation of foods as a colorant, flavorant, and for its preservative action. Turmeric powder is used in the preparations of curries, dhals, salad-based raitas, fish, meat products, etc. Curry powder formulations have turmeric as an ingredient and the details are given in Table 13.6, Table 13.7, and Table 13.8. Curry powder is a very general term used for spice mixtures made for specific dishes — vegetable curry, fish curry, chicken curry, mutton curry, etc. Especially, in the southern states of India, a number of curry powders are in vogue specific for *sambhar*, *rasam*, dry vegetables, dry powders for mixing with rice, and so on. Curcumin is insoluble in water, and hence when used in confectionary, agricultural products, and processed marine products, a solution in alcohol or propylene glycol is employed. However, it is relatively stable to heat but not to light. In acidic and neutral medium, it imparts an appealing yellow color to food, but in alkaline medium, the color changes to a dark reddish brown. Presence of certain metal compounds such as iron and boron can also influence the color. These factors are important when curcumin is used as a food color.

13.7 COMMERCIAL USE

Commercially, turmeric is used for:

- Manufacture of kumkum
- pH test paper because curcumin changes from yellow to brown red in alkali

TABLE 13.7
Curry Powder Blends as Per U.S. Specifications

Spice	U.S. Standard Formula No. 1 %	General Purpose Curry Formulae No. 2 %	No. 3 %	No. 4 %	No. 5 %
Coriander	32	37	40	35	25
Turmeric	38	10	10	25	25
Fenugreek	10	0	0	7	5
Cinnamon	7	2	10	0	0
Cumin	5	2	0	15	25
Cardamom	2	4	5	0	5
Ginger (Cochin)	3	2	5	5	5
Pepper (white)	3	5	15	5	0
Poppy seed	0	35	0	0	0
Cloves	0	2	3	0	0
Cayenne pepper	0	1	1	5	0
Bay leaf	0	0	5	0	0
Chili peppers	0	0	0	0	5
Allspice	0	0	3	0	0
Mustard seed	0	0	0	3	5
Lemon peel (dried)	0	0	3	0	0

Source: Farrel (1985). Formula 1 in the U.S. Military specification Mil-C-35042 A. Formula 2 is considered a mild curry, Formula 3 is sweet curry, and Formula 4 is a hot curry type. Formula 5 is a very hot pungent Indian-style curry.

TABLE 13.8
Curry Blends and Masala Mixes Having Turmeric as a Component

Spice	Ingredients
Indian curry blends	Basic curry blend consists of coriander, cumin, red pepper, and *turmeric*. Special blends for fish, meat, etc. contain in addition to the above, ginger, cardamom, clove, cinnamon, mustard, fenugreek, curry leaf, mint, coriander leaf, and celery seed, depending upon the particular blend.
Pickling masala blends	Many different types of pickling blends are in vogue. The important ingredients are mango/or lime pieces, Chili pepper, ginger, garlic, mustard oil, mustard seed paste, *turmeric*, sesame seeds, mint and cilantro. Mango, lime and mixed fruit pickles are the most common ones
Burmese curry blend	Onion, garlic, ginger, *turmeric*, fish sauce, chillies and tamarind
Malaysian curry blend	Lemongrass, atar anise, ginger, galangal, pandan leaf, tamarind, mint, coriander, *turmeric*, shallot
Mediterranean spice blend	Cardamom, ginger, cassia cinnamon, black pepper, cumin, fenugreek, lovage, mace, cubeb, long pepper, allspice, nutmeg, rose petals, lavender blossoms, orange blossoms, grains of paradise, chillies, nigella, onion, thyme, rosemary, and *turmeric*

Source: Compiled from various sources.

- Fluorescence test for detection of turmeric
- Manufacture of curcumin and its use in ice creams, gelatins, lemonades, and liquor
- Manufacture of oleoresin and their use in processed foods

The marketed products, which are branded, processed foods are listed in Table 13.9 where turmeric is used as one of the ingredients.

TABLE 13.9
List of Manufactured Products Containing Turmeric

Products

Powder and paste for vegetarian foods
    Turmeric powder
    Sambhar powder
    Rasam powder
    Curry powder
    Curry paste
    Amchoor paste
    Channa gravy
    Mustard paste
Paste for nonvegetarian foods
    Mutton curry paste
    Butter chicken curry paste
    Fish curry paste
    Tandoori curry paste
Masalas — vegetarian
    Instant upma masala
    Kharabhath masala
    Subzi masala
    Paw bhaji masala
    Chhole masala
    Garam masala
    Vangibhath masala
    Green masala
Masalas — nonvegetarian
    Tandoori chicken masala
    Chicken masala
    Fish masala
    Meat masala
    Biryani mix
    Kitchen king masala
    Khabab masala
Pickles
    Mixed pickle
    Mango pickle
    Garlic pickle
    Vadu mango pickle
    Green chillies pickle
    Lime pickle
    Mixed vegetable pickle
    Avakkai mango pickle
    Chillies pickle

Continued

TABLE 13.9 *(Continued)*
List of Manufactured Products Containing Turmeric

Ready-to-eat products
    Bisibele bath
    Tomato rice
    Lemon rice
    Rajma masala
    Sansar
    Kadi pakora
    Mixed vegetable curry
    Pineapple sweet and sour curry
    Navaratna kurma
    Alu methi
    Channa masala
    Dal fry
    Awadhi alu mutter
    Punjabi kadhi pakora
    Yellow dal tadka
    Potato sagu
    Hyderabadi biryani bhath

REFERENCES

Aggarwal, B.B., Kumar, A., Aggarwal, M.S. and Shishodia S. (2004) Curcumin derived from turmeric (*Curcuma longa* L): a spice for all seasons.. *In*: Bagchi, D., and Preuss, H.G. (eds.). *Phytopharmaceuticals in cancer chemoprevention.* CRC Press, LLC. pp. 349-387.

Asai, A. and Miyazawa, T. (2001) Dietary curcuminoids prevent high fat die –induced lipid accumulation in rat liver and epididymal adipose tissue. *J. Nutrition.,* 131(11), 2932-2935.

Awang, I.P.R., Chutan, H. and Ahmed, F.R.H. (1992) Curcumin for upgrading skin colour of broilers, *Pertanika.,* 15(1), 37-38.

Billing, J. and Sherman, P.W. (1998). Antimicrobial functions of spices: Why some like it hot. *The Quart. Rev. Biol.,* 73,3-49.

Breslin, P.A.S. (2001) Human gustation and flavor. *Flavor. Fragr. J.* 16, 439–456

Carterette, E. and Friedman, M.P. (eds.). (1989) *Flavor chemistry. trends and development*; American Chemical Society: Washington, DC, 1989.

Chatterjee, S., Padwal Desai, S.R. and Thomas, P. (1999) Effect of gamma-irradiation on the colour power of turmeric and red chillies during storage. *Food Res.Intern.,* 31(9), 625-628.

Chattopadhyay, I., Biswas, K., Bandyopadhyay, U. and Banerjee, R.K. (2004) Turmeric and curcumin: Biological actions and medicinal applications. *Current Sci.,* 87, 44-53.

Dattatreya, B.S., Kamath, A. and Bhat, K.K. (2002) Developments and challenges in flavor perception and measurement - a review. *Food Rev. Intern.,* 18, 223-242.

Downham, A. and Collins, P. (2000) Colouring our foods in the last and next millennium. *Internatl. J. Food. Sci. Technol.,* 35,5–22.

Dweck, A.C. (2002) Natural ingredients for colouring and styling. *Intern. J. Cosmetic Sci.,* 24, 287–302.

Farrel, K.T. (1985). *Spices, Condiments and Seasonings.* The AVI Publishing, CT.

Finley, J.W. (2005) Proposed criteria for assessing the efficacy of cancer reduction by plant foods enriched in carotenoids, glucosinolates, polyphenols and selenocompounds, *Ann. Botany.,* 95, 1075-1096.

Garg, S.N., Bansal, R.P., Gupta, M.M. and Kumar, S. (1999) Variation in the rhizome essential oil and curcumin contents and oil quality in the land races of turmeric *Curcuma longa* of North Indian plains. *Flavor Fragr. J.,* 14,315–318.

Garg, S.N., Mengi,N., Patra, N.K., Charles, R. and Kumar, S. (2002) Chemical examination of the leaf essential oil of *Curcuma longa* L. from the North Indian plains, *Flavor. Fragr. J.,* 17, 103–104.

Govindarajan, V.S. (1980) Turmeric–Chemistry, Technology and Quality. *Crit. Rev. Food. Sci. Nutrition,* 12, 199-301.

Guichard, E. (2002) Interactions between flavor compounds and food ingredients and their influence on flavor perception. *Food Rev. Intern.,* 18, 49–70.

Henry, B. (1998) Use of capsicum and turmeric as natural colors in the food industry. In: *Proceedings of the World Spice Congress – 1998.* Spices Board of India and All India Spices Exporters Forum, Cochin, Kerala, India. pp. 27-38.

Hirasa, K. and Takemasa, M. (1998) *Spice Science and Technology,* Marcel Dekker Inc., New York.

Joe, B., Vijayakumar, M. and Lokesh, B.R. (2004) Biological properties of curcumin – cellular and molecular mechanisms of action. *CRC Critical Rev. Food. Sci. Nutr.,* 44, 97-111.

Khanna, N.M. (1999) 'Turmeric-nature's precious gift', *Current Sci.,* 76(10), 1351-1356.

Lewis, W.H., and Lewis, M.P.F.E. (2003). *Medical botany — Plants affecting human health.* John Wiley & Sons, Inc. pp.xx+812.

Lim, D.K., Choi, U. and Shin, D.H. (1996) Antioxidant activity of ethanol extract from Korean medicinal plants. *Korean J. Fd. Sci. & Tech.,* 28(1),83-89.

Lokhande, H.T. and Doruugade, V.A. (1999) Dyeing nylon with natural dyes. *American Dyestuff Reporter.,* 88 (2), 29-34.

Magda, R.R. (1994) Turmeric: a seasoning, dye and medicine', Food *marketing and Technol.,* 8(5), 9-10.

Masuda, T., Maekawa, T., Hidaka, K., Bando, H., Takeda, Y. and Yamaguchi, H. (2001) Chemical studies on antioxidant mechanism of curcumin. *J. Agric. Food Chem.,* 49 (5), 2539-2547.

Masuda, T., Toi, Y., Bando, H., Maekawa, T., Takeda, Y. and Yamaguchi, H. (2002) Structural identification of new curcumin dimmers and their contribution to the antioxidant mechanism of curcumin. *J. Agric. Food Chem.,* 50(9), 19-23.

Negi, P.S., Jayaprakash, G.K., Jaganmohan rao, L. and Sakariah, K.K. (1999) Antibacterial activity of turmeric oil a by-product from curcumin manufacture. *J. Agric. Food Chem.,* 47 (10), 4297-4300.

Noguchi, N., Komuro, E., Niki, E. and Willson, R.L. (1994) Action of curcumin as an antioxidant against lipid peroxidation. J. *Japan oil Chem. Soc.,* 43(12), 1046-1051.

Okada, K., Wangpoengtrakuk, C., Tanaka, T., Toyokuni, S., Uchida, K. and Osawa, T. (2001) Curcumin and especially tetrahydrocurcumin ameliorate oxidative stress induced renal injury in mice. *J. Nutrition.,* 131(8), 2090-2095.

Osawa, T., Sugiyama, Y., Inayoshi, M. and Kawakishi, S. (1995) Antioxidative activity of tetrahydro curcumininoids. *Biosci. Biotech. Biochem.* 59(9), 1609-1612.

Patel, K. and Srinivasan, K. (2004) Digestive stimulant action of spices : A myth or reality?. *Indian J. Med. Res.,* 119, 167-179.

Pfeiffer, E., Hˆhle,S., Solyom, A.M. and Metzler, M. (2003) Studies on the stability of turmeric constituents.*J. Food Eng.*56, 257–259

Purseglove, J.W., Brown, E.G. and Green, C.L. (1981) *Spices.* Volume 1. Tropical agriculture series (UK). Longman, London, UK.

Raina, V.K., Srivastava, S.K., Jain, N., Ahmad, A., Syamasundar, K.V. and Aggarwal, K.K. (2002) Essential oil composition of *Curcuma longa* L. cv. Roma from the plains of northern India, *Flavor Fragr. J.* 17, 99–102

Ratnambal, M.J., Nirmal Babu, N., Nair, M.K. and Edision, S. (1992) PCT-13 and PCT-14 two high yielding varieties of turmeric. *J. of Plantation Crops,* 20 (2),79-84.

Roy, K., Gullapalli, S., Chaudhuri, U.R. and Chakraborty, R. (2004) The use of a natural colorant based on betalain in the manufacture of sweet products in India. *Intern. J.Food Sci. Technol.,* 39, 1087–1091.

Saavala, M. (2003) Auspicious Hindu Houses. The new middle classes in Hyderabad, India. *Social Anthropol.,* 11, 231–247.

Salzer, U.J. (1977) The analysis of essential oils from seasoning, a critical review. *CRC Critical Rev. Fd. Sci. Nutri.,* 9(4), 345.

Scalbert, A.,, Manach, C., Morand, C., Remesy, C. and Jimenez, L. (2005) Dietary polyphenols and the prevention of diseases. *CRC Crit. Rev.Food Sci.Nutr.,,* 45,287–306.

Sherman, P.W. and Billing, J. (1999) Darwinian gastronomy: Why we use spices. Bio Sci., 49 (6),453-463.

Sherman, P.W. and Hash G.A. (2001) Why vegetable recipes are not very spicy. *Evolution and human behavior.*, 22, 147-163.

Song, E.K., Cho, H., Kim, J.S., Kim, N.Y., An, N.H., Kim, J.A., Lee, S.H. and Kim, Y.C.L. (2001) Diaryl-heptanoids with free radical scavenging and hepatoprotective activity invitro from *Curcuma longa*. *Planta-Medica*, 67(9),876-877.

Srimal, R.C. and Dhawan, B.N. (1973) Pharmacology of diferuloyl methane (curcumin), a non-steroidal anti-inflammatory agent, *J. Pharm. Pharmacol.*, 25,447– 452,

Srinivasan, K. (2005 a) Spices as influencers of body metabolism: an overview of three decades of research. *Food Res. Intern.*, 38, 77–86.

Sultan, S.I.AL. (2003) The Effect of *Curcuma longa* L. (Tumeric) on overall performance of broiler chickens, Intern. *J. Poultry Sci.*2, 351-353.

Tainter, D.R. and Grenis, A.T. (2001) *Spices and Seasonings: a food technology handbook,* Wiley-VCH: New York, A John Wiley & Sons Inc. Publications. P. ix+249.

Tokita, F., Ishikawa, M., Shibuya, K., Kashimizu, M. and Abe, R. (1984) *Nippon Nogeikagaku Gakkaishi*, 58, 585 (Cited from Hirasa and Takemasa, 1998).

Vargas, F.D., Jiménez, A.R. and Lûpez, O.P. (2000) Natural pigments: Carotenoids, anthocyanins, and betalains — characteristics, biosynthesis, processing, and stability*Crit. Rev. Food Sci. Nutr.*, 40, 173–289.

Watanabe, S. and Fukui, T., (1997) Inhibitory effect of curcumin on the peroxisomes proliferation by trichlro ethylene'. *Japanese J. Toxicol. Environmental health*, 43(1), 3

Watnabe, S. and Fukui, T. (2000) Suppresive effect of curcumin on trichloroethylene-induced oxidative stress', J. *Nutri. Sci. Vitaminol.*, 46(5),230-234.

WHO (1999) *WHO monographs on selected medicinal plants*. Volume 1. World Health Organization Geneva. P. v+289.

14 Other Economically Important *Curcuma* Species

J. Skornickova, T. Rehse, and M. Sabu

CONTENTS

14.1 INTRODUCTION

Curcuma is an important genus in the family Zingiberaceae. Various species have been used medicinally, as a yellow dye for cloth, and for flavoring and coloring food since time immemorial. Its generic name originated from the Arabic word *kurkum* meaning "yellow," and most likely refers to the deep yellow rhizome color of the true turmeric (*Curcuma longa* L.). Besides *C. longa,* there are several species of economic importance as medicine, such as *C. aromatica* Salisb., *C. amada* Roxb., *C. caesia* Roxb., *C. aeruginosa* Roxb., and *C. zanthorrhiza* Roxb. Others are beautiful and splendid plants of great ornamental value, such as *C. alismatifolia* Gagnep., *C. elata* Roxb., and *C. roscoeana* Wall. Locals and tribal people in most Asian countries use *Curcuma* species in religious rituals, as a foodstuff, and as medicinal plants.

The true number of species in the genus is not known, although anywhere from 70 to over 110 have been reported, distributed throughout tropical Asia. Their greatest diversity is in India, Myanmar, and Thailand, with a few species extending to China, Australia, and the South Pacific. However, some popular species can be found cultivated and naturalized all over the tropics. In their native range, they are an important component of the understorey in shaded, wet areas, or as some of the first pioneers in disturbed areas such as roadsides and ditches.

Sadly, their true identity is often unclear. Many *Curcuma* species were described as early as 200 years ago, and poor descriptions, lack of type specimens, and difficulties in preserving useful specimens (making it necessary to work with living flowering material) make their study one of the most difficult among gingers. The appropriate scientific names rarely correspond to even some of the most commonly cultivated species. Many previous workers have mentioned that the genus is badly in need of revision: most species are quite variable, but many of them look alike. In addition, they may hybridize in the wild, and hybrids may get naturalized.

Much information about different *Curcuma* species is available not only in historical and recent literature, but also on the Internet and through the horticultural industry. Since much of this information use unreliable scientific and vernacular names, the information we present here may seem incomplete. Most of the information presented is from our own observations from the field in India, Thailand, and Myanmar, as well as from our collections and from helpful individuals from botanical gardens and nurseries. However, for the sake of providing an overall picture, some details on the chemistry and pharmacology are added from the literature, but the taxonomic identity in such reports may not be reliable.

14.2 GENERAL FEATURES

Curcumas are perennial rhizomatous herbs, the leafy shoot dying back during the dry period of monsoonal areas. Most are medium sized, about 0.5 to 1.5 m tall, but the smallest species are just 15 to 20 cm, while the stateliest one can reach up to 3.5 m. The rhizome can be simply ovoid without branches, or branched (branches often incorrectly called sessile or palmate tubers). The branches may be short and serve to simply increase storage area, or long creeping runners, which initiate new shoots well away from the previous year's growth. They are usually light brown externally, but internally they can be white, different shades of yellow, bluish to deep blue, light to deep orange, yellow with greenish borders, gray, etc. This character is considered quite specific, even though the deepness of the color may be influenced by age of the plant or perhaps also by ecological factors. Root tubers are present in all *Curcuma* species, placed on the ends of roots distanced about 5 to 20 cm from the rhizome, containing mostly starch. These are not capable of sprouting, and their function is exclusively for the purpose of sustaining the plant during its dormant period when the leafy shoot dries up, storing energy for the next season. In most species, a pseudostem (false stem) is formed by closely embracing leaf sheaths. It can be pure green or with a pink or deep red tinge, low to the ground or rather high up, depending on the general size and habit of a particular species. The leaves can be petiolate or sessile, but even species with petiolate leaves usually produce almost sessile leaves initially. The lamina is usually lanceolate, oblong or ovate, rarely linear; bright green to deep green above and usually paler green beneath. Venation is usually prominent, and in many species beautifully sulcate. The presence of an indumentum (fine hairs) is quite variable in the species that set seed,* and thus not a reliable specific character. Some species have a purple or violet patch on the upper side of the lamina, which in some species penetrates to the lower surface. Its size, shape, relative placement on the lamina (e.g., the upper half of the lamina near the midrib or the full length), color, and density may be helpful in the determination of some species, but is usually variable in the seed-setting species.*

The flowers are arranged in spike-like inflorescences, which are formed by bracts usually connate (fused) at their sides to the neighboring bracts, resulting in pouches. The bracts in the lower portion of the inflorescence are fertile (bearing one cincinnus of flowers per bract), usually less colored, often green, and more connate than the top series of bracts.

The top bracts are usually much larger, longer, and more brightly colored (called a coma). It is the coma that is usually the main attraction of *Curcuma* plants and probably serves the purpose of attracting pollinators. Several species have no distinctive coma bracts. When all bracts are of the same color, their spikes are fertile almost to the top but for a few uppermost bracts, which are usually smaller. Inflorescences may originate from two places relative to the foliage — laterally or centrally. Lateral inflorescences appear just after the premonsoon showers to soon after the rainy season starts; in some species, even sometimes before the leaves develop, appearing disconnected from them. Species with central inflorescences bloom later, between the leaf sheaths, as the monsoon progresses. A central, terminal inflorescence with a short peduncle may appear at ground level, which may seem lateral. Many of the species which set seeds are capable of blooming twice — once in the beginning of the season (laterally) and if the monsoon rains are sufficient also later in the summer. Recently, a central inflorescence protruding from the pseudostem through lateral slits has been observed in *Curcuma rubrobracteata* by Skornickova et al. (2003). Most species bloom in the beginning of the rainy season, other species flower a couple of weeks after the monsoon starts, but a few species flower during the autumn, e.g., *C. longa, C. amada*, and *C. roscoeana*.

14.2.1 EAST INDIAN ARROWROOT

East Indian arrowroot, also known as Travancore starch, is a starch prepared from multiple species of *Curcuma* L. (Zingiberaceae). East Indian arrowroot is used in much of Asia, including India, Bangladesh, Cambodia, and Indonesia. The main sources are *C. zedoaria, C. angustifolia*, and *C. aeruginosa*.

Their Sanskrit names (*nirbisha, apavisha, vishaha*) imply a very old usage as an antidote for poisoning. Now it is considered an easily digestible starch, traditionally used in India for making biscuits for feeding infants and as a health drink to support convalescence in the elderly. To improve health in adults, it is also taken in the case of stomach disorders, especially for dysentery. It has cooling and demulcent properties, used in soothing irritated mucous membranes. Arrowroot powder is also one of the major components of *chyavanaprash*, a kind of ayurvedic rejuvenating tonic, commonly used to improve immunity, memory, and general health. The arrowroot is also used in some traditional medicines for reducing swelling due to bone fracture and giving early relief.

---

* A distinction is made here between the seed-setting species and those species that appear to reproduce only by vegetative means through rhizome divisions. The seed-setting species generally exhibit a range of natural variation in many characters that would be expected in populations. The non seed-setting species are those for which seeds have rarely (if ever) been reported, and generally are consistent with clonal populations: little variation in vegetative or reproductive characters has been seen. The loss of ability to produce seeds is likely due to polyploidy, hybridization, or both, but this has not been consistently demonstrated beyond *C. longa*, a known sterile triploid. The characters therefore used to distinguish between species were often described from one individual, and it is unknown for most species, which characters vary and which do not. The material presented here attempts to distinguish between the species, which we have found to produce seeds and which do not, and the degree of variation found during our observations.

FIGURE 14.1 Extraction of starch from *C. aeruginosa* in Kerala, India. (Photo: M. Sabu.)

Arrowroot powder is also used generally as a starch for culinary purposes, in certain Indian traditional cuisines for the preparation of various sweets, ice creams, milkshakes, etc. In Java, the local people rasp, pound, and cook *Curcuma* rhizomes to prepare puddings (Burkill, 1966).

Starch is extracted by the simple method of rubbing the rhizome (and in some species also the root tubers) into a paste, which is then mixed with water (Figure 14.1). This is strained through a fine muslin cloth tied at the mouth of a vessel filled with water. Only fine particles of starch pass through the cloth into the water and settle at the bottom. Any bitter components and contaminants do not settle, and the water is decanted and the white residue is washed a few times until it is snow-white, and then sun dried to a white powder. One kilogram of fresh rhizomes will yield approximately 125 g of starch (Sabu and Skornickova, 2003).

Van Rheede (1678 to 1693), who wrote the Hortus Malabaricus, the first printed record of the plants of the Malabar Coast, mentioned the value of this powder as a part of the Indian diet. He also observed that locals extract this very esteemed starch, known as *kua-podi* in Malayalam, from *C. zedoaria* Roxb. Since then many others have described its use in a number of other *Curcuma* species, and there is evidence that at least 14 species are currently being used in India alone for "arrowroot," although other species are used in other countries. Some species are cultivated, such as *C. longa, C. zedoaria, C. aromatica,* and others, but many are simply collected by the locals from the wild. Some of these species are becoming rare and endangered due to over-exploitation of their rhizomes by local and tribal people.

This white powder has often been confused with the true arrowroot (or West Indian arrowroot) produced from *Maranta arundinacea* L. (Marantaceae) in the Americas, or Tahiti arrowroot produced from *Tacca* spp. (Taccaceae).

14.2.2 ORNAMENTAL SPECIES

Some *Curcuma* species are beautiful and splendid garden plants. Their flowers are usually not very showy, but the colorful coma is quite attractive. Most species grow in clumps, serving also as good foliage plants for garden landscaping. Some species are widely cultivated for their foliage, with white variegations or bronzed leaves. Curcumas are not only very popular as garden and pot plants, but some species have also become very popular as cut flowers. All *Curcuma* species have a natural dormancy, which makes them convenient to package and ship safely and which fortunately coincide with winter in the northern hemisphere. No matter where they are grown they should be given dry periods, otherwise they may not flower again. Many species are rather easy to grow and can flower within a couple of months, quickly rewarding its gardener. Even though hybridization for new cultivars is now common, there are still plenty of species with ornamental potential, which have not yet been selected for the horticultural trade, such as *Curcuma ecomata*.

14.2.3 MEDICINAL PROPERTIES

Many *Curcuma* species were recognized by local and tribal people all over Asia as valuable sources of medicine centuries ago. Perhaps the richest knowledge of the use of different gingers, curcumas in particular, can be found in India. Such wide usage of *Curcuma* species in India may be connected with the unique role of turmeric, which is an integral part of the religious rites and traditional medicine knowledge in the Hindu culture, since time immemorial. It was also used in different systems of folk medicines. There are also a few ethno-veterinary Indian records that natives sometimes use wild *Curcuma* species and various gingers as medicines for their animals. Chinese herbals dating back to the first century (A.D.) also report the use of turmeric. Currently, all countries from India to Philippines, China, and Japan to Indonesia use turmeric (Figure 14.2) and other *Curcuma* species as a source of herbal remedies. However, it is unknown how long these cultures have used their incredibly rich knowledge.

In the past 20 years, modern medicine and the pharmacological industry have turned towards *Curcuma* with the hope of finding remedies for serious diseases as well as natural remedies for common maladies. Thus, there are studies on the chemical contents, structure, and composition of curcuminoids and essential oils of various species, and they are being extensively tested for their medicinal properties. However, the identity of most taxa studied may be unclear, except *C. longa* L. and perhaps *C. zanthorrhiza* Roxb, both of which are major sources of curcuminoids possessing antioxidant and anti-inflammatory properties. Many of the recent findings are promising, as there are positive results from experiments carried out on curcuminoids to counteract cancer, diabetes, cataractogenesis, liver diseases, and HIV. *C. zanthorrhiza* has been reported for lipid reducing and sedative properties (Wichtl, 1998) and to protect against hepatotoxins (Lin et al., 1995), which is consistent with its use in Asia as cholagogue and traditional cure for jaundice. Among other species being tested for various effects are, e.g., *C. zedouria* with proven antiamoebic activity (Ansari and Ahmad, 1991) and hepatoprotective constituents (Matsuda et al., 1998, 2001), *C. caesia* possessing antifungal and antibacterial properties (Garg and Jain, 1998; Banerjee and Nigam, 1976), and from the recently described Chinese species *C. wenyujin*, which exhibits antitumor active compounds (Shukui et al., 1995).

14.3 *CURCUMA AERUGINOSA* ROXB.

Curcuma aeruginosa (Figure 14.2a) is one of the oldest named species of the genus, described by William Roxburgh almost 200 years ago. The specific epithet is derived from the striking greenish-blue color of the rhizome. It is quite a large and stately plant and can reach almost 2 m. The plant was first described from a collection in Myanmar (then Burma), where it is certainly native to Pegu and surrounding areas, and brought for cultivation to Calcutta Botanic Garden, India. However, it is probably a widespread species, as it grows native in South India, and potentially elsewhere in Asia. Its leaves have a feather-shaped dark maroon patch in the distal part of the lamina, along the midrib on the adaxial (upper) side. The leaves are glabrous on both sides. Inflorescences appear at the beginning of the rainy season during the first premonsoon showers, just after the leaves begin to sprout. In South India, the leaves are used for wrapping fresh fish. The leaves are also used to

FIGURE 14.2 Some useful and ornamental *Curcuma* species. a: *C. aeruginosa*; b: *C. aromatica*; c: *C. aurantiaca*; d: *C. alismatifolia*; e: *C. rhabdota*; f: *C. pseudomontana*; g: *C. roscoeana*; h: *C. parviflora*; i: *C. zanthorrhiza*; j: *C. harmandii*; k: *C. zedoaria*; l: *C. longa*. (Photo: J. Skornickova; a, e, g, h, i, j taken at RBG Edinburgh.)

cover inner surface of bamboo baskets where paddy seeds are kept for sprouting prior to sowing, because of the belief that it promotes germination. Also in South India, it is one of the most esteemed species for extracting arrowroot, since it is believed that species with the blue-colored rhizome is of the best value. In Malaysia, it is used medicinally for asthma, cough, and as a paste with coconut oil for dandruff, and in Indonesia as purgative during childbirth. The species is distributed from India and the Andaman Islands to Cambodia, Malaysia, and Java.

Taxonomically unverified reports: Several studies have investigated the chemical composition of *Curcuma aeruginosa* from different parts of Asia, namely in S. India (Jirovetz et al, 2000), Malaysia (Sirat et al., 1998) and Vietnam (Tuyet et al., 1995), and it remains a question if they all represent *C. aeruginosa*. Two compounds, namely, curzerenone and 1,8-cineole were found to be present among three major compounds in all three analysis, however, the relative percentage differed as well as the absolute content and amount of minor compounds which included camphor, furanogermenone, curcumenol, isocurcumenol, zedoarol, and many others. We are of the opinion that it is unlikely they represent the same species.

14.4 *C. ALISMATIFOLIA* GAGNEP.

The leaves of this plant are generally only 60 to 80 cm, but the inflorescence may reach up to 1.5 m. The fertile bracts are relatively small and green, but the coma bracts are much larger and spreading, much like a tulip. They are the main attraction of this beautiful plant. This similarity with the Dutch tulip has earned the name Siam Tulip, or Tulip ginger for this species.

This species is the most utilized ornamental species of *Curcuma*. It is the mainstay of the Thai ginger horticultural industry, mass-produced via tissue culture, but it is also cultivated elsewhere. Exported to countries all over the world, it is used for mass plantings in landscapes, as a home pot plant, and as an extremely long-lasting (2 to 3 weeks) cut flower. There are countless cultivars, the coma varying anywhere between white and dark pink, with streaks or patches of coloration. Among the best-known favorites are "Chiangmai Pink," "Chiangmai Ruby," "Tropic Snow," "Thai Beauty," and "Thai Supreme."

14.5 *C. AMADA* ROXB.

The name of this species comes from the peculiar smell of the rhizome, which resembles that of an unripe mango. The leaves are plain green and glabrous on both sides. It is a late-flowering species; its greenish-white (or sometimes slightly tinged with rose) inflorescence invariably appears from the center of leaves around mid-October to the beginning of December. Its origin was probably in Bengal, and it is nowadays found cultivated in many parts of India. Bengali people use it for making pickles and curries. It is cultivated in Bengal and Bihar, but cultivation declined drastically due to low market prizes in recent years. Its common names (recently employed) are *amada* in Bengali, and *amadi bihari* or *ama haldi* in Hindi. The rhizomes are considered cooling; they are used as carminative and to promote digestion. They are supposed to be useful against prurigo. Watt (1889) mentions several medicinal uses for external application, including rheumatism, sprains, and bruises.

Similar to *C. mangga* of Indonesia by virtue of the mango smell of the rhizome, it is readily recognizable by the central position of the inflorescence.

The ether extract of *Curcuma amada* lowered the cholesterol level of experimentally induced hypercholesterolemic rabbits (Pachauri and Mukherjee, 1970). The crude extract from rhizomes of *C. amada* showed strong antifungal activity against *Trichophyton rubrum*, common cause of skin infections in West Bengal and Eastern India, and also against *Aspergillus niger* (Gupta and Banerjee, 1972).

An improved cultivar "Amba" has been developed at the High Altitude Research Station at Pottangi, Orissa State, India.

14.6 *C. AMARISSIMA* ROSCOE

This is a stately species with a large rhizome, which is of a deep yellow color with a circle of a green-blue shade. Its specific epithet comes from its extremely bitter rhizome. The strikingly red-brown petioles and whitish coma with deep violet tips make this a species with great ornamental potential. However, its relationship with the similar species *Curcuma ferruginea* is yet to be investigated.

14.7 *C. ANGUSTIFOLIA* ROXB.

Reported as a source of arrowroot in early history, it has narrow, green, glabrous leaves. Flowers usually appear in the beginning of the rainy season before the leaves are fully developed, but it continues flowering some time even after the leaves are developed. It is not a very showy species, as the inflorescences are rather small with yellow flowers and pink coma bracts near to the ground, and leaves of old plants with a strong rhizome reach only to 1.5 m. In English, it is called Wild or East Indian arrowroot, or as narrow-leaved turmeric; *tikhur* and *tavakhira* are common names in Hindi.

Watt (1889) described the cultivation of this plant as a commercial crop and the extraction of starch. It was the favored diet of tribals in yesteryears; it was also used in the preparation of sweetmeats that were consumed with great relish by the hill tribes.

14.8 *C. AROMATICA* SALISB.

As is obvious from its specific epithet, this plant is esteemed for the warm, aromatic taste of its rhizomes. It is known to be a species under wide cultivation within the Asian tropics, but also has an obscure history. Anthony Salisbury described it in 1807 from material sent for cultivation to England and there is no evidence of its type locality, nor a reliable type specimen. There are several taxa fitting to the original description provided by Salisbury, and this name is now attached to different plants such as the one shown in Figure 14.2b. Thus, many of the uses mentioned in the literature for *C. aromatica* must be taken with a grain of salt, as most references to *C. aromatica* are likely that of *C. zedoaria* (Chritstm.) Roscoe, *C. zanthorrhiza* Roxb., or an unknown complex of species.

Taxonomically unverified reports: Uses for some of these plants attributed to *Curcuma aromatica* include an anti-inflamatory effect (Jangde et al., 1998) when the rhizome is applied to bruises and sprains; it is also a stimulant, tonic and carminative, and is useful for curing leucoderma and blood diseases. Kojima et al. (1998) analyzed by GC/MS the essential oils derived from two samples of Japanese and one sample of Indian *C. aromatica* rhizomes. While composition of the oils of two Japanese samples was similar despite the different colour of the volatile oil, the essential oil of Indian sample gave distinctly different chemical composition. The Japanese samples contained curdione, germacrone, 1,8-cineole, (4S, 5S)-germacrone-4,5-epoxide, β-elemene, and linalool as the major components whereas those in the oil from Indian sample contained β-curcumene, ar-curcumene, xanthorrhizol, germacrone, camphor, and curzerenone. Authors pointed out that the presence of the markedly different metabolites may not be related to the different soil and climatic conditions and classification of these plants needs re-evaluation. Bordoloi et al. (1999) analyzed the essential oils from the rhizome and leaves of *C. aromatica* from North East India. Both analyses included camphor as a major component, followed by another 4 major components: camphor, curzerenone, α-turmerone, ar-turmerone and 1,8-cineole. It is very likely that all samples examined in these two works represent three different species, however it is not clear if any of them represents the true *C. aromatica*.

Curcuma aromatica is often equated with *vanaharidra* or wild turmeric of Sanskrit writers (Dymock et al., 1893, Watt, 1889). However, identity of the plant described by Dymock et al.

(1893) and Watt (1889) in their works as *C. aromatica* has been recently established as *C. zanthorrhiza*, which is common species in the western part of South India (Skornickova & Sabu, 2005b).

14.9 *C. AURANTIACA* ZIJP

Commonly known as the "Rainbow *Curcuma*," it is one of the most variable species in the genus with high ornamental potential. Native to the extreme south of Thailand, extending to Java and South India, it is of medium size (usually ca. 25 to 70 cm) with pretty, sulcate leaves, the inflorescences borne from the center of the leaves. They are incredibly colorful and the inflorescence can vary from green to pink, orange with a red tinge to deep reddish brown. The coma is white to a deep pink. Its specific epithet comes from the unusual striking orange color of its flowers. The anther is always without spurs. In the horticultural trade, this name is occasionally applied to another superficially similar *Curcuma* species, in which the anther is clearly spurred. *C. aurantiaca* species has high ornamental potential (Figure 14.2c), is rather easy to grow, and flowers on a regular basis. Rhizomes are known to be used as a talisman in Peninsular Malaysia (Ibrahim, 1995), and young inflorescences can be eaten as a vegetable (Burkill, 1966).

14.10 *C. AUSTRALASICA* HOOK. F.

A pretty, large species (up to 1.3 m), this is the only native representative of the genus from Australia, where it grows in disturbed areas, generally near roadsides. One variety of this species is known in the horticultural trade as "Aussie Plume." It is suitable for gardens as well as a cut flower.

14.11 *C. BICOLOR* MOOD AND K. LARSEN

This wonderful species is a recently described one, which was already in the horticultural trade in the U.S. under the name "Candy Corn" but had no scientific name. Its natural range is unknown, although it has been found in the wild only in northern Thailand. It is one of the most floriferous and showy curcumas in cultivation, with a small but strikingly colored basal red inflorescence. Its flowering period can last for a couple of months; the flowers are attractively red–orange bicolored (Mood and Larsen, 2001).

14.12 *C. CAESIA* ROXB.

Curcuma caesia is a species with a blue rhizome (commonly called black turmeric), the color is much brighter and deeper blue than in *C. aeruginosa*. The leaves have a deep red-violet patch, which runs throughout the whole lamina. Usually, the upper side of the leaf is rough, shortly pubescent, but this character may vary. A paste made from the rhizome is used to cure blood dysentery and as poultice in rheumatic pain; the leaves are used as a wrapping material, and dry leaves as fuel by Bengalis in Bangladesh (Yusuf et al., 2002). Some northeastern Indian tribes use this plant as a talisman to keep evil spirits away. *C. caesia* is known especially from Bengal, North East India, and Bangladesh. It is rarely found in other parts of India, such as Madhya Pradesh, Jharkhand, Chattisgarh, and Orissa. Its common names are usually derived from the deep color of the rhizome and thus this plant is called by following vernacular names: black zedoary (English); *sati* (Sanskrit); *kali haldi* (Hindi); *kala haldi*, *nil kantha*, and *kalo halood* (Bengali); *kalihalada* (Maharati); and *kala hailla* (Chakma-Bangladesh). It is used in local medicine in Bengal, and the tribals in Madhya Pradesh and Chattisgarh use it against a variety of illnesses.

14.13 *C. CORDATA* WALL.

Curcuma cordata Wall. has a name that has not been historically verified, but is included here because it is one of the most common plants utilized in the Thai horticultural trade. The original plant was brought from Myanmar (Burma), near Rangoon, to the Botanical Garden in Calcutta in the early 1800s, a flourishing botanical garden at the time. It was originally assumed to be the same species as *C. petiolata* (see section 14.21), but finally considered different enough to warrant its own species name. A century later, however, it was determined again to likely be the same plant. Thus far, it is unknown what the original plants looked like and we here discuss only the horticultural species in use today with this name.

A beautiful Thai species, *C. cordata* is prized for its long-lasting inflorescence as a cut flower and as a reliable bloomer in tropical settings. Known as the "Pastel Hidden Ginger," its terminal inflorescence is large, cylindrical, with a strongly pastel pink coma. The lower bracts are broadly rounded, waxy, and range from white to green, often rimmed with the pink color of the coma. The flowering stem is seemingly resistant to many of the fungal pathogens of other horticultural *Curcuma* species. The leaves may be a lighter or darker green than most other curcumas, broadly ovate, and strongly sulcate with a cordate to acute leaf base. The plant may reach 1 m, but is generally slightly smaller.

14.14 *C. ELATA* ROXB.

A huge species, which can reach up to 2.5 m and thus makes a good landscaping element, is a native of Myanmar. The main attraction is the large inflorescence spike with a deep rosy or even crimson coma, which comes out before the leaves appear. In the horticultural trade *C. elata* is often called "Giant Plume Ginger."

14.15 *C. GLANS* K. LARSEN AND MOOD

This is another recently described, highly attractive species from Thailand, which flowers in the very beginning of the season and the first inflorescences appear before the leaves. Unlike all other *Curcuma* species, the flowers open in the late afternoon and will last the night until the next afternoon. This plant is of medium size, with a clumping growth habit. It makes multiple pseudo-stems with about three shiny dark green leaves, which are silvery hirsute beneath. The inflorescence is white with occasional red markings, and the flowers are a beautiful white and yellow with lavender "glands," presumably used for nocturnal insect pollination (Mood and Larsen, 2001).

14.16 *C. HARMANDII* GAGNEP.

The beauty of this species lies in the green, long, linear bracts and delicate white flowers with a pink and yellow-dotted labellum resembling that of an orchid (Figure 14.2j). The unusual shape of the inflorescence has drawn the attention of the horticultural trade, and the variety "Jade Pagoda" is sometimes available in the market.

14.17 *C. LATIFOLIA* ROSCOE

This is perhaps the largest species in the genus, reaching over 3 m. It differs from *C. elata* in having a red patch on the upper side of leaf. The spikes are showy with a huge and bright pink coma. As per the original description, the labellum of the flowers should have a few red-brown lines, but as we have observed, this is probably a variable character.

14.18 C. MANGGA VALETON AND ZIJP

This is another species in which the rhizome has a mango smell similar to *C. amada*. It was described from Java. *C. amada* is a late, centrally flowering species; while *C. mangga* is an early, laterally flowering species. In the Malay Peninsula and Indonesia, it is part of a mixture used for curing fevers and for abdominal problems, in some parts it is used externally against evil spirits (Perry, 1980).

14.19 C. MONTANA ROXB.

This species is the first out of the numerous Indian *Curcuma* species described by William Roxburgh at the end of the 18th century. Since then, the name has been quite neglected and there is not much information about this plant. It is, however, one of the common species in forest undergrowth in eastern India, especially the Eastern Ghats. By producing only a central inflorescence, superficially it resembles *C. longa* and *C. amada*. Like in those two species, the spike is greenish white to white and coma is usually also white or with a slight pink tinge. Its rhizome is however light yellow, and does not have any strong smell. Due to its abundance, local tribal people use it for the extraction of East Indian Arrowroot.

14.20 C. PARVIFLORA WALL.

This is a small green and white *Curcuma*, with white and violet flowers (Figure 14.2h). This species has been grown extensively as a potted plant. Originally described from Myanmar, this species is found in riparian areas throughout Myanmar and Thailand. Although it is generally a small plant, some larger cultivars have been produced. Often confused with *Curcuma thorelii*, this species has white lateral staminodes in contrast to the purple staminodes of the other species. Both species are morphologically related to *C. gracillima*, although how closely is unknown. In Thailand the inflorescence is eaten as a vegetable. The pulp of the rhizome is applied to cuts (Sirirugsa, 1998).

14.21 C. PETIOLATA ROXB.

Another species that has an established identity in the horticultural trade, but its historical identy is uncertain. *Curcuma petiolata* Roxb. is considered by some to be the same as the later-named *Curcuma cordata*, but the two horticultural taxa currently sold are quite distinct. The following discusses the horticultural plant of *C. petiolata* only.

A medium plant growing from 1 m to 1.5 m, *C petiolata* is often grown for its foliage as a landscape element, often in large numbers, rather than for its shortly pedunculate inflorescences. The leaves are narrowly elliptic, and in some cultivars are strikingly variegated with thick cream bands at the leaf margin. The terminal inflorescence has a white coma tinged with pink at the tips, and coma bracts that are only slightly longer than the lower green bracts.

14.22 C. PSEUDOMONTANA J. GRAHAM

This is a highly variable seed-setting species with a beautiful well-developed coma and deep yellow flowers (Figure 14.2f). The leaves are broadly ovate, and prominently sulcate with a bright green color. It is surely a potentially ornamental species. In Maharashtra, the plant is used during the Ganesh Puja for worshiping the goddess Gauri and thus is called "Gauri's phool" (Flower of Gauri). It is so far known only from India, particularly from Maharashtra. There, the plant is abundant on rocky, shrubby slopes and it flowers richly from June to October, first laterally and later again from the center of the leaves. Due to its habitat, this plant develops many rather big tubers, which have been boiled and eaten during a scarcity of grain in the past.

14.23 *C. RHABDOTA* SIRIRUGSA AND M. F. NEWMAN

This has been one of the species available for sale in Thailand due to its ornamental value, yet it was unknown to science until recently. A small plant, reaching only 0.5 m, it has a quite striking inflorescence. The fertile bracts are white or greenish, beautifully striped with a deep red-brownish color (see Figure 14.2e). The coma bracts vary from white to pink with brown-red streaking. This species is probably native to deciduous Dipterocarp forests in Laos, and therefore should be planted in a very free-draining soil containing gravel and sharp sand when in cultivation. The plants need rather high temperatures, and night temperatures should not drop below 14°C. A dormancy period, with minimum watering, is required for flowering (Sirirugsa and Newman, 2000). It is often known in the U.S. horticultural trade as *C. gracillima* "Candy Cane," its unusual and very ornamental bract design making it a perfect cut flower. Recently, new hybrids have appeared as a result of horticultural breeding between *C. rhabdota* and *C. alismatifolia*.

14.24 *C. RHOMBA* MOOD AND K. LARSEN

This highly attractive species is known from open and dry forests of Vietnam and northern Thailand. Its name derives from the diamond (rhomboid) shape of its lateral staminodes. The plant is over 1-m high and there are several pseudostems per plant. The inflorescence is terminal but arising very near to the ground and pushing the leaves of the pseudostem apart. The inflorescence consists of red fertile bracts with no coma, and the flowers are a deep orange. It is sold under the name "Sri Pak" or "Sri Pok" (Mood and Larsen, 2001).

14.25 *C. ROSCOEANA* WALL.

This is an extremely beautiful and splendid plant, and certainly one of the most ornamental not only among *Curcuma* species but of the whole family (Figure 14.2g). Sometimes called the "Jewel of Burma" or "Pride of Burma," it can be found also in the wild in Thailand and recently has been reported from the Andaman Islands, India (Skornickova and Sabu, 2005a). It is an important part of Thai horticultural exports, and is available in the European and the U.S. horticulture markets as well. It is suitable for garden as well as a cut flower because of the long-lasting nature of the inflorescence bracts. This species can be easily recognized by its bright orange to scarlet spike with no obvious coma, and especially by the peculiar arrangement of the bracts, in three, four, or five serial rows. Flowers are usually a creamy color with a yellow center on the labellum, but plants with yellow flowers can be rarely seen also. The pseudostem and petioles are often shaded with a dark reddish tinge; leaves are petiolate and cordate at the base.

14.26 *C. RUBESCENS* ROXB.

This is a species with beautiful, rather large foliage and thus good for landscaping purposes. The leaves are of a deep green color while the petioles and midribs are of a deep red wine color. Rarely, the whole leaf is shaded by deep red color. Its size is around 1 to 1.5 m, and inflorescence is rather small and nonshowy. It is a native of Bengal, can be found in North East India, and it has been reported from Bangladesh. *C. rubescens* is one of the many species known to be used for extraction of East Indian Arrowroot (Roxburgh 1810 and 1820, Dymock et al. 1893). Nowadays, it is quite a rare species in the wild, sometimes cultivated by locals for its ornamental foliage.

14.27 *C. SPARGANIFOLIA* GAGNEP.

A *Curcuma* species with narrow leaves, it is known to grow wild in Cambodia and Thailand. A few cultivars are available in the horticultural market (e.g., "Siam Ruby"), suitable for garden as well as pot planting. The inflorescence is small and globose, with a few dark-tipped bright pink bracts. Small but interestingly colored inflorescences on long stalks make it a great cut flower.

14.28 *C. THORELII* GAGNEP.

A pretty, small to mid-sized species not exceeding 1 m in height. The fertile bracts are green, embracing a flower with a violet labellum; the coma is pure white, sometimes streaked by green especially on the lower coma bracts. The inflorescence bracts are arranged in definite rows, similar to those of *C. roscoeana*. In the horticultural trade, it is one of the most often confused species, as the names *C. thorelii*, *C. gracillima*, and *C. parviflora* are often interchangeable, applied to any small *Curcuma* species with green fertile bracts, white and violet flowers, and a white coma. Since the variation of three above mentioned species is rather high, this group needs more investigation. All three of these species are small and can be successfully grown in pots.

14.29 *C. ZANTHORRHIZA* ROXB.

This is a stately species, which can reach over 2 m in height. It is one of the most extensively utilized species of the genus *Curcuma*, and is known as *temu lawak* or *temu lawas* in Malaysia and Indonesia, where it is also widely cultivated and earned the nickname Javanese turmeric. It has been reported also from Thailand, Vietnam, Philippines, and China. This species is also common in South India, but has been misidentified for a very long time as *C. aromatica* and *C. zedoaria* (Skornickova and Sabu, 2005b). The leaves have a red patch, conspicuous especially in the beginning of the season when it protrudes to the lower side of the leaf. Later in the season this patch fades. It is an early-flowering species and produces a beautiful spike and a pink coma (Figure 14.2i). This species got its specific epithet from the deep yellow orange color of its rhizomes, which are medicinally valued especially in Malaysia and Indonesia. The rhizome can be used to dye textiles and food, but it is not as bright or long lasting as the dye prepared from *C. longa*.

In Malaysia, the rhizomes of *C. zanthorrhiza* are employed in mixtures involving other plants that are used in healing rituals for diseases attributed to evil spirits. A paste made from the rhizome is rubbed on the abdomen, or a decoction made and the tonic administrated to women after giving birth. It is used to cure skin diseases, diseases of the digestive and urinary systems, and gallstones (Prana, 1977, 1978). It is also one of the common gingers included in the traditional herbal mixture *Jamu* originally from Java, but it is now easily available in Indonesia as well as Malaysia (Ibrahim, 1995). It has been also reported as a cure for acne (Ibu Moorati, 1989). It is used to bring on the period in amenorrhea and as a puerperal tonic. It is passed over the body following childbirth. In Holland, it is used in treating gallstones (Duke, 2003).

This species is also being used in the manufacture of starch used in certain special occasions. A beverage is made in Indonesia by boiling the rhizomes in water and sweetening. Cooked inflorescence is served with rice (Duke, 2003).

Taxonomically unverified reports, Bruneton (1999), reported that the plant is rich in sesquiterpenes such as bisacurol, bisacurone, ar-curcumene, turmerones, xanthorrhizol, and zingiberene. Curcuminoids present to the extent of 1 to 2% consist mainly of curcumin and its derivatives. Yasni et al. (1991) tested the beneficial effects of four plants, which are part of traditional Indonesian diet and folk medicine, on diabetic rats. They found that *C. zanthorrhiza* improved diabetic symptoms such as growth retardation, hyperphagia, polydipsia, elevation of glucose and triglyceride in the serum, and had other positive effects on liver phospholipid metabolism. Their further research,

Yasni et al. (1993, 1994), identified alpha-curcumene as an active principle exerting triglyceride-lowering activity. Lin et al. (1995) found that extracts of *C. zanthorrhiza* can protect the liver from various hepatoxins and suggested that *C. zanthorrhiza* could be useful in the treatment of liver injuries. Xanthorrhizol, a sesquiterpene phenol, is the putative characteristic constituent of this species, Prana (1977), and was reported to possess bactericidal activity against both Gram-negative and Gram-positive bacteria.

14.30 *C. ZEDOARIA* (CHRISTM.) ROSCOE

Curcuma zedoaria is a rather pretty species flowering in the beginning of the rainy season when the leaves start to develop. This species is found in the wild as well as in cultivation all over India and Bangladesh (see Figure 14.2j) and is reported from most other Asian countries. However, upon historical and taxonomic investigation based on flowering material from different parts of Asia, it is very clear that the name *C. zedoaria* is applied to several superficially similar species. These all share the presence of a nice pink coma and leaves with a red patch over the midrib on the upper side of the leaf, which is conspicuous especially in the young leaves.

The lectotype of *C. zedoaria* as established by Burtt (1977) is a species described by Rheede (1692) as *kua*. It is native to South India and can be found also in West Bengal, Assam, and Bangladesh. For that reason, the local uses reported by Rheede are the most accurate ones. Rheede mentioned that the fresh root arrests all inflammation of the intestines, purges the kidneys, cures gonorrhea, and purges the blood, and the juice of the leaves is a moderate laxative. Starch produced from this species is one of the most esteemed in India. Starch extracted from the rhizome as arrowroot is used for making cakes and as a substitute for barley starch given to children.

Taxonomically unverified reports: In Bangladesh, a decoction is used to cure painful bowel movements and is known as a diuretic, a stimulant, and to have carminative and antihelmintic properties. It is also applied to bruises and sprains. A decoction of the rhizome with black pepper, cinnamon, and honey is used to cure the common cold (Dymock et al., 1863). Fresh leaves are used in local market as wrapping material and dry leaves as fuel by the poor, (Rahman and Yusuf, 1996; Yusuf et al., 2002). In Thailand, the young rhizome is eaten as a vegetable used in soup and is used to relieve stomach ache, and the leaves are used for flavoring fish and other foods, Sirirugsa (1998).

The medicinal uses of *C. zedoaria* were summarized from various sources by Duke (2003) as antipyretic, aromatic, carminative, demulcent, expectorant, stomachic, stimulant, and tonic. Fresh rhizomes have diuretic properties and are used in checking leucorrhea and gonorrheal discharge and for purifying blood. Rhizomes are chewed to alleviate cough and clear throat. A decoction of rhizomes along with long pepper, cinnamon, and honey is said to be beneficial for colds, fevers, bronchitis, and coughs. Rhizomes are used in medicines given to women after childbirth. A paste of the rhizome mixed with alum is applied to sprains and bruises. It is also used in dermatitis, sprains, ulcer, and wounds in Asian countries. Duke (2003) also reported that young shoots, and fresh young leaves, scented like lemon grass, are used as a vegetable and in salads. Zedoary is used in the manufacture of liquors, various essences, bitters and in cosmetics and perfumes. Dried rhizomes are used as a spice, mainly in bitters (such as Swedish bitters) and liquors (such as Italy's Ramazzotti). However, information provided by Duke (2003) certainly indicates a mixture of uses for many different species.

Pharmacological research suffers the same problem of misidentification of plant material, and the following information, which appeared in the literature during the past decade, has probably very little to do with the chemical composition of the true C. zedoaria. Garg et al. (2005) reported the chemical composition of the leaf oil of *C. zedoaria* by GC-MS and identified 23 compounds. The major ones are α-terpenyl acetate (8.4%), iso-oborneol (7%), dihydrocurdione (9%), and selino-4 (15), 7(11)-dien-8-one (9.4%). Yoshioka et al. (1998) tested dehydrocurdione, a sesquiterpene derived from *C. zedoaria*, and demonstrated strong antiinflammatory potency related to its antioxidant effect. Curcumol and curdione are regarded as anticancer agents, especially for cervical cancer and lymphosarcoma (Duke, 2003). Chinese studies indicate usefulness in managing cervical cancer and in improving the efficacy of chemotherapy and radiation, Bown (2001). Matsuda et al. (1998) analyzed the sesquiterpenes from the rhizome of C. zedoaria and their action on mice with induced acute liver injury. The principal sesquiterpenoids identified were furanodiene, germacrone, curdione, neocurdione, curcuminol, isocurcuminol, acrugideol, zedoarondiol, curcumenone, and curcumin. They showed potent hepatoprotective effect. Kim et al. (2001) demonstrated strong dose-dependent lysosomal enzyme activity in the polysaccharides from the rhizomes. They hypothesized that C. zedoaria has macrophage-stimulating activity, and that it can be used as a biological response modifier. Hong et al (2002) reported that the sesquiterpenoids, β-turmerone and ar-turmerone inhibited LPS-induced prostaglandin E2 production in cultured mouse macrophage cell in a dose-dependent manner. These compounds exhibited inhibitory effects on LPS-induced nitric oxide production in the cell system. Curcumenol derived from C. zedoaria grown in Brazil exhibited very potent and dose-related analgesic activity, being much more potent than different reference analgesic drugs. The results confirm and justify the popular use of this plant for the treatment of aches, Navarro et al. (2002).

14.31 OTHER SPECIES OF *CURCUMA*

There are also many species that are probably used by local and tribal people. However, information about these species is not available. Much study needs to be done to bring out the chemical composition as well as their chemical properties. The authors of the chapter are currently working toward a taxonomic revision of the genus *Curcuma* to re-investigate identities of individual species, to provide tools for their identification, and to ensure correct usage of the scientific names.

REFERENCES

Ansari, M.H. and Ahmad, S. (1991) *Curcuma zedoaria* root extract: In vitro demonstration of antiamoebic activity. *Biomed. Res. (Aligarh).,* 2:192-196.

Banerjee, A. and Nigam, S.S. (1976) Antifungal activity of the essential oil of *Curcuma caesia* Roxb. *Indian J. Med. Res.,* 64:1318-1321.

Bordoloi, A.K., Sperkova, J. and Leclercq, P.A. (1999) Essential oil of *Curcuma aromatica* Salisb. from North East India. *J. Essent. Oil Res.,* 11:537-540.

Bown, D. (2001) *New Encyclopaedia of Herbs and Their Uses,* D.K. Pub., New York.

Bruneton, J. (1999) *Pharmacognosy, Phytochemistry, Medicinal Plants.* 2nd ed., Laviosier Pub., Paris.

Burkill I.H. (1966) *A Dictionary of the Economic Products of the Malay Peninsula.* Vol. I (A-H). Ministry of Agriculture and Co-operatives, Kuala Lumpur. Malaysia. Pp. 714-725. (Reprint).

Burtt, B.L. (1977) *Curcuma zedoaria. Gard. Bull. Singapore,* 30:59-62.

Duke, J.A. (2003) *CRC Handbook of Medicinal Spices.* CRC Press, Boca Raton, USA.

Dymock, W., Warden, C.J.H. and Hooper D. (1893) *Pharmacographia Indica.* London, Bombay, Calcutta.

Garg, S.C. and Jain, R.K. (1998) Antimicrobial efficacy of essential oil from *Curcuma caesia. Indian. J. Microbiol.,* 38:169-170.

Garg, S.N., Naquvi, A.A., Bansal, R.P., Bahl, J.R. and Kumar, S. (2005) Chemical composition of the essential oil from the leaves of *Curcuma zedoaria* Rosc. of Indian origin. *J. Essent. Oil Res.,* 17:29-31.

Gupta, S. K. and Banerjee, A.B. (1972) Screening of West Bengal plants for antifungal activity. *Econ. Bot.,* 26:255-259.

Hong, C.H., Noh, M.S., Lee, W.Y. and Lee, S.K. (2002) Inhibitory effects of natural sesquiterpenoids isolated from the rhizomes of *Curcuma zedoaria* on prosglandin E-2 and nitric oxide production. *Pl. Med.,* 68:545-547.

Ibrahim, H. (1995) Peninsular Malaysian Gingers: Their Traditional Uses. *Bull. Heliconia Soc. Int.,* 7(3):1-4.

Ibu Moorati, S. (1989) Temu lawak (*Curcuma xanthorrhiza* Roxb. - Zingiberaceae) in the treatment of acne vulgaris. Sixth Asian Symposium on Medicinal Plants and Spices. Bandung, Indonesia.

Jangde, C.R., Phadnaik, B.S. and Bisen, V.V. (1998) Anti-inflammatory activity of extracts of *Curcuma aromatica* Salisb. *Indian Vet. J.,* 75:76-77.

Jirovetz, L., Buchbauer, G., Puschmann, C., Shafi, M.P. and Nambiar, M.K.G. (2000) Essential oil analysis of *Curcuma aeruginosa* Roxb. leaves from South India. *J. Essent. Oil Res.,* 12:47-49.

Kim, K.I., Shin, K.S., and Jun, W.J. (2001) Effects of polysaccharides from rhizomes of *Curcuma zedoaria* on macrophage functions. *Biosci. Biotech. Biochem.*, 65:2369-2377.

Kojima, H., Yanai, T. and Toyota, A. (1998) Essential oil constituents from Japanese and Indian *Curcuma aromatica* rhizomes. *Pl. Med.*, 64:380-381.

Lin, S.C., Lin, C.C., Lin, Y.H., Supriyatna, S. and Teng, C. W. (1995) Protective and therapeutic effects of *Curcuma xanthorrhiza* on hepatotoxin-induced liver damage. *Am. J. Chin. Med.,* 23:243-254.

Matsuda, H., Morikawa, T., Ninomiya, K. and Yoshikawa, M. (2001) Hepatoprotective constituents from Zedoariae rhizoma: Absolute stereostructures of three new carabrane-type sesquiterpenes, curcumenolactones A, B, and C. *Bioorg. Med. Chem.,* 9:909-916.

Matsuda, H., Ninomiya, K., Morikawa, T. and Yoshikawa, M. (1998) Inhibitory effect and action mechanism of sesquiterpenes from Zedoariae rhizoma on D-galactosamine/lipopolysaccharide-induced liver injury. *Bioorg. Med. Chem. Lett.,* 8:339-344.

Mood, J. and Larsen, K. (2001) New Curcumas from South-east Asia. *New Plantsman*, 8:207-217.

Navarro, D. de F., Souza, M.M. de, Neto, R.A., Golin, V., Niero, R., Yunes, R.A., Monache, F. and Cechinel Filho, V. (2002) Phytochemical analysis and analgesic properties of *Curcuma zedoaria* grown in Brazil. *Phytomedic.,* 9:427-432.

Pachauri, S.P. and Mukherjee, S.K. (1970) Effect of *Curcuma longa* (Haridar) and *Curcuma amada* (Amragandhi) on the cholesterol level in experimental hypercholesterolemia of rabbits. *J. Res. Indian Med.*, 5:27-31.

Perry, L. M. (1980) Zingiberaceae. *Medicinal Plants of East and South East Asia*: Attributed properties and uses. MIT Press, Cambridge, Massachusetts & London. England. Pp. 436-444.

Prana, M. S. (1977). Studies on some Indonesian *Curcuma* species, Ph.D. Thesis. University of Birmingham, 1-166.

Prana, M.S. (1978) Temu Lawak (*Curcuma xanthorrhiza* Roxb.). *Buletin Kebun Raya,* 3(6):191-194.

Rahman, M.A. and Yusuf, M. (1996) Diversity, Ecology and Ethnobotany of the Zingiberaceae of Bangladesh. *J. Econ. Taxon. Bot.*, Addit. Ser., 12:13-19.

Rheede tot Drakenstein, H.A. van (1678-1693) *Hortus Indicus Malabaricus*. Vol.11, Amsterdam.

Roxburgh, W. (1810) Descriptions of several of the Monandrous Plants of India. *Asiat. Res.* 11:318-362.

Roxburgh, W. (1820) *Flora Indica*. Pp. 1-84. Serampore.

Sabu, M. and Skornickova, J. (2003) *Curcuma aeruginosa* Roxb. – A source of East Indian Arrowroot. Proceedings of the 3rd Symposium on the family Zingiberaceae, Khon Kaen, Thailand. Pp. 196-200.

Salisbury, A. (1807) *Curcuma aromatica*. The Paradisus Londinensis t. 96.

Shukui, Q., Lin, W. and Jun, Q. (1995) Elemene emulsion for advanced lung cancer (Meeting abstract); Cancer conference 12th Asia Pacific, Singapore. P. 298.

Sirat, H.M., Jamail, S. and Hussain, J. (1998) Essential oil of *Curcuma aeruginosa* Roxb. from Malaysia. *J. Essent. Oil Res.,* 10:453-458.

Sirirugusa, P. (1998) Thai Zingiberaceae: Species diversity and their uses. *Pure Appl. Chem.*, 70:2111-2118.

Sirirugsa, P., and Newman, M.F. (2000) A new species of *Curcuma* L. (Zingiberaceae) from S.E. Asia. *New Plantsman*, 6:196-197.

Skornickova, J., Sabu, M., and Prasanthkumar, M.G. (2003) New species of *Curcuma* from Mizoram. *Gardens Bull. Singapore*, 55:89-95.

Skornickova, J. and Sabu M. (2005a) *Curcuma roscoeana* Wall. In India. *Gard. Bull. Singapore*, 57:187-198.

Skornickova, J. and Sabu, M. (2005b) The identity and distribution of *Curcuma zanthorrhiza* Roxb. *Gard. Bull. Singapore*, 57:199-210.

Tuyet, N.T.B., Dûng, N.X. and Leclercq, P.A. (1995) Characterization of the leaf oil of *Curcuma aeruginosa* Roxb. from Vietnam. *J. Essent. Oil Res.,* 7:657-659.

Watt, G. (1889) *A Dictionary of the Economic Products of India*. Vol. II. Reprint ed.1972. Cosmo Publications. Delhi.

Wichtl, M. (1998) *Curcuma* (Turmeric): Biological activity and active compounds. In: Lawson, L.D. and Bauer, R., eds. Phytomedicines of Europe: Chemical and Biological Activity. *ACS Symp. Ser.,* Vol. 691. American Chemical Society: Washington, D.C. Pp. 133-139.

Yasni, S., Imaizumi, K. and Sugano, M. (1991) Effects of an Indonesian medicinal plant, *Curcuma xanthorrhiza* Roxb. on the levels of serum glucose and triglyceride, fatty-acid desaturation, and bile-acid excretion in streptozotocin-induced diabetic rats. *Agric. Biol. Chem.,* 55:3005-3010.

Yasni, S., Imaizumi, K., Nakamura, M., Aimoto, J. and Sugano, M. (1993) Effects of *Curcuma xanthorrhiza* Roxb. and curcuminoids on the level of serum and liver lipids, serum apoliprotein A-1 and lipogenic enzymes in rats. *Food Chem. Toxicol.,* 31:213-218.

Yasni, S., Imaizumi, K., Sin, K., Sugano, M., Nonaka, G. and Sidik. (1994) Identification of an active principle in essential oils and hexane-soluble fractions of *Curcuma xanthorrhiza* Roxb. showing triglyceride lowering action in rats. *Food Chem. Toxicol.,* 32:273-278.

Yoshioka, T., Fujii, E., Endo, M., Wada, K., Tokunaga, Y., Shiba, N., Hohsho, H., Shibuya, H., Muraki, T. (1988) Antiinflammatory potency of dehydrocurdione, a zedoary-derived sesquiterpene. *Inflamm. Res.,* 47:476-481.

Yusuf, M., Rahman, M.A., Chowdhury, J.U. and Begum, J. (2002) Indigenous knowledge about the use of Zingibers in Bangladesh. *J. Econ. Taxon. Bot.,* 26(3):566-570.

15 Turmeric Production — Constraints, Gaps, and Future Vision

P.A. Valsala and K.V. Peter

CONTENTS

15.1 INTRODUCTION

Turmeric is intensively cultivated in India, Sri Lanka, parts of China, Pakistan, Bangladesh, Taiwan, Indonesia, Malagasy Republic, and Vietnam. It is cultivated on a small scale in several countries in the Caribbean, Central, and South America among which Jamaica, Haiti, and Peru are important. The chemical and physical characteristics of turmeric differ from region to region, and preferences as to the origin and physical forms are expressed by users for certain applications. In the international market, turmeric is branded according to the geographical origin of the produce. The major types entering the international trade are Alleppey, Madras, and West Indian turmeric. Alleppey turmeric comes from Kerala, especially from the region of Thodupuzha and Muvattupuzha taluks and is characterized by deep yellow to orange yellow in color and has high curcumin, up to 6.5%. U.S. has a special preference to Alleppey Finger Turmeric, where it is used as food colorant. Madras turmeric, exported from Madras, comes from several regional cultivars of Tamil Nadu and Andhra Pradesh. This brand is mustard yellow in color and has a curcumin content of around 3.5%. U.K. has a special preference to Madras turmeric, where it is regarded as superior in quality and flavor. West Indian turmeric comes from the Caribbean, Central, and South American countries and the rhizomes are dull yellowish brown in color and are regarded inferior in quality to Indian turmeric (Purseglove et al., 1981). Other geographical races from India that are popular in trade are "Erode" and "Salem turmeric" (Tamil Nadu); "Duggirala," "Nizamabad," and "Cuddappah" (Andhra Pradesh); and "Rajpore" and "Sangli turmeric" (Maharashtra) (http.//finance.indianart. com/markets/commodity/turmeric.html).

TABLE 15.1
Area, Production, Productivity, and Export of Turmeric From India During 1970–1971 to 2001–2002

Period	Area (ha)	Production (tons)	Yield (kg/ha)	Export (t)	Value (Rs. million)
1970–1971	80,500	150,600	1871	11,109	38.35
1980–1981	101,500	216,900	2137	14,517	78.82
1990–1991	119,000	342,400	2877	13,624	154.85
2000–2001	187,430	719.600	3839	44,627	115.58
2001–2002	162,950	552,300	3389	37,778	907.40
2002–2003	149,880	526,430	3512	32,402	1033.80

15.2 PRODUCTION AND TRADE

India is the largest producer, exporter, and consumer of turmeric in the world, with 82% of world production and 45% of export (Peter, 1997). Turmeric occupies about 6% of total area under spices in India. Table 15.1 gives the production and export for the last three decades.

India produced 0.526 million tonnes of dry turmeric from an area of 0.150 million ha in 2002 to 2003, with a productivity of 3512 kg/ha. The export was 32,402 t that earned foreign exchange to the tune of Rs. 1033.8 millions. The area, production, and export showed a sharp upward trend from 1970–1971 to 2000–2001. The increase in export was remarkable, and it was 192% during 2002–2003 (Spices Board, 2004). Increased export may be due to ban of synthetic colors in food industry in developed countries coupled with the knowledge about the medicinal properties of this spice. Out of the total production in India, 92 to 95% is consumed within the country, and the remaining 5 to 8% only is exported.

In India, the main turmeric growing states are Andhra Pradesh, Tamil Nadu, Orissa, Karnataka, West Bengal, Maharashtra, Meghalaya, and Kerala. Andhra Pradesh, the lead state of turmeric cultivation, shares 38% of area and 45% of production (Spices Board, 2004). Other states in the order of ranking in production are Tamil Nadu (21%), Orissa (12%), Karnataka (6.4%), West Bengal (3.9%), Maharashtra (1.5%), Meghalaya (1.5%), and Kerala (1.4%). Production of turmeric in Kerala, which produces the renowned Alleppey turmeric, is only 7900 t (Spices Board, 2004). In these states 70 to 75% of the area is occupied by the local cultivars, and the rest is by improved varieties.

India and other turmeric-producing Asian countries supply 75% of the world's turmeric trade. Major importers are Middle East and North African countries, Iran, Japan, and Sri Lanka. Other importers, Europe and North America, share 15% of the world trade, and the supply is from India and Central and Latin American countries. Taiwan exports mostly to Japan. The U.S. imports 97% of the requirement from India, and the rest is met by the islands of the Pacific and Thailand (http://finance.indiamart.com/markets/commodity/turmeric.html).

India exports turmeric as dry whole rhizome, powder, oleoresin, and oil, and as an ingredient of curry powders and mixtures, to more than 100 countries. The major importers are UAE (4724 t), U.S. (3914 t), Japan (2614 t), U.K. (2005 t), Malaysia (1993 t), Srilanka (1907 t), and Netherlands (1742 t). Bangladesh (1580 t), South Africa (1254 t), Saudi Arabia (1070 t), Iran (948 t), Germany (934 t), Egypt (712 t), and Morocco (453 t). These importing countries represent 80% of the Indian export trade (Spices Board, 2004). Item-wise export of turmeric from India during 1998–1999 and 2002–2003 is given in Table 15.2.

In 2002–2003, India exported 12,428 t of dry turmeric rhizome, 19,975 t of turmeric powder, 265.22 t of oleoresin, and 1.84 t of oil. Export of curry powder/mixtures was 8492 t. The quantity of turmeric powder in curry powder/mixtures is around 5%. Curry powder export showed highest growth rate (62.88%), followed by oleoresin (10.2%) and turmeric powder

TABLE 15.2
Item-Wise Export of Turmeric from India During 1998 to 1999 and 2002 to 2003

Items	1998–1999		2002–2003		Growth Rate (%)
	Quantity	Value	Quantity	Value	
Turmeric dry rhizome	26281.43	810.61	12428.00	363.92	–5.27
Turmeric powder	11016.03	480.84	19975.00	669.88	8.13
Oil	2.84	2.18	1.84	2.21	–5.43
Oleoresin	240.63	319.96	265.22	275.13	10.2
Curry powder/mixtures	5213.45	359.79	8491.91	689.37	62.88

Quantity in t; value in Rs. Million

(8.13%). Export of turmeric rhizome showed a negative trend of –5.27% and oil of –5.43% (Spices Board, 2004).

15.2.1 PRODUCT UTILIZATION

In the East, turmeric is used daily and extensively by all classes of people in the preparation of various dishes. It not only adds flavor but also color. It was used earlier as a dye for cotton textiles. Now it is used in medicine and cosmetics to impart yellow hue to the product. Turmeric also finds use in pest and disease management of crops. In India, it is regarded sacred by the Hindus, and is used in ceremonial and religious functions.

Turmeric is extensively used in the traditional systems of medicine in India and China and as a household remedy for various ailments including biliary disorders, anorexia, cough, diabetic wounds, hepatic disorders, rheumatism, and sinusitis. It is also used for healing small pox and chicken pox lesions and in curing jaundice and leprosy, as a complexion promoter of body and as a brain tonic, etc. (Verghese, 1999). For the last few decades, extensive work has been done to establish the biological activities and pharmacological action of turmeric and its extracts and isolates. Curcumin, the main yellow, bioactive component of turmeric has a wide spectrum of biological actions. These include anti-inflammatory, antioxidant, anticarcinogenic, antimutagenic, anticoagulant, antifertility, antidiabetic, antibacterial, antifungal, antiprotozoal, antiviral, antifibrotic, antivenom, antiulcer, hypotensive, and hypocholestermic activities. Its anticancer effect is mainly mediated through the induction of apoptosis. It has chemotherapeutic potential against colo-rectal cancer. It is found that turmeric reduces fat deposit in broiler chicken diet. Its anti-inflammatory, anticancer, and antioxidant properties may be clinically exploited to control rheumatism, carcinogenesis, and oxidative stress-related pathogenesis. Clinically, curcumin is used to reduce postoperative inflammation. Safety evaluation studies indicate that both turmeric and curcumin are well tolerated at a very high dose without any toxic effects. Thus, both turmeric and curcumin have the potential for the development of modern medicine for the treatment of various diseases (Chattopadhyay et al., 2004).

In U.S., use of turmeric as complementary medicine is recommended for most of the disorders mentioned earlier and is available for use as capsule-containing powder, fluid extract, and tincture. Standardized powder 400 to 600 mg thrice per day is a recommended dose for adults (http://www.umm.edu/_attned/constterbs/Turmerich.html).

15.3 RESEARCH ATTAINMENTS

Attempts were made in India to formulate specific research and developmental programs by Indian Council of Agricultural Research, State Agricultural Universities, Spices Board at Cochin, Directorate of Arecanut and Spices Development at Calicut, and Council of Scientific and Industrial Research. All India Coordinated Research Project (AICRP) on spices is the largest spices research network in the country providing the much needed R and D base (Ghosh et al., 1999). The Indian Institute of Spices Research (IISR) and AICRP centers together hold 1299 germplasm accessions that show variation with respect to fresh yield (6.6 to 50 t/ha), dry yield (1504 to 8558 kg/ha), curing percentage (14.06 to 28.17%), oleoresin (12.1 to 21.1%), curcumin (2.3 to 7%) and oil (2.33 to 6.55%). The accessions also varied in their response to biotic stress, leaf spot (*Colletotrichum capsici*), and leaf blotch (*Taphrina maculens*) diseases and shoot borer infection (*Dichocrocis punctiferalis*).

Concerted efforts made during the last three decades resulted in the release of many high-yielding varieties such as Co-1, Krishna, Sugandham, BSR-1, BSR-2, Roma, Suroma, Ranga, Rasmi, Rajendra Sonia, Suvarna, Suguna, Sudharsana, Prabha, Prathibha, Megha turmeric-1, Kanthi, Sobha, Varna, Sona, etc., yielding 4000 to 6000 kg dry turmeric/ha (Ghosh et al., 1999; Kurian et al., 2004).

Refined agrotechniques for various agroclimatic zones were developed with respect to rapid multiplication of planting materials, optimum time of planting, spacing, manuring, earthing up, weed control, and disease and pest management. Integrated nutrient management alone influences crop yield, up to 50 to 60%. Being a rhizomatous crop, turmeric loves rich organic manure and needs higher amount of nitrogen and potassium for dry matter accumulation (Venkatesan and Gurumurthy, 2002).

In India, turmeric is cultivated as rainfed and irrigated. Trials conducted at Tamil Nadu revealed that drip irrigation with fertigation realizes higher returns with efficient water and nitrogen use (Selvaraj et al., 2000).

As turmeric responds well to organic inputs, sustainable ecofriendly organic farming technology with good returns has been worked out for states such as Tamil Nadu (Ravikumar, 2002). In organic farming, the use of biofertilizers with *Azospirillum* and phosphobacterium is practiced. In order to get additional income, short duration crops such as onion, maize, colocassia, and pigeon pea are intercropped with turmeric. Integrated pest and disease management systems were developed, and advances were made to control rhizome rot of turmeric with biocontrol agents such as *Trichoderma harzianum*, *Trichoderma hamatum*, and *Trichoderma viridae*.

Development of a power tiller-operated turmeric harvester by Tamil Nadu Agricultural University, Coimbatore, has increased the harvesting efficiency to 99% and has reduced the harvesting cost (Kathirvel and Manian, 1999). Improved technologies have been developed for curing and coloring in turmeric by the leading Food Processing Institute, Central Food Technological Research Institute, Mysore.

Salient achievements in the field of biotechnology in turmeric involved development of micropropagation techniques, plant regeneration from callus, and micropropagation of related species and genera. The technology will be useful for rapid multiplication of elite breeding materials.

The active principles responsible for flavor, color, nutritional, and therapeutic value were identified, and successful efforts were made to find nontraditional applications such as antioxidant and antiseptic and use in cosmetic and health-food industry. The coloring pigments include curcumin [1,7-bis(4-hydroxy-3-methoxy-phenyl)-1,6-heptadiene-3,5-dione]; demethoxy curcumin [4-hydroxycinnamoyl(4-hydroxy-3-methoxycinnamoyl) methane], and bis-demethoxy curcumin [bis-(4-hydroxycinnamoyl) methane], and they are collectively known as curcuminoids. Among them, curcumin is the main coloring constituent of turmeric rhizome (Verghese, 1999).

15.4 CONSTRAINTS AND GAPS

At present, turmeric is a commercial spice crop of the tropics, and its product is a commodity of international trade. World demand for turmeric in recent years is continuously on the rise. For overall development of the sector, the gaps and constraints in the sector have to be analyzed and solutions worked out.

Important threats and weaknesses identified are the following:

In India, turmeric production is unsteady, and quality varies from season to season and region to region. Adoption of appropriate production technology incorporating ecofriendly crop management practices is lacking. The national productivity of turmeric is only 3389 kg/ha, even though released varieties are capable of producing 6000 kg/ha. The climate and soil factors prevalent in Kerala are very much suitable for turmeric production, and the Alleppey Finger Turmeric exported from here has special demand in U.S. for use as food colorant. Even then, production and productivity are very low. Turmeric is mostly cultivated in smallholdings frequently as intercrop. Production is thus generally unplanned, affecting the quality. There are no measures that focus on organic farming, value addition of products, and product diversification. Major constraints in India in the spices sector are weak marketing infrastructure and lack of export-oriented production strategies, high cost of production owing to lower productivity, quality aspects such as high pesticide residue, and poor trading strategies, all of which have led to a decline in India's share in the global spice trade (Selvan and Manojkumar, 2003).

Quality assurance is one area that needs attention to push India's products in international market. Problems such as pesticide residues have deteriorated the quality of raw materials. Restrictions and levies imposed by both the exporting and the importing nations act as fiscal controls and hamper exports.

In spite of regular internal demand and growing export, trade in this sector is largely unorganized. Due to many intermediaries, producers get only about 65 to 70% of the consumers' price, the remaining being taken away by wholesalers and retailers as margin of profit and other marketing cost. There is normally no government intervention in the trade with regard to marketing procedures or in the maintenance of a reasonable price to the farmers.

Significant positive correlation between prices and production exists in turmeric. Being an annual crop, increase in prices has immediate effect on production. When prices start increasing, farmers put more area under the crop and adopt better farming practices, resulting in an increase in production. This occasionally leads to a glut in market and a slump in prices.

15.5 FUTURE VISION

Turmeric is a medicinal spice with a wide spectrum of biological properties. The demand for turmeric is continuously on the rise. This rise is mainly due to the growing awareness on health and nutritional aspects of turmeric and the market trend, which is moving away from synthetic colors to natural ones. Increase in the popularity of Indian curry powder and other spice blends has boosted the demand for turmeric. The international spice industry is lately going through a sea of changes with increase in number of competition leading to a glut in world market and stringent quality regulations imposed by developed nations. This has called for immediate measures to focus on organic spices, value addition, and improved productivity. Future developmental program for turmeric should focus on the exploitation of its use in traditional as well as nontraditional areas, with value addition and supply of clean quality products at competitive price. The following aspects should find a place in the future developmental programs.

Turmeric has been used in ayurvedic medicines since ancient times, with various biological applications. Crude extract also has increasing medicinal applications. Curcumin, the yellow coloring pigment, is a powerful antioxidant. Its efficiency is comparable to standard antioxidants such as Vitamin C and E. The bright yellow color and antioxidant properties against lipid peroxidation promote its use in butter, margarine, cheese, etc. (Suresh, 2003). Curcumin is also nonmutagenic and has shown to suppress mutagenicity of several mutagens. It also exhibits many direct anti-inflammatory effects for prevention and cure of rheumatic arthritis and bronchial asthma, and is also known for its cholesterol-lowering effect and for inhibition of platelet aggregation. Curcumin has exhibited hepatitic protection properties also. Daily consumption of turmeric as spice and increased use of curcumin in neutraceuticals industry should be campaigned owing to aforesaid medicinal properties.

Although the crude extract and curcumin have numerous medicinal applications, clinical applications can be made only after intensive research on its bioactivity, mechanism of action, and pharmacotherapeutic, pharmacokinetics, and toxicity studies. Recent years saw an increased enthusiasm in treating various diseases with natural products. Curcumin is now available in pure form. Novel drugs to control various diseases including inflammatory disorders, carcinogenesis, and oxidative stress-induced pathogenesis have to be developed from curcumin after extensive studies on its mechanism of action and pharmacological effects. The recent findings on the effect of curcumin in slowing down the Alzheimer's disease is yet another area that needs intensive studies.

World demand for organically produced foods are growing rapidly in developed countries such as Europe, U.S., Japan, Australia, and Middle East. India, with its "intrinsic quality spices" grown under wide agroecological regions, can definitely utilize this expanding organic spice sector. With one of the lowest per capita consumption of fertilizers and pesticides in the world, it is rather easy for Indian farmers to embrace upon organic spice farming to meet the growing global demand. It is forecasted that by 2006 to 2007, there could be a fivefold increase in organic market all over the world.

With the global food industry increasingly turning toward oils and oleoresins and natural flavor, value addition of products holds great potential for India. The market for oleoresin is increasing, especially in the food-processing sector, mainly due to bacteriological problems with other forms of spices (Ashalatha and Gangopadhyay, 2001). Value-added products are less susceptible to price competition compared with spices in the raw form. To achieve competitive advantage, India needs to resort to low volume, high cost trade through value addition.

Importing countries have regional preferences. U.K. prefers Madras turmeric and U.S. and Canada prefer Alleppey finger turmeric. Increasing productivity, quality, thrust on value addition with brand building initiatives, expanding production base to nonconventional areas, and exploring new markets with new uses and applications are a few of the major strategies that India has to pursue. For reducing cost of production, high-yielding varieties with excellent quality aspects should be cultivated with ecofriendly and sustainable production technology. In India, the crop shows much variability for economic characters. This conserved genetic variability may be utilized for evolving better varieties for higher production. Crop improvement programs should shift from mere clonal selection to planned breeding programs. The breeding strategy should utilize biotechnological tools such as anther culture, *in vitro* fertilization, micropropagation, and recombinant DNA technology (Renjith et al., 2000). Transgenics, which are resistant to rhizome rot disease, could be produced through recombinant DNA technology. Genes coding for resistance to rhizome rot disease caused by *Pythium* spp. may be searched in the conserved variability.

Appropriate production technology with emphasis on organic and ecofriendly farming should be developed in tune with Good Agricultural Practices. Cultivation practices with respect to planting, harvesting, and cleaning the produce have to be mechanized. The intrinsic quality of turmeric varies with respect to location and season. The factors influencing quality parameters in turmeric should be studied so as to schedule harvesting/processing to suit various end products.

Stringent measures need to be adopted to assure the quality of the produce. Presence of pesticide residues, mycotoxins, and microbial contaminations are the major reasons for the detention of consignments of spice/spice products exported from India. They are detained due to occurrence of residues of various pesticides such as chloropyriphos, ethion, quinalphos, cypermethrin, fenvalerate, phosphamidon, and phosalone by the major importing countries such as U.S., U.K., Germany, Spain, Italy, and Australia. Consignments are also detained for the incidence of Aflatoxins, presence of salmonella and filth content, and for improper labeling. Major reason for this is the lack of proper certification system of the consignment prior to its shipment. With the liberalization of economic policies, compulsory preshipment inspection and quality control of export consignments from India are withdrawn. Preshipment inspection/certification by any of the agencies approved by the Government of India namely Export Inspection Agency (EIA) under Export Inspection Council of India and Directorate of Marketing & Inspection (AGMARK) ensure decent quality for the spices exported from India.

Use of turmeric as a natural, cheap, cosmetic for body beautification and hygiene has been practiced from time immemorial. Its germicidal and cosmetic properties should be further exploited by value addition for development of products such as bathing soap, body cream, lotion, powder, etc.

Fluctuations in prices in national and international markets on account of variable supply and demand, which bring into conflict the interest of growers and user industries, have led to market instability. Wide fluctuation in production has to be checked by keeping the prices at reasonable remunerative levels. During glut period, price fluctuations have to be stabilized through market interventions. This will create confidence among farmers and encourage them to take up regular cultivation of this annual crop, leading to steady growth and production.

Active branding exercise in importing countries needs to be ensured. Since the consumer is becoming more and more health conscious, for long-term success, products with best quality should be sold. The Indian spice logo represented through the visible symbol of a "fresh green leaf in elliptical ring" as an identity for Indian spice, with quality, should be promoted. The consumer can pick up quality spices just looking at the logo, the "fresh green leaf in elliptical ring" (Menon, 1993). Increasing world consumption is essential. There is ample scope for increasing the consumption of turmeric, provided proper and well-planned promotional campaigns are undertaken.

The knowledge accumulated on various uses of turmeric shows that it is a gift of Mother Nature, which could be exploited in numerous ways for the benefit of humanity.

REFERENCES

Ashalatha and Gangopadhyay, S. (2001) SWOT Analysis of Indian Spice Industry. www. commodity India.com, Commodity of the month, June 2001.

Chattopadhyay, I., Biswas, K., Bandyopadhyay, U. and Banerjee, R.K. (2004) Turmeric and Curcumin. Biological actions and medicinal applications. *Curr. Sci.* 87(1): 44-53.

Ghosh, S.P., Pal, R.N., Peter, K.V. and Ravindran, P.N. (1999). Four decades of spices research and development - an overview. *Indian Spices,* 36(4), 11-17.

http://finance.indiamart.com/markets/commodity/turmeric.html.

http://www.umm.edu/attned/constterbs/Turmerich.html.

India Finance and Investment Guide (2005) Turmeric. Accessed from http://finance.indiamart.com/market/commodity/turmeric.htm. (Accessed Oct. 2, 2005).

Kathirvel, K. and Manian, R. (1999) Turmeric Harvester. *Spice India,* 12(8): 21-22.

Kurian, A., Nybe, E.V., Valsala, P.A. and Sankar, A. (2004) Curcumin rich turmeric varieties from Kerala Agricultural University. *Spice India,* 14(11): 38-39.

Menon, K.P.G. (1993) Indian Spices Logo - Heralding Excellence in Quality. *Indian Spices,* 30(4), 5-6.

Peter, K.V. (1997) Fifty years of research on major spices in India. *Indian Spices,* 34(1): 20-36.

Purseglove, J.W., Brown, E.G., Grew, C.L. and Robbins, S.R.J. (1981) *Spices, Vol.2.* Longman Inc. New York

Ravikumar, P. (2002) Production technology for organic turmeric. *Spice India,* 15(2): 2-6.

Renjith, D., Valsala, P.A. and Nybe, E.V. (2000) Response of Turmeric (*Curcuma domestica* Val.) to *in vivo* and *in vitro* pollination. *J. Spices Aromatic Crops,* 10(2): 135-139.

Selvan, M.T. and Manojkumar, K. (2003) Indian Spices. Challenges and Opportunities. *Indian J. Arecanut, Spices Medic. Plants,* 5(1): 1-4.

Selvaraj, P.K. Krishnamurthi, V.V. and Manickasundram, P. (2000) Drip irrigation is a boon for turmeric crop. *Spice India,* 13(6): 2-5.

Spices Board. (2004) Spices Statistics V ed. Ministry of Commerce & Industry, Government of India, Cochin - 682 025, pp.1-281.

Suresh, M.P. (2003) Medicinal applications of spices and herbs - An attractive segment. *Spice India,* 16(4): 11-12

Velayudhan, K.C., Muralidharan, V.K., Amalraj, V.A., Gautham, P.L., Mandal, S. and Dinesh Kumar. (1999) *Curcuma Genetic Resources*. Scientific Monograph. No.4. National Bureau of Plant Genetic Resources, New Delhi.

Venkatesan, B. and Gurumurthy, S. (2002) Integrated nutrient management in turmeric (*Curcuma longa* L.). *Spice India*, 15(3): 26-28.

Verghese, J. (1999) Curcuminoids, the magic dye of *C. longa* L. rhizome. Indian Spices, 36(4): 19-25.

Epilogue

"If I had only a single herb to depend upon for all possible help and dietary needs, I without much hesitation, chose the Indian spice turmeric" (Frowley, 1997). A spice, a medicine par excellence, a beauty aid, and, above all, a plant having considerable sociocultural and religious associations — all these make turmeric the real golden gift of Mother Nature, the spice of life, and "the earthy herb of the sun." In the Ayurvedic system of India, turmeric is "*tridoshara*," capable of correcting the imbalances of the three *doshas* (*vata, pitta,* and *kapha*), the three humors that keep the human body in a functional balance. Turmeric has been in use for the last 6000 years — the saga goes on and on. New medicinal properties are being unraveled, and new uses are being discovered. Now we know, in addition to its traditional medicinal uses, it is also anticarcinogenic, anti-Alzheimeric, antimutagenic, and so on. "It is an awesome antioxidant, five to eight times stronger than Vitamin E, three times more powerful than the grape and pine bark extract, more powerful than Vitamin C, eugenol, capsaicin, and BHT, and especially potent in scavenging the hydroxyl radicals, the most reactive of all oxidants" (Jager, 1997). Turmeric detoxifies the body system by promoting the production of glutathione–S–transferase, and in this way indirectly helps the body to cure itself.

At least preliminary studies indicate that a daily dose of turmeric can protect from the toxic chemicals in the automobile exhaust gases and thereby from the air pollution in cities. Some studies also indicate that turmeric offers protection from cigarette smoke. Turmeric is a great woman's herb and has helped the women of the Orient from many gynecological problems. It even guards the body against uterine and breast cancers. "It helps beautify the skin and improve the complexion, promoting circulation and nutrition to the surface of the body. It tones up the immune system and also improves the health of heart and liver" (Frowley, 1997). It is also an obesity fighter.

No wonder the ancients named turmeric *Oushadhi* (the medicinal herb); for them, ginger was the *Mahaoushadhi* — the great medicinal herb. Together they protected the health of the people of an ancient land from time immemorial. Turmeric is likened to the divine mother bestowing numerous blessings and helping us in all dangers, difficulties, and conditions of weaknesses and debility. It vitalizes the body's own natural healing energy through its action of strengthening, digestion, and circulation and aiding in the regulation of all bodily systems. Hundreds of publications and reports on the pharmacological properties of turmeric and its components stand testimony to the medicinal value of this plant and further vindicate its use in traditional Indian medicine.

"Outer beauty, inner purity," this is what turmeric gives us. Used on daily basis for thousands of years in India by most of its population, it is traditionally considered to have the essentials of beneficial properties. What Ayurveda has known for millennia, modern science is now starting to prove for itself in laboratories and clinics around the world. Turmeric improves liver and heart functions, helps with arthritis and diabetes, and attacks cancer and carcinogens. "You will also notice the difference it makes with your skin" (Jager, 1997).

Charaka, the founder of the Indian traditional medicine, compiled his text *Charakasamhita* at least about 4000 years BP and he recommends turmeric for skin diseases and to purify the "body mind" to help the lungs expel phlegm. Far away from India, in Hawaii, turmeric (known as *Olena*) juice is a medicine for earaches and purifies the sinuses via the nose. The root is eaten for curing bronchitis and asthma, and they use turmeric paste as skin ointment. They sprinkle turmeric juice mixed with seawater on people and places to remove negativity and restore harmony. In Unani medicine, which originated in Greece and was patronized by the Persian kings, turmeric is considered the safest herb of choice for all blood disorders since it purifies, stimulates, and builds the blood.

The saga of turmeric, which started about 8000 years BP goes on and on. The crop that was shrouded in mythology in the past has come out to the modern to help mankind cure many of its ailments. Indeed, turmeric is the golden spice of life. It carries the energy of life to our entire being and connects us to the beneficent forces of this conscious universe in which we live. It is also perhaps the most useful and certainly the most commonly used Ayurvedic herb. Turmeric is a good place to start studying and using Ayurveda and a good herb to take a new lease on life (Frowley, 1997).

The need of the hour is to verify and validate the traditional uses by subjecting them to proper experimental studies. The poor solubility and bioavailability of curcumin limits its uses, as well as that of turmeric as a phytoceutical. Soluble curcumin derivatives that are nontoxic and very effective clinically may have to be developed for efficient exploitation of the potentialities of turmeric as a medicinal plant. The bitter taste of turmeric contributed by some of the components of the essential oil limits its usefulness in health drinks. Debittered turmeric can form a good prebiotic/probiotic health drink for many ailments afflicting the middle-aged and older people of the modern world.

REFERENCES

Frowley, D. (1997). In: Jager, P. de. (1997) *Turmeric*, Vidyasagar Pub., California, U.S., pp. 67.
Jager, P. de. (1997) *Turmeric*, Vidyasagar Pub., California, U.S. pp. 67.

Index